A Modern Course in Statistical Physics

A Modern Course in Statistical Physics

by L. E. Reichl
Department of Physics and
Center for Statistical Mechanics and Thermodynamics
University of Texas at Austin

Foreword by Ilya Prigogine

UNIVERSITY OF TEXAS PRESS, AUSTIN

Library of Congress Cataloging in Publication Data

Reichl, L E
A modern course in statistical physics.

Includes index.
1. Statistical physics. I. Title.
QC174.8.R44 530.1'5'95 79-15287
ISBN 0-292-75051-X

Printed in the United States of America

Illustrations by Pamela Vesterby

Contents

Foreword

It is a pleasure for me to present this monograph in statistical physics to the public. In the last decade this field has undergone a real explosion. I remember quite vividly the first conferences which I organized after World War II in Brussels and Florence as the Secretary of the Commission for Statistical Mechanics and Thermodynamics of the International Union of Pure and Applied Physics. They were small and, I may add, friendly gatherings. Today the field has increased markedly in its scope, and the number of problems which have been studied has multiplied. More than ever a synthetic view introducing the student to these recent developments is needed. Many years ago this was the aim of Fowler's *Statistical Mechanics* (Cambridge University Press, 1929), which has been used successfully as an introduction to this field by many generations of students. I believe that in recent years no book which could perform such a service has been published. Of course, there are many outstanding monographs in statistical physics which have appeared, but most of them are either too elementary or too specialized. It is quite an achievement to have brought together so many modern aspects of statistical physics. *A Modern Course in Statistical Physics* is an excellent introduction to the field of statistical physics, and I am sure it will encourage many young people to explore further this fascinating subject. I might add that Dr. Reichl is very well prepared for this task, having made a series of interesting contributions to statistical physics which range over a variety of subjects, including the formal many-body theory, nonlinear response theory, transport theory, and low-temperature physics.

Ilya Prigogine, Nobel Laureate

Professor of Physics and Chemical Engineering,
University of Texas at Austin

Professor of Physical Chemistry,
Free University of Brussels

Acknowledgments

First and foremost, I wish to thank Professor Ilya Prigogine, Director of the Center for Statistical Mechanics. His deep understanding of many fields of science and his infectious enthusiasm for ideas have led me to explore many of the subjects discussed here. His continued support and encouragement have made this book possible. I am also grateful for a grant from the European Atomic Energy Commission and the hospitality of the Free University of Brussels during the fall of 1977 when part of this book was written.

This work is based on lecture notes from a series of special topics courses given at the University of Texas at Austin from 1974 to 1977. The active participation of the students has enriched this manuscript and made those courses a pleasure to teach. I wish to thank them all.

A Modern Course in Statistical Physics

1

Introduction

The field of statistical physics has seen many exciting developments in recent years. New results in ergodic theory, nonlinear chemical physics, stochastic theory, quantum fluids, critical phenomena, hydrodynamics, and transport theory have revolutionized the subject, and yet these results are rarely presented in a form that students who have little background in statistical physics can appreciate or understand. This book has been written in an effort to incorporate these exciting subjects into a basic course on statistical physics. It includes in a unified and integrated manner the foundations of statistical physics and develops from them most of the tools needed to understand modern research in all of the above fields. For example, in the field of ergodic theory the work of Fermi, Pasta, and Ulam; Kolmogorov, Arnold, and Moser; Henon and Heiles; and Ford, among others, has deepened our understanding of the structure and behavior of phase space for a variety of nonlinear dynamical systems and has made ergodic theory a modern field of research. In an effort to introduce this field to students, a careful discussion is given (Chaps. 7 and 8) of the behavior of reversible probability flows in phase space, including specific examples of ergodic and mixing flows. In addition, the results of Henon and Heiles and Ford are used to discuss the probable origin of the irreversibility observed in nature.

Nonlinear chemical physics is still in its infancy, but it has already given a conceptual framework within which we can understand the thermodynamic origin of life processes. The discovery of "dissipative structures" (nonlinear spatial and temporal structures) in nonlinear chemical systems and subsequent work in this area by Prigogine, Glansdorff, Nicolis, Lefever, and others have opened a new field in chemistry and biophysics. In this book material has been included on chemical thermodynamics (Chap. 4), chemical hydrodynamics (Chap. 14), chemical kinetics (Chap. 13), and nonlinear stability theory (App. D), all of which are necessary to understand the theory of dissipative structures. The simplest mathematical model of a dissipative structure (the Brusselator) is discussed in some detail (Chap. 17).

The use of stochastic theory to study fluctuation phenomena in nonlinear chemical systems and its growing use in population dynamics and ecology have brought new life to this field in recent years. In order to give the student some familiarity with the field of stochastic theory, the Chapman-Kolmogorov and master equations are derived in Chap. 6 and applied to Markov chains, diffusion, stationary Markov processes, and linear and nonlinear birth and death processes. Also included is the method of van Kampen for obtaining an approximation to the master equation for systems which have a large parameter.

The theory of superfluids rarely appears in general textbooks on statistical physics, but the theory of such systems is incorporated at appropriate places throughout this book (Chaps. 3, 4, 7, 9, 13, 14, and 15). The relation between phase space and field operators is discussed in a simple way (App. B), and the standard treatments of ideal quantum fluids are given (Chap. 9). In addition, the book contains discussions of the Ginzburg-Landau theory of superconductors (Chap. 4), the thermodynamics (Chaps. 3 and 4) and hydrodynamics (Chap. 14) of superfluids, the concept of off-diagonal, long-range order (Chap. 7), the finite temperature BCS treatment of superconductors (Chap. 9), Goldstone bosons (Chap. 14), Wick's theorem (Chaps. 10, 12), Dyson's equations (Chap. 12), and the use of exact propagators for finding excitations in interacting quantum fluids (Chap. 12). Also included is an extensive discussion of the classical theory of fluids (Chap. 11), including the Ornstein-Zernike equation, the hypernetted chain, and the Percus-Yevick equation, and we compare the predictions with experiment.

Since the work of Wilson on the renormalization group in the early 1970's, the field of critical phenomena has undergone a revolution, and various aspects of this field are treated extensively in this book. A thorough discussion of the thermodynamics of phase transitions is given and is used to study transitions in *PVT* systems, binary mixtures, ferromagnetic systems, and superconductors in Chap. 4. Critical exponents are introduced in Chap. 4, and all critical exponents for the van der Waals equation are computed and compared with experiment. A number of exactly soluble microscopic models are treated (Chap. 9) in addition to the Widom and Kadanoff scaling theories (Chap. 10), and the Wilson theory (Chap. 10) is used to obtain the critical exponents for the triangular lattice and the Gaussian and S^4-models. Examples of nonequilibrium phase transitions in chemical and hydrodynamic systems are also included (Chap. 17).

In recent years, the field of hydrodynamics has flourished as a tool for understanding long-wavelength phenomena in classical fluids, solids, liquid crystals, and superfluids. This book contains a thorough grounding in hydrodynamics. Completely general equations for fluid flow (App. A) and Onsager's relations (Chap. 14) are derived, and the Curie principle is used to obtain the hydrodynamic equations of simple classical fluids, reacting chemical fluids, and superfluids (App. C and Chap. 14). In addition the book contains an extensive discussion of correlation functions, causality, the fluctuation-dissipation theorem, the theory of light scattering, and the origin of hydrodynamics in terms of conserved quantities and broken symmetries (Chap. 15).

With the discovery by Alder and Wainwright of long time tails in the velocity autocorrelation function, the traditional picture of transport processes developed by Boltzmann was shown to be inadequate even for dilute gases. Subsequent work by Kawasaki and Oppenheim, Dorfmann and Cohen, Ernst, van Leeuven, Dufty, McLennan, and others has shown the origin of the tails to be hydrodynamic processes in the fluids. In this book, transport theory is discussed from many points of view. Both the mean free path expressions for transport coefficients (Chap. 13) and expressions for the transport coefficients in terms of correlation functions are given (Chap. 15). In addition, the Boltzmann and Lorentz-Boltzmann equations are derived, and microscopic expressions for the coefficients of diffusion, shear viscosity, and thermal conductivity are obtained (Chap. 13) using a method due to Resibois. Finally, the breakdown of the Boltzmann view of transport processes is discussed, and the long time tails are derived, first from hydrodynamics and then from microscopic theory (Chap. 16).

While the subject matter of the book is modern, the method of presentation is traditional and is designed to build in students an understanding of statistical physics beginning from well-defined foundations. The book is essentially divided into four parts. Chaps. 2 through 4 deal with thermodynamics, Chaps. 5 through 8 deal with probability theory and the foundations of statistical mechanics, Chaps. 9 through 12 deal with equilibrium statistical mechanics, and Chaps. 13 through 17 deal with nonequilibrium thermodynamics, nonequilibrium statistical mechanics, and hydrodynamics.

It has been my experience that many students, even at the graduate level, have difficulty using thermodynamics. This book assumes that the student has had some contact with thermodynamics at a more elementary level where the ideas are illustrated microscopically with the simple kinetic theory of ideal gases. My aim in the first few chapters is to develop thermodynamics as a subject in its own right and to develop in students an appreciation for its simplicity and internal beauty as well as a facility in extracting the huge wealth of information it contains. In Chap. 2 we review the foundations of thermodynamics and thermodynamic stability theory, while Chaps. 3 and 4 are devoted to a large variety of applications. In Chap. 3 we deal with applications of thermodynamics which do not directly involve phase transitions, such as the cooling of gases, mixing, osmosis, chemical thermodynamics, and the two-fluid model of superfluids. Many of the ideas introduced there we will see again in later chapters. Chap. 4 is devoted to the thermodynamics of phase transitions and the use of stability theory in analyzing phase transitions. We discuss in great detail first-order transitions in liquid-vapor-solid transitions, with particular emphasis on the liquid-vapor transition and its critical points, and we use the van der Waals equation to illustrate many ideas. We then introduce the Ginzburg-Landau theory of λ-points and discuss a variety of transitions which involve broken symmetry. We conclude the chapter with a discussion of critical point exponents and compute them for the vapor-liquid transition using the van der Waals equation. In Chap. 10 we will use Wilson theory to compute them microscopically.

Having developed some intuition concerning the macroscopic behavior of complex systems, we then turn to the microscopic foundations. Chaps. 5 through 8 are devoted to probability theory and the foundations of statistical mechanics. Chap. 5 contains a review of the basic concepts from probability theory, such as counting, distribution functions, cumulant expansions, the central limit theorem, and the law of large numbers. In Chap. 6 we develop the theory of discrete stochastic variables in the Markov approximation. We are able to obtain conditions under which the probability distribution for such systems decays to a unique probability distribution and to obtain equations of motion for moments of the stochastic variables. The theory developed here has many applications in chemical physics, laser physics, population dynamics, biophysics, and other subjects and prepares the way for the more complicated subject of statistical mechanics.

Chaps. 7 and 8 are devoted to the foundations of statistical mechanics. In Chap. 7 we derive the equation of motion for the probability density of classical dynamical systems and show that Hamiltonian dynamics places severe restrictions on the way the probability distribution for such systems evolves. We also obtain the BBGKY hierarchy of equations for the reduced probability densities and the microscopic balance equations. We then derive analogous equations for density matrix of quantum systems and for the reduced density matrices, and we introduce the concept of off-diagonal, long-range order (ODLRO). In Chap. 8 we look more closely at the type of flow that can occur in dynamical systems and introduce the concepts of ergodic and mixing flows which appear to be minimal requirements if a system is to decay to thermodynamic equilibrium. We also discuss recent work on anharmonic oscillator systems by Kolmogorov, Arnold, and Moser, by Henon and Heiles, and by Ford which shows that such systems cannot always decay to thermodynamic equilibrium.

Chaps. 9–12 are devoted entirely to equilibrium statistical mechanics. In Chap. 9 we derive the probability distribution of closed and open systems and relate them to thermodynamic quantities. We then present a variety of examples which are chosen to illustrate various techniques and physical ideas. These examples include the theory of solids, ideal quantum fluids, superconductors, phase transitions in order-disorder systems, the Lee-Yang theory of phase transitions, and a simple microscopic derivation of the van der Waals equation.

In Chap. 10 we introduce equilibrium fluctuation theory of fluid and spin systems and show qualitatively how the spatial extent of correlations between fluctuations increases as we approach the critical point. Then we introduce the idea of scaling and use Wilson theory to obtain microscopic expressions for the critical exponents of spin lattices, the Gaussian model, and the S^4 model.

Chaps. 11 and 12 are devoted to the equilibrium theory of classical and quantum fluids, respectively. These two chapters are placed side by side to make possible a comparison of the standard theoretical treatment of the two subjects. We find some similarities in the broad outline but huge differences in the basic tools used. These chapters are the only ones in which we use diagrammatic methods, but we are able to introduce them in a simple way and they help to make many aspects of the theory of fluids more understandable and physically meaningful. In Chap. 11 we derive the virial expansion for the grand potential and the radial distribution function of hard-core classical fluids using a binary expansion, and we compute the virial coefficients for several types of interparticle potential and compare them to experiment. We also derive the Ornstein-Zernike equation and obtain the Percus-Yevick and hypernetted chain approximations to the radial distribution function and compare them with experiment. To conclude Chap. 11, we derive quantum corrections to the virial coefficients for dilute systems. In Chap. 12 we study quantum fluids using a perturbation expansion rather than binary expansion and therefore can use second quantization. We use the temperature-dependent formalism to study quantum fluids and not the ground-state formalism so commonly used in the early days of the subject. The temperature-dependent formalism is far more powerful and is becoming more and more important because of its use in general relativity and in discussing phase transitions at finite temperatures. We give the cumulant expansion of the grand potential and the exact one-body propagator (a quantity closely related to the one-body reduced density matrix), and we show how to obtain information about the exact excitations in a quantum fluid. We illustrate the ideas of this chapter by studying electron gases, weakly coupled Fermi fluids with short-ranged interactions, and weakly coupled condensed boson systems.

The last part of the book, Chaps. 13–17, deals with nonequilibrium thermodynamics, statistical mechanics, hydrodynamics, and transport theory. In Chap. 13, we look at simple transport processes in dilute classical systems and derive mean free path expressions for the coefficients of diffusion, shear viscosity, and thermal conductivity, and we derive the Boltzmann and Lorentz-Boltzmann equations and use them to obtain microscopic expressions for the three transport coefficients. We conclude Chap. 13 with a derivation of the quantum kinetic equation for a weakly coupled quantum gas.

Chap. 14 is devoted to the macroscopic theory of coupled transport processes. We derive Onsager's relations for the phenomenological coefficients and illustrate them for the thermomechanical effect. We then derive the hydrodynamic equations for multicomponent chemically reacting systems and for superfluid systems using symmetry arguments and show how Onsager's relations reduce the number of independent transport coefficients in these equations. In Chap. 15 we derive general properties of the correlation functions for equilibrium fluctuations and use macroscopic linear response theory and causality to derive the fluctuation-dissipation theorem. We illustrate the use of the fluctuation-dissipation theorem for a Brownian particle and for the case of light scattering. Finally, we use microscopic linear response theory and projection operator methods to generalize the definition of hydrodynamic equations to include the case of modes (such as Goldstone bosons) which result from a broken symmetry.

In Chap. 16 we show that the correlation functions for currents in classical fluids have long time tails, rather than the exponential decay predicted by Boltzmann. These tails are due to hydrodynamic modes (many-body effects) in the fluid and can be obtained microscopically by resuming the virial expansion for the current correlation function. We derive the microscopic expressions for the correlation functions for the case of the velocity autocorrelation function and discuss the implications of the long time tails and the many-body effects for hydrodynamics in general.

Finally, in Chap. 17 we conclude with the fascinating subject of nonequilibrium phase transitions. We discuss the thermodynamic stability theory for systems far from equilibrium, and we show how nonlinearities on the rate equations for chemical systems can lead to chemical clocks and nonlinear waves, while in hydrodynamic systems they lead to spatial cells in the flow pattern of the fluids.

This book, while covering a large number of topics, has been written specifically for students. All concepts are illustrated with examples which are chosen to clarify physical content without introducing overly difficult mathematics. The students always know where they stand in relation to other areas of statistical physics. In addition, each chapter has problems to help the students test their understanding.

The book has too much material to cover in a one-semester course, but it is written in such a way that parts can be excluded, and the instructor can select areas most relevant to the needs of the students. For a one-semester course, I suggest that the following material be covered: Chap 2, a review; some examples from Chap. 3; Chap. 4, Secs. B–D, F, G, and J–L; Chap. 5, a review; Chap. 6, Secs. B–D; Chap. 7, Sec. B; Chap. 8, Sec. B; Chap. 9, Secs. B–D; Chap. 10, Secs. B–C; Chaps. 11–12, some parts, depending on the interests of the class; Chap. 13, Secs. B–G; Chap. 14, Secs. B–D; and Chap. 15, Secs. B–F. In a two-semester course, it is possible to cover a larger amount of the book, and the students can easily read the rest.

The book is intended to introduce the students to a variety of subjects and resource materials which they can then pursue in greater depth if they wish. We have tried to use standardized notation as much as possible. In writing a book which surveys the entire field of statistical physics, it is impossible to include or even to reference everyone's work. We have included references which are especially pertinent to the points of view we take in this book and which will lead students easily to other work in the same field. We apologize in advance to those whose work we have not been able to include.

2

Introduction to Thermodynamics

A. INTRODUCTORY REMARKS

The science of thermodynamics began with the observation that matter in the aggregate can exist in macroscopic states which are stable and do not change in time. These stable "equilibrium" states are characterized by definite mechanical properties, such as color, size, and texture, which change as the substance becomes hotter or colder (changes its temperature). However, any given equilibrium state can always be reproduced by bringing the substance back to the same temperature. Once a system reaches its equilibrium state, all changes cease and the system will remain forever in that state unless some external influence acts to change it. This inherent stability and reproducibility of the equilibrium states can be seen everywhere in the world around us.

Thermodynamics has been able to describe, with remarkable accuracy, the macroscopic behavior of a huge variety of systems over the entire range of experimentally accessible temperatures (10^{-4} K to 10^{6} K). It provides a truly universal theory of matter in the aggregate. And yet, the entire subject is based on only four laws, which may be stated rather simply as follows: Zeroth Law—it is possible to build a thermometer; First Law—energy is conserved; Second Law—not all heat energy can be converted into work; and Third Law—we can never reach the coldest temperature. However, even though these laws sound rather simple, their implications are vast and give us important tools for studying the behavior and stability of systems in equilibrium and, in some cases, of systems far from equilibrium.

This chapter will be devoted to a review of various aspects of thermodynamics that will be used throughout the remainder of the book. We shall begin by introducing the variables which are used in thermodynamics, and the mathematics needed to calculate changes in the thermodynamic state of a system. As we shall see, many different sets of mechanical variables can be used to describe thermodynamic systems. In order to become familiar with some of these mechanical variables, we shall write the experimentally observed equations of state for a variety of thermodynamic systems.

As we have mentioned above, thermodynamics is based on four laws. We shall discuss the content of these laws in some detail, with particular emphasis on the second law. The second law is extremely important both in equilibrium and out of equilibrium because it gives us a criterion for testing the stability of equilibrium systems and, in some cases, nonequilibrium systems.

There are a number of different thermodynamic potentials that can be used to describe the behavior and stability of thermodynamic systems, depending on the type of constraints imposed on the system. For a system which is completely isolated from the world, the internal energy will be a minimum for the equilibrium state. However, if we couple the system thermally, mechanically, or chemically to the outside world, other thermodynamic potentials will be minimized. We will introduce the five most commonly used thermodynamic potentials (internal energy, enthalpy, Helmholtz free energy, Gibbs free energy, and the grand potential), and we will discuss the conditions under which each one is minimized at equilibrium and why they are called potentials.

When experiments are performed on thermodynamic systems, the quantities which are easiest to measure are the response functions. Generally, we change one parameter in the system and see how another parameter responds to that change, under highly controlled conditions. The quantity that measures the way in which the system responds is called a response function. In this chapter, we shall introduce a variety of thermal and mechanical response functions and give relations between them.

Isolated equilibrium systems are systems in a state of maximum entropy. Any fluctuations which occur in such systems must cause a decrease in entropy if the equilibrium state is to be stable. We can use this fact to find relations between the intensive state variables for different parts of a system if those parts are to be in mechanical, thermal, and chemical equilibrium. In addition, we can find restrictions on the sign of the response functions which must be satisfied for stable equilibrium. We shall find these conditions and discuss the restrictions they place on the Helmholtz and Gibbs free energy.

Finally, we shall apply the ideas discussed in this chapter to the case of a classical ideal gas and obtain expressions for the various thermodynamic potentials, starting from experimentally accessible information about the system.

B. STATE VARIABLES AND EXACT DIFFERENTIALS

Thermodynamics describes the behavior of systems with many degrees of freedom after they have reached a state of thermal equilibrium—a state in which all past history is forgotten and all macroscopic quantities cease to change in time. The amazing feature of such systems is that, even though they contain many degrees of freedom ($\sim 10^{23}$) in chaotic motion, their thermodynamic state can be specified completely in terms of a few parameters—called *state variables.* In general, there are many state variables which can be used to specify the thermodynamic state of a system, but only a few (usually two or three) are independent. In practice, one chooses state variables which are accessible to experiment and obtains relations between them. Then, the "machinery" of thermodynamics enables one to obtain the values of any other state variables of interest.

State variables may be either *extensive* or *intensive.* Extensive variables always change in value when the size (spatial extent and number of degrees of freedom) of the system is changed, and intensive variables need not. Certain pairs of intensive and extensive state variables often occur together because they correspond to generalized forces and displacements which appear in expressions for thermodynamic work. Some examples of such extensive and intensive pairs are, respectively, N (number of particles) and μ' (the chemical potential per particle), V (volume) and P (pressure), $\mathbf{M}$ (magnetization) and $\mathscr{H}$ (magnetic field strength), L (length) and J (tension), A (area) and σ (surface tension), $\mathbf{P}$ (electric polarization) and $\mathbf{E}$ (electric field). The pair of state variables related to the heat content of a thermodynamic system are the temperature, T, which is intensive, and the entropy, S, which is extensive.

Other state variables used to describe the thermodynamic behavior of a system are the various response functions, such as heat capacity, C; compressibility, K; magnetic susceptibility, χ; and various thermodynamic potentials, such as the internal energy, U; enthalpy, H; Helmholtz free energy, A; Gibbs free energy, G; and the grand potential, Ω. We shall become thoroughly acquainted with these state variables in subsequent sections.

If we change the thermodynamic state of our system, the amount by which the state variables change must be independent of the path taken. If this were not so, the state variables would contain information about the history of the system. It is precisely this property of state variables which makes them so useful in studying changes in the equilibrium state of various systems. Mathematically, changes in state variables correspond to exact differentials;[1] therefore, before we begin our discussion of thermodynamics, it is useful to review the theory of exact differentials. This will be the subject of the remainder of this section.

Given a function $F = F(x_1, x_2)$ depending on two independent variables x_1 and x_2, the differential of F is defined

$$dF = \left(\frac{\partial F}{\partial x_1}\right)_{x_2} dx_1 + \left(\frac{\partial F}{\partial x_2}\right)_{x_1} dx_2, \tag{2.1}$$

where $(\partial F/\partial x_1)_{x_2}$ is the derivative of F with respect to x_1 holding x_2 fixed. If F and its derivatives are continuous and

$$\left[\frac{\partial}{\partial x_1}\left(\frac{\partial F}{\partial x_2}\right)_{x_1}\right]_{x_2} = \left[\frac{\partial}{\partial x_2}\left(\frac{\partial F}{\partial x_1}\right)_{x_2}\right]_{x_1}, \tag{2.2}$$

then dF is an exact differential. If we denote

$$c_1(x_1, x_2) \equiv \left(\frac{\partial F}{\partial x_1}\right)_{x_2} \quad \text{and} \quad c_2(x_1, x_2) \equiv \left(\frac{\partial F}{\partial x_2}\right)_{x_1},$$

then the variables c_1 and x_1 and variables c_2 and x_2 are called "conjugate" variables with respect to the function F.

The fact that dF is exact has the following consequences:

(*a*) The value of the integral

$$F(A) - F(B) = \int_A^B dF = \int_A^B (c_1\, dx_1 + c_2\, dx_2)$$

is independent of the path taken between A and B and depends only on the end points A and B.

(*b*) The integral of dF around a closed path is zero,

$$\oint_{\text{closed}} dF = \oint_{\text{closed}} (c_1\, dx_1 + c_2\, dx_2) \equiv 0.$$

(*c*) If one knows only the differential dF, then the function F can be found to within an additive constant.

If F depends on more than two variables, then the statements given above generalize in a simple way: Let $F = F(x_1, x_2, \ldots, x_n)$, then the differential, dF, may be written

$$dF = \sum_{i=1}^{n} \left(\frac{\partial F}{\partial x_i}\right)_{\{x_j \neq i\}} dx_i. \tag{2.3}$$

The notation $(\partial F/\partial x_i)_{\{x_j \neq i\}}$ means that the derivative of F is taken with respect to x_i holding *all* variables *but* x_i constant. For any pair of variables, the following relation holds:

$$\left[\frac{\partial}{\partial x_l}\left(\frac{\partial F}{\partial x_k}\right)_{\{x_j \neq k\}}\right]_{\{x_j \neq l\}} = \left[\frac{\partial}{\partial x_k}\left(\frac{\partial F}{\partial x_l}\right)_{\{x_j \neq l\}}\right]_{\{x_j \neq k\}}. \tag{2.4}$$

An example for the case of three independent variables is

$$dF = c_1\, dx_1 + c_2\, dx_2 + c_3\, dx_3.$$

Then Eq. (2.4) leads to the result

$$\left(\frac{\partial c_1}{\partial x_2}\right)_{x_1, x_3} = \left(\frac{\partial c_2}{\partial x_1}\right)_{x_2, x_3}, \qquad \left(\frac{\partial c_1}{\partial x_3}\right)_{x_1, x_2} = \left(\frac{\partial c_3}{\partial x_1}\right)_{x_2, x_3},$$

$$\left(\frac{\partial c_2}{\partial x_3}\right)_{x_1, x_2} = \left(\frac{\partial c_3}{\partial x_2}\right)_{x_1, x_3}.$$

Differentials of all state variables are exact and have the above properties.

Given four state variables x, y, z, and w, such that $F(x, y, z) = 0$ and w is a function of any two of the variables x, y, or z, one can obtain the following useful relations:

$$\left(\frac{\partial x}{\partial y}\right)_z = \frac{1}{\left(\frac{\partial y}{\partial x}\right)_z}. \tag{2.5}$$

$$\left(\frac{\partial x}{\partial y}\right)_z\left(\frac{\partial y}{\partial z}\right)_x\left(\frac{\partial z}{\partial x}\right)_y = -1. \tag{2.6}$$

$$\left(\frac{\partial x}{\partial w}\right)_z = \left(\frac{\partial x}{\partial y}\right)_z\left(\frac{\partial y}{\partial w}\right)_z. \tag{2.7}$$

$$\left(\frac{\partial x}{\partial y}\right)_z = \left(\frac{\partial x}{\partial y}\right)_w + \left(\frac{\partial x}{\partial w}\right)_y\left(\frac{\partial w}{\partial y}\right)_z. \tag{2.8}$$

It is a simple matter to derive Eqs. (2.5)–(2.8). We will first consider Eqs. (2.5) and (2.6). Let us choose variables y and z to be independent, $x = x(y, z)$, and then choose x and z to be independent, $y = y(x, z)$, and write the following differentials: $dx = (\partial x/\partial y)_z\, dy + (\partial x/\partial z)_y\, dz$ and $dy = (\partial y/\partial x)_z\, dx + (\partial y/\partial z)_x\, dz$. If we eliminate dy between these equations, we obtain

$$\left[\left(\frac{\partial x}{\partial y}\right)_z\left(\frac{\partial y}{\partial x}\right)_z - 1\right] dx + \left[\left(\frac{\partial x}{\partial y}\right)_z\left(\frac{\partial y}{\partial z}\right)_x + \left(\frac{\partial x}{\partial z}\right)_y\right] dz = 0.$$

Because dx and dz may be varied independently, their coefficients may be set equal to zero separately. The result is Eqs. (2.5) and (2.6).

To derive Eq. (2.7) we let y and z be independent so that $x = x(y, z)$ and write the differential for dx. If we then divide by dw, we obtain

$$\frac{dx}{dw} = \left(\frac{\partial x}{\partial y}\right)_z \frac{dy}{dw} + \left(\frac{\partial x}{\partial z}\right)_y \frac{dz}{dw}.$$

For constant z, $dz = 0$ and we find Eq. (2.7).

Finally, to derive Eq. (2.8), we let x be a function of y and w, $x = x(y, w)$. If we write the differential of x, divide it by dy, and restrict the entire equation to constant z, we obtain Eq. (2.8).

When integrating the exact differential, $dF = c_1(x_1, x_2)\, dx_1 + c_2(x_1, x_2)\, dx_2$, one must be careful. As an example, let us consider the differential $d\phi = (x^2 + y)\, dx + x\, dy$. Then $(\partial\phi/\partial x)_y = x^2 + y$ and $(\partial\phi/\partial y)_x = x$. Since

$$\left[\frac{\partial}{\partial y}\left(\frac{\partial \phi}{\partial x}\right)_y\right]_x = \left[\frac{\partial}{\partial x}\left(\frac{\partial \phi}{\partial y}\right)_x\right]_y = 1,$$

the differential is exact. There are several ways in which we can integrate the differential. Let us consider two of them:

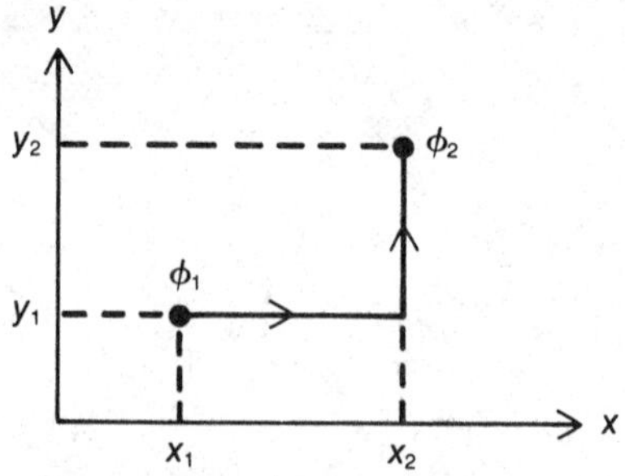

Fig. 2.1. Path of integration for the differential $d\phi = (x^2 + y)\,dx + x\,dy$.

(*a*) Choose a definite path (cf. Fig. 2.1) (the result is independent of path and depends only on the end points).

$$\begin{aligned}\phi_2 - \phi_1 &= \int_{x_1}^{x_2} (x^2 + y_1)\,dx + \int_{y_1}^{y_2} x_2\,dy \\ &= \tfrac{1}{3}(x_2^3 - x_1^3) + y_1(x_2 - x_1) + x_2(y_2 - y_1) \\ &= \tfrac{1}{3}(x_2^3 - x_1^3) + x_2y_2 - x_1y_1 = (\tfrac{1}{3}x^3 + xy)\Big|_{x_1,y_1}^{x_2,y_2},\end{aligned}$$

thus $\phi = \frac{1}{3}x^3 + xy$ plus a constant.

(*b*) Integrate by inspection. First do the indefinite integrals

$$\int \left(\frac{\partial \phi}{\partial x}\right)_y dx = \int (x^2 + y)\,dx = \frac{x^3}{3} + xy + K_1(y),$$

where $K_1(y)$ may depend only on y, and

$$\int \left(\frac{\partial \phi}{\partial y}\right)_x dy = xy + K_2(x),$$

where $K_2(x)$ may depend only on x. Comparing the two expressions, we see that $K_1(y) = K$ a constant and $K_2(x) = x^3/3 + K$. Thus, we find again $\phi = x^3/3 + xy + K$.

C. EQUATIONS OF STATE

An equation of state is a functional relation between the state variables for a system in equilibrium which reduces the number of independent degrees of freedom needed to describe the state of the system. It usually is an equation which relates the thermal state variables T of S to the mechanical state variables for that system and contains a great deal of information about the thermodynamic behavior of the system. It is useful to give examples of empirically obtained equations of state.

1. Ideal Gas Law

The best known equation of state is the ideal gas law

$$PV = nRT, \tag{2.9}$$

where n is the number of moles, T is temperature in degrees Kelvin, P is the pressure in Pascals, V is the volume in cubic meters, and $R = 8.314$ J/mol·K is the molar gas constant. The ideal gas law gives a good description of a gas which is so dilute that the effect of interaction between particles can be neglected.

If there are m different types of particles in the gas, then the ideal gas law takes the form

$$PV = \sum_{i=1}^{m} n_i RT, \tag{2.10}$$

where n_i is the number of moles of the ith constituent.

2. Virial Expansion[2]

The virial expansion

$$P = \left(\frac{nRT}{V}\right)\left[1 + \frac{n}{V}B(T) + \left(\frac{n}{V}\right)^2 C(T) + \cdots\right] \tag{2.11}$$

expresses the equation of state of a gas as a density expansion. The quantities $B(T)$ and $C(T)$ are called the second and third virial coefficients and are func-

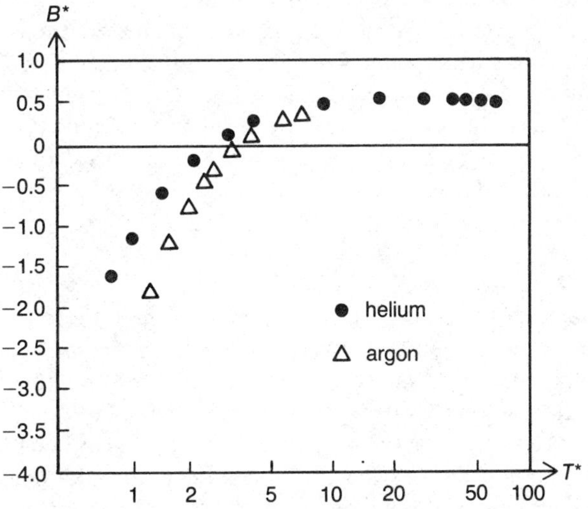

Fig. 2.2. Plot of the second virial coefficients for helium and argon: $B^* = B/b_0$ and $T^* = kT/\epsilon$; for helium $b_0 = 21.07 \times 10^{-6}$ m^3/mol and $\epsilon/k = 10.22$ K; for argon $b_0 = 49.8 \times 10^{-6}$ m^3/mol and $\epsilon/k = 119.8$ K (based on Ref. 2.2, pp. 164 and 1110).

tions of temperature only. As we shall see in Chap. 11, the virial coefficients may be computed in terms of the interparticle potential. Comparison between experimental and theoretical values for the virial coefficients is an important method for obtaining the force constants for various interparticle potentials. In Fig. 2.2, we have plotted the second virial coefficient for helium and argon. The curves are typical of most gases. At low temperatures $B(T)$ is negative because the kinetic energy is small and the attractive forces between particles reduce the pressure. At high temperatures the attractive forces have little effect and corrections to the pressure become positive. At high temperature the second virial coefficient has a maximum.

For an ideal classical gas all virial coefficients are zero, but for an ideal quantum gas (Bose-Einstein or Fermi-Dirac) the virial coefficients are nonzero. The "statistics" of quantum particles give rise to effective interactions.

3. Van der Waals Equation of State[3]

The van der Waals equation of state is of immense importance historically because it was the first equation of state which applies to both the gas and liquid phases and exhibits a phase transition. It contains most of the important qualitative features of the gas and liquid phases although it becomes less accurate as density increases. The van der Waals equation contains corrections to the ideal gas equation of state, which take into account the form of the interaction between real particles. The interaction potential between molecules in a gas contains both a strong repulsive region (hard core) and a weaker attractive tail. For an ideal gas, as the pressure is increased, the volume of the system can decrease without limit. For a real gas this cannot be true because the hard core limits the close-packed density to some finite value. Therefore, as pressure is increased the volume tends to some minimum value $V = V_{\min} = nb$ where b is an experimental constant. The ideal gas equation of state must be corrected and assumes the form

$$P = \frac{nRT}{(V - V_{\min})} = \frac{nRT}{(V - nb)}.$$

The attractive part of the potential causes the pressure to be decreased slightly relative to that of a noninteracting gas. The decrease in pressure will be proportional to the probability that two molecules interact and this, in turn, is proportional to the square of the density of particles $(N/V)^2$. We therefore correct the pressure by a factor proportional to the square of the density—which we write $a(n^2/V^2)$. The constant a is an experimental constant which depends on the type of molecule being considered. The equation of state can now be written

$$\left(P + \frac{an^2}{V^2}\right)(V - nb) = nRT. \tag{2.12}$$

In Table 2.1 we have given values of a and b for simple gases.

	a (Pa·m^6/mol)	b (m^3/mol)
H_2	0.02476	0.02661
He	0.003456	0.02370
CO_2	0.3639	0.04267
H_2O	0.5535	0.03049
O_2	0.1378	0.03183
N_2	0.1408	0.03913

Table 2.1. Van der Waals constants for some simple gases (based on Ref. 2.14).

The second virial coefficient for a van der Waals gas is easily found to be

$$B_{VW}(T) = \left(b - \frac{a}{RT}\right). \tag{2.13}$$

We see that $B_{VW}(T)$ will be negative at low temperatures and will become positive at high temperatures but does not exhibit the maximum observed in real gases. Thus, the van der Waals equation does not predict all the observed features of real gases. However, it describes enough of them to make it a worthwhile equation to study. In subsequent chapters, we will repeatedly use the van der Waals equation to study the thermodynamic properties of interacting fluids.

4. Solids

Solids have the property that their coefficient of thermal expansion, $\alpha = 1/V(\partial V/\partial T)_{P,N}$, and their isothermal compressibility, $\kappa_T = -1/V(\partial V/\partial P)_{T,N}$, are very small. Therefore, we can expand the volume of a solid in a Taylor series about its zero temperature, zero pressure value, V_0, and obtain the following equation of state:

$$V = V_0(1 + \alpha_P T - \kappa_T P), \tag{2.14}$$

where T is measured in degrees Kelvin. Typical values[4] of κ_T are of the order of 10^{-10}/Pa. or 10^{-5}/atm. For example, for solid Ag (silver) at room temperature, $\kappa_T = 1.3 \times 10^{-10}$/Pa. (for $P = 0$ Pa.), and for diamond at room temperature, $\kappa_T = 1.6 \times 10^{-10}$/Pa. (for $P = 4.0 \times 10^8$ Pa. $-$ 10^{10} Pa.). Typical values of α_P are of the order 10^{-4}/K. For example, for solid Na (sodium) at room temperature, $\alpha_P = 2 \times 10^{-4}$/K, and for solid K (potassium) $\alpha_P = 2 \times 10^{-4}$/K.

5. Stretched Wire

For a stretched wire in the elastic limit, Hook's law applies, and we can write

$$J = A(t)(L - L_0), \tag{2.15}$$

where J is the tension measured in Newtons per meter, $A(t)$ is a temperature-dependent force constant, L is the length of the stretched wire, and L_0 is the length of the wire when $J = 0$.

6. Surface Tension[5]

For pure liquids in equilibrium with their vapor phase, the surface tension between the liquid and vapor phases has an equation of state of the form

$$\sigma = \sigma_0\left(1 - \frac{t}{t'}\right)^n, \tag{2.16}$$

where t is temperature in degrees Celsius, σ_0 is the surface tension at $t = 0°\mathrm{C}$, t' is an experimentally determined temperature within a few degrees of the critical temperature, and n is an experimental constant which has a value between one and two.

7. Electric Polarization[5–7]

When an electric field $\mathbf{E}$ is applied to a dielectric material, the particles composing the dielectric will be distorted and an electric polarization field, $\mathbf{P}$ ($\mathbf{P}$ is the induced electric dipole moment per unit volume), will be set up by the material. The polarization is related to the electric field, $\mathbf{E}$, and the electric displacement, $\mathbf{D}$, by the equation

$$\mathbf{D} = \epsilon_0\mathbf{E} + \mathbf{P} \tag{2.17}$$

where ϵ_0 is the permittivity constant, $\epsilon_0 = 8.854 \times 10^{-12}\ \mathrm{C^2/N\cdot m^2}$. The electric field, $\mathbf{E}$, has units of Newtons per coulomb (N/C), and the electric displacement and electric polarization have units of coulomb per square meter ($\mathrm{C/m^2}$). $\mathbf{E}$ results from both external and surface charges. The magnitude of the polarization field, $\mathbf{P}$, will depend on the temperature. A typical equation of state for a homogeneous dielectric is

$$\mathbf{P} = \left(a + \frac{b}{T}\right)\mathbf{E} \tag{2.18}$$

for temperatures not too low. Here a and b are experimental constants and T is temperature in degrees Kelvin.

8. Curie's Law[5–7]

If we consider a paramagnetic solid at constant pressure, the volume changes very little as a function of temperature. We can then specify the state in terms of applied magnetic field and induced magnetization. When the external field is applied, the spins line up to produce a magnetization $\mathbf{M}$ (magnetic moment per unit volume). The magnetic induction field, $\mathbf{B}$ (measured in units of teslas,

1T = 1 Web/m²), the magnetic field strength, $\mathscr{H}$ (measured in units of ampere/meter), and the magnetization are related through the equation

$$\mathbf{B} = \mu_0 \mathscr{H} + \mu_0 \mathbf{M}, \tag{2.19}$$

where μ_0 is the permeability constant ($\mu_0 = 4\pi \times 10^{-7}$ T·m/A). The equation of state for such a system at room temperature is well approximated by Curie's law

$$\mathbf{M} = \frac{nD}{T} \mathscr{H}, \tag{2.20}$$

where n is the number of moles, D is an experimental constant dependent on the type of material used, and the temperature, T, is measured in degrees Kelvin.

D. LAWS OF THERMODYNAMICS[5]

Thermodynamics is based upon four laws. Before we can discuss these laws in a meaningful way, it is helpful to introduce some basic concepts.

A system is in *thermodynamic equilibrium* if the mechanical variables do not change in time and if there are no macroscopic flow processes present. Two systems are separated by an *insulating wall* (a wall that prevents transfer of matter between the systems) if the thermodynamic coordinates of one can be changed arbitrarily without causing changes in the thermodynamic coordinates of the other. Two systems are separated by a *conducting wall* if arbitrary changes in the coordinates of one cause changes in the coordinates of the other. An insulating wall prevents transfer of heat. A conducting wall allows transfer of heat.

It is useful to distinguish among three types of thermodynamic systems. An *isolated system* is one which is surrounded by an insulating wall, so that no heat or matter can be exchanged with the surrounding medium. A *closed system* is one which is surrounded by a conducting wall so that heat can be exchanged but matter cannot. An *open system* is one which allows both heat and matter exchange with the surrounding medium.

It is possible to change from one equilibrium state to another. Such changes can occur reversibly or irreversibly. A reversible change is one for which the system remains infinitesimally close to the thermodynamic equilibrium, that is, is performed quasi statically. Such changes can always be reversed and the system brought back to its original thermodynamic state without causing any changes in the thermodynamic state of the universe. For each step of a reversible process, the state variables have a well-defined meaning.

An irreversible or spontaneous change from one equilibrium state to another is one in which the system does not stay infinitesimally close to equilibrium during each step. Such changes often occur rapidly and give rise to flows and "friction" effects. After an irreversible change the system cannot be brought back to its original thermodynamic state without causing a change in the thermodynamic state of the universe. With these ideas in mind, we can now discuss the four laws of thermodynamics.

1. Zeroth Law: Two bodies, each in thermodynamic equilibrium with a third system, are in thermodynamic equilibrium with each other

The zeroth law is of fundamental importance to experimental thermodynamics because it enables us to introduce the concept of a thermometer and to measure temperatures of various systems in a reproducible manner. If we place a thermometer in contact with a given reference system, such as water at the triple point (where ice, water, and vapor coexist), then the mechanical variables describing the thermodynamic state of the thermometer (e.g., the height of a mercury column, the resistance of a resistor, or the pressure of a fixed volume container of gas) always take on the same values. If we then place the thermometer in contact with a third system and the mechanical variables do not change, then we say that the third system, the thermometer, and water at the triple point all have the same "temperature." Changes in the mechanical variables of the thermometer as it is cooled or heated are used as a measure of temperature change.

2. First Law: Energy is conserved

The first law tells us that there is a store of energy in the system, called the internal energy, U, which can be changed by causing the system to do mechanical work, $đW$, by adding heat energy, $đQ$, to the system, or by adding matter to the system (chemical work), $\mu' \, dN$. The change in the internal energy which results from these three processes is given by

$$dU = đQ - đW + \sum_j \mu'_j \, dN_j, \tag{2.21}$$

where the sum over j is over all types of particles in the system, and the work, $đW$, may be due to changes in any relevant extensive mechanical variable. In general,

$$đW = P\,dV - J\,dL - \sigma\,dA - \mathbf{E}\cdot d\mathbf{P} - \mathscr{H}\cdot d\mathbf{M}, \tag{2.22}$$

where dU, dV, dL, dA, $d\mathbf{P}$, $d\mathbf{M}$, and dN_j are exact differentials, but $đQ$ and $đW$ are not because they depend on the path taken (on the way in which heat is added or work done). We may think of P, J, σ, $\mathbf{E}$, $\mathscr{H}$, and μ'_j as generalized forces and dV, dL, dA, $d\mathbf{P}$, $d\mathbf{M}$, and dN_j as generalized displacements. Note that, if we have an isolated system (no heat or particle transfer), then any changes in internal energy will be equal to the mechanical work done on or by the system. This is a very useful fact because it allows us to measure differences in internal energy of two states of a given system by isolating the system and measuring the work needed to go from one state to another.

It is useful to introduce a generalized force, Y, which denotes any of the quantities, $-P, J, \sigma, \mathbf{E}, \mathscr{H}, \ldots$, and a generalized displacement, X, which denotes the corresponding displacements, $V, L, A, \mathbf{P}, \mathbf{M}, \ldots$, respectively. Then the first law can be written in the form

$$dU = đQ + Y\,dX + \sum_j \mu_j'\,dN_j. \tag{2.23}$$

Note that pressure has a different sign from the rest. If we increase the applied pressure the volume decreases, whereas if we increase the applied force, Y, for all other cases, the extensive variable X increases.

3. Second Law: Heat flows spontaneously from high temperatures to low temperatures

There are a number of ways to state the second law, the one given above being the simplest. Three alternative versions are[5]

(*a*) *The spontaneous tendency of a system to go toward thermodynamic equilibrium cannot be reversed without at the same time changing some organized energy, work, into disorganized energy, heat.*
(*b*) *In a cyclic process, it is not possible to convert heat from a hot reservoir into work without at the same time transferring some heat to a colder reservoir.*
(*c*) *The entropy change of any system and its surroundings, considered together, is positive and approaches zero for any process which approaches reversibility.*

The second law is of immense importance from many points of view. From it we can compute the maximum possible efficiency of an engine which transforms heat into work. It also enables us to introduce a new state variable, the entropy,

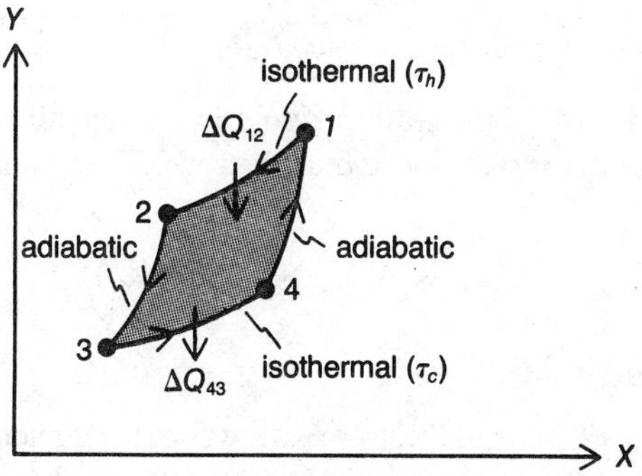

Fig. 2.3. A Carnot engine for arbitrary mechanical variables: processes $1 \to 2$ and $3 \to 4$ occur isothermally at temperatures τ_h and τ_c, respectively; processes $2 \to 3$ and $4 \to 1$ occur adiabatically; ΔQ_{12} is the heat absorbed and ΔQ_{43} is the heat emitted; the shaded area is equal to the work done during the cycle; the whole process takes place reversibly.

S, which is conjugate to the temperature. The entropy gives us a measure of the degree of disorder in a system, a means for determining the stability of equilibrium states and, in general, forms an important link between reversible and irreversible processes.

The second law is most easily discussed in terms of an ideal heat engine first introduced by Carnot. The construction of all heat engines is based on the observation that, if heat is allowed to flow from a high temperature to a lower temperature, part of the heat can be turned into work. Carnot observed that temperature differences can disappear spontaneously without producing work. Therefore, he proposed a very simple heat engine consisting only of reversible steps, thereby eliminating wasteful heat flows. The Carnot engine consists of four steps (cf. Fig. 2.3):

(*a*) Isothermal (constant temperature) absorption of heat ΔQ_{12} from a reservoir at a high temperature τ_h (we use Δ to indicate a finite rather than an infinitesimal amount of heat) (the process 1 to 2 in Fig. 2.3).
(*b*) Adiabatic (constant heat content) change in temperature to the lower value τ_c (the process 2 to 3 in Fig. 2.3).
(*c*) Isothermal expulsion of heat ΔQ_{43} into a reservoir at temperature τ_c (the process 3 to 4 in Fig. 2.3).
(*d*) Adiabatic return to the initial state at temperature τ_h (the process 4 to 1 in Fig. 2.3).

The work done by the engine during one complete cycle can be found by integrating the differential element of work $Y\,dX$ about the entire cycle. We see that the total work ΔW_{tot} done by the engine is given by the shaded area in Fig. 2.3.

The total efficiency η of the heat engine is given by the ratio of the work done to heat absorbed:

$$\eta = \frac{\Delta W_{tot}}{\Delta Q_{12}}. \tag{2.24}$$

Since the internal energy U is a state variable and independent of path, the total change ΔU_{tot} for one complete cycle must be zero. The first law then enables us to write

$$\Delta U_{tot} = \Delta Q_{tot} - \Delta W_{tot} = 0 \tag{2.25}$$

and thus

$$\Delta W_{tot} = \Delta Q_{tot} = \Delta Q_{12} + \Delta Q_{34} = \Delta Q_{12} - \Delta Q_{43}. \tag{2.26}$$

If we combine Eqs. (2.24) and (2.26) we can write the efficiency in the form

$$\eta = 1 - \frac{\Delta Q_{43}}{\Delta Q_{12}}. \tag{2.27}$$

A 100% efficient engine is one which converts all the heat it absorbs into work. However, as we shall see, no such engine can exist in nature.

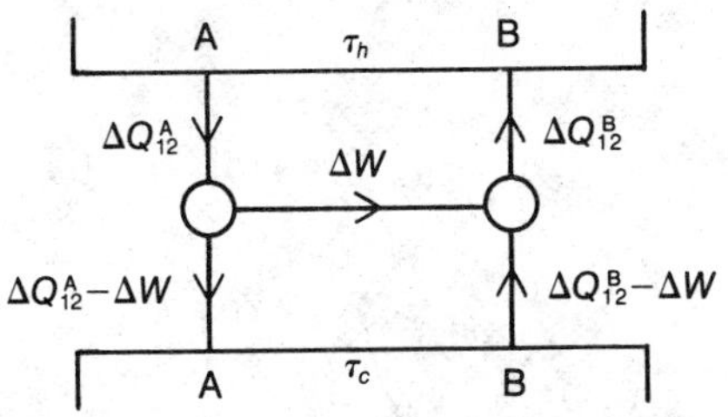

Fig. 2.4. Two heat engines working together: engine B is a Carnot engine acting as a heat pump; engine A is a heat engine with irreversible elements; engine A cannot have a greater efficiency than engine B without violating the second law.

The great beauty and utility of the Carnot engine lies in the fact that it is the most efficient of all heat engines operating between two heat reservoirs, each at a (different) fixed temperature. This is a consequence of the second law. To prove this let us consider two heat engines, A and B (cf. Fig. 2.4), which run between the same two reservoirs τ_h and τ_c. Let us assume that engine A is a heat engine with irreversible elements and B is a reversible Carnot engine. We will adjust the mechanical variables X and Y so that during one cycle both engines perform the same amount of work (note that X_A and Y_A need not be the same mechanical variables as X_B and Y_B):

$$\Delta W_{\text{tot}}^{A} = \Delta W_{\text{tot}}^{B} \equiv \Delta W. \tag{2.28}$$

Let us now assume that engine A is more efficient than engine B.

$$\eta_A > \eta_B \tag{2.29}$$

and thus

$$\frac{\Delta W}{\Delta Q_{12}^{A}} > \frac{\Delta W}{\Delta Q_{12}^{B}} \tag{2.30}$$

or

$$\Delta Q_{12}^{B} > \Delta Q_{12}^{A}. \tag{2.31}$$

We can use the work produced by engine A to drive the Carnot engine as a refrigerator. Since the Carnot engine is reversible, it will have the same efficiency whether it runs as a heat engine or as a heat pump. The work, ΔW, produced by A will be used to enable the Carnot engine B to pump heat from the low-temperature reservoir to the high-temperature reservoir. The net heat extracted from reservoir τ_c and delivered to reservoir τ_h is

$$\Delta Q_{12}^{B} - \Delta W - (\Delta Q_{12}^{A} - \Delta W) = \Delta Q_{12}^{B} - \Delta Q_{12}^{A}. \tag{2.32}$$

If engine A is more efficient than engine B, then the combined system has caused heat to flow from low temperature to high temperature without any work being expended by an outside source. This violates the second law and therefore engine A cannot be more efficient than the Carnot engine. If we now assume

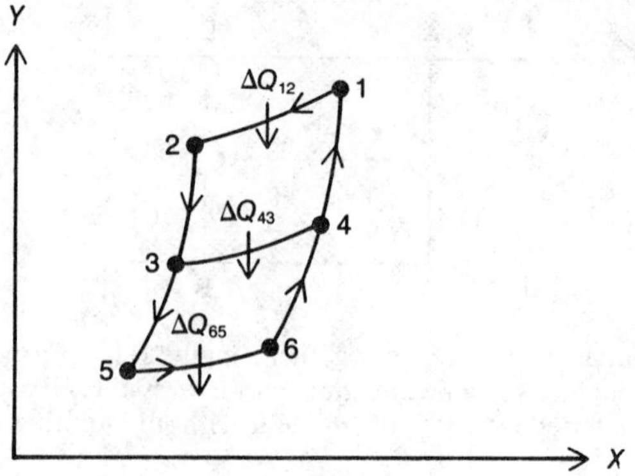

Fig. 2.5. Two Carnot engines running between three reservoirs: $\tau_h > \tau > \tau_c$ are equal to one Carnot engine running between reservoirs $\tau_h > \tau_c$.

that both engines are Carnot engines, we can show, by similar arguments, that they both must have the same efficiency. Thus, we reach the following conclusion: *No engine can be more efficient than a Carnot engine and all Carnot engines have the same efficiency.*

From the above discussion, we see that the efficiency of a Carnot engine is completely independent of the choice of mechanical variables X and Y and therefore can only depend on the temperatures τ_h and τ_c of the two reservoirs. This enables us to define an absolute temperature scale. From Eq. (2.27) we see that

$$\frac{\Delta Q_{43}}{\Delta Q_{12}} = f(\tau_h, \tau_c), \tag{2.33}$$

where $f(\tau_h, \tau_c)$ is some function of temperatures τ_h and τ_c. The function $f(\tau_h, \tau_c)$ has a very special form. Let us consider two heat engines running between three reservoirs $\tau_h > \tau' > \tau_c$ (cf. Fig. 2.5). We can write

$$\frac{\Delta Q_{43}}{\Delta Q_{12}} = f(\tau_h, \tau'), \tag{2.34}$$

$$\frac{\Delta Q_{65}}{\Delta Q_{43}} = f(\tau', \tau_c), \tag{2.35}$$

and

$$\frac{\Delta Q_{65}}{\Delta Q_{12}} = f(\tau_h, \tau_c), \tag{2.36}$$

so that

$$f(\tau_h, \tau_c) = f(\tau_h, \tau')f(\tau', \tau_c). \tag{2.37}$$

Thus, $f(\tau_h, \tau_c) = g(\tau_h)g^{-1}(\tau_c)$ where $g(\tau)$ is some function of temperature.

One of the first temperature scales proposed but not widely used is due to W. Thompson (Lord Kelvin) and is called the Thompson scale.[8] It has the form

$$\frac{\Delta Q_{43}}{\Delta Q_{12}} = \frac{e^{\tau_c{}^\circ}}{e^{\tau_h{}^\circ}}. \tag{2.38}$$

The Thompson scale is defined so that a given unit of heat ΔQ_{12} flowing between temperatures $\tau^\circ \rightarrow (\tau^\circ - 1)$ always produces the same amount of work, regardless of the value of τ°.

A more practical scale, the Kelvin scale, was also introduced by Thompson. It is defined as

$$\frac{\Delta Q_{43}}{\Delta Q_{12}} = \frac{\tau_c}{\tau_h}. \tag{2.39}$$

As we will see below, the Kelvin scale is identical to the temperature used in the ideal gas equation of state and is the temperature measured by a gas thermometer. For this reason, the Kelvin scale is the internationally accepted temperature scale at the present time.

As a specific example of a Carnot cycle, let us consider a heat engine composed of an ideal gas (or a very dilute real gas). The mechanical variables for an ideal gas are pressure $Y = -P$ and volume $X = V$. The equation of state for an ideal gas is $PV = nRT$ and the internal energy is $U = \frac{3}{2}nRT$, where temperature T is measured in degrees Kelvin. To find the efficiency of an ideal gas Carnot cycle we must compute the heat absorbed from the high-temperature reservoir and the heat expelled into the low-temperature reservoir. We assume the gas is in a container, one wall of which is a piston. Let us consider the four steps in the cycle (cf. Fig. 2.6):

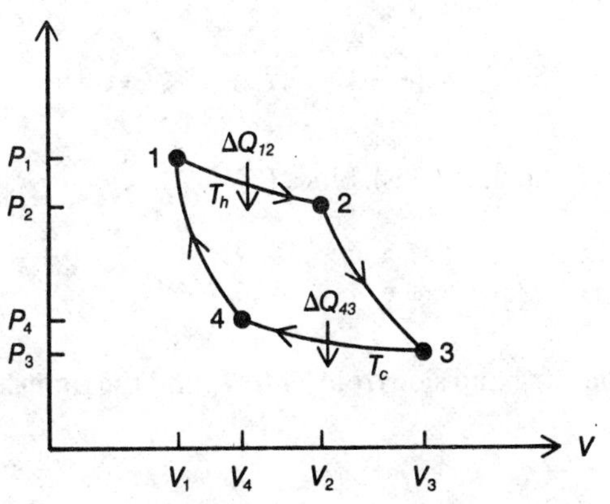

Fig. 2.6. A Carnot engine using an ideal gas as a running substance.

(*a*) *Isothermal Expansion.* Temperature is held fixed at $T = T_h$ and the volume is increased reversibly by allowing the gas to force the piston outward (we reduce the external force holding the piston in place). During such a process the temperature can only be kept fixed by adding heat. The heat absorbed can be found from the first law:

$$đQ = dU + đW. \tag{2.40}$$

If we choose V and T as independent variables and hold the number of particles fixed, we can write

$$đQ = \left(\frac{\partial U}{\partial T}\right)_{V,N} dT + \left(\frac{\partial U}{\partial V}\right)_{T,N} dV + P\,dV. \tag{2.41}$$

However, the temperature is held fixed at $T = T_h$ so that $dT = 0$. Furthermore, since $U = \frac{3}{2}nRT$, we find $(\partial U/\partial V)_{T,N} = 0$. Thus,

$$đQ = P\,dV = nRT_h \frac{dV}{V}. \tag{2.42}$$

We can now integrate from V_1 to V_2 to obtain

$$\Delta Q_{12} = \int_{V_1}^{V_2} nRT_h \frac{dV}{V} = nRT_h \ln \frac{V_2}{V_1} \tag{2.43}$$

for the heat absorbed in the first part of the cycle.

(*b*) *Adiabatic Expansion.* We now isolate the system and allow it to expand, thereby lowering the temperature. Since the process is adiabatic we have $\Delta Q_{23} = 0$. However, we still need a relation between P_2, V_2 and P_3, V_3. From the first law, we can write

$$đQ = \left(\frac{\partial U}{\partial T}\right)_{V,N} dT + P\,dV \equiv 0 \tag{2.44}$$

($(\partial U/\partial V)_{T,N} = 0$ for an ideal gas). Since $(\partial U/\partial T)_{V,N} = \frac{3}{2}nR$, we obtain

$$\frac{3}{2}\frac{dT}{T} = -\frac{dV}{V}. \tag{2.45}$$

If we integrate the left-hand side from T_h to T_c and the right-hand side from V_2 to V_3, we find

$$T_c V_3^{2/3} = T_h V_2^{2/3} \tag{2.46}$$

for an adiabatic process.

(*c*) *Isothermal Compression.* We now compress the gas by forcing the piston in, and we maintain the temperature at T_c. To do this we must draw heat out of the system. By a calculation similar to step (*a*) we find

$$\Delta Q_{43} = nRT_c \ln \frac{V_3}{V_4} = -\Delta Q_{34} \tag{2.47}$$

for the heat absorbed during the part C of the cycle. We see that ΔQ_{34} is negative and therefore heat is expelled from the system.

(*d*) *Adiabatic Compression.* Finally, we isolate the system and compress the gas so that it heats up and returns to its initial state. By a calculation similar to step (*b*) we obtain the relation

$$T_c V_4^{2/3} = T_h V_1^{2/3} \tag{2.48}$$

between the state variables at points 1 and 4.

If we now combine Eqs. (2.43) and (2.46)–(2.48), the efficiency of the Carnot cycle can be written

$$\eta = 1 - \frac{\Delta Q_{43}}{\Delta Q_{12}} = 1 - \frac{T_c \ln V_3/V_4}{T_h \ln V_2/V_1} = 1 - \frac{T_c}{T_h} \tag{2.49}$$

as we expected. We see then that the Kelvin temperature scale is the same as the ideal gas scale, and the efficiency of the Carnot engine depends only on the temperatures of the reservoirs.

The units of degrees Kelvin are the same as degrees Celsius. The ice point of water at atmospheric pressure is defined as 0°C and the boiling point as 100°C. The triple point of water is 0.01°C. To obtain a relation between degrees Kelvin and degrees Celsius we can measure pressure of a real *dilute* gas as a function of temperature at fixed volume. It is found experimentally that the pressure varies linearly with temperature and goes to zero at $t_c = -273.15$°C. Thus, from the ideal gas law, we see that degrees Kelvin, T, are related to degrees Celsius, t_c, by the equation

$$T = (t_c + 273.15). \tag{2.50}$$

The triple point of water is fixed at $T = 273.16$ K.

We can now use the Carnot cycle to define a new state variable called the entropy. Let us first note that from Eq. (2.49) we can write the following relation for a Carnot cycle:

$$\frac{\Delta Q_{12}}{T_h} + \frac{\Delta Q_{34}}{T_c} = 0. \tag{2.51}$$

Eq. (2.51) can be generalized to the case of an arbitrary *reversible* heat engine because we can consider such an engine as being composed of a sum of many infinitesimal Carnot cycles (cf. Fig. 2.7). Thus, for an arbitrary reversible heat engine we have

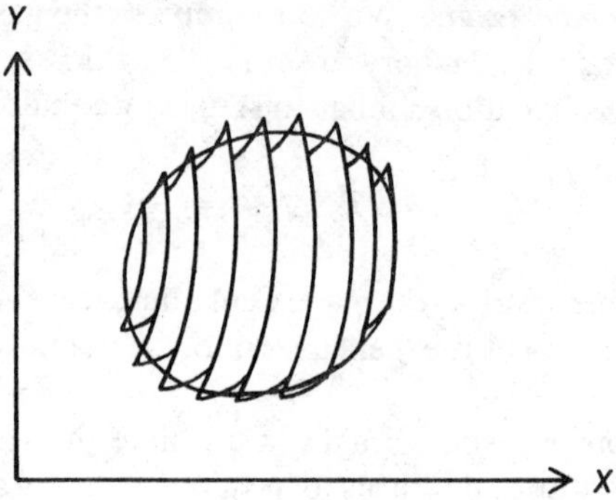

Fig. 2.7. An arbitrary reversible heat engine is composed of many infinitesimal Carnot engines; the area enclosed by the curve is equal to the work done by the heat engine.

$$\oint \frac{đQ}{T} = 0. \tag{2.52}$$

The quantity

$$dS \equiv \frac{đQ}{T} \tag{2.53}$$

is an exact differential and the quantity S, called the entropy, may be considered a new state variable since the integral of dS about a closed path gives zero.

No heat engine can be more efficient than a Carnot engine. Thus, an engine which runs between the same two reservoirs but contains spontaneous or irreversible processes in some part of the cycle will have a lower efficiency and we can write

$$\frac{\Delta Q_{43}}{\Delta Q_{12}} > \frac{T_c}{T_h} \tag{2.54}$$

or

$$\frac{\Delta Q_{12}}{T_h} - \frac{\Delta Q_{43}}{T_c} < 0. \tag{2.55}$$

For an arbitrary heat engine which contains an irreversible part, Eq. (2.55) gives the very important relation

$$\oint \frac{đQ}{T} < 0. \tag{2.56}$$

For an irreversible process, $đQ/T$ can no longer be considered an exact differential.

For any process, reversible or irreversible, the entropy change can be found by constructing a reversible path and computing the quantity $\Delta S = \int_{\text{rev}} dS \equiv \int_{\text{rev}} đQ/T$ because entropy change depends only on the end points (entropy is a state variable). However, for an irreversible process the integral $\int_{rev} đQ/T$ will be larger than the integral $\int_{irr} đQ/T$ evaluated over the actual irreversible path. Thus $\Delta S = \int_{rev} đQ/T > \int_{irr} đQ/T$. This result is usually written in the form

$$dS \geqslant \frac{đQ}{T}, \tag{2.57}$$

where the equality holds for a reversible process and the inequality holds for an irreversible process.

For an isolated system, $đQ \equiv 0$, we obtain the important relation

$$\Delta S \geqslant 0, \tag{2.58}$$

where the equality holds for a reversible process and the inequality holds for a spontaneous or irreversible process. Since the equilibrium state is, by definition, a state which is stable against spontaneous changes, Eq. (2.58) tells us that the equilibrium state is a state of maximum entropy. As we shall see, this fact gives an important criterion for determining the stability of the equilibrium state for an isolated system.

4. Third Law: The difference in entropy between states connected by a reversible process goes to zero in the limit $T \to 0$ K[8–10]

The third law was first proposed by Nernst in 1906 on the basis of experimental observations and is a consequence of quantum mechanics. Roughly speaking, a system at zero temperature drops into its lowest quantum state and in this sense becomes completely ordered. If entropy can be thought of as a measure of disorder, then at $T = 0$ K it must take its lowest value.

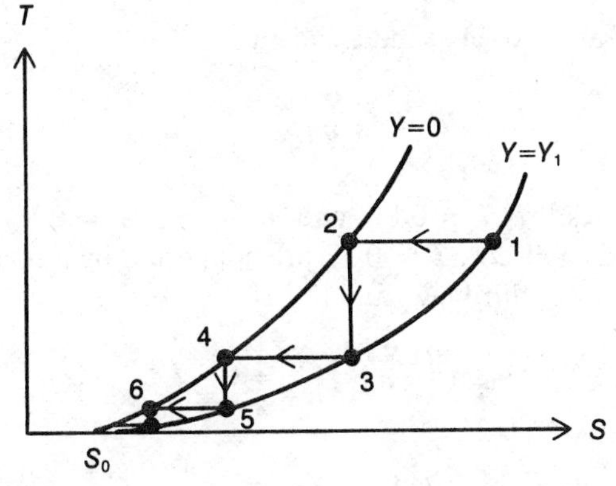

Fig. 2.8. The fact that the curves $Y = 0$ and $Y = Y_1$ must approach the same point (third law) makes it impossible to reach absolute zero by a finite number of reversible steps.

An alternative statement of the third law, and a direct consequence of the above statement, is, *It is impossible to reach absolute zero in a finite number of steps if a reversible process is used.* This alternative statement is easily demonstrated by means of a plot in the S–T plane. In Fig. 2.8 we have plotted the curves as a function of S and T for two states $Y = 0$ and $Y = Y_1$ for an arbitrary system. (A specific example might be a paramagnetic salt with $Y = H$.) We can cool the system by alternating between the two states, adiabatically and isothermally. From Eq. (2.6) we write

$$\left(\frac{\partial T}{\partial Y}\right)_S = -\left(\frac{\partial T}{\partial S}\right)_Y \left(\frac{\partial S}{\partial Y}\right)_T. \tag{2.59}$$

As we shall show in Sec. 2.I, thermal stability requires that $(\partial S/\partial T)_Y \geqslant 0$. Eq. (2.59) tells us that if T decreases as Y increases isentropically then S must decrease as Y decreases isothermally, as shown in Fig. 2.8. For the process $1 \to 2$ we change from state $Y = Y_1$ to state $Y = 0$ isothermally, thus squeezing out heat, and the entropy decreases. For process $2 \to 3$, we increase Y adiabatically from $Y = 0$ to $Y = Y_1$ and thus decrease the temperature. We can repeat these processes as many times as we wish. However, as we approach $T = 0$ K, we know by the third law that the two curves must approach the same point and must therefore begin to approach each other, thus making it impossible to reach $T = 0$ K in a finite number of steps.

Another consequence of the third law is that certain derivatives of the entropy must approach zero as $T \to 0$ K. Let us consider a process at $T = 0$ K such that $Y \to Y + dY$ and $X \to X + dX$. Then the change in entropy if Y, T, and N are chosen as independent variables is (assume $dN = 0$)

$$dS = \left(\frac{\partial S}{\partial Y}\right)_{N,T=0} dY, \tag{2.60}$$

or if X, T, and N are chosen as independent

$$dS = \left(\frac{\partial S}{\partial X}\right)_{N,T=0} dX. \tag{2.61}$$

Thus, if the states $(Y, T = 0\text{ K})$ and $(Y + dY, T = 0\text{ K})$ or the states $(X, T = 0\text{ K})$ and $(X + dX, T = 0\text{ K})$ are connected by a reversible process, we must have $dS = 0$ (third law) and therefore

$$\left(\frac{\partial S}{\partial Y}\right)_{N,T=0} = 0 \tag{2.62}$$

and

$$\left(\frac{\partial S}{\partial X}\right)_{N,T=0} = 0. \tag{2.63}$$

Eqs. (2.62) and (2.63) appear to be satisfied by real substances.

E. FUNDAMENTAL EQUATION OF THERMODYNAMICS[10]

The entropy plays a central role in both equilibrium and nonequilibrium thermodynamics. It can be thought of as a measure of the disorder in a system. As we shall see in Chap. 9, entropy is obtained microscopically by state counting. The entropy of an isolated system is proportional to the logarithm of the number of states available to the system. Thus, for example, a quantum system in a definite quantum state (pure state) has zero entropy. However, if the same system has finite probability of being in any of a number of quantum states, its entropy will be nonzero and may be quite large.

The entropy is an extensive, additive quantity. If a system is composed of a number of independent subsystems then the entropy of the whole system will be the sum of the entropies of the subsystems. This additive property of the entropy is expressed mathematically by the relation

$$S(\lambda U, \lambda X, \{\lambda N_i\}) = \lambda S(U, X, \{N_i\}). \tag{2.64}$$

That is, the entropy is a homogeneous first-order function of the extensive state variables of the system. If we increase *all* the extensive state variables by a factor λ, then the entropy must also increase by λ.

Differential changes in the entropy are related to differential changes in the extensive state variables through the combined first and second laws of thermodynamics:

$$T\,dS \geqslant dU - Y\,dX - \sum_j \mu_j'\,dN_j. \tag{2.65}$$

The equality holds if changes in the thermodynamic state are reversible. The inequality holds if they are spontaneous or irreversible. Eqs. (2.64) and (2.65) now enable us to define the fundamental equation of thermodynamics. Let us take the derivative of λS with respect to λ:

$$\begin{aligned}\frac{d}{d\lambda}(\lambda S) = S = \left(\frac{\partial S}{\partial \lambda U}\right)_{X,\{N_j\}} \frac{d}{d\lambda}(\lambda U) + \left(\frac{\partial S}{\partial \lambda X}\right)_{U,\{N_j\}} \frac{d}{d\lambda}(\lambda X) \\ + \sum_j \left(\frac{\partial S}{\partial \lambda N_j}\right)_{X,\{N_{i\neq j}\},U} \frac{d(\lambda N_j)}{d\lambda}.\end{aligned} \tag{2.66}$$

However, from Eq. (2.65) we see that

$$\left(\frac{\partial S}{\partial U}\right)_{X,\{N_j\}} = \frac{1}{T}, \tag{2.67a}$$

$$\left(\frac{\partial S}{\partial X}\right)_{U,\{N_j\}} = -\frac{Y}{T}, \tag{2.67b}$$

and

$$\left(\frac{\partial S}{\partial N_j}\right)_{U,X,\{N_{i\neq j}\}} = -\frac{\mu_j'}{T}. \tag{2.67c}$$

Thus, Eq. (2.66) becomes

Fundamental equation (Eulers equation)

$$S = \frac{U}{T} - \frac{YX}{T} - \sum_j \frac{\mu_j' N_j}{T}. \tag{2.68}$$

Eqs. (2.67*a*)–(2.67*c*) are called equations of state. Eq. (2.68) is called the fundamental equation (or Eulers equation) of thermodynamics because it contains all possible thermodynamic information about a thermodynamic system. Eqs. (2.64), (2.67), and (2.68) show that the internal energy, like the entropy, must be a first-order homogeneous function of the extensive state variables, while the intensive state variables, T, Y, and μ_j', are zeroth-order homogeneous functions of the extensive state variables. That is, the parameters T, Y, and μ_j' can only depend on densities.

If we take the differential of Eq. (2.68) and subtract Eq. (2.65) (we shall consider the reversible case), then we obtain another important equation:

$$0 = S\,dT + X\,dY + \sum_j N_j\,d\mu_j', \tag{2.69}$$

called the Gibbs-Duheim equation, which relates differentials of intensive state variables. For a system containing only one kind of particle the Gibbs-Duheim equation takes the form

$$d\mu' = -\frac{S}{N}\,dT - \frac{X}{N}\,dY. \tag{2.70}$$

It is often useful to obtain the chemical potential μ' by integrating the Gibbs-Duheim equation.

F. THERMODYNAMIC POTENTIALS[9]

In conservative mechanical systems, such as a spring or a mass raised in a gravitational field, work can be stored in the form of potential energy and subsequently retrieved. *Under certain circumstances* the same is true for thermodynamic systems. We can store energy in a thermodynamic system by doing work on it through a reversible process, and we can eventually retrieve that energy in the form of work. The energy which is stored and retrievable in the form of work is called the *free energy*. There are as many different forms of free energy in a thermodynamic system as there are combinations of constraints. In this section, we shall discuss the five most common ones: internal energy, U; the enthalpy, H; the Helmholtz free energy, A; the Gibbs free energy, G; and the grand potential, Ω. These quantities play a role analogous to that of the potential energy in a spring and for that reason are also called the thermodynamic potentials.

1. Internal Energy

From Eq. (2.68) the fundamental equation for the internal energy can be written

$$U = ST + YX + \sum_j \mu_j' N_j, \tag{2.71}$$

where T, Y, and μ_j' are considered to be functions of S, X, and $\{N_j\}$ (cf. Eqs. (2.67*a*–*c*)); from Eq. (2.65) the differential form of the internal energy can be written

$$dU \leqslant T\,dS + Y\,dX + \sum_j \mu_j'\,dN_j. \tag{2.72}$$

The equality holds for reversible changes and the inequality holds for changes which are spontaneous. From Eq. (2.72) we see that

$$T = \left(\frac{\partial U}{\partial S}\right)_{X,\{N_j\}}, \tag{2.73a}$$

$$Y = \left(\frac{\partial U}{\partial X}\right)_{S,\{N_j\}}, \tag{2.73b}$$

and

$$\mu_j' = \left(\frac{\partial U}{\partial N_j}\right)_{S,X,\{N_{l\neq j}\}}. \tag{2.73c}$$

We can use the fact that dU is an exact differential to find relations between derivatives of the intensive variables T, Y, and μ_j'. From Eq. (2.4) we know, for example, that

$$\left[\frac{\partial}{\partial X}\left(\frac{\partial U}{\partial S}\right)_{X,\{N_j\}}\right]_{S,\{N_j\}} = \left[\frac{\partial}{\partial S}\left(\frac{\partial U}{\partial X}\right)_{S,\{N_j\}}\right]_{X,\{N_j\}}. \tag{2.74}$$

Comparing Eqs. (2.73) and (2.74) we obtain

$$\left(\frac{\partial T}{\partial X}\right)_{S,\{N_j\}} = \left(\frac{\partial Y}{\partial S}\right)_{X,\{N_j\}}. \tag{2.75a}$$

$(i + 1)$ more relations like Eq. (2.75*a*) exist and lead to the identities

$$\left(\frac{\partial T}{\partial N_j}\right)_{S,X,\{N_{l\neq j}\}} = \left(\frac{\partial \mu_j'}{\partial S}\right)_{X,\{N_j\}}, \tag{2.75b}$$

$$\left(\frac{\partial Y}{\partial N_j}\right)_{S,X,\{N_{l\neq j}\}} = \left(\frac{\partial \mu_j'}{\partial X}\right)_{S,\{N_j\}}, \tag{2.75c}$$

and

$$\left(\frac{\partial \mu_j'}{\partial N_i}\right)_{S,X,\{N_{l\neq i}\}} = \left(\frac{\partial \mu_i'}{\partial N_j}\right)_{S,X,\{N_{l\neq j}\}}. \tag{2.75d}$$

Eqs. (2.75*a*)–(2.75*d*) are extremely important both theoretically and experimentally because they provide a relation between rates of change of seemingly diverse quantities. They are called Maxwell relations.

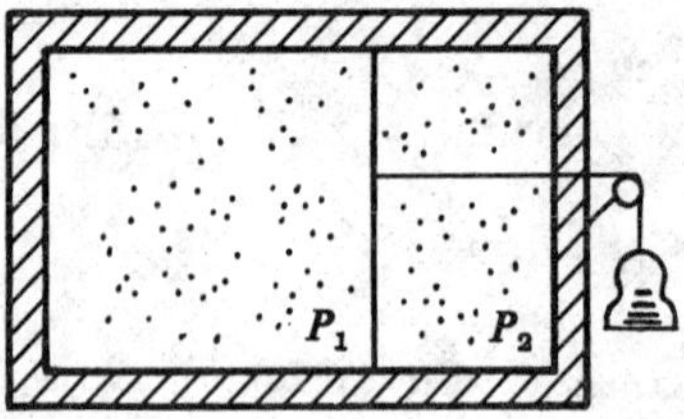

Fig. 2.9. For a reversible process in a closed, insulated box of fixed size ($\Delta S = 0, \Delta V = 0, \Delta N = 0$), the work done in lifting the weight will be equal to the change in the internal energy, $(\Delta U)_{S,V,N} = -\Delta W_{\text{free}}$.

The internal energy is a thermodynamic potential or free energy because, for processes carried out reversibly in an isolated, closed system at fixed X and $\{N_j\}$, the change in internal energy is equal to the maximum amount of work that can be done on or by the system. As a specific example let us consider a PVT system (cf. Fig. 2.9). We shall enclose a gas in an insulated box with fixed total volume and divide it into two parts by a movable conducting wall. We can do work on the gas or have the gas do work by attaching a mass in a gravitational field to the partition via a pulley and insulated string. To do work reversibly, we assume that the mass is composed of infinitesimal pieces which can be added or removed one by one. If $P_1A + mg > P_2A$ then work is done on the gas by the mass, and if $P_1A + mg < P_2A$ the gas does work on the mass. The first law can be written

$$\Delta U = \Delta Q - \Delta W, \tag{2.76}$$

where ΔU is the change in total internal energy of the gas, ΔQ is the heat flow through the walls, and ΔW can be divided into work done due to change in size of the box, $\int P\,dV$, and work done by the gas in raising the weight, ΔW_{free}:

$$\Delta W = \int P\,dV + \Delta W_{\text{free}}. \tag{2.77}$$

For a reversible process, $\Delta Q = \int T\,dS$. For the reversible process pictured in Fig. 2.9, $\Delta Q = 0, \Delta V = 0$, and $\Delta N_j = 0$ (if no chemical reactions take place). Therefore,

$$\begin{pmatrix}\text{Reversible}\\ \text{process}\end{pmatrix} \qquad (\Delta U)_{\{S,V,N_j\}} = -\Delta W_{\text{free}}. \tag{2.78}$$

Thus, for a reversible process at constant S, V, and N_j, work can be stored in the form of internal energy and can be recovered completely. Under these conditions, internal energy behaves like a potential energy.

For a spontaneous process work can only be done at constant S, V, and $\{N_i\}$

if we allow heat to leak through the walls. The first and second laws for a spontaneous process take the form (cf. Eq. (2.58))

$$\int dU = \Delta U < \int T\,dS - \int P\,dV - \Delta W_{\text{free}} + \sum_j \int \mu_j'\,dN_j, \tag{2.79}$$

where the integrals are taken over a reversible path between initial and final states and not the actual spontaneous path. We can do work on the gas spontaneously by allowing the mass to drop very fast. Then part of the work goes into stirring up the gas. In order for the process to occur at constant entropy, some heat must leak out since $\Delta Q < \int T\,dS = 0$. Thus, for a spontaneous process

$$\text{(Spontaneous)} \qquad (\Delta U)_{S,V,\{N_j\}} < -\Delta W_{\text{free}}. \tag{2.80}$$

Not all work is changed to internal energy and is retrievable. Some is wasted in stirring the gas. (Note that for this process the entropy of the universe has increased since heat has been added to the surrounding medium.)

For processes involving mechanical variables Y and X we can write Eqs. (2.78) and (2.80) in the form

$$(\Delta U)_{S,X,\{N_j\}} \leqslant (-\Delta W_{\text{free}}), \tag{2.81}$$

where ΔW_{free} is any work done by the system other than that required to change X. *For a reversible process at constant* S, X, *and* $\{N_j\}$, *work can be stored as internal energy and can be recovered completely.*

If a process takes place in which no work is done on or by the system, then Eq. (2.81) becomes

$$(\Delta U)_{S,X,\{N_j\}} \leqslant 0 \tag{2.82}$$

and the internal energy either does not change (reversible process) or decreases (spontaneous process). Since a system in equilibrium cannot change its state spontaneously, we see that *an equilibrium state at fixed* S, X, *and* $\{N_j\}$ *is a state of minimum internal energy.*

2. Enthalpy

The internal energy is the convenient potential to use for processes carried out at constant X, S, and $\{N_j\}$. However, it often happens that we wish to study the thermodynamics of processes which occur at constant S, Y, and $\{N_j\}$. Then it is more convenient to use the enthalpy.

The enthalpy, H, is useful for systems which are thermally isolated and closed but mechanically coupled to the outside world. It is obtained by adding to the internal energy an additional energy due to the mechanical coupling:

$$H \equiv U - XY = ST + \sum_j \mu_j' N_j. \tag{2.83}$$

The addition of the term $-XY$ has the effect of changing the independent variables from (S, X, N_j) to (S, Y, N_j) and is called a Legendre transformation. If we take the differential of Eq. (2.83) and combine it with Eq. (2.72), we obtain

$$dH \leqslant T\,dS - X\,dY + \sum_j \mu_j'\,dN_j \tag{2.84}$$

and, therefore,

$$T = \left(\frac{\partial H}{\partial S}\right)_{Y,\{N_j\}}, \tag{2.85a}$$

$$X = -\left(\frac{\partial H}{\partial Y}\right)_{S,\{N_j\}}, \tag{2.85b}$$

and

$$\mu_j' = \left(\frac{\partial H}{\partial N_j}\right)_{S,Y,\{N_{l\neq j}\}}. \tag{2.85c}$$

Since dH is an exact differential, we can use Eq. (2.4) to obtain a new set of Maxwell relations:

$$\left(\frac{\partial T}{\partial Y}\right)_{S,\{N_j\}} = -\left(\frac{\partial X}{\partial S}\right)_{Y,\{N_j\}}, \tag{2.86a}$$

$$\left(\frac{\partial T}{\partial N_j}\right)_{S,Y,\{N_{l\neq j}\}} = \left(\frac{\partial \mu'}{\partial S}\right)_{Y,\{N_j\}}, \tag{2.86b}$$

$$\left(\frac{\partial X}{\partial N_j}\right)_{S,Y,\{N_{l\neq j}\}} = -\left(\frac{\partial \mu_j'}{\partial Y}\right)_{S,\{N_j\}}, \tag{2.86c}$$

and

$$\left(\frac{\partial \mu_j'}{\partial N_i}\right)_{S,Y,\{N_{l\neq i}\}} = \left(\frac{\partial \mu_i'}{\partial N_j}\right)_{S,Y,\{N_{l\neq j}\}}, \tag{2.86d}$$

which relate seemingly diverse partial derivatives.

For a YXT system, the enthalpy is a thermodynamic potential for reversible processes carried out at constant Y. The discussion for the enthalpy is completely analogous to that for the internal energy except that now we allow the extensive variable X to change and maintain the system at constant Y. We then find

$$\Delta H \leqslant \int T\,dS - \int X\,dY - \Delta W_{\text{free}} + \sum_j \int \mu_j'\,dN_j, \tag{2.87}$$

where the equality holds for a reversible process and the inequality holds for a spontaneous process (ΔW_{free} is defined in Sec. 2.F.1). Therefore,

$$(\Delta H)_{S,Y,\{N_j\}} \leqslant (-\Delta W_{\text{free}}) \tag{2.88}$$

and we conclude that, *for a reversible process at constant* S, Y, *and* $\{N_j\}$, *work can be stored as enthalpy and can be recovered completely.*

If a process takes place at constant S, Y, and $\{N_j\}$ in which no work is done on or by the system, then

$$(\Delta H)_{S,Y,\{N_j\}} \leqslant 0. \tag{2.89}$$

Since the equilibrium state cannot change spontaneously, we find that *the equilibrium state at fixed* S, Y, *and* $\{N_j\}$ *is a state of minimum enthalpy.*

3. Helmholtz Free Energy

For processes carried out at constant T, X, and $\{N_j\}$, the Helmholtz free energy corresponds to a thermodynamic potential. The Helmholtz free energy, A, is useful for systems which are closed and thermally coupled to the outside world but are mechanically isolated (held at constant X). We obtain the Helmholtz free energy from the internal energy by adding a term due to the thermal coupling:

$$A = U - ST = YX + \sum_j \mu'_j N_j. \tag{2.90}$$

The addition of ST is a Legendre transformation which changes the independent variables from $(S, X, \{N_j\})$ to $(T, X, \{N_j\})$. If we take the differential of Eq. (2.90) and use Eq. (2.72), we find

$$dA \leqslant -S\,dT + Y\,dX + \sum_j \mu'_j\,dN_j. \tag{2.91}$$

Therefore,

$$S = -\left(\frac{\partial A}{\partial T}\right)_{X,\{N_j\}}, \tag{2.92a}$$

$$Y = \left(\frac{\partial A}{\partial X}\right)_{T,\{N_j\}}, \tag{2.92b}$$

and

$$\mu'_j = \left(\frac{\partial A}{\partial N_j}\right)_{T,X,\{N_{i\neq j}\}}. \tag{2.92c}$$

Again, from Eq. (2.4), we obtain Maxwell relations

$$\left(\frac{\partial S}{\partial X}\right)_{T,\{N_j\}} = -\left(\frac{\partial Y}{\partial T}\right)_{X,\{N_j\}}, \tag{2.93a}$$

$$\left(\frac{\partial S}{\partial N_j}\right)_{T,X,\{N_{i\neq j}\}} = -\left(\frac{\partial \mu'_j}{\partial T}\right)_{X,\{N_j\}}, \tag{2.93b}$$

$$\left(\frac{\partial Y}{\partial N_j}\right)_{T,X,\{N_{i\neq j}\}} = \left(\frac{\partial \mu'_j}{\partial X}\right)_{T,\{N_j\}}, \tag{2.93c}$$

and

$$\left(\frac{\partial \mu'_j}{\partial N_i}\right)_{T,X,\{N_{l\neq i}\}} = \left(\frac{\partial \mu'_i}{\partial N_j}\right)_{T,X,\{N_{l\neq j}\}} \tag{2.93d}$$

for the system.

For a YXT system, the Helmholtz free energy is a thermodynamic potential for reversible processes carried out at constant T, X, and $\{N_j\}$. For a change in the thermodynamic state of the system, the change in the Helmholtz free energy can be written

$$\Delta A \leqslant -\int S\,dT + \int Y\,dX - \Delta W_{\text{free}} + \sum_j \int \mu'_j\,dN_j, \tag{2.94}$$

where the inequality holds for spontaneous processes and the equality holds for reversible processes (ΔW_{free} is defined in Sec. 2.F.1). For a process carried out at fixed T, X, and $\{N_j\}$, we find

$$(\Delta A)_{T,X,\{N_j\}} \leqslant (-\Delta W_{\text{free}}), \tag{2.95}$$

and we conclude that *for a reversible process at constant* T, X, *and* $\{N_j\}$ *work can be stored as Helmholtz free energy and can be recovered completely.*

If no work is done for a process occurring at fixed T, X, and $\{N_j\}$, Eq. (2.95) becomes

$$(\Delta A)_{T,X,\{N_j\}} \leqslant 0. \tag{2.96}$$

Thus, *an equilibrium state at fixed* T, X, *and* $\{N_j\}$ *is a state of minimum Helmholtz free energy.*

4. Gibbs Free Energy

For processes carried out at constant Y, T, and $\{N_j\}$ the Gibbs free energy corresponds to the thermodynamic potential. Such a process is coupled both thermally and mechanically to the outside world. We obtain the Gibbs free energy, G, from the internal energy by adding terms due to the thermal and mechanical coupling,

$$G = U - TS - XY = \sum_j \mu'_j N_j. \tag{2.97}$$

In this way we change from independent variables (S, X, $\{N_j\}$) to variables (T, Y, $\{N_j\}$). If we add the differential of Eq. (2.97) to Eq. (2.72), we obtain

$$dG \leqslant -S\,dT - X\,dY + \sum_j \mu'_j\,dN_j, \tag{2.98}$$

so that

$$S = -\left(\frac{\partial G}{\partial T}\right)_{Y,\{N_j\}}, \tag{2.99a}$$

$$X = -\left(\frac{\partial G}{\partial Y}\right)_{T,\{N_j\}}, \tag{2.99b}$$

and

$$\mu_j' = \left(\frac{\partial G}{\partial N_j}\right)_{T,Y,\{N_{i\neq j}\}}. \tag{2.99c}$$

The Maxwell relations obtained from the Gibbs free energy are

$$\left(\frac{\partial S}{\partial Y}\right)_{T,\{N_j\}} = \left(\frac{\partial X}{\partial T}\right)_{Y,\{N_j\}}, \tag{2.100a}$$

$$\left(\frac{\partial S}{\partial N_j}\right)_{T,Y,\{N_{i\neq j}\}} = -\left(\frac{\partial \mu_j'}{\partial T}\right)_{Y,\{N_i\}}, \tag{2.100b}$$

$$\left(\frac{\partial X}{\partial N_j}\right)_{T,Y,\{N_{i\neq j}\}} = -\left(\frac{\partial \mu_j'}{\partial Y}\right)_{T,\{N_i\}}, \tag{2.100c}$$

and

$$\left(\frac{\partial \mu_j'}{\partial N_i}\right)_{T,Y,\{N_{l\neq i}\}} = \left(\frac{\partial \mu_i'}{\partial N_j}\right)_{T,Y,\{N_{l\neq j}\}} \tag{2.100d}$$

and again relate seemingly diverse partial derivatives.

For a YXT system, the Gibbs free energy is a thermodynamic potential for reversible processes carried out at constant T, Y, and $\{N_j\}$. For a change in the thermodynamic state of the system, the change in Gibbs free energy can be written

$$\Delta G \leqslant -\int S\,dT - \int X\,dY - \Delta W_{\text{free}} + \sum_j \int \mu_j'\,dN_j, \tag{2.101}$$

where the equality holds for reversible processes and the inequality holds for spontaneous processes (ΔW_{free} is defined in Sec. 2.F.1). For processes at fixed T, Y, and $\{N_j\}$,

$$(\Delta G)_{T,Y,\{N_j\}} \leqslant (-\Delta W_{\text{free}}). \tag{2.102}$$

Thus, *for a reversible process at constant* T, Y, *and* $\{N_j\}$, *work can be stored as Gibbs free energy and can be recovered completely.*

For a process at fixed T, Y, and $\{N_j\}$ for which no work is done,

$$(\Delta G)_{T,Y,\{N_j\}} \leqslant 0, \tag{2.103}$$

and we conclude that *an equilibrium state at fixed* T, Y, *and* $\{N_j\}$ *is a state of minimum Gibbs free energy.*

5. Grand Potential

A potential which is extremely useful for the study of quantum systems is the grand potential. It is a thermodynamic potential energy for processes carried out in open systems where particle number can vary but T, X, and $\{\mu_j'\}$ are kept fixed.

The grand potential, Ω, can be obtained from the internal energy by adding terms due to thermal and chemical coupling of the system to the outside world:

$$\Omega = U - TS - \sum_j \mu_j' N_j = XY. \tag{2.104}$$

The Legendre transformation in Eq. (2.104) changes the independent variables from $(S, X, \{N_i\})$ to $(T, X, \{\mu_j'\})$. If we add the differential of Eq. (2.104) to Eq. (2.72), we obtain

$$d\Omega \leqslant -S\,dT + Y\,dX - \sum_j N_j\,d\mu_j', \tag{2.105}$$

and thus

$$S = -\left(\frac{\partial \Omega}{\partial T}\right)_{X,\{\mu_j'\}}, \tag{2.106a}$$

$$Y = \left(\frac{\partial \Omega}{\partial X}\right)_{T,\{\mu_j'\}}, \tag{2.106b}$$

and

$$N_j = -\left(\frac{\partial \Omega}{\partial \mu_j'}\right)_{T,X,\{\mu_l' \neq j\}}. \tag{2.106c}$$

The Maxwell relations obtained from the grand potential are

$$\left(\frac{\partial S}{\partial X}\right)_{T,\{\mu_j'\}} = -\left(\frac{\partial Y}{\partial T}\right)_{X,\{\mu_j'\}}, \tag{2.107a}$$

$$\left(\frac{\partial S}{\partial \mu_j'}\right)_{T,X,\{\mu_l' \neq j\}} = -\left(\frac{\partial N_j}{\partial T}\right)_{X,\{\mu_j'\}}, \tag{2.107b}$$

$$\left(\frac{\partial Y}{\partial \mu_j'}\right)_{T,X,\{\mu_l' \neq j\}} = \left(\frac{\partial N_j}{\partial X}\right)_{T,\{\mu_j'\}}, \tag{2.107c}$$

and

$$\left(\frac{\partial N_i}{\partial \mu_j'}\right)_{T,V,\{\mu_l' \neq j\}} = \left(\frac{\partial N_j}{\partial \mu_i'}\right)_{T,V,\{\mu_l' \neq i\}} \tag{2.107d}$$

and are very useful in treating open systems. The grand potential is a thermodynamic potential energy for a reversible process carried out at constant T, X, and $\{\mu_j'\}$ (variable particle number). For a change in the thermodynamic state of the system, the change in the grand potential can be written

$$\Delta\Omega \leqslant -\int S\,dT + \int Y\,dX - \Delta W_{\text{free}} - \sum_j \int N_j\,d\mu_j', \tag{2.108}$$

where the equality holds for reversible changes and the inequality holds for spontaneous changes (ΔW_{free} is defined in Sec. 2.F.1). For a process at fixed T, X, and $\{\mu_j'\}$,

$$(\Delta\Omega)_{T,X,\{\mu_j\}} \leqslant (-\Delta W_{\text{free}}). \tag{2.109}$$

Thus, *for a reversible process at constant* T, X, *and* $\{\mu_j'\}$, *work can be stored as grand potential and can be recovered completely.*

For a process at fixed T, X, and $\{\mu_j'\}$ for which no work is done,

$$(\Delta\Omega)_{T,X,\{\mu_j'\}} \leqslant 0, \tag{2.110}$$

and we find that *an equilibrium state at fixed* T, X, *and* $\{\mu_j'\}$ *is a state of minimum grand potential.*

6. Thermodynamic Potential Densities

For *PVT* systems it is often useful to write the thermodynamic relations obtained in Secs. 2.F.1–2.F.5 in terms of densities. Let us denote the total mass of the system by

$$M = \sum_j m_j N_j = \sum_j M_j, \tag{2.111}$$

where m_j is the mass of one of the constituents of type j and M_j is the total mass of constituent j. The chemical potential per unit mass of constituent j is $\mu_j = \mu_j'/m_j$, and the internal energy, entropy, and volume per unit mass can be written $u = U/M$, $s = S/M$, and $v = V/M$, respectively. The fundamental equation in terms of these densities becomes

$$u = Ts - Pv + \sum_j \mu_j c_j, \tag{2.112}$$

where c_j is the concentration of the jth constituent, $c_j = M_j/M$.

The differential form of the first and second laws can be written in terms of densities in the following way. First write each of the differentials appearing in Eq. (2.72) in terms of the corresponding densities: that is, $du = (1/M)\,dU + U\,d(1/M)$ and substitute into Eq. (2.72). Then make use of Eq. (2.71) to set the coefficient of $d(1/M)$ to zero. The result is

$$du \leqslant T\,ds - P\,dv + \sum_j \mu_j\,dc_j. \tag{2.113}$$

Differential forms for densities of the enthalpy, Helmholtz free energy, Gibbs free energy and grand potential are easily obtained from Eqs. (2.112) and (2.113) and the fundamental equations for the various potentials.

We have now discussed the five thermodynamic potentials which are most commonly used. However, it is clear that we can always construct more thermodynamic potentials by varying combinations of parameters other than those considered here.

G. RESPONSE FUNCTIONS

The response functions are the thermodynamic quantities most accessible to experiment. They give us information about how a specific state variable changes as other independent state variables are changed under controlled conditions. The response functions can be divided into thermal response functions, such as the heat capacities, and mechanical response functions, such as compressibility and susceptibility. We shall introduce some of the more common response functions in this section.

1. Heat Capacity

The heat capacity, C, is a measure of the amount of heat needed to raise the temperature of a system by a given amount. In general, it is defined as the derivative $C = đQ/dT$. When we measure the heat capacity, we try to fix all independent variables except the temperature. Thus, there are as many different heat capacities as there are combinations of independent variables and they each contain different information about the system. We shall derive the heat capacity at constant X and N, $C_{X,N}$, and heat capacity at constant Y and N, $C_{Y,N}$, in two different ways: first from the first law and then from the definition of the entropy.

To obtain an expression for $C_{X,N}$, we shall assume that X and T are independent variables and hold particle numbers $\{N_j\}$ fixed (for simplicity, we shall denote $\{N_j\}$ as N). Then $dN_j = 0$ and the first law becomes

$$đQ = dU - Y\,dX = \left(\frac{\partial U}{\partial T}\right)_{X,N} dT + \left(\frac{\partial U}{\partial X}\right)_{T,N} dX - Y\,dX. \quad (2.114)$$

For constant X and N, $đQ = (\partial U/\partial T)_{X,N}\,dT$ and we find

$$C_{X,N} = \left(\frac{\partial U}{\partial T}\right)_{X,N} \quad (2.115)$$

for the heat capacity at constant X and N.

To obtain an expression for $C_{Y,N}$, we shall assume that Y and T are independent variables and hold particle numbers $\{N_j\}$ fixed. Then

$$dX = \left(\frac{\partial X}{\partial T}\right)_{Y,N} dT + \left(\frac{\partial X}{\partial Y}\right)_{T,N} dY. \quad (2.116)$$

If we substitute for dX in Eq. (2.114), we obtain

$$đQ = \left\{C_{X,N} + \left[\left(\frac{\partial U}{\partial X}\right)_{T,N} - Y\right]\left(\frac{\partial X}{\partial T}\right)_{Y,N}\right\} dT + \left[\left(\frac{\partial U}{\partial X}\right)_{T,N} - Y\right]\left(\frac{\partial X}{\partial Y}\right)_{T,N} dY. \quad (2.117)$$

For constant Y and N, $đQ = C_{Y,N}\,dT$ and we find

$$C_{Y,N} = C_{X,N} + \left[\left(\frac{\partial U}{\partial X}\right)_{T,N} - Y\right]\left(\frac{\partial X}{\partial T}\right)_{Y,N} \quad (2.118)$$

for the heat capacity at constant Y and N.

It is now useful to rederive expressions for $C_{X,N}$ and $C_{Y,N}$ from the entropy. Let us assume the T and X are independent and that $\{N_j\}$ are held fixed. Then, for a reversible process,

$$đQ = T\,dS = T\left(\frac{\partial S}{\partial T}\right)_{X,N} dT + T\left(\frac{\partial S}{\partial X}\right)_{T,N} dX. \tag{2.119}$$

For a change in state at constant X and $\{N_j\}$, Eq. (2.119) becomes

$$đQ = T\left(\frac{\partial S}{\partial T}\right)_{X,N} dT \tag{2.120}$$

and therefore

$$C_{X,N} = T\left(\frac{\partial S}{\partial T}\right)_{X,N} = -T\left(\frac{\partial^2 A}{\partial T^2}\right)_{X,N}. \tag{2.121}$$

The second term comes from Eq. (2.92*a*).

Let us now assume that T and Y are independent and that particle numbers $\{N_j\}$ are held fixed. Then for a reversible process we can write

$$đQ = T\,dS = T\left(\frac{\partial S}{\partial T}\right)_{Y,N} dT + T\left(\frac{\partial S}{\partial Y}\right)_{T,N} dY, \tag{2.122}$$

and from Eq. (2.119), we make a change of variables to write

$$đQ = \left[T\left(\frac{\partial S}{\partial T}\right)_{X,N} + T\left(\frac{\partial S}{\partial X}\right)_{T,N}\left(\frac{\partial X}{\partial T}\right)_{Y,N}\right] dT + T\left(\frac{\partial S}{\partial X}\right)_{T,N}\left(\frac{\partial X}{\partial Y}\right)_{T,N} dY. \tag{2.123}$$

Comparing Eqs. (2.122) and (2.123), we find

$$C_{Y,N} = T\left(\frac{\partial S}{\partial T}\right)_{Y,N} = C_{X,N} + T\left(\frac{\partial S}{\partial X}\right)_{T,N}\left(\frac{\partial X}{\partial T}\right)_{Y,N} = -T\left(\frac{\partial^2 G}{\partial T^2}\right)_{Y,N}. \tag{2.124}$$

The last term in Eq. (2.124) comes from Eq. (2.99*a*). Comparing Eqs. (2.118) and (2.124), we obtain the identity

$$\left(\frac{\partial S}{\partial X}\right)_{T,N} = \frac{1}{T}\left[\left(\frac{\partial U}{\partial X}\right)_{T,N} - Y\right] = -\left(\frac{\partial Y}{\partial T}\right)_{X,N}, \tag{2.125}$$

and therefore

$$-\left(\frac{\partial^2 Y}{\partial T^2}\right)_{X,N} = \frac{1}{T}\left(\frac{\partial C_{X,N}}{\partial X}\right)_{T,N}, \tag{2.126}$$

where we have used Eqs. (2.125) and (2.4).

2. Mechanical Response Functions for *PVT* Systems

For *PVT* systems, we often want to know how the volume changes as we change the pressure. If the change occurs at constant temperature and particle number, then the interesting response function is the isothermal compressibility

$$\kappa_{T,N} = -\frac{1}{V}\left(\frac{\partial V}{\partial P}\right)_{T,N} = \frac{1}{\rho}\left(\frac{\partial \rho}{\partial P}\right)_{T,N} = -\frac{1}{V}\left(\frac{\partial^2 G}{\partial P^2}\right)_{T,N}, \tag{2.127}$$

where $\rho = mN/V$ is the mass density and we have made use of Eq. (2.99*b*). If the change occurs at constant entropy and particle number (assuming a reversible change), then the interesting response function is the adiabatic compressibility

$$\kappa_{S,N} \equiv -\frac{1}{V}\left(\frac{\partial V}{\partial P}\right)_{S,N} = \frac{1}{\rho}\left(\frac{\partial \rho}{\partial P}\right)_{S,N} = -\frac{1}{V}\left(\frac{\partial^2 H}{\partial P^2}\right)_{S,N}, \tag{2.128}$$

where we have used Eq. (2.85*b*). If we wish to measure how the volume changes as we change the temperature, then the thermal expansivity is the response function of interest. It is defined

$$\alpha_{P,N} = \frac{1}{V}\left(\frac{\partial V}{\partial T}\right)_{P,N} = -\frac{1}{\rho}\left(\frac{\partial \rho}{\partial T}\right)_{P,N}. \tag{2.129}$$

Thus, we complete our expressions for the response functions.

The thermal and mechanical response functions are related. Using the identities in Sec. 2.B, we can show that

$$\kappa_{T,N}(C_{P,N} - C_{V,N}) = TV\alpha_{P,N}^2, \tag{2.130}$$

$$C_{P,N}(\kappa_{T,N} - \kappa_{S,N}) = TV\alpha_{P,N}^2, \tag{2.131}$$

and

$$\frac{C_{P,N}}{C_{V,N}} = \frac{\kappa_{T,N}}{\kappa_{S,N}}. \tag{2.132}$$

The derivation of these identities is left as a problem.

3. Mechanical Response Functions for Magnetic Systems

For magnetic systems, we can introduce a completely analogous set of response functions, but they have different names. If we wish to find how the magnetization of a system changes as we change the applied external field at fixed temperature and particle number, then the response function of interest is the isothermal susceptibility,

$$\chi_{T,N} = \left(\frac{\partial M}{\partial \mathscr{H}}\right)_{T,N} = -\left(\frac{\partial^2 G}{\partial \mathscr{H}^2}\right)_{T,N}. \tag{2.133}$$

For an anisotropic medium, the susceptibility will be a tensor. For an isotropic medium we can treat it as a scalar. If we keep the entropy fixed rather than the temperature, the interesting response function is the adiabatic susceptibility,

$$\chi_{S,N} = \left(\frac{\partial M}{\partial \mathscr{H}}\right)_{S,N} = -\left(\frac{\partial^2 H}{\partial \mathscr{H}^2}\right)_{S,N}. \tag{2.134}$$

The response function analogous to the thermal expansivity is defined

$$\alpha_{\mathscr{H},N} = \left(\frac{\partial M}{\partial T}\right)_{\mathscr{H},N} \tag{2.135}$$

and is a measure of how the magnetication changes as we change the temperature, holding the external field and particle number fixed.

If we use the identities in Sec. 2.B, we can show that

$$\chi_{T,N}(C_{\mathscr{H},N} - C_{M,N}) = T\alpha_{\mathscr{H}}^2, \tag{2.136}$$

$$C_{\mathscr{H}}(\chi_{T,N} - \chi_{S,N}) = T\alpha_{\mathscr{H}}^2, \tag{2.137}$$

and

$$\frac{C_{\mathscr{H},N}}{C_{M,N}} = \frac{\chi_{T,N}}{\chi_{S,N}} \tag{2.138}$$

in analogy to Eqs. (2.130)–(2.132).

The response functions are the quantities most directly related to experiment. For any choice of mechanical variables, we can define a set of mechanical response functions analogous to those we have given for *PVT* and $\mathscr{H}MT$ systems.

H. STABILITY OF THE EQUILIBRIUM STATE

The entropy of an isolated equilibrium system (Sec. 2.D) must be a maximum. Thus, any local fluctuations in thermodynamic quantities in the system must cause the entropy to decrease. If this were not so, the system could spontaneously move to a new equilibrium state with a higher entropy because of fluctuations. For a system in a stable equilibrium state this, by definition, cannot happen.

We can use the fact that the entropy must be maximum to obtain conditions for local equilibrium and for local stability of equilibrium systems. We will restrict ourselves to *PVT* systems. However, our arguments also apply to general *YXT* systems.

1. Conditions for Local Equilibrium in a *PVT* System

Let us consider a mixture of l types of particle in an isolated box of volume V_T divided into two parts, A and B, by a conducting porous wall which is free to move and through which particles can pass (cf. Fig. 2.10). We shall assume that no chemical reactions can occur. Since the box is closed and isolated, the total internal energy, U_T, is

$$U_T = \sum_{\alpha=\mathrm{A,B}} U_\alpha, \tag{2.139}$$

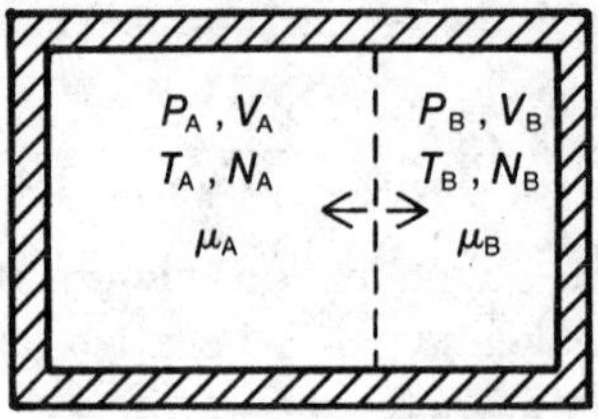

Fig. 2.10. An isolated, closed box containing gas separated into parts by a movable porous membrane: the total internal energy, the total particle number, and the total volume cannot change.

where U_α is the internal energy of compartment α (α = A,B): the total volume V_T, is

$$V_T = \sum_{\alpha=A,B} V_\alpha, \tag{2.140}$$

where V_α is the volume of compartment α; the total number of particles of type j is

$$N_{Tj} = \sum_{\alpha=A,B} N_{\alpha j}, \tag{2.141}$$

where $N_{\alpha j}$ is the number of particles of type j in compartment α; and the total entropy is

$$S_T = \sum_{\alpha=A,B} S_\alpha, \tag{2.142}$$

where S_α is the entropy in compartment α.

Let us now assume that spontaneous changes can occur in energy, volume, and particle number of each cell subject to the constraints

$$\Delta U_T = \Delta V_T = \Delta N_{Tj} = 0 \tag{2.143}$$

(we assume that no chemical reactions can occur) so that $\Delta U_A = -\Delta U_B$, $\Delta V_A = -\Delta V_B$, and $\Delta N_{Aj} = -\Delta N_{Bj}$. The entropy change for these spontaneous changes can be written

$$\Delta S_T = \sum_{\alpha=A,B}\left[\left(\frac{\partial S_\alpha}{\partial U_\alpha}\right)_{V_\alpha,\{N_{\alpha j}\}}\Delta U_\alpha + \left(\frac{\partial S_\alpha}{\partial V_\alpha}\right)_{U_\alpha,\{N_{\alpha j}\}}\Delta V_\alpha + \sum_{j=1}^{l}\left(\frac{\partial S_\alpha}{\partial N_{\alpha j}}\right)_{U_\alpha,V_\alpha}\Delta N_{\alpha j}\right] + \cdots \tag{2.144}$$

From Eqs. (2.67*a–c*) and Eq. (2.143), we can write Eq. (2.144) in the form

$$\Delta S_T = \left(\frac{1}{T_A} - \frac{1}{T_B}\right)\Delta U_A + \left(\frac{P_A}{T_A} - \frac{P_B}{T_B}\right)\Delta V_A + \sum_{j=1}^{l}\left(\frac{\mu'_{Bj}}{T_B} - \frac{\mu'_{Aj}}{T_A}\right)\Delta N_{Aj} + \cdots \tag{2.145}$$

where T_α, P_α, and $\mu'_{\alpha j}$ are the temperature, pressure, and chemical potential of compartment α.

For a system in equilibrium, the entropy is a maximum. Therefore, any spontaneous changes must cause the entropy to decrease. However, ΔU_A, ΔV_A, and ΔN_{Aj} can be positive or negative. Thus, in order to ensure that $\Delta S \leqslant 0$, we must have

$$T_A = T_B, \tag{2.146}$$

$$P_A = P_B, \tag{2.147}$$

and

$$\mu'_{Aj} = \mu'_{Bj} \qquad (j = 1, \ldots, l). \tag{2.148}$$

Eqs. (2.146)–(2.148) give the conditions for local equilibrium in a system in which no chemical reactions occur.

It is interesting to note that, if the partition is not porous, then $\Delta N_A = \Delta N_B = 0$ and we can have $\mu'_{Aj} \neq \mu'_{Bj}$. If the partition is nonporous and fixed in position, then we can have $P_A \neq P_B$ and $\mu'_{Aj} \neq \mu'_{Bj}$ and still have equilibrium.

2. Conditions for Local Stability[11]

Stability of the equilibrium state also places certain conditions on the sign of the response functions. Let us consider a closed isolated box with volume, V_T, total entropy, S_T, total energy, U_T, and total number of particles, N_T.

For simplicity, we assume that only one kind of particle is present and that the box is divided into cells, A and B. We will denote the equilibrium volume, entropy, internal energy, and number of particles of the αth cell by $V_\alpha{}^0$, $S_\alpha{}^0$, $U_\alpha{}^0$, and $N_\alpha{}^0$, respectively. The equilibrium pressure, temperature, and chemical potential of the whole system are denoted P^0, T^0, and μ'^0, respectively (they must be the same for the two cells).

Because there are a finite number of particles in the box, there will be spontaneous fluctuations of the thermodynamic variables of each cell about their respective equilibrium values. These spontaneous fluctuations must be such that V_T, U_T, and N_T remain fixed. However, since the equilibrium state is stable, the fluctuations must cause S_T to *decrease*. If it did not decrease, the equilibrium state would be unstable and spontaneous fluctuations would cause the system to move to a more stable equilibrium state of higher entropy.

We will assume that fluctuations about the equilibrium state are small and expand the entropy of the αth cell in a Taylor expansion about its equilibrium value:

$$\begin{aligned} S_\alpha(U_\alpha, V_\alpha, N_\alpha) = {} & S_\alpha{}^0(U_\alpha{}^0, V_\alpha{}^0, N_\alpha{}^0) + \left(\frac{\partial S}{\partial U}\right)^0_{V,N} \Delta U_\alpha + \left(\frac{\partial S}{\partial V}\right)^0_{U,N} \Delta V_\alpha \\ & + \left(\frac{\partial S}{\partial N}\right)^0_{U,V} \Delta N_\alpha + \frac{1}{2}\left\{\Delta\left(\frac{\partial S_\alpha}{\partial U_\alpha}\right)_{V,N} \Delta U_\alpha + \Delta\left(\frac{\partial S_\alpha}{\partial V_\alpha}\right)_{U,N} \Delta V_\alpha \right. \\ & \left. + \Delta\left(\frac{\partial S_\alpha}{\partial N_\alpha}\right)_{U,N} \Delta N_\alpha\right\} + \cdots, \end{aligned} \tag{2.149}$$

where

$$\Delta\left(\frac{\partial S_\alpha}{\partial U_\alpha}\right)_{V,N} \equiv \left(\frac{\partial^2 S}{\partial U^2}\right)^0_{V,N} \Delta U_\alpha + \left[\frac{\partial}{\partial V}\left(\frac{\partial S}{\partial U}\right)_{V,N}\right]^0_{U,N} \Delta V_\alpha + \left[\frac{\partial}{\partial N}\left(\frac{\partial S}{\partial U}\right)_{V,N}\right]^0_{U,V} \Delta N_\alpha \tag{2.150}$$

and similar expressions hold for $\Delta(\partial S_\alpha/\partial V_\alpha)_{U,N}$ and $\Delta(\partial S_\alpha/\partial N_\alpha)_{U,V}$. In Eqs. (2.149) and (2.150) the superscript "0" on the partial derivatives means they are evaluated at equilibrium. The fluctuations ΔU_α, ΔV_α, and ΔN_α are defined $\Delta U_\alpha = U_\alpha - U_\alpha{}^0$, $\Delta V_\alpha = V_\alpha - V_\alpha{}^0$, and $\Delta N_\alpha = N_\alpha - N_\alpha{}^0$ and denote the deviation of the quantities U_α, V_α, and N_α from their absolute equilibrium values. We obtain the total entropy by adding the entropy in the two cells. Because of the local equilibrium conditions in Eqs. (2.146)–(2.148) the first-order terms in the expression for the total entropy drop out and we find

$$\Delta S_T = \frac{1}{2}\sum_{\alpha=\mathrm{A,B}} \left\{\Delta\left(\frac{\partial S_\alpha}{\partial U_\alpha}\right)_{V,N} \Delta U_\alpha + \Delta\left(\frac{\partial S_\alpha}{\partial V_\alpha}\right)_{U,N} \Delta V_\alpha + \Delta\left(\frac{\partial S_\alpha}{\partial N_\alpha}\right)_{U,V} \Delta N_\alpha\right\}. \tag{2.151}$$

Eq. (2.151) can be written in simpler form if we make use of Eqs. (2.67*a*–*c*). We then find

$$\Delta S_T = \frac{1}{2}\sum_{\alpha=\mathrm{A,B}} \left[\Delta\left(\frac{1}{T}\right)_\alpha \Delta U_\alpha + \Delta\left(\frac{P}{T}\right)_\alpha \Delta V_\alpha - \Delta\left(\frac{\mu'}{T}\right)_\alpha \Delta N_\alpha\right] \tag{2.152}$$

or

$$\Delta S_T = \frac{1}{2T}\sum_{\alpha=\mathrm{A,B}} [-\Delta T_\alpha \Delta S_\alpha + \Delta P_\alpha \Delta V_\alpha - \Delta\mu'_\alpha \Delta N_\alpha]. \tag{2.153}$$

(In Eq. (2.153), we have used the relation $T\Delta S = \Delta U + P\Delta V - \mu'\Delta N$.) Eq. (2.153) gives the entropy change, due to the fluctuations, in a completely general form. We can now expand ΔS_T in terms of any independent variables we choose.

Let us choose the variables ΔT_α, ΔP_α, and ΔN_α. Then

$$\Delta\mu_\alpha = \left(\frac{\partial\mu'}{\partial T}\right)^0_{P,N} \Delta T_\alpha + \left(\frac{\partial\mu'}{\partial P}\right)^0_{T,N} \Delta P_\alpha + \left(\frac{\partial\mu'}{\partial N}\right)^0_{T,P} \Delta N_\alpha, \tag{2.154}$$

$$\Delta S_\alpha = \left(\frac{\partial S}{\partial T}\right)^0_{P,N} \Delta T_\alpha + \left(\frac{\partial S}{\partial P}\right)^0_{T,N} \Delta P_\alpha + \left(\frac{\partial S}{\partial N}\right)^0_{T,P} \Delta N_\alpha, \tag{2.155}$$

and

$$\Delta V_\alpha = \left(\frac{\partial V}{\partial T}\right)^0_{P,N} \Delta T_\alpha + \left(\frac{\partial V}{\partial P}\right)^0_{T,N} \Delta P_\alpha + \left(\frac{\partial V}{\partial N}\right)^0_{T,P} \Delta N_\alpha. \tag{2.156}$$

We now use Maxwell relations in Eqs. (2.100*a*)–(2.100*d*) to obtain

$$\Delta S_T = \frac{1}{2T} \sum_{\alpha=\mathrm{A,B}} \left\{ \frac{-C_P}{T} (\Delta T)_\alpha^2 + 2\left(\frac{\partial V}{\partial T}\right)^0_{P,N} \Delta P_\alpha \Delta T_\alpha + \left(\frac{\partial V}{\partial P}\right)^0_{T,N} (\Delta P_\alpha)^2 - \left(\frac{\partial \mu'}{\partial N}\right)^0_{P,T} (\Delta N_\alpha)^2 \right\}, \quad (2.157)$$

where C_P is the heat capacity at constant pressure. We can make one final rearrangement of Eq. (2.157). If we use the identity in Eq. (2.130), we obtain

$$\Delta S_T = -\frac{1}{2T} \sum_{\alpha=\mathrm{A,B}} \left\{ \frac{C_V}{T} (\Delta T_\alpha)^2 + \frac{1}{\kappa_T V} [(\Delta V_\alpha)^2_{N_\alpha}] + \left(\frac{\partial \mu'}{\partial N}\right)^0_{P,T} (\Delta N_\alpha)^2 \right\}, \quad (2.158)$$

where $(\Delta V_\alpha)_{N_\alpha}$ denotes volume fluctuations at constant particle number,

$$(\Delta V_\alpha)_{N_\alpha} \equiv \left(\frac{\partial V}{\partial T}\right)^0_{N,P} \Delta T_\alpha + \left(\frac{\partial V}{\partial P}\right)^0_{N,T} \Delta P_\alpha. \quad (2.159)$$

The fluctuations (ΔT_α), (ΔV_α), and (ΔN_α) can occur independently of one another. As we have already noted, the equilibrium state will be stable if $\Delta S_T \leqslant 0$. From Eq. (2.158) we see that this will be true if

$$C_V \geqslant 0, \qquad \kappa_T \geqslant 0, \quad \text{and} \quad \left(\frac{\partial \mu'}{\partial N}\right)^0_{P,T} \geqslant 0. \quad (2.160)$$

Conditions (2.160) are another statement of Le Châtelier's famous principle: *If a system is in stable equilibrium, then any spontaneous change in its parameters must bring about processes which tend to restore the system to equilibrium.*

The first condition, $C_V \geqslant 0$, is a condition for thermal stability. If a small excess of heat is added to a volume element of fluid, then the temperature of the volume element must increase relative to its surroundings so that some of the heat will flow out again. This requires that the heat capacity be positive, $C > 0$. If $C < 0$, the temperature would decrease and even more heat would flow in, thus leading to an instability.

The second condition, $\kappa_T \geqslant 0$, is a condition for mechanical stability: If a small volume element of fluid spontaneously increases, the pressure of the fluid inside the fluid element must decrease relative to its surroundings so that the larger pressure of the surroundings will stop the growth of the volume element. This requires that the compressibility be positive, $\kappa \geqslant 0$. If $\kappa < 0$, the pressure would increase and the volume element would continue to grow, thus leading to an instability.

The third condition, $(\partial\mu/\partial N)^0$, is a condition for chemical stability. If particles are added to a system, its chemical potential, and therefore its total energy, must increase. If it did not, the system could act as a sink for matter.

The identities in Eqs. (2.130)–(2.132) plus the stability conditions in Eq. (2.160) lead to the relations

$$C_P > C_V > 0 \tag{2.161}$$

and

$$\kappa_T > \kappa_S > 0. \tag{2.162}$$

Thus, we have derived quite general conditions which must be satisfied by the response functions for thermodynamically stable systems.

3. Implications of Stability Requirements for the Free Energies

The stability conditions place restrictions on the derivatives of the thermodynamic potentials. Before we show this it is useful to introduce the concept of concave and convex functions.[12]

(*a*) A function $f(x)$ is a *convex* function if for any x_1 and x_2 the chord joining the points $f(x_1)$ and $f(x_2)$ lies above or on the curve $f(x)$ for all x in the interval $x_1 < x < x_2$. If $f'(x)$ exists at a given point, the tangent at that point always lies below the function except at the point of tangency; $f''(x) \geqslant 0$ for all x (cf. Fig. 2.11).

(*b*) A function $f(x)$ is a *concave* function of x if the function $-f(x)$ is convex.

We can now consider the effect of the stability requirements on the Helmholtz and Gibbs free energies.

From Eq. (2.92*a*) and the stability condition (2.161), we can write

$$\left(\frac{\partial^2 A}{\partial T^2}\right)_{V,N} = -\left(\frac{\partial S}{\partial T}\right)_{V,N} = -\frac{C_V}{T} < 0, \tag{2.163}$$

and from Eq. (2.92*b*) and the stability condition (2.162), we can write

$$\left(\frac{\partial^2 A}{\partial V^2}\right)_{T,N} = -\left(\frac{\partial P}{\partial V}\right)_{T,N} = \frac{1}{V\kappa_{T,N}} > 0. \tag{2.164}$$

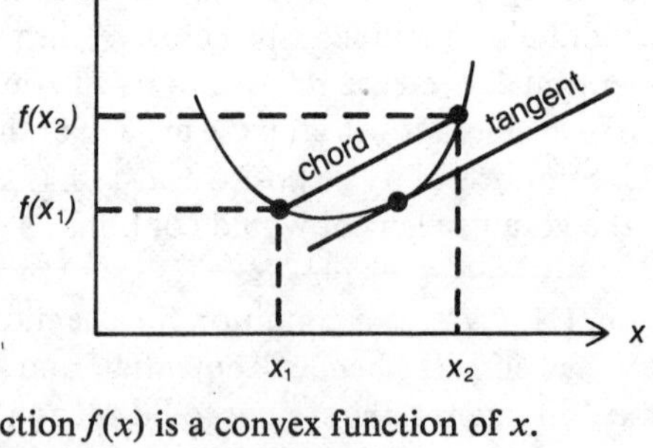

Fig. 2.11. The function $f(x)$ is a convex function of x.

Thus, the Helmholtz free energy is a concave function of temperature and a convex function of volume.

From Eq. (2.99*a*) and the stability condition (2.161), we find that

$$\left(\frac{\partial^2 G}{\partial T^2}\right)_{P,N} = -\left(\frac{\partial S}{\partial T}\right)_{P,N} = -\frac{C_{P,N}}{T} < 0, \tag{2.165}$$

while from Eq. (2.99*b*) and Eq. (2.163), we find

$$\left(\frac{\partial^2 G}{\partial P^2}\right)_{T,N} = \left(\frac{\partial V}{\partial P}\right)_{T,N} = -V\kappa_{T,N} < 0. \tag{2.166}$$

Thus, the Gibbs free energy is a concave function of temperature and a concave function of pressure.

It is interesting to sketch the free energy and its slope as functions of pressure and temperature. A sketch of the Gibbs free energy, for a range of pressure and temperature for which no phase transition occurs, is given in Fig 2.12. The case of phase transitions is given in Chap. 4.

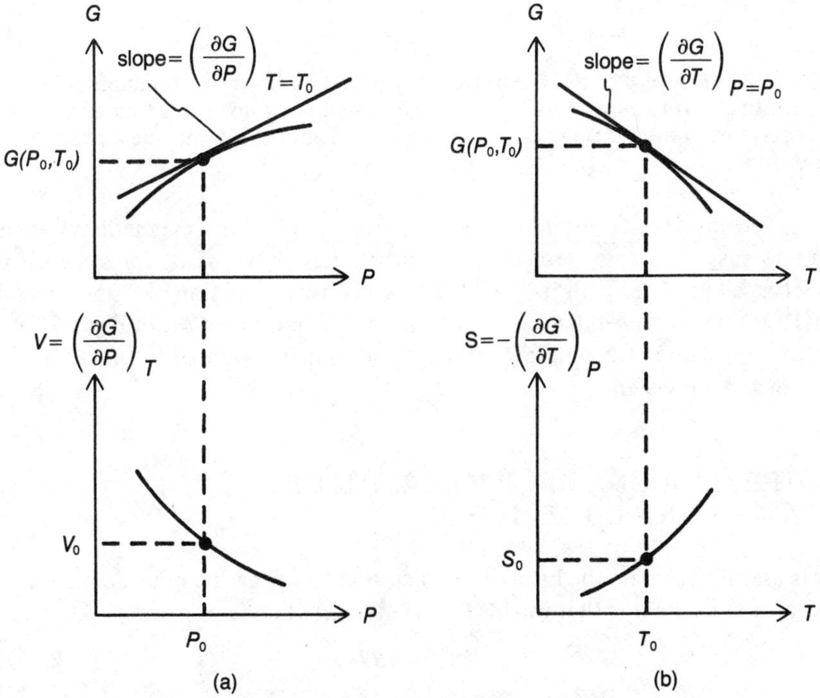

Fig. 2.12. (*a*) A plot of the Gibbs free energy and its slope as a function of pressure; (*b*) a plot of the Gibbs free energy and its slope as a function of temperature; both plots are done in a region which does not include a phase transition.

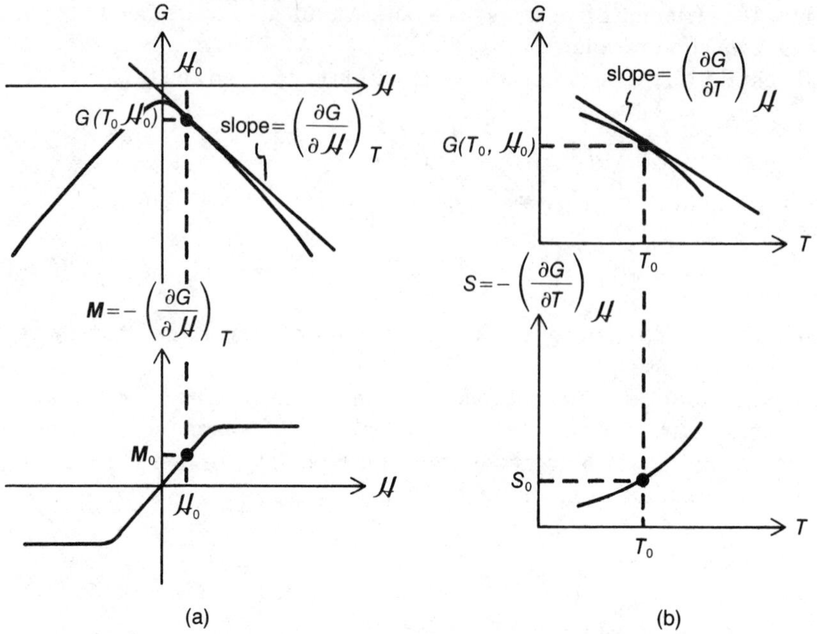

Fig. 2.13. (*a*) A plot of the Gibbs free energy and its slope as a function of applied field; (*b*) a plot of the Gibbs free energy and its slope as a function of temperature; both plots are done in a region which does not include a phase transition.

The form of the Gibbs and Helmholtz free energies for a magnetic system is not so easy to obtain. However, Griffiths[13] has shown that, for systems of uncharged particles with spin, $G(T, \mathscr{H})$ is a concave function of T and $\mathscr{H}$ and $A(T, M)$ is a concave function of T and a convex function of M. In Fig. 2.13, we sketch the Gibbs free energy and its slope as a function of T and $\mathscr{H}$ for a paramagnetic system.

I. THERMODYNAMIC PROPERTIES OF A CLASSICAL IDEAL GAS

It is useful to discuss the behavior of a classical ideal gas in some detail at this point. Let us assume that the ideal gas equation of state

$$PV = nRT \tag{2.167}$$

and constant volume heat capacity

$$C_V = \alpha nR \tag{2.168}$$

have been obtained from experiment ($\alpha = \frac{3}{2}$ for monatomic gases, $\alpha = \frac{5}{2}$ for diatomic gases, and $\alpha = 3$ for polyatomic gases). Given Eqs. (2.167) and (2.168), we can derive all the thermodynamic potentials. We shall begin with the internal energy and entropy.

1. Internal Energy and Entropy

From Eqs. (2.115) and (2.168) we can write

$$\left(\frac{\partial U}{\partial T}\right)_{V,N} = \alpha nR \tag{2.169}$$

and, from Eq. (2.125),

$$\left(\frac{\partial U}{\partial V}\right)_{T,N} = T\left(\frac{\partial P}{\partial T}\right)_{V,N} - P = 0. \tag{2.170}$$

It is a simple matter to integrate Eqs. (2.169) and (2.170) to obtain

$$U(T) = U_0 + \alpha nR(T - T_0) \tag{2.171}$$

(U_0 and T_0 are integration constants). The internal energy of an ideal gas depends only on temperature.

The entropy can be obtained from Eqs. (2.121) and (2.125). We find

$$\left(\frac{\partial S}{\partial T}\right)_{V,N} = \frac{C_V}{T} = \frac{\alpha nR}{T} \tag{2.172}$$

and

$$\left(\frac{\partial S}{\partial V}\right)_{T,N} = \left(\frac{\partial P}{\partial T}\right)_{V,N} = \frac{nR}{V}. \tag{2.173}$$

We can now integrate to obtain

$$S(T, V) = S_0 + nR \ln\left[\left(\frac{T}{T_0}\right)^{\alpha} \frac{V}{V_0}\right]. \tag{2.174}$$

If we wish to find the dependence of the internal energy on S and V, we simply combine Eqs. (2.171) and (2.174) to obtain

$$U(S, V) = U_0 + \alpha nRT_0\left[\left(\frac{V_0}{V}\right)^{1/\alpha} \exp\left(\frac{S - S_0}{\alpha nR}\right) - 1\right]. \tag{2.175}$$

Note that T_0 is just a constant. Eq. (2.175) gives the functional dependence of U on S and V and is the fundamental equation for the internal energy of an ideal gas.

2. Enthalpy

We can obtain the enthalpy from the internal energy by using the equation $H = U + PV$. If we wish to write the enthalpy in terms of its natural variables

S and P, we must use Eqs. (2.167) and (2.174) to write the volume and temperature in terms of S and P. Then we obtain

$$H(S, P) = H_0 - (\alpha + 1)nRT_0\left[1 - \left(\frac{P}{P_0}\right)^{1/\alpha+1} \exp\left(\frac{S - S_0}{nR(\alpha + 1)}\right)\right]. \tag{2.176}$$

Eq. (2.176) is the fundamental equation for the enthalpy of an ideal gas.

3. Helmholtz and Gibbs Free Energies

The Helmholtz free energy is obtained from the fundamental equation $A = U - TS$. From Eqs. (2.171) and (2.174) we find

$$A(T, V) = A_0 + (\alpha nR - S_0)(T - T_0) - nRT \ln\left[\left(\frac{T}{T_0}\right)^{\alpha} \frac{V}{V_0}\right]. \tag{2.177}$$

Eq. (2.177) is the fundamental equation for the Helmholtz free energy of an ideal gas.

The Gibbs free energy may be obtained from the equation $G = A + PV$. From Eqs. (2.167) and (2.177) we find

$$G(T, P) = G_0 + [(\alpha + 1)nR - S_0](T - T_0) - nRT \ln\left[\left(\frac{T}{T_0}\right)^{\alpha+1} \frac{P_0}{P}\right]. \tag{2.178}$$

Eq. (2.178) is the fundamental equation for the Gibbs free energy of an ideal gas.

The grand potential may be obtained in a similar manner from the fundamental equations $\Omega = -PV$ and $G = N\mu'$ and the Gibbs-Duheim equation. It is a worthwhile exercise to check that the Helmholtz free energy for an ideal gas is a concave function of T and a convex function of V; and that the Gibbs free energy is a concave function of T and P.

REFERENCES

1. Most books on advanced calculus contain thorough discussions of exact differentials.
2. J. O. Hirschfelder, C. F. Curtiss, and R. B. Bird, *Molecular Theory of Gases and Liquids* (John Wiley & Sons, New York, 1954).
3. J. B. Partington, *An Advanced Treatise on Physical Chemistry*, Vol. I (Longmans, Green & Co., London, 1949).
4. *International Critical Tables*, ed. E. W. Washburn (McGraw-Hill Book Co., New York, 1928).
5. M. W. Zemansky, *Heat and Thermodynamics* (McGraw-Hill Book Co., New York, 1957).
6. O. D. Jefimenko, *Electricity and Magnetism* (Appleton-Century-Crofts, New York, 1966).

7. P. M. Morse, *Thermal Physics* (W. A. Benjamin, New York, 1965).
8. J. S. Dugdale, *Entropy and Low Temperature Physics* (Hutchinson University Library, London, 1966).
9. D. ter Haar and H. Wergeland, *Elements of Thermodynamics* (Addison-Wesley Publishing Co., Reading, Mass., 1966).
10. H. B. Callen, *Thermodynamics* (John Wiley & Sons, New York, 1960).
11. P. Glansdorff and I. Prigogine, *Thermodynamic Theory of Structure, Stability and Fluctuations* (Wiley-Interscience, New York, 1971).
12. H. E. Stanley, *Introduction to Phase Transitions and Critical Phenomena* (Oxford University Press, Oxford, 1971).
13. R. K. Griffiths, J. Math. Phys. *5* 1215 (1964).
14. D. Hodgeman, ed., *Handbook of Chemistry and Physics* (Chemical Rubber Publishing Co., Cleveland, 1962).

PROBLEMS

1. Test the following differentials for exactness. For those cases in which the differential is exact, find the function $u(x, y)$.

(*a*) $du = \dfrac{-y}{x^2 + y^2}\,dx + \dfrac{x}{x^2 + y^2}\,dy.$

(*b*) $du = (y - x^2)\,dx + (x + y^2)\,dy.$

(*c*) $du = (2y^2 - 3x)\,dx - 4xy\,dy.$

2. An isotropic magnetic material satisfying Curie's law, $M = nD\mathscr{H}/T$, has a constant heat capacity $C_M = C$ and an internal energy which is independent of magnetization, M. Assume that this substance is used to create a Carnot engine and that the engine operates between temperatures T_h and T_c such that $T_h > T_c$.

(*a*) Give a qualitative plot of the curve for one complete cycle of the engine.
(*b*) How much work is done by the engine?
(*c*) Prove that the engine has efficiency $\eta = (T_h - T_c)/T_h$.

3. Find the efficiency of the heat engine shown in Fig. 2.14. Draw the diagram for a Carnot cycle in the *S*-*T* plane.

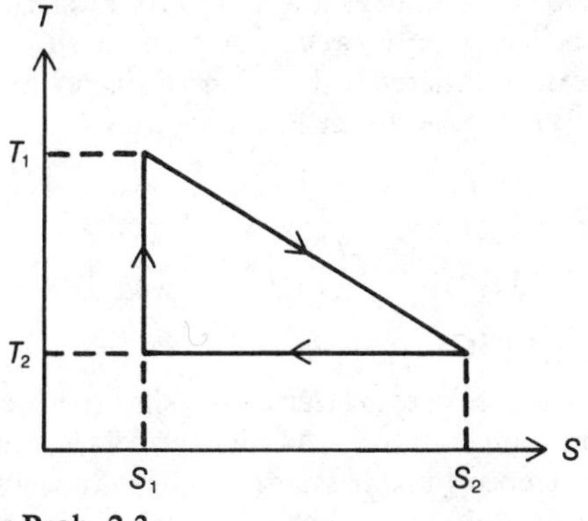

Fig. 2.14. Figure for Prob. 2.3.

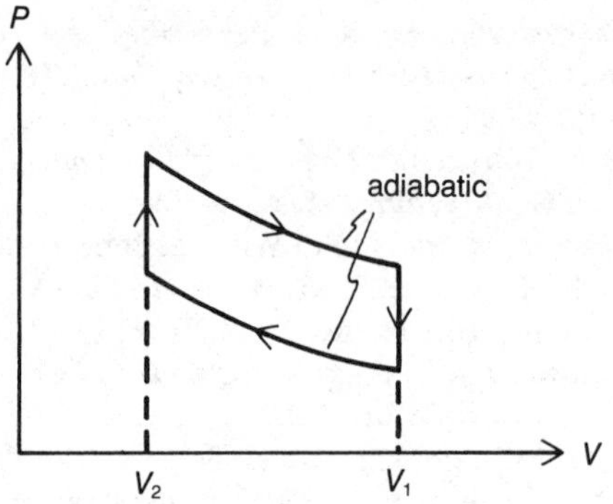

Fig. 2.15. Figure for Prob. 2.4.

4. A gasoline engine (Otto cycle) may be represented by the cycle in Fig. 2.15. Find the efficiency in terms of the compression ratio, (V_1/V_2), and compute its value for an ideal gas assuming $(V_1/V_2) \sim 9$.

5. Two equations of state often used in place of the van der Waals equation are the Berthelot equation

$$\left(P + \frac{n^2 a_B}{TV^2}\right)(V - nb_B) = nRT$$

and the Dieterici equation

$$P\, e^{na_D/VRT}(V - b_D) = nRT,$$

where a_B, b_B, a_D, and b_D are experimental constants. Find the second virial coefficient for each of these equations of state.

6. Experimentally one finds that for a rubber band

$$\left(\frac{\partial J}{\partial L}\right)_T = \frac{aT}{L_0}\left(1 + 2\left(\frac{L_0}{L}\right)^3\right) \quad \text{and} \quad \left(\frac{\partial J}{\partial T}\right)_L = \frac{aL}{L_0}\left(1 - \left(\frac{L_0}{L}\right)^3\right),$$

where J is the tension, $a = 1.0 \times 10^3$ dyne/K, and $L_0 = 0.5$ m is the length of the band when no tension is applied.

(*a*) Compute $(\partial L/\partial T)_J$ and discuss its physical meaning.

(*b*) Find the equation of state and show that dJ is an exact differential.

(*c*) Assume that the heat capacity at constant length is $C_L = 1.0$ joules/K. Find the work necessary to stretch the band reversibly and adiabatically to a length of 1 m. What is the change in temperature?

7. Prove that

$$\left(\frac{\partial U}{\partial Y}\right)_T = T\left(\frac{\partial X}{\partial T}\right)_Y + Y\left(\frac{\partial X}{\partial Y}\right)_T,$$

where U is the internal energy.

8. A paramagnetic substance obeying Curie's law sits in a magnetic field $\mathscr{H}_0$ and has magnetization M_0 and temperature T_0. Assume the field is turned off adiabatically. Find the change in temperature. (Assume the heat capacity is independent of temperature, C_M = constant.) Explain physically what is happening.

9. For a low-density gas the virial expansion can be terminated at first order and the equation of state is

$$P = \frac{nRT}{V}\left[1 + \frac{n}{V}B(T)\right].$$

The heat capacity will have corrections to its ideal gas value. We can write it in the form

$$C_V = \tfrac{3}{2}nR - \frac{n^2R}{V}F(T).$$

(*a*) Find the form that $F(T)$ must have in order for the two equations to be thermodynamically consistent.
(*b*) Find C_p.
(*c*) Find the entropy and internal energy.

10. Show that

$$T\,dS = C_x\left(\frac{\partial T}{\partial Y}\right)_x dY + C_y\left(\frac{\partial T}{\partial X}\right)_y dX$$

where C_x is heat capacity at constant x and C_y is heat capacity at constant y.

11. Prove that for a substance obeying Curie's law, $M = nD\mathscr{H}/T$, the internal energy and heat capacity depend only on temperature.

12. The equation of state for one mole of a gas is $PV = RT - aP/T$, and the heat capacity at constant pressure is $C_p = \tfrac{5}{2}R + 2aPF(T)$ where $F(T)$ is some function of temperature. Find $F(T)$.

13. Given an isotropic magnetic substance satisfying Curie's law, $M = nD\mathscr{H}/T$, with a constant heat capacity $C_M = C$, find the internal energy, entropy, enthalpy, Helmholtz free energy, and Gibbs free energy as a function of M and T.

14. Prove that $\kappa_T(C_P - C_V) = TV\alpha_T^2$ and $C_P/C_V = \kappa_T/\kappa_S$.

15. Compute $C_P - C_V$ for a gas obeying the Dieterici equation of state (cf. Prob. 2.5).

16. Obtain an expression for the grand potential as a function of V, T, and μ' for an ideal gas.

17. Prove that

$$C_{Y,N} = \left(\frac{\partial H}{\partial T}\right)_{Y,N} \quad \text{and} \quad \left(\frac{\partial H}{\partial Y}\right)_{T,N} = T\left(\frac{\partial X}{\partial T}\right)_{Y,N} + X.$$

18. Sketch the Helmholtz free energy and its slope as a function of V and T for a PVT system and as a function of M and T for a paramagnetic system. Assume that you are well away from a phase transition.

19. If there are l different kinds of particles in a closed isolated box, show from stability arguments that

$$\sum_{i=1}^{l}\sum_{j=1}^{l}\frac{\partial \mu_i'}{\partial N_j}\Delta N_i\,\Delta N_j > 0.$$

20. Black body radiation is electromagnetic radiation kept in a closed cavity in equilibrium with the walls at some temperature T. One finds that, if the volume of the cavity is increased for fixed T, more radiation is generated but the energy density (energy/volume) $e(T)$ remains constant. Thus, the internal energy has the form $U = Ve(T)$, where $e(T)$ depends only on temperature and V is the volume. Furthermore, one finds that the radiation exerts a pressure $P = \frac{1}{3}e(T)$ on the walls. Find the temperature dependence of U, P, and the entropy S.

3 Applications of Thermodynamics

A. INTRODUCTORY REMARKS

For a physicist, thermodynamics becomes most interesting when it is applied to real systems. In order to demonstrate its versatility, we shall apply it to a number of systems which have been selected for their practical importance or conceptual importance. In this chapter, we restrict ourselves to examples which do not directly involve the concept of a phase transition. In Chap. 4, we will discuss the thermodynamics of phase transitions.

We begin with a subject of great practical and historic importance—the cooling of gases. It is often necessary to cool substances below the temperature of their surroundings. The refrigerator most commonly used for this purpose is based on the so-called Joule-Thompson effect. There are two important ways to cool gases. We can let them do work against their own intermolecular forces by letting them expand freely (Joule effect); or we can force them through a small constriction, thus causing cooling at low temperatures or heating at high temperatures (Joule-Thompson effect). The Joule-Thompson effect is by far the more effective of the two methods. We shall discuss both methods in this chapter and use the van der Waals equation of state to obtain estimates of the cooling effects for some real gases.

As we have seen in Chap. 2, for reversible processes, changes in entropy content are proportional to changes in heat content of thermodynamic systems. For irreversible processes, this is no longer true. We can have entropy increase in an isolated system, even though no heat has been added. Therefore, it is often useful to think of an increase in entropy as being related to an increase in disorder in a system. One of the most convincing illustrations of this is the entropy change which occurs when two substances which have the same temperature and pressure, but different identities, are mixed. Thermodynamics predicts that the entropy will increase solely due to mixing of the substances.

When the entropy of a system changes due to mixing, so will all other thermodynamic quantities. One of the most interesting examples of this is osmosis. We can fill a container with water and separate it into two parts by a membrane permeable to water but not salt, for example. If we put a small amount of salt into one side, the pressure of the resulting salt solution will increase markedly because of mixing.

Chemical reactions can be characterized in a rather simple way in terms of a thermodynamic quantity called the affinity. The affinity gives a measure of the distance of a chemical reaction from thermodynamic equilibrium and will be useful in later chapters when we discuss chemical systems out of equilibrium. We can obtain an expression for the affinity by using the conditions for thermodynamic equilibrium and stability introduced in Chap. 2, and at the same time we can learn a number of interesting facts about the thermodynamic behavior of chemical reactions.

A rather spectacular phenomenon which occurs in superfluid systems is the thermomechanical, or fountain, effect. As we shall see, it is possible to use a heat source to drive a superfluid fountain. Even though superfluids are highly degenerate quantum systems, the fountain effect surprisingly can be described completely in terms of classical thermodynamics. All we need is a system composed of two interpenetrating fluids, one of which carries no entropy. Then simple arguments give us the fountain effect.

We will discuss all of the above phenomena in detail in this chapter.

B. COOLING AND LIQUEFACTION OF GASES[1]

All neutral gases (if we exclude gravitational effects) interact via a short-ranged potential which has a repulsive core and a short-ranged attractive region. If such a gas is allowed to expand, it must do work against the attractive forces and its temperature will decrease. This effect can be used to cool a gas, although the amount of cooling that occurs via this mechanism alone is very small.

We shall study two different methods for cooling, one based solely on free expansion and one which involves throttling of the gas through a porous plug or construction. The second method is the basis for gas liquifiers commonly used in the laboratory.

1. Joule Effect: Free Expansion

Experiments which attempted to measure cooling due to free expansion were first performed by Gay-Lussac in 1807 and were improved upon by Joule in 1843. The free expansion process is shown schematically in Fig. 3.1. The gas is initially confined to an insulated chamber with volume V_i, at pressure P_i and temperature T_i. It is then allowed to expand suddenly into an insulated evacuated chamber with volume V_f. Since the gas expands freely, no work will be done and $\Delta W = 0$. Furthermore, since both chambers are insulated, no heat will be added and we find $\Delta Q \equiv 0$. Thus, from the first law, the internal energy must remain constant, $\Delta U \equiv 0$, during the process. The only effect of the free expansion is a transfer of energy between the potential energy and kinetic energy of the particles.

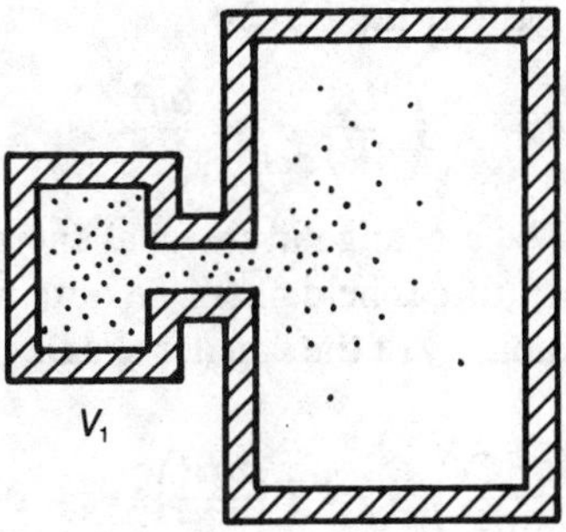

Fig. 3.1. The Joule effect: free expansion of a gas from an insulated chamber of volume V_1 into an insulated evacuated chamber of volume V_2 causes cooling.

Because free expansion takes place spontaneously, the entropy of the gas will increase even though no heat is added. During the expansion we cannot use thermodynamics to describe the state of the system because it will not be in equilibrium, even locally. However, after the system has settled down and reached a final equilibrium state, we can use thermodynamics to relate the initial and final states by finding a reversible path between them.

During the expansion the internal energy does not change. Thus, the internal energies of the initial and final states must be the same, and for our imaginary reversible path we can write

$$dU = 0 = \left(\frac{\partial U}{\partial T}\right)_{V,N} dT + \left(\frac{\partial U}{\partial V}\right)_{T,N} dV \tag{3.1}$$

(particle number will also remain constant during the expansion) or

$$dT = -\frac{\left(\frac{\partial U}{\partial V}\right)_{T,N}}{\left(\frac{\partial U}{\partial T}\right)_{V,N}} dV = \left(\frac{\partial T}{\partial V}\right)_{U,N} dV. \tag{3.2}$$

In Eq. (3.2) we have used the chain rule (cf. Eq. (2.6)). The quantity $(\partial T/\partial V)_{U,N}$ is called the differential Joule coefficient.

Let us compute the Joule coefficient for various gases. From Eq. (2.125) we know that

$$\left(\frac{\partial U}{\partial V}\right)_{T,N} = T\left(\frac{\partial P}{\partial T}\right)_{V,N} - P. \tag{3.3}$$

For an ideal gas we have the equation of state $PV = nRT$. Thus, $(\partial U/\partial V)_{T,N} = 0$ and $(\partial T/\partial V)_{U,N} = 0$, and we conclude that the temperature of an ideal gas does not change during free expansion.

For a van der Waals gas (cf. Eq. (2.12)),

$$\left(\frac{\partial U}{\partial V}\right)_{T,N} = \frac{an^2}{V^2}. \tag{3.4}$$

For this case, there will be a change in internal energy due to interactions as volume changes if temperature is held fixed. To obtain the Joule coefficient, we must also find the heat capacity of the van der Waals gas. From Eq. (2.126), we obtain

$$\left(\frac{\partial C_V}{\partial V}\right)_{T,N} = T\left(\frac{\partial^2 P}{\partial T^2}\right)_{V,N} = 0, \tag{3.5}$$

and therefore C_V is independent of volume,

$$C_V = C_V(T), \tag{3.6}$$

for a van der Waals gas. Since the heat capacity C_V contains no volume corrections due to interactions, it is thermodynamically consistent to choose its value to be equal to that of an ideal gas; $C_V = \frac{3}{2}nR$ (above the critical point). The Joule coefficient for a van der Waals gas then becomes

$$\left(\frac{\partial T}{\partial V}\right)_{U,N} = -\frac{2}{3}\frac{an}{RTV^2}. \tag{3.7}$$

Using Eq. (3.7), we can integrate Eq. (3.2) between initial and final states. We find

$$T_f^{\,2} = \frac{4an}{3R}\left(\frac{1}{V_f} - \frac{1}{V_i}\right) + T_i^2 \tag{3.8}$$

and

$$\Delta T = T_f - T_i \approx \frac{2an}{3RT_i}\left(\frac{1}{V_f} - \frac{1}{V_i}\right) + \cdots \tag{3.9}$$

If $V_f > V_i$ the temperature will always decrease. We can use the values of a given in Table 2.1 to estimate the change in temperature for some simple cases. Let us assume that $V_i = 10^{-3}\ \text{m}^3$, $T_i = 300$ K, $n = 1$ mole, and $V_f = \infty$ (this will give maximum cooling). For oxygen $\Delta T \sim -0.4 \times 10^{-3}$ K and for carbon dioxide $\Delta T \sim -1.0 \times 10^{-3}$ K. We must conclude that free expansion alone is not a very effective way to cool a gas.

2. Joule-Thompson Effect: Throttling

Throttling of a gas through a porous plug or small constriction provides a much more efficient means of cooling and is the basis of most liquefaction machines. The throttling process in its simplest form is pictured in Fig. 3.2. A gas initially at a pressure P_i, temperature T_i, and volume V_i is forced through a porous plug into another chamber, maintained at pressure $P_f < P_i$. All chambers and the plug are insulated, $\Delta Q = 0$, for the process. The gas inside the plug is forced

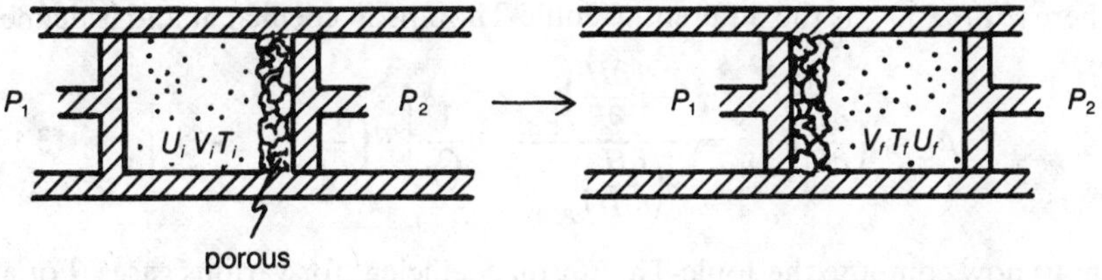

Fig. 3.2. The Joule-Thompson effect: throttling of a gas through a porous plug can cause cooling or heating.

through narrow twisting chambers irreversibly (the chamber walls are insulated). Work must be done to force it through. Even though the entire process is irreversible, we can use thermodynamics to relate the initial and final states.

The net work done by the gas is

$$\Delta W = \int_0^{V_f} P_f \, dV_f + \int_{V_i}^0 P_i \, dV_i = P_f V_f - P_i V_i. \tag{3.10}$$

From the first law $\Delta U = -\Delta W$ since $\Delta Q \equiv 0$. Thus,

$$U_f + P_f V_f = U_i + P_i V_i \tag{3.11}$$

or, in terms of enthalpy,

$$H_f = H_i. \tag{3.12}$$

Thus the throttling process is one which takes place at constant enthalpy.

Let us now construct a hypothetical reversible path to describe the process. For each differential change we have (assuming total particle number remains constant)

$$dH = 0 = T\,dS + V\,dP. \tag{3.13}$$

We see that the increase in entropy due to the throttling process is accompanied by a decrease in pressure. It is convenient to use temperature and pressure as independent variables rather than entropy and pressure. We therefore expand the entropy

$$dS = \left(\frac{\partial S}{\partial T}\right)_{P,N} dT + \left(\frac{\partial S}{\partial P}\right)_{T,N} dP \tag{3.14}$$

and obtain

$$dH = 0 = C_p \, dT + \left[V - T\left(\frac{\partial V}{\partial T}\right)_{P,N}\right] dP. \tag{3.15}$$

In Eq. (3.15) we have used Eqs. (2.124) and (2.100*a*); C_p is the constant pressure heat capacity. Eq. (3.15) can be rewritten in the form

$$dT = \left(\frac{\partial T}{\partial P}\right)_{H,N} dP, \tag{3.16}$$

where $(\partial T/\partial P)_{H,N}$ is the differential Joule-Thompson coefficient and is defined

$$\mu_J = \left(\frac{\partial T}{\partial P}\right)_{H,N} = \frac{-\left(\frac{\partial H}{\partial P}\right)_{T,N}}{\left(\frac{\partial H}{\partial T}\right)_{P,N}} = \frac{1}{C_p}\left[T\left(\frac{\partial V}{\partial T}\right)_{P,N} - V\right]. \tag{3.17}$$

Let us now compute the Joule-Thompson coefficient for various gases. For an ideal gas $(\partial V/\partial T)_{P,N} = V/T$ and therefore $(\partial T/\partial P)_{H,N} = 0$. There will be no temperature change during the throttling process for an ideal gas. Furthermore, since $T_i = T_f$ for the ideal gas, $P_iV_i = P_fV_f$ and no net work will be done ($\Delta W = 0$). If work had been done on or by the gas we would expect a temperature change since the process is adiabatic.

For a van der Waals gas, assuming that $C_V = \frac{3}{2}nR$, we find

$$\left(\frac{\partial T}{\partial P}\right)_{H,N} = \frac{1}{R}\left[\frac{2a}{RT}\left(\frac{\tilde{v} - b}{\tilde{v}}\right)^2 - b\right] \Bigg/ \left[\frac{5}{2} - \frac{3a}{RT\tilde{v}}\left(\frac{\tilde{v} - b}{\tilde{v}}\right)^2\right], \tag{3.18}$$

where $\tilde{v} = V/n$ is the volume per mole. (Remember that the van der Waals equation becomes meaningless for $\tilde{v} < b$.) For an interacting gas the Joule-Thompson coefficient can change sign. This is easier to see if we consider low densities so that $RT\tilde{v} \gg a$ and $\tilde{v} \gg b$. Then

$$\left(\frac{\partial T}{\partial P}\right)_{H,N} \approx \frac{2}{5R}\left[\frac{2a}{RT} - b\right]. \tag{3.19}$$

For low temperatures $(\partial T/\partial P)_{H,N} > 0$ and gases cool in the throttling process, but at high temperature $(\partial T/\partial P)_{H,N} < 0$ and they heat up. Two effects determine the behavior of the Joule-Thompson coefficient: one is expansion of the gas, which will always cause cooling, and the other is work done on or by the gas. If $P_iV_i > P_fV_f$ then net work is done on the gas, which causes heating. If $P_iV_i < P_fV_f$ then work is done by the gas, which causes cooling.

The inversion temperature (temperature at which the sign of $(\partial T/\partial P)_{H,N}$ changes) for the Joule-Thompson coefficient will be a function of pressure. Since $C_p > 0$, the condition for inversion is

$$\left(\frac{\partial V}{\partial T}\right)_{P,N} = \frac{V}{T} \tag{3.20}$$

or, for a van der Waals gas,

$$\frac{2a}{RT}\left(\frac{\tilde{v} - b}{\tilde{v}}\right)^2 = b. \tag{3.21}$$

We can use van der Waals equation (2.12) to write Eq. (3.21) in terms of pressure and temperature:

$$P = \frac{2}{b}\sqrt{\frac{2aRT}{b}} - \frac{3}{2}\frac{RT}{b} - \frac{a}{b^2}. \tag{3.22}$$

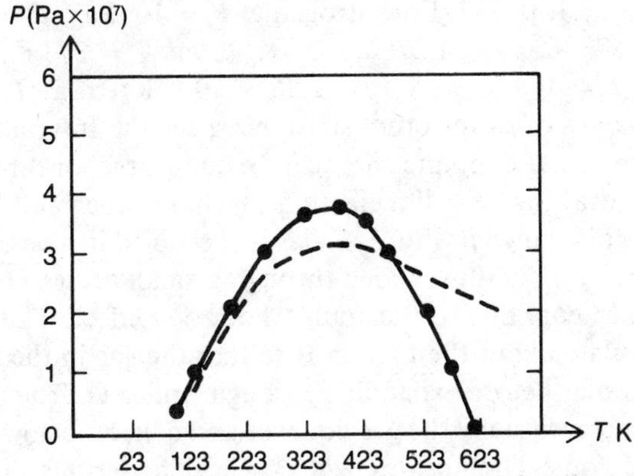

Fig. 3.3. Plot of the inversion temperature versus pressure for the Joule-Thompson coefficient of N_2: the solid line is the experimental curve (based on Ref. 3.1, p. 282); the dashed line is the curve predicted by the van der Waals equation for $a = 0.1408\ \mathrm{Pa \cdot m^6/mol}$ and $b = 0.03913\ \mathrm{m^3/mol}$.

The inversion curve predicted by the van der Waals equation has the shape of a parabola with a maximum at $T_{\max}^{VW} = 8a/9Rb$. For CO_2, $T_{\max}^{VW} = 911$ K while the experimental[1] value is $T_{\max}^{\text{ex}} = 1500$ K. For H_2, $T_{\max}^{VW} = 99$ K while the experimental[1] value is $T_{\max}^{\text{ex}} = 202$ K. In Fig. 3.3 we plot the van der Waals and the experimental inversion curves for N_2. The van der Waals equation predicts an inversion curve which lies below the experimental curve but qualitatively has the correct shape. Some experimental values of the Joule-Kelvin coefficient

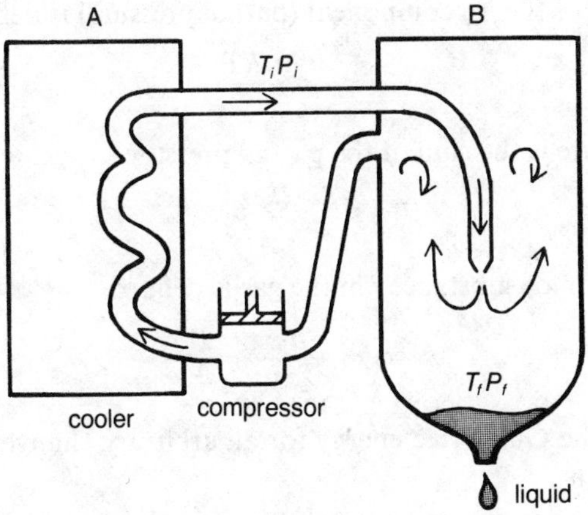

Fig. 3.4. Schematic drawing of a gas liquifier using the Joule-Thompson effect.

are given in Ref. 3.1, p. 282. For nitrogen at $P = 10^5$ Pa., $\mu_J = 1.37 \times 10^{-7}$ K/Pa. at $T = 573$ K, $\mu_J = 1.27 \times 10^{-6}$ K/Pa. at $T = 373$ K, $\mu_J = 6.40 \times 10^{-6}$ K/Pa. at $T = 173$ K, and $\mu_J = 2.36 \times 10^{-5}$ K/Pa. at $T = 93$ K. (For experimental values of μ_J for other substances, see the International Critical Tables.)[2] We see that the cooling effect can be quite large for throttling.

A schematic drawing of a liquefaction machine which utilizes the Joule-Thompson effect is shown in Fig. 3.4. Gas is precooled in a vessel A below its inversion temperature and expanded through a small orifice O into vessel B, thus causing it to cool due to the Joule-Thompson effect. The cooled gas is allowed to circulate about the tube in B so that the gas in the tube becomes progressively cooler before expanding through orifice O. The process is run continuously and eventually the gas liquefies and collects below.

At times the Joule-Thompson effect can lead to serious difficulties. For example,[3] highly compressed H_2, which has a low inversion temperature, can ignite spontaneously when leaking from a damaged container, because of Joule-Thompson heating.

C. ENTROPY OF MIXING AND THE GIBBS PARADOX[1]

If entropy is a measure of disorder in a system, then we expect that an entropy increase will be associated with the mixing of substances, since this causes an increase in disorder.

Let us consider an ideal gas containing n_1 moles of A_1, n_2 moles of $A_2, \ldots, n_m$ moles of A_m. The equation of state for the gas is

$$P = \frac{1}{V}\sum_{j=1}^{m} n_j RT = \frac{n_1 RT}{V} + \frac{n_2 RT}{V} + \cdots + \frac{n_m RT}{V}. \tag{3.23}$$

The pressure P_j of the jth component (partial pressure) is defined

$$P_j = \frac{n_j RT}{V}. \tag{3.24}$$

The total pressure is the sum of the partial pressures,

$$P = \sum_{j=1}^{m} P_j. \tag{3.25}$$

The mole fraction of substance j in the gas is defined

$$x_j = \frac{n_j}{\sum_{j=1} n_j} = \frac{P_j}{P}. \tag{3.26}$$

The change in the Gibbs free energy for an arbitrary change in the variables $P, T, n_1, \ldots, n_m$ is

$$dG = -S\,dT + V\,dP + \sum_{j=1}^{m} \tilde{\mu}_j\,dn_j, \tag{3.27}$$

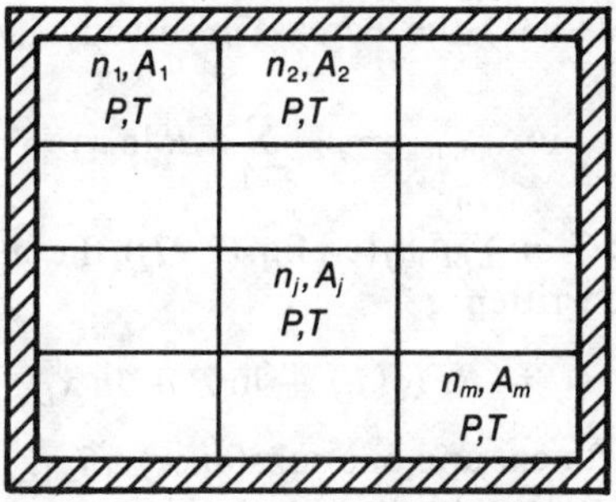

Fig. 3.5. Insulated box containing m different ideal gases separated by partitions, each maintained at pressure P and temperature T.

where $\tilde{\mu}_j$ is the chemical potential per mole and S, V, and $\tilde{\mu}_j$ are given by Eqs. (2.99*a*)–(2.99*c*), respectively, but with mole number n_j replacing particle number N_j.

The chemical potential $\tilde{\mu}_j$ is an intensive quantity. It must depend on the mole numbers n_j in such a way that, if *all* n_j are multiplied by the same factor, then $\tilde{\mu}_j$ does not change. This will be true if $\tilde{\mu}_j$ depends only on the mole fractions.

We wish to find the entropy of mixing for a mixture of ideal gases. Let us first find the Gibbs free energy for a single component system. If the equation of state is $P = nRT/V$ and the heat capacity at constant volume is $C_V = \frac{3}{2}nR$, then the Gibbs free energy is

$$G(P, T, n) = nRT(\phi(T) + \ln P), \tag{3.28}$$

where $\phi(T)$ is a function of temperature only (cf. Eq. (2.178)).

Let us now consider a box at temperature T and pressure P partitioned into m compartments and let us assume that compartment 1 contains n_1 moles of A_1 at pressure P and temperature T, compartment 2 contains n_2 moles of A_2 at pressure P and temperature T, etc. (cf. Fig. 3.5). The Gibbs free energy of the system is the sum of the free energies of each compartment:

$$G_i(P, T, n_1, \ldots, n_m) = \sum_{j=1}^{m} n_j RT(\phi(T) + \ln P). \tag{3.29}$$

If we now remove the partitions and let the gases mix so that the final temperature and pressure are T and P, the Gibbs free energy of the mixture will be

$$G_f(P, T, n_1, \ldots, n_m) = \sum_{j=1}^{m} n_j RT(\phi(T) + \ln P + \ln x_j), \tag{3.30}$$

where we have used the relation $P_j = Px_j$. The change in the Gibbs free energy during the mixing process is

$$G_f - G_i = \sum_{j=1}^{m} n_j RT \ln x_j. \tag{3.31}$$

From Eqs. (2.99*a*) and (3.31), the increase in entropy due to mixing is

$$\Delta S_{\text{mixing}} = -\sum_{j=1}^{m} n_j R \ln x_j, \tag{3.32}$$

and from the equation $G = \sum_j \tilde{\mu}_j n_j$ (cf. Eq. (2.97)), the chemical potential of the jth component can be written:

$$\tilde{\mu}_j = RT(\phi(T) + \ln P + \ln x_j). \tag{3.33}$$

If $x_j = 1$ (one compartment and one substance) $\Delta S_{\text{mixing}} \equiv 0$, as it should be. If there are two compartments, each containing one mole, then $x_1 = \frac{1}{2}$ and $x_2 = \frac{1}{2}$ and $\Delta S_{\text{mixing}} = 2R \ln 2$ and the entropy increases during mixing.

It is important to note that Eq. (3.32) contains no explicit reference to the type of particles in the various compartments. As long as they are different, mixing increases the entropy. However, if they are identical, Eq. (3.32) tells us that there will still be an increase in entropy when the partitions are removed, even though the concept of mixing loses its meaning. Clearly, Eq. (3.32) does not work for identical particles. This was first noticed by Gibbs and is called the *Gibbs paradox*. The resolution of the Gibbs paradox lies in quantum mechanics, as we shall see when we come to statistical mechanics (cf. Chap. 9). Identical particles must be counted in a different way from distinguishable particles (they have different "statistics"). This difference between identical and distinguishable particles persists even in the classical limit and leads to a resolution of the Gibbs paradox.

D. OSMOTIC PRESSURE IN DILUTE SOLUTIONS

Each spring, when the weather begins to warm, sap rises in the trees, leaves grow, and the yearly cycle of life starts again. The rising of sap is one of many examples in biological systems of the phenomenon called osmosis. We can easily demonstrate the effect in the laboratory. Let us take a beaker of water and immerse in it a long tube open at both ends. The water levels of the tube and of the beaker will be the same. Next, close off the bottom end of the tube with a membrane which is permeable to water but not sugar. The water levels will still be the same in the tube and the beaker. Now add a bit of sugar to the water in the tube. Water will begin to enter the tube through the membrane and the level of the sugar solution will rise a distance h above the level of the water in the beaker (cf. Fig. 3.6). The excess pressure created in the tube, $\pi = \rho_s hg$, is called the osmotic pressure (ρ_s is the density of the sugar solution and g is the acceleration of gravity). After equilibrium is reached, the pressure on the side of the membrane with sugar solution will be greater by a factor π than on the water side. The membrane must sustain the unbalanced force.

It is instructive to show the same phenomenon schematically, as in Fig. 3.7. Equilibrium for a two-part system at temperature T consisting of water (A) and

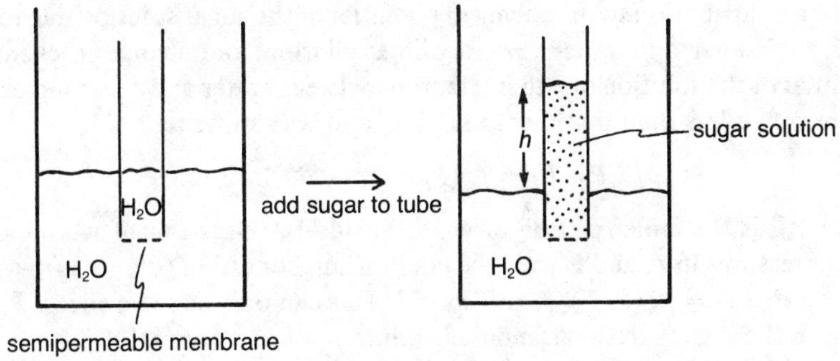

Fig. 3.6. The osmotic pressure of the sugar solution is $\pi = \rho_s hg$, where ρ_s is the density of the sugar solution and g is the acceleration of gravity.

sugar solution (B) separated by a membrane permeable only to water occurs when there is an imbalance of pressure. However, since water is free to move through the membrane, for equilibrium we must have

$$\tilde{\mu}_w{}^{\mathrm{A}}(P_{\mathrm{A}}, T) = \tilde{\mu}_w{}^{\mathrm{B}}(P_{\mathrm{B}}, T), \tag{3.34}$$

where $\tilde{\mu}_w{}^{\mathrm{A}}$ and $\tilde{\mu}_w{}^{\mathrm{B}}$ are the chemical potentials of water in sides A and B, respectively. (Note that since the membrane is fixed we need not have equal pressures.)

We can easily write the chemical potentials for water in A and B. In side A we have

$$\tilde{\mu}_w{}^{\mathrm{A}}(P_{\mathrm{A}}, T) = \tilde{\mu}_w(P_{\mathrm{A}}, T)_0, \tag{3.35}$$

where $\tilde{\mu}_w(P_{\mathrm{A}}, T)_0$ is the chemical potential of pure water. In side B, we have a sugar solution. From the preceding section (cf. Eq. (3.33)) we know that there

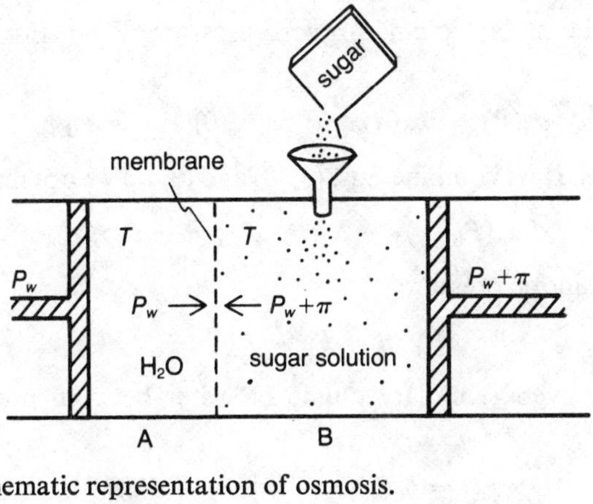

Fig. 3.7. A schematic representation of osmosis.

will be a contribution to the chemical potential of the sugar solution due to the mixing of sugar and water. For a dilute solution, the change in chemical potential of the solution due to interaction between water and sugar molecules will be much less than that due to mixing, and we can write

$$\tilde{\mu}_w{}^{\mathrm{B}}(P_{\mathrm{B}}, T) = \tilde{\mu}_w(P_{\mathrm{B}}, T)_0 + RT \ln x_w{}^{\mathrm{B}}, \tag{3.36}$$

where $x_w{}^{\mathrm{B}}$ is the mole fraction of water in side B. (Note that if we imposed equal pressures in A and B then the equilibrium condition (Eq. (3.34)) would read $\tilde{\mu}_w(P, T)_0 = \tilde{\mu}_w(P, T)_0 + RT \ln x_w{}^{\mathrm{B}}$. This can only be satisfied if $x_w{}^{\mathrm{B}} = 1$; that is, if the sugar solution is infinitely dilute.)

Let us now obtain an expression for $\tilde{\mu}_w(P_{\mathrm{B}}, T)_0 - \tilde{\mu}_w(P_{\mathrm{A}}, T)_0$ from a different point of view. Since the Gibbs free energy of water in B is $G_w{}^{\mathrm{B}}(P_{\mathrm{B}}, T) = n_w{}^{\mathrm{B}}\tilde{\mu}_w{}^{\mathrm{B}}(P_{\mathrm{B}}, T)$, we can use Eq. (2.99*b*) to write

$$\left(\frac{\partial \tilde{\mu}_w}{\partial P}\right)_{T, n_w} = \frac{V}{n_w} = \tilde{v}_w, \tag{3.37}$$

where $\tilde{v}_w$ is the molar volume of water. Thus, for fixed T and n_w, we find

$$\tilde{\mu}_w(P_{\mathrm{B}}, T)_0 - \tilde{\mu}_w(P_{\mathrm{A}}, T)_0 = \int_{P_{\mathrm{A}}}^{P_{\mathrm{B}}} \tilde{v}_w \, dP. \tag{3.38}$$

In general, the molar volume will be a function of pressure and is related to the isothermal compressibility through the relation $\kappa_T = -(1/\tilde{v}_w)(\partial \tilde{v}_w/\partial P)_T$. It is a fairly good approximation to assume that κ_T is independent of pressure. Then we find

$$d \ln \tilde{v}_w(P) = -\kappa_T \, dP \tag{3.39}$$

and

$$\ln \frac{\tilde{v}_w(P)}{\tilde{v}_w(0)} = -\kappa_T P, \tag{3.40}$$

where $\tilde{v}_w(0)$ is the molar volume for zero pressure. From Eq. (3.40) we can write

$$\tilde{v}_w(P) = \tilde{v}_w(0)\, e^{-\kappa_T P} \approx \tilde{v}_w(0)(1 - \kappa_T P) \tag{3.41}$$

for low pressures. If we combine Eqs. (3.38) and (3.41) we obtain

$$\tilde{\mu}_w(P_{\mathrm{B}}, T)_0 - \tilde{\mu}_w(P_{\mathrm{A}}, T)_0 = \pi\langle\tilde{v}_w\rangle, \tag{3.42}$$

where π is the osmotic pressure

$$\pi = P_{\mathrm{B}} - P_{\mathrm{A}} \tag{3.43}$$

and $\langle\tilde{v}_w\rangle$ is the average molar volume of water between pressure P_{A} and P_{B}:

$$\langle\tilde{v}_w\rangle = \tilde{v}_w(0)[1 - \tfrac{1}{2}\kappa_T(P_{\mathrm{A}} + P_{\mathrm{B}})]. \tag{3.44}$$

From Eqs. (3.34)–(3.36) and (3.42) we find the following equation for the osmotic pressure:

$$-RT \ln x_w{}^{\mathrm{B}} = \pi\langle \tilde{v}_w \rangle. \tag{3.45}$$

For very dilute solutions $x_w{}^{\mathrm{B}} \sim 1$ and $x_s{}^{\mathrm{B}} \sim 0$ so that $n_s{}^{\mathrm{B}}/n_w{}^{\mathrm{B}} \ll 1$ and Eq. (3.45) takes the form

$$\pi V_{\mathrm{B}} = n_s{}^{\mathrm{B}} RT. \tag{3.46}$$

(We have let $\langle \tilde{v}_w \rangle = V_{\mathrm{B}}/n_w^{\mathrm{B}}$ and we have used the approximation $\ln x = (x - 1)/x$ when $x > \frac{1}{2}$.) Eq. (3.46) is called van't Hoff's law and, surprisingly, looks very much like the ideal gas law, although we are by no means dealing with a mixture of ideal gases. Eq. (3.46) is well verified for dilute solutions. Although we have considered the special case of a dilute water-sugar system, the arguments and Eq. (3.46) hold for any dilute neutral solvent-solute system.

It is useful to give some experimental values of π. For solutions containing M moles of sucrose ($C_{12}H_{22}O_{11}$) for each kilogram of water at 303 K, we have $\pi = 2.53 \times 10^5$ Pa. for $M = 0.1$, $\pi = 5.17 \times 10^5$ Pa. for $M = 0.2$, and $\pi = 7.81 \times 10^5$ Pa. for $M = 0.3$ (cf. Ref. 3.2).

E. THE THERMODYNAMICS OF CHEMICAL REACTIONS[1,4,5]

Chemical reactions occur in systems containing several species of molecules (which we will call A, B, C, and D), which can transform into one another through inelastic collisions. A typical case might be one in which molecules A and B can collide inelastically to form molecules C and D; conversely, molecules C and D can collide inelastically to form molecules A and B (cf. Fig. 3.8). As we shall see in Chap. 13, the collisions occur at random and can be either elastic or inelastic. To be inelastic and result in a reaction the two molecules must have sufficient energy to overcome any potential barriers to the reaction

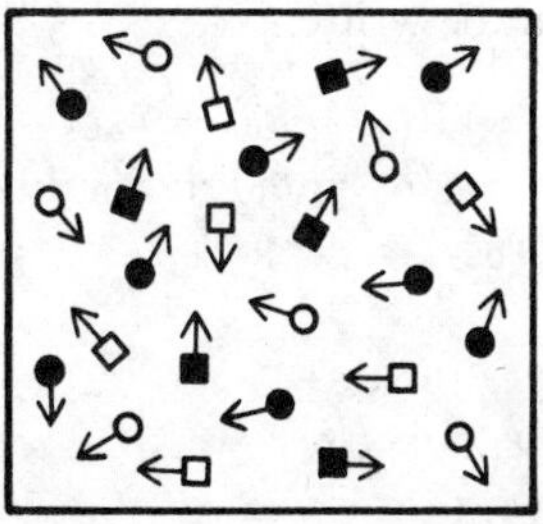

Fig. 3.8. A mixture of molecules which move and collide at random: collisions between circles can lead to squares and vice versa if the relative energy of the colliding particles is sufficiently high.

which might exist. Chemical equilibrium is a dynamical state of the system. It occurs when the rate of production of each chemical species is equal to its rate of depletion through chemical reactions. The chemical reactions themselves never stop, even at equilibrium.

In the early part of this century a Belgian scientist, de Donder, found that it was possible to characterize each chemical reaction by a single variable, ξ, called the degree of reaction. In terms of ξ, it is then possible to determine when the Gibbs free energy has reached its minimum value (chemical reactions usually take place in systems with fixed temperature and pressure) and therefore when the chemical system reaches chemical equilibrium.

1. The Affinity

Let us consider a chemical reaction of the form

$$-\nu_A A - \nu_B B \rightleftharpoons \nu_C C + \nu_D D. \tag{3.47}$$

The quantities ν_A, ν_B, ν_C, and ν_D are called stoichiometric coefficients; ν_j is the number of molecules of type j needed for the reaction to take place. By convention, ν_A and ν_B are negative. Let us assume that initially there are $n_A = -\nu_A n_0$ moles of A, $n_C = \nu_C n_0'$ moles of C, $n_B = -\nu_B n_0 + N_B$ moles of B, and $n_D = \nu_D n_0' + N_D$ moles of D. The reaction to the right will be complete when

$$n_A = 0, \qquad n_B = N_B, \qquad n_C = \nu_C(n_0 + n_0'), \qquad n_D = \nu_D(n_0 + n_0') + N_D.$$

The reaction to the left will be complete when

$$n_A = -\nu_A(n_0 + n_0'), \qquad n_B = -\nu_B(n_0 + n_0') + N_B, \qquad n_C = 0, \qquad n_D = N_D.$$

We next define the degree of reaction by the equation

$$\xi = (n_0 + n_0') + \frac{n_A}{\nu_A}. \tag{3.48}$$

As we have defined it, ξ has the units of moles. In terms of ξ, the number of moles of each substance can be written

$$\begin{aligned} n_A &= -\nu_A(n_0 + n_0') + \nu_A \xi, \\ n_B &= -\nu_B(n_0 + n_0') + N_B + \nu_B \xi, \\ n_C &= \nu_C \xi, \end{aligned} \tag{3.49}$$

and

$$n_D = \nu_D \xi + N_D.$$

Any changes in the concentrations due to the reaction can therefore be written

$$\begin{aligned} dn_A &= \nu_A\, d\xi \qquad & dn_C &= \nu_C\, d\xi \\ dn_B &= \nu_B\, d\xi \qquad & dn_D &= \nu_D\, d\xi \end{aligned} \tag{3.50}$$

or

$$\frac{dn_A}{\nu_A} = \frac{dn_B}{\nu_B} = \frac{dn_C}{\nu_C} = \frac{dn_D}{\nu_D} = d\xi. \tag{3.51}$$

Eqs. (3.50) and (3.51) are very important because they tell us that any changes in the thermodynamic properties of a system due to a given reaction can be characterized by a single variable.

For systems at constant pressure and temperature, the change in the Gibbs free energy due to a reaction may be written

$$dG = \sum_{j=1}^{m} \tilde{\mu}_j \, dn_j = \sum_{j=1}^{m} \tilde{\mu}_j \nu_j \, d\xi, \tag{3.52}$$

where the sum is over the species which participate in the reaction. Therefore,

$$\left(\frac{\partial G}{\partial \xi}\right)_{P,T} = \sum_{j=1}^{m} \tilde{\mu}_j \nu_j. \tag{3.53}$$

The quantity

$$\mathscr{A} \equiv \sum_{j=1}^{m} \tilde{\mu}_j \nu_j \tag{3.54}$$

is called the affinity (in some books affinity is defined with an opposite sign). At chemical equilibrium, the Gibbs free energy must be a minimum,

$$\left(\frac{\partial G}{\partial \xi}\right)_{P,T}^{\text{equil}} = \mathscr{A}^{\text{equil}} = 0, \tag{3.55}$$

and, therefore, at chemical equilibrium, the affinity must be zero.

We can easily find the sign of the affinity as the system moves toward chemical equilibrium from the left or right. At constant P and T, the Gibbs free energy, G, must always decrease as the system moves toward chemical equilibrium (at equilibrium G is a minimum). Therefore,

$$dG = \left(\frac{\partial G}{\partial \xi}\right)_{P,T} d\xi < 0. \tag{3.56}$$

If the reaction goes to the right, then $d\xi > 0$ and $\mathscr{A} < 0$. If the reaction goes to the left, then $d\xi < 0$ and $\mathscr{A} > 0$.

We may use the ideal gas equations of Sec. 3.C to get some idea of how the Gibbs free energy behaves as a function of ξ. From Eq. (3.33) we may write

$$\mathscr{A} = \left(\frac{\partial G}{\partial \xi}\right)_{P,T} = \sum_{j=A}^{D} \tilde{\mu}_j \nu_j = \sum_{j=A}^{D} \nu_j RT(\phi(T) + \ln P + \ln x_j)$$

$$= \sum_{j=A}^{D} \nu_j RT(\phi(T) + \ln P) + RT \ln \frac{x_C^{\nu_C} x_D^{\nu_D}}{x_A^{|\nu_A|} x_B^{|\nu_B|}}. \tag{3.57}$$

(The sum ranges over the four types of molecules A, B, C, D.) From Eqs. (3.49)

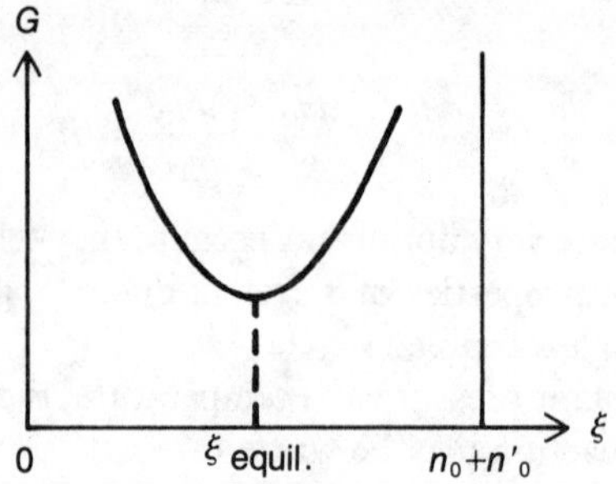

Fig. 3.9. A plot of the Gibbs free energy versus the degree of reaction for a typical chemical reaction.

and (3.57), we see that, when $n_C = 0$, $\xi = 0$ and $(\partial G/\partial \xi)_{P,T} = -\infty$. When $n_A = 0$, $\xi = (n_0 + n_0')$ and $(\partial G/\partial \xi)_{P,T} = +\infty$. Therefore, G is a convex function of ξ with a minimum at chemical equilibrium (cf. Fig. 3.9). The affinity $\mathscr{A} = (\partial G/\partial \xi)_{P,T}$ has values ranging from $-\infty$ to ∞.

If there are r chemical reactions in the system involving species j, then there will be r parameters ξ_k needed to describe the rate of change of the number of moles n_j,

$$dn_j = \sum_{k=1}^{r} \nu_{jk}\, d\xi_k. \tag{3.58}$$

The sum over k is over all chemical reactions in which molecules of type j participate.

2. Stability

Given the fact that the Gibbs free energy for fixed P and T is minimum at equilibrium, we can deduce a number of interesting general properties of chemical reactions. First, let us note that at equilibrium

$$\left(\frac{\partial G}{\partial \xi}\right)_{T,P} \equiv \mathscr{A} = 0 \tag{3.59}$$

and

$$\left(\frac{\partial^2 G}{\partial \xi^2}\right)_{T,P} = \left(\frac{\partial \mathscr{A}}{\partial \xi}\right)_{T,P} > 0. \tag{3.60}$$

Eqs. (3.59) and (3.60) are statements of the fact that the Gibbs free energy is minimum at equilibrium for fixed T and P.

From the fundamental equation, $H = G + TS$, we obtain several important relations. First, let us note that at equilibrium

$$\left(\frac{\partial H}{\partial \xi}\right)_{P,T} = T\left(\frac{\partial S}{\partial \xi}\right)_{P,T} \tag{3.61}$$

(we have used Eq. (3.59)). Thus, changes in enthalpy during the reaction are proportional to the change in heat content or entropy. The left-hand side of Eq. (3.61) is called the heat of reaction. It is the heat absorbed per unit reaction in the neighborhood of equilibrium. For an exothermic reaction $(\partial H/\partial \xi)_{P,T}$ is negative. For an endothermic reaction $(\partial H/\partial \xi)_{P,T}$ is positive. From Eq. (2.99*a*), Eq. (3.61) can be written

$$\left(\frac{\partial H}{\partial \xi}\right)_{P,T} = -T\left[\frac{\partial}{\partial \xi}\left(\frac{\partial G}{\partial T}\right)_{P,\xi}\right]_{P,T} = -T\left[\frac{\partial}{\partial T}\left(\frac{\partial G}{\partial \xi}\right)_{P,T}\right]_{P,\xi} = -T\left(\frac{\partial \mathscr{A}}{\partial T}\right)_{P,\xi}. \tag{3.62}$$

Eq. (3.62) will be useful below.

We can now obtain some general properties of chemical reactions. From the chain rule (Eq. (2.6)) we can write

$$\left(\frac{\partial \xi}{\partial T}\right)_{P,\mathscr{A}} = -\frac{\left(\frac{\partial \mathscr{A}}{\partial T}\right)_{P,\xi}}{\left(\frac{\partial \mathscr{A}}{\partial \xi}\right)_{P,T}} = \frac{1}{T}\frac{\left(\frac{\partial H}{\partial \xi}\right)_{P,T}}{\left(\frac{\partial \mathscr{A}}{\partial \xi}\right)_{P,T}}. \tag{3.63}$$

The denominator in Eq. (3.63) is always positive. Thus, at equilibrium any small increase in temperature causes the reaction to shift in a direction in which heat is absorbed.

Let us next note that

$$\left(\frac{\partial \mathscr{A}}{\partial P}\right)_{T,\xi} = \left(\frac{\partial V}{\partial \xi}\right)_{P,T}. \tag{3.64}$$

In Eq. (3.64) we have used Eqs. (3.53), (3.54), (2.99*b*), and (2.4). Thus,

$$\left(\frac{\partial \xi}{\partial P}\right)_{T,\mathscr{A}} = -\frac{\left(\frac{\partial \mathscr{A}}{\partial P}\right)_{T,\xi}}{\left(\frac{\partial \mathscr{A}}{\partial \xi}\right)_{T,P}} = -\frac{\left(\frac{\partial V}{\partial \xi}\right)_{P,T}}{\left(\frac{\partial \mathscr{A}}{\partial \xi}\right)_{P,T}}. \tag{3.65}$$

At equilibrium an increase in pressure at fixed temperature will cause the reaction to shift in a direction which decreases the total volume.

3. Law of Mass Action and Heat of Reaction

For a system in chemical equilibrium at fixed temperature and pressure, the ratio of mole fractions $x_C^{\nu_C}x_D^{\nu_D}/x_A^{|\nu_A|}x_B^{|\nu_B|}$ must remain constant. This is easily seen from Eq. (3.57). At chemical equilibrium the affinity $\mathscr{A} = 0$ and we have

$$\begin{aligned}\ln\left(\frac{x_C^{\nu_C}x_D^{\nu_D}}{x_A^{|\nu_A|}x_B^{|\nu_B|}}P^{\nu_C+\nu_D+\nu_A+\nu_B}\right) &= -(\nu_C\phi_C(T) + \nu_D\phi_D(T) + \nu_A\phi_A(T) + \nu_B\phi_B(T)) \\ &= \ln K(T),\end{aligned} \tag{3.66}$$

where $K(T)$ is the equilibrium constant for the reaction $-\nu_A A - \nu_B B \rightleftharpoons \nu_C C + \nu_D D$. Eq. (3.66) is called the law of mass action. From it we can compute the degree of reaction and therefore the mole fractions at equilibrium as a function of pressure and temperature (this is left as a problem).

From Eq. (3.66), we can also obtain the heat of reaction of a chemical reaction. We shall consider systems which are dilute enough that we can approximate the thermodynamic quantities by those of an ideal gas. We must first find a relation between the enthalpy and the Gibbs free energy for an ideal gas. From Eq. (2.171) and the fact that $C_p = (\alpha + 1)nR$, we can write the enthalpy in the form

$$H(T) = H_0 + (\alpha + 1)nR(T - T_0) = H_0 + \int_{T_0}^{T} C_p\, dT'. \tag{3.67}$$

From Eqs. (2.167) and (2.174) the entropy can be written

$$S(T, P) = S_0 + \int_{T_0}^{T} \frac{C_p\, dT'}{T'} - nR \ln \frac{P}{P_0}. \tag{3.68}$$

The Gibbs free energy, which is related to the enthalpy by the fundamental equation $G = H - TS$, takes the form

$$G(T, P) = H_0 + \int_{T_0}^{T} C_p\, dT' - TS_0 - T \int_{T_0}^{T} C_p \frac{dT'}{T'} + nRT \ln \frac{P}{P_0} \tag{3.69}$$

or, after integration by parts,

$$G(T, P) = H_0 - T \int_{T_0}^{T} \frac{\int_{T_0}^{T'} C_p\, dT''}{T'^2}\, dT' - TS_0 + nRT \ln \frac{P}{P_0}. \tag{3.70}$$

From the fundamental equation, $G = n\tilde{\mu}$, we can write the chemical potential as

$$\tilde{\mu}(P, T) = \tilde{h}_0 - T \int_{T_0}^{T} \frac{\int_{T_0}^{T'} \tilde{c}_p\, dT''}{T'^2}\, dT' - \tilde{s}_0 T + RT \ln \frac{P}{P_0}, \tag{3.71}$$

where $\tilde{h}_0 = H_0/n$, $\tilde{c}_p = C_p/n$, and $\tilde{s}_0 = S_0/n$. If we now compare Eqs. (3.28) and (3.71), we find

$$\phi(T) = \frac{\tilde{h}_0}{RT} - \frac{1}{R} \int_{T_0}^{T} \frac{\int_{T_0}^{T'} \tilde{c}_p\, dT}{T'^2}\, dT' - \frac{\tilde{s}_0}{R}. \tag{3.72}$$

Since $\phi(T)$ depends only on temperature, we can write

$$\frac{d\phi}{dT} = -\frac{\tilde{h}_0}{RT^2} - \frac{\int_{T_0}^{T'} \tilde{c}_p\, dT'}{RT^2} = -\frac{\tilde{h}}{RT^2} \tag{3.73}$$

where h is the total enthalpy per mole (cf. Eq. (3.67)).

We now can relate the equilibrium constant to the heat of reaction. If we take the temperature derivative of Eq. (3.66), we find

$$\frac{d \ln K(T)}{dT} = +\frac{\Delta \tilde{h}}{RT^2}, \tag{3.74}$$

where

$$\Delta \tilde{h} = \nu_C \tilde{h}_C + \nu_D \tilde{h}_D + \nu_A \tilde{h}_A + \nu_B \tilde{h}_B \tag{3.75}$$

is the heat of reaction and is equal to the heat given off or absorbed during the chemical reaction. Eq. (3.74) is called van't Hoff's relation. We can write van't Hoff's relation in another form if we note that

$$\left(\frac{\partial H}{\partial \xi}\right)_{P,T} = -T\left(\frac{\partial \mathscr{A}}{\partial T}\right)_{P,\xi} = -\mathscr{A} - RT^2\left[\frac{\partial}{\partial T}\sum_j \nu_j \phi_j(T)\right]_{P,\xi} \tag{3.76}$$

However, at equilibrium $\mathscr{A} = 0$ and from Eq. (3.66) we find

$$\left(\frac{\partial H}{\partial \xi}\right)_{P,T} = -RT^2 \frac{dK(T)}{dT} = \Delta h \tag{3.77}$$

The affinity and the degree of reaction are extremely important because they enable us to determine the distance of a chemical reaction from its equilibrium state. We shall see them again when we write the hydrodynamic equations for systems undergoing chemical reactions and when we treat nonequilibrium phase transitions in chemical systems.

F. THERMOMECHANICAL EFFECT[6]

Liquid He^4, below 2.19 K, appears to be composed of two interpenetrating fluids. One fluid (superfluid) can flow through cracks so small that even He^4 gas cannot leak through, and it appears to carry no entropy. The other fluid (normal fluid) behaves normally. The fact that the superfluid carries no entropy leads to some very interesting effects. Let us consider two vessels filled with liquid He^4 at a temperature below 2.19 K, and let us assume that they are connected by a

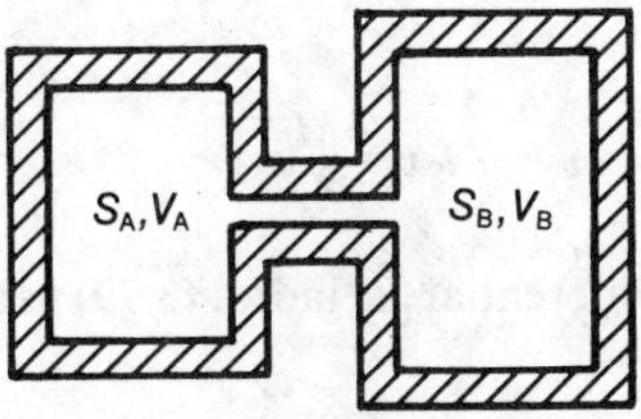

Fig. 3.10. Two vessels containing liquid He^4 below 2.19 K and connected by a very fine capillary; only superfluid can pass between the two vessels.

capillary so thin that only the superfluid can flow through it (cf. Fig. 3.10). Let us further assume that the vessels are insulated from the outside world. This means that the total entropy must remain constant if no irreversible processes take place. Let us also assume that the total mass and total volume of the system remain constant. Under these conditions, the equilibrium state is a state of minimum internal energy.

We can obtain the condition for equilibrium between the vessels if we assume that matter can flow between them but entropy cannot. The total internal energy will be denoted U_T, and u_l will denote internal energy per kilogram in vessel l. The total internal energy is given by

$$U_T = \sum_{l=\mathrm{A,B}} M_l u_l, \tag{3.78}$$

where M_l is the total mass in vessel l. In equilibrium, the total internal energy must be a minimum. Thus,

$$\delta U_T = 0 = \sum_{l=\mathrm{A,B}} (u_l\, \delta M_l + M_l\, \delta u_l). \tag{3.79}$$

Let us now assume that the total entropy and volume of each vessel is a constant: S_l = constant and V_l = constant. The entropy of the fluid in vessel l can be written in terms of the mass of fluid in l, M_l, and the entropy per kilogram, s_l (specific entropy), as

$$S_l = M_l s_l. \tag{3.80}$$

Similarly,

$$V_l = M_l v_l \tag{3.81}$$

where v_l is the volume per kilogram (specific volume) of fluid in vessel l. Since S_l and V_l are constants, we find

$$\delta S_l = M_l\, \delta s_l + s_l\, \delta M_l = 0 \tag{3.82}$$

and

$$\delta V_l = M_l\, \delta v_l + v_l\, \delta M_l = 0 \tag{3.83}$$

or

$$\delta s_l = -s_l \frac{\delta M_l}{M_l} \tag{3.84}$$

and

$$\delta v_l = -v_l \frac{\delta M_l}{M_l}. \tag{3.85}$$

Let us now expand the differential δu_l in Eq. (3.79) in terms of specific entropy and specific volume. Thus,

$$0 = \sum_{l=\mathrm{A,B}} \left(u_l\, \delta M_l + M_l\left[\left(\frac{\partial u_l}{\partial s_l}\right)_v \delta s_l + \left(\frac{\partial u_l}{\partial v_l}\right)_s \delta v_l\right]\right). \tag{3.86}$$

If we make use of Eqs. (2.73*a*) and (2.73*b*) and Eqs. (3.84) and (3.85), we find

$$\sum_{l=\mathrm{A,B}} (u_l - s_l T_l + v_l P_l)\,\delta M_l = 0. \tag{3.87}$$

Since total mass is conserved, we can write $\delta M_{\mathrm{A}} = -\delta M_{\mathrm{B}}$. This relation, together with the fundamental equation for the Gibbs free energy, yields the equilibrium condition

$$\mu_{\mathrm{A}}(T_{\mathrm{A}}, P_{\mathrm{A}}) = \mu_{\mathrm{B}}(T_{\mathrm{B}}, P_{\mathrm{B}}), \tag{3.88}$$

as we would expect (μ is the chemical potential per kilogram). Thus, equilibrium occurs when the chemical potentials are equal, but the pressures and temperatures need not be equal.

We can now vary the temperature and pressure in one of the vessels in such a way that the two vessels remain in equilibrium. That is, we change the pressure and temperature in such a way that the chemical potentials remain equal. Then we find

$$\Delta\mu = 0 = -s\,\Delta T + v\,\Delta P, \tag{3.89}$$

where $s = -(\partial\mu/\partial T)_P$ and $v = (\partial\mu/\partial P)_T$. Eq. (3.89) yields the relation

$$\Delta P = \frac{s}{v}\Delta T, \tag{3.90}$$

which must be satisfied if the two vessels are to remain in equilibrium. Thus, if there is a temperature difference between the two vessels, there will also be a pressure difference. The vessel with higher temperature will also have higher pressure. This is called the thermomechanical effect. If initially we connect the two vessels when $T_{\mathrm{B}} > T_{\mathrm{A}}$ and $P_{\mathrm{B}} = P_{\mathrm{A}}$, matter will flow from the colder vessel A to the warmer one until Eq. (3.90) is satisfied. Notice that this appears to contradict the second law of thermodynamics, which states that heat must flow spontaneously from high temperature to low temperature. However, if the

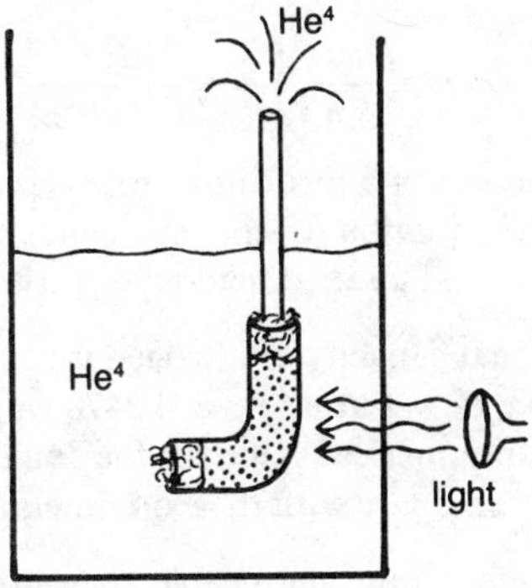

Fig. 3.11. The fountain effect.

helium that flows through the capillary carries no entropy (no heat), the second law is not violated.

The thermomechanical effect is most dramatically demonstrated in terms of the so-called fountain effect, first discovered by Allen and Jones.[7] Imagine a small elbow tube filled with very fine powder, with cotton stuffed in each end. Assume that a long, thin capillary tube is put in one end and the elbow tube is immersed in liquid He^4 at a temperature below 2.19 K. If we now irradiate the elbow tube with a burst of light, helium will be drawn into the elbow tube and will spurt out of the capillary tube (cf. Fig. 3.11). This is called the fountain effect and is a consequence of Eq. (3.90). When the helium in the elbow tube is heated by the radiation, superfluid will flow in to equalize the chemical potentials (and increase the pressure in the elbow tube), overshoot, and spurt out of the capillary tube.

REFERENCES

1. M. W. Zemansky, *Heat and Thermodynamics* (McGraw-Hill Book Co., New York, 1957).
2. *International Critical Tables*, ed. E. W. Washburn (McGraw-Hill Book Co., New York, 1928).
3. A. Sommerfeld, *Thermodynamics and Statistical Mechanics* (Academic Press, New York, 1964).
4. H. B. Callen, *Thermodynamics* (John Wiley & Sons, New York, 1960).
5. I. Prigogine and R. Defay, *Chemical Thermodynamics* (Longmans, Green & Co., London, 1954).
6. F. London, *Superfluids II: Macroscopic Theory of Superfluid Helium* (Dover Publications, New York, 1964).
7. J. F. Allen and J. Jones, Nature *141* 243 (1938).

PROBLEMS

1. For a gas obeying the Dieterici equation of state,

$$P \exp\left(\frac{na_D}{VRT}\right) = \frac{nRT}{(V - nb_D)},$$

compute the Joule coefficient. For a throttling process find the maximum inversion temperature and obtain the equation for the inversion curve. Make a rough sketch of the inversion curve predicted by the Dieterici equation of state.

2. The constant pressure heat capacity of nitrogen is roughly constant over a wide range of temperatures. At $P = 1$ atm. $c_p \approx 0.2470$ cal/gm K. Use the van der Waals equation to obtain a numerical value for the Joule-Thompson coefficient at $T = 373$ K. How does this compare with the experimental value given in Sec. 3.B?

3. A tiny sack made of membrane permeable to water but not NaCl (sodium chloride) is filled with a 1% solution of NaCl and water immersed in an open

beaker of pure water at 38°C at the depth of 1 ft. What osmotic pressure is experienced by the sack? What is the total pressure of the solution in the sack? Assume the sack is small enough that the pressure of the surrounding water can be assumed constant. (An example of such a sack is a human blood cell.)

4. Use Eq. (3.46) to compute the osmotic pressure for a solution of M moles of sucrose ($C_{12}H_{22}O_{11}$) and 1000 grams of water at a temperature of 30°C. Find the osmotic pressure for $M = .100$, $M = .200$, and $M = .300$ and compare your results to the experimental values given at the end of Sec. 3.D.

5. An insulated box is partitioned into m insulated compartments, each containing an ideal gas of a different molecular species. Assume that each compartment has the same pressure but a different number of moles, a different temperature, and a different volume. (The thermodynamic variables for the ith compartment are (P, n_i, T_i, V_i).) If all partitions are suddenly removed, and the system is allowed to reach equilibrium:

(*a*) Find the final temperature and pressure, and the entropy of mixing. (Assume the particles are monatomic.)

(*b*) For the special case of $m = 2$ and parameters $n_1 = 1$ mole, $T_1 = 300$ K, $V_1 = 1$ liter, $n_2 = 2$ moles, and $V_2 = 2$ liters, obtain numerical values for all parameters in part (*a*). What are the units of entropy?

6. A box is partitioned into m compartments, each containing an ideal gas of a different molecular species. Assume that each compartment has the same temperature but a different number of moles, a different pressure, and a different volume. (The thermodynamic variables for the ith compartment are (P_i, T, n_i, V_i).) If the partitions are all suddenly removed and the system is allowed to reach equilibrium:

(*a*) Find the final temperature, pressure, and entropy of mixing.

(*b*) For the special case of $m = 2$ and parameters $n_1 = 1$ mole, $P_1 = 1$ atm, $V_1 = 1$ liter, $n_2 = 2$ moles, and $V_2 = 2$ liters, obtain numerical values for all parameters in part (*a*).

7. The mixing of ideal gases by suddenly removing partitions between them is a spontaneous process and leads to an increase of entropy of the system due to mixing. It is possible to devise a reversible, isothermal process using semipermeable membranes whereby a mixture of two ideal gases at temperature T can be separated into two compartments, each containing a different gas, and such that the entropies of the initial and final states are the same. Explain how this might be done.

8. One mole of H_2S and 2 moles of H_2O in the gas phase are mixed together at pressure P_0 and temperature T_0. As a result of collisions they undergo the reaction

$$H_2S + 2H_2O \rightleftharpoons 3H_2 + SO_2.$$

Find the number of mole fractions of each substance and the affinity as a function of ξ (ξ is the degree of reaction). For what value of ξ is chemical equilibrium reached? Find the equilibrium constant $K(T_0)$ and the volume of the system at equilibrium. (Determine all quantities as a function of P_0 and T_0 and assume the system is dilute enough that the ideal gas equations can be used.)

9. For a mixture of reacting ideal gases in equilibrium, compute $(\partial V/\partial P)_{T,\{n_i\}}$ and

$(\partial V/\partial T)_{P,\{n_i\}}$ as function concentrations, pressure, and temperature. Compute the same quantities for a mixture of ideal gases which cannot react. How do they differ?

10. Work Prob. 3.8, but use the chemical reaction $N_2 + 3H_2 \rightleftharpoons 2NH_3$.

11. Show that $(\partial U/\partial \xi)_{T,V} = T(\partial S/\partial \xi)_{T,V}$ and that

$$\mathscr{A} = \left(\frac{\partial U}{\partial \xi}\right)_{S,V} = \left(\frac{\partial H}{\partial \xi}\right)_{S,P} = \left(\frac{\partial A}{\partial \xi}\right)_{T,V}.$$

12. Assume that the two vessels of liquid He^4 in Sec. 3.F are maintained under isothermal conditions; that is, vessel A is held at temperature T_A and vessel B is held at temperature T_B. If an amount of mass, ΔM, is transferred from vessel A to vessel B, how much heat must flow out of (or into) each vessel? (Assume $T_A > T_B$.)

4

Thermodynamics of Phase Transitions

A. INTRODUCTORY REMARKS

Thermodynamic systems can exist in a number of phases, each of which can exhibit dramatically different macroscopic behavior. Generally, systems become more ordered as temperature is lowered. Forces of cohesion tend to overcome thermal motion, and atoms rearrange themselves in a more ordered state. Phase changes occur abruptly at some critical temperature although evidence that one will occur can be found on a microscopic scale as the critical temperature is approached. The study of the transition region between phases is one of the most interesting fields of statistical physics and one that we shall return to often throughout this book. In this chapter we will be concerned solely with the thermodynamics of phase transitions, that is, the description of phase transitions in terms of macroscopic variables. In later chapters we shall study them from a microscopic point of view.

The first step in trying to understand the phase changes that occur in a system is to map out the phase diagram for the system. At a transition point, two (or more) phases can coexist in equilibrium with each other. The condition for equilibrium between phases is obtained from the equilibrium conditions derived in Chap. 2. Since phases can exchange matter, equilibrium occurs when the chemical potentials of the phases become equal for given values of Y and T. From the equilibrium condition, we can determine the maximum number of phases that can coexist and, in principle, find equations for the regions of coexistence (e.g., the Clausius-Clapeyron equation).

At phase transitions the chemical potentials of the phases, and therefore the Gibbs free energy, must change continuously. However, phase transitions can be divided into two classes according to the behavior of derivatives of the Gibbs free energy. Phase transitions which are accompanied by a discontinuous change of state (discontinuous first derivatives of the Gibbs free energy) are called first-order phase transitions. Phase transitions which are accompanied by a continuous change of state (higher order derivatives of the Gibbs free energy will be discontinuous) are called continuous phase transitions. We give examples of both in this chapter.

Classical fluids provide some of the most familiar examples of first-order phase transitions. The vapor-liquid, vapor-solid, and liquid-solid transitions are all first order. We shall discuss the phase transitions in classical fluids in some detail. For the vapor-solid and vapor-liquid transitions, we can use the Clausius-Clapeyron equation to find explicit equations for the coexistence curves. Since the vapor-liquid transition terminates in a critical point, we will focus on it and compare the observed behavior of the vapor-liquid coexistence region to that predicted by the van der Waals equation.

A binary mixture of molecules provides a different example of a phase transition. For that system below a certain critical temperature, we can have a physical separation of the mixture into two fluids, each rich in one of the types of molecules. Following Hildebrand, it is possible to write a rather simple expression for the Gibbs free energy of the mixture (on the level of the van der Waals equation) which contains the essential features of the phase transition. We then can use stability arguments to study the phase transition in some detail.

Most phase transitions have associated with them a critical point (the liquid-solid transition is one of the few which does not). That is, there is a well-defined temperature above which one phase exists, and as the temperature is lowered a new phase appears. When a new phase appears as we lower the temperature, it often has different symmetry properties and some new variable, called the order parameter, appears which characterizes the new phase. For first-order phase transitions, there need not be a connection between the symmetries of the high-temperature phase and of the low-temperature phase. For a continuous transition, since the state changes continuously, there generally will be a well-defined connection between symmetry properties of the two phases. Ginzburg and Landau were able to construct a completely general theory of continuous symmetry-breaking phase transitions which involves a power series expansion of the free energy in terms of the order parameter. We shall discuss the Ginzburg-Landau theory in this chapter and show how it can be applied to magnetic systems at the Curie point and to superfluid systems.

Superconductors are especially interesting from the standpoint of thermodynamics because they give us quite a different application of the Clausius-Clapeyron equation and they provide a very clear example of the uses of the Ginzburg-Landau expansion. Many features of the Ginzburg-Landau expansion for superconductors can be taken over directly to superfluid He^4 and, with some major complications, to superfluid He^3. We therefore discuss superconductors in some detail in this chapter.

Since a great deal of this chapter is devoted to classical fluids, we are including a brief discussion of the phase diagrams for liquid He^4 and liquid He^3, which are quantum fluids. These systems give a good illustration of the third law of thermodynamics.

The critical point plays a unique role in the theory of phase transitions. As a system approaches its critical point, from high temperatures, it begins to adjust itself on a microscopic level. Large fluctuations occur which signal the emergence of a new order parameter which finally does appear at the critical point itself. At the critical point, some thermodynamic variables can become infinite. Critical points occur in a huge variety of systems, but regardless of the particular substance or mechanical variable involved, there appears to be a great similarity in the behavior of all systems as they approach their critical points. For this reason, there is a whole field of physics devoted to the study of critical phenomena.

One of the best ways to characterize the behavior of systems as they approach the critical point is by means of critical exponents. We shall define critical exponents in this chapter and give explicit examples of some of them for the liquid-vapor transition in simple fluids and for the Curie point. We shall also compute critical exponents for the liquid-vapor transition using the van der Waals equation. This will give us another means of comparing the van der Waals equation with the experimentally observed behavior of real fluids.

In this chapter we will discuss several equations which describe the behavior of systems around the critical point. Among them are the van der Waals equation, the equation for regular binary mixtures, and the Ginzburg-Landau expansion. All these theories are known as mean field theories because they can be derived by assuming that each particle moves in the mean field of all other particles. Mean field theories do not give the correct value for the critical exponents, because they do not correctly take into account short-ranged correlations which are important near the critical point. However, they give qualitatively correct behavior near critical points and therefore are useful in building our intuition.

B. COEXISTENCE OF PHASES: GIBBS PHASE RULE

Most systems can exist in a number of different phases, each of which can exhibit quite different macroscopic behavior. The particular phase that is realized in nature for a given set of independent variables is the one with the lowest free energy. It can happen that, for certain values of the independent variables, two or more phases of a system can coexist. There is a simple rule, called the Gibbs phase rule, which tells us the number of phases that can coexist. Generally, coexisting phases are in thermal and mechanical equilibrium and can exchange matter. Under these conditions, the temperature and chemical potentials of the phases must be equal (cf. Sec. 2.H) and there will be another condition between mechanical variables expressing mechanical equilibrium. For example, for a simple *PVT* system, the pressures of the two phases may be equal (this need not be the case if they are separated by a surface which is not free to move).

For simplicity, let us first consider a *YXT* system which is pure (composed of one kind of particle). For a pure system, two phases, I and II, can coexist at a fixed value of Y and T if their respective chemical potentials are equal:

$$\mu^{I}(Y, T) = \mu^{II}(Y, T). \tag{4.1}$$

(The chemical potentials are functions only of intensive variables.) Eq. (4.1) gives a relation between the values of Y and T for which the phases can coexist,

$$Y = Y(T), \tag{4.2}$$

and in the Y-T plane defines a coexistence curve for the two phases. If the pure system has three phases, I, II, and III, they can only coexist at a single point in the Y-T plane (the triple point). Three coexisting phases must satisfy the equations

$$\mu^{I}(Y, T) = \mu^{II}(Y, T) = \mu^{III}(Y, T). \tag{4.3}$$

Since we have two equations and two unknowns, the triple point is uniquely determined. For a pure system, four phases cannot coexist, because we would then have three equations and two unknowns and there would be no solution.

For a mixture of l different types of particles, $l + 2$ phases can coexist. To show this, we note that if there are l types of particles in each phase then there will be $l + 1$ independent variables for each phase, namely, $(Y, T, x_i, \ldots, x_{l-1})$ where x_i is the mole fraction of particles of type i. If we have several phases coexisting, the chemical potentials for a given type of particle must be equal in the various phases. Thus, if there are r coexisting phases at a given value of Y

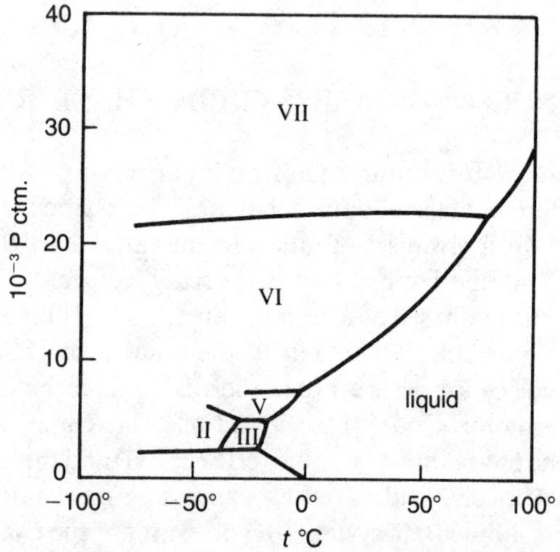

Fig. 4.1. Coexistence curves for the solid and liquid phases of water: in accordance with the Gibbs phase rule, no more than three phases can coexist (based on Ref. 4.25).

and T, the condition for equilibrium is

$$\mu_1{}^{\mathrm{I}}(Y, T, x_1{}^{\mathrm{I}}, x_2{}^{\mathrm{I}}, \ldots, x_{l-1}^{\mathrm{I}}) = \mu_1^{\mathrm{II}}(Y, T, x_1^{\mathrm{II}}, x_2^{\mathrm{II}}, \ldots, x_{l-1}^{\mathrm{II}}) = \cdots$$
$$= \mu_1{}^{r}(Y, T, x_1{}^{r}, \ldots, x_{l-1}^{r}). \tag{4.4}$$

$$\mu_2{}^{\mathrm{I}}(Y, T, x_1{}^{\mathrm{I}}, \ldots, x_{l-1}^{\mathrm{I}}) = \mu_2^{\mathrm{II}}(Y, T, x_1^{\mathrm{II}}, \ldots, x_{l-1}^{\mathrm{II}}) = \cdots$$
$$= \mu_2{}^{r}(Y, T, x_1{}^{r}, \ldots, x_{l-1}^{r}). \tag{4.5}$$

$$\vdots \qquad\qquad \vdots$$

$$\mu_l{}^{\mathrm{I}}(Y, T, x_1{}^{\mathrm{I}}, \ldots, x_{l-1}^{\mathrm{I}}) = \mu_l^{\mathrm{II}}(Y, T, x_1^{\mathrm{II}}, \ldots, x_{l-1}^{\mathrm{II}}) = \cdots$$
$$= \mu_l{}^{r}(Y, T, x_1{}^{r}, \ldots, x_{l-1}^{r}). \tag{4.6}$$

Eqs. (4.4)–(4.6) give $l(r - 1)$ equations to determine $2 + r(l - 1)$ unknowns. For a solution, the number of equations cannot be greater than the number of unknowns. Thus, we must have $l(r - 1) \leqslant 2 + r(l - 1)$ or $r \leqslant l + 2$. The number of coexisting phases must be less than or equal to $l + 2$, where l is the number of different types of particle. For a pure state $l = 1$ and $r \leqslant 3$ as we found before. For a binary mixture $l = 2$ and $r \leqslant 4$, and, at most, four different phases can coexist.

As an example of the Gibbs phase rule, we show the coexistence curves for various solid phases of water (cf. Fig. 4.1). We see that, although water exists in many solid phases, no more than three phases can coexist at a given temperature and pressure.

C. CLASSIFICATION OF PHASE TRANSITIONS

As we change the independent intensive variables ($Y, T, x_1, \ldots, x_l$) of a system, we reach values of the variables for which a phase change can occur. At such points the chemical potentials (which are functions only of intensive variables) of the phases must be equal and the phases can coexist.

In Sec. 2.F we found that the Gibbs free energy is closely related to the chemical potential. The fundamental equation for the Gibbs free energy is

$$G = \sum_j \tilde{\mu}_j n_j, \tag{4.7}$$

where $\tilde{\mu}_j$ is the chemical potential per mole, and for processes which occur at constant Y and T

$$(dG)_{YT} = \sum_j \tilde{\mu}_j \, dn_j. \tag{4.8}$$

Thus, at a phase transition, the Gibbs free energy of each phase must have the same value, and the derivatives $\tilde{\mu}_j = (\partial G/\partial n_j)_{Y,T,\{n_i \neq j\}}$ must be equal. However, no restriction is placed on the derivatives $X = -(\partial G/\partial Y)_{T,\{n_j\}}$ and $S = (\partial G/\partial T)_{Y,\{n_j\}}$. The behavior of these derivatives is used to classify phase transitions. If the derivatives $(\partial G/\partial Y)_{T,\{n_j\}}$ and $(\partial G/\partial T)_{Y,\{n_j\}}$ are discontinuous at the transition point (that is, if the extensive variable X and the entropy S have different values in the two phases), the transition is called "first-order." If the

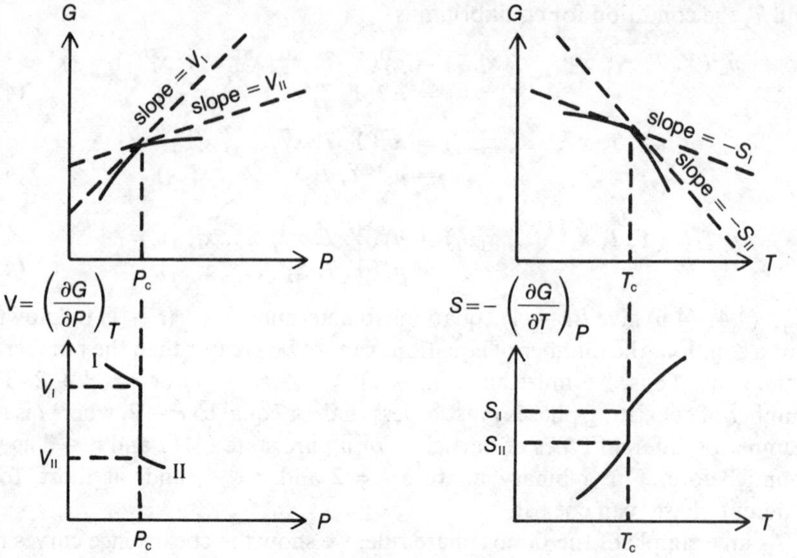

Fig. 4.2. Typical behavior for the Gibbs free energy at a first-order phase transition.

derivatives $(\partial G/\partial T)_{Y,\{n_j\}}$ and $(\partial G/\partial Y)_{T,\{n_j\}}$ are continuous at the transition but higher order derivatives are discontinuous, then the phase transition is continuous. (The terminology "*n*th-order phase transition" was introduced by Ehrenfest to indicate a phase transition for which the *n*th derivative of *G* was the first discontinuous derivative. However, for some systems higher order derivatives are infinite, and the theory proposed by Ehrenfest breaks down for those cases.)

Let us now plot the Gibbs free energy for first-order and continuous transitions in a *PVT* system. For such a system the Gibbs free energy must be a concave function of *P* and *T* (cf. Sec. 2.H). The Gibbs free energy and its first derivatives are plotted in Fig. 4.2 for a first-order phase transition. A discontinuity in $(\partial G/\partial P)_{T,\{n_j\}}$ means that there is the discontinuity in the volume of the phases,

$$\Delta V = V^{\mathrm{I}} - V^{\mathrm{II}} = \left(\frac{\partial G}{\partial P}\right)^{\mathrm{I}}_{T,\{n_j\}} - \left(\frac{\partial G}{\partial P}\right)^{\mathrm{II}}_{T,\{n_j\}}, \tag{4.9}$$

and a discontinuity in $(\partial G/\partial T)_{P,\{n_j\}}$ means there is a discontinuity in the entropy of the two phases,

$$\Delta S = S^{\mathrm{I}} - S^{\mathrm{II}} = \left(\frac{\partial G}{\partial T}\right)^{\mathrm{II}}_{P,\{n_j\}} - \left(\frac{\partial G}{\partial T}\right)^{\mathrm{I}}_{P,\{n_j\}}. \tag{4.10}$$

Since the Gibbs free energy is the same for both phases at the transition, the

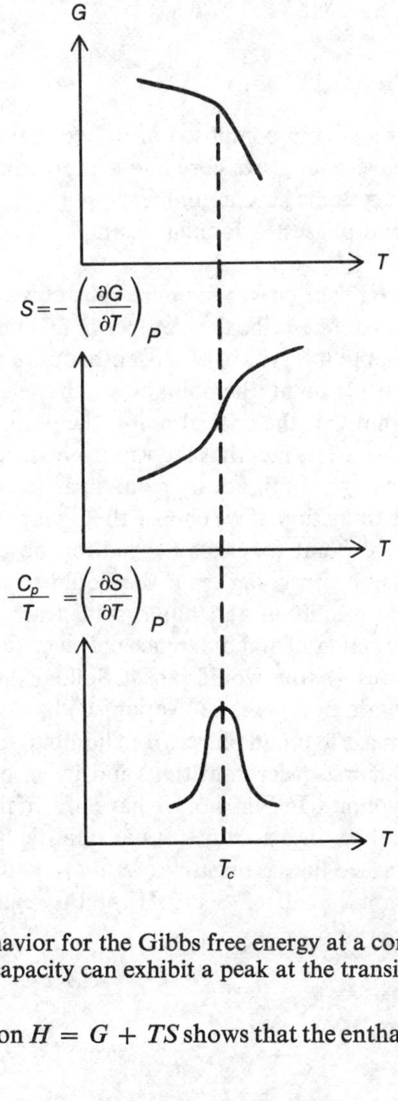

Fig. 4.3. Typical behavior for the Gibbs free energy at a continuous phase transition; the heat capacity can exhibit a peak at the transition.

fundamental equation $H = G + TS$ shows that the enthalpy of the two phases is different,

$$\Delta H = T\Delta S = H^{\mathrm{I}} - H^{\mathrm{II}}, \tag{4.11}$$

for a first-order phase transition; ΔH is called latent heat of the transition.

For a continuous phase transition, the Gibbs free energy is continuous but its slope changes rapidly. This in turn leads to a peaking in the heat capacity at the transition point. An example is given in Fig. 4.3. For a continuous transition, there is no abrupt change in the entropy or the extensive variable (as a function of Y and T) at the transition.

In the subsequent sections we shall give examples of first-order and continuous phase transitions.

D. PURE *PVT* SYSTEMS[1–3]

1. Phase Diagrams

A pure *PVT* system is a system composed of only one type of molecule. The molecules generally have a repulsive core and a short-range attractive region outside the core. Such systems have a number of phases: a gas phase, a liquid phase, and various solid phases. A familiar example of a pure *PVT* system is water.

A typical set of coexistence curves for pure substances is given in Fig. 4.4. (Note that Fig. 4.4 does not describe the isotopes of helium, He^3 or He^4, which have superfluid phases, but it is typical of most other pure substances.) Point A on the diagram is the triple point, the point at which the gas, liquid, and solid phases can coexist. Point C is the critical point, the point at which the vapor pressure curve terminates. The fact that the vapor pressure curve has a critical point means that we can go continuously from a gas to a liquid without ever going through a phase transition, if we choose the right path. The fusion curve does not have a critical point (one has never been observed). We must go through a phase transition in going from the liquid to the solid state. This difference between the gas-liquid and liquid-solid transitions indicates that there is a much greater fundamental difference between liquids and solids than between liquids and gases—as one would expect. Solids exhibit spatial ordering, while liquids and gases do not. (We use "vapor" and "gas" interchangeably.)

The transitions from gas to liquid phase, from liquid to solid phase, and from gas to solid phase are all first-order transitions and are accompanied by a latent heat and a change in volume. In Fig. 4.5, we have drawn the phase diagram in the *P-V* plane. The dashed lines are lines of constant temperature. We notice that the slope of the dashed lines is negative $(\partial P/\partial V)_T \leqslant 0$. This is a statement of the stability condition $\kappa_T > 0$ (cf. Sec. 2.H). In the region of coexistence of

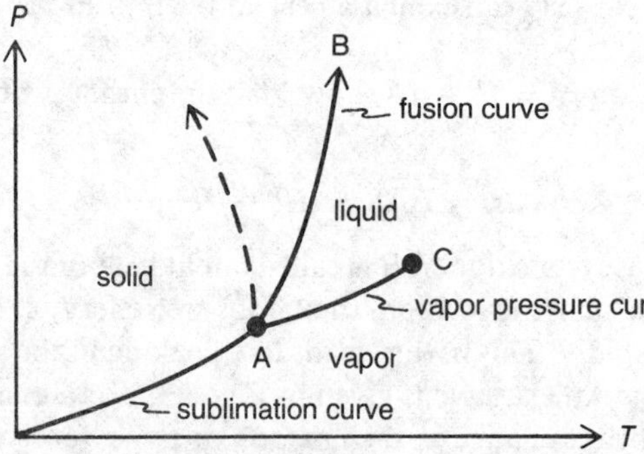

Fig. 4.4. Coexistence curves for a typical pure *PVT* system: point A is the triple point and point C is the critical point; the dashed line is an example of a fusion curve with negative slope.

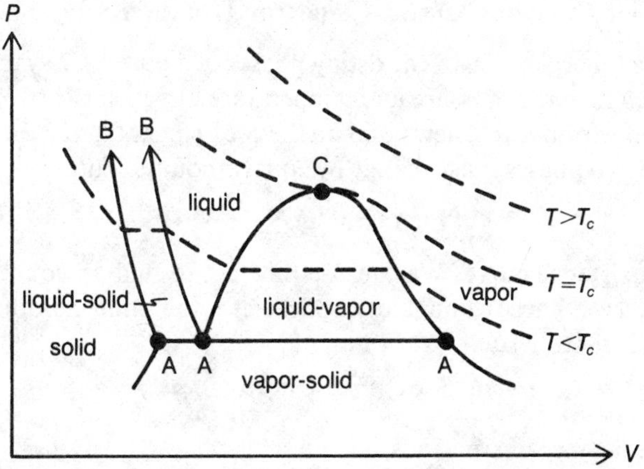

Fig. 4.5. A plot of the coexistence regions for a typical *PVT* system: all the phase transitions here are first order; the dashed lines represent isotherms.

phases, the isotherms (dashed lines) are always flat, indicating that in these regions the change in volume occurs for constant P and T.

It is interesting to plot the pressure as a function of volume and temperature in a three-dimensional figure. As we can see in Fig. 4.6, the result is similar to that of a mountain with ski slopes. The height of the mountain at any given value of V and T (latitude and longitude) is the pressure. Fig. 4.6 actually corresponds to a plot of the equation of state for the pure system. The shaded region is the region of coexistence of more than one phase. Fig. 4.4 is a projection of Fig. 4.6 on the P-T plane and Fig. 4.5 is a projection of Fig. 4.6 on the V-P plane.

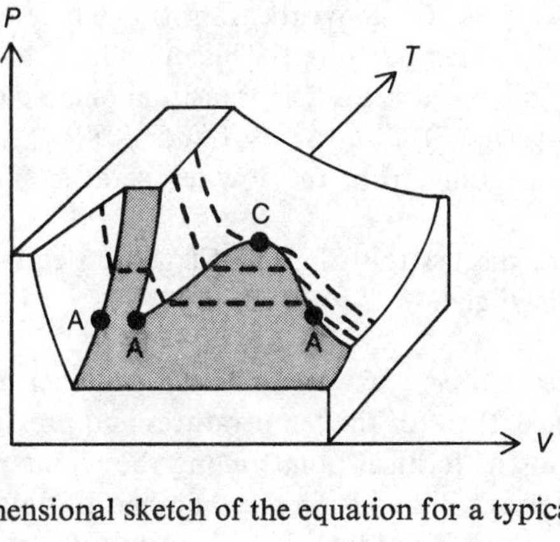

Fig. 4.6. Three-dimensional sketch of the equation for a typical pure *PVT* system.

2. Coexistence Curves: Clausius-Clapeyron Equation

The Gibbs free energies of two coexisting phases of a pure *PVT* system must be equal. If we change the pressure and temperature at which the two phases coexist (that is, if we move to a new point on the coexistence curve), the Gibbs free energy of the two phases must change by equal amounts. Thus,

$$dG^{\text{I}} = dG^{\text{II}} \tag{4.12}$$

along the coexistence curve. We can use this fact to find an equation for the coexistence curve. Since the number of particles is the same in both phases, we can use Eq. (2.98) to write

$$V^{\text{I}}\, dP - S^{\text{I}}\, dT = V^{\text{II}}\, dP - S^{\text{II}}\, dT \tag{4.13}$$

along the coexistence curve. Thus,

$$\left(\frac{dP}{dT}\right)_{\text{coex}} = \frac{\Delta S}{\Delta V} = \frac{S^{\text{I}} - S^{\text{II}}}{V^{\text{I}} - V^{\text{II}}} \tag{4.14}$$

along the coexistence curve. S and V are functions of P and T. Eq. (4.14) can also be written in terms of the latent heat $\Delta H = T\,\Delta S$ (cf. Eq. (4.11))

$$\left(\frac{dP}{dT}\right)_{\text{coex}} = \frac{\Delta H}{T\,\Delta V}. \tag{4.15}$$

Eq. (4.15) is called the Clausius-Clapeyron equation; ΔH is the heat absorbed in the transition from phase II to phase I, and ΔV is the change in volume between the two phases.

In going from a low-temperature phase to a high-temperature phase, the latent heat must always be positive. That is, heat must be absorbed. To show this, we assume that phase I is the high-temperature phase and phase II is the low-temperature phase. Since the equilibrium state is a state of minimum free energy, we must have $G^{\text{I}} < G^{\text{II}}$ above the transition temperature and $G^{\text{I}} > G^{\text{II}}$ below the transition temperature. This implies that $(\partial G^{\text{I}}/\partial T)_{P\{n_j\}} < (\partial G^{\text{II}}/\partial T)_{P\{n_j\}}$ both above and below the transition (simple drawing illustrates this). Thus, $S^{\text{I}} = -(\partial G^{\text{I}}/\partial T)_{P\{n_j\}} > S^{\text{II}} = -(\partial G^{\text{II}}/\partial T)_{P,\{n_j\}}$ and $\Delta S = (1/T)\Delta H$ is always positive in going from the low-temperature phase to the high-temperature phase.

It is of interest to discuss the Clausius-Clapeyron equation for the three coexistence curves in Fig. 4.4.

(*a*) *Vapor Pressure Curve.* If we evacuate a chamber and partially fill it with a pure substance, then for the temperatures and pressures from point A (triple point) to point C (critical point) along the vapor pressure curve the vapor and liquid phases will coexist in the chamber. For a given temperature, T, the pressure of the vapor and liquid is completely determined and is called the saturated vapor pressure. As we change the temperature of the system the vapor pressure will also change. The Clausius-Clapeyron equation tells us how the vapor pressure changes as a function of temperature along the coexistence curve.

We can obtain a rather simple equation for the vapor pressure curve. Let us assume that volume changes of the liquid may be neglected with respect to volume changes of the vapor as we move along the coexistence curve, and let us assume that the vapor obeys the ideal gas law. Then

$$\Delta V \approx V_g = \frac{nRT}{P}. \tag{4.16}$$

Furthermore, we assume that the latent heat of vaporization is roughly constant over the range of temperatures considered. Then the Clausius-Clapeyron equation for the vapor pressure curve takes the form

$$\frac{dP}{dT} = \frac{\Delta\tilde{h}_{gl}P}{RT^2}, \tag{4.17}$$

where $\Delta\tilde{h}_{gl}$ is the latent heat of vaporization per mole. We can integrate Eq. (4.17) to obtain

$$P = P_0\, e^{-\Delta\tilde{h}_{gl}/RT}. \tag{4.18}$$

Thus, as the temperature is increased, the vapor pressure increases exponentially. Conversely, if we increase the pressure, the temperature of coexistence (boiling point) increases.

It is of interest to determine the heat capacity of a vapor in equilibrium with its liquid phase. Along the vapor pressure curve there is only one independent variable, which we choose to be temperature. The entropy for a vapor is a function of both pressure and temperature, but along the vapor pressure curve the pressure is related to the temperature by the Clausius-Clapeyron equation. We define the heat capacity as

$$\begin{aligned} C_{\text{coex}} &= \left(\frac{đQ}{dT}\right)_{\text{coex}} = T\left(\frac{dS}{dT}\right)_{\text{coex}} = T\left(\frac{\partial S}{\partial T}\right)_P + T\left(\frac{\partial S}{\partial P}\right)_T\left(\frac{dP}{dT}\right)_{\text{coex}} \\ &= C_P - T\left(\frac{\partial V}{\partial T}\right)_P\left(\frac{dP}{dT}\right)_{\text{coex}}, \end{aligned} \tag{4.19}$$

where we have used Eqs. (2.8) and (2.100*a*). If we use the ideal gas equation of state to describe the properties of the vapor phase, then we obtain the following expression for the heat capacity along the coexistence curve:

$$C_{\text{coex}} = C_P - \frac{n\,\Delta\tilde{h}_{gl}}{T}. \tag{4.20}$$

At low temperatures, the heat capacity of the vapor can be negative. Thus, if the temperature of the vapor is raised and it is maintained in equilibrium with the liquid phase, the vapor gives off heat.

(*b*) *Fusion Curve.* The fusion curve does not terminate at a critical point but can have either positive or negative slope. In Fig. 4.4 we have used a solid line for the fusion curve with a positive slope and a dashed line for the case of negative slope. The Clausius-Clapeyron equation for the liquid-solid transition is

$$\frac{dP}{dT} = \frac{\Delta \tilde{h}_{ls}}{T \Delta \tilde{v}_{ls}}, \tag{4.21}$$

where $\Delta \tilde{v}_{ls}$ is the change in molar volume in going from the solid to the liquid phase and $\Delta \tilde{h}_{ls}$ is the latent heat of fusion per mole. If the volume of the solid is greater than that of the liquid, then $\Delta \tilde{v}_{ls}$ will be negative and the slope $(dP/dT)_v$ will be negative. For the case of a fusion curve with positive slope, if we increase the pressure at a fixed temperature, we simply drive the system deeper into the solid phase. However, if the fusion curve has a negative slope, then increasing the pressure at fixed temperature can drive the system into the liquid phase. Water is an example of a system whose fusion curve has negative slope. The negative slope of the fusion curve for water makes ice skating possible. As the skate blades exert pressure on the ice, it turns to water and the skater floats along on a narrow puddle.

(*c*) *Sublimation Curve.*[1] If a solid is placed in an evacuated chamber and maintained at some pressure and temperature along the sublimation curve, a vapor will coexist in equilibrium with the solid phase.

We can obtain a fairly simple equation for the sublimation curve. Let us first note that infinitesimal changes in the molar enthalpy of the solid can be written

$$d\tilde{h} = T\, d\tilde{s} + \tilde{v}\, dP = \tilde{c}_p\, dT + \tilde{v}(1 - T\alpha_p)\, dP, \tag{4.22}$$

where α_p is the thermal expansivity. We can integrate Eq. (4.22) to find the difference between the enthalpy of the solid at zero pressure and temperature and that of a point on the sublimation curve. The vapor pressure along the sublimation curve is generally very small. Thus we can neglect pressure variations in Eq. (4.22) and write

$$\tilde{h}_s = \tilde{h}_s^{\,0} + \int_0^T \tilde{c}_p^{\,s}\, dT, \tag{4.23}$$

where $\tilde{h}_s^{\,0}$ is the molar enthalpy of the solid at zero pressure and temperature. From Eq. (4.22) the molar enthalpy of the vapor can be written

$$\tilde{h}_g = \tilde{h}_g^{\,0} + \int_0^T \tilde{c}_p^{\,g}\, dT, \tag{4.24}$$

where $\tilde{h}_g{}^0$ is the molar enthalpy of the vapor at zero pressure and temperature and $\tilde{h}_g$ is its value at a point on the sublimation curve with temperature T. If we now use Eqs. (4.23) and (4.24) and assume that the vapor can be described by an ideal gas, the Clausius-Clapeyron equation for the sublimation curve can be written

$$\frac{dP}{dT} = \frac{P\,\Delta\tilde{h}_{gs}}{RT^2}, \tag{4.25}$$

where $\Delta\tilde{h}_{gs} \approx \tilde{h}_g - \tilde{h}_s$. In Eq. (4.25) we have assumed that volume changes of the solid can be neglected relative to those of the vapor. If we integrate (4.25) for the case of a monatomic vapor $\tilde{c}_P = \frac{5}{2}R$, we find

$$\ln P = -\frac{\Delta\tilde{h}^0_{gs}}{RT} + \tfrac{5}{2}\ln T - \frac{1}{R}\int \frac{\int_0^T \tilde{c}_P{}^s\, dT'}{T^2}\, dT + \text{constant}. \tag{4.26}$$

The quantity $\Delta\tilde{h}^0_{gs}$ is the latent heat of sublimation at zero temperature and pressure. Eq. (4.26) is the equation for the sublimation curve for small pressures. It can be used to obtain values of $\Delta\tilde{h}^0_{gs}$ from experimental data.[1]

3. Liquid-Vapor Coexistence Region[2,3]

The liquid-vapor coexistence region culminates in a critical point and will be of special interest later. Therefore, it is useful at this point to examine the coexistence region under the critical point more closely. Let us redraw the coexistence curve for the liquid-vapor transition in the *P-V* plane (cf. Fig. 4.7).

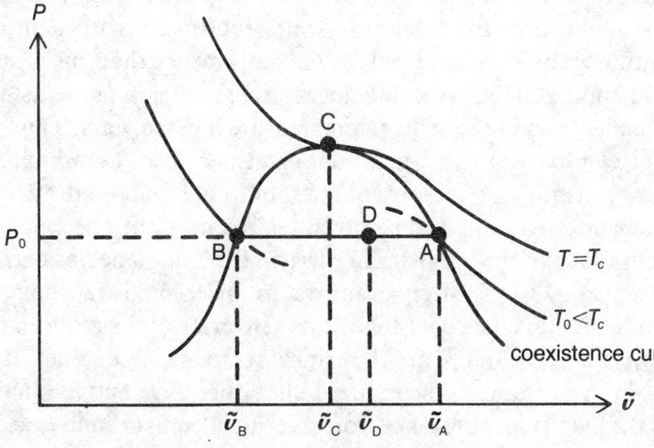

Fig. 4.7. Coexistence curve for the vapor-liquid coexistence region for a pure *PVT* system: plot is of pressure versus molar volume.

The isotherms for stable thermodynamic states are indicated by the solid lines. As we decrease the molar volume of a gas with temperature fixed at $T_0 < T_c$, the pressure increases until we reach the coexistence curve (point A). At point A the vapor starts to condense and the pressure remains fixed until all vapor has changed to liquid (point B). Then the pressure begins to rise again.

The amounts of liquid and vapor which coexist are given by the lever rule. Let us consider a system with temperature $T_0 < T_c$, pressure P_0, and total molar volume $\tilde{v}_d$. The system is then in a state in which the liquid and vapor phases coexist. The total molar volume, $\tilde{v}_d$, is given in terms of the molar volume of the liquid at point B, $\tilde{v}_l$, and the molar volume of vapor at point A, $\tilde{v}_g$, by

$$\tilde{v}_d = x_l\tilde{v}_l + x_g\tilde{v}_g, \tag{4.27}$$

where x_l is the mole fraction of liquid at point D and x_g is the mole fraction of gas at point D. If we multiply Eq. (4.27) by $(x_l + x_g) \equiv 1$, we find

$$\frac{x_l}{x_g} = \frac{(\tilde{v}_g - \tilde{v}_d)}{(\tilde{v}_d - \tilde{v}_l)}. \tag{4.28}$$

Thus the ratio of the mole fractions of liquid to gas at point D is equal to the inverse ratio of the distance between point D and points A and B.

As long as $(\partial V/\partial P)_T < 0$ the system is mechanically stable (cf. Eq. (2.160)). If we continue the isotherm T_0 past the points A and B (the dashed line) we obtain curves which are mechanically stable but no longer correspond to a minimum of free energy. States along the dashed line at point A correspond to super-cooled vapor states, while those along the dashed line at point B correspond to super-heated liquid states. Such states are metastable and can be produced in the laboratory for very pure samples. It is possible that the super-heated liquid curve can extend into the region of negative pressure. Such states can also be realized in the laboratory but require walls to maintain them. Systems with negative pressure pull in on walls rather than push out against them. As we approach the critical temperature, the region of metastable states becomes smaller, and at the critical temperature it disappears. Thus, no metastable states can exist at the critical temperature. Also, as we approach the critical temperature, the molar volumes of the liquid and vapor phases approach one another, and at the critical temperature they become equal.

The actual shape of the coexistence curve in the T-V plane has been given by E. A. Guggenheim[4] for a variety of pure substances and is reproduced in Fig. 4.8. Guggenheim plots the coexistence curve in terms of the reduced quantities T/T_c and ρ/ρ_c where T_c and ρ_c are the critical temperature and density, respectively, of a given substance. The reduced quantities T/T_c and ρ/ρ_c measure the distance of the particular substance from its critical point (T_c and ρ_c are different for each substance). Most substances, when plotted in terms of reduced temperature and density, lie in approximately the same curve. This is an example of

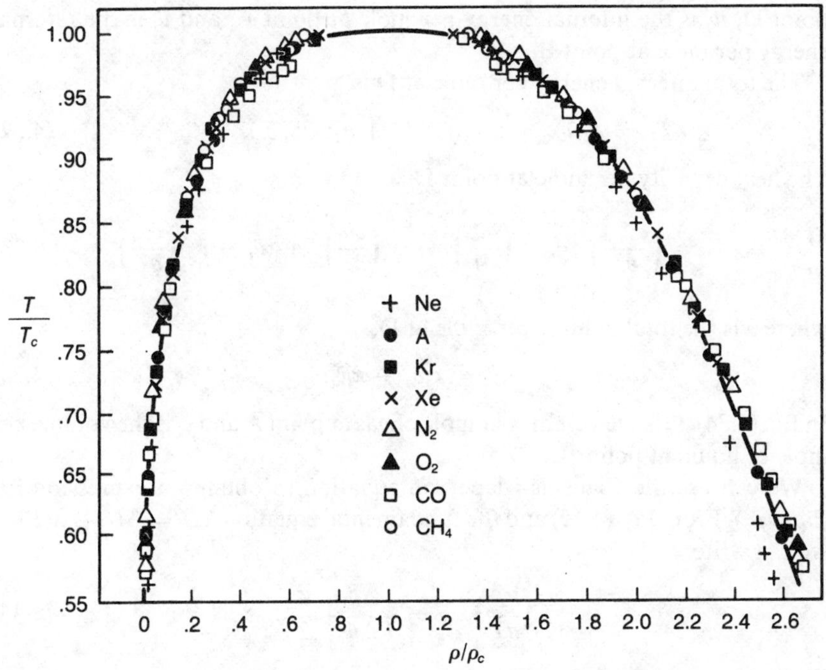

Fig. 4.8. Experimental vapor-liquid coexistence curve for a variety of substances: the plot is of reduced temperature versus reduced density (based on Ref. 4.4).

the so-called law of corresponding states, which states that all pure classical fluids, when described in terms of reduced quantities, obey the same equation of state.[5] We shall see this again when we return to the van der Waals equation. The reduced densities of the liquid and gas phases along the coexistence curves obey the following equations:[4]

$$\frac{\rho_l + \rho_g}{2\rho_c} = 1 + \frac{3}{4}\left(1 - \frac{T}{T_c}\right) \tag{4.29}$$

and

$$\frac{\rho_l - \rho_g}{\rho_c} = \frac{7}{2}\left(1 - \frac{T}{T_c}\right)^{1/3}. \tag{4.30}$$

These equations will be useful later.

We can find an expression for the heat capacity per mole $\tilde{c}_v$ inside the coexistence curve.[2] Let us again consider Fig. 4.7. The total internal energy at point D is

$$U_{\text{tot}} = n_g\tilde{u}_g(v_g, T) + n_l\tilde{u}_l(v_l, T), \tag{4.31}$$

where n_g and n_l are the number of moles of gas and liquid, respectively, at point D, $\tilde{u}_g$ is the internal energy per mole at point A, and $\tilde{u}_l$ is the internal energy per mole at point B.

The total internal energy per mole at D is

$$\tilde{u}_{\text{tot}} = x_g \tilde{u}_g(v_g, T) + x_l \tilde{u}_l(v_l, T). \tag{4.32}$$

The heat capacity per mole at point D is

$$\tilde{c}_v = \left(\frac{\partial \tilde{u}_{\text{tot}}}{\partial T}\right)_{\tilde{v}} = x_g \left(\frac{\partial \tilde{u}_g}{\partial T}\right)_{\tilde{v}} + x_l \left(\frac{\partial \tilde{u}_l}{\partial T}\right)_{\tilde{v}} + (\tilde{u}_l - \tilde{u}_g)\left(\frac{\partial x_l}{\partial T}\right)_{\tilde{v}}, \tag{4.33}$$

where $\tilde{v}$ is the total volume per mole at D,

$$\tilde{v} = x_g \tilde{v}_g + x_l \tilde{v}_l. \tag{4.34}$$

In Eq. (4.34), $\tilde{v}_g$ is the volume per mole of gas at point A and $\tilde{v}_l$ is the volume per mole of liquid at point B.

We can use the Clausius-Clapeyron equation to obtain an expression for $(\tilde{u}_l - \tilde{u}_g)$. From Eq. (4.15) and the fundamental equation $\Delta H = \Delta U + \Delta(PV)$, we can write

$$\left(\frac{dP}{dT}\right)_{\text{coex}} = \frac{1}{T}\frac{\Delta \tilde{u}}{\Delta \tilde{v}} + \frac{P}{T}. \tag{4.35}$$

Thus,

$$\tilde{u}_g - \tilde{u}_l = \left[\left(T\left(\frac{dP}{dT}\right) - P\right)(\tilde{v}_g - \tilde{v}_l)\right]_{\text{coex}}. \tag{4.36}$$

Let us next obtain an expression for $(\partial x_l/\partial T)_{\tilde{v}}$. If we take the derivative of $\tilde{v}$ in Eq. (4.34) with respect to T holding $\tilde{v}$ fixed, we find

$$(\tilde{v}_g - \tilde{v}_l)\left(\frac{\partial x_l}{\partial T}\right)_{\tilde{v}} = x_l \left(\frac{\partial \tilde{v}_l}{\partial T}\right)_{\tilde{v}} + x_g \left(\frac{\partial \tilde{v}_g}{\partial T}\right)_{\tilde{v}}, \tag{4.37}$$

where we have used the identity $x_l + x_g = 1$. Let us now note that

$$\left(\frac{\partial \tilde{v}_l}{\partial T}\right)_{\tilde{v}} = \left(\frac{\partial \tilde{v}_l}{\partial T}\right)_{\text{coex}}. \tag{4.38}$$

Since $\tilde{v}$ by definition lies on the coexistence curve, Eq. (4.37) becomes

$$\left(\frac{\partial x_l}{\partial T}\right)_{\tilde{v}} = \frac{\left[x_l \left(\frac{\partial \tilde{v}_l}{\partial T}\right) + x_g \left(\frac{\partial \tilde{v}_g}{\partial T}\right)\right]_{\text{coex}}}{(\tilde{v}_g - \tilde{v}_l)}. \tag{4.39}$$

In order to obtain the desired expression for $\tilde{c}_v$, we still must write the first two terms in Eq. (4.33) in a slightly different form. From Eq. (2.8), we note that

$$\left(\frac{\partial \tilde{u}_l}{\partial T}\right)_{\tilde{v}} = \left(\frac{\partial \tilde{u}_l}{\partial T}\right)_{\tilde{v}_l} + \left(\frac{\partial \tilde{u}_l}{\partial \tilde{v}_l}\right)_T \left(\frac{\partial \tilde{v}_l}{\partial T}\right)_{\text{coex}}, \tag{4.40}$$

where $\tilde{c}_{\tilde{v}_l} = (\partial \tilde{u}_l/\partial T)_{\tilde{v}_l}$ is the heat capacity per mole of the liquid phase. We can write a similar equation for $(\partial \tilde{u}_g/\partial T)_{\tilde{v}_g}$. If we now combine Eqs. (4.33), (4.36), (4.39), and (4.40), we obtain

$$\tilde{c}_{\tilde{v}} = x_g\tilde{c}_{\tilde{v}_g} + x_l\tilde{c}_{\tilde{v}_l} + x_g\left(\frac{\partial \tilde{u}_g}{\partial \tilde{v}_g}\right)_T\left(\frac{\partial \tilde{v}_g}{\partial T}\right)_{\text{coex}} + x_l\left(\frac{\partial \tilde{u}_l}{\partial \tilde{v}_l}\right)_T\left(\frac{\partial \tilde{v}_l}{\partial T}\right)_{\text{coex}}$$
$$- \left[T\left(\frac{dP}{dT}\right) - P\right]_{\text{coex}}\left[x_l\left(\frac{\partial \tilde{v}_l}{\partial T}\right) + x_g\left(\frac{\partial \tilde{v}_g}{\partial T}\right)\right]_{\text{coex}}. \tag{4.41}$$

However, from Eq. (2.125)

$$\left(\frac{\partial \tilde{u}_g}{\partial \tilde{v}_g}\right)_T = T\left(\frac{\partial P_g}{\partial T}\right)_{\tilde{v}_g} - P_g, \tag{4.42}$$

and from Eq. (2.8)

$$\left(\frac{\partial P_g}{\partial T}\right)_{\tilde{v}_g} = \left(\frac{dP}{dT}\right)_{\text{coex}} - \left(\frac{\partial P_g}{\partial \tilde{v}_g}\right)_T\left(\frac{\partial \tilde{v}_g}{\partial T}\right)_{\text{coex}} \tag{4.43}$$

with similar expressions for the liquid phase. If we combine terms, the heat capacity per mole at point D becomes

$$\tilde{c}_{\tilde{v}} = x_g\left[\tilde{c}_{\tilde{v}_g} - T\left(\frac{\partial P_g}{\partial \tilde{v}_g}\right)_T\left(\frac{\partial \tilde{v}_g}{\partial T}\right)^2_{\text{coex}}\right] + x_l\left[\tilde{c}_{\tilde{v}_l} - T\left(\frac{\partial P_l}{\partial \tilde{v}_l}\right)_T\left(\frac{\partial \tilde{v}_l}{\partial T}\right)^2_{\text{coex}}\right]. \tag{4.44}$$

This expression will be useful later. Let us note that the heat capacity at constant pressure is always infinite below the coexistence curve. If we add heat, liquid will turn to vapor but the temperature will not change. Thus, $c_p = (đQ/dT)_p = \infty$ in the two-phase region, while c_v can remain finite.

4. Van der Waals Equation

The van der Waals equation was first derived by van der Waals in his doctoral dissertation in 1873. It was the first, and to this day is the simplest, equation of state which exhibits many of the essential features of the liquid-vapor phase transition. The van der Waals equation is cubic in the molar volume and can be written in the form

$$\tilde{v}^3 - \left(b + \frac{RT}{P}\right)\tilde{v}^2 + \frac{a}{P}\tilde{v} - \frac{ab}{P} = 0. \tag{4.45}$$

An isotherm of the van der Waals equation is plotted in Fig. 4.9. For small values of T and P, the cubic equation will have three distinct real roots (three values of $\tilde{v}$) for each value of P and T. As T increases, the roots will coalesce at a critical temperature, T_c, and above T two of the roots will become imaginary and therefore unphysical. As $T \to \infty$ Eq. (4.45) reduces to the ideal gas equation of state, $\tilde{v} = RT/P$. The critical point is the point at which the roots of Eq.

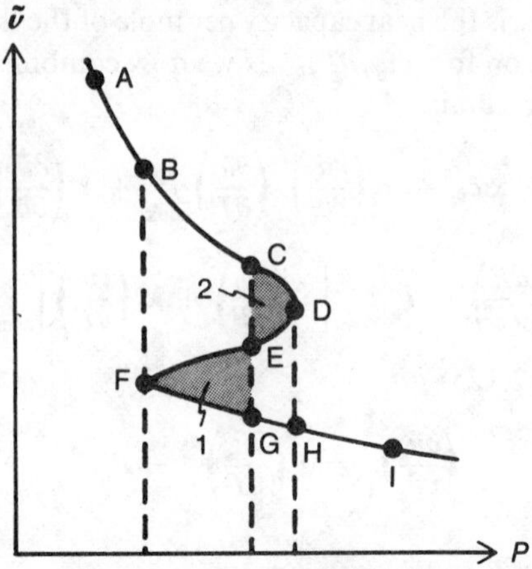

Fig. 4.9. Sketch of a typical van der Waals isotherm: the line from D to F corresponds to mechanically unstable states; the area CDE is labeled 2, and the area EFG is labeled 1; the area under the curve $\tilde{v}(P)$, between any two points, is equal to the difference in molar Gibbs free energy between the points.

(4.45) coalesce. It is also the point at which the critical isotherm ($T = T_c$) has a vanishing slope $(\partial P/\partial \tilde{v})_{T=T_c} = 0$ and an inflection point $(\partial^2 P/\partial \tilde{v}^2)_{T=T_c} = 0$. (An inflection point is a point where the curve changes from convex to concave and $(\partial^2 P/\partial \tilde{v}^2)_T$ changes sign.) If one uses the fact that

$$\left(\frac{\partial P}{\partial \tilde{v}}\right)_{T=T_c} = 0 \quad \text{and} \quad \left(\frac{\partial^2 P}{\partial \tilde{v}^2}\right)_{T=T_c} = 0 \tag{4.46}$$

at the critical point, then we obtain the following values for the temperature, T_c, pressure, P_c, and molar volume, $\tilde{v}_c$, at the critical point:

$$P_c = \frac{a}{27b^2}, \qquad \tilde{v}_c = 3b, \qquad T_c = \frac{8a}{27bR}. \tag{4.47}$$

If we introduce reduced variables $\bar{P} = P/P_c$, $\bar{T} = T/T_c$, and $\bar{V} = \tilde{v}/\tilde{v}_c$, then we may write the van der Waals equation in the form

$$\left(\bar{P} + \frac{3}{\bar{V}^2}\right)(3\bar{V} - 1) = 8\bar{T}. \tag{4.48}$$

It is important to note that Eq. (4.48) is independent of a and b. We are now measuring pressure, volume, and temperature in terms of their distance from the critical point. The values of $\tilde{v}_c$, T_c, and P_c will differ for different gases, but *all gases obey the same equation* if they are the same distance from their respective critical points, that is, have the same values of T/T_c, P/P_c, and $\tilde{v}/\tilde{v}_c$. Thus, we see again the *law of corresponding states*.

An unphysical aspect of the van der Waals equation is its prediction of positive slope $(\partial P/\partial \tilde{v})_T$ for certain segments of the isotherms below T_c (the

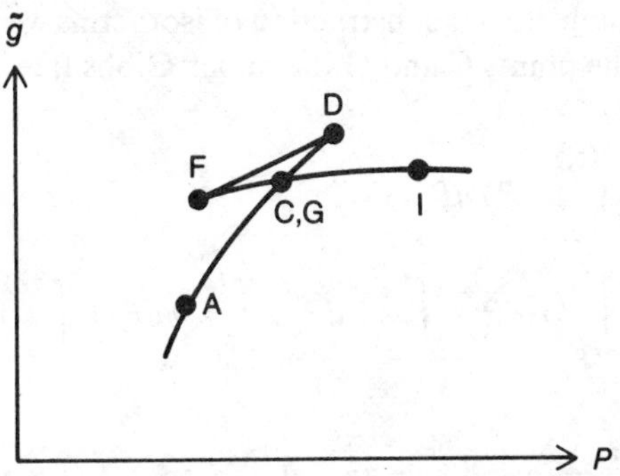

Fig. 4.10. A plot of the molar Gibbs free energy as a function of pressure for the isotherm in Fig. 4.9.

segment between D and F in Fig. 4.9). This region corresponds to mechanically unstable thermodynamic states. However, the unphysical parts of the *P-V* curve can be removed by use of the so-called Maxwell construction.

From Eq. (2.98) we can write the equation for infinitesimal changes in the molar Gibbs free energy in the form

$$d\tilde{g} = -\tilde{s}\,dT + \tilde{v}\,dP. \tag{4.49}$$

If we now restrict ourselves to one of the van der Waals isotherms so $dT = 0$, we can determine how $\tilde{g}$ varies with pressure along that isotherm. In Fig. 4.9 we plot the molar volume as a function of pressure along a typical van der Waals isotherm, and in Fig. 4.10 we plot the molar Gibbs free energy as a function of pressure for the isotherm in Fig. 4.9. Along the isotherm the difference in molar Gibbs free energy between any two points is equal to the area under the curve $\tilde{v}(P)$ between those two points,

$$\tilde{g}_2 - \tilde{g}_1 = \int_{P_1}^{P_2} \tilde{v}(P)\,dP. \tag{4.50}$$

The Gibbs free energy increases and is concave between A and D. Between D and F it decreases and is convex. Then between F and I it becomes concave again and increases. We see that between D and F the states are mechanically unstable since mechanical stability requires that $\tilde{g}$ be concave (cf. Sec. 2.H). The regions from A to D and from F to I are both mechanically stable since $\tilde{g}$ is concave. However, only the curve ACI corresponds to states in thermodynamic equilibrium because for these states the Gibbs free energy is a minimum. The states FCD are metastable. The equilibrium states thus correspond to those states whose Gibbs free energy has values lying along the curve ACI. To obtain the equilibrium states on the isotherm between C and G we must draw a straight line (line of constant pressure) between them, since this is the only way the Gibbs free energy will remain constant in going from C to G. The physical isotherm (isotherm containing equilibrium states) is the line ABCEGHI in Fig. 4.9.

Before we can complete our construction of isotherms we must decide where C and G lie. For the points C and G the molar Gibbs free energies are equal. Thus,

$$0 = \int_{P_C}^{P_G} \tilde{v}(P)\, dP$$

$$= \int_{P_C}^{P_D} \tilde{v}\, dP + \int_{P_D}^{P_E} \tilde{v}\, dP + \int_{P_E}^{P_F} \tilde{v}\, dP + \int_{P_F}^{P_G} \tilde{v}\, dP \tag{4.51}$$

or, rearranging,

$$\int_{P_C}^{P_D} \tilde{v}\, dP - \int_{P_E}^{P_D} \tilde{v}\, dP = \int_{P_F}^{P_E} \tilde{v}\, dP - \int_{P_F}^{P_G} \tilde{v}\, dP. \tag{4.52}$$

The left-hand side is equal to area 2 in Fig. 4.9 and the right-hand side is equal to area 1. Thus, the line from G to C must be drawn so that the areas 1 and 2 are equal:

$$\text{Area } 1 = \text{Area } 2. \tag{4.53}$$

If this is done, the curve ACEGI then gives the equilibrium states of the system. The condition given in Eq. (4.53) is called the Maxwell construction. Thus, with the Maxwell construction, we obtain the equilibrium isotherms from the van der Waals equation and the curves for metastable states.

E. REGULAR BINARY SOLUTIONS[6–8]

In Sec. 4.D, we considered phase transitions that can occur in pure *PVT* systems. If we now consider mixtures of different types of particles, a phase transition can occur in which we get a physical separation of the fluid into regions containing different concentrations of the various types of particles. The simplest example of this type of phase transition occurs for binary mixtures.

Let us consider a container filled with a mixture of particles of types A and B. If the particles do not interact the total Gibbs free energy of the system will be

$$G = n_A\tilde{\mu}_A{}^0(P, T) + n_B\tilde{\mu}_B{}^0(P, T) + RTn_A \ln x_A + RTn_B \ln x_B \tag{4.54}$$

(cf. Eq. (3.36)), where n_A and n_B are the number of moles of A and B, respectively, x_A and x_B are their mole fractions, and $\tilde{\mu}_A{}^0$ and $\tilde{\mu}_B{}^0$ are the chemical potentials per mole of A and B, respectively, in the absence of mixing. Let us now assume that particles A and B interact. This will add a term to the internal energy which is proportional to the number of moles of each type of particle; $U_{\text{int}} = \lambda n_A n_B/n$ where $n = n_A + n_B$. Thus, the Gibbs free energy per mole for the interacting mixture takes the form

$$\tilde{g} = \frac{G}{n} = x_A\tilde{\mu}_A{}^0 + x_B\tilde{\mu}_B{}^0 + RTx_A \ln x_A + RTx_B \ln x_B + \lambda x_A x_B. \tag{4.55}$$

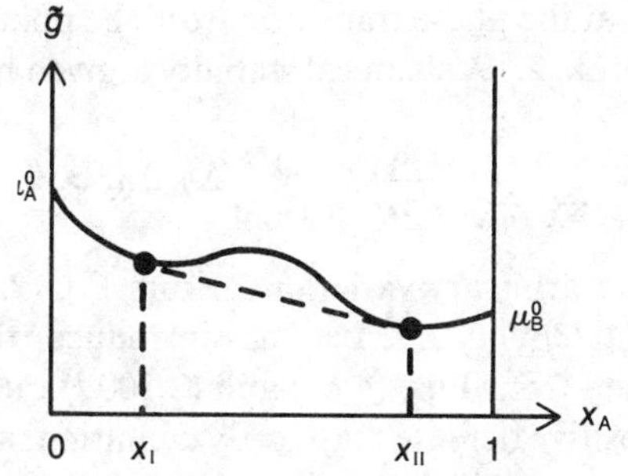

Fig. 4.11. For $T < \lambda/2R$, the molar Gibbs free energy of a binary mixture can have two local minima as a function of x_A for fixed P and T.

A binary system of the type given in Eq. (4.55) is called regular and was first introduced by Hildebrand. The coefficient λ is a measure of the strength of interaction between A and B. If λ is negative, A and B attract. If λ is positive, A and B repel. As we shall see, if λ is sufficiently large and positive and if the temperature is sufficiently low, the repulsive interaction energy will win out over the thermal energy which tends to mix the particles, and the system will separate into two phases, one rich in A and the other rich in B.

From Eq. (4.55), it is easy to see that a phase transition can occur. Let us first make note of the identities

$$(a)\quad \frac{\partial x_A}{\partial n_A} = \frac{x_B}{n} \quad \text{and} \quad (b)\quad \frac{\partial x_A}{\partial n_B} = -\frac{x_A}{n} \tag{4.56}$$

and remember that $x_B = 1 - x_A$. Thus, the molar Gibbs free energy, $\tilde{g}$, depends on three independent variables P, T, and x_A. If we now consider the derivatives

$$\left(\frac{\partial \tilde{g}}{\partial x_A}\right)_{P,T} = \tilde{\mu}_A{}^0 - \tilde{\mu}_B{}^0 + \lambda(1 - 2x_A) + RT \ln \frac{x_A}{(1 - x_A)} \tag{4.57}$$

and

$$\left(\frac{\partial^2 \tilde{g}}{\partial x_A{}^2}\right)_{P,T} = -2\lambda + \frac{RT}{x_A(1 - x_A)} \geqslant 4RT - 2\lambda, \tag{4.58}$$

we find that for temperatures $T < (\lambda/2R)$, there is a range of values of x_A centered at $x_A = \frac{1}{2}$ for which $\tilde{g}$ can be a concave function of x_A while outside this range $\tilde{g}$ is convex. This means that for $T < (\lambda/2R)$, $\tilde{g}$ can have two local minima and one local maximum (cf. Fig. 4.11). Each of the minima corresponds to a stable thermodynamic state. These extrema occur for values of x_A such that

$$\lambda(1 - 2x_A) + RT \ln \frac{x_A}{(1 - x_A)} = 0. \tag{4.59}$$

One extremum occurs for $x_A = \frac{1}{2}$. This is the local maximum. The two minima can be found graphically.

It is useful to look at the phase transition from the point of view of stability. As we have seen in Prob. 2.19, chemical stability is given by the condition

$$\sum_{i=A}^{B}\sum_{j=A}^{B}\left(\frac{\partial\tilde{\mu}_i}{\partial n_j}\right)_{P,T,\{n_{l\neq j}\}}\Delta n_i\,\Delta n_j > 0 \tag{4.60}$$

where Δn_i and Δn_j are arbitrary variations. From Eq. (2.100*d*) we know that $(\partial\tilde{\mu}_A/\partial n_B)_{T,P,n_A} = (\partial\tilde{\mu}_B/\partial n_A)_{T,P,n_B}$. Let us introduce the notation $\tilde{\mu}_{i,j} \equiv (\partial\tilde{\mu}_i/\partial n_j)_{P,T,\{n_l\neq j\}}$. Then from Eqs. (4.60) and (2.100*d*) the quantities $\tilde{\mu}_{i,j}$ must form a symmetric positive definite matrix. By definition, a symmetric matrix is positive definite if every principal minor is positive. Thus, for the binary solution we have the conditions

$$(a)\quad \tilde{\mu}_{A;A} > 0;\qquad (b)\quad \tilde{\mu}_{B;B} > 0;\qquad \text{and}\qquad (c)\quad \det\begin{vmatrix}\tilde{\mu}_{A;A} & \tilde{\mu}_{A;B}\\ \tilde{\mu}_{B;A} & \tilde{\mu}_{B;B}\end{vmatrix} > 0 \tag{4.61}$$

if the mixture is to be chemically stable.

Let us now note that for processes at constant P and T,

$$\sum_{ij} n_j\left(\frac{\partial\tilde{\mu}_j}{\partial n_i}\right)_{P,T,\{n_{l\neq i}\}} dn_i = 0 \tag{4.62}$$

(cf. Eq. (2.69)). Since the changes dn_i are arbitrary, this gives

$$\sum_j n_j\left(\frac{\partial\tilde{\mu}_j}{\partial n_i}\right)_{P,T,\{n_{l\neq i}\}} = 0. \tag{4.63}$$

For the binary system Eq. (4.63) takes the form

$$n_A\tilde{\mu}_{A;A} + n_B\tilde{\mu}_{B;A} = 0 \tag{4.64}$$

and

$$n_B\tilde{\mu}_{B;B} + n_A\tilde{\mu}_{A;B} = 0 \tag{4.65}$$

and thus

$$\tilde{\mu}_{A;B} = \tilde{\mu}_{B;A} < 0 \tag{4.66}$$

for chemical stability.

Let us now determine what Eqs. (4.61) and (4.66) imply about the stability of the binary system. The chemical potentials for components A and B can be found from Eq. (4.55):

$$\tilde{\mu}_A = \left(\frac{\partial G}{\partial n_A}\right)_{P,T,n_B} = \tilde{\mu}_A{}^0(P, T) + RT\ln x_A + \lambda x_B{}^2 \tag{4.67}$$

and

$$\tilde{\mu}_B = \left(\frac{\partial G}{\partial n_B}\right)_{P,T,n_A} = \tilde{\mu}_B{}^0(P, T) + RT\ln x_B + \lambda x_A{}^2. \tag{4.68}$$

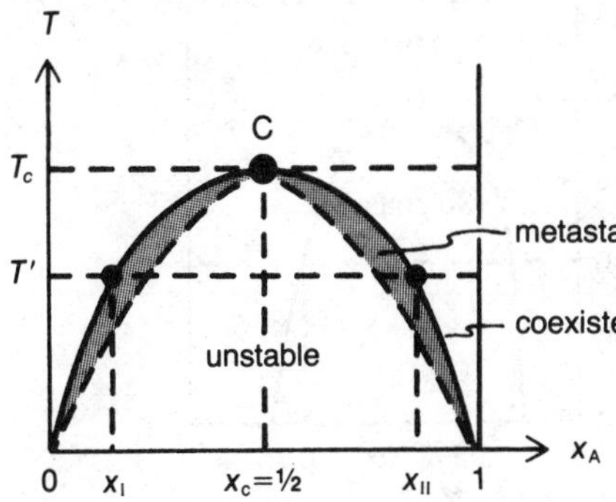

Fig. 4.12. Stability curve for a binary mixture.

Thus,

$$\tilde{\mu}_{A;A} = \left(\frac{\partial \tilde{\mu}_A}{\partial n_A}\right)_{P,T,n_B} = \frac{RTx_B}{x_A n} - \frac{2\lambda x_B^2}{n} \tag{4.69}$$

and

$$\tilde{\mu}_{A;B} = \left(\frac{\partial \tilde{\mu}_A}{\partial n_B}\right)_{P,T,n_B} = -\frac{RT}{n} + \frac{2\lambda x_B x_A}{n}. \tag{4.70}$$

Eqs. (4.69) and (4.70) give identical conditions for the stability of the binary solution; namely,

$$\frac{RT}{2\lambda} \geqslant (1 - x_A)x_A. \tag{4.71}$$

If we plot Eq. (4.71) in a T-x_A plane, we find a curve (cf. dashed line in Fig. 4.12) symmetric about $x_A = \frac{1}{2}$ which has its maximum value at $T_c = \lambda/2R$ and $x_A{}^c = \frac{1}{2}$. This is the critical point for the binary system. The region outside the dashed curve in Fig. 4.12 is chemically stable. To the left of the unstable region is a B-rich phase (phase I). To the right of the unstable region is an A-rich phase (phase II). However, not all states outside the unstable region are equilibrium states. Some may be metastable. At the critical point itself,

$$\left(\frac{\partial^2 \tilde{g}}{\partial x_A{}^2}\right)_{P,T} = \left(\frac{\partial^3 \tilde{g}}{\partial x_A{}^3}\right)_{P,T} = 0 \quad \text{and} \quad \left(\frac{\partial^4 \tilde{g}}{\partial x_A{}^4}\right)_{P,T} > 0.$$

The coexistence curve for equilibrium states is given by the condition that the chemical potentials of A in phases I and II be equal, $\tilde{\mu}_A{}^{\mathrm{I}} = \tilde{\mu}_A^{\mathrm{II}}$, and the chemical potentials of B in phases I and II be equal, $\tilde{\mu}_B{}^{\mathrm{I}} = \tilde{\mu}_B^{\mathrm{II}}$ (cf. Sec. 4.B). Thus, for coexistence, we must have

$$RT \ln x_A{}^{\mathrm{I}} + \lambda(1 - x_A{}^{\mathrm{I}})^2 = RT \ln x_A^{\mathrm{II}} + \lambda(1 - x_A^{\mathrm{II}})^2 \tag{4.72}$$

and

$$RT \ln x_B{}^{\mathrm{I}} + \lambda(1 - x_B{}^{\mathrm{I}})^2 = RT \ln x_B^{\mathrm{II}} + \lambda(1 - x_B^{\mathrm{II}})^2. \tag{4.73}$$

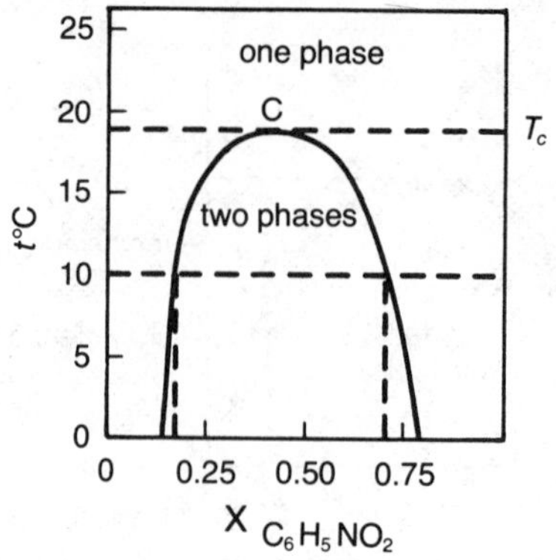

Fig. 4.13. Phase diagram for *n*-hexane-nitrobenzene at atmospheric pressure (based on Ref. 4.7, p. 238). Solid line is coexistence curve.

Since the curve is symmetric about $x_A = \frac{1}{2}$, we have $x_A^{II} = 1 - x_A{}^I$ and Eq. (4.72) reduces to Eq. (4.59). Thus, the two minima of $\tilde{g}$ lie on the coexistence curve.

The region under the coexistence curve is the region in which two phases coexist. In the metastable region a single phase can exist, but it is not in a stable equilibrium state. We see that there are many analogies between the liquid-gas transition and the separation of a binary system into two phases. In Fig. 4.12, let us follow the horizontal line at temperature $T' < T_c$. At $x_A = 0$, we have a system consisting only of B-particles. As we start adding A-particles, the concentration of A increases until we reach the coexistence curve at I. At this point the system separates into two phases, one in which A has concentration $x_A{}^I$ and another in which A has concentration x_A^{II}. As we increase the number of A particles relative to B, the amount of phase x_A^{II} increases and $x_A{}^I$ decreases until we reach the coexistence curve at II. At II the $x_A{}^I$ phase has disappeared and we again have a single phase of concentration x_A^{II}.

An example of a system exhibiting this type of behavior is an *n*-hexane–nitrobenzene system at atmospheric pressure. The phase diagram for this system is given in Fig. 4.13.

F. GINZBURG-LANDAU THEORY: λ-POINTS[3]

In Sec. 4.C, we discussed three types of first-order phase transitions in a pure *PVT* system. In two of them, the liquid-solid and vapor-solid transitions, the translational symmetry of the high-temperature phase is broken at the transition point, whereas for the vapor-liquid transition no symmetry of the system is broken.

The average density of both the liquid and the gas phases is independent of position and therefore is invariant under all elements of the translation group, whereas the solid phase has a periodic average density and is translationally invariant only with respect to a subgroup of the translation group. Thus, in going from the gas or liquid phase to the solid phase, translational symmetry is broken.

Both the liquid-solid and vapor-solid transitions are first order; that is, the slope of the free energy curve changes discontinuously as a function of Y and T at the transition. However, we can also have symmetry-breaking transitions in which the slope of the free energy curve changes continuously. Such transitions are called λ-points, because the heat capacity always exhibits a λ-shaped peak at the transition point. As we shall see, there are different types of symmetry that can be broken at a phase transition. For the liquid-solid transition, translational symmetry is broken. In the transition from a paramagnetic to a ferromagnetic system, rotational symmetry is broken because a spontaneous magnetization occurs which defines a unique direction in space. In the transition from normal liquid He^4 to superfluid liquid He^4, gauge symmetry is broken. All transitions which involve a broken symmetry and a continuous change in the slope of the free energy curve can be described within the framework of a single theory due to Ginzburg and Landau.

In every symmetry-breaking transition, a new macroscopic parameter (called the order parameter) appears in the less symmetric phase. The order parameter may be a scalar, a vector, a tensor, a complex number, or some other quantity. In a first-order symmetry-breaking transition there is a discontinuous jump in the state of the system (the density or entropy, for example) at the transition, and therefore there need be no relation between the symmetry properties of the two phases. For a continuous transition, the state changes continuously, and the symmetry properties of the two phases are closely related. Indeed, in a continuous transition one phase will always have a lower symmetry than the other. Usually the lower temperature phase is less symmetric, but this need not always be true.

For simplicity let us assume that the order parameter is a vector $\boldsymbol{\eta}$ and discuss the general form of the free energy in the region of the phase transition. The free energy must be such that it will be minimized for $\boldsymbol{\eta} = 0$ above the transition and for $\boldsymbol{\eta} \neq 0$ below the transition. Furthermore, the free energy, which is a number, must be a scalar function of the order parameter. Thus, if the order parameter is a vector, the free energy can only depend on scalar products of the order parameter. In general, near the phase transition we can expand the free energy (we will denote the free energy as $\Phi(T)$; the particular free energy we use will depend on the system) in the form

$$\Phi(T, \boldsymbol{\eta}) = \Phi_0(T) + \alpha_2(T)|\eta|^2 + \alpha_4(T)|\eta|^4 + \cdots, \tag{4.74}$$

where $|\eta|^2 = \boldsymbol{\eta}\cdot\boldsymbol{\eta}$. No first-order or third-order terms will appear because we cannot construct a scalar from them.

The form of $\alpha_2(T)$ is chosen so that at the critical temperature and above it the free energy will only be minimized for $|\eta| = 0$ and below the critical

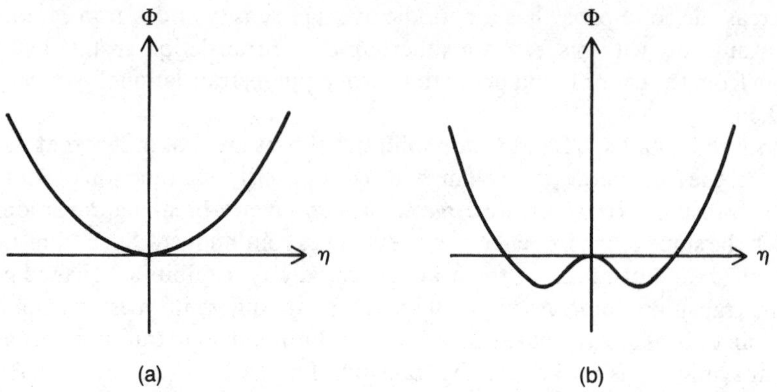

Fig. 4.14. Ginzburg-Landau free energy above and below the critical point: figure (*a*) corresponds to $\alpha_2 > 0$ and figure (*b*) corresponds to $\alpha_2 < 0$; for both cases, $\alpha_3 \equiv 0$.

temperature it will be minimized for $|\eta| > 0$. In general, the free energy will be minimum if $(\partial\Phi/\partial\boldsymbol{\eta})_{T,Y} = 0$ and $(\partial^2\Phi/\partial^2\boldsymbol{\eta})_{Y,T} > 0$. If we choose $\alpha_2(T) > 0$ for $T > T_c$ and $\alpha_2(T) < 0$ for $T < T_c$, the above condition is satisfied. $\Phi(T, \boldsymbol{\eta})$ will have its minimum value for $T > T_c$ if $\boldsymbol{\eta} = 0$, while for $T < T_c$ we can have $\boldsymbol{\eta} \neq 0$ and have a minimum. Since the free energy must vary continuously through the transition point, at $T = T_c$ we must have $\alpha_2(T_c) = 0$. We can combine all this information if we write $\alpha_2(T)$ in the form

$$\alpha_2(T) = \alpha_0(T - T_c), \tag{4.75}$$

where α_0 is some function of T.

In order to have global stability, we must have

$$\alpha_4(T) > 0. \tag{4.76}$$

This will ensure that as we increase $|\eta|$ to very large values the free energy will increase and not decrease. In Fig. 4.14, we sketch the free energy for $\alpha_2 > 0$ and $\alpha_2 < 0$. The free energy has extrema when

$$\frac{\partial\Phi}{\partial\boldsymbol{\eta}} = 2\alpha_2\boldsymbol{\eta} + 4\alpha_4\boldsymbol{\eta}|\eta|^2 = 0; \tag{4.77}$$

that is, when

$$\boldsymbol{\eta} = 0 \quad \text{or} \quad \pm\sqrt{\frac{-\alpha_2}{2\alpha_4}}\,\hat{\boldsymbol{\eta}}. \tag{4.78}$$

When $\alpha_2 > 0$ the minimum occurs for $\boldsymbol{\eta} = 0$. When $\alpha_2 < 0$ the minimum occurs for $\boldsymbol{\eta} = \pm\sqrt{(-\alpha_2/2\alpha_4)}\hat{\boldsymbol{\eta}}$. Thus, below the critical temperature, the order parameter is nonzero and increases as $(|T - T_c|)^{1/2}$. The vector $\boldsymbol{\eta}$ can be pointed in either the positive or the negative direction along a given axis.

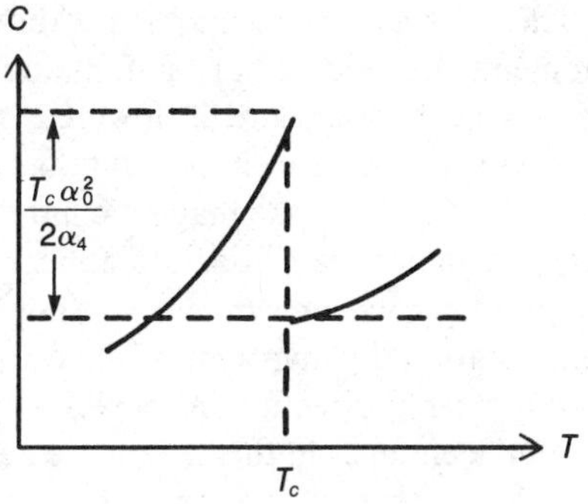

Fig. 4.15. The jump in the heat capacity at the λ-point as predicted by the Ginzburg-Landau theory.

The free energy takes the following form: above the transition,

$$\Phi(T, \eta) = \Phi_0(T) \qquad T > T_c, \tag{4.79}$$

while below it,

$$\Phi(T, \eta) = \Phi_0(T) - \frac{\alpha_0^2(T - T_c)^2}{4\alpha_4} \qquad T < T_c. \tag{4.80}$$

From Eqs. (4.79) and (4.80), the jump in the heat capacity at the critical point is easy to obtain. From Eq. (2.124),

$$C = -T\left(\frac{\partial^2\Phi}{\partial T^2}\right). \tag{4.81}$$

If we neglect derivatives of α_0 and α_4 (we assume they vary slowly with temperature), we find

$$C(T < T_c)_{T=T_c} - C(T > T_c)_{T=T_c} = T_c\frac{\alpha_0^2}{2\alpha_4}. \tag{4.82}$$

Thus, the jump in the heat capacity has the shape of a λ, as in Fig. 4.15.

The Ginzburg-Landau theory applies to all continuous symmetry-breaking transitions, although the form of the expansion varies considerably for different physical systems. A good theoretical discussion of its application to phase transitions resulting in structural changes in crystal lattices can be found in Ref. 4.3. Experimental curves for structural phase transitions are given in Ref. 4.9. In the next sections we shall apply the Ginzburg-Landau theory to magnetic systems and superconductors.

G. CURIE POINT[10]

The transition from a paramagnetic to a ferromagnetic system is one of the simplest examples of a λ-point. A common type of paramagnetic system is a lattice containing particles with a magnetic moment. Above the Curie temperature, the magnetic moments are oriented at random and there is no net magnetization. However, as the temperature is lowered, magnetic interaction energy between lattice sites becomes more important than randomizing thermal energy. Below the Curie temperature, the magnetic moments become ordered on the average and a spontaneous magnetization appears. The symmetry that is broken at the Curie point is rotation symmetry. Above the Curie point, the paramagnetic system is rotationally invariant, while below it the spontaneous magnetization selects a preferred direction in space.

If we assume that volume changes in the system are negligible at the transition, we can write the following expression for the Helmholtz free energy of an isotropic system:

$$A(\mathbf{M},T) = A(0, T) + \alpha_2(T)|M^2| + \alpha_4(T)|M^4| + \cdots, \tag{4.83}$$

where $A(0, T)$ is a magnetization-independent contribution to the free energy above the Curie temperature, $|M^2| = \mathbf{M}\cdot\mathbf{M}$, and $\mathbf{M}$ is the average magnetization of the system. If we apply an external magnetic field $\mathscr{H}$ to the system, then for fixed $\mathscr{H}$ the Gibbs free energy,

$$\begin{aligned} G(\mathscr{H}, T) &= A(\mathbf{M}, T) - \mathscr{H}\cdot\mathbf{M} \\ &= G_0(\mathscr{H}, T) + \alpha_2(T, \mathscr{H})|M|^2 + \alpha_4(T, \mathscr{H})|M|^4 + \cdots, \end{aligned} \tag{4.84}$$

will be a minimum. Here $G_0(\mathscr{H}, T) = A(0, T) - \mathscr{H}\cdot\mathbf{M}$ and we assume that the dependence of α_2 and α_4 on $\mathscr{H}$ can be neglected for small $\mathscr{H}$. From our discussion of the Ginzburg-Landau theory in Sec. 4.F, we know that

$$\alpha_2(T) = \alpha_0(T - T_c) \quad \text{and} \quad \alpha_4(T) > 0. \tag{4.85}$$

To lowest order in $\mathbf{M}$, the Gibbs free energy is minimized when

$$\left(\frac{\partial G}{\partial \mathbf{M}}\right)_{T,\mathscr{H}} = -\mathscr{H} + 2\alpha_2(T)\mathbf{M} = 0 \tag{4.86}$$

or

$$\mathbf{M} = \frac{1}{2\alpha_2}\mathscr{H}. \tag{4.87}$$

Thus, above the Curie point we have the possibility of a nonzero magnetization when an external field is present. (Below the Curie point we must retain higher order terms to get a meaningful result.) From Eq. (4.87), we obtain the isothermal susceptibility

$$\chi_T = \left(\frac{\partial \mathbf{M}}{\partial \mathscr{H}}\right)_T = \frac{1}{2\alpha_2(T)} = \frac{1}{2\alpha_0(T - T_c)}. \tag{4.88}$$

Because the system is isotropic, the susceptibility is a scalar. Note that the susceptibility becomes infinite at the critical point and, therefore, at the

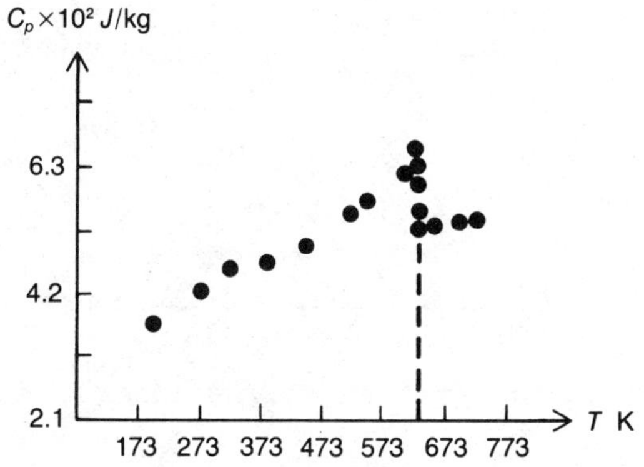

Fig. 4.16. Specific heat of nickel in the neighborhood of the Curie point: the dashed line gives the Curie point (based on Ref. 4.26).

critical point a very small external field can have a large effect on the magnetization. Below the critical point, in the absence of an external field, the magnetization behaves as

$$M = \sqrt{\frac{\alpha_0(T_c - T)}{2\alpha_4}}. \tag{4.89}$$

Thus,

$$\left(\frac{\partial M}{\partial T}\right)_{\mathscr{H}} = -\frac{1}{2}\sqrt{\frac{\alpha_0}{2\alpha_4(T_c - T)}} \tag{4.90}$$

and the magnetization increases abruptly below the critical point. The heat capacity at the Curie point exhibits the characteristic λ-shaped peak. An example of a λ-point in nickel is given in Fig. 4.16.

H. SUPERCONDUCTORS[11,12]

1. Experimental Properties

Superconductivity was first observed in 1911 by Kamerlingh Onnes. He found that the resistance to current flow in mercury drops to zero at about 4.2 K (cf. Fig. 4.17). At first this was interpreted as a transition to a state with infinite conductivity. However, infinite conductivity imposes certain conditions on the magnetic field which were not subsequently observed. The relation between the electric current, **J**, and the applied electric field, **E**, in a metal is given by Ohm's law,

$$\mathbf{J} = \sigma\mathbf{E}, \tag{4.91}$$

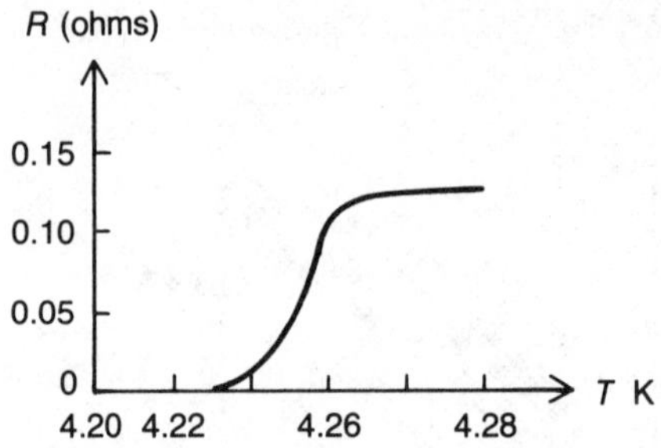

Fig. 4.17. The resistance of mercury drops to zero at about 4.2 K (based on Ref. 4.27).

where σ is the conductivity. The electric field **E** is related to the magnetic field **B** by Ampère's law,

$$\nabla_{\mathbf{r}} \times \mathbf{E} = -\frac{\partial \mathbf{B}}{\partial t}. \tag{4.92}$$

If we substitute Eq. (4.91) into Eq. (4.92), we see that for infinite conductivity $\sigma \to \infty$ and $\partial \mathbf{B}/\partial t = 0$. This in turn implies that the state of the system depends on its history. If we first cool the sample below the transition temperature and then apply an external magnetic field, $\mathscr{H}$, surface currents must be created in the sample to keep any field from entering the sample, since **B** must remain zero inside (cf. Fig. 4.18). However, if we place the sample in the $\mathscr{H}$-field before

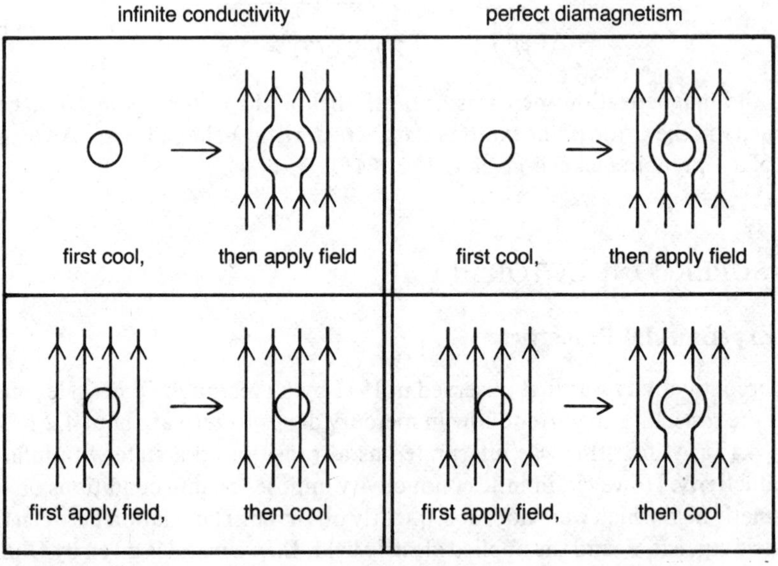

Fig. 4.18. A superconducting sample has different behavior if it is a perfect conductor or a perfect diamagnet (circle is sample).

cooling, a **B**-field is created inside. Then, if we cool the sample, the **B**-field must stay inside. Thus, the final states depend on how we prepare the sample. With the hypothesis of infinite conductivity, the state below the transition temperature cannot be a thermodynamic state since it depends on history.

In 1933, Meissner and Ochsenfeld[13] cooled a monocrystal of tin in a magnetic field and found that the field inside the sample was expelled below the transition point for tin. This is contrary to what is expected if the transition is to a state with infinite conductivity; it instead implies a transition to a state of perfect diamagnetism, $\mathbf{B} = 0$. It is now known that superconductors are perfect diamagnets. When superconducting metals are cooled below their transition point in the presence of a magnetic field, currents are set up on the surface of the sample in such a way that the magnetic fields created by the currents cancel any magnetic fields initially inside the medium. Thus $\mathbf{B} = 0$ inside a superconducting sample regardless of the history of its preparation.

No electric field is necessary to cause a current to flow in a superconductor. A magnetic field is sufficient. In a normal conductor, an electric field causes electrons to move at a constant average velocity because interaction with lattice impurities acts as a friction which removes energy from the electron current. In a superconductor, an electric field accelerates part of the electrons in the metal. No significant frictional effects act to slow them down. This behavior is reminiscent of the frictionless superflow observed in liquid He^4 below 2.19 K (cf. Sec. 3.F). Indeed, the superfluid flow in He^4 and the supercurrents in superconductors are related phenomena. The origin of the apparently frictionless flow in both cases lies in quantum mechanics. It is now known that electrons in a superconducting metal can experience an effective attractive interaction due to interaction with lattice phonons. Because of this attraction, a fraction of the electrons (we never know which ones) can form "bound pairs." The state of minimum free energy is the one in which the bound pairs all have the same quantum numbers. Thus, the bound pairs form a single macroscopically occupied quantum state which acts coherently and forms the condensed phase. As we shall see, the order parameter of the condensed phase behaves like an effective wave function of the pairs. Because the pairs in the condensed phase act coherently (as one state), any friction effects due to lattice impurities must act on the entire phase (which will contain many pairs and have a large mass) and not on a single pair. Thus, when an electric field is applied, the condensed phase moves as a whole and is not slowed significantly by frictional effects.

The condensed phase flows at a steady rate when a magnetic field is present and is accelerated when an electric field is applied. Just below the transition temperature, only a small fraction of electrons in the metal are condensed and participate in superflow. As temperature is lowered, thermal effects which tend to destroy the condensed phase become less important and a larger fraction of the electrons condense.

If a superconductor is placed in a large enough external magnetic field, the superconducting state can be destroyed. A plot of magnetic induction, B, versus

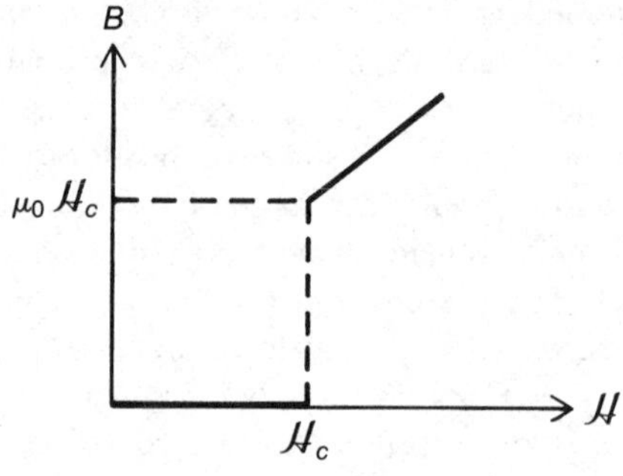

Fig. 4.19. A plot of the magnetic induction, **B**, versus the applied magnetic field, $\mathscr{H}$, in a superconductor.

applied field, $\mathscr{H}$, appears as in Fig. 4.19. For applied field, $\mathscr{H}$, with a value less than some temperature-dependent critical value $\mathscr{H}_c(T)$, the system is a perfect diamagnet; that is, the permeability $\mu = 0$ and therefore $B = 0$. However, for $\mathscr{H} > \mathscr{H}_c(T)$ the system becomes normal and $B = \mu\mathscr{H}$. (For normal metals $\mu \approx \mu_0$, the permeability of the vacuum.) Thus,

$$B = \begin{cases} 0 & \mathscr{H} < \mathscr{H}_c(T) \\ \mu_0\mathscr{H} & \mathscr{H} > \mathscr{H}_c(T) \end{cases}. \tag{4.93}$$

The critical field, $\mathscr{H}_c(T)$, has been measured as a function of the temperature and has roughly the same behavior for most metals (cf. Fig. 4.20). The co-existence curve for the normal and super-conducting phases is well approximated by the equation

$$\mathscr{H}_c(T) = \mathscr{H}_0\left(1 - \frac{T^2}{T_c^2}\right), \tag{4.94}$$

where T_c is the critical temperature when no external fields are present. The slope $d\mathscr{H}_c/dT = 0$ at $T = 0$ K and is negative at $T = T_c$. The phase diagram for

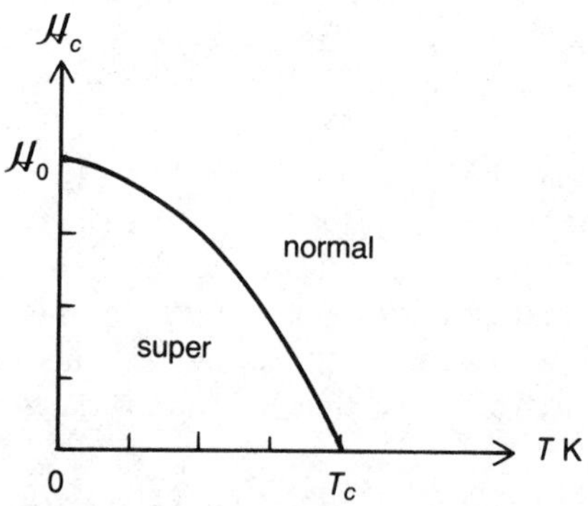

Fig. 4.20. The coexistence curve for normal and superconducting states.

a superconducting metal has analogies to the vapor-liquid transition in a *PVT* system, if we let H_c replace the specific volume. Inside the coexistence curve, condensate begins to appear.

Along the coexistence curve, the chemical potentials of the superconducting and normal phases must be equal and, therefore, any changes in the chemical potentials must be equal. Thus, along the coexistence curve

$$-s_n\, dT - B_n\, d\mathscr{H} = -s_s\, dT - B_s\, d\mathscr{H} \tag{4.95}$$

or

$$s_n - s_s = -\mu_0 \mathscr{H}_c(T)\left(\frac{d\mathscr{H}}{dT}\right)_{\text{coex}}. \tag{4.96}$$

Eq. (4.96) is the Clausius-Clapeyron equation for superconductors. We have used the fact that $B_s = 0$ and $B_n = \mu_0 \mathscr{H}_c(T)$ on the coexistence curve. Here $s_{n(s)}$ is the entropy per unit volume of the normal (superconducting) phase. We see that the transition has a latent heat (is first order) for all temperatures except $T = T_c$ where $\mathscr{H}_c = 0$. When no external magnetic fields are present, the transition is second order.

The change in the heat capacity per unit volume at the transition is

$$(c_n - c_s)_{\text{coex}} = \left[T\frac{\partial}{\partial T}(s_n - s_s)\right]_{\text{coex}} = 2\mu_0 \frac{\mathscr{H}_c^2}{T_c}\left(\frac{T}{T_c} - \frac{3T^3}{T_c^3}\right). \tag{4.97}$$

We have used Eq. (4.94) to evaluate the derivatives $(d\mathscr{H}_c/dT)$. At low temperatures the heat capacity of the normal phase is higher than that of the superconducting phase. At $T = T_c$ (the critical point) the heat capacity is higher in the superconductor and has a finite jump, $(c_s - c_n)_{T=T_c} = (4\mu_0/T_c)\mathscr{H}_0^2$. Thus, in the transition to a superconducting state, the critical point appears to be a λ-point. It is worthwhile noting that as $T \to 0$, $(s_s - s_n) \to 0$ since $(d\mathscr{H}_c/dT)_{T\to 0} \to 0$. This is in agreement with the third law of thermodynamics.

It is useful to obtain the difference between the Gibbs free energies of the normal and superconducting phases for $\mathscr{H} = 0$. The differential of the Gibbs free energy per unit volume, g, is

$$dg = -s\, dT - B\, d\mathscr{H}. \tag{4.98}$$

If we integrate Eq. (4.98) at a fixed temperature, we can write

$$g(T, \mathscr{H}) - g(T, 0) = -\int_0^{\mathscr{H}} B\, d\mathscr{H}. \tag{4.99}$$

For the normal phase we have

$$g_n(T, \mathscr{H}) - g_n(T, 0) = -\frac{\mu_0}{2}\mathscr{H}^2. \tag{4.100}$$

For the superconductor ($\mathscr{H} < \mathscr{H}_c$) we have

$$g_s(T, \mathscr{H}) - g_s(T, 0) = 0 \tag{4.101}$$

(cf. Fig. 4.21) since $\mathbf{B} = 0$ in the superconductor. If we next use the fact that

$$g_n(T, \mathscr{H}_c) = g_s(T, \mathscr{H}_c), \tag{4.102}$$

we obtain the desired result,

$$g_s(T, 0) = g_n(T, 0) - \frac{\mu_0}{2} \mathscr{H}_c^2(T). \tag{4.103}$$

Thus, the Gibbs free energy per unit volume of the superconductor in the absence of a magnetic field is lower than that of the normal phase by a factor $(\mu_0/2)\mathscr{H}_c^2(T)$ (the so-called condensation energy) at a temperature T. Since the Gibbs free energy must be a minimum for fixed $\mathscr{H}$ and T, the condensed phase is the physically realized state. Note that $g_s(T_c, 0) = g_n(T_c, 0)$, as it should.

2. Ginzburg-Landau Theory of Superconductors[3,12]

The phase transition from normal to superconducting phase is a transition in which gauge symmetry is broken. The condensed phase in a superconductor corresponds to a macroscopically occupied quantum state and therefore is described by a macroscopic "wave function," Ψ. The "wave function," Ψ, is the order parameter of the system and, in general, is a complex function. A gauge transformation is a transformation which changes the phase of all wave functions in the system. It is generated by the number operator.

Under a gauge transformation the order parameter Ψ changes its phase. However, the free energy must remain invariant under the gauge transformation. Therefore, if no magnetic fields are present, the Ginzburg-Landau expression for the Helmholtz free energy per unit volume must have the form

$$a_s(T) = a_n(T) + \alpha_2(T)|\Psi|^2 + \alpha_4(T)|\Psi|^4, \tag{4.104}$$

where $|\Psi|^2 = \Psi^*\Psi$. The quantity $a_n(T)$ is the free energy per unit volume of the normal phase, $\alpha_2 = \alpha_0(T - T_c)$ and $\alpha_4(T) > 0$; $|\Psi|^2$ can be interpreted as the number density of condensed pairs of electrons $|\Psi|^2 = n_s$. In writing Eq. (4.104), we have implicitly assumed that the superconductor is infinitely large so the order parameter will be independent of position. For $T < T_c$, $|\Psi| = (|\alpha_2|/2\alpha_4)^{1/2}$ and

$$a_s(T) = a_n(T) - \frac{\alpha_2^2}{4\alpha_4} \tag{4.105}$$

(cf. Eq. (4.80)). Thus, from Eq. (4.103) we obtain the relation

$$\frac{\alpha_2^2}{4\alpha_4} = \frac{\mu_0}{2} \mathscr{H}_c^2(T) \tag{4.106}$$

between the critical field and the Ginzburg-Landau parameters.

One of the most interesting applications of the Ginzburg-Landau theory is to the case of a sample which may contain both normal and condensed regions or is of finite size. Then we must allow the order parameter to vary in position.

Let us assume that a constant external field, $\mathscr{H}$, is applied to the sample, but let us allow the possibility of a spatially varying order parameter $\Psi(\mathbf{r})$. If the sample contains both normal and condensed regions, then there will also be a local spatially varying induction field $\mathbf{B}(\mathbf{r})$. The order parameter must be treated as if it were the wave function of a particle. Gradients in the order parameter (spatial variations) will give contributions to the free energy of a form similar to the kinetic energy of a quantum particle. Furthermore, if a local induction field $\mathbf{B}(\mathbf{r})$ is present, the canonical momentum of the condensed phase will have a contribution from the vector potential $\mathscr{A}(\mathbf{r})$ associated with $\mathbf{B}(\mathbf{r})$. The local vector potential $\mathscr{A}(\mathbf{r})$ is related to the local induction field, $\mathbf{B}(\mathbf{r})$, through the relation

$$\mathbf{B}(\mathbf{r}) = \nabla_{\mathbf{r}} \times \mathscr{A}(\mathbf{r}). \tag{4.107}$$

The Helmholtz free energy $a_s(\mathbf{r}, \mathbf{B}, T)$ per unit volume will be given by

$$\begin{aligned} a_s(\mathbf{r}, \mathbf{B}, T) &= a_n(T) + \alpha_2(T)|\Psi(\mathbf{r})|^2 + \alpha_4(T)|\Psi(r)|^4 \\ &\quad + \frac{1}{2m}|-i\hbar\nabla_{\mathbf{r}}\Psi - e\mathscr{A}\Psi|^2 + \frac{B^2}{2\mu_0}, \end{aligned} \tag{4.108}$$

where e is the charge of the electron pairs and $\hbar$ is Planck's constant. The quantity $-i\hbar\nabla_{\mathbf{r}}\Psi - e\mathscr{A}\Psi$ is the canonical momentum associated with the condensate. It has exactly the same form as the canonical momentum of a charged quantum particle in a magnetic field: m is the mass of the pairs. The Gibbs free energy per unit volume is given by a Legendre transformation,

$$g(\mathbf{r}, \mathscr{H}, T) = a(\mathbf{r}, \mathbf{B}, T) - \mathbf{B}\cdot\mathscr{H}. \tag{4.109}$$

The total Gibbs free energy is found by integrating $g(\mathbf{r}, \mathscr{H}, T)$ over the entire volume. Thus,

$$\begin{aligned} G_s(\mathscr{H}, T) = \int d\mathbf{r}\Big[& a_n(T) + \alpha_2(T)|\Psi(\mathbf{r})|^2 + \alpha_4(T)|\Psi(\mathbf{r})|^4 \\ & + \frac{1}{2m}|-i\hbar\nabla_{\mathbf{r}}\Psi - e\mathscr{A}\Psi|^2 + \frac{B^2}{2\mu_0} - \mathbf{B}\cdot\mathscr{H}\Big]. \end{aligned} \tag{4.110}$$

If we now extremize $G_s(\mathscr{H}, T)$ with respect to variations in $\Psi^*(\mathbf{r})$ and assume that on the boundary of the sample the normal component of the canonical momentum $((\hbar/i)\nabla_{\mathbf{r}}\Psi - e\mathscr{A}\Psi)$ is zero, we find

$$\alpha_2\Psi + \alpha_4\Psi|\Psi|^2 + \frac{1}{2m}\left(\frac{\hbar}{i}\nabla_{\mathbf{r}} - e\mathscr{A}(\mathbf{r})\right)^2\Psi = 0. \tag{4.111}$$

To obtain Eq. (4.111) we have had to integrate by parts. If we next extremize $G_s(\mathscr{H}, T)$ with respect to variations in $\mathscr{A}(\mathbf{r})$, we obtain

$$\nabla_{\mathbf{r}} \times \mathbf{B} = -\frac{e\hbar}{2mi}(\Psi^*\nabla_{\mathbf{r}}\Psi - \Psi\nabla_{\mathbf{r}}\Psi^*) - \frac{e^2}{m}\mathscr{A}|\Psi|^2. \tag{4.112}$$

To obtain Eq. (4.112), we have used the vector identities

$$\int dV\mathbf{C}\cdot(\nabla_{\mathbf{r}} \times \mathbf{A}) = \oint d\mathbf{S}\cdot(\mathbf{A} \times \mathbf{C}) + \int dV\mathbf{A}\cdot\nabla_{\mathbf{r}} \times \mathbf{C} \tag{4.113}$$

and

$$\oint d\mathbf{S}\cdot(\mathbf{E} \times \mathbf{F}) = \oint \mathbf{E}\cdot(\mathbf{F} \times d\mathbf{S}), \tag{4.114}$$

where $\int dV$ is an integral over the total volume V of the sample and $\oint d\mathbf{S}$ is an integral over the surface enclosing the volume V. We have assumed that $(\mathbf{B} - \mu_0\mathscr{H}) \times \hat{\mathbf{n}} = 0$, where $\hat{\mathbf{n}}$ is the unit vector normal to the surface of the sample. Thus, right on the surface the tangent component of the induction field is the same as that of a normal metal. However, it will go to zero in the sample.

The local super current in a superconductor is driven by the local magnetic induction field and is defined

$$\mathbf{J}_s(\mathbf{r}) = \nabla_{\mathbf{r}} \times \mathbf{B}(\mathbf{r}) \tag{4.115}$$

as we would expect. It is a combination of the current induced by $\mathscr{H}$ and the magnetization current. If we compare Eqs. (4.112) and (4.115) we see that the super current $\mathbf{J}_s(\mathbf{r})$ has the same functional form as the probability current of a free particle. Inside a sample, where $\mathbf{B}(\mathbf{r}) = 0$, there is no super current. The super current is confined to the surface of the sample.

We can use Eqs. (4.111) and (4.112) to determine how the order parameter, $\Psi(\mathbf{r})$, varies in space at the edge of the sample.

Let us assume that $\mathscr{A}(\mathbf{r}) \equiv 0$, but that Ψ can vary in space in the z-direction, $\Psi = \Psi(z)$. Under these conditions, we can assume that $\Psi(z)$ is a real function and, therefore, there is no super current, $\mathbf{J}_s = 0$. Let us next introduce a dimensionless function,

$$f(z) = \Psi(z)\sqrt{\frac{\alpha_4}{|\alpha_2|}}. \tag{4.116}$$

Then Eq. (4.111) takes the form

$$-\xi^2(T)\frac{d^2f}{dz^2} - f + f^3 = 0, \tag{4.117}$$

where $\xi(T)$ is called the Ginzburg-Landau coherence length and is defined

$$\xi(T) = \sqrt{\frac{\hbar^2}{2m|\alpha_2|}}. \tag{4.118}$$

Eq. (4.117) can be used to find how the order parameter varies in space on the boundary of a superconducting sample.

Let us consider a sample that is infinitely large in the x- and y-directions but extends only from $z = 0$ to $z = \infty$ in the z-direction. The region from $z = -\infty$ to $z = 0$ is empty. We will assume that at $z = 0$ there is no condensate,

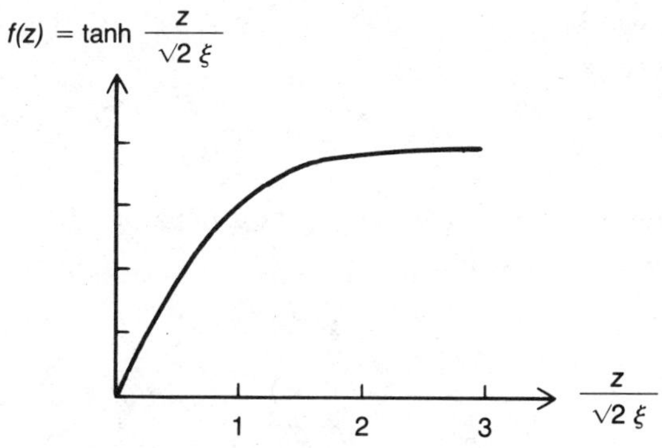

Fig. 4.21. A plot of $f(z) = \tanh(z/\sqrt{2}\xi)$ versus $z/\sqrt{2}\xi$.

$f(z = 0) = 0$; but deep in the interior of the sample it takes its maximum value $f(z = \infty) = 1$ (that is, $\Psi(z) = (|\alpha_2|/\alpha_4)^{1/2}$ deep in the interior of the sample). Eq. (4.117) is a nonlinear equation for f, but it can be solved. To solve it we must find its first integral. Let us multiply Eq. (4.117) by df/dz and rearrange and integrate terms. We then find

$$-\xi^2\left(\frac{df}{dz}\right)^2 = f^2 - \tfrac{1}{2}f^4 + C \tag{4.119}$$

where C is an integration constant. We will choose C so that the boundary conditions

$$\frac{df}{dz} \xrightarrow[z\to\infty]{} 0 \quad \text{and} \quad f \xrightarrow[z\to\infty]{} 1$$

are satisfied. This gives $C = -\frac{1}{2}$ and Eq. (4.119) takes the form

$$\xi^2\left(\frac{df}{dz}\right)^2 = \tfrac{1}{2}(1 - f^2)^2. \tag{4.120}$$

We can now solve Eq. (4.120) and find

$$f(z) = \tanh\frac{z}{\sqrt{2}\xi} \tag{4.121}$$

($f(z)$ is plotted in Fig. 4.21). Most of the variation of the order parameter occurs within a distance $z = 2\xi(T)$ from the boundary. Thus, the order parameter is zero on the surface but increases to its maximum size within a distance $2\xi(T)$ of the surface. Near a critical point, $\xi(T)$ will become very large.

It is also useful to introduce the penetration depth for the magnetic field and determine the manner in which the induction field $\mathbf{B}(\mathbf{r})$ dies away at the surface of the sample. Let us assume that the order parameter is constant throughout the superconductor (we neglect variations at the surface, although this is often a bad assumption). Let us also assume that the vector potential is pointed in the

y-direction and can vary in the z-direction $\mathscr{A} = \mathscr{A}_y(z)\hat{\mathbf{y}}$. The super current then takes the form

$$\mathbf{J}_s = -\frac{e^2 n_s}{m} \mathscr{A}_y(z)\hat{\mathbf{y}}. \tag{4.122}$$

The super current is proportional to the vector potential. If we now use Eqs. (4.107) and (4.115), we find

$$\mathscr{A}_y(z) = \lambda^2 \frac{d^2 \mathscr{A}_y}{dz^2} \tag{4.123}$$

where

$$\lambda = \sqrt{\frac{2m\alpha_4}{|\alpha_2| e^2}} \tag{4.124}$$

is the penetration depth. Thus the vector potential drops off exponentially inside the superconductor,

$$\mathscr{A}_y = \mathscr{A}_0\, e^{-z/\lambda}, \tag{4.125}$$

and we obtain the result that all super currents are confined to within a distance λ of the surface. Note that the penetration depth also becomes very large near the critical point.

There are many more applications of the Ginzburg-Landau theory of superconductors than can be presented here. The interested reader should see Ref. 4.12.

I. THE HELIUM LIQUIDS

From the standpoint of statistical physics, helium has proven to be one of the most unique and interesting elements in nature. Because of its small atomic mass and weak attractive interaction, helium remains in the liquid state for a wide range of pressures all the way down to zero degrees Kelvin (as far as we can tell).

The helium atom occurs in nature in two stable isotopic forms, He^3 and He^4. He^3, with nuclear spin $\frac{1}{2}$, obeys Fermi-Dirac statistics; while He^4, with nuclear spin 0, obeys Bose-Einstein statistics. At very low temperatures, where quantum effects become important, He^3 and He^4 provide two of the few examples in nature of quantum liquids.

Chemically, He^3 and He^4 are virtually identical. The only difference between them is a difference in mass. However, at low temperatures, the two systems exhibit very different behavior due to the difference in their statistics. Liquid He^4, which is a boson liquid, exhibits a rather straightforward transition to a superfluid state at 2.19 K. This can be understood as a condensation of particles into a single quantum state. Liquid He^3 also undergoes a transition to a superfluid state, but at a much lower temperature (2.7×10^{-3} K). The mechanism

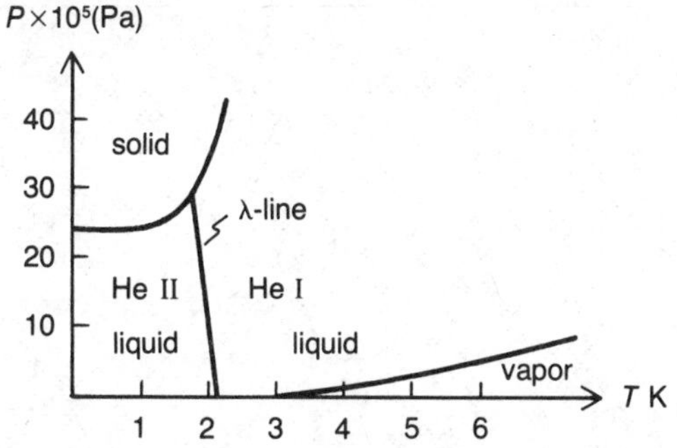

Fig. 4.22. Coexistence curves for He^4 (based on Ref. 4.17, p. 64).

for the superfluid transition in liquid He^3 is quite different from that of liquid He^4. In liquid He^3, particles (more accurately, quasiparticles) form bound pairs with a spin $s = 1$ and relative angular momentum, $\ell = 1$. The mechanism is similar to the formation of bound pairs in a superconductor except that the pairs in a superconductor are formed with spin 0 and angular momentum 0. Thus, the bound pairs in a superconductor are spherical with no magnetic moment, while those in liquid He^3 are flatter along one axis, carry angular momentum, and have a net magnetic moment. The fact that the bound pairs in liquid He^3 have structure leads to many fascinating effects never before observed in any other physical system.

While we cannot discuss the theory of these systems at this point, it is worthwhile to look at their phase diagrams since they present such a contrast to those of classical fluids and they tend to confirm the third law.

1. Liquid He^4 [14,15]

He^4 was first liquefied in 1908 by Kamerlingh Onnes at a temperature of 4.215 K at a pressure of 1 atm. Unlike the classical liquids we described in Sec. 4.C, it has two triple points. The coexistence curves for liquid He^4 are shown in Fig. 4.22 (compare them with the coexistence curve for a classical liquid in Fig. 4.4). He^4 at low temperature has four phases. The solid phase only appears for pressures above 25 atm and the transition between the liquid and solid phases is first order. The liquid phase continues down to $T = 0$ K. However there are in fact two liquid phases. As the normal liquid (liquid He(I)) is cooled, a line of λ-points occurs at about $T = 2$ K (the exact temperature depends on the pressure), indicating that a continuous symmetry-breaking phase transition has occurred. There is a triple point at each end of the line. The symmetry that is broken is gauge symmetry. Below the λ-line, the liquid phase (which was called liquid He(II) by Keesom and Wolfke)[16] begins to exhibit very strange properties. The first experimenters who worked with liquid He^4 found that it was able to leak out of their containers through cracks so tiny that even He^4 gas could

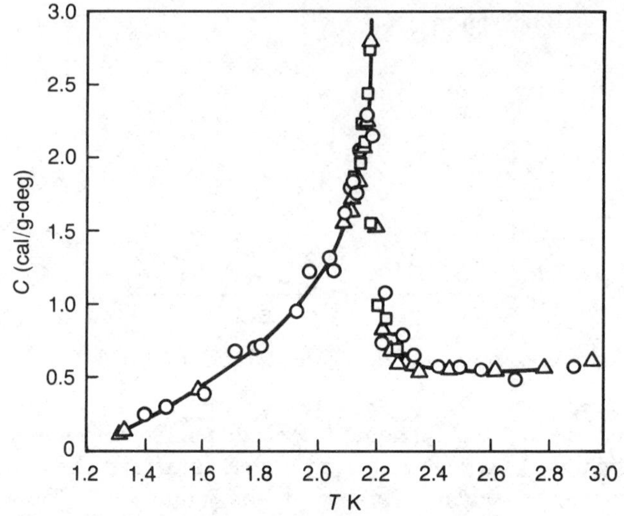

Fig. 4.23. Specific heat of He^4 at vapor pressure at the λ-point (based on Ref. 4.14).

not leak through. This apparently frictionless flow is a consequence of the fact that the condensed phase is a highly coherent macroscopic quantum state. It is analogous to the apparent frictionless flow of the condensed phase in superconductors. The order parameter for the condensed phase in liquid He^4 is a macroscopic "wave function" and the Ginzburg-Landau theory for the condensed phase of He^4 is very similar to that of the condensed phase in superconductors, except that in liquid He^4 the particles are not charged. The specific heat of liquid He^4 along the λ-line is shown in Fig. 4.23. We can see that it has the lambda shape characteristic of a continuous phase transition.

The phase diagram of He^4 provides a good example of the third law. The vapor-liquid and solid-liquid coexistence curves approach the P-axis with zero slope. From Eqs. (2.63) and (2.93a) we see that this is a consequence of the third law.

2. Liquid He^3 [17–20]

The He^3 atom is the rarer of the two helium isotopes. Its relative abundance in natural helium gas is one part in a million. Therefore, in order to obtain it in large quantities, it must be "grown" artificially from tritium solutions through β-decay of the tritium atom. Thus, He^3 was not obtainable in large enough quantities to study until the late 1940's and it was first liquefied in 1948 by Sydoriack, Grilly, and Hammel.[21] Since the He^3 atom has only $\frac{3}{4}$ the mass of a He^4 atom, it has a larger zero point energy than the He^4 atom. As a result, He^3

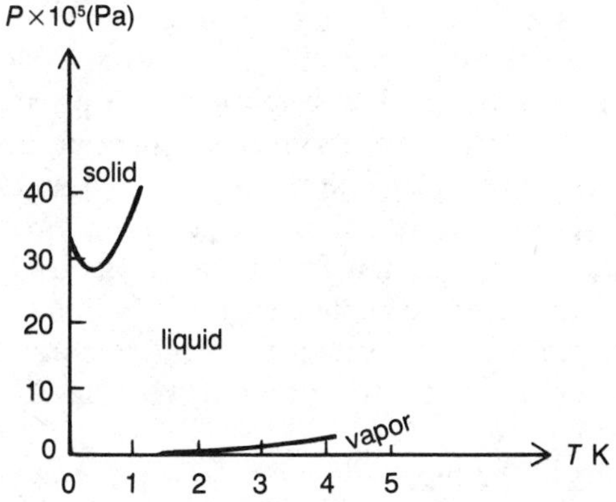

Fig. 4.24. Coexistence curves for He^3 (based on Ref. 4.17, p. 64).

boils at temperature about 25% lower than He^4 and it requires a pressure about 25% greater than that of He^4 to solidify.

The phase diagram for He^3 (on the same scale as that for He^4) is given in Fig. 4.24. On this scale there is no transition to a superfluid state. There is, however, a minimum in the liquid-solid coexistence curve. This is attributed to the spin of the He^3 atom. At low temperature the spin lattice of the He^3 solid has a higher entropy than the liquid. The entropy difference, $\Delta S = S_{\text{liquid}} - S_{\text{solid}}$, is positive at high temperature, vanishes at about $T = 0.3$ K, and becomes negative below 0.3 K. Since volume differences remain virtually unchanged, the Clausius-Clapeyron equation $dP/dT = \Delta S/\Delta V$ leads to a positive slope at high temperature and a negative slope at low temperature. At low temperature, if the third law is to be satisfied, the slope of the liquid-solid coexistence curve must become flat as $T \to 0$ K.

Superfluidity was first observed in liquid He^3 in 1971 by Osheroff, Richardson, and Lee.[22] The transition occurs at 2.7×10^{-3} K at a pressure of about

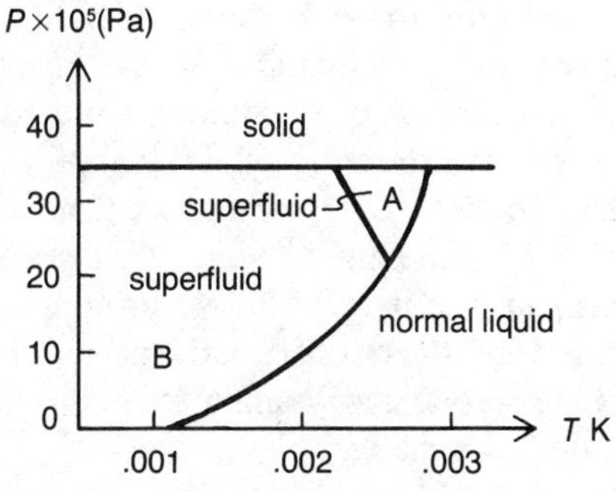

Fig. 4.25. Coexistence curves for superfluid phases of He^3 when no magnetic field is applied (based on Ref. 4.17, p. 68).

34 atm. The phase diagram for a small temperature interval is shown in Fig. 4.25. There are, in fact, several superfluid phases in liquid He^3, depending on how the bound pairs orient themselves. The so-called A-phase is an anisotropic phase. The bound pairs (which have shape) all orient on the average in the same direction. This defines a unique axis in the fluid and all macroscopic properties depend on their orientation with respect to that axis. The B-phase is a more isotropic phase and has many features in common with the superfluid phase of a superconductor. If we apply a magnetic field to liquid He^3, a third superfluid phase appears. The transition between the normal and superfluid phases appears to be continuous, while that between the A and B superfluid phases appears to be first order.

The present theory of the superfluid phases of liquid He^3 is again based on Ginzburg-Landau expansions. However, the transition is more complicated than that of a superconductor because rotational symmetry is broken in addition to gauge symmetry.

J. CRITICAL EXPONENTS[10,23,24]

The critical point is the point at which the order parameter of a new phase begins to grow continuously from zero. We have seen many examples of critical points. In the liquid-vapor transition, a critical point terminated the liquid-vapor coexistence curve and was the one point for which the Gibbs free energy changed continuously. For the binary mixture, the critical point marked the temperature at which phase separation could first take place as we lowered the temperature. In the spin system the critical point for transition from a paramagnetic state to ferromagnetic state was the Curie point. In the superconductor, the critical point was the point at which the condensate first appeared as we lowered the temperature in the absence of an external field. In liquid He^4, the condensed and normal phases were separated by a line of critical points.

Systems exhibit dramatically new behavior below the critical point. As we approach the critical point from above (higher temperature), the system anticipates its new behavior by making "adjustments" on a microscopic scale. These "adjustments" appear in the form of fluctuations in density, magnetization, etc., which become very large as the critical point is approached. Just below the critical point, the order parameter of the new phase first becomes nonzero.

A subject of great interest is the way in which various systems approach the critical point. For this purpose it is useful to introduce the idea of critical exponents. As one approaches the critical point, various thermodynamic functions may diverge or go to zero or even remain finite. It is therefore convenient to introduce an expansion parameter,

$$\epsilon = \frac{T - T_c}{T_c}, \tag{4.126}$$

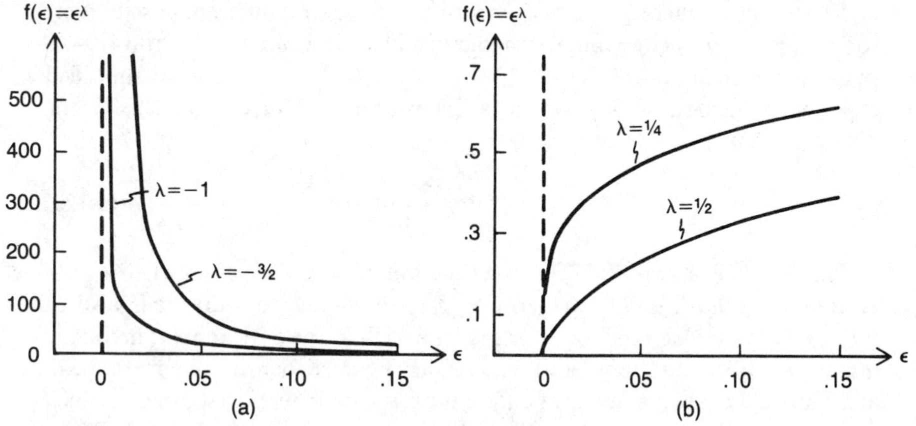

Fig. 4.26. Plot of $f(\epsilon) = \epsilon^\lambda$ for $\lambda \neq 0$: (*a*) plot for $\lambda = -1$ and $\lambda = -\frac{3}{2}$; (*b*) plot for $\lambda = \frac{1}{4}$ and $\lambda = \frac{1}{2}$.

where T_c is the critical temperature, which is a measure of the distance from the critical point in terms of reduced variables.

Near the critical point all thermodynamic functions can be written in the form

$$f(\epsilon) = A\epsilon^\lambda(1 + B\epsilon^y + \cdots) \tag{4.127}$$

where $y > 0$. The critical exponent, for the function $f(\epsilon)$, is defined

$$\lambda = \lim_{\epsilon \to 0} \frac{\ln f(\epsilon)}{\ln \epsilon}. \tag{4.128}$$

If λ is negative, $f(\epsilon)$ diverges at the critical point. If λ is positive, $f(\epsilon)$ goes to zero at the critical point. In Fig. 4.26, we have plotted $f(\epsilon) = \epsilon^\lambda$ for $\lambda = -1, -\frac{3}{2}, \frac{1}{2},$

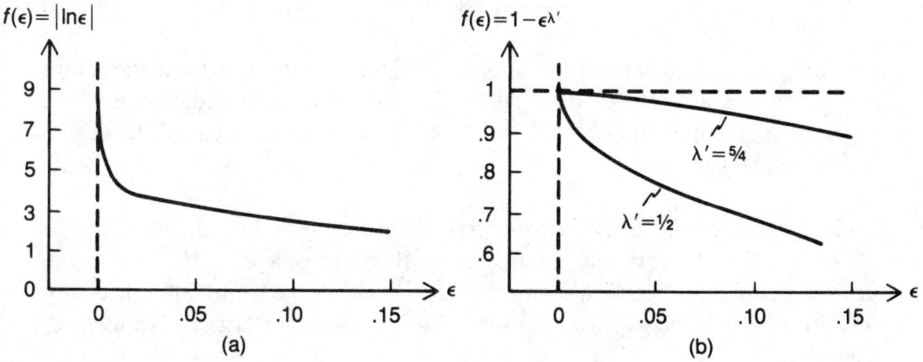

Fig. 4.27. Plot of $f(\epsilon)$ for cases when $\lambda = 0$: (*a*) plot of $f(\epsilon) = |\ln \epsilon|$; (*b*) plot of $f(\epsilon) = 1 - \epsilon^{\lambda'}$.

and $\frac{1}{4}$. The case where $\lambda = 0$ may correspond to several different possibilities; for example, it may correspond to a logarithmic divergence $f(\epsilon) = A|\ln \epsilon| + B$ or to a dependence on ϵ of the form $f(\epsilon) = A + B\epsilon^{1/2}$. For such cases a modified exponent is introduced. If j is the smallest integer, such that $d^jf(\epsilon)/d\epsilon^j = f^{(j)}(\epsilon)$ diverges, then

$$\lambda' = j + \lim_{\epsilon \to 0} \frac{\ln|f^{(j)}(\epsilon)|}{\ln \epsilon}. \tag{4.129}$$

In Fig. 4.27, we have plotted $f(\epsilon)$ as a function of ϵ for some cases when $\lambda = 0$. Although we have chosen to write ϵ in terms of temperature, it is also of interest to introduce exponents for the approach to the critical point involving quantities other than temperature, such as pressure, density, magnetic field, etc. Thus, there are a number of different critical exponents that can be defined for a system depending on how the critical point is approached. We shall give some examples in the next section.

K. THE CRITICAL EXPONENTS FOR PURE *PVT* SYSTEMS

1. Experimental Values

It is found experimentally that many fluids have the following behavior as the critical point is approached.

(*a*) *Degree of the Critical Isotherm.* The deviation of the pressure $(P - P_c)$ from its critical value varies *at least* as the fourth power of $(V - V_c)$ as the critical point is approached along the critical isotherm. It is convenient to express this fact by introducing a critical exponent δ, such that

$$\frac{P - P_c}{P_c^0} \equiv A_\delta \left| \frac{\rho - \rho_c}{\rho_c} \right|^\delta \operatorname{sign}(\rho - \rho_c) \qquad (T = T_c) \tag{4.130}$$

where P_c is the critical pressure, ρ_c is the critical density, A_δ is a constant, and P_c^0 is the pressure of an ideal gas at the critical density and temperature. Experimentally it is found that $6 > \delta \geqslant 4$. The exponent δ is called the degree of the critical isotherm.

(*b*) *Degree of the Coexistence Curve.* Guggenheim has shown that the deviation $(T - T_c)$ varies approximately as the third power of $(V - V_c)$ as the critical point is approached along the coexistence curve from either direction (cf. Eq. (4.30)). One expresses this fact by introducing a critical exponent β, such that

$$\frac{\rho_l - \rho_g}{\rho_c} = A_\beta(-\epsilon)^\beta \tag{4.131}$$

where ρ_l is the density of liquid at temperature, $T < T_c$, ρ_g is the density of gas at temperature $T < T_c$, each evaluated on the coexistence curve, and A_β is a constant. The quantity $\rho_l - \rho_g$ is the order parameter of system. It is zero above the critical point and nonzero below it. The exponent β is called the degree of the coexistence curve and is found from experiment to have values $\beta \approx 0.34$.

(*c*) *Heat Capacity.* The heat capacity at constant volume appears to have a logarithmic divergence for $T \rightarrow T_c$ along the critical isochore ($V = V_c$). The critical exponent for heat capacity is denoted α and is defined

$$C_V = \begin{cases} A'_\alpha(-\epsilon)^{-\alpha'} & (T < T_c) \\ A_\alpha(+\epsilon)^{-\alpha} & (T > T_c) \end{cases} \quad (\rho = \rho_c) \tag{4.132}$$

where A'_α and A_α are constants. The exponents α and α' are found experimentally to have values $\alpha \sim 0.1$ and $\alpha' \sim 0.1$.

(*d*) *Isothermal Compressibility.* The isothermal compressibility diverges approximately as a simple pole:

$$\frac{\kappa_T}{\kappa_T{}^0} = \begin{cases} A'_\gamma(-\epsilon)^{-\gamma'} & T < T_c \quad \rho = \rho_L(T) \text{ or } \rho_G(T) \\ A_\gamma(\epsilon)^{-\gamma} & T > T_c \quad \rho = \rho_c \end{cases} \tag{4.133}$$

where A'_γ and A_γ are constants. For $T < T_c$ one approaches the critical point along the coexistence curve; for $T > T_c$ one approaches it along the critical isochore. Typical experimental values of γ' and γ are $\gamma' \sim 1.2$ and $\gamma \sim 1.3$.

(*e*) *Exponent Inequalities.* It is possible to obtain inequalities between the critical exponents using thermodynamic arguments. We shall give an example here. Eq. (4.44) can be rewritten in terms of the mass density as

$$c_v = x_g c_{v_g} + x_l c_{v_l} + \frac{x_g T}{\rho_g \kappa_T{}^g}\left(\frac{\partial \rho_g}{\partial T}\right)^2_{\text{coex}} + \frac{x_l T}{\rho_l \kappa_T{}^l}\left(\frac{\partial \rho_l}{\partial T}\right)^2_{\text{coex}} \tag{4.134}$$

where c_v, c_{v_g}, and c_{v_l} are now specific heats (heat capacity per kilogram), and κ_T is the isothermal compressibility. All terms on the right-hand side of Eq. (4.134) are positive. Thus, we can write

$$c_v \geqslant \frac{x_g T}{\rho_g \kappa_T{}^g}\left(\frac{\partial \rho_g}{\partial T}\right)^2_{\text{coex}}. \tag{4.135}$$

As the critical point is approached for fixed volume, $x_g \rightarrow \frac{1}{2}$, $\rho_g \rightarrow \rho_c$ (ρ_c is the critical density), κ_T diverges as $(T_c - T)^{-\gamma'}$ (cf. Eq. (4.133)), and $(\partial \rho_g/\partial T)_{\text{coex}}$ diverges as $(T_c - T)^{\beta-1}$ if we assume that $[\frac{1}{2}(\rho_l + \rho_g) - \rho_c]$ goes to zero more slowly than $(\rho_l - \rho_g)$ (cf. Eqs. (4.29) and (4.30)). Thus,

$$c_v \geqslant \frac{1}{2}\frac{T_c B(T_c - T)^{\gamma' + 2\beta - 2}}{\rho_c} \tag{4.136}$$

where B is a constant, and

$$\ln c_v \geqslant (2 - \gamma' - 2\beta)|\ln(-\epsilon)|. \tag{4.137}$$

If we next divide by $|\ln(-\epsilon)|$ and take the limit $T \to T_c^-$, we find

$$\alpha' + 2\beta + \gamma' \geqslant 2. \tag{4.138}$$

The inequality in Eq. (4.138) is roughly satisfied by real fluids. If we choose $\alpha' = 0.1, \beta = \frac{1}{3}$, and $\gamma' = 1.3$, then $\alpha' + 2\beta + \gamma' \approx 2$. Eq. (4.138) is called the Rushbrook inequality.

2. Van der Waals Equation

It is interesting to find values of the critical exponents predicted by the van der Waals equation. Let us introduce the expansion parameters:

$$\epsilon = \frac{T - T_c}{T_c}, \qquad \omega = \frac{V - V_c}{V_c}, \qquad \pi = \frac{P - P_c}{P_c}. \tag{4.139}$$

We can then rewrite the van der Waals equation in the form

$$\left[(1 + \pi) + \frac{3}{(1 + \omega)^2}\right][3(\omega + 1) - 1] = 8(1 + \epsilon) \tag{4.140}$$

or in the form

$$2\pi\left(1 + \frac{7\omega}{2} + 4\omega^2 + \frac{3\omega^3}{2}\right) = -3\omega^3 + 8\epsilon(1 + 2\omega + \omega^2). \tag{4.141}$$

We now can find various exponents.

(*a*) *Degree of the Critical Isotherm.* If we set $\epsilon = 0 (T = T_c)$ we may expand π in powers of ω,

$$\pi = (-\tfrac{3}{2})\omega^3\left(1 - \frac{7\omega}{2} + \cdots\right) \qquad (T = T_c). \tag{4.142}$$

Hence, $\delta = 3$ for the van der Waals theory. This does *not* agree with experiment where typically $\delta > 4$.

(*b*) *Degree of the Coexistence Curve.*[2] To find the degree of the coexistence curve, let us first expand π in powers of ϵ and ω. We find

$$\pi \approx -\tfrac{3}{2}\omega^3 - 6\omega\epsilon + 4\epsilon + 9\omega^2\epsilon + \cdots. \tag{4.143}$$

The values of ω on the coexistence curve can be found from the condition that along the isotherm

$$\int_{\bar{V}_g}^{\bar{V}_l} \bar{V}\, d\bar{P} = 0 \tag{4.144}$$

and

$$\bar{P}(\bar{V}_g) = \bar{P}(\bar{V}_l) \tag{4.145}$$

where $\bar{P} = P/P_c$ is the reduced pressure and $\bar{V}_g = V_g/V_c$ and $\bar{V}_l = V_l/V_c$ are the so-called reduced or orthobaric volumes, the values of the reduced volumes of the gas and liquid, respectively, on the coexistence curve. If we note that $\bar{V}_g = 1 + \omega_g$ and $\bar{V}_l = 1 - \omega_l$, we find

$$4\epsilon(\omega_g + \omega_l) + 4\epsilon(\omega_l^2 - \omega_g^2) + \omega_l^3 + \omega_g^3 = 0 \tag{4.146}$$

and

$$4\epsilon(\omega_g + \omega_l) - 6\epsilon(\omega_l^2 - \omega_g^2) + \omega_l^3 + \omega_g^3 = 0. \tag{4.147}$$

Near the critical point $\omega_l \sim \omega_g$ and Eqs. (4.146) and (4.147) have the solution

$$\omega_g \sim \omega_l = 2|\epsilon|^{1/2}. \tag{4.148}$$

Thus, the coexistence curve behaves as a parabola,

$$(\bar{V} - 1)^2 = 4|\epsilon| \tag{4.149}$$

and

$$\bar{V}_g - \bar{V}_l = \omega_g + \omega_l = 4|\epsilon|^{1/2} \tag{4.150}$$

From Eq. (4.150), we can easily show that $\beta = \frac{1}{2}$ for the van der Waals equation, while experiment gives $\beta = \frac{1}{3}$. Thus the van der Waals equation fails again to give good predictions at the critical point.

(c) *Heat Capacity.* The jump in the heat capacity at the critical point can be obtained from Eq. (4.44). Let us approach the critical point along the critical isochore $V = V_c$. There, $x_l \approx x_g \to \frac{1}{2}$ and $\tilde{c}_{v_g} \to \tilde{c}_{v_l}$ as $T \to T_c$. Thus the jump in the heat capacity per mole is given by

$$\tilde{c}_{\tilde{v}_c}(T_c^-) - \tilde{c}_{\tilde{v}_c}(T_c^+) = \lim_{T \to T_{c-}} -\tfrac{1}{2}T\left[\left(\frac{\partial P_g}{\partial \tilde{v}_g}\right)_T\left(\frac{\partial \tilde{v}_g}{\partial T}\right)^2_{\text{coex}} + \left(\frac{\partial P_l}{\partial \tilde{v}_l}\right)_T\left(\frac{\partial \tilde{v}_l}{\partial T}\right)^2_{\text{coex}}\right]. \tag{4.151}$$

Along the coexistence curve,

$$\left(\frac{\partial \bar{V}}{\partial T}\right)_{\text{coex}} = \mp|\epsilon|^{-1/2} \tag{4.152}$$

where the minus sign applies to the liquid and the plus sign applies to the gas (cf. Eq. (4.148)). From Eqs. (4.143) and (4.148), we find

$$\left(\frac{d\bar{P}_l}{d\bar{V}_l}\right)_{\bar{T}} \equiv \left(\frac{d\bar{P}_g}{d\bar{V}_g}\right)_{\bar{T}} = 6|\epsilon| - \tfrac{9}{2}\omega^2 - 18\omega|\epsilon| + \cdots = -12|\epsilon| \pm 0(|\epsilon|^{3/2}). \tag{4.153}$$

If we next note that $P_cV_c/RT_c = \frac{3}{8}$, we find

$$\tilde{c}_{\tilde{v}_c}(T_c^-) - \tilde{c}_{\tilde{v}_c}(T_c^+) = \tfrac{9}{2}R + 0(\epsilon). \tag{4.154}$$

Above the critical point the heat capacity behaves like an ideal gas (cf. Sec. 3.B) and we obtain[8]

$$\tilde{c}_{\tilde{v}} = \tfrac{3}{2}R \qquad (T > T_c)$$
$$\tilde{c}_{\tilde{v}} = \tfrac{3}{2}R + \tfrac{9}{2}R\left[1 - \left(\frac{28}{25}\right)\epsilon + \cdots\right] \qquad (T < T_c). \tag{4.155}$$

Thus the van der Waals equation predicts a finite jump in the heat capacity at the critical point, $\alpha' = \alpha = 0$, while in real gases there is a logarithmic singularity in $\tilde{c}_{\tilde{v}}$ at the critical point, $\alpha = \alpha' = 0.1$.

(*d*) *Isothermal Compressibility.* We can easily compute the critical exponent γ for a van der Waals gas. If the critical point is approached along the critical isochore, we find

$$\frac{-1}{V\kappa_T} = \left(\frac{\partial P}{\partial V}\right)_T = \frac{P_c}{V_c}\left(\frac{\partial \pi}{\partial \omega}\right)_T = \frac{P_c}{V_c}(-6\epsilon) \qquad (T > T_c). \tag{4.156}$$

Therefore, the van der Waals equation predicts $\gamma = 1$. The isothermal compressibility diverges as a simple pole. One obtains the same result for γ'. In that case, the critical point is approached along the coexistence curves, for $T < T_c$, $\gamma' = 1$ but the coefficient differs by a factor of 2. These results are in qualitative agreement with experiment, where values of γ and γ' are $\gamma = 1.3$ and $\gamma' = 1.2$.

The critical exponents obtained for van der Waals theory (and Ginzburg-Landau theory) are similar to those obtained for all mean field theories. The common feature of these theories is that they can be derived assuming that the particles move in a mean field due to all other particles (we shall derive the van der Waals equation in Chap. 9). The mean field theories do not properly take into account the effects of short-ranged correlations at the critical point and do not give the correct results for the critical exponents. We shall return to this point when we discuss the Wilson theory in Chap. 10.

L. CRITICAL EXPONENTS FOR THE CURIE POINT

For magnetic systems, exponents β, γ, δ, and α can be defined in analogy with pure fluids. The phase diagrams for simple magnetic systems are given in Figs. 4.28–4.30. In Fig. 4.28 we sketch the coexistence curve for a ferromagnetic system: below some critical temperature the spins begin to order spontaneously; the coexistence curve separates the two directions of magnetization. In Fig. 4.29 we plot some isotherms of the magnetic system, and in Fig. 4.30 we plot the magnetization as a function of temperature. It is helpful to refer to these curves when defining the various exponents.

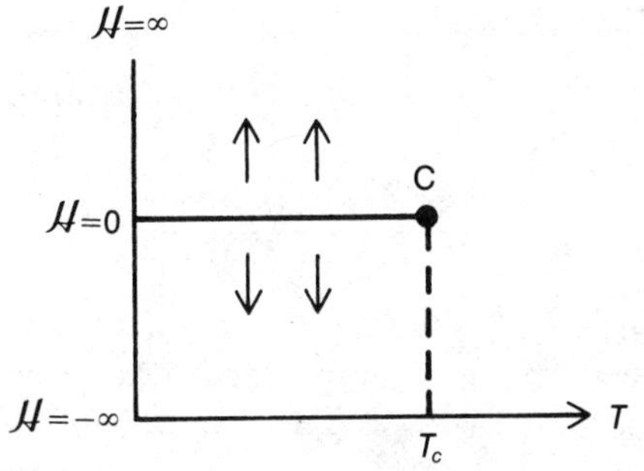

Fig. 4.28. Coexistence curve for a typical magnetic system: below the Curie point the magnetization occurs spontaneously; the curve $\mathcal{H} = 0$ separates the two possible orientations of the magnetization (based on Ref. 4.23).

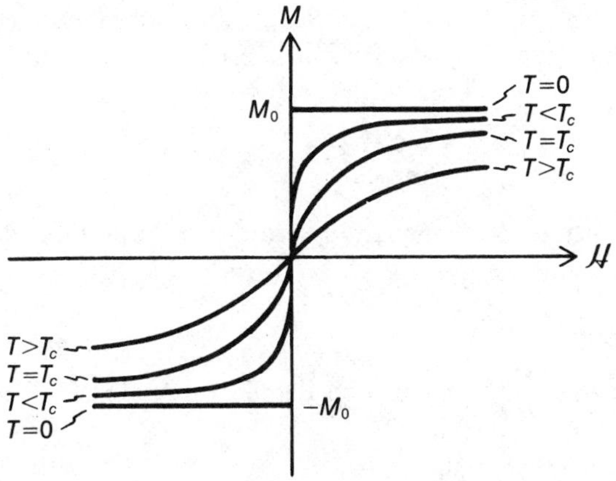

Fig. 4.29. A sketch of the isotherms for a ferromagnetic system (based on Ref. 4.10).

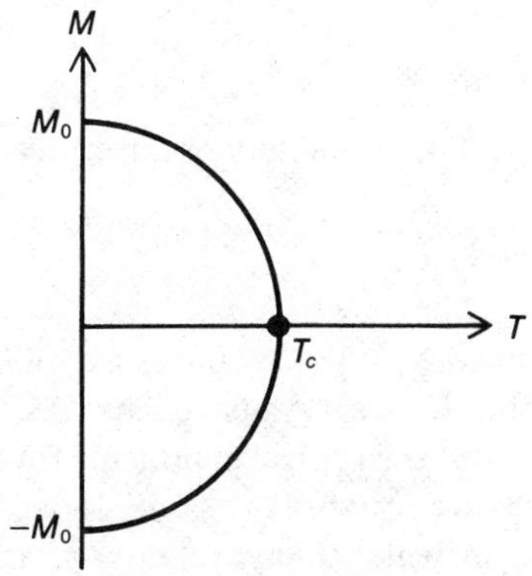

Fig. 4.30. A sketch of the magnetization for a simple ferromagnetic system (based on Ref. 4.10).

1. Degree of Critical Isotherm

The exponent δ describes the variation of magnetization with magnetic field along the critical isotherm

$$\frac{\mathscr{H}}{\mathscr{H}_c^0} = B_\delta \left| \frac{M_{\mathscr{H}}(T_c)}{M_0(0)} \right|^\delta \tag{4.157}$$

where $\mathscr{H}_c^0 \equiv kT_c/m_0$, $M_0(0)$ is the magnetization in zero field at zero temperature, m_0 is the magnetic moment per spin, and B_δ is a proportionality constant. Experimentally, δ has values $4 \leqslant \delta \leqslant 6$ in agreement with the values of δ for pure fluids.

2. Magnetic Exponent

In a magnetic system, the exponent β describes how the magnetization approaches its value at the critical point when no external field is present. It is defined

$$\frac{M_0(T)}{M_0(0)} = B_\beta(-\epsilon)^\beta \tag{4.158}$$

where B_β is a constant. For magnetic systems, $\beta \approx \frac{1}{3}$ as it is for fluids.

3. Heat Capacity

For magnetic systems, the coefficients α and α' are defined

$$C_{\mathscr{H}}(\mathscr{H} = 0) = \begin{cases} B_\alpha'(-\epsilon)^{-\alpha'} & T < T_c \\ B_\alpha \epsilon^{-\alpha} & T > T_c \end{cases} \tag{4.159}$$

where B_α and B_α' are constants. Experimentally, one finds $\alpha \sim \alpha' \sim 0$.

4. Magnetic Susceptibility

The magnetic susceptibility in the vicinity of the critical point can be written

$$\frac{\chi_T}{\chi_T{}^0} = \begin{cases} B_\gamma'(-\epsilon)^{-\gamma'} & T < T_c \quad \mathscr{H} = 0 \\ B_\gamma \epsilon^{-\gamma} & T > T_c \quad \mathscr{H} = 0 \end{cases} \tag{4.160}$$

where B_γ' and B_γ are constants and $\chi_T{}^0$ is the susceptibility of a noninteracting system at the critical point. For real systems, γ has been found to be $\gamma \sim 1.3$.

The striking feature about the critical exponents for fluids and for magnetic systems is that the values are roughly the same. Indeed, there appears to be a great similarity in the way in which many systems approach their critical points. In this chapter, we have only been able to discuss two systems. We will consider some other systems in Chap. 10.

We have introduced four critical exponents, α, β, γ, and δ; in Chap. 10 we shall introduce another. However, in this book we do not have space to discuss all the exponents that now exist to characterize the way in which various quantities approach the critical point. More specialized discussions can be found in Ref. 4.10.

REFERENCES

1. M. W. Zemansky, *Heat and Thermodynamics* (McGraw-Hill Book Co., New York, 1957).
2. D. ter Haar and H. Wergeland, *Elements of Thermodynamics* (Addison-Wesley Publishing Co., Reading, Mass., 1966).
3. L. D. Landau and E. M. Lifshitz, *Statistical Physics* (Pergamon Press, Oxford, 1958).
4. E. A. Guggenheim, J. Chem. Phys. *13* 253 (1945).
5. J. de Boer and A. Michels, Physica *5* 945 (1938).
6. J. H. Hildebrand, *Solubility of Non-electrolytes* (Reinhold, New York, 1936).
7. I. Prigogine and R. Defay, *Chemical Thermodynamics* (Longmans, Green & Co., London, 1954).
8. J. S. Rowlinson, *Liquids and Liquid Mixtures* (Butterworth, London, 1969).
9. F. C. Nix and W. Shockley, Rev. Mod. Phys. *10* 1 (1938).
10. H. E. Stanley, *Introduction to Phase Transitions and Critical Phenomena* (Oxford University Press, Oxford, 1971).
11. F. London, *Superfluids I: Macroscopic Theory of Superconductivity* (Dover Publications, New York, 1961).
12. M. Tinkham, *Introduction to Superconductivity* (McGraw-Hill Book Co., New York, 1975).
13. N. Meissner and R. Ochsenfeld, Naturwissenschaften *21* 787 (1933).
14. F. London, *Superfluids II: Macroscopic Theory of Superfluid Helium* (Dover Publications, New York, 1964).
15. W. E. Keller, *Helium-3 and Helium-4* (Plenum Press, New York, 1969).
16. W. H. Keesom and M. Wolfke, Proc. Roy. Acad. Amsterdam *31* 90 (1928).
17. N. D. Mermin and D. M. Lee, Scientific American Dec. 1976, p. 56.
18. A. J. Leggett, Rev. Mod. Phys. *47* 331 (1975).
19. J. Wheatley, Rev. Mod. Phys. *47* 415 (1975).
20. P. W. Anderson and W. F. Brinkman in *The Helium Liquids*, ed. Armitage and Farquhar (Academic Press, New York, 1975).
21. S. G. Sydoriack, E. R. Grilly, and E. F. Hammel, Phys. Rev. *75* 303 (1949).
22. D. D. Osheroff, R. C. Richardson, and D. M. Lee, Phys. Rev. Lett. *28* 885 (1972).
23. M. E. Fisher, Repts. Prog. Phys. *30* 615 (1967).
24. P. Heller, Repts. Prog. Phys. *30* 731 (1967).
25. P. N. Bridgeman, J. Chem. Phys. *5* 964 (1937).
26. E. Lapp, Ann. Physique (10) *12* 442 (1929).
27. H. K. Onnes, Leiden Comm. *122b*, *124c* (1911).

PROBLEMS

1. Plot the Helmholtz free energy A and its derivatives $(\partial A/\partial V)_{T,\{n_j\}}$ and $(\partial A/\partial T)_{V,\{n_j\}}$ as a function of V and T, respectively, for the transition in Fig. 4.2.

2. A spherical liquid drop floats in equilibrium with its saturated vapor. The drop has radius r and surface tension σ (assumed constant).
   (*a*) Find the pressure difference between the liquid and the saturated vapor. Assume that the thickness of the surface is small compared to the radius.
   (*b*) Show that the saturated vapor pressure of the droplet is given by the approximate formula

$$P_r = P_\infty \exp\left(\frac{2v_l\sigma}{rRT}\right)$$

   where P_∞ is the vapor pressure of a bulk quantity of liquid in equilibrium with its vapor, v_l is the molar volume of the liquid, and R is the gas constant. For given vapor pressure P_r, are droplets of radius $r' < r$ stable?

3. Find the coefficient of thermal expansion $\alpha_{\text{coex}} = (1/V)(\partial V/\partial T)_{\text{coex}}$ for a vapor maintained in equilibrium with its liquid phase. Discuss its behavior.

4. A pot of soup boils at 103°C at the bottom of a hill and boils at 98°C at the top. If the hill is 1000 ft high, what is the latent heat of vaporization of the soup?

5. Deduce the Maxwell construction using the stability properties of the Helmholtz free energy rather than the Gibbs free energy.

6. For a van der Waals gas, plot the isotherms $\bar{T} = .5$, $\bar{T} = 1$, and $\bar{T} = 1.5$ in the $\bar{P}$-$\bar{V}$ plane.

7. Find the critical temperature, pressure, and volume for a gas described by the Dieterici equation of state. Does this equation of state yield a law of corresponding states?

8. Find the local minima of the Gibbs free energy for a binary solution for $T = 1.5T_c$, $T = T_c$, and $T = .5T_c$. Sketch the Gibbs free energy for these temperatures.

9. For a binary system, sketch $\tilde{\mu}_A$ as a function of x_B for $T = 1.5T_c$, $T = T_c$, and $T = .5T_c$. In what regions is the system stable? How can you construct the equilibrium states below T_c?

10. Assume that a *small* amount of a third substance C is added to the binary solution of A and B and assume that the internal energy due to interactions has the form

$$\Delta U_{\text{int}} = \lambda_{AB}\frac{n_A n_B}{n} + \lambda_{AC}\frac{n_A n_C}{n} + \lambda_{BC}\frac{n_B n_C}{n}$$

where $n = n_A + n_B + n_C$ and $n_C \ll n_A$ or n_B. If $\lambda_{AC} \approx \lambda_{BC}$, is the critical temperature of the system raised or lowered? Consider the case of positive $\lambda_{AC} \approx \lambda_{BC}$ and assume that the entropy is given entirely by mixing.

11. For a binary mixture of particles A and B in equilibrium with its vapor, find an equation relating the vapor pressure to the concentration of A in the solution and the concentration of A in the vapor. Assume that the vapor can be treated as an ideal gas.

12. Consider the case in which the order parameter, η, is a scalar which can be positive or negative. Then a cubic term can contribute to the free energy. For the free energy,

$$G(Y, T, \eta) = G_0(Y, T) + \sum_{i=1}^{4} \alpha_i \eta^i,$$

where $\alpha_i = \alpha_i(Y, T)$, find the form of α_i, for $i = 1, \ldots, 4$, at the critical point and on either side of the critical point for a continuous transition. Plot $G(Y, T, \eta)$ as a function of η for $T > T_c$, $T = T_c$, and $T < T_c$ and describe physically what is happening.

13. The order parameter, $\Psi(\mathbf{r})$, for a superconductor must be a single valued function of position. For this to be true, the phase, $\phi(\mathbf{r})$, of the order parameter (defined $\Psi(\mathbf{r}) = |\Psi(\mathbf{r})| e^{i\phi(\mathbf{r})}$) must change only by $2\pi n$ (where n is an integer) when traversing a closed path in the sample. Use this fact to show that the total flux enclosed by path is quantized and can only change in integer amounts.

14. Derive Eqs. (4.111) and (4.112).

15. Consider again the superconducting sample discussed in Sec. 4.H. We assume the sample fills an infinite region of space, $-\infty \leqslant x \leqslant \infty$, $-\infty \leqslant y \leqslant \infty$, and $0 \leqslant z \leqslant \infty$. Apply an external field, $\mathscr{H}_0(z) = \mathscr{H}_0(z)\hat{\mathbf{y}}$, and assume that the order parameter depends only on z and is real (for this case it is consistent to assume that $\Psi(z)$ is real but we cannot always do it). Find the equations for $\Psi(z)$, $\mathbf{J}_s(z)$, and $\mathscr{A}(z)$. Consider the boundary conditions $\mathbf{B}(0) = \mu_0\mathscr{H}_0$ and $d\Psi(z)/dz|_{z=0} = 0$ at $z = 0$; and $\mathbf{B}(\infty) = 0$, $\Psi(\infty) = \sqrt{|\alpha_2|/\alpha_4}$ and $d\Psi(z)/dz|_{z=\infty} = 0$ at $z = \infty$. Assume that deviations from the uniform solution, $\Psi(\infty) = \sqrt{|\alpha_2|/\alpha_4}$, are small and write $\Psi(z) = \Psi(\infty) + \phi(z)$ where $\phi(z)$ is small. Linearize the equations for $\Psi(z)$ and $\mathscr{A}(z)$ in the parameter $\phi(z)$ and solve for $\Psi(z)$ and $\mathscr{A}(z)$. What are $\Psi(0)$ and $\mathscr{A}(0)$? Under what conditions is our assumption that $\phi(z)$ is small valid?

16. In the region of the critical point of a pure PVT system, the deviation of the pressure $\Delta P(V, T) \equiv P(V, T) - P_c$ from its critical value can be expanded in a Taylor series about $V = V_c$ and $T = T_c$. Expand ΔP to terms quadratic in ΔV and ΔT ($\Delta V = V - V_c$ and $\Delta T = T - T_c$). From what we know about the isotherm of fluids at the critical point, which terms in the Taylor expansion are identically zero? Show that the exponent $\gamma = 1$.

17. For a system whose equation of state is given by the Berthelot equation (cf. Prob. 2.5), find the critical temperature, pressure, and volume in terms of the experimental constants a and b. Does the Berthelot equation yield a law of corresponding states? Find the critical exponents δ, β, α, and γ from the Berthelot equation.

5 Elementary Probability Theory

A. INTRODUCTORY REMARKS

Thermodynamics is a theory which relates the average values of physical quantities, such as energy, magnetization, particle number, etc., to one another. It tells us nothing about processes occurring at the microscopic level. The microscopic theory of systems with a large number of degrees of freedom is based on probabilistic concepts. Therefore, it is useful at this point to review some ideas from elementary probability theory. These ideas will appear again and again throughout the remainder of this book.

When we deal with a large number of objects, it is often necessary to count and classify them. Two classifications which naturally arise are those of permutations and combinations. A permutation is an arrangement of objects in a definite order. A combination is a collection without regard to order. We shall learn how to use these classifications in this chapter.

At the beginning of this chapter we will give an intuitive definition of probability, and at the end we will justify it with the law of large numbers. In order to facilitate analysis of problems, we shall introduce some concepts from set theory and show how they may be used in probability theory.

Stochastic variables are, by definition, variables whose values are determined by the outcome of experiments. The most we can know about a stochastic variable is the probability that a particular value of it will be realized in an experiment. We shall introduce the concept of probability distributions for both discrete and continuous stochastic variables and show how to obtain the moments of a stochastic variable by using characteristic functions. In addition, we shall introduce the idea of a cumulant expansion for characteristic functions.

For concreteness, we shall discuss in some detail the binomial distribution and its two limiting cases, the Gaussian and Poisson distributions. We shall then apply the Gaussian distribution to the case of random walk on a one-dimensional lattice.

Finally, we shall derive the central limit theorem and the related law of large numbers. The central limit theorem tells us that if we have a large number of experiments which measure some stochastic variable, X, then the probability distribution of the average of all the measurements approaches a Gaussian, regardless of the form (there are a few restrictions) of the distribution for X itself. The law of large numbers gives a quantitative justification for the concept of probability; it is included here for completeness.

B. PERMUTATIONS AND COMBINATIONS[1,2]

When applying probability theory to real situations, we are often faced with counting problems which are complex. On such occasions it is useful to keep in mind two very important principles:

(*a*) *Addition principle*: If two operations are mutually exclusive and the first can be done in m ways while the second can be done in n ways, then one or the other can be done in $m + n$ ways.

(*b*) *Multiplication principle*: If an operation can be performed in n ways, and after it is performed in any one of these ways a second operation is performed which can be performed in any one of m ways, then the two operations can be performed in $n \times m$ ways.

These two principles will underly much of our discussion in the remainder of this chapter.

When dealing with large numbers of objects it is often necessary to find the number of permutations and/or combinations of the objects. A *permutation* is any arrangement of a set of N distinct objects in a definite order. The number of different permutations of N distinct objects is $N!$ The number of different permutations of any R objects taken from the set of N objects is $N!/(N-R)!$

Proof: Let us assume that we have N ordered spaces and N distinct objects with which to fill them. Then the first space can be filled N ways, and after it is filled, the second can be filled in $(N-1)$ ways, etc. Thus the N spaces can be filled in

$$N(N-1)(N-2) \times \cdots \times 1 = N!$$

ways.

To find the number of permutations, P_R^N, of N distinct objects taken R at a time, let us assume we have R ordered spaces to fill. Then the first can be filled in N ways, the second in $(N-1)$ ways, . . ., and the Rth in $(N-R+1)$ ways. The total number of ways that R ordered spaces can be filled using N distinct objects is

$$P_R^N = N(N-1) \times \cdots \times (N-R+1) = \frac{N!}{(N-R)!}.$$

A *combination* is a selection of N distinct objects without regard to order. The number of different combinations of any R objects taken from the set of N objects is $N!/(N-R)!\,R!$.

Proof: R distinct objects have R permutations. If we let C_R^N denote the number of combinations of N distinct objects taken R at a time, then $R!C_R^N = P_R^N$ and $C_R^N = N!/(N - R)!R!$.

The number of permutations of a set of N objects which contains n_1 identical elements of one kind, n_2 identical elements of another kind, ..., and n_k identical elements of a kth kind is $N!/n_1!\,n_2!\cdots n_k!$, where $n_1 + n_2 + \cdots + n_k = N$. The proof of this statement is left as a problem.

C. DEFINITION OF PROBABILITY[1-3]

Probability is a quantization of our *expectation* of the outcome of an event or experiment. Suppose that one possible outcome of an experiment is A. Then, the probability of A occurring is $P(A)$ if, out of N identical experiments, we *expect* that $NP(A)$ will result in the outcome A. As N becomes very large ($N \to \infty$) we *expect* that the fraction of experiments which result in A will approach $P(A)$. An important special case is one in which an experiment can result in any of n different equally likely outcomes. If exactly m of these outcomes corresponds to event A, then $P(A) = m/n$.

The concept of a sample space is often useful for obtaining relations between probabilities and for analyzing experiments. A *sample space* of an experiment is a set, S, of elements such that any outcome of the experiment corresponds to one or more elements of the set. An *event* is a subset of a sample space S of an experiment. The probability of an event A can be found by using the following procedure:

(*a*) Set up a sample space S of all possible outcomes.
(*b*) Assign probabilities to the elements of the sample space (the sample points). For the special case of a sample space of N equally likely outcomes, assign a probability $1/N$ to each point.
(*c*) To obtain the probability of an event A, add the probabilities assigned to elements of the subset of S that corresponds to A.

In working with probabilities, some ideas from set theory are useful. The *union* of two events A and B is denoted $A \cup B$. $A \cup B$ is the set of all points belonging to A or B or both (cf. Fig. 5.1*a*). The *intersection* of two events is denoted $A \cap B$. $A \cap B$ is the set of all points belonging to both A and B (cf. Fig. 5.1*b*). If the events A and B are *mutually exclusive*, then $A \cap B = \varnothing$ where $\varnothing$ is the empty set ($A \cap B$ = contains no points) (cf. Fig. 5.1*c*).

We can obtain some useful relations between the probabilities of different events. We shall let $P(A)$ denote the probability that event A is the outcome of an experiment ($P(\varnothing) = 0$, $P(S) = 1$); we shall let $P(A \cap B)$ denote the probability that *both* events A and B occur as the result of an experiment; and finally we shall let $P(A \cup B)$ denote the probability that event A or event

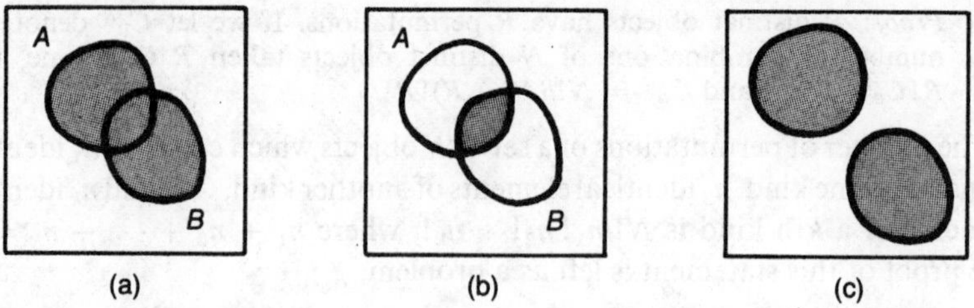

Fig. 5.1. (*a*) The shaded area is the union of A and B, $A \cup B$; (*b*) the shaded area is the intersection of A and B, $A \cap B$; (*c*) when A and B are mutually exclusive, there is no overlap.

B or both occur as the outcome of an experiment. Then the probability $P(A \cup B)$ may be written

$$P(A \cup B) = P(A) + P(B) - P(A \cap B). \tag{5.1}$$

In writing $P(A) + P(B)$, we take the region $A \cap B$ into account twice. Therefore, we have to subtract a factor $P(A \cap B)$.

If the two events A and B are mutually exclusive, then they have no points in common and

$$P(A \cup B) = P(A) + P(B). \tag{5.2}$$

If events $A_1, A_2, \ldots, A_m$ are *mutually exclusive* and *exhaustive*, then $A_1 \cup A_2 \cup \cdots \cup A_m = S$ and the m events form a *partition* of the sample space S into m subsets. If $A_1, A_2, \ldots, A_m$ form a partition, then

$$P(A_1) + P(A_2) + \cdots + P(A_m) = 1. \tag{5.3}$$

We shall see Eq. (5.3) often in this book.

The events A and B are *independent if and only if*

$$P(A \cap B) = P(A)P(B). \tag{5.4}$$

Note that since $P(A \cap B) \neq 0$, A and B have some points in common. Therefore, independent events are not mutually exclusive events. They are completely different concepts. For mutually exclusive events, $P(A \cap B) = 0$.

The *conditional probability* $P(B|A)$ gives us the probability that event A occurs as the result of an experiment *if* B also occurs. $P(B|A)$ is defined by the equation

$$P(B|A) = \frac{P(A \cap B)}{P(B)}. \tag{5.5}$$

Since $P(A \cap B) = P(B \cap A)$, we find also that

$$P(A)P(A|B) = P(B)P(B|A). \tag{5.6}$$

From Eq. (5.4), we see that, if A and B are independent, then

$$P(B|A) = P(A). \tag{5.7}$$

The conditional probability $P(B|A)$ is essentially the probability of event A if we use the set B as the sample space rather than S. An example may be helpful. Consider the case where $P(A) = \frac{2}{3}$ and $P(B) = \frac{2}{3}$, and the events A and B exhaust all possibilities so that $P(A \cup B) = 1$. Then from Eq. (5.1) we find $P(A \cap B) = \frac{1}{3}$. Since $P(A \cap B) \neq P(A)P(B)$ for this case, the events A and B are *not* independent. The conditional probability is $P(B|A) = P(A \cap B)/P(B) = \frac{1}{2}$. Thus, $\frac{1}{2}$ of the points in B will be in A also, and vice versa.

D. DISTRIBUTION FUNCTIONS[1-3]

Before we can understand the meaning of a distribution function, we must introduce the concept of a stochastic, or random, variable (the two words are interchangeable, but we shall refer to them as stochastic variables). A quantity whose value is a number determined by the outcome of an experiment is called a *stochastic variable*. A stochastic variable, X, on a sample space, S, is a function which maps elements of S into the set of real numbers $\{R\}$ in such a way that the inverse mapping of every interval in $\{R\}$ corresponds to an event of S (in other words, a stochastic variable is a function which assigns a real number to each sample point).

In a given experiment, a stochastic variable may have any one of a number of values. Therefore, one must be careful to distinguish a stochastic variable (usually denoted by a capital letter X) from its possible values $\{x_j\}$. Some examples of stochastic variables are

(i) The number of heads which appear each time three coins are tossed
(ii) The maximum number which appears when four dice are tossed

(*Note*: the statement "select at random" means that all selections are equally possible.)

1. Discrete Stochastic Variables

Let X be a stochastic variable on S which can take on a countable (finite or infinite) set of values $X(S) = \{x_1, x_2, \ldots\}$. One can make $X(S)$ a probability space by defining a probability for each value of x_i. The set of values $f(x_i)$ is called the probability distribution of S. The probability distribution, $f(x_i)$, must satisfy the conditions

$$f(x_i) \geqslant 0 \tag{5.8}$$

and

$$\sum_i f(x_i) = 1. \tag{5.9}$$

In Eq. (5.9), the sum is taken over all values $\{x_i\}$ of the stochastic variable X.

If we can determine the distribution function, $f(x_i)$, for the stochastic

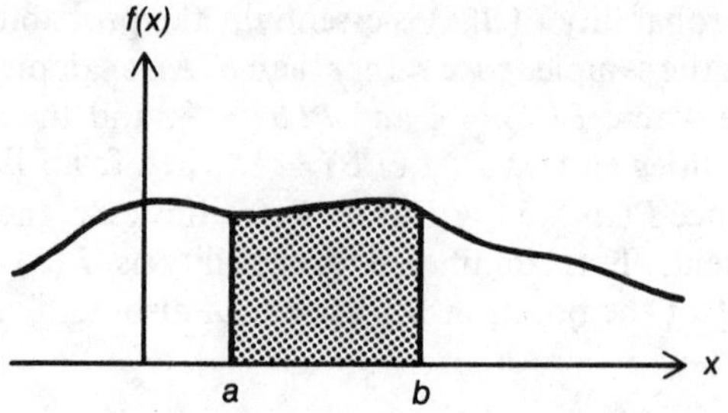

Fig. 5.2. The shaded area is the probability of finding the stochastic variable, x, in the interval $a \leqslant x \leqslant b$.

variable, X, then we have obtained all possible information about it. In practice, this is often difficult. We usually cannot determine $f(x_i)$, but we can often obtain information about the moments of X. The nth moment of X is defined

$$\langle X^n \rangle = \sum_i x_i^n f(x_i). \qquad (5.10)$$

Some of the moments have special names. The moment $\langle X \rangle$ is called the mean, or expected, value of X. The combination $\langle X^2 \rangle - \langle X \rangle^2$ is called the variance of X, and the standard deviation of X is defined

$$\sigma_X \equiv (\langle X^2 \rangle - \langle X \rangle^2)^{1/2}. \qquad (5.11)$$

The moments of X give us information about the spread and shape of the distribution function, $f(x_i)$. The most important moments are the lower order ones since they contain information about the overall behavior of the probability distribution. The first moment $\langle X \rangle$ is important because it gives the mean value of X. The standard deviation is important because it gives us a measure of the width of the distribution $f(x_i)$. A small standard deviation means that $f(x_i)$ is sharply peaked about $\langle X \rangle$ and we can be fairly certain that X will have a value close to $\langle X \rangle$.

2. Continuous Stochastic Variables

Let X be a stochastic variable which can take on a continuous set of values, such as an interval on the real axis. From the definition of a random variable, we know that an interval $a \leqslant X \leqslant b$ corresponds to an event. Let us assume that there exists a piecewise continuous function $f(x)$ such that the probability, $P(a \leqslant X \leqslant b)$, that X has a value in the interval $a \leqslant X \leqslant b$ is given by the area under the curve $f(x)$ between $x = a$ and $x = b$ (cf. Fig. 5.2):

$$P(a \leqslant X \leqslant b) = \int_a^b f_X(x)\, dx. \qquad (5.12)$$

Then X is a continuous stochastic variable and $f_X(x)$ is the probability density of X. The probability density must satisfy the conditions

$$f_X(x) \geqslant 0 \tag{5.13}$$

and

$$\int f_X(x)\, dx = 1. \tag{5.14}$$

In Eq. (5.14) the integral is over the entire range of X.

As before, it is useful to introduce the moments of X. The nth moment of X is defined

$$\langle X^n \rangle = \int dx x^n f_X(x) \tag{5.15}$$

where $\langle X \rangle$ is the mean, or expected, value of X, and the variance and standard deviation are defined as they were above. If all the moments, $\langle X^n \rangle$, are known, then the probability density is completely specified. This is easy to see if we introduce the characteristic function $\phi_X(k)$, defined

$$\phi_X(k) = \langle e^{ikX} \rangle = \int dx\, e^{ikx} f_X(x) = \sum_{n=0}^{\infty} \frac{(ik)^n \langle X^n \rangle}{n!}. \tag{5.16}$$

The series expansion in Eq. (5.16) is meaningful only if the higher moments, $\langle X^n \rangle$, are small so that the series converges. The probability density, $f_X(x)$, is the Fourier transform of the characteristic function

$$f_X(x) = \frac{1}{2\pi} \int dk\, e^{-ikx} \phi_X(k). \tag{5.17}$$

Thus, if all the moments are known, the probability density is completely specified through Eqs. (5.16) and (5.17). If we are given the characteristic function, we can obtain the moments by differentiation through the equation

$$\langle X^n \rangle = \frac{1}{i^n} \left. \frac{d^n \phi_X(k)}{dk^n} \right|_{k=0}. \tag{5.18}$$

Eq. (5.18) provides a simple way of obtaining the moments if we know $f_X(x)$.

As we shall see in subsequent sections, it is useful to write the characteristic function in terms of a cumulant expansion rather than to expand it directly in terms of moments. The cumulant expansion is defined

$$\phi_X(k) = \exp\left[\sum_{n=1}^{\infty} \frac{(ik)^n}{n!} C_n(X) \right], \tag{5.19}$$

where $C_n(X)$ is the nth-order cumulant. If we expand Eqs. (5.16) and (5.19) in powers of k and equate terms of the same order in k, we find the following expressions for the first four cumulants:

$$C_1(X) = \langle X \rangle, \tag{5.20a}$$

$$C_2(X) = \langle X^2 \rangle - \langle X \rangle^2, \tag{5.20b}$$

$$C_3(X) = \langle X^3 \rangle - 3\langle X \rangle \langle X^2 \rangle + 2\langle X \rangle^3, \tag{5.20c}$$

and

$$C_4(X) = \langle X^4 \rangle - 3\langle X^2 \rangle^2 - 4\langle X \rangle \langle X^3 \rangle + 12\langle X \rangle^2 \langle X^2 \rangle - 6\langle X \rangle^4. \tag{5.20d}$$

If higher order cumulants rapidly go to zero, we can often obtain a good approximation to $\phi_X(k)$ by retaining only the first few cumulants in Eq. (5.19). We see that $C_1(X)$ is just the mean value of X and $C_2(X)$ is the variance.

Often we wish to find the probability density, not of X but of some new stochastic variable, $Y = H(X)$, where $H(X)$ is a known function of X. The probability density of $Y, f_Y(y)$ is defined

$$f_Y(y) = \int dx\, \delta(y - H(x)) f_X(x) \tag{5.21}$$

where $\delta(y - H(x))$ is the Dirac delta function.

3. Joint Probability Distributions

Let X and Y be stochastic variables on a sample space S with values $X(S) = \{x_1, x_2, \ldots\}$ and $Y(S) = \{y_1, y_2, \ldots\}$, respectively. We can make the product set $X(S) \times Y(S) = \{(x_1, y_1), (x_1, y_2), \ldots, (x_i, y_j), \ldots\}$ into a probability space by defining the probability of the ordered pair $\{x_i, y_j\}$ to be $P(X = x_i, Y = y_j) = f(x_i, y_j)$. The function $f(x_i, y_j)$ is the joint probability distribution of X and Y. If the stochastic variables X and Y are continuous, then we simply write the probability density as $f(x, y)$ and integrate rather than sum over the variables x and y. We shall work with continuous stochastic variables below, but it is easy to change to the case of discrete variables if the need arises.

If we know the joint probability density, $f(x, y)$, we can obtain the probability density for X by integrating (summing if Y is discrete) over y,

$$f_X(x) = \int dy f(x, y). \tag{5.22}$$

The joint probability density must always be positive, $f(x, y) \geqslant 0$, and is normalized. Thus,

$$\iint dx\, dy f(x, y) = 1. \tag{5.23}$$

The covariance of X and Y is defined

$$\begin{aligned} \operatorname{cov}(X, Y) &= \iint dx\, dy (x - \langle X \rangle)(y - \langle Y \rangle) f(x, y) \\ &= \iint dx\, dy\, xy f(x, y) - \langle X \rangle \langle Y \rangle \\ &= \langle XY \rangle - \langle X \rangle \langle Y \rangle. \end{aligned} \tag{5.24}$$

The *correlation* of X and Y is defined

$$\mathrm{cor}(X, Y) = \frac{\mathrm{cov}(X, Y)}{\sigma_X \sigma_Y}. \tag{5.25}$$

The correlation cor(X, Y) is dimensionless and has properties:

(i) $\mathrm{cor}(X, Y) = \mathrm{cor}(Y, X)$.
(ii) $-1 \leqslant \mathrm{cor}(X, Y) \leqslant 1$.
(iii) $\mathrm{cor}(X, X) = 1, \mathrm{cor}(X, -X) = -1$.
(iv) $\mathrm{cor}(aX + b, cY + d) = \mathrm{cor}(X, Y)$ if $a, c \neq 0$.

Pairs of stochastic variables with identical distributions can have different correlations.

For two random variables X and Y which are *independent*, the following relations hold:

(i′) $f(x, y) = f_X(x) f_Y(y)$.
(ii′) $\langle XY \rangle = \langle X \rangle \langle Y \rangle$.
(iii′) $\langle (X + Y)^2 \rangle - \langle X + Y \rangle^2 = \langle X^2 \rangle - \langle X \rangle^2 + \langle Y^2 \rangle - \langle Y \rangle^2$.
(iv′) $\mathrm{cov}(X, Y) = 0$.

Note that the converse of (iv′) does not necessarily hold. If $\mathrm{cov}(X, Y) = 0$, it does not always mean that X and Y are independent. The notion of a joint distribution function can be extended to any finite number of stochastic variables.

When we deal with a number of stochastic variables, we often wish to find the probability density for a stochastic variable, z, which is a function of the old stochastic variables. For example, if we know the joint probability density, $f(x, y)$, we may wish to find the probability density for a variable $z = G(x, y)$, where $G(x, y)$ is a known function of x and y. The probability density, $f_Z(z)$, for the stochastic variable Z is defined

$$f_Z(z) = \iint dx\, dy\, \delta(z - G(x, y)) f(x, y). \tag{5.26}$$

For the case when X and Y are independent, we can write $f(x, y) = f_X(x) f_Y(y)$ where $f_X(x)$ and $f_Y(y)$ are the probability densities for X and Y, respectively. Then Eq. (5.26) becomes

$$f_Z(z) = \iint dx\, dy\, \delta(z - G(x, y)) f_X(x) f_Y(y). \tag{5.27}$$

If we write the delta function in terms of a Fourier integral, the characteristic function of $f_Z(z)$ is easily found to be

$$\phi_Z(k) = \iint e^{ikG(x,y)} f_X(x) f_Y(y)\, dx\, dy \tag{5.28}$$

if X and Y are independent. If $G(x, y) = x + y$, then

$$\phi_Z(k) = \iint e^{ik(x+y)} f_X(x) f_Y(y)\, dx\, dy = \phi_X(k)\phi_Y(k). \tag{5.29}$$

Thus, the characteristic function of the sum of independent stochastic variables is the product of the characteristic functions of the individual stochastic variables.

E. BINOMIAL DISTRIBUTION[1-4]

One of the most common applications of probability theory is to the case of a large number, N, of experiments, each having two possible outcomes. The probability distribution for one of the outcomes is called the binomial distribution. In the limit of large N, the binomial distribution can be approximated by either the Gaussian or the Poisson distribution, depending of the size of the probability of a given outcome. We shall consider all three of these distributions in this section.

1. Binomial Distribution

Let us carry out a sequence of N statistically independent trials and assume that each trial can have only one of two outcomes, $+1$ or -1. Let us denote the probability of outcome $+1$ by p and the probability of outcome -1 by q. Then $(p + q) = 1$.

In a given sequence of N trials, the outcome $+1$ can occur n_1 times and the outcome -1 can occur n_2 times, where $N = n_1 + n_2$. The probability for a given *permutation* of n_1 outcomes $+1$ and n_2 outcomes -1 is $p^{n_1}q^{n_2}$ since the N trials are statistically independent. The probability for any *combination* of n_1 outcomes $+1$ and n_2 outcomes -1 is

$$P_N(n_1) \equiv \frac{N!}{n_1!\, n_2!} p^{n_1} q^{n_2} \tag{5.30}$$

since a combination of n_1 outcomes $+1$ and n_2 outcomes -1 contains $N!/n_1!\, n_2! \cdots$ permutations. Eq. (5.30) is called the *binomial distribution*.

By the binomial theorem, we have the normalization condition

$$\sum_{n_1=0}^{N} P_N(n_1) = \sum_{n_1=0}^{N} \frac{N!}{n_1!\,(N - n_1)!} p^{n_1} q^{N-n_1} = (p + q)^N = 1. \tag{5.31}$$

The first moment, or mean value, of outcome $+1$ is

$$\begin{aligned} \langle n_1 \rangle &= \sum_{n_1=0}^{\infty} n_1 P_N(n_1) = \sum_{n_1=0}^{\infty} \frac{N!\, n_1}{n_1!\,(N - n_1)!} p^{n_1} q^{N-n_1} \\ &= p \frac{\partial}{\partial p} \left[\sum_{n_1=0}^{\infty} P_N(n_1) \right] = p \frac{\partial}{\partial p} (p + q)^N = pN. \end{aligned} \tag{5.32}$$

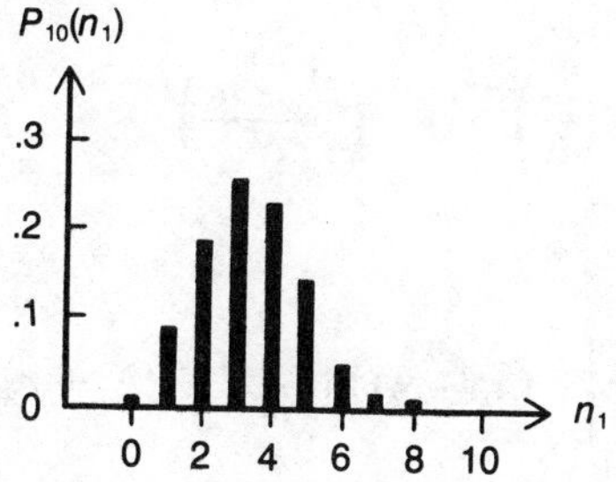

Fig. 5.3. The binomial distribution for $p = \frac{1}{3}$ and $N = 10$: for this case, $\langle n_1 \rangle = \frac{10}{3}$.

In a similar manner, we obtain for the second moment

$$\langle n_1{}^2 \rangle = \sum_{n_1=0}^{\infty} n_1{}^2 P_N(n_1) = (Np)^2 + Npq. \tag{5.33}$$

The variance therefore becomes

$$\sigma_N{}^2 = \langle n_1{}^2 \rangle - \langle n_1 \rangle^2 = Npq, \tag{5.34}$$

and the standard deviation is

$$\sigma_N = \sqrt{Npq}. \tag{5.35}$$

The fractional deviation is

$$\frac{\sigma_N}{\langle n_1 \rangle} = \sqrt{\left(\frac{q}{p}\right)} \frac{1}{\sqrt{N}}. \tag{5.36}$$

The fractional deviation is a measure of the deviation of the fraction of $+1$ outcomes, n_1/N, from its expected value p, in any single sequence of N trials. A small value of $\sigma_N/\langle n_1 \rangle$ means that n_1/N will be close to p. For $N \to \infty$, $\sigma_N/\langle n_1 \rangle \to 0$ so that $n_1/N \to p$. The binomial distribution for the case $N = 10$ and $p = \frac{1}{3}$ is plotted in Fig. 5.3.

2. Gaussian (or Normal) Distribution

In the limit of large N and large pN (i.e., p not very small) the binomial distribution approaches a Gaussian distribution. If we make use of Stirling's formula,

$$n_1! \approx \sqrt{2\pi n_1} \left(\frac{n_1}{e}\right)^{n_1} \tag{5.37}$$

for $n_1 > 10$, we can write the binomial distribution in the following forms:

$$P_N(n_1) \equiv \frac{\sqrt{2\pi N}\left(\dfrac{N}{e}\right)^N p^{n_1}(1-p)^{N-n_1}}{\sqrt{2\pi n_1}\sqrt{2\pi(N-n_1)}\left(\dfrac{n_1}{e}\right)^{n_1}\left(\dfrac{N-n_1}{e}\right)^{N-n_1}} \tag{5.38}$$

or

$$P_N(n_1) = \frac{1}{\sqrt{2\pi N}} \left(\frac{n_1}{N}\right)^{-n_1-1/2} \left(\frac{N-n_1}{N}\right)^{n_1-N-1/2} p^{n_1}(1-p)^{N-n_1}. \tag{5.39}$$

This can be rewritten

$$P_N(n_1) = \frac{1}{\sqrt{2\pi N}} \exp\left[-(n_1+\tfrac{1}{2})\ln\left(\frac{n_1}{N}\right) - (N-n_1+\tfrac{1}{2})\ln\left(\frac{N-n_1}{N}\right) + n_1 \ln p + (N-n_1)\ln(1-p)\right]. \tag{5.40}$$

The binomial distribution $P_N(n_1)$ exhibits a maximum at $n_1 = \langle n_1 \rangle = Np$. Thus, the average value $\langle n_1 \rangle$ grows with N. For $N \gg 1$, the values of n_1 in the region of $\langle n_1 \rangle$ will be very large relative to integer changes in n_1 which can occur, and in the region of the maximum we can treat n_1 as a continuous variable.

We wish to examine Eq. (5.40) in the region of the maximum. Since $n_1 \sim Np$ in this region we can approximate Eq. (5.40) by the expression

$$P_N(n_1) = \frac{1}{\sqrt{2\pi N}} \exp\left[-n_1\ln\left(\frac{n_1}{N}\right) - (N-n_1)\ln\left(\frac{N-n_1}{N}\right) + n_1 \ln p + (N-n_1)\ln(1-p)\right]. \tag{5.41}$$

It is simple to show that the condition for a maximum, $(dP_N/dn_1)_{n_1=\langle n_1\rangle} = 0$, yields the result $\langle n_1 \rangle = Np$.

If we expand the argument of the exponent in a Taylor series about $n_1 = \langle n_1 \rangle$, we obtain the following expression for $P_N(n_1)$:

$$P_N(n_1) = P_N(\langle n_1 \rangle)\exp[\tfrac{1}{2}B_2\epsilon^2 + \tfrac{1}{6}B_3\epsilon^3 + \cdots] \tag{5.42}$$

where $\epsilon = n_1 - \langle n_1 \rangle$ and

$$B_k = \left(\frac{d^k \ln[\sqrt{2\pi N}P_N(n_1)]}{d{n_1}^k}\right)_{n_1=\langle n_1\rangle}. \tag{5.43}$$

We can evaluate B_2 and B_3 from Eqs. (5.41) and (5.43). We first note that

$$\frac{d^2 \ln[\sqrt{2\pi N}P_N(n_1)]}{d{n_1}^2} = -\frac{1}{n_1} - \frac{1}{N-n_1}$$

so that

$$B_2 = -\frac{1}{Npq}, \tag{5.44}$$

and

$$\frac{d^3 \ln[\sqrt{2\pi N}P_N(n_1)]}{d{n_1}^3} = \frac{1}{{n_1}^2} - \frac{1}{(N-n_1)^2},$$

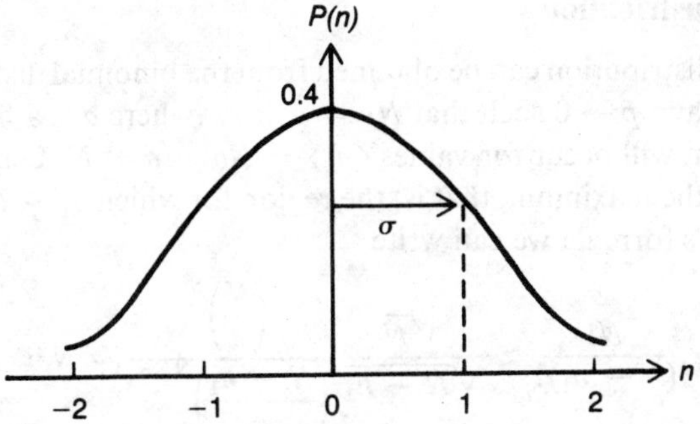

Fig. 5.4. The Gaussian distribution for a mean value $\langle n\rangle = 0$ and a standard deviation $\sigma = 1$.

so that

$$B_3 = \frac{1}{N^2p^2q^2}(q^2 - p^2). \tag{5.45}$$

Thus,

$$|B_3| < \frac{1}{N^2p^2q^2}. \tag{5.46}$$

Similarly, one can show that

$$|B_k| < \frac{1}{(Npq)^{k-1}}. \tag{5.47}$$

Therefore, terms of higher order in ϵ^2 can be neglected if $\epsilon \ll Npq$, and we obtain

$$P_N(n_1) \approx P_N(\langle n_1\rangle)\, e^{-(1/2)|B_2|\epsilon^2}. \tag{5.48}$$

We note further that if $|B_2|\epsilon^2 \gg 1$ or $\sqrt{Npq} \ll \epsilon$ the probability $P_N(n_1)$ will go to zero very fast as one moves away from the maximum value. Thus, for $Npq \gg 1$, the Gaussian will give a very good approximation to the binomial distribution since the quantities $\sqrt{Npq}$ and Npq will be well separated.

We can use the normalization condition to determine the coefficient in Eq. (5.48). We find

$$P_N(\langle n_1\rangle) = \frac{1}{\sqrt{2\pi}\sqrt{Npq}}. \tag{5.49}$$

If we note that the standard deviation is given by $\sigma_N = \sqrt{Npq}$ (cf. Eq. (5.34)) we may write the expression for $P_N(n_1)$ in the form

$$P_N(n_1) = \frac{1}{\sigma_N\sqrt{2\pi}} \exp\left\{-\frac{1}{2}\frac{(n_1 - \langle n_1\rangle)^2}{\sigma_N{}^2}\right\}. \tag{5.50}$$

Eq. (5.50) is the Gaussian distribution for outcomes $+1$. It is important to note that the Gaussian distribution is entirely determined in terms of the first and second moments, $\langle n_1\rangle$ and $\langle n_1{}^2\rangle$. In Fig. 5.4 we plot the Gaussian distribution.

3. Poisson Distribution

The Poisson distribution can be obtained from the binomial distribution in the limit $N \to \infty$ and $p \to 0$ such that $Np = a \ll N$ (where a is a finite constant). The maximum will occur for values $\langle n_1 \rangle = Np = a \ll N$. Let us look at the region about the maximum, that is, the region for which $n_1 \sim Np \ll N$. Then, using Stirling's formula we can write

$$\frac{N!}{(N-n_1)!} \approx \frac{\sqrt{N}}{\sqrt{N-n_1}} \frac{\left(\frac{N}{e}\right)^N}{\left(\frac{N-n_1}{e}\right)^{N-n_1}} \sim N^{n_1}. \tag{5.51}$$

Furthermore, for $p \to 0$

$$(1-p)^{N-n_1} \approx (1-p)^N = (1-p)^{a/p} \to e^{-a}, \tag{5.52}$$

where we have used the definition $e^z = \lim_{n\to\infty}(1 + (z/n))^n$. We may now combine the above results to obtain

$$P_N(n_1) = \frac{a^{n_1} e^{-a}}{n_1!} \tag{5.53}$$

in the limit $p \to 0$ and $N \to \infty$ such that the product $Np \to a$. Eq. (5.53) is the Poisson distribution. It applies when many experiments are carried out but the result $+1$ has only a small probability of occurring. The expected number of outcomes $+1$ will be a.

The Poisson distribution is normalized to one

$$\sum_{n_1=0}^{\infty} \frac{a^{n_1} e^{-a}}{n_1!} = e^a e^{-a} = 1. \tag{5.54}$$

We note that the Poisson distribution for n_1 is determined entirely in terms of the first moment, $\langle n_1 \rangle = a$. Thus, it is sufficient to know only the first moment in order to find the probability density for a Poisson process. In Fig. 5.5 we plot the Poisson distribution for $\langle n_1 \rangle = 2 = a$.

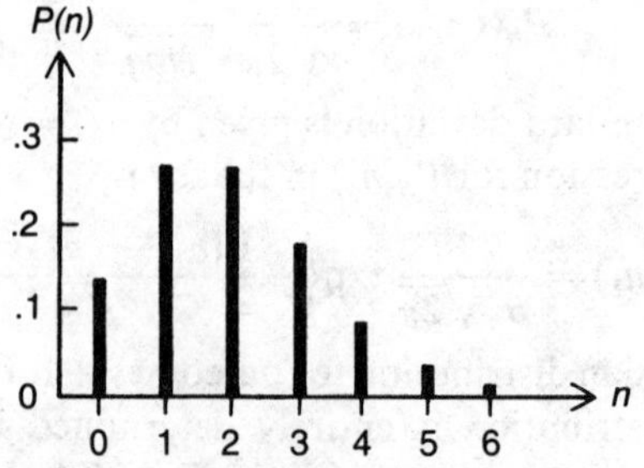

Fig. 5.5. The Poisson distribution for $\langle n \rangle = 2$.

F. RANDOM WALK[5]

The problem of random walk is an example of the application of the binomial distribution to a problem in physics. Consider a particle which is constrained to move along a line. It has a probability $p = \frac{1}{2}$ of taking a step to the right and a probability $q = \frac{1}{2}$ of taking a step to the left. Let us assume that the particle takes N steps and that the outcome of each step is independent of previous steps (the steps are statistically independent). Let n_1 be the number of steps to the right and n_2 the number of steps to the left. The net displacement after N steps is given by $m = n_1 - n_2$, where $n_1 + n_2 = N$ and therefore $m = 2n_1 - N$. The net displacement, m, can have values $-N \leqslant m \leqslant N$. For a large number of steps, N, the probability of a net displacement m is found by substituting $n_1 = (m + N)/2$ in Eq. (5.50). The result is

$$P_N(m) = \left[\frac{2}{\pi N}\right]^{1/2} \exp\left\{-\frac{m^2}{2N}\right\}. \tag{5.55}$$

Thus, the net displacement obeys a Gaussian distribution.

If each step length is l, then the net displacement (in terms of position) is $x = ml$. If we consider intervals Δx, which are large compared to l, then the probability that the particle lies in the interval $x \to x + \Delta x$ after N steps is $P_N(x)\,\Delta x = P_N(m)(\Delta x/2l)$. Therefore,

$$P_N(x) = \frac{1}{(2\pi N l^2)^{1/2}} \exp\left(\frac{-x^2}{2Nl^2}\right). \tag{5.56}$$

If the particle takes n steps per unit time, then the probability that the particle lies in the interval $x \to x + \Delta x$ at time t is

$$P(x, t)\,\Delta x = \frac{1}{2(\pi D t)^{1/2}} \exp\left[\frac{-x^2}{4Dt}\right] \Delta x \tag{5.57}$$

where $N = nt$ and $D = \frac{1}{2}nl^2$ is the diffusion coefficient. The probability density as a function of time is plotted in Fig. 5.6. Initially $P(x, t)$ has the form

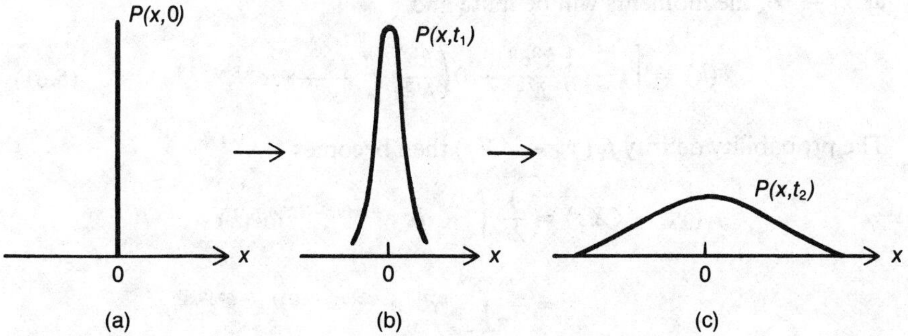

Fig. 5.6. A plot of the probability density for the displacement of a particle by random walk as a function of time: here, $0 < t_1 < t_2 < \infty$.

of a delta function $\delta(x)$. As time passes, it spreads in such a way that the total area underneath the curve is equal to one, but the probability of finding the particles at some distance along the x-axis grows.

G. CENTRAL LIMIT THEOREM[3,4]

Given a random variable X with a probability density $f_X(x)$, we wish to find the distribution of the random variable Y, where the values of Y correspond to the average of N measurements of X:

$$y_N = \frac{x_1 + x_2 + \cdots + x_N}{N}. \tag{5.58}$$

(Note that X could correspond to the displacement of a particle after one step and $f_X(x)\,dx$ could correspond to the probability that it is displaced a distance $x \to x + dx$ in one step.) Let us consider the probability density $f_Y(y_N - \langle X\rangle)$. Its characteristic function can be written

$$\begin{aligned}\Phi(k) &= \int e^{ik(y_N - \langle X\rangle)} f_Y(y_N - \langle X\rangle)\,dy_N \\ &= \int e^{i(k/N)((x_1 - \langle X\rangle) + (x_2 - \langle X\rangle) + \cdots + (x_N - \langle X\rangle))} f_X(x_1) f_X(x_2) \cdots f_X(x_N)\,dx_1 \cdots dx_N \\ &= [\phi(k/N)]^N. \end{aligned} \tag{5.59}$$

If $\sigma^2 = \langle X^2\rangle - \langle X\rangle^2$, then

$$\phi\left(\frac{k}{N}\right) = \int e^{i(k/N)(x_1 - \langle X\rangle)} f_X(x_1)\,dx_1 = 1 - \frac{1}{2}\frac{k^2}{N^2}\sigma^2 + \cdots \tag{5.60}$$

The function $\phi(k/N)$ will decrease with increasing k because of the oscillatory character of $e^{i(k/N)x_1}$. The function $[\phi(k/N)]^N$ will decrease even more rapidly with increasing k. Furthermore, if the function $f_X(x_1)$ goes to zero fast enough as $x_1 \to \infty$, the moments will be finite and

$$\Phi(k) = \left[1 - \frac{1}{2}\frac{k^2\sigma^2}{N^2} + 0\left(\frac{k^3}{N^3}\right)\right]^N \xrightarrow[N\to\infty]{} e^{-k^2\sigma^2/2N}. \tag{5.61}$$

The probability density $f_Y(y_N - \langle X\rangle)$ then becomes

$$\begin{aligned} f_Y(y_N - \langle X\rangle) &= \frac{1}{2\pi}\int_{-\infty}^{\infty} dk\, e^{-ik(y_N - \langle X\rangle)}\Phi(k) \\ &= \frac{1}{2\pi}\int_{-\infty}^{\infty} dk\, e^{-ik(y_N - \langle X\rangle)}\, e^{-k^2\sigma^2/2N} \\ &= \sqrt{\frac{N}{2\pi}}\frac{1}{\sigma}\, e^{-N(y_N - \langle X\rangle)^2/2\sigma^2}. \end{aligned} \tag{5.62}$$

Thus, regardless of the form of $f_X(x)$, the average of a large number of measurements of X will be a Gaussian centered at $\langle X \rangle$, with a standard deviation $N^{-1/2}$ times the standard deviation of the probability density of X. The only requirements are that $f(x)$ have finite moments, that the measurements of X be statistically independent, and that N be large. This result is called the *central limit theorem* and helps explain why the Gaussian distribution describes so many phenomena that occur in nature.

H. LAW OF LARGE NUMBERS[2,3]

The law of large numbers underlies the intuitive concept of probability that we introduced in Sec. 5.C. Much of the content of the law of large numbers is contained in the central limit theorem. However, the law of large numbers is more general than the central limit theorem so it is worthwhile to discuss it briefly here.

The law of large numbers applies to N independent experiments and may be stated as follows: If an event, A, has a probability, p, of occurring, then the fraction of outcomes, A, approaches p in the limit $N \to \infty$. The law of large numbers has been proven in varying degrees of generality. We shall consider a rather specialized version of it. The proof we give here has two steps. The first step involves the derivation of the so-called Tchebycheff inequality; in the second step this inequality is used to derive the law of large numbers.

The Tchebycheff inequality establishes a relation between the variance and the probability that a stochastic variable can deviate by an arbitrary amount ϵ (ϵ is positive) from its average value. The variance, σ_X^2, of a random variable, X, is written

$$\sigma_X^2 = \int_{-\infty}^{\infty} dx(x - \langle x \rangle)^2 f_X(x). \tag{5.63}$$

If we now delete that range of the variable, x, for which $|x - \langle x \rangle| \leqslant \epsilon$, we can write

$$\sigma_X^2 \geqslant \int_{-\infty}^{\langle x \rangle - \epsilon} dx(x - \langle x \rangle)^2 f_X(x) + \int_{\langle x \rangle + \epsilon}^{\infty} dx(x - \langle x \rangle)^2 f_X(x). \tag{5.64}$$

Since $|x - \langle x \rangle|^2 \geqslant \epsilon^2$, we can replace $(x - \langle x \rangle)^2$ by ϵ^2 in Eq. (5.64) and write

$$\sigma_X^2 \geqslant \epsilon^2 \left[\int_{-\infty}^{\langle x \rangle - \epsilon} dx f_X(x) + \int_{\langle x \rangle + \epsilon}^{\infty} dx f_X(x) \right] = \epsilon^2 P(|x - \langle x \rangle| \geqslant \epsilon) \tag{5.65}$$

where $P(|x - \langle x \rangle| \geqslant \epsilon)$ is the probability that the stochastic variable, X, deviates from $\langle x \rangle$ by more than $\pm\epsilon$. From Eq. (5.64) we obtain the Tchebycheff inequality

$$P(|x - \langle x \rangle| \geqslant \epsilon) \leqslant \frac{\sigma_X^2}{\epsilon^2}. \tag{5.66}$$

Thus, for fixed variance, $\sigma_X{}^2$, the probability that x can differ from its mean by more than $\pm\epsilon$ decreases as ϵ^{-2} for increasing ϵ.

We now come to the law of large numbers. Let us consider N independent experiments on the random variable X. We define again the mean value of the outcomes:

$$y_N = \frac{x_1 + x_2 + \cdots + x_N}{N}. \tag{5.67}$$

The law of large numbers states that the probability that y_N deviates from $\langle x\rangle$ goes to zero as $N \to \infty$. Thus, $\lim_{N\to\infty} P(|y_N - \langle x\rangle| \geqslant \epsilon) = 0$. To prove this, let us first note that $\langle y_N\rangle = \langle x\rangle$; from relation 5.iii′ (p. 145), since we have independent events, the variance $\sigma_{y_N}^2$ behaves as

$$\sigma_{y_N}^2 = \frac{1}{N}\sigma_X{}^2. \tag{5.68}$$

We now use Tchebycheff inequality to write

$$P(|y_N - \langle x\rangle| \geqslant \epsilon) \leqslant \frac{\sigma_{y_N}^2}{\epsilon^2} = \frac{\sigma_X{}^2}{N\epsilon^2}. \tag{5.69}$$

Thus, in the limit $N \to \infty$ we find

$$\lim_{N\to\infty} P(|y_N - \langle x\rangle| \geqslant \epsilon) \to 0, \tag{5.70}$$

provided that σ_X is finite. In deriving Eq. (5.70), we have only assumed that $\sigma_X{}^2$ is finite. The law of large numbers, which we have derived here, is a rather special form of a more general law due to Khintchine.[3]

REFERENCES

1. F. Mosteller, R. E. K. Rourke, and G. B. Thomas, *Probability and Statistics* (Addison-Wesley Publishing Co., Reading, Mass., 1967).
2. S. Lipschutz, *Probability*, Schaum's Outline Series (McGraw-Hill Book Co., New York, 1965).
3. W. Feller, *An Introduction to Probability Theory and Its Applications*, Vol. I (John Wiley & Sons, New York, 1968).
4. F. Reif, *Fundamentals of Statistical and Thermal Physics* (McGraw-Hill Book Co., New York, 1965).
5. S. Chandrasekhar, Rev. Mod. Phys. *15* 1 (1943).

PROBLEMS

1. Prove that the number of permutations of N objects which contain n_1 identical elements of one kind, n_2 identical elements of another kind, and n_K identical elements of a Kth kind is $\left(\dfrac{N!}{n_1!\, n_2! \cdots n_K!}\right)$.

2. A bus has nine seats facing forward and eight facing backward. In how many ways can seven passengers be seated if two refuse to ride facing forward and three refuse to ride facing backward?

3. Find the number of permutations of the letters in the word *monotonous*. In how many ways are four *o*'s together? In how many ways are three *o*'s together?

4. In how many ways can six red balls, five blue balls, and five white balls be placed in a row so that the balls on the ends of the row have the same color?

5. Three coins are tossed. Find the probability of getting

(*a*) No heads
(*b*) At least one head

Show that events "heads on first coin" and "tails on last two coins" are independent. Show that events "two coins heads" and event "three coins heads" are dependent.

6. Various six-digit numbers can be formed by permuting the digits 666655. All arrangements are equally likely. Given that a number is even, what is the probability that two fives are together? (*Hint:* you must find a conditional probability.)

7. An urn contains seven red marbles, four white marbles, and five blue marbles. If three marbles are drawn in succession (each being replaced before the next is drawn), what is the probability that the first marble is red, the second is white, and the third is blue?

8. Fifteen boys go hiking. Five get lost, eight get sunburned, and six return home without any problems. What is the probability that a sunburned boy got lost? What is the probability that a lost boy got sunburned?

9. A stochastic variable, X, can have values, $X_1 = 1$ and $X_2 = 2$. The stochastic variable Y can have values $Y_1 = 2$ and $Y_2 = 3$. Consider the following two cases:

(*a*) Let the joint probability be given by $f(x_1, y_1) = f(x_1, y_2) = f(x_2, y_1) = f(x_2, y_2) = \frac{1}{4}$.
(*b*) Let the joint probability be given by $f(x_1, y_1) = f(x_2, y_2) = 0$ and $f(x_1, y_2) = f(x_2, y_1) = \frac{1}{2}$.

Find the covariance of X and Y for the two cases. For either case, are the random variables X and Y independent?

10. The probability of an archer hitting his target is one in four.

(*a*) If he shoots five times, what is the probability of hitting the target at least three times?
(*b*) How many times must he shoot so that the probability of hitting the target at least once is more than 80%?

11. A die is loaded so that even numbers occur three times as often as odd numbers.

(*a*) If the die is thrown nine times, what is the probability that an odd number occurs five times?
(*b*) If the die is thrown one thousand times, what is the probability that more than one-half the throws will be odd?

12. A book with seven hundred misprints contains fourteen hundred pages. What is the probability that one page contains two misprints?

13. A thin sheet of gold foil (assume it to be one atom thick) is fired upon by a beam of neutrons. The neutrons are assumed equally likely to hit any part of the foil but only "see" the gold nuclei. Assume that enough neutrons are fired so that each nucleus is hit an average of two times. What fraction of nuclei suffers no hits. What fraction suffers two hits.

14. Consider a box of volume V_T which contains N_T particles. Assume that the N_T particles have an equally likely chance of being anywhere in the box. Let us observe a small subvolume V of the box.

(*a*) What is the average number of particles in V?
(*b*) What are the variance and standard deviation?
(*c*) Sketch the probability distribution for number of particles in V when $V = V_T$, $V = \frac{1}{2}V_T$, and $V = 10^{-6}V_T$. Assume that $N = 10^{23}$.

15. A stochastic variable X has a continuous Gaussian distribution with mean value $\langle x \rangle = 0$ and variance $\langle x^2 \rangle = 1$. Two independent values are chosen at random. Find the distribution function for the stochastic variable $\xi = x_1^2 + x_2^2$ by using the method of characteristic functions.

16. An intoxicated man trying to walk a straight line has an equally likely chance of going forward or backward a distance $-a \leqslant x \leqslant a$ with each step. After N steps, what is his mean displacement $\langle x \rangle$ and what is the variance $\langle x^2 \rangle - \langle x \rangle^2$? What is the meaning of variance?

17. Assume that the probability of taking a step of length $x \to x + dx$ is given by

$$f(x)\,dx = \frac{1}{\pi}\frac{a}{x^2 + a^2}\,dx.$$

Find the probability distribution for the total displacement after N steps. Does it satisfy the central limit theorem? Should it? Find the first four cumulants for this distribution.

18. Consider the random walk problem in one dimension. The probability of a displacement between x and $x + dx$ is $f(x)\,dx = (1/(2\pi\sigma^2)^{1/2})\,e^{-(x-l)^2/2\sigma^2}dx$. After N steps:

(*a*) What is the mean displacement?
(*b*) What is the variance?

19. The moments of random variable X are given by $\langle x^n \rangle = (1/A)^n$ where A is a constant. What is the probability distribution of X?

6
Master Equation

A. INTRODUCTORY REMARKS

Now that we have reviewed some of the basic ideas of probability theory (cf. Chap. 5), we will begin to study how probability distributions evolve in time. In this chapter we will derive equations for the evolution of probability distributions for processes in which most of the memory effects in the evolution can be neglected (Markov processes), and we shall apply these equations to a variety of problems involving discrete stochastic variables. The level at which we discuss the evolution of probability distributions in this chapter is semiphenomenological because we do not concern ourselves with the underlying dynamics of the stochastic variables. We approximate the dynamics by a judicious choice of the transition matrix. In the next chapter we will find equations of motion for the probability distributions of systems governed by Hamiltonian dynamics.

The equation which governs the evolution of the probability distribution for Markov processes is the master equation. It is one of the most important equations in statistical physics because of its almost universal applicability. It has been applied to problems in chemistry, biology, population dynamics, laser physics, Brownian motion, fluids, and semiconductors, to name only a few cases. As a system of stochastic variables evolves, there are transitions between various values of the stochastic variables. Through transitions, the probability of finding the system in a given state changes until the system reaches a final equilibrium steady state in which transitions cannot cause further changes in the probability distribution (it can happen that it never reaches a steady state, but we will be most concerned with cases for which it can).

To derive the master equation, we must assume that the probability of each transition depends only on the preceding step and not on any previous history. This assumption applies to many systems, although in Chap. 16 we shall see that it breaks down for fluid systems. If the transitions between values of the stochastic variable only occur in small steps, then the master equation reduces approximately to a partial differential equation for the probability density (one example of such a partial differential equation is the Fokker-Planck equation).

One of the simplest applications of the master equation is to the case of Markov chains. These are processes which involve transitions between discrete stochastic variables at discrete times. We will get a very good picture of the decay to a unique equilibrium state for the case in which the transition matrix is "regular," and we can introduce the concept of ergodicity, which we shall come back to in Chap. 8 for dynamical systems.

Another simple application of the master equation is to the random walk problem. We shall show that by solving the master equation we obtain the same result for the spread of the probability density for a particle undergoing random walk that we found in Chap. 5, and we shall derive the diffusion equation.

It is possible sometimes to solve the master equation if we can find the eigenvectors of the transition matrix. Since the transition matrix is not symmetric in general, the solution involves the introduction of left and right eigenvectors. We shall obtain the general form of solution for the case of stationary stochastic processes and write the solution in terms of the left and right eigenvectors of the transition matrix. We leave as a homework problem the solution of some typical Fokker-Planck equations in terms of the eigenvectors of the differential operators appearing in the Fokker-Planck equation.

Birth and death processes correspond to one of the widest applications of the master equation. These are processes in which the transition occurs only a single step at a time. They received their name from their use in population dynamics. For some linear birth and death processes, the master equation can be solved exactly by introducing a generating function (similar to a characteristic function). We can obtain a partial differential equation for the generating function which may be solved in simple cases. We shall treat one example of an exactly solvable problem and give some other examples as homework problems.

In many cases the master equation cannot be solved exactly. However, if the system has a large parameter, such as total volume or total number of particles, a systematic expansion procedure has been developed by van Kampen to find approximate solutions. We shall discuss this method and an example of its use. The example we choose is a problem from nonlinear population dynamics which leads to the Malthus-Verhulst equation, a nonlinear equation for the evolution of the average number of individuals in a society. It is an interesting example because it exhibits a phase transition.

B. DERIVATION OF THE MASTER EQUATION[1–3]

Let us consider a system whose properties can be described in terms of a single stochastic variable Y. Y could denote the velocity of a Brownian particle, the number of particles in a box, or the number of people in a queue, to name a few of many possibilities.

We will use the following notation for the probability density for the stochastic variable Y:

$$P_1(y_1, t) \equiv \text{(the probability density that the stochastic variable } Y \text{ has value } y_1 \text{ at time } t_1\text{)}; \tag{6.1}$$

$$P_2(y_1, t_1; y_2, t_2) \equiv \text{(the joint probability density that the stochastic variable } Y \text{ has value } y_1 \text{ at time } t_1 \text{ and } y_2 \text{ at time } t_2\text{)}; \tag{6.2}$$

$$P_n(y_1, t_1; y_2, t_2; \ldots; y_n, t_n) \equiv \text{(the joint probability density that the stochastic variable } Y \text{ has value } y_1 \text{ at time } t_1, y_2 \text{ at time } t_2, \ldots, y_n \text{ at time } t_n. \tag{6.3}$$

The joint probability densities are positive:

$$P_n \geqslant 0; \tag{6.4}$$

they can be reduced:

$$\int P_n(y_1, t_1; y_2, t_2; \ldots; y_n, t_n)\, dy_n = P_{n-1}(y_1, t_1; y_2, t_2; \ldots; y_{n-1}, t_{n-1}); \tag{6.5}$$

and they are normalized:

$$\int P_1(y_1, t_1)\, dy_1 = 1. \tag{6.6}$$

In Eqs. (6.5) and (6.6), we have assumed that Y is a continuous stochastic variable. However, if Y is discrete we simply replace the integrations by summations. We can introduce time-dependent moments of the stochastic variables, $\langle y_1(t_1) y_2(t_2) \times \cdots \times y_n(t_n)\rangle$. They are defined

$$\langle y_1(t_1), y_2(t_2), \ldots, y_n(t_n)\rangle = \int \cdots \int y_1 y_2 \cdots y_n P_n(y_1, t_1; \ldots; y_n, t_n)\, dy_1 \cdots dy_n \tag{6.7}$$

and give the correlation between values of the stochastic variable at different times.

A process is called *stationary* if

$$P_n(y_1, t_1; y_2, t_2; \ldots; y_n, t_n) = P_n(y_1, t_1 + \tau; y_2, t_2 + \tau; \ldots; y_n, t_n + \tau) \tag{6.8}$$

for all n and τ. Thus, for a stationary process

$$P_1(y_1, t_1) = P_1(y_1), \tag{6.9}$$

and $\langle y_1(t_1) y_2(t_2)\rangle$ depends only on $|t_1 - t_2|$—the absolute value of the difference in times. All physical processes in equilibrium are stationary.

We shall also introduce a conditional probability:

$$P_{1|1}(y_1, t_1 \mid y_2, t_2) = \text{(the conditional probability density for the stochastic variable } Y \text{ to have value } y_2 \text{ at time } t_2 \text{ given that it had value } y_1 \text{ at time } t_1). \tag{6.10}$$

It is defined by the identity

$$P_1(y_1, t_1)P_{1|1}(y_1, t_1 | y_2, t_2) = P_2(y_1, t_1; y_2, t_2). \tag{6.11}$$

Combining Eqs. (6.5) and (6.11) we obtain the following relation between the probability densities at different times:

$$P_1(y_2, t_2) = \int P_1(y_1, t_1)P_{1|1}(y_1, t_1 | y_2, t_2)\, dy_1 \tag{6.12}$$

where the conditional probability $P_{1|1}(y_1, t_1 | y_2, t_2)$ has the property

$$\int P_{1|1}(y_1, t_1 | y_2, t_2)\, dy_2 = 1, \tag{6.13}$$

as can be demonstrated easily.

We can also introduce a joint conditional probability density as follows:

$$P_{k|l}(y_1, t_1; \ldots; y_k, t_k | y_{k+1}, t_{k+1}; \ldots; y_{k+l}, t_{k+l})$$
$$= \text{(the joint conditional probability density that the stochastic variable } Y \text{ has values } (y_{k+1}, t_{k+1}; \ldots; y_{k+l}, t_{k+l}) \text{ given that } (y_1, t_1; \ldots; y_k, t_k) \text{ are fixed).} \tag{6.14}$$

The joint conditional probability density is defined

$$P_{k|l}(y_1, t_1; \ldots; y_k, t_k | y_{k+1}, t_{k+1}; \ldots; y_{k+l}, t_{k+l})$$
$$= \frac{P_{k+l}(y_1, t_1; \ldots; y_k, t_k; y_{k+1}, t_{k+1}; \ldots; y_{k+l}, t_{k+l})}{P_k(y_1, t_1; \ldots; y_k, t_k)}. \tag{6.15}$$

The joint probability densities are important when there are correlations between values of the stochastic variable at different times—that is, if the stochastic variable has some memory of its past. However, if the stochastic variable has no memory of its past, then the expressions for the joint probability densities and the joint conditional probability densities simplify considerably.

If the stochastic variable has memory only of its immediate past, the joint conditional probability density $P_{n-1|1}(y_1, t_1; \ldots; y_{n-1}, t_{n-1} | y_n, t_n)$, where $t_1 < t_2 < \cdots < t_n$, must have the form

$$P_{n-1|1}(y_1, t_1; \ldots; y_{n-1}, t_{n-1} | y_n, t_n) = P_{1|1}(y_{n-1}, t_{n-1} | y_n, t_n). \tag{6.16}$$

That is, the conditional probability density for y_n at t_n is fully determined by the value of y_{n-1} at t_{n-1} and is not affected by any knowledge of the stochastic variable Y at earlier times. The conditional probability density $P_{1|1}(y_1, t_1 | y_2, t_2)$

is called the *transition probability*. A process for which Eq. (6.16) is satisfied is called a Markov process. A Markov process is fully determined by the two functions $P_1(y, t)$ and $P_{1|1}(y_1, t_1|y_2, t_2)$. The whole hierarchy of probability densities can be constructed from them. For example,

$$\begin{aligned} P_3(y_1, t_1; y_2, t_2; y_3, t_3) &= P_2(y_1, t_1; y_2, t_2)P_{2|1}(y_1, t_1; y_2, t_2|y_3, t_3) \\ &= P_1(y_1, t_1)P_{1|1}(y_1, t_1|y_2, t_2)P_{1|1}(y_2, t_2|y_3, t_3). \end{aligned} \tag{6.17}$$

If we integrate Eq. (6.17) over y_2 assuming $t_1 < t_2 < t_3$, we obtain

$$P(y_1, t_1; y_3, t_3) = P_1(y_1, t_1) \int P_{1|1}(y_1, t_1|y_2, t_2)P_{1|1}(y_2, t_2|y_3, t_3)\, dy_2. \tag{6.18}$$

If we now divide Eq. (6.18) by $P_1(y_1, t_1)$, we obtain

$$P_{1|1}(y_1, t_1|y_3, t_3) = \int P_{1|1}(y_1, t_1|y_2, t_2)P_{1|1}(y_2, t_2|y_3, t_3)\, dy_2. \tag{6.19}$$

Eq. (6.19) is called the Chapman-Kolmogorov equation. Notice that we have broken the probability of a transition from y_1, t_1 to y_3, t_3 into a process involving two successive steps, first from y_1, t_1 to y_2, t_2 and then from y_2, t_2 to y_3, t_3. The Markov character is exhibited by the fact that the probability of the two successive steps is the product of the probability of the individual steps. The successive steps are statistically *independent*. The probability of the transition $y_2, t_2 \rightarrow y_3, t_3$ is not affected by the fact that it was preceded by a transition $y_1, t_1 \rightarrow y_2, t_2$.

If we integrate Eq. (6.18) over y_1 and use Eq. (6.5), we find

$$P_1(y_2, t_2) = \int P_1(y_1, t_1)P_{1|1}(y_1, t_1|y_2, t_2)\, dy_1 \tag{6.20}$$

(where we have relabeled indices), as we expect.

From Eq. (6.20) we can find the differential equation for the probability density $P_1(y_1, t_1)$. Let us rewrite Eq. (6.20) in the form

$$P_1(y_2, t_1 + \tau) = \int P_1(y_1, t_1)P_{1|1}(y_1, t_1|y_2, t_1 + \tau)\, dy_1. \tag{6.21}$$

The time derivative of $P_1(y, t)$ is given by

$$\frac{dP_1(y, t)}{dt} = \lim_{\tau \to 0} \frac{P_1(y, t + \tau) - P_1(y, t)}{\tau}, \tag{6.22}$$

and, therefore, to take the time derivative of Eq. (6.21), we must evaluate the quantity

$$\lim_{\tau \to 0} \frac{P_{1|1}(y_1, t|y_2, t + \tau) - P_{1|1}(y_1, t|y_2, t)}{\tau}.$$

Let us expand the conditional probability density $P_{1|1}(y_1, t_1|y_2, t_1 + \tau)$ in a

Taylor series for small τ, in such a way that to each order in τ the normalization of $P_{1|1}(y_1, t_1|y_2, t_1 + \tau)$ (cf. Eq. 6.13) is preserved. If we note that $P_{1|1}(y_1, t_1|y_2, t_1) = \delta(y_1 - y_2)$ (cf. Eq. 6.12), we obtain

$$P_{1|1}(y_1, t_1|y_2, t_1 + \tau) = \delta(y_1 - y_2) - \tau \int dy W_{t_1}(y_1, y)\, \delta(y_1 - y_2) + \tau W_{t_1}(y_1, y_2). \quad (6.23)$$

The quantity $W_{t_1}(y_1, y_2)$ is the conditional (or transition) probability density per unit time that the system changes from state y_1 to y_2 in the time interval $t_1 \rightarrow t_1 + \tau$. In other words, it is a transition rate. The quantity $\tau \int dy W_{t_1}(y_1, y)$ is the total probability for a transition out of state y_1 (into any other state) in time $t_1 \rightarrow t_1 + \tau$. Thus $(1 - \tau \int dy W_{t_1}(y_1, y))$ is the total probability that no transition occurs in time τ, $(1 - \tau \int dy W_{t_1}(y_1, y))\, \delta(y_1 - y_2)$ is the probability density that no transition occurs in time τ, and $\tau W_{t_1}(y_1, y_2)$ is the probability density for a transition from state y_1 to state y_2 in time $t_1 \rightarrow t_1 + \tau$. Eq. (6.23) has the correct normalization.

If we now combine Eqs. (6.21), (6.22), and (6.23) we obtain

$$\frac{\partial P_1(y_2, t)}{\partial t} = \int dy_1 \{W(y_1, y_2)P_1(y_1, t) - W(y_2, y_1)P_1(y_2, t)\}. \quad (6.24)$$

Eq. (6.24) is called the *master equation*. It gives the rate of change of the probability density $P_1(y_2, t)$ due to transitions into the state y_2 from all other states y_1 (first term on the right) and transitions out of state y_2 into all other states y_1 (second term on right). There are many applications of the master equation, and we shall consider some of them in subsequent sections.

For the case when y is a continuous variable and changes in y take place in small jumps, we can derive a partial differential equation for $P_1(y, t)$ called the Fokker-Planck equation. Let us assume that the changes in the coordinate y can only take place in small jumps. Then $W(y', y)$, the transition probability per unit time, decreases rapidly with increasing $|y - y'|$. We can write $W(y', y) \equiv W(y'; y - y') \equiv W(y'; \xi)$, where $\xi = y - y'$ is the size of the jump. The master equation then becomes

$$\frac{\partial P_1(y, t)}{\partial t} = \int d\xi W(y - \xi; \xi)P_1(y - \xi, t) - P_1(y, t) \int d\xi W(y; -\xi). \quad (6.25)$$

We next expand the product $W(y - \xi, \xi)P_1(y - \xi, t)$ in powers of ξ. Eq. (6.25) then becomes

$$\frac{\partial P_1(y, t)}{\partial t} = \int W(y; \xi)P_1(y, t)\, d\xi - \int d\xi \xi \frac{\partial}{\partial y} W(y; \xi)P_1(y, t) + \frac{1}{2}\int d\xi \xi^2 \frac{\partial^2}{\partial y^2} W(y; \xi)P_1(y; t) + \cdots - P_1(y, t) \int d\xi W(y; -\xi). \quad (6.26)$$

The first and last terms on the right side of Eq. (6.26) cancel and we obtain

$$\frac{\partial P_1(y, t)}{\partial t} = -\frac{\partial}{\partial y}[\alpha_1(y)P_1(y, t)] + \frac{1}{2}\frac{\partial^2}{\partial y^2}[\alpha_2(y)P_1(y, t)], \tag{6.27}$$

where $\alpha_n(y)$ is the nth-order jump moment,

$$\alpha_n(y) = \int d\xi\, \xi^n W(y; \xi). \tag{6.28}$$

Eq. (6.27) is called the Fokker-Planck equation. It is important to note that in many physical problems the jump size is determined by some discrete finite parameter characteristic of the system and the assumptions inherent in the derivation of the Fokker-Planck equation do not apply.[4] Then an expansion procedure similar to that given in Sec. 6.G can be used to find an approximate partial differential equation to describe the system.

C. MARKOV CHAINS[5,6]

One of the simplest examples of a Markov process is that of a Markov chain. This involves transitions between values of a discrete stochastic variable Y occurring at discrete times. Let us assume that Y can take on values $Y = (y_1, y_2, \ldots, y_l)$, and that time is measured in discrete steps $t = s\tau$, where s is an integer and τ is a fundamental time interval. Then, for one step from $t = 0$ to $t = \tau$, Eq. (6.20) can be written

$$P_1(y_i, 1) = \sum_{j=1}^{l} P_1(y_j, 0)P_{1|1}(y_j, 0 | y_i, 1) \tag{6.29}$$

where $P_{1|1}(y_j, 0 | y_i, 1)$ gives the probability for transition from value y_j to y_i in one step. It contains all possible information about the basic transition mechanism in the system. Let us introduce the notation

$$Q_{ji} \equiv P_{1|1}(y_j, 0 | y_i, 1) \tag{6.30}$$

where $\sum_i Q_{ji} = 1$ (cf. Eq. (6.13)). Then we can write Eq. (6.29) in the form of a matrix equation

$$\mathbf{P}(1) = \mathbf{P}(0)\overline{\overline{\mathbf{Q}}} \tag{6.31}$$

where $\mathbf{P}(s)$ is an l-dimension vector at time s whose ith component is $P_1(y_i, s)$; $\overline{\overline{\mathbf{Q}}}$ is the $l \times l$ dimensional transition matrix whose (ji)th component is Q_{ji}. In general, $\overline{\overline{\mathbf{Q}}}$ is not a symmetric matrix. After s transitions, the system will be in the state

$$\mathbf{P}(s) = \mathbf{P}(0)(\overline{\overline{\mathbf{Q}}})^s. \tag{6.32}$$

One often wishes to know the state of the system after a large number of transitions ($s \to \infty$). Does the system retain some memory of its initial state $\mathbf{P}(0)$, or after many transitions does the system proceed to some unique final state independent of the initial state?

The behavior of the probability vector, $\mathbf{P}(s)$, for large s depends on the structure of the transition matrix $\bar{\bar{\mathbf{Q}}}$. We will discuss, briefly, a few of the more interesting cases.

1. Regular Transition Matrix

A stochastic matrix, $\bar{\bar{\mathbf{Q}}}$, is called *regular* if all elements of some power $\bar{\bar{\mathbf{Q}}}^k$ are positive. If $\bar{\bar{\mathbf{Q}}}$ is regular then $\mathbf{P}(s)$ tends to a unique fixed probability vector $\boldsymbol{\pi}$ (stationary state) such that the components of $\boldsymbol{\pi}$ are all positive. The sequence of powers of $\bar{\bar{\mathbf{Q}}}$ ($\bar{\bar{\mathbf{Q}}}$, $\bar{\bar{\mathbf{Q}}}^2$, $\bar{\bar{\mathbf{Q}}}^3$, etc.) tends to a unique stationary transition matrix, $\bar{\bar{\mathbf{M}}}$, whose rows are each the stationary vector $\boldsymbol{\pi}$:

$$\mathbf{P}(s) \to \boldsymbol{\pi} = (\pi_1, \pi_2, \ldots, \pi_l) \tag{6.33}$$

and

$$(\bar{\bar{\mathbf{Q}}})^s \to \bar{\bar{\mathbf{M}}} = \begin{pmatrix} \pi_1, \pi_2, \ldots, \pi_l \\ \pi_1, \pi_2, \ldots, \pi_l \\ \vdots \\ \pi_1, \pi_2, \ldots, \pi_l \end{pmatrix}. \tag{6.34}$$

To show that powers of a regular stochastic matrix, $\bar{\bar{\mathbf{Q}}}$, approach a steady-state matrix, $\bar{\bar{\mathbf{M}}}$, one must show that for each column of $(\bar{\bar{\mathbf{Q}}})^s$ the difference between the maximum and minimum elements in the column tends to zero as $s \to \infty$. (We will not give the proof here. See, for example, Ref. 6.5.)

Once one shows that $(\bar{\bar{\mathbf{Q}}})^s \underset{s\to\infty}{\to} \bar{\bar{\mathbf{M}}}$, it is easy to show that $\bar{\bar{\mathbf{M}}} = \bar{\bar{\mathbf{Q}}}\bar{\bar{\mathbf{M}}}$ and $\boldsymbol{\pi} = \mathbf{P}(0)\bar{\bar{\mathbf{M}}}$. For example, consider the case $n = 2$. Then

$$\bar{\bar{\mathbf{Q}}}\bar{\bar{\mathbf{M}}} = \begin{pmatrix} Q_{11} & Q_{12} \\ Q_{21} & Q_{22} \end{pmatrix}\begin{pmatrix} \pi_1 & \pi_2 \\ \pi_1 & \pi_2 \end{pmatrix} = \begin{pmatrix} \pi_1 \sum_{j=1}^{2} Q_{1j} & \pi_2 \sum_{j=1}^{2} Q_{1j} \\ \pi_1 \sum_{j=1}^{2} Q_{2j} & \pi_2 \sum_{j=1}^{2} Q_{2j} \end{pmatrix} = \bar{\bar{\mathbf{M}}}$$

and

$$\mathbf{P}(0)\bar{\bar{\mathbf{M}}} = (P_1(y_1, 0), P_1(y_2, 0))\begin{pmatrix} \pi_1 & \pi_2 \\ \pi_1 & \pi_2 \end{pmatrix} = \boldsymbol{\pi}$$

because rows of stochastic vectors and matrices must add to one.

For simple cases one can easily compute the steady-state vector $\boldsymbol{\pi}$. An example will illustrate this. Consider two pots, A and B, with three red balls and two white balls distributed between them so that A always has two balls and B always has three balls. There are three different configurations for the pots as shown in Fig. 6.1. We obtain transitions between these three configurations by

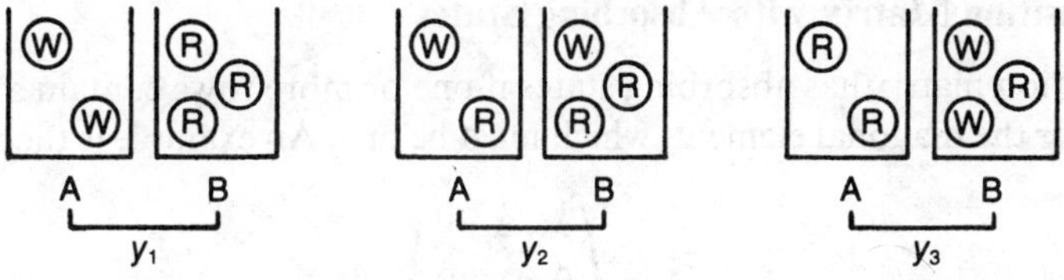

Fig. 6.1. Three possible configurations for two white balls and three red balls in two pots, if one pot must have two balls and the other must have three balls.

picking a ball out of A at random and one out of B at random and interchanging them. Inspection shows that we can make the following transitions from y_i to y_j with transition probability Q_{ij} such that $Q_{11} = 0$, $Q_{12} = 1$, $Q_{13} = 0$, $Q_{21} = \frac{1}{6}$, $Q_{22} = \frac{1}{2}$, $Q_{23} = \frac{1}{3}$, $Q_{31} = 0$, $Q_{32} = \frac{2}{3}$, and $Q_{33} = \frac{1}{3}$. The transition matrix is therefore given by

$$\bar{\bar{\mathbf{Q}}} = \begin{pmatrix} 0 & 1 & 0 \\ \frac{1}{6} & \frac{1}{2} & \frac{1}{3} \\ 0 & \frac{2}{3} & \frac{1}{3} \end{pmatrix}.$$

Since

$$\bar{\bar{\mathbf{Q}}}^2 = \begin{pmatrix} \frac{1}{6} & \frac{1}{2} & \frac{1}{3} \\ \frac{1}{12} & \frac{23}{36} & \frac{10}{36} \\ \frac{1}{9} & \frac{22}{36} & \frac{1}{3} \end{pmatrix}$$

the transition matrix is regular and we expect to find a unique stationary state $\boldsymbol{\pi}$, independent of our initial state.

Let $\boldsymbol{\pi} = (x, y, z)$ denote the stationary state. Then, since $\boldsymbol{\pi}$ must satisfy the equation $\boldsymbol{\pi} = \boldsymbol{\pi}\bar{\bar{\mathbf{Q}}}$, we obtain the following equations for x, y, and z: $(\frac{1}{6}y = x)$, $(x + \frac{1}{2}y + \frac{2}{3}z = y)$, and $(\frac{1}{3}(y + z) = z)$. Since $x + y + z = 1$, we obtain the following solutions: $x = \frac{1}{10}$, $y = \frac{6}{10}$, and $z = \frac{3}{10}$. Hence, for an arbitrary initial state, $\mathbf{P}(0)$, the state of the system tends to $\boldsymbol{\pi} = (\frac{1}{10}, \frac{6}{10}, \frac{3}{10})$ after many transitions, and we have lost all knowledge of the initial state.

We can say this in another way. Let us assume that, initially, the system is in configuration y_1. We are certain of this. The initial state (of our knowledge) is $\mathbf{P}(0) = (1, 0, 0)$. Now perform s transitions (s large). What configuration is the system in? We do not know! The state is now $\mathbf{P}(s) = (\frac{1}{10}, \frac{6}{10}, \frac{3}{10})$ and will remain that way. There is a 10% chance the system is in configuration y_1, a 60% to be in y_2, and a 30% chance to be in y_3. If we could perform the experiment many times, that is, start in y_1 and make s transitions and then look at the configuration, we would find that 10% of the time it would end up in configuration y_1, 60% of the time it would end up in configuration y_2, and 30% of the time it would end up in configuration y_3. But for each experiment, we could never be certain of the outcome.

2. Transition Matrix with Absorbing States

A transition matrix has absorbing states if one or more rows contain all zeros except for the diagonal element, which must be one. An example is the matrix

$$\bar{\bar{\mathbf{Q}}} = \begin{pmatrix} \frac{1}{2} & \frac{1}{2} & 0 \\ 0 & 1 & 0 \\ \frac{1}{3} & \frac{1}{3} & \frac{1}{3} \end{pmatrix}.$$

In this case, the value y_2 of the stochastic variable, Y, is absorbing. Once the stochastic variable reaches the value y_2 it can never change because all powers of $\bar{\bar{\mathbf{Q}}}$ will also have (0, 1, 0) in the second row.

After s steps the probability vector is

$$\mathbf{P}(s) = (P_1(y_1, s), P_1(y_2, s), P_1(y_3, s)) = \mathbf{P}(0)(\bar{\bar{\mathbf{Q}}})^s.$$

The entry $P_1(y_2, s)$ is the probability that the system reaches the value y_2 on or before s transitions. A transition matrix with an absorbing state cannot be regular.

The example given in Sec. 6.C.1 (two pots and five balls) would become an example of a system with an absorbing state if the red balls were made of plutonium and three balls together exceeded the critical mass. Then when configuration y_1 was reached, the whole experiment would end with a bang; $P_1(y_1, s)$ would be the probability that the bang occurred on or before the sth transition. For that case,

$$\bar{\bar{\mathbf{Q}}} = \begin{pmatrix} 1 & 0 & 0 \\ \frac{1}{6} & \frac{1}{2} & \frac{1}{3} \\ 0 & \frac{2}{3} & \frac{1}{3} \end{pmatrix}.$$

It is of interest to mention the possibility of "ergodicity" in Markov chains: A Markov chain is "ergodic" if it is possible to move from any state to every other state during the transitions. In general a Markov chain will not be ergodic if the transition matrix is block diagonal. Consider the matrix

$$\bar{\bar{\mathbf{Q}}} = \begin{pmatrix} Q_{11} & Q_{12} & 0 & 0 \\ Q_{21} & Q_{22} & 0 & 0 \\ 0 & 0 & Q_{33} & Q_{34} \\ 0 & 0 & Q_{43} & Q_{44} \end{pmatrix}.$$

In this case, one can never make transitions from values y_1 or y_2 to values y_3 or y_4. A transition matrix with absorbing states is a special case of a block diagonal matrix, so it cannot be ergodic. A regular transition matrix is always ergodic, because regardless of the initial state one always approaches a final state in which there is a finite probability of having any of the values of the random variable Y.

D. RANDOM WALK AND THE DIFFUSION EQUATION[2]

Let us consider again the problem of random walk on a line (cf. Sec. 5.F). We will assume that a particle moves along the x-axis with step size l and that the time between steps is τ. Then Eq. (6.20) can be written

$$P(n_2l, s\tau) = \sum_{n_1} P(n_1l, (s-1)\tau)P_{1|1}(n_1l, (s-1)\tau|n_2l, s\tau) \tag{6.35}$$

where $n = 0, \pm 1, \pm 2, \ldots$ indicates the absolute position of the particle. The quantity $P_{1|1}(n_1l, (s-1)\tau|n_2l, s\tau)$ is the conditional probability that in one step it will go from n_1 to n_2. If there is an equal probability of taking a step to the left or right,

$$P_{1|1}(n_1l, (s-1)\tau|n_2l, s\tau) = \tfrac{1}{2}\delta_{n_2,n_1+1} + \tfrac{1}{2}\delta_{n_2,n_1-1}, \tag{6.36}$$

and Eq. (6.35) becomes

$$P_1(nl, s\tau) = \tfrac{1}{2}P_1((n-1)l, (s-1)\tau) + \tfrac{1}{2}P_1((n+1)l, (s-1)\tau). \tag{6.37}$$

We can obtain a differential equation for the probability distribution if we rewrite Eq. (6.37) in the form

$$\frac{P_1(nl, s\tau) - P_1(nl, (s-1)\tau)}{\tau} = \frac{l^2}{2\tau}\left[\frac{P_1((n+1)l, (s-1)\tau) + P_1((n-1)l, (s-1)\tau) - 2P_1(nl, (s-1)\tau)}{l^2}\right]. \tag{6.38}$$

If we now let $x = nl$ and $t = s\tau$ and take the limit $l \to 0$, $\tau \to 0$, in such a way that $D \equiv l^2/2\tau$ is finite, we obtain the diffusion equation

$$\frac{\partial P_1(x, t)}{\partial t} = D\frac{\partial^2 P_1(x, t)}{\partial x^2}. \tag{6.39}$$

Eq. (6.39) is a Fokker-Planck equation.

Let us now solve Eq. (6.39) assuming that, initially, $P_1(x, 0) = \delta(x)$ where $\delta(x)$ is a Dirac delta function. We first introduce the Fourier transform of $P_1(x, t)$:

$$\tilde{P}_1(k, t) = \int_{-\infty}^{\infty} dx P_1(x, t)\, e^{ikx}. \tag{6.40}$$

Then Eq. (6.39) takes the form

$$\frac{\partial \tilde{P}_1(k, t)}{\partial t} = -Dk^2\tilde{P}_1(k, t). \tag{6.41}$$

We can solve Eq. (6.41) to obtain

$$\tilde{P}_1(k, t) = A\, e^{-Dk^2t}. \tag{6.42}$$

The integration constant, A, can be determined from the initial condition. Since $\tilde{P}_1(k, 0) = 1$, we have $A = 1$. We can now take the inverse transform to obtain

$$P_1(x, t) = \frac{1}{2\pi}\int_{-\infty}^{\infty} dk\, e^{-ikx}\, e^{-Dk^2t} = \frac{1}{\sqrt{4\pi Dt}}\, e^{-x^2/4Dt}. \tag{6.43}$$

Eq. (6.43) gives the probability of finding the particle at point x at time t if it started at $x = 0$ at time $t = 0$. It is the same equation that we obtained in Sec. 5.F.

It is interesting to obtain the equation of motion of the moments $\langle x(t)\rangle$ and $\langle x^2(t)\rangle$. The moment $\langle x(t)\rangle = \int_{-\infty}^{\infty} dx x P_1(x, t)$ gives the average position of the particle as a function of time. If we multiply Eq. (6.39) by x and integrate over x, we find

$$\frac{d\langle x(t)\rangle}{dt} = 0. \tag{6.44}$$

Thus, the average position of the particle does not change with time. Let us now consider the moment $\langle x^2(t)\rangle$. If we multiply Eq. (6.39) by x^2 and integrate over position, we find

$$\frac{d\langle x^2(t)\rangle}{dt} = 2D \tag{6.45}$$

or

$$\langle x^2(t)\rangle = 2Dt. \tag{6.46}$$

Eq. (6.46) is characteristic of a diffusion process. The moment $\langle x^2(t)\rangle$ is a measure of the width of $P_1(x, t)$ at time t. We see that $P_1(x, t)$ spreads with time, as we expect.

E. DISCRETE STATIONARY MARKOV PROCESSES: GENERAL SOLUTION[2]

For stationary processes the probability densities must obey the condition in Eq. (6.8), and therefore the Chapman-Kolmogorov equation reduces to the form

$$P_{1|1}(y_1|y_3; t) = \int dy P_{1|1}(y_1|y; t_0) P_{1|1}(y|y_3; t - t_0) \tag{6.47}$$

where $t = t_3 - t_1$ and $t_0 = t_2 - t_1$.

If we restrict ourselves to discrete stochastic variables $y = n\Delta$ and discrete times $t = s\tau$, where n and s are integers, we can write

$$P_{1|1}(m|n; s) > 0, \tag{6.48}$$

$$P_1(n) = \sum_m P_1(m) P_{1|1}(m|n; s), \tag{6.49}$$

$$\sum_n P_{1|1}(m|n; s) = 1, \tag{6.50}$$

and

$$P_{1|1}(m|n; s) = \sum_k P_{1|1}(m|k; s - 1)P_{1|1}(k|n; 1). \tag{6.51}$$

In Eq. (6.51), the quantity $P_{1|1}(k|n; 1)$ is the conditional probability that, if the system is in state k, it will jump to state n in the next step. It therefore contains all necessary information about the basic transition mechanism in the system. The elements $P_{1|1}(k|n; 1)$ form a matrix which we shall denote

$$P_{1|1}(k|n; 1) \equiv Q(k, n). \tag{6.52}$$

The solution of Eq. (6.51) is given formally by

$$P_{1|1}(m|n; s) = (\bar{\bar{\mathbf{Q}}}^s)_{mn} \tag{6.53}$$

where the right-hand side denotes the mnth element of the matrix $\bar{\bar{\mathbf{Q}}}$ raised to the sth power. We shall now obtain an expression for $P_{1|1}(m|n; s)$ in terms of the eigenvalues of the matrix $\bar{\bar{\mathbf{Q}}}$.

The matrix $\bar{\bar{\mathbf{Q}}}$ in general is not a symmetric matrix. Therefore, the right and left eigenvectors of $\bar{\bar{\mathbf{Q}}}$ will be different. The left eigenvector problem can be written

$$\lambda_i x_{im} = \sum_{n=1}^{l} x_{in} Q_{nm} \tag{6.54}$$

and the right eigenvector problem can be written

$$\lambda_j y_{mj} = \sum_{n=1}^{l} Q_{mn} y_{nj}. \tag{6.55}$$

The eigenvalues of $\bar{\bar{\mathbf{Q}}}$ are given by values of λ which satisfy the condition that the determinant of the matrix $\bar{\bar{\mathbf{Q}}} - \lambda\bar{\bar{\mathbf{I}}}$ be zero:

$$\det|\bar{\bar{\mathbf{Q}}} - \bar{\bar{\mathbf{I}}}\lambda| = 0 \tag{6.56}$$

($\bar{\bar{\mathbf{I}}}$ is the unit matrix). If $\bar{\bar{\mathbf{Q}}}$ is an $l \times l$ matrix, it will have l eigenvalues which may or may not be real. Corresponding to each eigenvalue λ_j there will be a left and a right eigenvector, $\mathbf{x}_j$ and $\mathbf{y}_j$, respectively.

We can show that the left and right eigenvectors given in Eqs. (6.54) and (6.55) are orthogonal. Let us first assume that the eigenvectors are normalized to one:

$$\sum_n x_{in} y_{ni} = \sum_i y_{ni} x_{in} = 1. \tag{6.57}$$

To show that they are orthogonal, let us multiply Eq. (6.54) on the right by y_{mj}, multiply Eq. (6.55) on the left by x_{im}, and subtract (6.55) from (6.54). We find

$$(\lambda_i - \lambda_j) \sum_m x_{im} y_{mj} = \sum_{m,n} [x_{in} Q_{nm} y_{mj} - x_{im} Q_{mn} y_{nj}] \equiv 0; \tag{6.58}$$

thus

$$\sum_m x_{im} y_{mj} = 0 \tag{6.59}$$

for $i \neq j$. Let us next expand an arbitrary vector in our space in terms of the complete set of left eigenvectors $\mathbf{x}_i$:

$$\mathbf{A} = \sum_i a_i \mathbf{x}_i \tag{6.60}$$

where $\mathbf{x}_i$ is the left eigenvector corresponding to eigenvalue λ_i. In terms of components, Eq. (6.60) reads

$$A_n = \sum_i a_i x_{in}. \tag{6.61}$$

Let us now multiply on the right by y_{nj} and sum over n:

$$\sum_n A_n y_{nj} = \sum_i \sum_n a_i x_{in} y_{nj} = a_j. \tag{6.62}$$

If we substitute Eq. (6.62) back into Eq. (6.61) and use the fact that **A** is an arbitrary vector, we find that

$$\sum_i y_{ni} x_{in'} = 0 \qquad \text{for} \quad n \neq n'. \tag{6.63}$$

Eqs. (6.57), (6.59), and (6.63) can be combined in the forms

$$\sum_m x_{im} y_{mj} = \delta_{ij} \tag{6.64}$$

and

$$\sum_i y_{mi} x_{in} = \delta_{nm} \tag{6.65}$$

which shows the orthonormality of the eigenvectors.

We next obtain some properties of the eigenvalues λ_i. First we prove that $|\lambda_i| \leqslant 1$. Let us take the absolute value of Eq. (6.55) and remember that all elements of the matrix $\overline{\overline{\mathbf{Q}}}$ are positive. If we use the triangle inequality ($|x + y| \leqslant |x| + |y|$) and the fact that $|\lambda_i y_{mi}| = |\lambda_i|\,|y_{mi}|$, we find

$$|\lambda_i|\,|y_{mi}| \leqslant \sum_n Q_{mn}|y_{ni}|. \tag{6.66}$$

Let us next assume that $|y_{ni}| \leqslant C$ (for all n), where C is some positive number. Then

$$|\lambda_i| C \leqslant C \sum_n Q_{mn} = C \tag{6.67}$$

(cf. Eqs. (6.50) and (6.52)). Thus,

$$|\lambda_i| \leqslant 1 \tag{6.68}$$

and all eigenvalues have a magnitude less than or equal to one.

Now that we have established Eq. (6.68), we can also prove that $\lambda = 1$ is always an eigenvalue. Let us first note that the right eigenvector, $y_{in} = 1$ (for all n), is an eigenvector with eigenvalue $\lambda_i = 1$. A left eigenvector with eigenvalue, $\lambda_i = 1$, must satisfy the equation

$$x_{im} = \sum_{n=1} x_{in} Q_{nm}. \tag{6.69}$$

From Eqs. (6.49) and (6.52) we see that $x_{im} = P_1(m)$ always satisfies Eq. (6.69). Thus, the right eigenvector, $y_{in} = 1$ (for all n), and the left eigenvector, $x_{im} = P_1(m)$, have eigenvalue $\lambda = 1$, and $\lambda = 1$ is always an eigenvalue of $\bar{\bar{\mathbf{Q}}}$. Furthermore, the choice $x_{im} = P_1(m)$ has the correct normalization.

We can expand the matrix $\bar{\bar{\mathbf{Q}}}$ in terms of its left and right eigenvectors. If we multiply Eq. (6.54) by y_{ni}, sum over i, and use Eq. (6.65), we obtain

$$Q_{nm} = \sum_i \lambda_i y_{ni} x_{im}. \tag{6.70}$$

If we now combine Eqs. (6.70) and (6.53) and use the orthonormality of the left and right eigenvectors, we can write

$$P_{1|1}(m|n; s) = \sum_i \lambda_i^s y_{mi} x_{in}. \tag{6.71}$$

Thus we obtain the general solution for the discrete stationary Markov process in terms of the eigenvalues and left and right eigenvectors of the transition matrix $\bar{\bar{\mathbf{Q}}}$.

For the special case of a "regular" transition matrix $\bar{\bar{\mathbf{Q}}}$, there will be only one eigenvalue of $\bar{\bar{\mathbf{Q}}}$ with value $\lambda = 1$. Let us denote it $\lambda_1 = 1$. Then Eq. (6.71) can be written

$$\begin{aligned} P_{1|1}(m|n; s) &= y_{m1} x_{1n} + \sum_{j \neq 1} \lambda_j^s y_{mj} x_{jn} \\ &= P_1(n) + \sum_{j \neq 1} \lambda_j^s y_{mj} x_{jn}. \end{aligned} \tag{6.72}$$

Since $\lambda_j^s < 1$ for $j \neq 1$, we can take the limit $s \to \infty$ to obtain

$$\lim_{s \to \infty} P_{1|1}(m|n; s) = P_1(n). \tag{6.73}$$

Thus, for regular transition matrices, the conditional probability tends to a unique value after long times.

F. BIRTH AND DEATH PROCESSES[1,7]

Birth and death processes are processes in which transitions can only take place one step at a time. Linear and nonlinear birth and death processes occur very commonly in chemistry and population dynamics and the master equation describing them can sometimes be solved exactly using a generating function. We shall treat an exactly soluble case in this section.

Let us consider a population of m bacteria at time t such that:

(i) The probability of a bacterium dying in time $t \to t + \Delta t$ is $\mu_m \Delta t$.
(ii) The probability of a bacterium being created in time $t \to t + \Delta t$ is $\lambda_m \Delta t$.
(iii) The probability of no change in the number of bacteria in time $t \to t + \Delta t$ is $(1 - (\lambda_m + \mu_m)\Delta t)$.
(iv) The probability of more than one birth or death in time $t \to t + \Delta t$ is zero.

Then we can write

$$P_{1|1}(m, t|n, t + \Delta t) = (1 - (\lambda_m + \mu_m)\Delta t)\,\delta_{n,m} + (\lambda_m\,\delta_{n,m+1} + \mu_m\,\delta_{n,m-1})\Delta t + \cdots \tag{6.74}$$

The conditional probability in Eq. (6.74) has the correct normalization. Let us now assume that the probability of a birth or death is proportional to the number of bacteria present so that $\lambda_m = m\lambda$ and $\mu_m = m\mu$. If we follow the procedure in Sec. 6.B, we obtain

$$\frac{\partial P_1}{\partial t}(n, t) = (n - 1)\lambda P_1(n - 1, t) + (n + 1)\mu P_1(n + 1, t) - (n\lambda + n\mu)P_1(n, t). \tag{6.75}$$

Eq. (6.75) is the master equation for the linear birth-death process. It is linear because the coefficients on the right-hand side are linear in n; $\lambda(\mu)$ is the probability per unit time that a single bacterium has a birth (death).

Master equations depending on discrete stochastic variables are most easily solved through use of a generating function. For the process considered here, we write the generating function as

$$F(z, t) = \sum_{n=-\infty}^{\infty} P(n, t)z^n. \tag{6.76}$$

Various moments of the stochastic variable n are obtained by taking derivatives of $F(z, t)$ with respect to z and allowing $z \to 1$. For example,

$$\langle n \rangle = \lim_{z\to 1} \frac{\partial}{\partial z} F(z, t) = \sum_{n=-\infty}^{\infty} nP(n, t) \tag{6.77}$$

and

$$\langle n^2 \rangle - \langle n \rangle = \lim_{z\to 1} \left(\frac{\partial}{\partial z}\right)^2 F(z, t) = \sum_{n=-\infty}^{\infty} (n^2 - n)P(n, t), \tag{6.78}$$

etc. Note that if $P(n, 0) = 0$ for $n < 0$ then $P(n, t) = 0$ for $n < 0$.

We can use Eq. (6.75) to obtain an equation for the generating function for a birth-death process. If we substitute Eq. (6.76) into the following differential equation,

$$\frac{\partial F}{\partial t} = (z - 1)(\lambda z - \mu)\frac{\partial F}{\partial z}, \tag{6.79}$$

and equate powers of z^n, we retrieve Eq. (6.75). Thus Eqs. (6.75) and (6.79) are equivalent. Eq. (6.79) is a first-order linear partial differential equation and may be solved using the method of characteristics.

To find a general solution to Eq. (6.79) we must first solve the two equations

$$\frac{dt}{1} = \frac{-dz}{(z-1)(\lambda z - \mu)} = \frac{dF}{0}. \tag{6.80}$$

If we integrate the equation on the left, we obtain

$$C_1 = \frac{(z-1)}{(\lambda z - \mu)} e^{-(\mu - \lambda)t} \tag{6.81}$$

where C_1 is an integration constant. If we integrate the equation on the right, we find

$$F(z, t) = C_2 \tag{6.82}$$

where C_2 is an integration constant.

Since $F(z, t)$ must remain constant, we can write

$$F\left(\frac{(\lambda z - \mu)}{(z-1)} e^{(\mu - \lambda)t}\right) = C_2, \tag{6.83}$$

since this is the most general form of F which satisfies Eqs. (6.79) and (6.82). The functional form of F can be obtained from Eq. (6.76). At time $t = 0$

$$F\left(\frac{(\lambda z - \mu)}{(z-1)}\right) = F(z, 0) = \sum_n P_1(n, 0)z^n. \tag{6.84}$$

Let us assume that, initially, the number of bacteria was m; that is, $P_1(n, 0) = \delta_{n,m}$. Then Eq. (6.84) becomes

$$F\left(\frac{(\lambda z - \mu)}{(z-1)}\right) = z^m. \tag{6.85}$$

Thus, if $u = (z-1)^{-1}(\lambda z - \mu)$, then

$$F(u) = \left(\frac{\mu - u}{\lambda - u}\right)^m, \tag{6.86}$$

and we find

$$F(z, t) = \left[\frac{(\mu z - \mu)\, e^{(\lambda - \mu)t} - \lambda z + \mu}{(\lambda z - \lambda)\, e^{(\lambda - \mu)t} - \lambda z + \mu}\right]^m. \tag{6.87}$$

Eq. (6.87) is exact and contains all possible information about the birth-death process. If we expand $F(z, t)$ in a power series in z, the coefficient of z^n is $P_1(n, t)$. If we take derivatives of $F(z, t)$, we obtain the moments of the population number. If $\lambda \leqslant \mu$ (probability of deaths greater than that of births) then $F(z, t) \underset{t\to\infty}{\to} 1$. Thus, $P_1(0, \infty) = 1$ and $P_1(n, \infty) = 0$ $(n \geqslant 1)$, and the population becomes extinct. If $\lambda > \mu$ then $F(z, t) \underset{t\to\infty}{\to} (\mu/\lambda)^m$.

The mean number of bacteria is given by Eq. (6.77). We find

$$\langle n \rangle = m\, e^{(\lambda - \mu)t}. \tag{6.88}$$

The variation in time of the average number of bacteria is exponential. The variance can easily be found from Eqs. (6.77) and (6.78).

G. EXPANSION OF THE MASTER EQUATION[8,9]

In previous sections we have applied the master equation to exactly soluble problems. However, most problems cannot be solved exactly and it is necessary to find an approximation to the master equation. Van Kampen has given a general procedure for "large systems" which we shall discuss in this section.

Let us consider the master equation for continuous times and discrete stochastic variables. For simplicity we shall assume that the transition rate is independent of time. The master equation can then be written in the form

$$\frac{\partial}{\partial t} P_1(n, t) = \sum_m \{P_1(m, t)W(m, n) - P_1(n, t)W(n, m)\}. \tag{6.89}$$

From Eq. (6.89) we can find the equation of motion for the average value of the stochastic variable n and its various moments. Before we do this, it is convenient to introduce the jump moments:

$$a_p(n) \equiv \sum_m (m - n)^p W(n, m) \tag{6.90}$$

where $a_p(n)$ is a function of the stochastic variable n. To find the equation of motion for the mean value, $\langle n(t) \rangle$, we multiply Eq. (6.89) by n and sum over n. If we relabel dummy variables, we obtain

$$\frac{\partial \langle n(t) \rangle}{\partial t} = \sum_{m,n} (m - n)W(n, m)P_1(n, t) = \langle a_1(n) \rangle. \tag{6.91}$$

We note that the right-hand side of Eq. (6.91) is not in general a linear function of $\langle n(t) \rangle$. It will depend on higher order moments of $n(t)$ and, therefore, we must also obtain an equation for higher order moments. If we multiply Eq. (6.89) by n^2 we obtain

$$\frac{\partial}{\partial t} \langle n^2(t) \rangle = \sum_{m,n} (m^2 - n^2)W(n, m)P_1(n, t) = \langle a_2(n) \rangle + 2\langle n a_1(n) \rangle, \tag{6.92}$$

etc. Thus, knowledge of the basic transition rate, $W(n, m)$, enables us to obtain a great deal of information about a system without actually solving the master equation.

The first step in obtaining an approximation to the master equation for "large systems" is to note that any system we consider will have some parameter, Ω, which characterizes its size. The parameter, Ω, might be particle

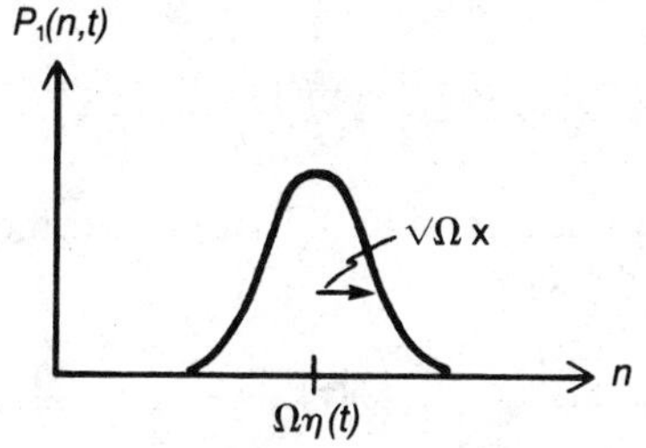

Fig. 6.2. Probability distribution peaked at $\Omega\eta(t)$: a given value of $n(t)$ can be written $n(t) = \Omega\eta(t) + \sqrt{\Omega}x(t)$.

number, volume, or any other suitable parameter. The approximation scheme involves expanding the master equation in powers of $1/\sqrt{\Omega}$.

The important parameters in the transition rate are the density, m/Ω, and the step size, Δn. Let us therefore expand the transition rate, $W(m, n)$, in the form

$$W(m, n) = f(\Omega)\left[\omega_0\left(\frac{m}{\Omega}, \Delta n\right) + \frac{1}{\Omega}\omega_1\left(\frac{m}{\Omega}, \Delta n\right) + \cdots\right] \tag{6.93}$$

where $\Delta n = n - m$ is the step size in the transition from $m \to n$, and $f(\Omega)$ is an arbitrary function of the size parameter Ω. For large Ω, we can neglect higher order terms in Eq. (6.93) and write the master equation as

$$\frac{\partial P_1(n, t)}{\partial t} = f(\Omega)\sum_{\Delta n}\left\{\omega_0\left(\frac{n - \Delta n}{\Omega}, \Delta n\right)P_1(n - \Delta n, t) - \omega_0\left(\frac{n}{\Omega}, -\Delta n\right)P_1(n, t)\right\}. \tag{6.94}$$

We now assume that $P_1(n, t)$ is sharply peaked at $n = \langle n(t)\rangle \equiv \Omega\eta(t)$ with a width proportional to $\sqrt{\Omega}$ (cf. Fig. 6.2). According to the central limit theorem, this is the type of behavior we expect for a large number of independent objects. The values of n can be expanded about their average value:

$$n(t) = \Omega\eta(t) + \sqrt{\Omega}x(t), \tag{6.95}$$

where $\sqrt{\Omega}x$ is the displacement of n from the average value $\Omega\eta(t)$. We can rewrite the master equation in terms of x rather than n. The probability of finding the system with value of the stochastic variable n in range $n \to n + \Delta n$ is

$$P_1(n, t)\,\Delta n \equiv \pi(x, t)\,\Delta x \tag{6.96}$$

where

$$\pi(x, t) \equiv \sqrt{\Omega}P_1(\Omega\eta(t) + \sqrt{\Omega}x, t). \tag{6.97}$$

We next note that

$$\frac{1}{\sqrt{\Omega}}\frac{\partial \pi}{\partial x} = \sqrt{\Omega}\frac{\partial P_1}{\partial n} \tag{6.98}$$

and

$$\frac{\partial \pi}{\partial t} = \sqrt{\Omega}\left\{\Omega \frac{d\eta}{dt}\frac{\partial P_1}{\partial n} + \frac{\partial P_1}{\partial t}\right\}. \tag{6.99}$$

Combining Eqs. (6.94), (6.95), and (6.97)–(6.99), we find the following expression for the master equation:

$$\frac{\partial \pi}{\partial t} - \sqrt{\Omega}\frac{d\eta}{dt}\frac{\partial \pi}{\partial x} = f(\Omega)\left\{\sum_{\Delta n}\left[\omega_0\left(\eta(t) + \frac{x}{\sqrt{\Omega}} - \frac{\Delta n}{\Omega}; \Delta n\right)\pi\left(x - \frac{\Delta n}{\sqrt{\Omega}}, t\right)\right.\right.$$
$$\left.\left. - \omega_0\left(\eta(t) + \frac{x}{\sqrt{\Omega}}; -\Delta n\right)\pi(x, t)\right]\right\}. \tag{6.100}$$

We now can expand the first term on the right-hand side in a Taylor series about $\Delta n/\sqrt{\Omega} = 0$:

$$\omega_0\left(\eta(t) + \frac{x}{\sqrt{\Omega}} - \frac{\Delta n}{\Omega}; \Delta n\right)\pi\left(x - \frac{\Delta n}{\sqrt{\Omega}}, t\right)$$
$$= \left(1 - \frac{\Delta n}{\sqrt{\Omega}}\frac{\partial}{\partial x} + \frac{1}{2}\left(\frac{\Delta n}{\sqrt{\Omega}}\right)^2\frac{\partial^2}{\partial x^2} + \cdots\right)\omega_0\left(\eta(t) + \frac{x}{\sqrt{\Omega}}; \Delta n\right)\pi(x, t), \tag{6.101}$$

and Eq. (6.100) takes the form

$$\frac{\partial \pi}{\partial t} - \sqrt{\Omega}\frac{d\eta}{dt}\frac{\partial \pi}{\partial x} = f(\Omega)\left[\frac{-\Delta n}{\sqrt{\Omega}}\frac{\partial}{\partial x} + \frac{1}{2}\left(\frac{\Delta n}{\sqrt{\Omega}}\right)^2\frac{\partial^2}{\partial x^2} + \cdots\right]$$
$$\times\ \omega_0\left(\eta(t) + \frac{x}{\sqrt{\Omega}}; \Delta n\right)\pi(x, t). \tag{6.102}$$

We will make one further simplification of Eq. (6.102). Let us rescale the time so that

$$f(\Omega)t = \Omega\tau. \tag{6.103}$$

Then the master equation takes the form

$$\frac{\partial \pi'}{\partial \tau} - \sqrt{\Omega}\frac{d\eta'}{d\tau}\frac{\partial \pi'}{\partial x} = \sum_{\Delta n}\Omega\left[-\left(\frac{\Delta n}{\sqrt{\Omega}}\right)\frac{\partial}{\partial x} + \frac{1}{2}\left(\frac{\Delta n}{\sqrt{\Omega}}\right)^2\frac{\partial^2}{\partial x^2} + \cdots\right]$$
$$\times\ \omega_0\left(\eta'(\tau) + \frac{x}{\sqrt{\Omega}}; \Delta n\right)\ \pi'(x, \tau) \tag{6.104}$$

where $\pi'(x, \tau) = \pi(x, \Omega\tau/f(\Omega))$ and $\eta'(\tau) = \eta(\Omega\tau/f(\Omega))$.

If we keep only terms of order $\sqrt{\Omega}$ in Eq. (6.104), we obtain

$$\frac{d\eta'}{d\tau}\frac{\partial \pi'}{\partial x} = \sum_{\Delta n}\Delta n\frac{\partial}{\partial x}\omega_0(\eta'(\tau), \Delta n)\pi'(x, \tau)$$
$$= \sum_{\Delta n}\Delta n\omega_0(\eta'(\tau), \Delta n)\frac{\partial \pi'}{\partial x}. \tag{6.105}$$

(We have performed a Taylor series expansion of $\omega_0(\eta'(\tau) + (x/\sqrt{\Omega}); \Delta n)$ about $x/\sqrt{\Omega} = 0$ and have kept only the lowest order term.) Eq. (6.105) can be satisfied if we choose

$$\frac{d\eta'}{d\tau} = \sum_{\Delta n} \Delta n \omega_0(\eta'(\tau), \Delta n) = a_1'(\eta'(\tau)) \tag{6.106}$$

where $a_1'(\eta'(\tau))$ is related to the original jump moments through the relation $f(\Omega)a_p'(n/\Omega) = a_p(n)$.

If we keep terms to zeroth order in Ω, the master equation can be written

$$\frac{\partial \pi'}{\partial \tau} = -(\partial_{\eta'} a_1') \frac{\partial}{\partial x} x\pi' + \tfrac{1}{2} a_2' \frac{\partial^2 \pi'}{\partial x^2}. \tag{6.107}$$

Eq. (6.107) is a Fokker-Planck equation for the probability distribution $\pi'(x, \tau)$. Note that the coefficients on the right-hand side depend on time because they contain the time-varying average value $\eta(t)$. For a stationary process, $\eta(t) = \text{constant}$, and the coefficients will be independent of time.

From Eq. (6.107), we can obtain the equation of motion for the mean values of the fluctuations $\langle x \rangle$ and its moments $\langle x^2 \rangle$, etc. If we multiply Eq. (6.107) by x and integrate over x, we find

$$\int_{-\infty}^{\infty} x \frac{\partial \pi'}{\partial \tau} dx = -(\partial_{\eta'} a_1') \int_{-\infty}^{\infty} x \frac{\partial}{\partial x} (x\pi')\, dx + \tfrac{1}{2} a_2' \int_{-\infty}^{\infty} x \frac{\partial^2 \pi'}{\partial x^2} dx.$$

We now integrate the terms on the right by parts to obtain

$$\frac{d\langle x \rangle}{d\tau} = (\partial_{\eta'} a_1')\langle x \rangle. \tag{6.108}$$

We have assumed that π' and its derivatives go to zero as $x \to \pm\infty$. Similarly,

$$\frac{d}{d\tau} \langle x^2 \rangle = 2(\partial_{\eta'} a_1')\langle x^2 \rangle + a_2'. \tag{6.109}$$

In Eqs. (6.108) and (6.109) terms of order $\Omega^{-1/2}$ have been neglected since those terms were neglected in the derivation of the Fokker-Planck equation. In order that the Fokker-Planck equation derived in Eq. (6.107) give an adequate description of the system, the average value of the fluctuation $\langle x(t) \rangle$ must be of order unity. Otherwise the decomposition in Eq. (6.95) becomes invalid and the expansion outlined here may no longer be used. One must always check to be sure that the fluctuations $\langle x(t) \rangle$ remain small. We can easily see that the behavior of $\langle x(t) \rangle$ is tied to the behavior of $\eta(t)$ through Eq. (6.108). Indeed, if $\langle x(t) \rangle$ is to remain small, $\partial_{\eta'} a_1'$ must have no real positive part. Otherwise $\langle x(t) \rangle$ would grow exponentially.

It is possible to write down a general solution for the Fokker-Planck equation (6.107). Following van Kampen[8] let us assume that at time $\tau = 0$, $\pi'(x, 0) = \delta(x - x_0)$, and let us introduce a function $s(\tau)$ defined

$$s(\tau) = \ln \frac{a_1'\{\eta'(0)\}}{a_1'\{\eta'(\tau)\}}. \tag{6.110}$$

Then $s(0) = 0$, $ds/d\tau = -\partial_{\eta'} a_1'\{\eta'\}$, and Eq. (6.107) takes the form

$$\frac{\partial \pi'}{\partial s} = \frac{\partial}{\partial x} x\pi' - \frac{a_2'}{2(\partial_{\eta'} a_1')} \frac{\partial^2 \pi'}{\partial x^2}. \tag{6.111}$$

Next let $x = y\, e^{-s}$ and $\pi'(x, s) = e^s Q(y, s)$ to obtain

$$\frac{\partial Q}{\partial s} = -\frac{a_2'}{2(\partial_{\eta'} a_1')} e^{2s} \frac{\partial^2 Q}{\partial y^2}. \tag{6.112}$$

Eq. (6.112) has the form of a generalized diffusion equation. The solution was given by Chandrasekhar:[10]

$$Q(y,s) = \frac{1}{[4\pi\gamma(s)]^{1/2}} \exp\left[-\frac{(y - y_0)^2}{4\gamma(s)}\right], \tag{6.113}$$

where

$$\gamma(s) = \frac{1}{2}\int_0^\tau a_2'\{\eta'(\tau)\}\, e^{2s(\tau)}\, d\tau \tag{6.114}$$

and $y_0 = x_0$. Thus,

$$\pi'(x, \tau) = \frac{1}{[2\pi\sigma^2]^{1/2}} \exp\left[-\frac{(x - x_0\, e^{-s})^2}{2\sigma^2}\right], \tag{6.115}$$

where

$$\sigma^2 = e^{-2s(\tau)} \int_0^\tau a_2'\{\eta'(\tau)\}\, e^{2s(\tau)}\, d\tau. \tag{6.116}$$

We see that processes described by the Fokker-Planck equation in Eq. (6.107) have a Gaussian distribution. This is the standard form of the solution for a Fokker-Planck equation with $\alpha_1(y) = y + C_1$ and $\alpha_2(y) = C_2$, where C_1 and C_2 are constants (cf. Eq. (6.27)). However, if higher order terms are considered or the Ω-expansion breaks down, the probability distribution no longer need be Gaussian in shape.

H. MALTHUS-VERHULST EQUATION[9]

Nonlinear birth and death processes occur when the individuals in a society are allowed to interact. A very simple example occurs when competition between members of a society leads to an increased death rate. Let us assume that the transition probability is identical to that in Sec. 6.D except that there is an additional contribution to the death rate proportional to the density of other individuals $(n-1)/\Omega$, where Ω is the size of the system. Then the transition rate takes the form

$$W(m, n) = \left(m\mu + \gamma\frac{m(m-1)}{\Omega}\right)\delta_{n,m-1} + m\lambda\,\delta_{n,m+1} \tag{6.117}$$

and the master equation can be written

$$\frac{\partial P_1(n, t)}{\partial t} = (n-1)\lambda P_1(n-1, t) + \left[(n+1)\mu + \frac{\gamma}{\Omega}n(n+1)\right]P_1(n+1, t)$$
$$- \left(n\lambda + n\mu + \frac{\gamma}{\Omega}n(n-1)\right)P_1(n, t). \tag{6.118}$$

Note that in the notation of Sec. 6.G we have added a term $f(\Omega)(1/\Omega)\omega_1(m/\Omega, \Delta n)$ to the transition rate. For our problem $f(\Omega) = \Omega$ and thus $t = \tau$ (cf. Eq. (6.103)).

From Eq. (6.118), the equation for the average number of individuals at time t is

$$\frac{d\langle n(t)\rangle}{dt} = (\lambda - \mu)\langle n(t)\rangle - \frac{\gamma}{\Omega}\langle n^2(t)\rangle + \frac{\gamma\langle n(t)\rangle}{\Omega} \equiv \langle a_1(n)\rangle. \tag{6.119}$$

The time evolution of the first-order moment depends on the second-order moment. This is a general feature of nonlinear problems. Eq. (6.119) is exact. If we now use Eq. (6.95) and retain terms of order Ω, we find

$$\frac{d\eta(t)}{dt} = (\lambda - \mu)\eta(t) - \gamma\eta^2(t), \tag{6.120}$$

and we have reduced Eq. (6.119) to a nonlinear equation involving only first-order moments. Eq. (6.120) is called the Malthus-Verhulst equation and gives the equation of motion for the average number of individuals in the society. Terms of order $1/\sqrt{\Omega}$ and smaller have been neglected. The solution to Eq. (6.120) is

$$\eta(t) = \frac{\eta(0)\,e^{(\lambda-\mu)t}}{1 + \eta(0)\dfrac{\gamma}{\lambda - \mu}\{e^{(\lambda-\mu)t} - 1\}}. \tag{6.121}$$

If $\lambda < \mu$, $\eta(t) \underset{t\to\infty}{\to} 0$ and the population dies away. If $\lambda > \mu$, $\eta(t) \underset{t\to\infty}{\to} (\lambda - \mu)/\gamma$ and the population tends to a finite constant value. Note that if $\gamma = 0$, so that there is no nonlinear death rate, the population would become infinite. The nonlinear term tends to stabilize the population. We see that, as a function of $(\lambda - \mu)$, there is a bifurcation. For $(\lambda - \mu) < 0$ the system approaches the state $\eta = 0$ and for $(\lambda - \mu) > 0$ it approaches the state $\eta = (\lambda - \mu)/\gamma$. As the transition rates change, there is an abrupt "phase transition" from one state to another.

The equation of motion of the average fluctuation $\langle x(t)\rangle$ is easily obtained from Eqs. (6.119) and (6.108). If we retain terms of order $\sqrt{\Omega}$ in the equation for $\langle x(t)\rangle$, we find

$$\frac{d\langle x(t)\rangle}{dt} = (\lambda - \mu - 2\gamma\eta(t))\langle x(t)\rangle. \tag{6.122}$$

Similarly, we can obtain the equation for the moment, $\langle x^2(t)\rangle$, of the fluctuations,

$$\frac{d}{dt}\langle x^2(t)\rangle = 2(\lambda - \mu - 2\gamma\eta(t))\langle x^2(t)\rangle + \mu\eta(t) + \lambda\eta(t) + \gamma\eta^2(t). \tag{6.123}$$

Eqs. (6.122) and (6.123) are valid if Ω is a good expansion parameter. For this to be so, $\langle x(t)\rangle$ must remain of order one for all times. We can check the equations to see if this is true.

Let us consider Eqs. (6.122) and (6.123) after the system has reached a steady state, that is, for large t. For $\lambda - \mu < 0$, $\eta(\infty) = 0$ and Eqs. (6.122) and (6.123) become, respectively,

$$\frac{d\langle x(t)\rangle}{dt} = (\lambda - \mu)\langle x(t)\rangle \tag{6.124}$$

and

$$\frac{d}{dt}\langle x^2(t)\rangle = 2(\lambda - \mu)\langle x^2(t)\rangle. \tag{6.125}$$

Both $\langle x(t)\rangle$ and $\langle x^2(t)\rangle$ approach zero for long times if $(\lambda - \mu) < 0$. Note that if $(\lambda - \mu) > 0$ the state $\eta(\infty) = 0$ is unstable. The average fluctuations $\langle x(t)\rangle$ grow.

Let us next consider the case $(\lambda - \mu) > 0$. Then $\eta(\infty) = (\lambda - \mu)/\gamma$ and Eqs. (6.122) and (6.123) become, respectively,

$$\frac{d\langle x(t)\rangle}{dt} = -(\lambda - \mu)\langle x(t)\rangle \tag{6.126}$$

and

$$\frac{d\langle x^2(t)\rangle}{dt} = -2(\lambda - \mu)\langle x^2(t)\rangle + \frac{2\lambda(\lambda - \mu)}{\gamma}. \tag{6.127}$$

Again after long times the fluctuations will die away, and for $(\lambda - \mu) > 0$, the state $\eta(\infty) = (\lambda - \mu)/\gamma$ is stable.

The Malthus-Verhulst problem is interesting because it gives us a rather simple example of a nonequilibrium phase transition which can arise because of nonlinearities in the transition rates. The macroscopic state of the system can undergo an abrupt change as the parameter $(\lambda - \mu)$ changes from positive to negative values. We shall consider more examples of nonequilibrium phase transitions in Chap. 17.

REFERENCES

1. N. U. Prabhu, *Stochastic Processes* (Macmillan Co., New York, 1965).
2. G. E. Uhlenbeck, unpublished lecture notes, University of Colorado, 1965.
3. N. G. van Kampen, unpublished lecture notes, University of Utrecht, 1970.
4. N. G. van Kampen, "The Diffusion Approximation for Markov Processes," preprint (1978).
5. A. Scheerer, *Probability on Discrete Sample Spaces* (International Textbook Co., Scranton, 1969), p. 227.
6. S. Lipschutz, *Probability*, Schaum's Outline Series (McGraw-Hill Book Co., New York, 1965).
7. I. N. Sneddon, *Elements of Partial Differential Equations* (McGraw-Hill Book Co., New York, 1957).
8. N. G. van Kampen, Can. J. Phys. *39* 551 (1961).
9. N. G. van Kampen, Adv. Chem. Phys. *34* 245 (1976).
10. S. Chandrasekhar, Rev. Mod. Phys. *15* 1 (1943).

PROBLEMS

1. There are two white marbles in urn A and four red marbles in urn B which can be interchanged. At each step of the process a marble is selected at random from each urn and the two marbles selected are interchanged.

(*a*) Find the transition matrix $\bar{\bar{Q}}$, its eigenvalues and eigenvectors.
(*b*) What is the probability that there are two red marbles in urn A after three steps? After many steps?
(*c*) Express the probability vector, $P(s)$, in terms of the left and right eigenvectors for this problem.

2. A trained mouse lives in the house shown in Fig. 6.3. A bell rings at regular intervals (very short compared to the mouse's lifetime). Each time it rings the mouse changes rooms. When he changes rooms, he is equally likely to pass through any of the doors in the room he is in. Approximately what fraction of his life does he spend in each room? Express $P(s)$ in terms of left and right eigenvectors for this problem.

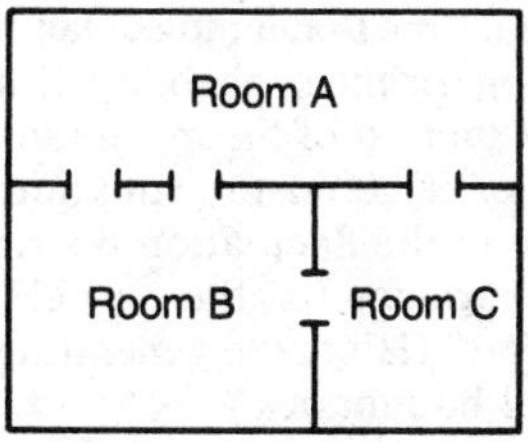

Fig. 6.3. Floor plan of mouse's house.

3. The Fokker-Planck equation,

$$\frac{\partial P}{\partial t} = \frac{\partial}{\partial x}(\alpha x P) + \frac{\beta}{2}\frac{\partial^2 P}{\partial x^2}.$$

can be solved by separation of variables, $P(x, t) = T(t)X(t)$. Assume that

$P(\pm\infty, t) = 0$ and that $P(x, 0) = \delta(x - x_0)$. Find the solution of the Fokker-Planck equation. (*Hint:* the solution involves Hermite polynomials.) After an infinite length of time the solution approaches a unique steady state. What is it? Find the first moment as a function of time.

4. The Fokker-Planck equation,

$$\frac{\partial P}{\partial t} = \frac{\partial}{\partial x}\left[\left(\alpha x - \frac{\beta}{2x}\right)P\right] + \frac{\beta}{2}\frac{\partial^2 P}{\partial x^2}.$$

can be solved by separation of variables. Find the solution and first moment as a function of time if $P(\pm\infty, t) = 0$ and $P(x, 0) = \delta(x - x_0)$. What is the steady-state solution? (*Hint:* the solution involves Laguerre polynomials.)

5. Consider a linear birth and death process which includes the possibility of a change in population due to emigration in addition to the change that occurs due to birth and death of individuals. Let $\alpha\,\Delta t$ be the probability that individuals enter the society and $\beta\,\Delta t$ be the probability that they leave the society in time $t \to t + \Delta t$ due to emigration. In addition, let $n\lambda\,\Delta t$ be the probability of a birth in time $t \to t + \Delta t$ and $n\mu\,\Delta t$ the probability of a death in time $t \to t + \Delta t$. Write down the master equation for this process. Find the equation of motion for the generating function $F(z, t)$, and solve it. Obtain expressions for the mean number of individuals $\langle n(t)\rangle$ in the population at time t and the variance. Discuss the meaning of your results.

6. A radio-active sample consists of n identical nuclei, each with a probability K per unit time of decaying. Since there are n nuclei, the probability per unit time of a decay is nK. Write down and solve the master equation for this process. Calculate the mean number of undecayed nuclei and the variance as a function of time. Assume that at time $t = 0$ there are n_0 nuclei present. Discuss the meaning of your results.

7. Consider the chemical reaction

$$A + X \underset{K_2}{\overset{K_1}{\rightleftharpoons}} A + Y,$$

where molecule A is a catalyst whose concentration is maintained constant. Assume that the total number of molecules X and Y is constant and equal to N. K_1 is the probability per unit time that a molecule X interacts with an A to produce Y. K_2 is the probability per unit time that a molecular Y interacts with A to produce X. Find the equation of motion for the probability distribution of molecules X and solve it. Find the equation of motion of the mean number of molecules X and the variance in the distribution of X. Find the equation of motion for the average fluctuations and the variance in the fluctuation distribution (cf. Sec. 6.G). In the limit $t \to \infty$, how does the variance for the particle distribution compare with that of a Gaussian distribution? (Hint: the generating function can be used since $n = 0$ and $n = N$ are natural boundaries.)

8. Consider the chemical reaction

$$X + Y \underset{K_2}{\overset{K_1}{\rightleftharpoons}} 2X$$

and assume that the total number of molecules is held fixed and is equal to N. That is, $n_x + n_y = N$ where n_x and n_y are the number of x and y molecules respectively. Let $K_1(n_y/\Omega)$ be the probability per unit time that a *single* molecule X undergoes the forward reaction (Ω is the volume), and let $K_2(n_x/\Omega)$ be the probability per unit time that a single X molecule undergoes the reverse reaction. Write down the master equation for this process, the equation of motion for the average number of X molecules, the equation of motion for the variance in the distribution of X particles, the equation of motion for the average fluctuation and for the variance in the fluctuation distribution. Use the Ω-expansion method. Is the Ω-expansion applicable? In the limit $t \to \infty$, how does the variance of the particle distribution compare with that of a Gaussian distribution?

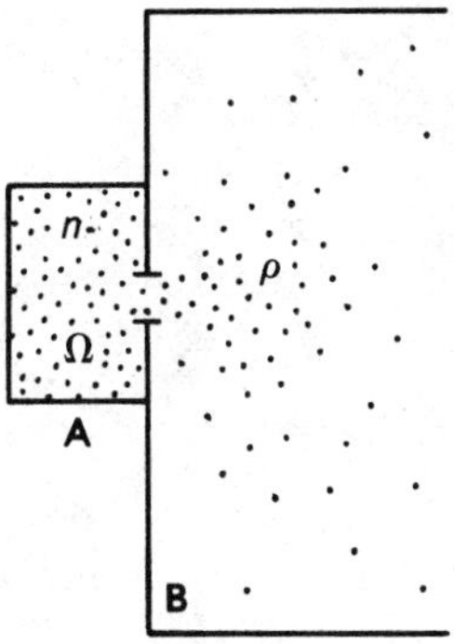

Fig. 6.4. Figure for Prob. 6.9.

9. Consider a box of volume Ω connected to another box of infinite volume via a small hole (cf. Fig. 6.4). Assume that the probability that a particle moves from box A to box B in time Δt is $(n/\Omega)\,\Delta t$ and the probability that a particle moves from box B to box A in time Δt is $\rho\,\Delta t$. Write down the master equation for the probability distribution of particles in A and solve it assuming that $t = 0$ and $n = n_0$ (where n is number of particles in box A). Find the average number of particles in A and the variance as a function of time.

10. A Brownian particle moving in one dimension in a fluid at temperature T has the equation of motion

$$m\dot{v}(t) + \gamma v(t) = f(t),$$

where m is the mass of the particle, $v(t)$ is its velocity, γ is a friction constant, and $f(t)$ is a random fluctuating force. Assume that $\langle f(t)\rangle = 0$ and $\langle f(t)f(t')\rangle = C\delta(t - t')$, where C is an unknown constant. Assume that the particle is given an initial velocity $v(0) = v_0$.

(*a*) Find $\langle v(t)\rangle$ and $\langle v^2(t)\rangle$. (Assume $\langle v(t)f(t)\rangle = 0$.)

(*b*) After a long time the particle will come to equilibrium with the fluid and $\frac{1}{2}m\langle v^2(\infty)\rangle = \frac{1}{2}k_BT$, where k_B is Boltzmann's constant. Use this information and the results of part (*a*) to find the constant C.

(*c*) Write the Fokker-Planck equation for the conditional velocity distribution function $P(v_0, 0|v(t), t)$. (Hint: you must find the jump moments $\alpha_n(v(t)) = \lim_{\tau\to\infty}\tau^{-1}\langle(\Delta v(\tau))^n\rangle$, where $\Delta v(\tau) = v(t + \tau) - v(t)$, from the equation of motion.) Write the solution to the Fokker-Planck equation.

7
Probability Distributions in Dynamical Systems

A. INTRODUCTORY REMARKS

In Chap. 6 we studied the time evolution of probability distributions in the Markov approximation, where the dynamics of the process was determined in terms of a single transition probability. The transition probability itself is usually determined phenomenologically, and thus the equations we derived, the master equation and the Fokker-Planck equation, may be considered as phenomenological equations. However, they exhibit exactly the type of behavior we need to describe the observed irreversible decay of systems to a unique equilibrium state. We must now ask the following question: can we derive similar irreversible equations from a rigorous microscopic theory of matter starting from Newton's reversible laws of mechanics? The answer to this question is not simple and will occupy us for much of the remainder of this book.

In this chapter we will set up the machinery for a microscopic probabilistic description of matter for both classical and quantum mechanical systems. That is, we shall lay the foundations for statistical mechanics. In subsequent chapters, we will apply this theory to a variety of problems involving systems both in and out of equilibrium and we will learn how thermodynamics and irreversible hydrodynamics are thought to arise from the reversible laws of dynamics.

This chapter is divided into two parts. The first part deals with classical systems and the second part deals with quantum systems. We shall generally be concerned with systems with a large number of degrees of freedom, such as N interacting particles in a box or N interacting objects on a lattice. The motion of such objects is described by Newton's laws or, equivalently, by Hamiltonian dynamics. In three dimensions, such a system has $3N$ degrees of freedom (if we neglect internal degrees of freedom) and classically is specified by $6N$ independent position and momentum coordinates whose motion is uniquely determined from Hamiltonian dynamics. If we set up a $6N$-dimensional phase space, whose $6N$ coordinates consist of the $3N$ momentum and $3N$-position variables of the particles, then the state of the system is given by a single point in the phase space, which moves according to Hamiltonian dynamics as the state of the system changes. If we are given a real N-particle system, we never know exactly what its state is. We only know with a certain probability that it is one of the points in the phase space. Thus, the state point can be regarded as a stochastic variable and we can assign a probability distribution to the points in phase space in accordance with our knowledge of the state of the system. We then can view the phase space as a probability fluid which flows according to Hamiltonian dynamics. In this way, we obtain a connection between the mechanical description and a probabilistic description of the system. The problem of finding an equation of motion for the probability density reduces to a problem in fluid dynamics. We shall find that the equation of motion for the N-body probability density is reversible, like the laws of dynamics, and, on the level at which we discuss it in this chapter, does not exhibit the irreversible behavior we hoped for. Something more is needed to obtain irreversibility. However, irreversible behavior can and has been obtained based on the equations we will derive in this chapter.

The N-body probability density for a classical system contains more information about the system than we need. In practice, the main use of the probability density is to find expectation values or correlation functions for various observables, since those are what we measure experimentally and what we deal with in thermodynamics. The observables we deal with in physics are generally one- or two-body operators, and to find their expectation values we only need reduced one- or two-body probability densities and not the full N-body probability density. The reduced n-body probability density can be obtained from the N-body probability density by integrating out all but n of the N variables. In this way we can also find the equations of motion for the reduced probability density given the equation of motion for the N-body probability density. We shall find that the equations of motion for the reduced probability densities form a hierarchy of equations called the BBGKY hierarchy (after Born, Bogoliubov, Green, Kirkwood, and Yvon who discovered it), which makes them impossible to solve without some approximation which terminates the hierarchy. This, in fact, is a general feature of all reduced descriptions.

In classical and quantum statistical mechanics, as in quantum mechanics, we can introduce the idea of a "Schrödinger picture" and a "Heisenberg picture" of the evolution of the system. In taking expectation values, either we let the probability density (state of the system) carry the time dependence (Schrödinger picture), or we let the phase functions carry the time dependence (Heisenberg picture). Both pictures are useful. We shall obtain the equation of motion for phase functions and use it to obtain the microscopic balance equations for the number density, momentum density, and energy density of the system. These are especially important because they are densities of quantities which are conserved during the evolution of the system and give rise to the hydrodynamic equations.

When we deal with quantum systems the phase space variables no longer commute and it is often useful to use representations other than the coordinate representation to describe the state of the system. Thus, we introduce the idea of a probability density operator (a positive definite Hermitian operator), which can be used to find the probability distribution in any desired representation. We then can use the Schrödinger equation to find the equation of motion for the probability density operator.

In quantum systems, just as for the classical case, it is useful to introduce a reduced description for the system. Thus, we shall introduce the reduced density operators and discuss the behavior of their eigenvalues. Of particular interest is the case of a transition to a superfluid state. When this happens, one of the eigenvalues of the reduced density operators can become macroscopic in size and the system is said to exhibit off-diagonal, long-range order (ODLRO).

In quantum systems, we no longer can introduce probability densities which specify both the position and momentum of the particle, because these quantities no longer commute. However, we can introduce quantities which are formally analogous, namely, the Wigner functions. The Wigner functions are not probability densities, because they can become negative. However, they can be used to obtain expectation values in a manner formally analogous to that of classical systems, and the reduced Wigner functions form a hierarchy which in the classical limit reduces to the classical BBGKY hierarchy.

We will conclude this chapter by giving the quantum version of the balance equations for the number density, momentum density, and energy density. These equations will be useful later when we discuss hydrodynamics.

In dealing with the quantum systems in this and subsequent chapters, we have assumed that the reader is familiar with the use of field operators and the number representation, since they vastly simplify all calculations. For those who have never used the number representation or who are a bit rusty as regards its use, we have discussed the relation between the coordinate and number representations in App. B.

B. PROBABILITY DENSITY AS A FLUID[1–3]

Let us consider a closed classical system with $3N$ degrees of freedom (for example, N particles in a three-dimensional box). The state of such a system is completely specified in terms of a set of $6N$ independent real variables $(\mathbf{p}^N, \mathbf{q}^N)$ ($\mathbf{p}^N$ and $\mathbf{q}^N$ denote the set of vectors $\mathbf{p}^N = (\mathbf{p}_1, \mathbf{p}_2, \ldots, \mathbf{p}_N)$ and $\mathbf{q}^N = (\mathbf{q}_1, \mathbf{q}_2, \ldots, \mathbf{q}_N)$, respectively; $\mathbf{p}_l$ and $\mathbf{q}_l$ are the momentum and position of the lth particle). If the state vector $\mathbf{X}^N = \mathbf{X}^N(\mathbf{p}^N, \mathbf{q}^N)$ is known at one time, then it is completely determined for any other time from Newton's laws.

If we can define a Hamiltonian, $H(\mathbf{X}^N, t)$, for the system, then the time evolution of the quantities $\mathbf{p}_l$ and $\mathbf{q}_l$ ($l = 1, \ldots, N$) is given by Hamilton's equations,

$$\dot{\mathbf{p}}_l \equiv \frac{d\mathbf{p}_l}{dt} = -\frac{\partial H^N}{\partial \mathbf{q}_l} \tag{7.1}$$

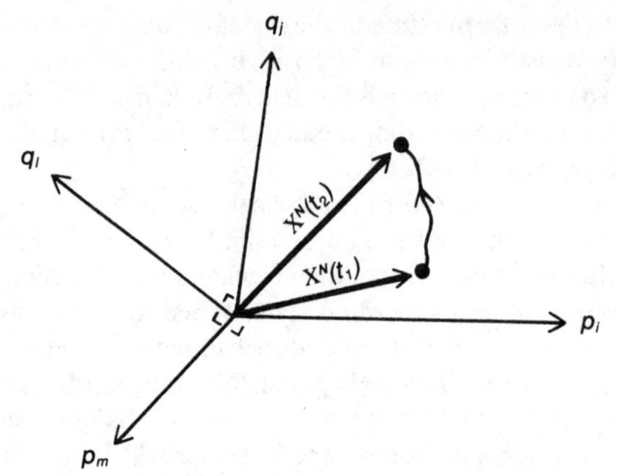

Fig. 7.1. Movement of a system point, $\mathbf{X}^N(t)$, in a $6N$-dimensional phase space ($t_2 > t_1$): we show only four of the $6N$ coordinate directions.

and

$$\dot{\mathbf{q}}_l \equiv \frac{d\mathbf{q}_l}{dt} = \frac{\partial H^N}{\partial \mathbf{p}_l}. \tag{7.2}$$

If the Hamiltonian does not depend explicitly on time, then it is a constant of the motion

$$H^N(\mathbf{X}^N) = E, \tag{7.3}$$

where E is the total energy of the system. In this case, the system is called conservative.

Let us now associate to the system a $6N$-dimensional phase space, Γ. The state vector $\mathbf{X}^N(\mathbf{p}^N, \mathbf{q}^N)$ then specifies a point in the phase space. As the system evolves in time and its state changes, the system point $\mathbf{X}^N$ traces out a trajectory in Γ-space (cf. Fig. 7.1). Since the subsequent motion of a classical system is uniquely determined from the initial conditions, it follows that no two trajectories in phase space can cross. If they could, one could not uniquely determine the subsequent motion of the trajectory.

When we deal with real physical systems, we can never specify exactly the state of the system. There will always be some uncertainty in the initial conditions. Therefore, it is useful to consider $\mathbf{X}^N$ as a stochastic variable and to introduce a probability density $\rho(\mathbf{X}^N, t)$ on the phase space, where $\rho(\mathbf{X}^N, t)\, d\mathbf{X}^N$ is the probability that the state point, $\mathbf{X}^N$, lies in the volume element $\mathbf{X}^N \to \mathbf{X}^N + d\mathbf{X}^N$ at time t. (Here $d\mathbf{X}^N = d\mathbf{q}_1 \times \cdots \times d\mathbf{q}_N\, d\mathbf{p}_1 \times \cdots \times d\mathbf{p}_N$.) In so doing we introduce a picture of phase space filled with a continuum (or fluid) of state points. If the fluid were composed of discrete points, then each point would be assigned a probability in accordance with our initial knowledge of the system and would carry this probability for all time (probability is conserved). The change in our knowledge of the state of the system with time is determined by the way in which the fluid flows. Since for real systems the state points form a continuum, we must introduce a probability density, $\rho(\mathbf{X}^N, t)$, on the phase space rather than a probability distribution, but the ideas are the same.

Because state points must always lie somewhere in the phase space, we have the normalization condition

$$\int_{\Gamma} \rho(\mathbf{X}^N, t)\, d\mathbf{X}^N = 1 \tag{7.4}$$

where the integration is taken over the entire phase space. If we want the probability of finding the state point in a small finite region R of Γ-space at time t, then we simply integrate the probability density over that region. If we let $P(R)$ denote the probability of finding the system in region R, then

$$P(R) = \int_{R} \rho(\mathbf{X}^N, t)\, d\mathbf{X}^N. \tag{7.5}$$

If at some time there is only a small uncertainty in the state of the system, the probability density will be sharply peaked in the region where the state is known to be located, and zero elsewhere. As time passes, the probability density may remain sharply peaked (although the peaked region can move through phase space) and we do not lose any knowledge about the state of the system. On the other hand, it might spread and become rather uniformly distributed, in which case all knowledge of the state of the system becomes lost (see Chap. 8 for some explicit examples of flow in phase space).

Probability behaves like a fluid in phase space. We can therefore use arguments from fluid mechanics to obtain the equation of motion for the probability density (cf. App. A). We will let $\dot{\mathbf{X}}^N = (\dot{\mathbf{q}}^N, \dot{\mathbf{p}}^N)$ denote the velocity of a state point, and we will consider a small volume element, V_0, at a fixed point in phase space. Since probability is conserved, the total decrease in the amount of probability in V_0 per unit time is entirely due to the flow of probability through the surface of V_0. Thus,

$$\frac{d}{dt} P(V_0) = \frac{\partial}{\partial t} \int_{V_0} \rho(\mathbf{X}^N, t)\, d\mathbf{X}^N = -\oint_{S_0} \rho(\mathbf{X}^N, t) \dot{\mathbf{X}}^N \cdot d\mathbf{S}^N, \tag{7.6}$$

where S_0 denotes the surface of volume element V_0, and $d\mathbf{S}^N$ is a differential area element on S_0. If we use Gauss's theorem and change the surface integral to a volume integral, we find

$$\frac{\partial}{\partial t} \int_{V_0} \rho(\mathbf{X}^N, t)\, d\mathbf{X}^N = -\int_{V_0} \nabla_{\mathbf{X}^N} \cdot (\rho(\mathbf{X}^N, t) \dot{\mathbf{X}}^N)\, d\mathbf{X}^N, \tag{7.7}$$

where $\nabla_{\mathbf{X}^N}$ denotes the gradient with respect to phase space variables $\nabla_{\mathbf{X}^N} = \left(\frac{\partial}{\partial \mathbf{q}_1}, \frac{\partial}{\partial \mathbf{q}_2}, \ldots, \frac{\partial}{\partial \mathbf{q}_N}; \frac{\partial}{\partial \mathbf{p}_1}, \ldots, \frac{\partial}{\partial \mathbf{p}_N}\right)$. We can take the derivative inside the integral because V_0 is fixed in space. If we equate arguments of the two integrals in Eq. (7.7), we obtain

$$\frac{\partial}{\partial t}\rho(\mathbf{X}^N, t) + \nabla_{\mathbf{X}^N}\cdot(\rho(\mathbf{X}^N, t)\dot{\mathbf{X}}^N) = 0. \tag{7.8}$$

Eq. (7.8) is the balance equation for the probability density in the $6N$-dimensional phase space.

We can use Hamilton's equations to show that the probability behaves like an incompressible fluid. A volume element in phase space changes in time according to the equation

$$d\mathbf{X}_t^N = \mathscr{J}^N(t, t_0)\, d\mathbf{X}_{t_0}^N, \tag{7.9}$$

where $\mathscr{J}^N(t, t_0)$, the Jacobian of the transformation, is the determinant of a $6N \times 6N$-dimensional matrix which we write symbolically as

$$\mathscr{J}^N(t, t_0) = \det\begin{vmatrix} \dfrac{\partial \mathbf{p}_t^N}{\partial \mathbf{p}_{t_0}^N} & \dfrac{\partial \mathbf{p}_t^N}{\partial \mathbf{q}_{t_0}^N} \\ \dfrac{\partial \mathbf{q}_t^N}{\partial \mathbf{p}_{t_0}^N} & \dfrac{\partial \mathbf{q}_t^N}{\partial \mathbf{q}_{t_0}^N} \end{vmatrix}. \tag{7.10}$$

The Jacobian $\mathscr{J}^N(t, t_0)$ can be shown to satisfy the relation

$$\mathscr{J}^N(t, t_0) = \mathscr{J}^N(t, t_1)\mathscr{J}^N(t_1, t_0) \tag{7.11}$$

if we remember that the product of the determinant of two matrices is equal to the determinant of the products.

Let us now assume that the system evolves for a short time interval $\Delta t = t - t_0$. Then the coordinates of a state point can be written

$$\mathbf{p}_t^N = \mathbf{p}_{t_0}^N + \dot{\mathbf{p}}_{t_0}^N \Delta t + 0(\Delta t)^2 \tag{7.12}$$

and

$$\mathbf{q}_t^N = \mathbf{q}_{t_0}^N + \dot{\mathbf{q}}_{t_0}^N \Delta t + 0(\Delta t)^2. \tag{7.13}$$

(Again Eqs. (7.12) and (7.13) have been written symbolically to denote the set of $6N$ equations for the components of $\mathbf{X}^N$.) If we combine Eqs. (7.10), (7.12), and (7.13), we find

$$\mathscr{J}^N(t, t_0) = \begin{vmatrix} 1 + \dfrac{\partial \dot{\mathbf{p}}_{t_0}^N}{\partial \mathbf{p}_{t_0}^N}\Delta t & \dfrac{\partial \dot{\mathbf{p}}_{t_0}^N}{\partial \mathbf{q}_{t_0}^N}\Delta t \\ \dfrac{\partial \dot{\mathbf{q}}_{t_0}^N}{\partial \mathbf{p}_{t_0}^N}\Delta t & 1 + \dfrac{\partial \dot{\mathbf{q}}_{t_0}^N}{\partial \mathbf{q}_{t_0}^N}\Delta t \end{vmatrix} = 1 + \left(\frac{\partial \dot{\mathbf{q}}_{t_0}^N}{\partial \mathbf{q}_{t_0}^N} + \frac{\partial \dot{\mathbf{p}}_{t_0}^N}{\partial \mathbf{p}_{t_0}^N}\right)\Delta t + 0(\Delta t)^2. \tag{7.14}$$

However, from Hamilton's equations (cf. Eqs. (7.1) and (7.2)) we obtain

$$\left(\frac{\partial \dot{\mathbf{q}}_{t_0}^N}{\partial \mathbf{q}_{t_0}^N} + \frac{\partial \dot{\mathbf{p}}_{t_0}^N}{\partial \mathbf{p}_{t_0}^N}\right) \equiv 0 \tag{7.15}$$

and, therefore,

$$\mathcal{J}^N(t, t_0) = 1 + 0(\Delta t)^2. \tag{7.16}$$

From Eq. (7.11), we can write

$$\mathcal{J}^N(t, 0) = \mathcal{J}^N(t, t_0)\mathcal{J}^N(t_0, 0) = \mathcal{J}^N(t_0, 0)(1 + 0(\Delta t)^2), \tag{7.17}$$

and the time derivative of the Jacobian becomes

$$\frac{d\mathcal{J}}{dt} = \lim_{\Delta t \to 0} \frac{\mathcal{J}^N(t_0 + \Delta t, 0) - \mathcal{J}^N(t_0, 0)}{\Delta t} = 0. \tag{7.18}$$

Thus, for a system whose dynamics is determined by Hamilton's equations, the Jacobian does not change in time and

$$\mathcal{J}^N(t, 0) = \mathcal{J}^N(0, 0) = 1. \tag{7.19}$$

Eq. (7.19) is extremely important for several reasons. First, it tells us that volume elements in phase space do not change in size during the flow (although they can change in shape):

$$d\mathbf{X}_t^N = d\mathbf{X}_{t_0}^N. \tag{7.20}$$

Second, it tells us that the probability behaves like an incompressible fluid since

$$\nabla_{\mathbf{X}^N} \cdot \dot{\mathbf{X}}^N \equiv 0 \tag{7.21}$$

(cf. Eq. (7.15)). If we combine Eqs. (7.8) and (7.21), the equation of motion for the probability density takes the form

$$\frac{\partial \rho^N}{\partial t} = -\dot{\mathbf{X}}^N \cdot \nabla_{\mathbf{X}^N} \rho^N. \tag{7.22}$$

Note that Eq. (7.22) gives the time rate of change of $\rho(\mathbf{p}^N, \mathbf{q}^N, t)$ at a *fixed point* in phase space. If we want the time rate of change as seen by an observer moving with the probability fluid, then we must find the equation for the total time derivative of $\rho(\mathbf{p}^N, \mathbf{q}^N, t)$. The total or convective time derivative is defined

$$\frac{d}{dt} = \left(\frac{\partial}{\partial t} + \dot{\mathbf{X}}^N \cdot \nabla_{\mathbf{X}^N}\right) \tag{7.23}$$

(cf. App. A, Eq. (A.5)) and, therefore, from Eq. (7.22) we obtain

$$\frac{d\rho^N}{dt} = 0. \tag{7.24}$$

Thus, the probability density remains constant in the neighborhood of a point moving with the probability fluid.

If we use Hamilton's equations, we can write Eq. (7.22) in the form

$$\frac{\partial \rho^N}{\partial t} = -\hat{\mathscr{H}}^N \rho^N, \tag{7.25}$$

where the operator $\hat{\mathscr{H}}^N$ is the Poisson bracket:

$$\hat{\mathscr{H}}^N = \sum_{j=1}^{N} \left(\frac{\partial H^N}{\partial \mathbf{p}_j} \cdot \frac{\partial}{\partial \mathbf{q}_j} - \frac{\partial H^N}{\partial \mathbf{q}_j} \cdot \frac{\partial}{\partial \mathbf{p}_j} \right). \tag{7.26}$$

We put a hat on $\hat{\mathscr{H}}^N$ to indicate that it is an operator. Eq. (7.22) is often written in another form,

$$i \frac{\partial \rho^N}{\partial t} = \hat{L}^N \rho^N \tag{7.27}$$

where $\hat{L}^N = -i\hat{\mathscr{H}}^N$. Eq. (7.27) is called the Liouville equation and $\hat{L}^N$, called the Liouville operator, is a Hermitian operator. If we know the probability density at some time $t = 0$, then we may solve Eq. (7.27) to find the probability density at time t:

$$\rho^N(\mathbf{q}^N, \mathbf{p}^N, t) = e^{-i\hat{L}^N t} \rho^N(\mathbf{q}^N, \mathbf{p}^N, 0). \tag{7.28}$$

A probability density which remains constant in time must satisfy the condition

$$\hat{L}^N \rho^N_{\text{stat}} = 0 \tag{7.29}$$

and is called a stationary state of the Liouville equation.

Following W. Gibbs, the probability density is often interpreted in terms of the concept of an "ensemble" of systems. Let us consider an ensemble of η identical systems (η very large). If we look at each system at a given time, it will be represented by a point in the $6N$-dimensional phase space. The distribution of points representing our ensemble of systems will be proportional to $\rho(\mathbf{X}^N, t)$. That is, the density of system points in phase space will be given by $\eta\rho(\mathbf{X}^N, t)$.

Eq. (7.28) gives the general equation for the evolution of the probability density of a classical dynamical system. However, it differs in an important respect from the behavior we found in Chap. 6 for the master equation. Since the Liouville operator is Hermitian (cf. Prob. 7.1) it will have real eigenvalues. Thus, the solution to Eq. (7.28) will oscillate and not decay to a unique equilibrium state. Furthermore, if we reverse the time in Eq. (7.28), we do not change the equation of motion for the probability density since the Liouville operator changes sign under time reversal. This is different from the Fokker-Planck equation, which changes into a new equation under time reversal. We see that Eq. (7.28) at first sight does not admit an irreversible decay of the system to a unique equilibrium state and thus cannot describe the decay to equilibrium that we observe so commonly in nature. And yet if we believe that the dynamics of systems is governed by Newton's laws, it is all we have. The problem of obtaining irreversible decay from the Liouville equation is one of the central problems of statistical physics and one which we will return to in this book.

C. THE BBGKY HIERARCHY[2,4]

The N-particle probability density contains much more information than we would ever need or want. Most quantities we measure experimentally can be expressed in terms of one-body or two-body phase functions. To find the expectation value of a one-body phase function we only need to know the one-body probability density. Similarly, to find the expectation value of a two-body phase function we only need to know the two-body probability density.

Let us denote the pair of phase space coordinates $\mathbf{q}_i$ and $\mathbf{p}_i$ by $\mathbf{X}_i$. Then the one-particle probability density is given by

$$\rho_1(\mathbf{X}_1, t) = \int \cdots \int d\mathbf{X}_2 \cdots d\mathbf{X}_N \rho^N(\mathbf{X}_1, \ldots, \mathbf{X}_N, t) \tag{7.30}$$

and the s-particle probability density is given by

$$\rho_s(\mathbf{X}_1, \ldots, \mathbf{X}_s, t) = \int \cdots \int d\mathbf{X}_{s+1} \cdots d\mathbf{X}_N \rho^N(\mathbf{X}_1, \ldots, \mathbf{X}_N, t). \tag{7.31}$$

One-body phase functions are usually written in the form

$$O_{(1)}{}^N(\mathbf{p}^N, \mathbf{q}^N) = \sum_{i=1}^{N} O(\mathbf{X}_i), \tag{7.32}$$

and two-body phase functions are written in the form

$$O_{(2)}{}^N(\mathbf{p}^N, \mathbf{q}^N) = \sum_{i<j}^{N(N-1)/2} O(\mathbf{X}_i, \mathbf{X}_j). \tag{7.33}$$

If the probability density, $\rho^N(\mathbf{X}_1, \ldots, \mathbf{X}_N, t)$, is known at time t, then the expectation value of the one-body phase function at time t is given by

$$\begin{aligned}\langle O_{(1)}(t)\rangle &= \mathrm{Tr}\, O_{(1)}{}^N \rho^N(t) \equiv \int \cdots \int d\mathbf{X}_1 \cdots d\mathbf{X}_N \sum_{i=1}^{N} O(\mathbf{X}_i)\rho^N(\mathbf{X}_1, \ldots, \mathbf{X}_N, t) \\ &= N \int d\mathbf{X}_1 O(\mathbf{X}_1)\rho_1(\mathbf{X}_1, t),\end{aligned} \tag{7.34}$$

where Tr denotes the trace. Similarly, the expectation value of the two-body phase function at time t is

$$\begin{aligned}\langle O_{(2)}(t)\rangle &= \mathrm{Tr}\, O_{(2)}{}^N \rho^N(t) \\ &= \sum_{i<j}^{N(N-1)/2} \int \cdots \int d\mathbf{X}_1 \cdots d\mathbf{X}_N O(\mathbf{X}_i, \mathbf{X}_j)\rho^N(\mathbf{X}_1, \ldots, \mathbf{X}_N, t) \\ &= \frac{N(N-1)}{2} \iint d\mathbf{X}_1\, d\mathbf{X}_2 O(\mathbf{X}_1, \mathbf{X}_2)\rho_2(\mathbf{X}_1, \mathbf{X}_2, t).\end{aligned} \tag{7.35}$$

In Eqs. (7.34) and (7.35), we have assumed that the probability density is symmetric under interchange of particle labels if the Hamiltonian is symmetric.

The equation of motion of $\rho^N(\mathbf{X}_1, \ldots, \mathbf{X}_N, t)$ is given by Eq. (7.22). From it we can find the equation of motion of $\rho_s(\mathbf{X}_1, \ldots, \mathbf{X}_s, t)$. It is convenient to introduce another quantity, $F_s(\mathbf{X}_1, \ldots, \mathbf{X}_s, t)$, defined

$$F_s(\mathbf{X}_1, \ldots, \mathbf{X}_s, t) \equiv V^s \int \cdots \int \rho^N(\mathbf{X}_1, \ldots, \mathbf{X}_N, t)\, d\mathbf{X}_{s+1} \cdots d\mathbf{X}_N \quad (7.36)$$

and

$$F_N(\mathbf{X}_1, \ldots, \mathbf{X}_N, t) \equiv V^N \rho^N(\mathbf{X}_1, \ldots, \mathbf{X}_N, t). \quad (7.37)$$

Let us assume that the evolution of the system is governed by a Hamiltonian of the form

$$H^N = \sum_{i=1}^{N} \frac{|\mathbf{p}_i|^2}{2m} + \sum_{i<j}^{N(N-1)/2} \phi(|\mathbf{q}_i - \mathbf{q}_j|), \quad (7.38)$$

where $\phi(|\mathbf{q}_i - \mathbf{q}_j|)$ is a two-body spherically symmetric interaction potential between particles i and j. The equation of motion of $\rho^N(\mathbf{X}_1, \ldots, \mathbf{X}_N, t)$ can be written

$$\frac{\partial \rho^N}{\partial t} = -\mathscr{H}^N \rho^N, \quad (7.39)$$

where

$$\mathscr{H}^N(\mathbf{X}_1, \ldots, \mathbf{X}_N) = \sum_{i=1}^{N} \frac{\mathbf{p}_i}{m} \cdot \frac{\partial}{\partial \mathbf{q}_i} - \sum_{i<j} \widehat{\Theta}_{ij} \quad (7.40)$$

and

$$\widehat{\Theta}_{ij} = \frac{\partial \phi_{ij}}{\partial \mathbf{q}_i} \cdot \frac{\partial}{\partial \mathbf{p}_i} + \frac{\partial \phi_{ij}}{\partial \mathbf{q}_j} \cdot \frac{\partial}{\partial \mathbf{p}_j} \quad (7.41)$$

(cf. Eqs. (7.25) and (7.26)).

If we integrate Eq. (7.39) over $\mathbf{X}_{s+1}, \ldots, \mathbf{X}_N$ and multiply by V^s, we obtain

$$\begin{aligned} \frac{\partial F_s}{\partial t} + \mathscr{H}^s F_s = V^s \int \cdots \int d\mathbf{X}_{s+1} \cdots d\mathbf{X}_N \\ \times \left\{ -\sum_{i=s+1}^{N} \frac{\mathbf{p}_i}{m} \cdot \frac{\partial}{\partial \mathbf{q}_i} + \sum_{i \le s; s+1 \le j \le N} \widehat{\Theta}_{ij} + \sum_{s+1 \le k < l} \widehat{\Theta}_{kl} \right\} \\ \times \rho^N(\mathbf{X}_1, \ldots, \mathbf{X}_N, t). \quad (7.42) \end{aligned}$$

If we assume that $\rho^N(\mathbf{X}_1, \ldots, \mathbf{X}_N, t) \to 0$ for large values of $\mathbf{X}_i$, then the first and third terms on the right-hand side of Eq. (7.42) go to zero. One can see this by using Gauss's theorem and changing the volume integration to surface integration. For a large system the contribution from $\rho^N(\mathbf{X}_1, \ldots, \mathbf{X}_N, t)$ on the surface goes to zero. The second term on the right-hand side can be written in the form

$$\begin{aligned} V^s \int \cdots \int d\mathbf{X}_{s+1} \cdots d\mathbf{X}_N \sum_{i \le s, s+1 \le j \le N} \widehat{\Theta}_{ij} \rho^N(\mathbf{X}_1, \ldots, \mathbf{X}_N, t) \\ = V^s(N-s) \sum_{i=1}^{s} \int d\mathbf{X}_{s+1} \widehat{\Theta}_{i,s+1} \int \cdots \int d\mathbf{X}_{s+2} \cdots d\mathbf{X}_N \rho^N(\mathbf{X}_1, \ldots, \mathbf{X}_N, t) \\ = \frac{(N-s)}{V} \sum_{i=1}^{s} \int d\mathbf{X}_{s+1} \widehat{\Theta}_{i,s+1} F_{s+1}(\mathbf{X}_1, \ldots, \mathbf{X}_{s+1}, t). \end{aligned}$$

Eq. (7.42) then becomes

$$\frac{\partial F_s}{\partial t} + \mathscr{H}^s F_s = \frac{(N-s)}{V} \sum_{i=1}^{s} \int d\mathbf{X}_{s+1} \hat{\Theta}_{i,s+1} F_{s+1}(\mathbf{X}_1, \ldots, \mathbf{X}_{s+1}, t). \tag{7.43}$$

For a fixed value of s we may take the limit $N \to \infty$, $V \to \infty$, such that $v \equiv V/N$ remains constant (this is called the thermodynamic limit) and Eq. (7.43) becomes

$$\frac{\partial F_s}{\partial t} + \mathscr{H}^s F_s = \frac{1}{v} \sum_{i=1}^{s} \int d\mathbf{X}_{s+1} \hat{\Theta}_{i,s+1} F_{s+1}(\mathbf{X}_1, \ldots, \mathbf{X}_{s+1}, t). \tag{7.44}$$

Eq. (7.44) gives a hierarchy of equations of motion for the reduced probability densities $F_s(\mathbf{X}_1, \ldots, \mathbf{X}_s, t)$. It is called the BBGKY hierarchy. The most useful equations in the hierarchy are those for $F_1(\mathbf{X}_1, t)$ and $F_2(\mathbf{X}_1, \mathbf{X}_2, t)$:

$$\frac{\partial F_1}{\partial t} + \frac{\mathbf{p}_1}{m} \cdot \frac{\partial F_1}{\partial \mathbf{q}_1} = \frac{1}{v} \int d\mathbf{X}_2 \hat{\Theta}_{1,2} F_2(\mathbf{X}_1, \mathbf{X}_2, t) \tag{7.45}$$

and

$$\frac{\partial F_2}{\partial t} + \left(\frac{\mathbf{p}_1}{m} \cdot \frac{\partial}{\partial \mathbf{q}_1} + \frac{\mathbf{p}_2}{m} \cdot \frac{\partial}{\partial \mathbf{q}_2} - \hat{\Theta}_{12} \right) F_2$$
$$= \frac{1}{v} \int d\mathbf{X}_3 (\hat{\Theta}_{13} + \hat{\Theta}_{23}) F_3(\mathbf{X}_1, \mathbf{X}_2, \mathbf{X}_3, t). \tag{7.46}$$

Notice that the equation of motion of F_1 depends on F_2, the equation of motion for F_2 depends on F_3, etc. This makes the equations of the hierarchy impossible to solve unless some way can be found to truncate it. For example, if we could find some way to write $F_2(\mathbf{X}_1, \mathbf{X}_2, t)$ in terms of $F_1(\mathbf{X}_1, t)$ and $F_1(\mathbf{X}_2, t)$, then we could in principle solve Eq. (7.45) for the reduced probability density $F_1(\mathbf{X}_1, t)$. Eq. (7.45) is called the *kinetic equation*.

D. MICROSCOPIC BALANCE EQUATIONS (CLASSICAL FLUIDS)

If we consider systems of particles with short-range interactions and long-wavelength inhomogeneities, we can derive microscopic balance equations (cf. App. A) for the particle density, momentum density, and energy density. These quantities are of particular interest because they are the densities of quantities that are conserved during the evolution of a system whose dynamics is governed by a Hamiltonian of the type given in Eq. (7.38) and they give rise to the hydrodynamic equations.

As we have seen, the expectation value of an arbitrary phase function $O^N(\mathbf{X}^N)$ at time t is given by

$$\langle O(t)\rangle = \mathrm{Tr}_N\, O^N\rho^N(t) \equiv \int\cdots\int d\mathbf{X}_1\cdots d\mathbf{X}_N O^N(\mathbf{X}^N)\rho^N(\mathbf{X}^N, t)$$

$$= \int\cdots\int d\mathbf{X}_1, \ldots, d\mathbf{X}_N O^N(\mathbf{X}^N)\, e^{-\hat{\mathscr{H}}^N(\mathbf{X}^N)t}\rho^N(\mathbf{X}^N, 0). \tag{7.47}$$

We have written the expectation value in the Schrödinger picture. That is, we have allowed the state function $\rho^N(\mathbf{X}^N, t)$ to evolve in time and have kept the phase function $O^N(\mathbf{X}^N)$ fixed at time $t = 0$. We could equally well allow the phase function to vary in time, keeping the state function fixed. We note that the evolution operator $\hat{\mathscr{H}}^N(\mathbf{X}^N)$ contains derivatives with respect to $\mathbf{p}^N$ and $\mathbf{q}^N$. If we expand the exponential in a power series and integrate by parts, there will be a change of sign for each partial derivative and we find

$$\langle O(t)\rangle = \mathrm{Tr}_N\, O^N = \int\cdots\int d\mathbf{X}_1\cdots d\mathbf{X}_N O^N(\mathbf{X}^N, t)\rho^N(\mathbf{X}^N)$$

$$= \int\cdots\int d\mathbf{X}_1\cdots d\mathbf{X}_N [e^{\hat{\mathscr{H}}^N(\mathbf{X}^N)t}O^N(\mathbf{X}^N)]\rho^N(\mathbf{X}^N). \tag{7.48}$$

Thus, we obtain the classical version of the Heisenberg picture. We see that phase functions and state functions evolve according to different laws. The equation of motion of a phase function is given by

$$\frac{\partial O^N(\mathbf{X}^N, t)}{\partial t} = \hat{\mathscr{H}}^N O^N(\mathbf{X}^N, t) = \sum_{i=1}^{N}\left(\dot{\mathbf{q}}_i\cdot\frac{\partial}{\partial \mathbf{q}_i} + \dot{\mathbf{p}}_i\cdot\frac{\partial}{\partial \mathbf{p}_i}\right)O^N(\mathbf{X}^N, t), \tag{7.49}$$

where Eq. (7.49) gives the evolution of $O^N(\mathbf{X}^N, t)$ at a fixed point.

We can use Eq. (7.49) to find microscopic balance equations for the number density, momentum density, and energy density. Before we begin, it is useful to write explicit expressions for the equations of motion of the phase space coordinates. For the Hamiltonian in Eq. (7.38) we find

$$\dot{\mathbf{q}}_i = \frac{\mathbf{p}_i}{m} \tag{7.50}$$

and

$$\dot{\mathbf{p}}_i = -\sum_{l\neq i}\frac{\partial\phi(|\mathbf{q}_i - \mathbf{q}_l|)}{\partial \mathbf{q}_i} = \sum_{l\neq i}\mathbf{F}_{il}, \tag{7.51}$$

where the force $\mathbf{F}_{il}$ has the property

$$\mathbf{F}_{il} \equiv -\frac{\partial\phi(|\mathbf{q}_i - \mathbf{q}_l|)}{\partial \mathbf{q}_i} = \frac{\partial\phi(|\mathbf{q}_i - \mathbf{q}_l|)}{\partial \mathbf{q}_l} = -\mathbf{F}_{li}. \tag{7.52}$$

Now we can obtain the balance equations.

1. Balance Equation for the Particle Density

The microscopic operator which gives the particle density at position **R** is defined

$$n(\mathbf{q}^N, \mathbf{R}) \equiv \sum_{i=1}^{N} \delta(\mathbf{q}_i - \mathbf{R}). \tag{7.53}$$

The equation of motion for $n(\mathbf{q}^N; \mathbf{R})$ can be obtained using Eqs. (7.49) and (7.53):

$$\frac{\partial n}{\partial t} = \sum_{i=1}^{N} \dot{\mathbf{q}}_i \cdot \frac{\partial}{\partial \mathbf{q}_i} \delta(\mathbf{q}_i - \mathbf{R}) = -\nabla_{\mathbf{R}} \cdot \sum_{i=1}^{N} \frac{\mathbf{p}_i}{m} \delta(\mathbf{q}_i - \mathbf{R}) \tag{7.54}$$

or

$$\frac{\partial}{\partial t} n(\mathbf{q}^N; \mathbf{R}) = -\nabla_{\mathbf{R}} \cdot \mathbf{J}^n(\mathbf{p}^N, \mathbf{q}^N; \mathbf{R}) \tag{7.55}$$

where

$$\mathbf{J}^n(\mathbf{p}^N, \mathbf{q}^N; \mathbf{R}) = \sum_{i=1}^{N} \frac{\mathbf{p}_i}{m} \delta(\mathbf{q}_i - \mathbf{R}). \tag{7.56}$$

The quantity $\mathbf{J}^n(\mathbf{p}^N, \mathbf{q}^N; \mathbf{R})$ is the microscopic particle current. Eq. (7.55) is a microscopic balance equation which expresses the conservation of particle number. Notice that Eq. (7.55) is invariant under time reversal, as it should be since Newton's laws are reversible.

2. Balance Equation for the Momentum Density

The momentum density is given by the operator

$$m\mathbf{J}^n(\mathbf{p}^N, \mathbf{q}^N; \mathbf{R}) \equiv \sum_{i=1}^{N} \mathbf{p}_i \, \delta(\mathbf{q}_i - \mathbf{R}). \tag{7.57}$$

The equation of motion of $m\mathbf{J}^n(\mathbf{p}^N, \mathbf{q}^N; \mathbf{R})$ is given by the equation

$$\frac{\partial}{\partial t} m\mathbf{J}^n(\mathbf{p}^N, \mathbf{q}^N; \mathbf{R}) = \sum_{i=1}^{N} \left[\dot{\mathbf{p}}_i \, \delta(\mathbf{q}_i - \mathbf{R}) + \mathbf{p}_i \dot{\mathbf{q}}_i \cdot \frac{\partial}{\partial \mathbf{q}_i} \delta(\mathbf{q}_i - \mathbf{R}) \right]. \tag{7.58}$$

Let us consider the terms on the right-hand side of Eq. (7.58), one at a time. The first term on the right can be written

$$\sum_{i=1}^{N} \dot{\mathbf{p}}_i \, \delta(\mathbf{q}_i - \mathbf{R}) = \sum_{i \neq l}^{N} \sum^{N} \mathbf{F}_{il} \, \delta(\mathbf{q}_i - \mathbf{R}) = \frac{1}{2} \sum_{i \neq l}^{N} \sum^{N} [\mathbf{F}_{il}(\delta(\mathbf{q}_i - \mathbf{R}) - \delta(\mathbf{q}_l - \mathbf{R}))]. \tag{7.59}$$

In the third step in Eq. (7.59) we have interchanged dummy indices and we have used $\mathbf{F}_{il} = -\mathbf{F}_{li}$.

In Eq. (7.59) we now restrict ourselves to short-ranged interactions and long-wavelength inhomogeneities. Let us assume that the particles are contained in a cubic box with sides of length L and volume $V = L^3$. We first introduce the Fourier expansion of the particle density:

$$n(\mathbf{q}^N, \mathbf{R}) = \sum_{i=1}^{N} \delta(\mathbf{q}_i - \mathbf{R}) = \frac{1}{V}\sum_{\mathbf{k}}\sum_{i=1}^{N} e^{i\mathbf{k}\cdot(\mathbf{q}_i - \mathbf{R})} = \frac{1}{V}\sum_{\mathbf{k}} n_{\mathbf{k}}(\mathbf{q}^N)\, e^{-i\mathbf{k}\cdot\mathbf{R}}. \tag{7.60}$$

The vector $\mathbf{k}$ has components $(2\pi l/L, 2\pi m/L, 2\pi n/L)$ where l, m, and n are integers. For long-wavelength disturbances, small integers will give the major contribution; that is, small $|\mathbf{k}|$ will give the major contribution. We can therefore make the following approximation:

$$\begin{aligned}(\delta(\mathbf{q}_i - \mathbf{R}) - \delta(\mathbf{q}_l - \mathbf{R})) &= \frac{1}{V}\sum_{\mathbf{k}} e^{-i\mathbf{k}\cdot\mathbf{R}}(e^{i\mathbf{k}\cdot\mathbf{q}_i} - e^{i\mathbf{k}\cdot\mathbf{q}_l}) \\ &= \frac{1}{V}\sum_{\mathbf{k}} e^{-i\mathbf{k}\cdot\mathbf{R}}\, e^{i\mathbf{k}\cdot\mathbf{q}_i}(i\mathbf{k}\cdot(\mathbf{q}_i - \mathbf{q}_l)) + O(k^2) \\ &= -\nabla_{\mathbf{R}}\cdot(\mathbf{q}_i - \mathbf{q}_l)\,\delta(\mathbf{q}_i - \mathbf{R}) + O(k^2).\end{aligned} \tag{7.61}$$

We may now write Eq. (7.59) as

$$\sum_{i=1}^{N} [\dot{\mathbf{p}}_i\, \delta(\mathbf{q}_i - \mathbf{R})] \simeq -\tfrac{1}{2}\nabla_{\mathbf{R}}\cdot\left[\sum_{i\neq l}^{N}\sum^{N} (\mathbf{q}_i - \mathbf{q}_l)\cdot\mathbf{F}_{il}\,\delta(\mathbf{q}_i - \mathbf{R})\right]. \tag{7.62}$$

Let us next consider the second term on the right-hand side of Eq. (7.58). We find

$$\sum_{i=1}^{N} \mathbf{p}_i\dot{\mathbf{q}}_i\cdot\frac{\partial}{\partial\mathbf{q}_i}\,\delta(\mathbf{q}_i - \mathbf{R}) = -\frac{1}{m}\,\nabla_{\mathbf{R}}\cdot\sum_{i=1}^{N} \mathbf{p}_i\mathbf{p}_i\,\delta(\mathbf{q}_i - \mathbf{R}). \tag{7.63}$$

If we combine Eqs. (7.58), (7.62), and (7.63), we obtain the following microscopic equation for the momentum density:

$$m\frac{\partial}{\partial t}\mathbf{J}^n(\mathbf{q}^N; \mathbf{R}) = -\nabla_{\mathbf{R}}\cdot\mathbf{J}^p(\mathbf{p}^N, \mathbf{q}^N; \mathbf{R}), \tag{7.64}$$

where the momentum current tensor is defined

$$\mathbf{J}^p(\mathbf{q}^N, \mathbf{p}^N; \mathbf{R}) = \sum_{i=1}^{N} \frac{\mathbf{p}_i\mathbf{p}_i}{m}\,\delta(\mathbf{q}_i - \mathbf{R}) + \frac{1}{2}\sum_{i\neq l}^{N}\sum^{N} (\mathbf{q}_i - \mathbf{q}_l)\cdot\mathbf{F}_{il}\,\delta(\mathbf{q}_i - \mathbf{R}). \tag{7.65}$$

Thus, we obtain a microscopic balance equation for the momentum density which is valid for long wavelengths. Again, the equation for momentum density is reversible. However, if we are to describe the viscous behavior of fluids actually observed in nature, it must somehow change to an irreversible equation when we take its expectation value and obtain the hydrodynamic equations.

3. Balance Equation for the Energy Density

We define the energy density by

$$h(\mathbf{p}^N, \mathbf{q}^N; \mathbf{R}) \equiv \sum_{i=1}^{N} h_i\, \delta(\mathbf{q}_i - \mathbf{R}), \tag{7.66}$$

where

$$h_i = \frac{|\mathbf{p}_i|^2}{2m} + \frac{1}{2}\sum_{j \neq i}^{N} \phi(|\mathbf{q}_i - \mathbf{q}_j|). \tag{7.67}$$

By arguments similar to those given for the momentum density, we obtain the following balance equation for the energy density for systems with long-wavelength inhomogeneities:

$$\frac{\partial}{\partial t} h(\mathbf{p}^N, \mathbf{q}^N; \mathbf{R}) = -\nabla_{\mathbf{R}} \cdot \mathbf{J}^h(\mathbf{p}^N, \mathbf{q}^N; \mathbf{R}), \tag{7.68}$$

where the energy flux, or current, is given by

$$\mathbf{J}^h(\mathbf{p}^N, \mathbf{q}^N; \mathbf{R}) = \sum_{i=1}^{N} \left[h_i \frac{\mathbf{p}_i}{m} \delta(\mathbf{q}_i - \mathbf{R}) \right] + \frac{1}{2}\sum_{i \neq j}^{N}\sum^{N} \left[\frac{(\mathbf{p}_i + \mathbf{p}_j)}{m} \cdot \mathbf{F}_{ij}(\mathbf{q}_i - \mathbf{q}_j)\, \delta(\mathbf{q}_i - \mathbf{R}) \right]. \tag{7.69}$$

As we shall see in later chapters, these microscopic balance equations form the microscopic basis for the hydrodynamic equations.

E. PROBABILITY DENSITY OPERATOR[7,8]

For quantum systems the phase space coordinates of particles do not commute. Therefore, rather than introduce a probability density on the phase space of the system, it is convenient to work with a more general object called the probability density operator, $\hat{\rho}(t)$. The density operator is a positive definite Hermitian operator and can be used to obtain expectation values of observables in any desired representation. Given a density operator for a system, the expectation value of any observable $\hat{O}$ at time t is defined

$$\langle O(t) \rangle = \mathrm{Tr}\, \hat{O}\hat{\rho}(t) = \mathrm{Tr}\, \hat{O}(t)\hat{\rho}, \tag{7.70}$$

where the density operator is normalized so that

$$\mathrm{Tr}\, \hat{\rho}(t) = 1. \tag{7.71}$$

In Eqs. (7.70) and (7.71) the trace can be evaluated using any convenient complete set of basis states.

Since the density operator is a positive definite Hermitian operator, it can be diagonalized and it will have positive real eigenvalues. If the set $\{|\pi_i\rangle\}$ denotes the eigenstates of the density operator and $\{p_i\}$ its eigenvalues ($p_i \geqslant 0$ since $\hat{\rho}$ is positive definite), then the density operator can be written in the form

$$\hat{\rho}(t) = \sum_i p_i|\pi_i(t)\rangle\langle\pi_i(t)|, \tag{7.72}$$

where p_i is the probability of finding the system in state $|\pi_i\rangle$ and

$$\sum_i p_i = 1. \tag{7.73}$$

The expectation value of any operator becomes

$$\langle O(t)\rangle = \text{Tr}\,\hat{O}\hat{\rho}(t) = \sum_i p_i\langle\pi_i(t)|O|\pi_i(t)\rangle. \tag{7.74}$$

In Eq. (7.74), $\langle\pi_i(t)|\hat{O}|\pi_i(t)\rangle$ is the expectation value of $\hat{O}$ in the state $|\pi_i(t)\rangle$ and p_i is the probability of finding the system in that state.

Let us assume that we have an arbitrary complete orthonormal set of states $\{|n\rangle\}$ which are not eigenstates of $\hat{\rho}(t)$. Then the probability $P_n(t)$ that the system is in the state $|n\rangle$ at time t is given by the expectation value of the density operator in that state:

$$P_n(t) = \langle n|\hat{\rho}(t)|n\rangle = \sum_i p_i\langle n|\pi_i(t)\rangle\langle\pi_i(t)|n\rangle. \tag{7.75}$$

The expectation value of an arbitrary operator $\hat{O}$ may be evaluated with respect to any representation. Thus, in the representation, $\{|n\rangle\}$, we have

$$\langle O(t)\rangle = \text{Tr}\,\hat{O}\hat{\rho}(t) = \sum_{n',n}\langle n|\hat{O}|n'\rangle\langle n'|\hat{\rho}(t)|n\rangle. \tag{7.76}$$

The quantity $\langle n'|\hat{\rho}(t)|n\rangle$ is called the density matrix.

If the dynamics of the system is described by a Hamiltonian, $\hat{H}$, then the states $|\pi_i(t)\rangle$ will obey the Schrödinger equation,

$$i\hbar\frac{\partial}{\partial t}|\pi_i(t)\rangle = \hat{H}|\pi_i(t)\rangle, \tag{7.77}$$

where $\hbar$ is Planck's constant. If we take the time derivative of $\hat{\rho}(t)$ and make use of the Schrödinger equation, we find an equation for $\hat{\rho}(t)$ in the Schrödinger picture

$$i\hbar\frac{\partial\hat{\rho}}{\partial t} = \hat{H}\hat{\rho} - \hat{\rho}\hat{H} \tag{7.78}$$

or

$$i\frac{\partial\hat{\rho}}{\partial t} = \hat{L}\hat{\rho}, \tag{7.79}$$

where

$$\hat{L} = \frac{1}{\hbar}[\hat{H},\ ] \tag{7.80}$$

is the quantum version of the Liouville operator and is Hermitian, and Eq.

(7.79) is the Liouville equation. If the density operator is known at time $t = 0$, then its value at time t is given by

$$\hat{\rho}(t) = e^{-i\hat{L}t}\hat{\rho}(0) = e^{-i\hat{H}t/\hbar}\hat{\rho}(0)\, e^{i\hat{H}t/\hbar}. \tag{7.81}$$

From Eqs. (7.70) and (7.71), we can use the invariance of the trace under cyclic rotation of operators to show that the operator $\hat{O}(t)$ evolves in time as

$$\hat{O}(t) = e^{i\hat{H}t/\hbar}\hat{O}(0)\, e^{-i\hat{H}t/\hbar} \tag{7.82}$$

and obeys the equation of motion,

$$-i\frac{\partial \hat{O}}{\partial t} = \frac{1}{\hbar}[\hat{H}, \hat{O}] = \hat{L}\hat{O}. \tag{7.83}$$

Eq. (7.83) describes the evolution of the system in the Heisenberg picture.

It is often convenient to expand the density operator in terms of a complete orthonormal set of eigenstates of the Hamiltonian, $\{|E_k\rangle\}$, where k ranges over all eigenstates. If we note that

$$1 = \sum_k |E_k\rangle\langle E_k|, \tag{7.84}$$

then Eq. (7.81) takes the form

$$\hat{\rho}(t) = \sum_{k,l} \langle E_k|\hat{\rho}(0)|E_l\rangle\, e^{(-i/\hbar)(E_k - E_l)t}|E_k\rangle\langle E_l|. \tag{7.85}$$

From Eq. (7.85), we see that a stationary state occurs when all off-diagonal matrix elements of $\hat{\rho}(0)$, for which $E_k \neq E_l$, vanish. Thus, for a system with no degenerate energy levels, the stationary state must be diagonal in the energy representation:

$$\hat{\rho}_{\text{stat}} = f(\hat{H}); \tag{7.86}$$

that is, it must be a function of the Hamiltonian. For a system with degenerate energy levels, one may still diagonalize both $\hat{\rho}$ and $\hat{H}$ simultaneously by introducing additional invariants of the motion, $\hat{I}$, which commute with $\hat{H}$. Thus, in general, a stationary state will be a function of $\hat{H}$ and all operators which commute with $\hat{H}$:

$$\hat{\rho}_{\text{stat}} = f(\hat{H}, \hat{I}). \tag{7.87}$$

For systems which approach thermodynamic equilibrium, the stationary state is the equilibrium state.

As a simple application of the density matrix, let us consider a beam of photons with various polarizations. Let us assume that the beam travels along the z-axis, and let us represent the state with polarization in the x-direction by $|\alpha\rangle = \begin{pmatrix}1\\0\end{pmatrix}$ and the state with polarization in the y-direction by $|\beta\rangle = \begin{pmatrix}0\\1\end{pmatrix}$. The state with polarization at some angle ϕ with respect to the x-axis is given by

$$|\psi\rangle = e^{i\alpha}\left[\cos\phi\begin{pmatrix}1\\0\end{pmatrix} + \sin\phi\begin{pmatrix}0\\1\end{pmatrix}\right], \tag{7.88}$$

where α is an overall phase factor. The density matrix describing this state is a pure state and is given by

$$\hat{\rho}_{\text{pure}} = \begin{pmatrix} \cos\phi \\ \sin\phi \end{pmatrix} (\cos\phi, \sin\phi) = \begin{pmatrix} \cos^2\phi & \cos\phi\sin\phi \\ \sin\phi\cos\phi & \sin^2\phi \end{pmatrix}. \tag{7.89}$$

The probability of finding the photon with polarization along the x-direction is $\langle\alpha|\hat{\rho}_{\text{pure}}|\alpha\rangle = \cos^2\phi$. Similarly, the probability of finding it polarized in the y-direction is $\langle\beta|\hat{\rho}_{\text{pure}}|\beta\rangle = \sin^2\phi$.

Eq. (7.89) gives the density matrix for a pure state. If we wish to describe a beam which is a mixture of pure states, then the density matrix of the mixture becomes

$$\hat{\rho}_{\text{mix}} = \sum_i p_i \begin{pmatrix} \cos^2\phi_i & \cos\phi_i\sin\phi_i \\ \sin\phi_i\cos\phi_i & \sin^2\phi_i \end{pmatrix}, \tag{7.90}$$

where p_i is the probability of finding a beam with polarization at angle ϕ_i.

F. REDUCED DENSITY OPERATOR[9]

For quantum systems, as for classical systems, we generally only need to find expectation values for one- and two-body operators. In view of this, it is useful to introduce the concept of one- and two-body reduced density matrices.

Let us first consider the expectation value of the one-body operator $\hat{O}_{(1)}$ (cf. App. B for a discussion of notation used in this section). This takes the form

$$\langle O_{(1)}(t)\rangle = \text{Tr}\, \hat{O}_{(1)}\hat{\rho}(t). \tag{7.91}$$

In the number representation (cf. App. B) we can write

$$\begin{aligned} \langle O_{(1)}(t)\rangle &= \sum_{\substack{\{n_\alpha\} \\ \sum_\alpha n_\alpha = N}} \sum_{k_1 k_1'} \langle k_1|\hat{O}_1|k_1'\rangle \langle\{n_\alpha\}|\hat{a}^+_{k_1}\hat{a}_{k_1'}\hat{\rho}^{nb}(t)|\{n_\alpha\}\rangle \\ &= \sum_{k_1 k_1'} \langle k_1|\hat{O}_1|k_1'\rangle\langle k_1'|\hat{\rho}_{(1)}(t)|k_1\rangle, \end{aligned} \tag{7.92}$$

where $\langle k_1'|\hat{\rho}_{(1)}(t)|k_1\rangle$ is the one-particle reduced density matrix and is defined

$$\langle k_1'|\hat{\rho}_{(1)}(t)|k_1\rangle = \text{Tr}\, \hat{a}^+_{k_1}\hat{a}_{k_1'}\hat{\rho}^{nb}(t) = \sum_{\substack{\{n_\alpha\} \\ \sum_\alpha n_\alpha = N}} \langle\{n_\alpha\}|\hat{a}^+_{k_1}\hat{a}_{k_1'}\hat{\rho}^{nb}(t)|\{n_\alpha\}\rangle. \tag{7.93}$$

Similarly, in terms of position variables we can write

$$\langle O_{(1)}(t)\rangle = \iint dr_1\, dr_1' \langle r_1|\hat{O}_1|r_1'\rangle\langle r_1'|\hat{\rho}_{(1)}(t)|r_1\rangle, \tag{7.94}$$

where

$$\begin{aligned} \langle r_1'|\hat{\rho}_{(1)}(t)|r_1\rangle &= \text{Tr}\, \hat{\psi}^+(r_1)\hat{\psi}(r_1')\hat{\rho}^{nb}(t) \\ &= \sum_{\substack{\{n_\alpha\} \\ \sum_\alpha n_\alpha = N}} \langle\{n_\alpha\}|\hat{\psi}^+(r_1)\hat{\psi}(r_1')\hat{\rho}^{nb}(t)|\{n_\alpha\}\rangle. \end{aligned} \tag{7.95}$$

The operators $\hat{\psi}^+(r)$ and $\hat{\psi}(r)$ are field operators.

The one-body reduced density matrices can also be written in terms of symmetrized or antisymmetrized matrix elements. For simplicity, we shall consider the case of fermions. Then we obtain

$$\begin{aligned}\langle O_{(1)}(t)\rangle &= \left(\frac{1}{N!}\right)^2 \sum_{\substack{k_1\cdots k_N\\ k_1'\cdots k_N'}} {}^{(-)}\langle k_1,\ldots,k_N|O_{(1)}^N|k_1',\ldots,k_N'\rangle^{(-)}\\ &\quad\times {}^{(-)}\langle k_1',\ldots,k_N'|\hat{\rho}^N(t)|k_1,\ldots,k_N\rangle^{(-)}\\ &= \frac{1}{(N-1)!}\sum_{k_1',k_1,\ldots,k_N}\langle k_1|O_1|k_1'\rangle\\ &\quad\times {}^{(-)}\langle k_1',k_2,\ldots,k_N|\hat{\rho}^N(t)|k_1,k_2,\ldots,k_N\rangle^{(-)}\end{aligned} \tag{7.96}$$

(cf. Eq. B.52). Comparison with Eq. (7.92) shows that

$$\langle k_1'|\hat{\rho}_{(1)}(t)|k_1\rangle = \frac{1}{(N-1)!}\sum_{k_2,\ldots,k_N} {}^{(-)}\langle k_1',k_2,\ldots,k_N|\rho^N(t)|k_1,k_2,\ldots,k_N\rangle^{(-)}. \tag{7.97}$$

Similarly, in terms of position variables we find

$$\begin{aligned}&\langle r_1'|\hat{\rho}_{(1)}(t)|r_1\rangle\\ &= \frac{1}{(N-1)!}\int\cdots\int dr_2\cdots dr_N\, {}^{(-)}\langle r_1',r_2,\ldots,r_N|\rho^N(t)|r_1,r_2,\ldots,r_N\rangle^{(-)}.\end{aligned} \tag{7.98}$$

The factors $(1/N!)^2$ and $1/(N-1)!$ in Eq. (7.96) go with the summations and prevent overcounting of states.

When we deal with two-body operators, we get similar expressions. Thus,

$$\begin{aligned}\langle O_{(2)}(t)\rangle &= \mathrm{Tr}\, \hat{O}_{(2)}\hat{\rho}(t) = \frac{1}{2}\sum_{k_1,k_2}\sum_{k_1',k_2'}\langle k_1,k_2|\hat{O}_{12}|k_1',k_2'\rangle\langle k_1',k_2'|\hat{\rho}_{(2)}(t)|k_1,k_2\rangle\\ &= \frac{1}{2}\int\cdots\int dr_1\, dr_2\, dr_1'\, dr_2'\langle r_1,r_2|\hat{O}_{12}|r_1',r_2'\rangle\langle r_1',r_2'|\hat{\rho}_{(2)}(t)|r_1,r_2\rangle,\end{aligned} \tag{7.99}$$

where

$$\begin{aligned}&\langle k_1',k_2'|\hat{\rho}_{(2)}(t)|k_1,k_2\rangle\\ &= \sum_{\substack{\{n_\alpha\}\\ \sum_\alpha n_\alpha = N}}\langle\{n_\alpha\}|\hat{a}_{k_1}^+\hat{a}_{k_2}^+\hat{a}_{k_2'}\hat{a}_{k_1'}\hat{\rho}^{nb}(t)|\{n_\alpha\}\rangle\\ &= \frac{1}{(N-2)!}\sum_{k_3,\ldots,k_N} {}^{(-)}\langle k_1',k_2',k_3,\ldots,k_N|\hat{\rho}^N(t)|k_1,k_2,\ldots,k_N\rangle^{(-)}\end{aligned} \tag{7.100}$$

and

$$\langle r_1', r_2'|\hat{\rho}_{(2)}(t)|r_1, r_2\rangle$$
$$= \sum_{\substack{\{n_\alpha\} \\ \Sigma_\alpha n_\alpha = N}} \langle\{n_\alpha\}|\hat{\psi}^+(r_1)\hat{\psi}^+(r_2)\hat{\psi}(r_2)\hat{\psi}(r_1)\hat{\rho}^{nb}(t)|\{n_\alpha\}\rangle$$
$$= \frac{1}{(N-2)!}\int\cdots\int dr_3\cdots dr_N{}^{(-)}\langle r_1', r_2', r_3, \ldots, r_N|\hat{\rho}^N(t)|r_1, r_2, \ldots, r_N\rangle^{(-)}. \tag{7.101}$$

In Eqs. (7.100) and (7.101), the expressions for the reduced two-body density matrices in terms of the number representation hold for both fermions and bosons. The expressions in terms of antisymmetrized matrix elements hold only for fermions. The expressions we have obtained for the one- and two-body reduced density matrices can be extended to the case of n-particle reduced density matrices.

It is of interest to study the properties of the operators $\hat{\rho}_{(1)}$ and $\hat{\rho}_{(2)}$. We first note that diagonal matrix elements of $\hat{\rho}_{(1)}$ with respect to a given state give the average occupation number of that state. Thus,

$$\langle k_1|\hat{\rho}_{(1)}(t)|k_1\rangle = \mathrm{Tr}\, \hat{a}_{k_1}^+\hat{a}_{k_1}\hat{\rho}^{nb}(t) = \langle n(k_1, t)\rangle \tag{7.102}$$

and

$$\langle r_1|\hat{\rho}_{(1)}(t)|r_1\rangle = \mathrm{Tr}\, \hat{\psi}^+(r_1)\hat{\psi}(r_1)\hat{\rho}^{nb}(t) = \langle n(r_1, t)\rangle, \tag{7.103}$$

where $\langle n(k_1, t)\rangle$ is the average number of particles with momentum k_1 at time t and $\langle n(r_1, t)\rangle$ is the average number of particles at position r_1 at time t. (Note that the momentum and position variables, k_1 and r_1, respectively, can also include spin variables; cf. App. B.) The one-particle reduced density operator is normalized to the total number of particles. Thus,

$$\sum_{k_1} \langle k_1|\rho_{(1)}(t)|k_1\rangle = \int dr_1\langle r_1|\hat{\rho}_{(1)}(t)|r_1\rangle = N. \tag{7.104}$$

Similarly, the two-particle reduced density operator is normalized to $N(N-1)$:

$$\sum_{k_1,k_2} \langle k_1, k_2|\rho_{(2)}(t)|k_1, k_2\rangle = \iint dr_1\, dr_2\langle r_1, r_2|\rho_2(t)|r_1, r_2\rangle = N(N-1). \tag{7.105}$$

From Eq. (7.102), we see that the operator $\hat{\rho}_{(1)}$ must have eigenvalues $\lambda_j \leqslant N$ and $\hat{\rho}_{(2)}$ must have eigenvalues $\lambda_j \leqslant N(N-1)$. It is useful to expand $\hat{\rho}_{(1)}$ in terms of its complete set of orthonormal eigenstates which we shall denote $\{|\pi_j^{(1)}\rangle\}$. Thus,

$$\hat{\rho}_{(1)} = \sum_j \lambda_j|\pi_j^{(1)}\rangle\langle\pi_j^{(1)}| \tag{7.106}$$

and the matrix element $\langle\pi_j^{(1)}|\hat{\rho}_{(1)}|\pi_j^{(1)}\rangle = \lambda_j$ is just the occupation number of the state $|\pi_j^{(1)}\rangle$. If we now take the trace of $\hat{\rho}_{(1)}$ in the position representation and assume that the states $\langle r|\pi_j^{(1)}\rangle = \pi_j^{(1)}(r)$ and $\langle\pi_j^{(1)}|r\rangle = \pi_j^{(1)*}(r)$ are orthonormal, we find

$$\int dr\langle r|\hat{\rho}_{(1)}|r\rangle = \sum_j \lambda_j \int dr\langle r|\pi_j^{(1)}\rangle\langle\pi_j^{(1)}|r\rangle$$

$$= \sum_j \lambda_j \int dr\pi_j^{(1)*}(r)\pi_j^{(1)}(r) = \sum_j \lambda_j = N. \qquad (7.107)$$

To have this normalization, $\pi_j^{(1)}(r)$ must have the volume dependence $\pi_j^{(1)}(r) \sim (V)^{-1/2}$.

Let us now consider the off-diagonal matrix element

$$\langle r_1|\hat{\rho}_{(1)}|r_2\rangle = \sum_j \lambda_j\langle r_1|\pi_j^{(1)}\rangle\langle\pi_j^{(1)}|r_2\rangle = \sum_j \lambda_j\pi_j^{(1)*}(r_2)\pi_j^{(1)}(r_1) \qquad (7.108)$$

and let us take the so-called thermodynamic limit $N \to \infty$ and $V \to \infty$ such that the number density N/V remains constant. Then we can consider what happens to $\langle r_1|\hat{\rho}_{(1)}|r_2\rangle$ when we let $|\mathbf{r}_1 - \mathbf{r}_2| \to \infty$. If the eigenvalues λ_j all remain finite as we take the thermodynamic limit, then

$$\lim_{|\mathbf{r}_1-\mathbf{r}_2|\to\infty} \langle r_1|\hat{\rho}_{(1)}|r_2\rangle = 0 \qquad (7.109)$$

since the product $\pi_j^{(1)*}(r_2)\pi_j^{(1)}(r_1) \sim 1/V$. However, if one of the eigenvalues, say λ_0, is proportional to the number of particles ($\lambda_0 = N\alpha$ where α is a finite fraction), then the limit $|\mathbf{r}_1 - \mathbf{r}_2| \to \infty$ will be nonzero since

$$\lim_{|\mathbf{r}_1-\mathbf{r}_2|\to\infty} \langle r_1|\hat{\rho}_{(1)}|r_2\rangle = \frac{N\alpha}{V} f(r_1, r_2) \qquad (7.110)$$

where $f(r_1, r_2)$ as a function of r_1 and r_2. A system which behaves according to Eq. (7.110) is said to exhibit off-diagonal, long-range order (ODLRO) in the one-body reduced density matrix. Fermi systems cannot exhibit ODLRO in the one-body reduced density matrices because of the Pauli exclusion principle, which restricts each eigenvalue to the values $\lambda_j \leqslant 1$. However, boson systems can exhibit ODLRO in $\langle r_1|\hat{\rho}_{(1)}|r_2\rangle$. This is precisely what happens when gauge symmetry is broken in boson systems and they become superfluid. An example of a boson system with ODLRO in $\langle r_1|\rho_{(1)}|r_2\rangle$ is superfluid liquid He^4. Fermi systems can exhibit ODLRO in higher order reduced density matrices (usually $\hat{\rho}_{(2)}$). Examples of ODLRO in Fermi systems are superconductivity in metals and superfluidity in liquid He^3.

G. WIGNER FUNCTION[10–15]

For quantum mechanical systems, the phase space coordinates of particles do not commute and, therefore, it is impossible to specify simultaneously the position and momentum of the particles. As a result, it is also not possible to define a distribution function on the phase space which can be interpreted as a probability density. However, Wigner was first to show that it is possible to introduce a function which is formally analogous to the classical probability density and which reduces to it in the classical limit.

We introduce one- and two-particle reduced Wigner functions in the following way. We first write the off-diagonal density matrices $\langle \mathbf{r}_1 | \rho_{(1)} | \mathbf{r}_1' \rangle$ and $\langle \mathbf{r}_1\mathbf{r}_2 | \rho_{(2)} | \mathbf{r}_1'\mathbf{r}_2' \rangle$ in terms of relative and center of mass coordinates (for simplicity, we neglect spin dependence). Thus,

$$\left\langle \mathbf{R} + \frac{\mathbf{r}}{2} \middle| \hat{\rho}_{(1)}(t) \middle| \mathbf{R} - \frac{\mathbf{r}}{2} \right\rangle = \operatorname{Tr} \hat{\psi}^{+}\left(\mathbf{R} + \frac{\mathbf{r}}{2}\right) \hat{\psi}\left(\mathbf{R} - \frac{\mathbf{r}}{2}\right) \hat{\rho}(t) \quad (7.111)$$

and

$$\left\langle \mathbf{R}_1 + \frac{\mathbf{r}_1}{2}, \mathbf{R}_2 + \frac{\mathbf{r}_2}{2} \middle| \hat{\rho}_{(2)}(t) \middle| \mathbf{R}_1 - \frac{\mathbf{r}_1}{2}, \mathbf{R}_2 - \frac{\mathbf{r}_2}{2} \right\rangle$$
$$= \operatorname{Tr} \hat{\psi}^{+}\left(\mathbf{R}_1 + \frac{\mathbf{r}_1}{2}\right) \hat{\psi}^{+}\left(\mathbf{R}_2 + \frac{\mathbf{r}_2}{2}\right) \hat{\psi}\left(\mathbf{R}_1 - \frac{\mathbf{r}_1}{2}\right) \hat{\psi}\left(\mathbf{R}_2 - \frac{\mathbf{r}_2}{2}\right) \hat{\rho}(t), \quad (7.112)$$

where in Eq. (7.111), for example, $\mathbf{r} = \mathbf{r}_1 - \mathbf{r}_1'$ and $\mathbf{R} = \frac{1}{2}(\mathbf{r}_1 + \mathbf{r}_1')$. The one-particle reduced Wigner function is then defined

$$f_1(\mathbf{k}, \mathbf{R}, t) \equiv \int d\mathbf{r}\, e^{i\mathbf{k}\cdot\mathbf{r}} \left\langle \mathbf{R} + \frac{\mathbf{r}}{2} \middle| \hat{\rho}_{(1)}(t) \middle| \mathbf{R} - \frac{\mathbf{r}}{2} \right\rangle, \quad (7.113)$$

and the two-particle reduced Wigner function is defined

$$f_2(\mathbf{k}_1, \mathbf{k}_2, \mathbf{R}_1, \mathbf{R}_2; t) = \iint d\mathbf{r}_1\, d\mathbf{r}_2\, e^{i\mathbf{k}_1\cdot\mathbf{r}_1}\, e^{i\cdot\mathbf{k}_2\mathbf{r}_2}$$
$$\times \left\langle \mathbf{R}_1 + \frac{\mathbf{r}_1}{2}; \mathbf{R}_2 + \frac{\mathbf{r}_2}{2} \middle| \rho_{(2)}(t) \middle| \mathbf{R}_1 - \frac{\mathbf{r}_1}{2}, \mathbf{R}_2 - \frac{\mathbf{r}_2}{2} \right\rangle. \quad (7.114)$$

Higher order Wigner functions can be defined in a similar manner.

In analogy to the classical distribution function, the one-particle Wigner function obeys the relations

$$\int \frac{d\mathbf{k}}{(2\pi)^3} f_1(\mathbf{k}, \mathbf{R}, t) = \langle \mathbf{R} | \hat{\rho}_{(1)}(t) | \mathbf{R} \rangle = n(\mathbf{R}, t) \quad (7.115)$$

and

$$\int d\mathbf{R} f_1(\mathbf{k}, \mathbf{R}, t) = \langle \mathbf{k} | \hat{\rho}_{(1)}(t) | \mathbf{k} \rangle = n(\mathbf{k}, t) \quad (7.116)$$

where $\mathbf{k}$ is the wave vector of the particles. The Wigner function can be used to take phase space averages in the same way as the classical distribution functions. For example, the average current is defined

$$\langle \mathbf{j}(\mathbf{R}, t) \rangle = \int \frac{d\mathbf{k}}{(2\pi)^3} \hbar \mathbf{k} f_1(\mathbf{k}, \mathbf{R}, t), \quad (7.117)$$

where $\hbar$ is Planck's constant. However, the Wigner function can become negative[8] and therefore cannot always be interpreted as a probability density.

We can derive the equation of motion for $f_1(\mathbf{k}, \mathbf{R}, t)$ in the following way. Let us first take the time derivative of the one-particle reduced density matrix,

$$\begin{aligned}\frac{\partial}{\partial t}\langle \mathbf{r}_1|\hat{\rho}_{(1)}(t)|\mathbf{r}_2\rangle &= \mathrm{Tr}\,\hat{\psi}^+(\mathbf{r}_1)\hat{\psi}(\mathbf{r}_2)\frac{\partial\hat{\rho}(t)}{\partial t}\\ &= -\frac{i}{\hbar}\mathrm{Tr}\,\hat{\psi}^+(\mathbf{r}_1)\hat{\psi}(\mathbf{r}_2)[\hat{H}, \hat{\rho}(t)]_-\\ &= -\frac{i}{\hbar}\mathrm{Tr}[\hat{\psi}^+(\mathbf{r}_1)\hat{\psi}(\mathbf{r}_2), \hat{H}]_-\hat{\rho}(t), \end{aligned} \tag{7.118}$$

where $\hat{H}$ is the Hamiltonian of the system, and we have made use of Eq. (7.78). The commutator in Eq. (7.118) can be expanded if we make use of the following identity for the commutator of the operator products:

$$[\hat{B}\hat{C}, \hat{A}]_- = \hat{B}[\hat{C}, \hat{A}]_\pm - [\hat{A}, \hat{B}]_\pm \hat{C} \tag{7.119}$$

where the minus denotes commutator and the plus denotes an anticommutator. If we also use the commutation relations in App. B, we can write Eq. (7.118) in the following form:

$$\begin{aligned}\frac{\partial}{\partial t}\langle \mathbf{r}_1|\hat{\rho}_{(1)}(t)|\mathbf{r}_2\rangle = &-\frac{i\hbar}{2m}(\nabla_{\mathbf{r}_1} + \nabla_{\mathbf{r}_2})\cdot(\nabla_{\mathbf{r}_1} - \nabla_{\mathbf{r}_2})\langle \mathbf{r}_1|\hat{\rho}_{(1)}(t)|\mathbf{r}_2\rangle\\ &- \frac{i}{\hbar}\int d\mathbf{r}'[V(\mathbf{r}_2 - \mathbf{r}') - V(\mathbf{r}_1 - \mathbf{r}')]\langle \mathbf{r}_1, \mathbf{r}'|\hat{\rho}_{(2)}(t)|\mathbf{r}_2, \mathbf{r}'\rangle, \end{aligned} \tag{7.120}$$

where we have used the Hamiltonian $\hat{H} = \hat{T} + \hat{V}$ with $\hat{T}$ and $\hat{V}$ defined as in App. B. For simplicity, we have dropped spin indices and we have assumed a spherically symmetric potential so that $V(\mathbf{r}_1 - \mathbf{r}_2) = \langle \mathbf{r}_1, \mathbf{r}_2|V|\mathbf{r}_1, \mathbf{r}_2\rangle$. The details of the derivation are left as a problem. Let us now change to center of mass and relative coordinates, multiply by $e^{i\mathbf{k}\cdot\mathbf{r}}$, and integrate over $d\mathbf{r}$. We then find

$$\begin{aligned}&\int d\mathbf{r}\, e^{i\mathbf{k}\cdot\mathbf{r}}\frac{\partial}{\partial t}\left\langle \mathbf{R} + \frac{\mathbf{r}}{2}\middle|\hat{\rho}_{(1)}(t)\middle|\mathbf{R} - \frac{\mathbf{r}}{2}\right\rangle\\ &= -\frac{i\hbar}{m}\int d\mathbf{r}\, e^{i\mathbf{k}\cdot\mathbf{r}}\nabla_{\mathbf{R}}\cdot\nabla_{\mathbf{r}}\left\langle \mathbf{R} + \frac{\mathbf{r}}{2}\middle|\hat{\rho}_{(1)}(t)\middle|\mathbf{R} - \frac{\mathbf{r}}{2}\right\rangle\\ &\quad - \frac{i}{\hbar}\int d\mathbf{r}'\int d\mathbf{r}\, e^{i\mathbf{k}\cdot\mathbf{r}}\left[V\left(\mathbf{R} - \frac{\mathbf{r}}{2} - \mathbf{r}'\right) - V\left(\mathbf{R} + \frac{\mathbf{r}}{2} - \mathbf{r}'\right)\right]\\ &\quad \times \left\langle \mathbf{R} + \frac{\mathbf{r}}{2}, \mathbf{r}'\middle|\hat{\rho}_{(2)}(t)\middle|\mathbf{R} - \frac{\mathbf{r}}{2}, \mathbf{r}'\right\rangle. \end{aligned} \tag{7.121}$$

In Eq. (7.121) we can integrate the first term on the right by parts to remove the derivative with respect to **r**. We can also introduce dummy variables into the second term on the right and make use of definitions in Eqs. (7.113) and (7.114) to obtain

$$\frac{\partial}{\partial t} f_1(\mathbf{k}, \mathbf{R}, t) = -\frac{\hbar}{m} \mathbf{k} \cdot \nabla_{\mathbf{R}} f_1(\mathbf{k}, \mathbf{R}, t)$$
$$- \frac{i}{\hbar} \int \frac{d\mathbf{k}'}{(2\pi)^3} \int d\mathbf{r}' \left[V\left(\mathbf{R} - \mathbf{r}' - \frac{1}{2i}\frac{\partial}{\partial \mathbf{k}}\right) - V\left(\mathbf{R} - \mathbf{r}' + \frac{1}{2i}\frac{\partial}{\partial \mathbf{k}}\right)\right]$$
$$\times f_2(\mathbf{k}, \mathbf{R}; \mathbf{k}', \mathbf{r}'; t). \tag{7.122}$$

Eq. (7.122) is the quantum kinetic equation for the one-particle reduced Wigner function. If we rewrite it in terms of momenta $\mathbf{p} = \hbar\mathbf{k}$ and $\mathbf{p}' = \hbar\mathbf{k}'$, it takes the form

$$\frac{\partial}{\partial t} f_1'(\mathbf{p}, \mathbf{R}, t) + \frac{\mathbf{p}}{m} \cdot \nabla_{\mathbf{R}} f_1'(\mathbf{p}, \mathbf{R}, t)$$
$$= -\frac{i}{\hbar} \int \frac{d\mathbf{p}}{(2\pi)^3} \int d\mathbf{r}' \left[V\left(\mathbf{R} - \mathbf{r}' - \frac{\hbar}{2i}\frac{\partial}{\partial \mathbf{p}}\right) - V\left(\mathbf{R} - \mathbf{r}' + \frac{\hbar}{2i}\frac{\partial}{\partial \mathbf{p}}\right)\right]$$
$$\times f_2'(\mathbf{p}, \mathbf{R}; \mathbf{p}', \mathbf{r}'; t), \tag{7.123}$$

where

$$f_n'(\mathbf{p}_1, \mathbf{r}_1; \ldots; \mathbf{p}_n, \mathbf{r}_n) = \frac{1}{\hbar^{3n}} f_n\left(\frac{1}{\hbar}\mathbf{p}_1, \mathbf{r}_1; \ldots; \frac{1}{\hbar}\mathbf{p}_n, \mathbf{r}_n\right). \tag{7.124}$$

We can now expand the potential on the right-hand side in powers of $\hbar$ and take the limit $\hbar \to 0$. We then retrieve the classical kinetic equation.

The Wigner function can be used to take the average value of a large class of ordinary functions of momentum and position but in some cases it will give the wrong answer. The average value of any quantity which is only a function of position or only a function of momentum can always be taken (this is easily seen from Eqs. (7.115) and (7.116)). However, only those functions which involve both position and momentum can be used for which the Weyl correspondence between the quantum and classical version of the operators holds. To see this, let us consider the classical function $O(\mathbf{p}, \mathbf{q})$ of phase space variables $\mathbf{p}$ and $\mathbf{q}$. We can find the quantum version of this function as follows. We first introduce its Fourier transform, $\tilde{O}(\boldsymbol{\sigma}, \boldsymbol{\eta})$, with the equation

$$O(\mathbf{p}, \mathbf{q}) = \iint d\boldsymbol{\sigma}\, d\boldsymbol{\eta}\, \tilde{O}(\boldsymbol{\sigma}, \boldsymbol{\eta})\, e^{i(\boldsymbol{\sigma}\cdot\mathbf{p} + \boldsymbol{\eta}\cdot\mathbf{q})}. \tag{7.125}$$

Matrix elements of the quantum operator corresponding to $O(\mathbf{p}, \mathbf{q})$ can be obtained from the Fourier transform, $\tilde{O}(\boldsymbol{\sigma}, \boldsymbol{\eta})$, via the equation

$$\langle \mathbf{r}' | \hat{O} | \mathbf{r}'' \rangle = \iint d\boldsymbol{\sigma}\, d\boldsymbol{\eta}\, \tilde{O}(\boldsymbol{\sigma}, \boldsymbol{\eta}) \langle \mathbf{r}' |\, e^{i(\boldsymbol{\sigma}\cdot\hat{\mathbf{p}} - \boldsymbol{\eta}\cdot\hat{\mathbf{q}})} | \mathbf{r}'' \rangle \tag{7.126}$$

where $\hat{\mathbf{p}}$ and $\hat{\mathbf{q}}$ are momentum and position operators. If the classical function $O(\mathbf{p}, \mathbf{q})$ and matrix elements of the corresponding operator $\hat{O}$ are related by the above procedure, then the expectation value of $\hat{O}$ is given by

$$\langle O(t)\rangle = \iint d\mathbf{p}\, d\mathbf{r} O(\mathbf{p}, \mathbf{r}) f_1^{\prime}(\mathbf{p}, \mathbf{r}, t). \tag{7.127}$$

There are some cases for which the Weyl procedure does not give the correct correspondence between classical and quantum operators (such as the commutator $[\hat{\mathbf{p}}, \hat{\mathbf{r}}]_-$, the square of the Hamiltonian $\hat{H}^2$, the square of the angular momentum $\mathbf{L}^2$, etc.) and the Wigner function gives the wrong result. Then it is necessary to introduce more general quantum phase space distributions which may in general be complex functions.

H. MICROSCOPIC BALANCE EQUATIONS (QUANTUM FLUIDS)[16]

For quantum systems with short-ranged interactions and long-wavelength inhomogeneities, we can derive microscopic balance equations for the particle density, momentum density, and energy density in a manner analogous to the derivation for classical systems. We first note, however, that the position and momentum operators satisfy the commutation relations,

$$[\hat{\mathbf{p}}_i, \hat{\mathbf{p}}_j]_- = 0, \qquad [\hat{\mathbf{q}}_i, \hat{\mathbf{q}}_j]_- = 0, \quad \text{and} \quad [\hat{\mathbf{q}}_i, \hat{\mathbf{p}}_j]_- = i\hbar\, \delta_{ij}. \tag{7.128}$$

The commutator of the momentum operator $\hat{\mathbf{p}}_i$ with an arbitrary function of coordinate operators is

$$[\hat{\mathbf{p}}_i, F(\hat{\mathbf{q}}_1, \ldots, \hat{\mathbf{q}}_N)]_- = -i\hbar \frac{\partial F}{\partial \hat{\mathbf{q}}_i}, \tag{7.129}$$

while the commutator of $\hat{\mathbf{q}}_i$ with an arbitrary function of momenta $G(\hat{\mathbf{p}}_1, \ldots, \hat{\mathbf{p}}_N)$ is

$$[\hat{\mathbf{q}}_i, G(\hat{\mathbf{p}}_1, \ldots, \hat{\mathbf{p}}_N)]_- = i\hbar \frac{\partial G}{\partial \hat{\mathbf{p}}_i}. \tag{7.130}$$

Let us assume that the dynamics of the system is governed by a Hamiltonian of the form

$$\hat{H}^N = \sum_{i=1}^{N} \frac{|\hat{\mathbf{p}}_i|^2}{2m} + \sum_{i<j}^{(1/2)N(N-1)} V(|\hat{\mathbf{q}}_i - \hat{\mathbf{q}}_j|). \tag{7.131}$$

Then from Eq. (7.83), the equation of motion for the operator $\hat{\mathbf{q}}_i$ is

$$\frac{\partial \hat{\mathbf{q}}_i}{\partial t} = \frac{i}{\hbar}[\hat{H}, \hat{\mathbf{q}}_i]_- = \frac{\hat{\mathbf{p}}_i}{m}. \tag{7.132}$$

The equation of motion of $\hat{\mathbf{p}}_i$ is given by

$$\frac{\partial \hat{\mathbf{p}}_i}{\partial t} = -\sum_{l\neq i}\left[\frac{\partial V(|\hat{\mathbf{q}}_i - \hat{\mathbf{q}}_l|)}{\partial \hat{\mathbf{q}}_i}\right] = \sum_{l\neq i} \hat{\mathbf{F}}_{il}. \tag{7.133}$$

These equations have the same form as the classical equations.

For quantum systems, the microscopic expressions for the densities must be Hermitian in order to be observable. Thus, we must have symmetrized expressions for operators which involve both momentum and position. Using Eq. (7.83) and the above equations, we can show that the balance equation for the particle number density is given by

$$\frac{\partial}{\partial t}\hat{n}(\hat{\mathbf{q}}^N; \mathbf{R}) = -\nabla_{\mathbf{R}} \cdot \hat{\mathbf{J}}^n(\hat{\mathbf{p}}^N, \hat{\mathbf{q}}^N; \mathbf{R}), \tag{7.134}$$

where the particle density is defined

$$\hat{n}(\hat{\mathbf{q}}^N; \mathbf{R}) = \sum_{i=1}^{N} \delta(\hat{\mathbf{q}}_i - \mathbf{R}) \tag{7.135}$$

and the particle current density is defined

$$\hat{\mathbf{J}}^n(\hat{\mathbf{p}}^N, \hat{\mathbf{q}}^N; \mathbf{R}) = \frac{1}{2}\sum_{i=1}^{N}\left[\frac{\hat{\mathbf{p}}_i}{m}\,\delta(\hat{\mathbf{q}}_i - \mathbf{R}) + \delta(\hat{\mathbf{q}}_i - \mathbf{R})\frac{\hat{\mathbf{p}}_i}{m}\right]. \tag{7.136}$$

As usual, we let $\hat{\mathbf{p}}^N$ denote the set of momenta $\hat{\mathbf{p}}^N = (\hat{\mathbf{p}}_1, \ldots, \hat{\mathbf{p}}_N)$ and we let $\hat{\mathbf{q}}^N$ denote the set of positions $\hat{\mathbf{q}}^N = (\hat{\mathbf{q}}_1, \ldots, \hat{\mathbf{q}}_N)$.

The balance equation for the momentum density takes the form

$$m\frac{\partial}{\partial t}\hat{\mathbf{J}}^n(\hat{\mathbf{p}}^N, \hat{\mathbf{q}}^N; \mathbf{R}) = -\nabla_{\mathbf{R}} \cdot \hat{\mathbf{J}}^p(\hat{\mathbf{p}}^N, \hat{\mathbf{q}}^N; \mathbf{R}) \tag{7.137}$$

where the momentum current tensor is defined

$$\begin{aligned}\hat{\mathbf{J}}^p(\hat{\mathbf{p}}^N, \hat{\mathbf{q}}^N; \mathbf{R}) = \frac{1}{4m}\sum_{i=1}^{N}&[\hat{\mathbf{p}}_i\hat{\mathbf{p}}_i\,\delta(\hat{\mathbf{q}}_i - \mathbf{R}) + (\hat{\mathbf{p}}_i\,\delta(\hat{\mathbf{q}}_i - \mathbf{R})\hat{\mathbf{p}}_i)^T \\ &+ \hat{\mathbf{p}}_i\,\delta(\hat{\mathbf{q}}_i - \mathbf{R})\hat{\mathbf{p}}_i + \delta(\hat{\mathbf{q}}_i - \mathbf{R})\hat{\mathbf{p}}_i\hat{\mathbf{p}}_i] \\ &+ \frac{1}{2}\sum_{i \neq l}^{N}\sum^{N}(\hat{\mathbf{q}}_i - \hat{\mathbf{q}}_l)\cdot\mathbf{F}_{il}\,\delta(\mathbf{q}_i - \mathbf{R}).\end{aligned} \tag{7.138}$$

In Eqs. (7.137) and (7.138), the notation

$$\nabla_{\mathbf{R}} \cdot (\hat{\mathbf{p}}_i\,\delta(\hat{\mathbf{q}}_i - \mathbf{R})\hat{\mathbf{p}}_i)^T \equiv \hat{\mathbf{p}}_i\nabla_{\mathbf{R}} \cdot \delta(\hat{\mathbf{q}}_i - \mathbf{R})\hat{\mathbf{p}}_i \tag{7.139}$$

has been used.

Finally, the balance equation for the energy density can be written

$$\frac{\partial}{\partial t}\hat{h}(\hat{\mathbf{p}}^N, \hat{\mathbf{q}}^N; \mathbf{R}) = -\nabla_{\mathbf{R}} \cdot \hat{\mathbf{J}}^h(\hat{\mathbf{p}}^N, \hat{\mathbf{q}}^N; \mathbf{R}), \tag{7.140}$$

where the energy density is defined

$$h(\hat{\mathbf{p}}^N, \hat{\mathbf{q}}^N; \mathbf{R}) \equiv \frac{1}{2}\sum_{i=1}^{N}[\hat{h}_i\,\delta(\hat{\mathbf{q}}_i - \mathbf{R}) + \delta(\hat{\mathbf{q}}_i - \mathbf{R})\hat{h}_i] \tag{7.141}$$

with

$$\hat{h}_i = \frac{|\hat{\mathbf{p}}_i|^2}{2m} + \frac{1}{2}\sum_{j \neq i} V(|\hat{\mathbf{q}}_i - \hat{\mathbf{q}}_j|) \tag{7.142}$$

and the energy current density is defined

$$\hat{\mathbf{J}}^h(\hat{\mathbf{p}}^N, \hat{\mathbf{q}}^N; \mathbf{R}) \equiv \frac{1}{4}\sum_{i=1}^{N}\left[\hat{h}_i \frac{\hat{\mathbf{p}}_i}{m}\,\delta(\hat{\mathbf{q}}_i - \mathbf{R}) + \hat{h}_i\,\delta(\hat{\mathbf{q}}_i - \mathbf{R})\frac{\hat{\mathbf{p}}_i}{m} + \frac{\hat{\mathbf{p}}_i}{m}\,\delta(\hat{\mathbf{q}}_i - \mathbf{R})\hat{h}_i + \delta(\hat{\mathbf{q}}_i - \mathbf{R})\frac{\hat{\mathbf{p}}_i}{m}\hat{h}_i\right]$$
$$+ \frac{1}{4}\sum_{i\neq j}^{N}\sum^{N}\left[\frac{(\hat{\mathbf{p}}_i + \hat{\mathbf{p}}_j)}{m}\cdot\hat{\mathbf{F}}_{ij}(\hat{\mathbf{q}}_i - \hat{\mathbf{q}}_j)\,\delta(\hat{\mathbf{q}}_i - \mathbf{R}) + \delta(\hat{\mathbf{q}}_i - \mathbf{R})(\hat{\mathbf{q}}_i - \hat{\mathbf{q}}_j)\hat{\mathbf{F}}_{ij}\cdot\frac{(\hat{\mathbf{p}}_i + \hat{\mathbf{p}}_j)}{m}\right]. \tag{7.143}$$

To obtain Eq. (7.143), one must use the fact that the center of mass coordinates commute with the relative coordinates.

REFERENCES

1. H. Goldstein, *Classical Mechanics* (Addison-Wesley Publishing Co., Reading, Mass., 1950).
2. R. L. Liboff, *Introduction to the Theory of Kinetic Equations* (John Wiley & Sons, New York, 1969).
3. I. Prigogine, *Nonequilibrium Statistical Mechanics* (Wiley-Interscience, New York, 1962).
4. W. N. Bogoliubov, "Problems of a Dynamical Theory in Statistical Physics," in *Studies in Statistical Mechanics*, ed. J. de Boer and G. E. Uhenbeck (North-Holland Publishing Co., Amsterdam, 1962).
5. M. S. Green, J. Chem. Phys. *22* 398 (1954).
6. A. Messiah, *Quantum Mechanics*, Vol. I (North-Holland Publishing Co., Amsterdam, 1961).
7. U. Fano, Rev. Mod. Phys. *29* 74 (1957).
8. J. von Neumann, *Mathematical Foundations of Quantum Mechanics* (Princeton University Press, Princeton, 1955).
9. C. N. Yang, Rev. Mod. Phys. *34* 694 (1962).
10. E. Wigner, Phys. Rev. *40* 749 (1932).
11. J. H. Irving and R. W. Zwanzig, J. Chem. Phys. *19* 1173 (1951).
12. J. Ross and J. G. Kirkwood, J. Chem. Phys. *22* 1094 (1956).
13. J. E. Moyal, Proc. Cambridge Phil. Soc. *45* 99 (1949).
14. T. Takabayasi, Progr. Theor. Phys. (Kyoto) *11* 341 (1954).
15. A. O. Barut, Phys. Rev. *108* 565 (1957).
16. H. Mori, Phys. Rev. *112* 1829 (1958).

PROBLEMS

1. Prove that the Jacobian, $\mathcal{J}(t, t_0)$, satisfies Eq. (7.11).

2. Show that the classical Liouville operator is Hermitian. Write the solution of the Liouville equation in terms of eigenfunctions and eigenvalues of the Liouville

operator. Do these equations show any tendency to decay to a unique equilibrium state?

3. For a noninteracting gas of N particles in a cubic box of volume $V = L^3$, where L is the length of the side of the box, find the solution of the Liouville equation at time t, assuming periodic boundary conditions. Discuss its behavior.

4. Obtain the solution of the Liouville equation at time t for a system of N independent classical harmonic oscillators.

5. Show that the total particle number, the total linear momentum, the total angular momentum, and the total energy are constants of the motion for a system governed by the Hamiltonian in Eq. (7.38). Use the Liouville operator to establish this.

6. For a gas of N particles in a gravitational field, derive the kinetic equation. Assume that the particles interact via a potential of the form $\phi(|\mathbf{q}_i - \mathbf{q}_j|)$ and are contained in a box of volume V.

7. Derive the balance equation for the energy density of a classical system in the long-wavelength limit.

8. The ground state of a system of noninteracting fermions is one for which the particles occupy all the lowest energy quantum states—one particle per state.

(*a*) Find a relation between the maximum momentum a particle can have (the so-called Fermi momentum, $|p_f|$) and the density of particles in the ground state.

(*b*) Assume that the fermions interact via a weak repulsive potential of the form

$$V(\mathbf{r}) = \begin{cases} \Delta & \text{for } |\mathbf{r}| < a \\ 0 & \text{for } |\mathbf{r}| > a \end{cases}$$

where $\mathbf{r}$ is the distance between particles. Use the Rayleigh-Schrödinger perturbation theory to compute the total ground state energy per unit volume to first order in Δ. (Assume that volume is large enough so that summations over wave number $\mathbf{k}$ can be replaced by integrations according to the standard procedure $\sum_{\mathbf{k}} \to V/(2\pi)^3 \int d\mathbf{k}$.) For the case when Δ is negative, make a qualitative sketch of the ground state energy as a function of p_f. Describe physically what is happening. The perturbation expansion of the ground state energy is given by

$$E_0 = E_0{}^0 + \langle\Psi_0{}^0|V|\Psi_0{}^0\rangle + \langle\Psi_0{}^0|(\hat{V} - E_0^{(1)})\frac{1}{\hat{H}_0 - E_0{}^0}(\hat{V} - E_0^{(1)})|\Psi_0{}^0\rangle + \cdots$$

where $|\Psi_0{}^0\rangle$ is the ground state wave function of the noninteracting system, $E_0^{(1)} = \langle\Psi_0{}^0|\hat{V}|\Psi_0{}^0\rangle$, and $E_0{}^0$ is the ground state energy of the noninteracting system.

9. The ground state of a noninteracting boson system (a pure state) is one for which all particles are in the momentum state with lowest energy $\mathbf{k} = 0$. Assume that the bosons interact via the same potential as in Prob. 7.8. Use the Rayleigh-Schrödinger perturbation theory to compute the ground state energy to second order in Δ. What will happen if Δ is negative?

10. Solve Prob. 7.4, but for the case of N noninteracting quantum harmonic oscillators.

11. Assume that the probability density operator of an N-particle system of Maxwell-Boltzmann particles is given by

$$\hat{\rho}^N(t) = \sum_j p_j |\Psi_j(t)\rangle\langle\Psi_j(t)| \qquad \text{(where } p_j \text{ is the probability of state } j\text{)}$$

and that the dynamics of the system is governed by a Hamiltonian of the form

$$H^N(\mathbf{p}_1, \ldots, \mathbf{q}_N) = \sum_{j=1}^{N} \frac{|\mathbf{p}_j|^2}{2m} + \sum_{i<j}^{(1/2)N(N-1)} V(|\mathbf{q}_i - \mathbf{q}_j|).$$

Find the equation of motion of the N-particle Wigner function

$$f^N(\mathbf{k}_1, \ldots, \mathbf{k}_N; \mathbf{r}_1, \ldots, \mathbf{r}_N; t) = \int \cdots \int d\mathbf{y}_1 \cdots d\mathbf{y}_N \exp\left\{i \sum_{j=1}^{N} \mathbf{k}_i \cdot \mathbf{y}_i\right\}$$
$$\times \langle \mathbf{r}_1 + \mathbf{y}_1, \ldots, \mathbf{r}_N + \mathbf{y}_N | \hat{\rho}(t) | \mathbf{r}_1 - \mathbf{y}_1, \ldots, \mathbf{r}_N - \mathbf{y}_N\rangle$$

and write it in a form in which each term is written explicitly in terms of $f^N(\mathbf{k}_1, \ldots, \mathbf{k}_N; \mathbf{r}_1, \ldots, \mathbf{r}_N; t)$.

12. Prove that for operators $\hat{O}_{(1)}$ satisfying the Weyl correspondence we have

$$\langle O_{(1)}(t)\rangle = \iint d\mathbf{r}'\, d\mathbf{r}'' \langle \mathbf{r}' | \hat{O}_1 | \mathbf{r}''\rangle \langle \mathbf{r}'' | \hat{\rho}_{(1)} | \mathbf{r}'\rangle$$
$$= \iint d\mathbf{p}\, d\mathbf{r} O(\mathbf{p}, \mathbf{r}) f'_{(1)}(\mathbf{p}, \mathbf{r}, t),$$

where $O(\mathbf{p}, \mathbf{r})$ is the classical version of the operator $\hat{O}_{(1)}$ and $f_1(\mathbf{p}, \mathbf{r}, t)$ is the one-body reduced Wigner function. As a special case, show that

$$\left\langle \frac{\hat{r}_i \hat{p}_i + \hat{p}_i \hat{r}_i}{2} \right\rangle = \iint d\mathbf{r}\, d\mathbf{p} r_i p_i f'_1(\mathbf{r}, \mathbf{p}).$$

13. Write the microscopic balance equations (7.134) and (7.137) in terms of field operators.

8
Ergodic Theory

A. INTRODUCTORY REMARKS

One of the oldest and most intriguing problems in statistical physics has been the origin of the irreversibility observed in nature, given that the laws of dynamics, and consequently the equation of motion for the probability distribution for dynamical systems, are reversible. In the nineteenth century, Boltzmann was able to write an equation of motion for the one-body reduced probability density for gases (cf. Chap. 13) which has a form similar to a master equation and is irreversible. The Boltzmann equation has proven to be one of the most useful equations in statistical physics and yet, in his day, Boltzmann was heavily attacked by mathematicians and physicists because he made some probabilistic assumptions in deriving it. In an effort to place his equation on firm theoretical ground, Boltzmann founded the subject of ergodic theory. The aim of ergodic theory is to understand the origin of the irreversibility observed in nature in terms of the types of flow that occur in the underlying phase space.

The subject of ergodic theory was primarily the domain of mathematicians until the advent of modern computers. However, in recent years it has become one of the more exciting subjects of research because of its importance in such diverse fields as celestial mechanics (stability of the solar system) and chemistry (stability of isolated excited molecules) and because it asks questions which lie at the very foundations of statistical mechanics.

As we shall see, the flow of probability in phase space is of a very special type. It is a little like the flow of oil through water. There are absolutely no diffusion processes present. Historically, two types of probability flow have been important in understanding the behavior of phase space, namely, ergodic flow and mixing flow. For systems with ergodic flow, we obtain a unique stationary probability density (a constant on the energy surface) which characterizes systems with a fixed energy at equilibrium. However, a system with ergodic flow cannot necessarily reach this equilibrium state if it does not start out there. For decay to equilibrium, we must have at least the additional property of mixing. Mixing systems are ergodic (the converse is not always true, however) and can exhibit random behavior in a coarse-grained sense. In addition, reduced distribution functions can be defined which decay to an equilibrium state. We shall give examples of both types of flow in this chapter.

Ergodic and mixing behavior for real systems is difficult to establish in general. It has been done only for a few model systems. However, there is a large class of conservative systems: the anharmonic oscillators which are of great importance in mechanics, chemistry, and the theory of solids. These systems are neither ergodic nor mixing but exhibit behavior reminiscent of both in local regions of their phase space. They have been studied extensively with computers in recent years and give great insight into the behavior of flows in phase space and the possible mechanism behind the irreversibility we observe in nature. We shall discuss these systems in some detail in this chapter.

B. ERGODIC FLOW[1–7]

Let us consider the Hamiltonian system discussed in Sec. 7.B and write Hamilton's equations in the form

$$\frac{dq_1}{\dfrac{\partial H^N}{\partial p_1}} = \frac{dq_2}{\dfrac{\partial H^N}{\partial p_2}} = \cdots = \frac{dp_{3N}}{-\dfrac{\partial H^N}{\partial q_{3N}}} = dt. \tag{8.1}$$

Eq. (8.1) provides us with $6N - 1$ equations between phase space coordinates which, when solved, give us $6N - 1$ constants or integrals of the motion

$$f_i(\mathbf{X}^N) = C_i, \tag{8.2}$$

where C_i is a constant. However, these integrals of the motion can be divided into two kinds, isolating and nonisolating. Isolating integrals define a whole surface in the phase space and are important in ergodic theory, while nonisolating integrals do not define a surface and are unimportant (cf. Ref. 8.1, p. 91, for a discussion and Ref. 8.8 for an example). One of the main problems of ergodic theory is to determine how many isolating integrals a given system has. An example of an isolating integral is the total energy. For N particles in a box it is probably the only isolating integral (at least for hard spheres).

Let us consider a system for which the only isolating integral of the motion is the total energy and assume that the system has total energy E. Then trajectories in Γ-space (the $6N$-dimensional phase space) which have energy E will be restricted to the energy surface S_E. The energy surface S_E is a $(6N - 1)$-dimensional "surface" in phase space defined by Eq. (7.3). The flow of state points on the energy surface is defined to be ergodic if almost all points $\mathbf{X}^N(\mathbf{p}^N, \mathbf{q}^N)$ on the surface move in such a way that they pass through every small finite neighborhood R_E on the energy surface. Or, in other words, each point samples small neighborhoods over the entire surface during the course of its motion (a given point $\mathbf{X}^N(\mathbf{p}^N, \mathbf{q}^N)$ cannot pass through every point on the surface, because a line which cannot intersect itself cannot fill a surface of two or more dimensions). Note that not all points need sample the surface, only "almost all." We can exclude a set of measure zero from this requirement.

A criterion for determining if a system is ergodic was established by Birkhoff[9] and is called the *ergodic theorem*. Let us consider an integrable phase function $f(\mathbf{X}^N)$ of the state point $\mathbf{X}^N$. We may define a phase average of the function $f(\mathbf{X}^N)$ on the energy surface by the equation

$$\langle f \rangle_S = \frac{1}{\Sigma(E)} \int_{S_E} f(\mathbf{X}^N)\, dS_E = \frac{1}{\Sigma(E)} \int_{\Gamma} \delta(H^N(\mathbf{X}^N) - E) f(\mathbf{X}^N)\, d\mathbf{X}^N \quad (8.3)$$

where dS_E is an area element of the energy surface which is invariant (does not change size) during the evolution of the system and $\Sigma(E)$ is the area of the energy surface and is defined

$$\Sigma(E) = \int_{S_E} dS_E = \int_{\Gamma} \delta(H^N(\mathbf{X}^N) - E)\, d\mathbf{X}^N \quad (8.4)$$

(we are using the notation of Sec. 7.B). We may define a time average of the function $f(\mathbf{X}^N)$ by the equation

$$\langle f \rangle_T = \lim_{T \to \infty} \frac{1}{T} \int_{t_0}^{t_0+T} f(\mathbf{X}^N(t))\, dt \quad (8.5)$$

for all trajectories for which the time average exists. Birkhoff showed that the time average in Eq. (8.5) exists for all integrable phase functions of physical interest (that is, for smooth functions).

In terms of the averages, the ergodic theorem may be stated as follows: A system is ergodic if for all phase functions $f(\mathbf{X}^N)$ (i) the time average f_T exists for almost all $\mathbf{X}^N$ (all but a set of measure zero) and (ii) when it exists it is equal to the phase average $\langle f \rangle_T = \langle f \rangle_S$.

To find the form of the invariant area element, dS_E, let us first write an expression for the volume of phase space, $\Omega(E)$, with energy less than E, that is, the region of phase space for which $0 < H^N(\mathbf{X}^N) < E$. We shall assume that the phase space can be divided into layers, each with different energy, and that the layers can be arranged in the order of increasing energy. (This is possible for all systems that we will consider.) The volume, $\Omega(E)$, can then be written

$$\Omega(E) = \int_{0 < H^N(\mathbf{X}^N) < E} d\mathbf{X}^N = \int_{0 < H^N(\mathbf{X}^N) < E} dA_H\, dn_H, \quad (8.6)$$

where dA_H is an area element on a surface of constant energy and dn_H is normal to that surface. Since $\nabla_{\mathbf{X}} H^N$ is a vector perpendicular to the surface $H^N(\mathbf{X}^N) =$ constant, we can write $dH^N = |\nabla_{\mathbf{X}} H^N|\, dn_H$ and the volume becomes

$$\Omega(E) = \int_0^E dH^N \Sigma(H^N), \quad (8.7)$$

where

$$\Sigma(H^N) = \int_{S_H} \frac{dA_H}{|\nabla_{\mathbf{X}} H^N|} \tag{8.8}$$

is a function of H^N and is an invariant measure of the surface area for a given value of H^N. If we take the derivative of $\Omega(E)$, we find

$$\frac{d\Omega(E)}{dE} = \Sigma(E) = \int_{S_E} \frac{dA_E}{|\nabla_{\mathbf{X}} H^N|_{H=E}} \tag{8.9}$$

The area, $\Sigma(E)$, is called the *structure function*. By the same argument, if we wish to take the average value of a function $f(\mathbf{X}^N)$ over the surface, we can write

$$\langle f \rangle_S = \frac{1}{\Sigma(E)} \int_{S_E} f(\mathbf{X}^N) \frac{dA_E}{|\nabla_{\mathbf{X}} H^N|_{H^N=E}} = \frac{1}{\Sigma(E)} \frac{d}{dE} \int_{0<H^N(\mathbf{X}^N)<E} f(\mathbf{X}^N)\, d\mathbf{X}^N. \tag{8.10}$$

Thus, the differential

$$dS_E = \frac{dA_E}{|\nabla_{\mathbf{X}} H^N|_{H^N=E}} \tag{8.11}$$

is the invariant surface area element.

If a system is ergodic, the fraction of time that its state, $\mathbf{X}^N(\mathbf{p}^N, \mathbf{q}^N)$, spends in a given region R_E of the energy surface will be equal to the fraction of the surface S_E occupied by R_E. Let us consider a function $\phi(R_E)$ such that $\phi(R_E) = 1$ when $\mathbf{X}^N$ is in R_E and $\phi(R_E) = 0$ otherwise. Then it is easy to see that

$$\lim_{T\to\infty} \frac{\tau_{R_E}}{T} = \frac{\Sigma(R_E)}{\Sigma(E)}, \tag{8.12}$$

where τ_{R_E} is the time the trajectory spends in R_E and $\Sigma(R_E)$ is the area occupied by R_E.

A system can exhibit ergodic flow on the energy surface only if there are no other isolating integrals of the motion which prevent trajectories from moving freely on the energy surface. If no other isolating integrals exist, the system is said to be metrically transitive (trajectories move freely on the energy surface).

If a system is ergodic, it will spend equal times in equal areas of the energy surface. If we perform measurements to decide where on the surface the system point is, we should find that result. We can also ask for the probability of finding the system in a given region R_E of the energy surface. Since we have nothing to distinguish one region from another, the best choice we can make is to assume that the probability $P(R_E)$ of finding the system in R_E is equal to the fraction of the energy surface occupied by R_E. Thus,

$$P(R_E) = \frac{1}{\Sigma(E)} \int_{R_E} dS_E = \frac{\Sigma(R_E)}{\Sigma(E)}. \tag{8.13}$$

From Eq. (8.13) it is a simple matter to write down a normalized probability density for the energy surface, namely,

$$\rho(\mathbf{X}^N, S_E) = \frac{1}{\Sigma(E)}. \tag{8.14}$$

Eq. (8.14) is called the *fundamental distribution law* by Khinchin and called the *microcanonical ensemble* by Gibbs. Since it is a function only of the energy, it is a stationary state of the Liouville equation (7.25). It says that all states on the energy surface are equally probable. Eq. (8.14) forms the foundation upon which all of equilibrium and most of nonequilibrium statistical mechanics are built. Its importance cannot be overemphasized.

We can give a simple example[4,3] of ergodic flow if we take as our energy surface the two-dimensional unit square $0 < p < 1$ and $0 < q < 1$. We shall assume that the equations of motion are given by

$$\dot{p} = \alpha \tag{8.15}$$

and

$$\dot{q} = 1 \tag{8.16}$$

and we shall impose periodic boundary conditions on the system. The equations of motion are easily solved to give

$$p(t) = p_0 + \alpha t \tag{8.17}$$

and

$$q(t) = q_0 + t. \tag{8.18}$$

The phase trajectory on the surface is given by

$$p = p_0 + \alpha(q - q_0). \tag{8.19}$$

If α is a rational number, $\alpha = m/n$, then the trajectory will be periodic and will repeat itself after a period $t = n$. If α is irrational, then the trajectory will be dense on the unit square (but will not fill it). When α is irrational, the system is ergodic (cf. Fig. 8.1). Let us show this explicitly. Since the phase space is periodic, any integrable function $f(p, q)$ can be expanded in a Fourier series,

$$f(p, q) = \sum_{l,m=-\infty}^{\infty} A_{lm}\, e^{2\pi i(lq + mp)}. \tag{8.20}$$

We wish to show that the time average and phase average of $f(p, q)$ are equal for α irrational. The time average is given by

$$\begin{aligned}\langle f\rangle_T &= \lim_{T\to\infty} \frac{1}{T}\int_{t_0}^{t_0+T} dt \sum_{l,m} A_{l,m}\, e^{2\pi i[l(q_0+t)+m(p_0+\alpha t)]}\, dt \\ &= A_{0,0} + \lim_{T\to\infty}\frac{1}{T}\sum_{l,m\neq 0} A_{l,m}\, e^{2\pi i[l(q_0+t_0)+m(p_0+\alpha t_0)]}\left(\frac{e^{2\pi i(l+m)T} - 1}{2\pi i(l+\alpha m)}\right).\end{aligned} \tag{8.21}$$

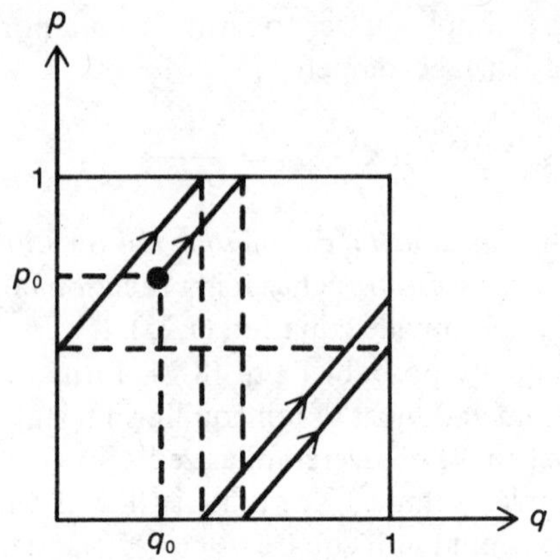

Fig. 8.1. For α irrational, the trajectory is dense on the unit square; for α rational, the trajectory is periodic.

For irrational α, the denominator can never equal zero. Therefore,

$$\langle f \rangle_T = A_{0,0}. \tag{8.22}$$

Similarly, we can show that

$$\langle f \rangle_s = \int_0^1 \int_0^1 dp\, dq f(p, q) = A_{0,0}. \tag{8.23}$$

Hence the system is ergodic (note that $dp\, dq = dp_0\, dq_0$ so that area is preserved).

Ergodicity is not sufficient to cause a system which initially has a nonstationary distribution to approach the stationary state in Eq. (8.16). Let us consider a probability density which is not constant over the unit square. For the sake of definiteness, assume that at time $t = 0$ the probability density is given by

$$\rho(p_0, q_0, 0) = \sin(\pi p_0) \sin(\pi q_0); \tag{8.24}$$

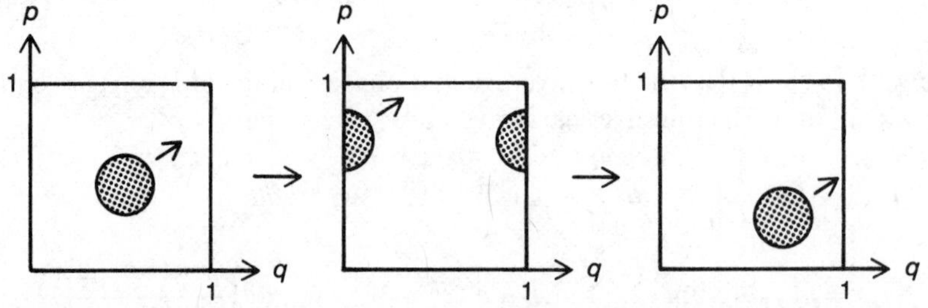

Fig. 8.2. Ergodicity alone is insufficient to change the shape of the probability density; the circle covers the square but does not change shape.

then at time t it will be given by

$$\rho(p_0, q_0, t) = \sin[\pi(p_0 - \alpha t)] \sin[\pi(q_0 - t)]. \tag{8.25}$$

The probability is not changed in shape, it is only displaced (cf. Fig. 8.2). After an infinite length of time it will have wandered uniformly over the entire energy surface.

In this section, we have discussed ergodic theory for classical systems. However, it is also possible to formulate analogous definitions for quantum systems. In fact, the criterion is rather simple. A quantum system is ergodic if and only if the system has a nondegenerate energy spectrum.[10] This means, of course, that there are no other observables which commute with the Hamiltonian.

C. MIXING FLOW[2–4]

A type of flow which begins to exhibit some features of irreversibility is mixing flow. Mixing flow is chaotic and causes any initial probability distribution to spread throughout the phase space. Mixing flow is ergodic, but ergodic flows are not always mixing.

A system is mixing if, for all square integrable functions, $f(\mathbf{X}^N)$ and $g(\mathbf{X}^N)$, on S_E,

$$\lim_{t\to\pm\infty} \frac{1}{\Sigma(E)} \int_{S_E} f(\mathbf{X}^N) g(\mathbf{X}^N(t))\, dS_E = \frac{\int_{S_E} f(\mathbf{X}^N)\, dS_E \int_{S_E} g(\mathbf{X}^N)\, dS_E}{(\Sigma(E))^2}. \tag{8.26}$$

Eq. (8.26) ensures that the average value of a dynamical function $f(\mathbf{X}^N)$ will approach a stationary value in the limit $t \to \pm\infty$. Let $g(\mathbf{X}^N) = \rho(\mathbf{X}^N)$, where $\rho(\mathbf{X}^N)$ is a nonstationary probability density. Then

$$\langle f(t) \rangle = \int_{S_E} f(\mathbf{X}^N) \rho(\mathbf{X}^N(t))\, dS_E \xrightarrow[t\to\pm\infty]{} \frac{1}{\Sigma(E)} \int_{S_E} f(\mathbf{X}^N)\, dS_E. \tag{8.27}$$

Thus, $f(t)$ approaches an average with respect to the stationary state $\rho_{St} = [\Sigma(E)]^{-1}$.

It is important to emphasize that mixing gives a coarse-grained and not a fine-grained approach to a stationary state. The average of the probability density becomes uniform, but the probability density itself cannot because of Eq. (7.24).

The probability density does not change in a neighborhood of a moving phase point, but it can change at a given point in space. We can visualize this if we consider a beaker containing oil and water. We add the oil and water carefully so that they are initially separated, and we assume that they cannot diffuse into one another. We then stir them together (cf. Fig. 8.3). The local density and the total volume of the oil remain constant, but the oil will get stretched into filaments throughout the water. Therefore, on the average, the density of the oil will become uniform throughout the beaker. If we are careful enough, we can also stir the oil back into its original shape. Therefore, while we get an approach to uniformity, the whole process can be reversed. However, as we shall see, mixing does lead to the appearance of random behavior in deterministic systems and coarse-grained irreversibility.

Fig. 8.3. Stirring oil and water together leads to uniformity, *on the average.*

The meaning of Eq. (8.26) may therefore be summarized as follows. Let A and B be two finite arbitrary regions on the surface S_E. Let us assume that all phase points initially lie in A. If the system is mixing, and we let it evolve in time, the fraction of points which lie in A or B in the limit $t \to \pm\infty$ will equal the fraction of area S_E occupied by A or B, respectively.

An important example of a classical system which is mixing and exhibits the type of behavior discussed in our oil and water example is the Bernoulli scheme[3,11,12]—the so-called Baker's transformation. The state points $X(p, q)$ for the Baker's transformation form a discrete set. The energy surface is two dimensional. The Baker's transformation consists of an alphabet with two "letters," 0 and 1, and a set of infinite sequences of elements,

$$S = (\ldots, S_{-2}, S_{-1}, S_0, S_1, S_2, \ldots) \tag{8.28}$$

where $S_k = (0 \text{ or } 1)$ and $k = (0, \pm 1, \pm 2, \ldots)$. Each different infinite sequence, S, corresponds to a state point in Γ-space, as we shall see. Some sequences S will contain completely random orderings of 0 and 1, while others will depend periodically on 0 and 1.

We can map each sequence, S, onto the unit square if we assign to S two numbers:

$$p = \sum_{k=0}^{-\infty} S_k 2^{k-1} \tag{8.29}$$

and

$$q = \sum_{k=1}^{\infty} S_k 2^{-k}. \tag{8.30}$$

We note that if all $S_k = 0$, then $p = q = 0$, and if all $S_k = 1$, then $p = q = 1$ since $\sum_{k=1}^{\infty} (\frac{1}{2})^k = 1$. All other cases fall in between.

We can introduce a dynamics into the phase space by means of the Bernoulli shift, which is defined so that

$$US_k = S_{k+1}. \tag{8.31}$$

That is, the operator U acting on a sequence S shifts each element to the right one place. One may verify that the shift acting on the sequence, S, is equivalent to the following transformation (Baker's transformation) on the unit square:

$$U(p, q) = \begin{cases} (2p, \frac{1}{2}q) & 0 \leqslant p < \frac{1}{2} \\ (2p - 1, \frac{1}{2}q + \frac{1}{2}) & \frac{1}{2} \leqslant p \leqslant 1. \end{cases} \tag{8.32}$$

The inverse transformation is

$$U^{-1}(p, q) = \begin{cases} (\frac{1}{2}p + \frac{1}{2}, 2q - 1) & \frac{1}{2} \leqslant q \leqslant \frac{1}{2} \\ (\frac{1}{2}p, 2q) & 0 < q < \frac{1}{2}. \end{cases} \tag{8.33}$$

The Jacobian of the transformation is easily shown to be equal to one. Therefore, the mapping U is area preserving.

The effect of the Baker's transformation is to stretch an initial area element into filaments throughout the unit square (cf. Fig. 8.4) much as a baker does in kneading dough. We note that whenever $S_0 = 0$ the point corresponding to the sequence S will lie to the left of $p = \frac{1}{2}$ and when $S_0 = 1$ it will lie to the right of $p = \frac{1}{2}$. Therefore, points corresponding to sequences with 0 and 1 distributed at random in the positions S_k will be shifted to the right or left of $p = \frac{1}{2}$ by U at random.

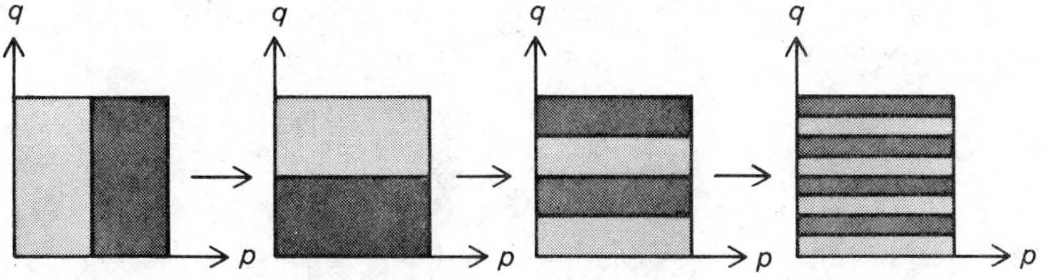

Fig. 8.4. The Baker's transformation stretches an initial area element into filaments which finally fill the phase space uniformly, *on the average.*

Let us now look at how a probability density evolves under the action of the shift U^{-1}. Let $\rho_0(p, q)$ be the initial probability density and let us assume that it is continuous and smooth. After we allow U^{-1} to act n times, we obtain

$$\rho_n(p, q) = U^{-n}\rho_0(p, q) = \rho_0(U^{-n}p, U^{-n}q). \tag{8.34}$$

More explicitly,

$$\rho_{n+1}(p, q) = \begin{cases} \rho_n(\frac{1}{2}p, 2q) & 0 \leqslant q \leqslant \frac{1}{2} \\ \rho_n(\frac{1}{2}p + \frac{1}{2}, 2q - 1) & \frac{1}{2} \leqslant q \leqslant 1. \end{cases} \tag{8.35}$$

We do not expect that the probability density $\rho_n(p, q)$ will become uniform as $n \to \pm\infty$. However, a coarse-grained probability will. As we have seen, there is an element of randomness in the position of a point in the p-direction. Let us therefore look at the reduced probability density $\phi(p)$. The quantity $\phi(p)\,dp$ is the probability of finding a point in the interval $p \to p + dp$. It is defined

$$\phi(p) \equiv \int_0^1 dq\rho(p, q). \tag{8.36}$$

Using Eq. (8.35), $\phi_{n+1}(p)$ becomes

$$\phi_{n+1}(p) = \tfrac{1}{2}\phi_n(\tfrac{1}{2}p) + \tfrac{1}{2}\phi_n(\tfrac{1}{2}p + \tfrac{1}{2}). \tag{8.37}$$

The reduced probability evolves in a Markovian manner.

We will now show that $\lim_{n\to\infty} \phi_n(p) = 1$ and, therefore, that the reduced or coarse-grained probability density approaches a constant.[13] If we iterate Eq. (8.37), we obtain

$$\phi_n(p) = \frac{1}{2^n} \sum_{k=0}^{2^n-1} \phi_0\left(\frac{1}{2^n} p + \frac{k}{2^n}\right). \tag{8.38}$$

For a continuous and smooth function $\phi_0(p)$, we obtain

$$\lim_{n\to\infty} \phi_n(p) = \lim_{n\to\infty} \int_0^{1-1/2^n} dy\phi_0\left(\frac{p}{2^n} + y\right) = \int_0^1 dy\phi_0(y) = 1, \tag{8.39}$$

where we have let $y = k2^{-n}$. Notice that for $n \to -\infty$, $\phi_n(p)$ will *not* approach a constant. However, a reduced probability density defined in terms of the variable q does. Therefore, the Baker's transformation exhibits irreversibility in a coarse-grained sense.

D. ANHARMONIC OSCILLATOR SYSTEMS

The study of anharmonic oscillator systems has long been important in statistical mechanics in connection with the theory of heat transport in solids.[14] One would like to know the mechanism of heat conduction. If heat is added to a harmonic lattice and divided unequally between the normal modes, there is no way for the system to reach equilibrium because the normal modes do not interact. However, if slight anharmonicities exist in the lattice, it was expected that equipartition of energy would occur and the system would thus reach equilibrium.

In 1955, Fermi, Pasta, and Ulam[15] conducted a computer experiment intending to show this. They studied a system of sixty-four oscillators with cubic and broken linear coupling. They found that when energy was added to a few of the lower modes there was no tendency for the energy to spread to the other modes. This behavior is quite different from what one would expect if the anharmonic oscillator system were an ergodic system. Then one expects the system to reach a stationary state in which all states with the same energy would be equally probable, and one expects to see energy sharing among the modes.

The type of behavior that Fermi, Pasta, and Ulam observed is now fairly well understood in terms of a theorem due to Kolmogorov,[16] Arnold,[17] and Moser[18] (commonly called the KAM theorem). The theorem states that for a system with weak anharmonic coupling (which satisfies the conditions of the KAM theorem) most of the energy surface will be composed of invariant tori and the system will exhibit behavior in many respects similar to that of an unperturbed harmonic oscillator system. Thus, the energy surface will not be metrically transitive. As the coupling is increased, however, the invariant regions of phase space break down and at some point one expects to see a sharp transition to chaotic behavior and something similar to equipartition of energy between the modes. (True equipartition requires ergodicity and it is not clear that anharmonic oscillator systems are ergodic above the transition energy.)

There is now a variety of nonlinear oscillator systems which have been studied and which exhibit a transition from stable to chaotic behavior as certain parameters are changed. A particularly clear method of visualizing the onset of instability in anharmonic systems is due to Henon and Heiles.[19] Henon and Heiles studied the bounded motion of orbits for a system with two degrees of freedom governed by the Hamiltonian

$$H = \tfrac{1}{2}(p_1^2 + p_2^2 + q_1^2 + q_2^2) + q_1 q_2^2 - \tfrac{1}{3}q_1^3. \tag{8.40}$$

The trajectories move in a four-dimensional phase space but are restricted to a three-dimensional surface because the total energy is a constant of the motion. It is possible to study a two-dimensional cross-section of the three-dimensional energy surface. For example, we can consider the surface $q_2 = 0$ and look at a trajectory each time it passes through the surface with positive velocity $p_2 > 0$. It is then possible to plot successive points of the trajectory ($q_2 = 0, p_2 > 0$) in the p_1, q_1 plane. If the only constant of motion is the total energy, E, then the points should be free to wander through the region of the p_1, q_1 plane corresponding to the energy surface. The motion we see will appear to be quite similar to ergodic motion. If there is an additional constant of the motion, the points will lie on a smooth curve in the p_1, q_1 plane.

Henon and Heiles studied trajectories whose motion was governed by Eq. (8.40) for a variety of energies. The results are sketched in Fig. 8.5. For an energy $E = 0.08333$, they found only smooth curves, indicating that to computer accuracy there was an additional constant of the motion. Each closed curve in Fig. 8.5 corresponds to one trajectory. The three points of intersection

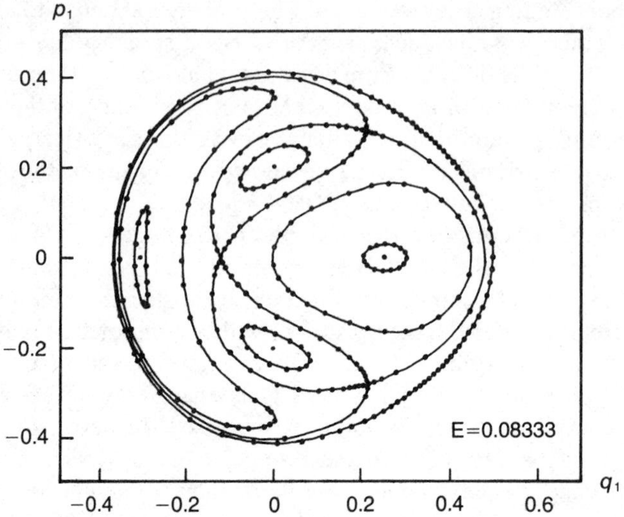

Fig. 8.5. Henon-Heiles result for $E = 0.08333$ (based on Ref. 8.19).

of lines are hyperbolic fixed points and the four points surrounded by curves are elliptic fixed points (cf. App. D). However, at an energy $E = 0.12500$, the picture begins to break down (cf. Fig. 8.6). Each closed curve in Fig. 8.6 corresponds to one trajectory. The five islands correspond to one trajectory and the random dots outside the closed curve correspond to one trajectory. At an

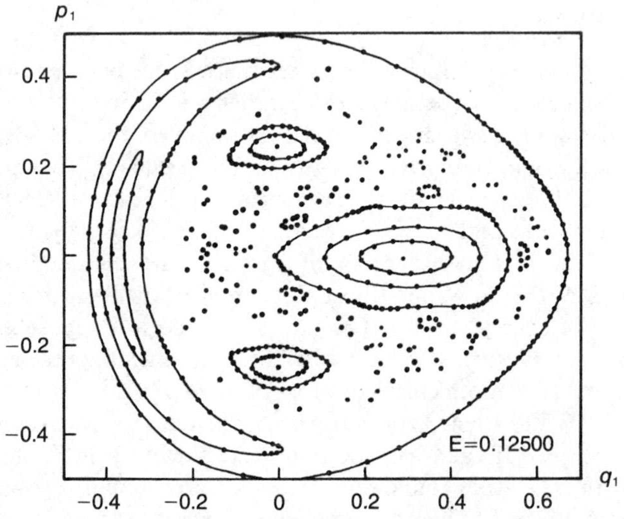

Fig. 8.6. Henon-Heiles result for $E = 0.12500$ (based on Ref. 8.19).

energy of $E = 0.16667$, almost no stable motion remains (cf. Fig. 8.7). A single trajectory is free to wander over almost the entire energy surface. In a very small energy range the system has undergone a transition from stable to chaotic behavior. Additional studies of the Henon-Heiles system have been made by Lunsford and Ford.[20] They found that trajectory points move apart linearly in the stable regions, whereas they move apart exponentially in the chaotic region.

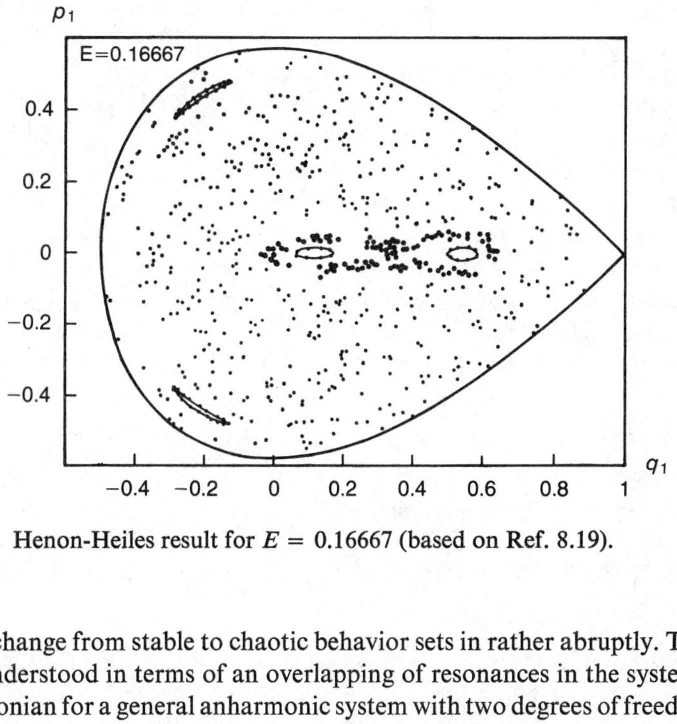

Fig. 8.7. Henon-Heiles result for $E = 0.16667$ (based on Ref. 8.19).

The change from stable to chaotic behavior sets in rather abruptly. This has been understood in terms of an overlapping of resonances in the system. The Hamiltonian for a general anharmonic system with two degrees of freedom can be written in terms of action angle variables in the form

$$H = H_0(J_1, J_2) + \lambda V(J_1, J_2, \phi_1, \phi_2) \tag{8.41}$$

by means of the transformation

$$p_i = -(2m\omega J_i)^{1/2} \sin \phi_i \quad \text{and} \quad q_i = \left(\frac{2J_i}{m\omega}\right)^{1/2} \cos \phi_i.$$

The function, $H_0(J_1, J_2)$, has a polynomial dependence on the action variables J_1 and J_2 (not merely a linear dependence as would be the case for a harmonic system) and no angle dependence, while $V(J_1, J_2, \phi_1, \phi_2)$ depends on both action and angle variables and is a periodic function of the angles. When

$\lambda = 0$ the action variables will be constants of the motion and the angles ϕ_1 and ϕ_2 will change in time according to the equations

$$\phi_i = \omega_i(J_1, J_2)t + \phi_{i,0} \tag{8.42}$$

for ($i = 1, 2$) where

$$\omega_i(J_1, J_2) = \frac{\partial H_0}{\partial J_i}. \tag{8.43}$$

For the anharmonic case, the frequencies $\omega_i(J_1, J_2)$ will be continuous because they depend on the action variables, even for two degrees of freedom. This continuous dependence on the action variables is quite different from a harmonic oscillator system where the frequencies, ω_i, are constant.

For systems which satisfy the conditions of the KAM theorem (namely, small λ and nonzero Hessian, $\det|\partial^2 H_0/\partial_{J_i}\partial_{J_j}| \neq 0$ for $(i, j) = 1$ and 2), only a very small region of phase space, the resonance regions, will exhibit chaotic behavior. The rest of the phase space will correspond to stable motion. If one tries to construct new action variables $\mathscr{J}_i$ which are constants of the motion when $\lambda \neq 0$ but small, one finds that this can be done for most of the phase space, except for the resonance zones. In the resonance zones, perturbation expansions for $\mathscr{J}_i$ diverge. Let us construct a perturbation expansion[21] for the action variables $\mathscr{J}_i$ to lowest order in λ. Let us consider a Hamiltonian of the form

$$H = H_0(J_1, J_2) + \lambda \sum_{n_1, n_2} V_{n_1, n_2}(J_1, J_2) \cos(n_1\phi_1 + n_2\phi_2) \tag{8.44}$$

and let us introduce the generator

$$F(\mathscr{J}_1, \mathscr{J}_2, \phi_1, \phi_2) = \mathscr{J}_1\phi_1 + \mathscr{J}_2\phi_2 + \sum_{n_1, n_2} B_{n_1, n_2} \sin(n_1\phi_1 + n_2\phi_2) \tag{8.45}$$

of a canonical transformation from variables J_i, ϕ_i to variables $\mathscr{J}_i, \Phi_i$, such that the variables $\mathscr{J}_i$ are constants of the motion. Then

$$J_i = \frac{\partial F}{\partial \phi_i} = \mathscr{J}_i + \sum_{n_1, n_2} n_i B_{n_1, n_2} \cos(n_1\phi_1 + n_2\phi_2) \tag{8.46}$$

for $i = 1, 2$ and

$$\Phi_i = \frac{\partial F}{\partial \mathscr{J}_i} \tag{8.47}$$

for $i = 1, 2$. If we substitute Eq. (8.46) into Eq. (8.44) and keep terms to lowest order in λ (this requires a Taylor series expansion of H_0), we obtain

$$H = H_0(\mathscr{J}_1, \mathscr{J}_2) + \sum_{n_1, n_2} \{(n_1\omega_1 + n_2\omega_2)B_{n_1, n_2} + \lambda V_{n_1, n_2}\} \cos(n_1\phi_1 + n_2\phi_2). \tag{8.48}$$

To lowest order in λ, $\mathscr{J}_1$ and $\mathscr{J}_2$ will be constants of motion if we choose

$$B_{n_1 n_2} = \frac{-\lambda V_{n_1,n_2}}{(n_1\omega_1 + n_2\omega_2)}. \tag{8.49}$$

Then

$$H = H_0(\mathscr{J}_1\mathscr{J}_2) + O(\lambda^2) \tag{8.50}$$

and

$$\mathscr{J}_1 = J_1 + \sum_{n_1,n_2} \frac{\lambda V_{n_1,n_2}}{(n_1\omega_1 + n_2\omega_2)} \cos(n_1\phi_1 + n_2\phi_2) + O(\lambda^2). \tag{8.51}$$

Note, however, that since ω_1 and ω_2 are functions of J_1 and J_2, there are values of J_1 and J_2 for which the denominator $(n_1\omega_1 + n_2\omega_2)$ can be zero, and the perturbation expansion becomes meaningless. Indeed, as long as

$$|n_1\omega_1 + n_2\omega_2| \leqslant \lambda V_{n_1 n_2}, \tag{8.52}$$

the perturbation expansion will diverge and $\mathscr{J}_i$ is not a well-behaved invariant. This region of phase space is called the resonance zone and $(n_1\omega_1 + n_2\omega_2) = 0$ is the resonance condition. It is in the resonance zones that one observes chaotic behavior.

If the regions of phase space which contain resonances, and a small region around each resonance, are excluded from the expansion for $\mathscr{J}_1$, then one can have a well-behaved expression for $\mathscr{J}_1$. Thus, one can exclude regions which satisfy the condition

$$[n_1\omega_1(J_1, J_2) + n_2\omega_2(J_1, J_2)] \ll \lambda V_{n_1,n_2}.$$

For smooth potentials, V_{n_1,n_2} decreases rapidly for increasing n_1 and n_2. Thus, for increasing n_1 and n_2, ever smaller regions of the phase space are excluded.

Kolmogorov, Arnold, and Moser proved that as $\lambda \to 0$ the amount of excluded phase space approaches zero. The idea behind their proof is easily seen in terms of a simple example.[22] Consider the unit line (a line of length one). It contains an infinite number of rational fractions, but they form a set of measure zero on the line. If we exclude a region

$$\left(\frac{m}{n} - \frac{\epsilon}{n^3}\right) \leqslant \frac{m}{n} \leqslant \left(\frac{m}{n} + \frac{\epsilon}{n^3}\right)$$

around each rational fraction, the total length of the unit line that is excluded is

$$\sum_{n=1}^{\infty}\sum_{m=1}^{n}\left(\frac{2\epsilon}{n^3}\right) = 2\epsilon \sum_{n=1}^{\infty}\frac{1}{n^2} = \frac{\epsilon\pi^2}{3} \xrightarrow[\epsilon\to 0]{} 0.$$

Thus, for small λ, we can exclude the resonance regions in the expansion of $\mathscr{J}_1$ and still have a large part of the phase space in which $\mathscr{J}_1$ is well defined and invariant tori can exist.

Walker and Ford[21] give a simple exactly soluble example of the type of distortion that a periodic potential can create in phase space. It is worth repeating here. They consider a Hamiltonian of the type

$$H = H_0(J_1, J_2) + \lambda J_1 J_2 \cos(2\phi_1 - 2\phi_2) = E \tag{8.53}$$

where

$$H_0(J_1, J_2) = J_1 + J_2 - J_1^2 - 3J_1J_2 + J_2^2. \tag{8.54}$$

For this model, there are two constants of the motion, the total energy $H = E$ and

$$I = J_1 + J_2. \tag{8.55}$$

Therefore, we do not expect to see any chaotic behavior for this system. However, the unperturbed phase space will still be distorted when $\lambda \neq 0$. The frequencies ω_i for this model are given by

$$\omega_1 = \frac{\partial H_0}{\partial J_1} = 1 - 2J_1 - 3J_2 \tag{8.56}$$

and

$$\omega_2 = \frac{\partial H_0}{\partial J_2} = 1 - 3J_1 + 2J_2. \tag{8.57}$$

If we want the frequencies to remain positive, we must choose $0 \leqslant J_1 \leqslant \frac{5}{13}$ and $0 \leqslant J_2 \leqslant \frac{1}{13}$ and, therefore, $E \leqslant \frac{3}{13}$.

Let us plot trajectories for the Walker-Ford case ($q_2 = 0, p_2 > 0$) (note that $q_i = (2J_i)^{1/2} \cos \phi_i$ and $p_i = -(2J_i)^{1/2} \sin \phi_i$). We find that for $\lambda = 0$ the trajectories trace out concentric circles in the p_1, q_1 plane. When the perturbation is turned on ($\lambda \neq 0$), the phase space becomes highly distorted. If we set $\phi_2 = \frac{3}{2}\pi$ (this means $q_2 = 0, p_2 > 0$) and substitute Eq. (8.55) into Eq. (8.53), we obtain the following equation for the perturbed level curves:

$$(3 + \lambda \cos 2\phi_1)J_1^2 - (5I + \lambda I \cos 2\phi_1)J_1 + I + I^2 - E = 0. \tag{8.58}$$

They are sketched in Fig. 8.8. Most of the phase space is only slightly distorted from the unperturbed case. However, there is a region which is highly distorted and in which two elliptic fixed points (surrounded by orbits) and two hyperbolic fixed points appear (cf. App. D). The fixed points occur for values of J_i and ϕ_i such that $\dot{J}_1 = \dot{J}_2 = (\dot{\phi}_1 - \dot{\phi}_2) = 0$. If we use the fact that $\dot{J}_i = -\partial H/\partial\phi_i$ and $\dot{\phi}_i = \partial H/\partial J_i$ and condition $\phi_2 = 3\pi/2$, we find that the hyperbolic orbits occur when

$$J_1 = \frac{(5 - \lambda)}{(1 - \lambda)} J_2 \quad \text{and} \quad (\phi_1 - \phi_2) = 0 \text{ and } \pi, \tag{8.59}$$

while the elliptic orbits occur for

$$J_1 = \frac{(5+\lambda)}{(1+\lambda)} J_2 \qquad (\phi_1 - \phi_2) = \frac{\pi}{2} \text{ and } \frac{3\pi}{2}. \tag{8.60}$$

The first-order resonance condition for this model (cf. Eq. (8.52)) is $2\omega_1 - 2\omega_2 = 0$ or, from Eqs. (8.56) and (8.57), $J_1 = 5J_2$. Therefore, from Eqs. (8.59) and (8.60) we see that the distorted region of phase space lies in the resonance zone.

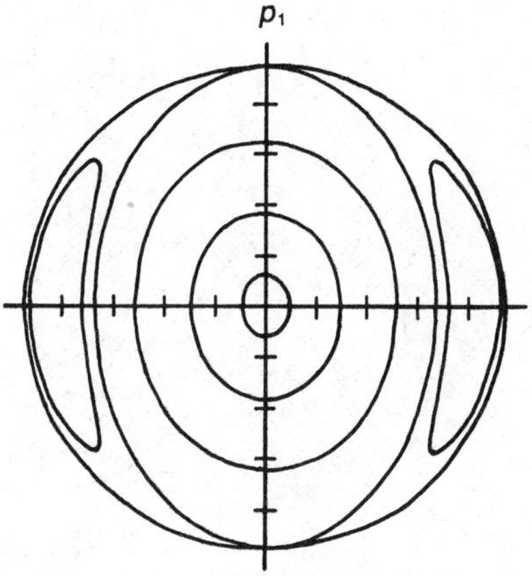

Fig. 8.8. Cross-section of the energy surface for the Hamiltonian $H(J_1, J_2, \phi_1, \phi_2) = J_1 + J_2 - J_1^2 - 3J_1J_2 + J_2^2 + \lambda J_1 J_2 \cos(2\phi_1 - 2\phi_2) = E$: there is no chaotic behavior (based on Ref. 8.21).

In general, for a Hamiltonian of the form

$$H = H_0(J_1, J_2) + \lambda V(J_1, J_2)\cos(n_1\phi_1 + n_2\phi_2) \tag{8.61}$$

there will be no chaotic behavior because there is always an extra constant of motion,

$$I = n_2 J_1 - n_1 J_2. \tag{8.62}$$

However, when the Hamiltonian is of the more general form given in Eq. (8.44), the extra constant of motion is destroyed and the resonance zones become more complicated and begin to overlap. When this occurs one begins to see chaotic behavior.

Walker and Ford study the example

$$H = H_0(J_1, J_2) + \lambda_1 J_1 J_2 \cos(2\phi_1 - 2\phi_2) + \lambda_2 J_1 J_2^{3/2} \cos(2\phi_1 - 3\phi_2), \tag{8.63}$$

where an extra cosine term has been added to Eq. (8.53). For this model there is no longer an extra constant of motion. There are two resonance zones which grow as λ_1 and λ_2 are increased. In Fig. 8.9, we sketch their results. For low energies there is no chaotic behavior (to computer accuracy). However, as the resonance zones grow and begin to overlap, the trajectories in the regions of overlap become unstable and begin to exhibit chaotic behavior. In Fig. 8.9*b*, the dots correspond to a single trajectory.

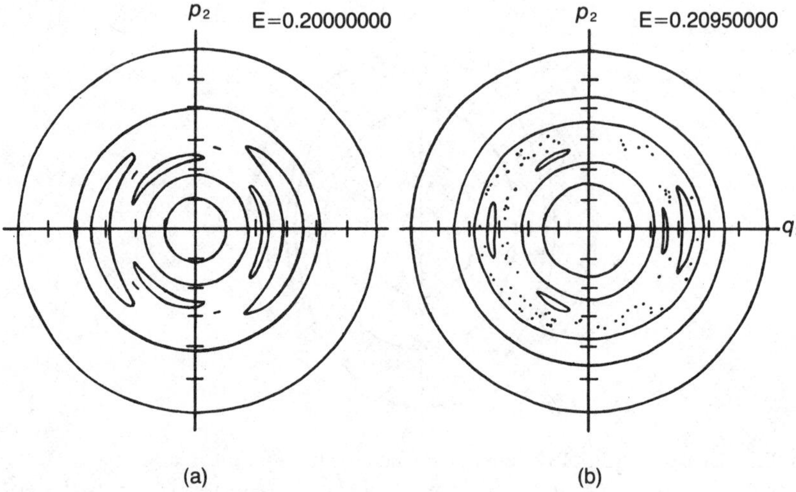

Fig. 8.9. Cross-section of the energy surface for the Hamiltonian $H(J_1, J_2, \phi_1, \phi_2) = J_1 + J_2 - J_1^2 - 3J_1J_2 + \lambda_1 J_1 J_2 \cos(2\phi_1 - 2\phi_2) + \lambda_2 J_1 J_2 \cos(2\phi_1 - 3\phi_2) = E$: (*a*) the flow in phase space below the transition; (*b*) the flow in phase space above the transition energy; when resonances begin to overlap, the motion becomes chaotic (based on Ref. 8.21).

Thus, from these simple examples we see that the chaotic, or ergodiclike, behavior of phase space for the anharmonic oscillator systems appears to be caused by the overlapping of resonances. If the energy surface is filled with resonance zones, as is often the case, then we expect chaotic behavior to set in at very low energy.

Anharmonic oscillator systems are a rather special type of system and their ergodicity has never been established, for obvious reasons. A completely different type of system is a system of hard spheres. For systems of hard spheres, ergodicity and mixing behavior have been established.[23] A proof that systems of particles with Lennard-Jones types of potential are ergodic has never been given. And yet, statistical mechanics, which is built on the assumption of ergodicity, appears to work perfectly for those systems.

The chaotic behavior illustrated in this section is indicative of unstable flow in phase space. Orbits in the chaotic region which initially neighbor one another move apart exponentially and may move to completely different parts of the energy surface. If we start with an ensemble of orbits in some region of phase space and assign a probability distribution to them, the probability distribution will spread on the energy surface and we will become less certain about the actual state of the system.

Systems with unstable flow have the potential of exhibiting decay to thermodynamic equilibrium: an initially localized probability distribution can spread and, in a coarse-grained sense, can fill the energy surface. Indeed, Prigogine, Grecos, and George[24] have given a criterion for decay to equilibrium which is based directly on the properties of resonances in a system.

REFERENCES

1. I. E. Farquhar, *Ergodic Theory in Statistical Mechanics* (Wiley-Interscience, New York, 1964).
2. J. L. Lebowitz and O. Penrose, Physics Today, Feb. 1973.
3. V. I. Arnold and A. Avez, *Ergodic Problems of Classical Mechanics* (W. A. Benjamin, New York, 1968).
4. I. E. Farquhar in *Irreversibility in the Many-Body Problem*, ed. J. Biel and J. Rae (Plenum Press, New York, 1972).
5. L. E. Reichl in *Long Time Predictions in Dynamics*, ed. V. Szebehely and B. Tapley (Reidel Publishing Co., Boston, 1976).
6. P. R. Halmos, *Lectures on Ergodic Theory* (Chelsea Publishing Co., New York, 1955).
7. A. I. Khinchin, *Mathematical Foundations of Statistical Mechanics* (Dover Publications, New York, 1949).
8. N. G. van Kampen, Physica *53* 98 (1971).
9. G. D. Birkhoff, Proc. Natl. Acad. Sci. (U.S.) *17* 656 (1931).
10. P. J. M. Bongaarts and Th. J. Siskens, Physica *68* 315 (1973); *71* 529 (1974).
11. J. Moser, *Stable and Random Motions in Dynamical Systems* (Princeton University Press, Princeton, 1973).
12. O. Penrose, *Foundations of Statistical Mechanics* (Pergamon Press, Oxford, 1970).
13. This proof is due to Professor M. Kac.
14. H. Wergeland in *Irreversibility in the Many-Body Problem*, ed. J. Biel and J. Rae (Plenum Press, New York, 1972).
15. *E. Fermi: Collected Papers*, Vol. II (University of Chicago Press, Chicago, 1965), p. 978.
16. A. N. Kolmogorov in R. Abraham, *Foundations of Mechanics* (W. A. Benjamin, New York, 1967), app. D.
17. V. I. Arnold, Russian Math. Surv. *18* 9 (1963); *18* 85 (1963).
18. J. Moser, Nachr. Akad. Wiss. Göttingen II, Math. Physik Kl. *1* (1962).
19. M. Henon and C. Heiles, Astron. J. *69* 73 (1964).
20. G. H. Lunsford and J. Ford, J. Math. Phys. *13* 700 (1972).
21. C. H. Walker and J. Ford, Phys. Rev. *188* 416 (1969).
22. J. Ford in *Fundamental Problems in Statistical Mechanics III*, ed. E. G. D. Cohen (North-Holland Publishing Co., Amsterdam, 1975).

23. Ya. G. Sinai in *The Boltzmann Equation*, ed. E. G. D. Cohen and W. Thirring (Springer-Verlag, Vienna, 1973).
24. I. Prigogine, A. Grecos, and Cl. George, Celestial Mech. *16* 489 (1977).

PROBLEMS

1. Show that a mixing system is also ergodic.

2. Prove that a single classical harmonic oscillator is ergodic. Draw the phase space and show how a small segment of it evolves. Show that a single quantum harmonic oscillator is ergodic.

3. A bead is free to slide on a straight frictionless wire of length, l, which is attached at each end ($x = 0$ and $x = l$) to infinitely massive walls. Show that the system is ergodic. Assume that the bead initially has momentum with magnitude in the interval $0 \le p \le p_{\max}$ and is located in the interval $0 < x < l/4$ along the wire. Draw the phase space for the system and show how the probability density evolves. How does the probability density look after long times?

4. Compute the structure function for a gas of N noninteracting particles in a box of volume V. Assume that the system has a total energy E.

5. Consider a bound system of fixed energy. Such a system occupies a finite volume, Ω, in phase space. Prove that if a system point, $\mathbf{X}(t)$, is in a volume element R_E of Ω at time t ($R_E \in \Omega$) then there exists a time T (the Poincaré recurrence time) such that $\mathbf{X}(t + T) \in R_E$ for most trajectories $\mathbf{X}(t)$ (all but a set of measure zero).

6. Given a two-dimensional phase space of unit area, assume that at time $n = 0$, the probability density $\phi_0(p, q)$ is defined

$$\phi_0(p, q) = 1 \quad \text{for} \quad \begin{cases} 0 < p < \frac{1}{2} \\ \frac{1}{4} < q < \frac{3}{4} \end{cases}$$

and

$$\phi_0(p, q) = 0 \quad \text{for} \quad \begin{cases} \frac{1}{2} < p < 1 \\ 0 < q < \frac{1}{4} \\ \frac{3}{4} < q < 1 \end{cases}$$

and that the dynamics of the system is that of the Baker's transformation. Sketch $\phi_n(p, q)$ for $n = 0, 1, 2, 3,$ and 4.

7. Consider a simple pendulum whose equation of motion is $\ddot{q} + k\sin(q) = 0$. Find the Hamiltonian for the system and sketch the flow in the phase space of the pendulum. Locate at least one elliptic fixed point and two hyperbolic fixed points for the system (cf. App. D) and prove that they are indeed hyperbolic and elliptic fixed points.

9

Equilibrium Statistical Mechanics: Soluble Models

A. INTRODUCTORY REMARKS

The state of thermodynamic equilibrium is a state which does not change in time and, therefore, is a stationary state of the Liouville equation. We have seen (Chap. 7) that for a system with a nondegenerate energy spectrum (ergodic) the stationary state is a function only of the total Hamiltonian. For a system with a degenerate energy spectrum the stationary state will depend on the total Hamiltonian and those operators which commute with it. For a closed isolated system, that is, one with fixed energy E, the stationary state is a constant on the energy surface. Thus, even though we do not yet know how to obtain irreversible decay from the Liouville equation, we do know the form of its stationary states. In this chapter, we shall establish a connection between the stationary probability densities for dynamical systems and thermodynamic variables.

The "fundamental" probability density that we obtained in Chap. 8 for ergodic systems applies to closed, isolated systems, that is, systems with fixed energy. As we have seen in the chapters on thermodynamics, we often want to deal with systems which are closed but not isolated or with systems that are opened. Closed nonisolated systems can have fluctuating total energy and open systems can have fluctuating energy and particle number. We must find the probability densities for these more general situations. At the same time, if we wish the probability density we obtain to describe the state of thermodynamic equilibrium, it must yield an extremum for the entropy of the system. We thus come to a stumbling block. How is the entropy expressed in terms of the probability density? There are many possibilities, but one, first introduced by Gibbs, yields the correct results for the thermodynamic properties of equilibrium systems and is now almost universally used. In this chapter we shall obtain the probability densities which extremize the Gibbs entropy for closed and open systems and we will show that they become equivalent to each other and to the fundamental probability density in the thermodynamic limit (the limit $N \rightarrow \infty$, $V \rightarrow \infty$, such that $N/V =$ constant where N is the number of degrees of freedom and V is the volume). The probability density we obtain for closed isolated systems is called the microcanonical ensemble, that for closed systems is called the canonical ensemble, and that for open systems is called the grand canonical ensemble.

Once we have obtained the probability density for an equilibrium system, we can begin to compute thermodynamic quantities in terms of the microscopic properties of the system, such as the mass of the constituent particles, their interaction potentials, their statistics, etc. In this chapter we have selected a variety of fairly simple examples to illustrate how the equilibrium probability densities can be applied to real systems. We have selected the examples for their historical and practical importance or because they illustrate important concepts that we shall need later. The systems we will study in this chapter are all exactly soluble (after we make suitable approximations). In subsequent chapters, we will introduce a variety of approximation schemes for systems which cannot be solved exactly.

We begin with a system which historically has been of great importance because it gave one of the early confirmations of quantum mechanics, it laid the foundation for the modern theory of solids, and it gives us an important insight into how we must view many-body systems. In the very crudest approximation solids can be considered as a collection of harmonic oscillators. If we treat the harmonic oscillators classically and assume that they do not interact, then statistical mechanics predicts a constant heat capacity for all temperatures. However, experimentally the heat capacity of solids is found to go to zero as $T \to 0$ K. Einstein, in 1906, showed that the experimental behavior could be reproduced fairly well, but not quite correctly, by assuming that the harmonic oscillators were quantum mechanical. Debye improved on the work of Einstein and obtained almost perfect agreement with experiment by assuming that the harmonic oscillators were coupled harmonically to their nearest neighbors. For the system considered by Debye, the Hamiltonian can be diagonalized and yields a picture of the solid as a gas of noninteracting collective modes called phonons (phonons are again noninteracting harmonic oscillators). The idea of describing a many-body system in terms of its noninteracting or weakly coupled collective modes, rather than its basic particle properties, is one of the most important ideas in many-body physics.

In the Einstein and Debye theory of solids, the particles are assumed to be distinguishable; that is, we do not allow exchange among particles on the lattice and therefore we always know where each one is (at very low temperature, exchange effects in solid He are important). The next example we consider in this chapter is that of the ideal gases. In this case, particles are free to move and, therefore, if all particles are identical, they must be considered as indistinguishable. There are two basic types of ideal gases, Bose-Einstein and Fermi-Dirac. A Bose-Einstein gas is composed of identical bosons and at very low temperatures can exhibit a phase transition (even though the particles do not interact) in which a macroscopic number of particles begin to occupy a single momentum state. (Momentum is a "good" quantum number for noninteracting particles in an infinite box.) The phase transition is a consequence of Bose-Einstein statistics which tends to favor accumulation of particles in a single quantum state and, therefore, causes an effective attraction between particles.

Noninteracting fermions, on the other hand, do not exhibit a phase transition but, because of the Pauli exclusion principle, have an interesting behavior of their own. No two fermions can occupy the same quantum state. Thus, no two fermions can have the same momentum and spin components and, at low temperature, the fermions begin to fill all the lower momentum states. For sufficiently dense systems, some particles in the gas can have very high momentum even at 0 K. Thus, a Fermi gas, even at $T = 0$ K, will have a large zero point pressure although no thermal energy is left in the system.

For an ideal Fermi gas, there is no condensation into a single momentum state. However, if we allow an attraction to exist between the fermions, pairs of fermions (we never know which two fermions they are) can form bound states and can condense in momentum space. This is what happens to electrons in a superconductor. In a superconducting solid, electrons interact with lattice phonons and with one another through a phonon-mediated interaction. For superconductors, the phonon-mediated interaction is attractive in a small neighborhood of the Fermi surface. The fermion pairs condense in momentum space into a single quantum state. Since all pairs have the same quantum numbers (this is the state of lowest free energy), they act coherently, thus giving rise to the unusual superfluid properties observed in such systems. In this chapter, we shall consider the simplest Hamiltonian which gives rise to condensation in a Fermi system and corresponds to a mean field description. In many respects, the Hamiltonian we shall use is reminiscent of the Debye Hamiltonian. In the mean field approximation, it contains only pairs of creation and annihilation operators and thus mathematically is like a harmonically coupled system which depends only on pairs of phase space variables. As Bogoliubov showed, the mean field Hamiltonian can be diagonalized and leads to the picture of a superconductor as an ideal Fermi gas of "bogolons." Bogolons are the collective modes of a condensed Fermi system. From the mean field Hamiltonian, we can find microscopic expressions for the order parameter of the condensed phase and the thermodynamic properties of the condensed phase.

There is a completely different class of systems which are now widely used to study phase transitions. These are the order-disorder transitions that can occur on lattices. Such systems are nontrivial but are still mathematically tractable. In a few notable cases, the thermodynamic properties of such systems can be obtained exactly even in the neighborhood of the phase transition. Thus, they yield a great deal of insight into the behavior of systems near a phase transition. The prototype of such systems is the Ising model, which was introduced by Ising to study the transition from a paramagnetic state to a ferromagnetic state on a magnetic lattice. However, it has since been applied to a large variety of systems, such as antiferromagnets, lattice gases, and DNA molecules. In this chapter we will discuss the physical ideas behind the Ising model and solve it using a mean field approximation (called the Bragg-Williams approximation). The Ising model can be solved exactly in one and two dimensions (in two dimensions it was first done by Onsager). In two dimensions it exhibits a phase transition, while in one dimension it does not. We shall solve the Ising model exactly in one dimension and then discuss Onsager's solution for the two-dimensional case and compare it with the solution in the mean field approximation.

A very interesting mathematical picture of equilibrium phase transitions has been given by Lee and Yang, who were able to relate phase transitions to the analytic properties of the grand partition function. We shall discuss the Lee-Yang theory in this chapter and give a simple example.

Finally, we shall use a model due to Ornstein and stability arguments due to van Kampen to derive the van der Waals equation from a mean field theory. The ideas we use in deriving the van der Waals equation are instructive and will be useful in later chapters.

B. EQUILIBRIUM ENSEMBLES

Most systems in nature, if they are isolated, will tend to a time-independent (stationary) state which may be described in terms of only a few state variables (cf. Chap. 2). For such stationary states (the states of thermodynamic equilibrium) the entropy is a maximum. However, a system with fixed internal energy, $U = E$, could be in any one of the many microscopic states on the energy surface E. If we only know the total energy, we have no way of distinguishing one microscopic state from another. Therefore, it is reasonable that we give all microscopic states on the energy surface an equal probability. As we have seen in Chap. 7, for ergodic systems, the underlying dynamics leads naturally to this choice of stationary state. Thus, for a system with fixed total energy, E, in thermal equilibrium, the best choice we can make for the equilibrium probability density is the fundamental distribution

$$\rho(\mathbf{X}^N, E) = \begin{cases} \dfrac{1}{\Sigma(E)} & \text{for } H(\mathbf{X}^N) = E \\ 0 & \text{otherwise} \end{cases} \tag{9.1}$$

where $\Sigma(E)$ is the area of the energy surface (structure function). This choice of probability density for an isolated system forms the foundation upon which statistical mechanics is built.

Once the distribution function is given, then it is a straightforward matter to compute the expectation values of various quantities, such as energy, magnetization, etc. However, there is one quantity which still eludes us. And that is the entropy. We know that the entropy must be additive and positive and must have a maximum value at equilibrium. There are many possible choices for such a quantity. They are

$$S = k_B \int d\mathbf{X}^N (\rho(\mathbf{X}^N))^{2n} C^N, \tag{9.2}$$

where n is some positive integer, and

$$S = -k_B \int d\mathbf{X}^N \rho(\mathbf{X}^N) \log [C^N \rho(\mathbf{X}^N)]. \tag{9.3}$$

The constant C^N is inserted to give the correct units. Both forms of entropy given in Eqs. (9.2) and (9.3) are concave functions and give a maximum entropy when the probability density is constant on the energy surface. In addition, they maximize the entropy when the probability density is constant along any energy surface in phase space. However, the later form of entropy, Eq. (9.3), was chosen by Gibbs[1] (and is the one we shall use) because it gives the correct expression for the temperature for systems which are closed but not isolated (not at fixed energy). For quantum systems, the Gibbs entropy takes the more general form

$$S = -k_B \operatorname{Tr} \hat{\rho} \ln \hat{\rho}, \tag{9.4}$$

where $\hat{\rho}$ is the density operator and the trace is taken over any complete orthonormal set of basis states.

As we have seen in Chap. 2, we often want to consider systems under a variety of external constraints. Closed, isolated systems have fixed total energy and fixed particle number. Closed systems have fixed particle number but varying energy so that only the average energy is specified. Open systems can have varying particle number and energy. A closed isolated system has an equilibrium probability density of the form given in Eq. (9.1). However, the probability densities for closed and open systems must still be determined. Once the equilibrium probability density for a system is known, the problem of computing thermodynamic quantities is straightforward.

1. Closed Isolated Systems: Microcanonical Ensemble[1–3]

We already know the form of the equilibrium probability density for closed isolated systems (systems with fixed energy E): it is simply a constant on the energy surface (cf. Eq. (9.1)). However, for the sake of completeness we will now rederive it using the Gibbs entropy. We shall restrict ourselves to classical systems and will indicate how the corresponding proof is done for the quantum case.

Let us consider a closed isolated system with a configuration space volume, V, and a fixed number of particles, N, which is constrained to the energy shell $E \to E + \Delta E$. We now consider an energy shell rather than just the energy surface because it simplifies calculations and because the Heisenberg uncertainty principle tells us that we can never determine the energy exactly. We can make ΔE as small as we like. To obtain the equilibrium probability density we must find an extremum of the Gibbs entropy subject to the normalization condition

$$\int_{E<H(\mathbf{X}^N)<E+\Delta E} d\mathbf{X}^N \rho(\mathbf{X}^N) = 1, \tag{9.5}$$

where the integration is restricted to the energy shell.

The simplest way of finding an extremum of the entropy subject to a constraint is to use the method of Lagrange multipliers. Since we have one constraint, we need one Lagrange multiplier, which we call α_0. We then require that the following variation be zero:

$$\delta\left[\int_{E\leq H(\mathbf{X}^N)\leq E+\Delta E} d\mathbf{X}^N(\alpha_0\rho(\mathbf{X}^N) - k_B\rho(\mathbf{X}^N)\ln[C^N\rho(\mathbf{X}^N)])\right] = 0 \tag{9.6}$$

or

$$\int_{E<H(\mathbf{X}^N)\leq E+\Delta E} d\mathbf{X}^N(\alpha_0 - k_B \ln[C^N\rho(\mathbf{X}^N)] - k_B)\,\delta\rho(\mathbf{X}^N) = 0. \tag{9.7}$$

Since the variation $\delta\rho(\mathbf{X}^N)$ is arbitrary, the integral will be zero, in general, if the integrand is zero. Thus,

$$(\alpha_0 - k_B \ln[C^N\rho(\mathbf{X}^N)] - k_B) = 0 \tag{9.8}$$

or

$$\rho(\mathbf{X}^N) = \frac{1}{C^N} e^{(\alpha_0/k_B - 1)} = \begin{cases} K & \text{for } E \leqslant H(\mathbf{X}^N) \leqslant E + \Delta E \\ 0 & \text{otherwise} \end{cases} \tag{9.9}$$

where K is a constant. We can now determine K (and therefore α_0) from the normalization condition in Eq. (9.5). We find

$$\rho(\mathbf{X}^N) = \begin{cases} \dfrac{1}{\Omega_{\Delta E}(E, V, N)} & \text{for } E \leqslant H(\mathbf{X}^N) \leqslant E + \Delta E \\ 0 & \text{otherwise} \end{cases} \tag{9.10}$$

where $\Omega_{\Delta E}(E, V, N) = \Delta E\Sigma(E, V, N)$ and $\Sigma(E, V, N)$ is the structure function for the surface of energy E (cf. Sec. 7.B). The quantity $\Omega_{\Delta E}(E, V, N)$ is the volume of the energy shell. If we substitute Eq. (9.10) into Eq. (9.3), we obtain

$$S(E, V, N) = k_B \ln\left(\frac{\Omega_{\Delta E}(E, V, N)}{C^N}\right). \tag{9.11}$$

The constant, C^N, has the same units as $\Omega_{\Delta E}(E, V, N)$ and cannot be determined classically. However, it can be determined from quantum mechanics. Then we find that $C^N = (h)^{3N}$ for distinguishable particles and $C^N = N!\,(h)^{3N}$ for indistinguishable particles, where h is Planck's constant. The insertion of the factor $N!$ is called correct Boltzmann counting. It is purely quantum mechanical in origin and cannot be justified classically except that it solves Gibbs paradox. We shall leave it as a homework problem to show that with the factor $N!$ there is no longer an entropy of mixing of identical particles. From the Heisenberg uncertainty principle, we know that h is the volume of a single state in phase space. Thus $\Omega_{\Delta E}(E, V, N)/C^N$ is the total number of states in the energy shell $E \to E + \Delta E$.

Eq. (9.11) is the "fundamental equation" for the system (cf. Sec. 2.E). From it we can derive all thermodynamic properties. For example, the temperature is given by $T = (\partial E/\partial S)_{V,N}$; the pressure, by $P = (\partial E/\partial V)_{S,N}$; and the chemical potential, by $\mu = (\partial E/\partial N)_{S,V}$.

For quantum systems the proof is similar. We find the set of states $|E, n\rangle$ with respect to which the density matrix is diagonal. Then the entropy can be written

$$S = -k_B \operatorname{Tr} \hat{\rho} \ln \hat{\rho} = -k_B \sum_{n=1}^{N(E)} P_n \ln P_n, \tag{9.12}$$

where $P_n = \langle E, n|\hat{\rho}|E, n\rangle$ is the probability of finding the system in state $|E, n\rangle$. The normalization condition takes the form

$$\operatorname{Tr} \hat{\rho} = \sum_{n=1}^{N(E)} P_n = 1. \tag{9.13}$$

The upper limit, $N(E)$, in the summation is the total number of states with energy E. If we now maximize the entropy subject to the normalization condition in Eq. (9.13) in the same manner as for the classical case, we obtain for the probability P_n

$$P_n = \frac{1}{N(E)} \tag{9.14}$$

and for the entropy

$$P_n = k_B \ln N(E). \tag{9.15}$$

Thus, again we find that the entropy is proportional to the logarithm of the number of states with energy E.

The probabilities in Eqs. (9.10) and (9.14) can be interpreted as describing a single system. Then they give the fraction of times the system can be found in a given microscopic state if one looks at the system many times. The probabilities can also be interpreted as describing an ensemble of identical systems, in which case they give the fraction of systems that can be found in a given microscopic state. An ensemble of systems with energy E, which is described by probability distribution of the type given in Eq. (9.10) or (9.14), is called a microcanonical ensemble.

It is worthwhile to show that the entropy we have obtained in Eq. (9.11) is extensive in the thermodynamic limit, that is, in the limit $N \to \infty$ and $V \to \infty$ such that $N/V =$ constant. Let us consider a system of N distinguishable particles with a separable Hamiltonian of the form

$$H(\mathbf{X}^N) = H(\mathbf{X}^\alpha) + H(\mathbf{X}^\beta), \tag{9.16}$$

where $\alpha + \beta = N$. An example of such a system would be a gas of noninteracting particles, in which case $H(\mathbf{X}^\alpha)$ and $H(\mathbf{X}^\beta)$ would also be separable. The total phase space will be the direct product of the phase spaces for each of the subsystems. Let us assume that the total system has energy E and occupies a volume $\Omega(E, N, V)$. We can write $\Omega(E, N, V)$ in the form

$$\Omega(E, N, V) = \int_{0<H(\mathbf{X}^N)<E} d\mathbf{X}^N = \int_{0<H(\mathbf{X}^\alpha)<E_\alpha<E} d\mathbf{X}^\alpha \int_{0<H(\mathbf{X}^\beta)<E-E_\alpha} d\mathbf{X}^\beta. \tag{9.17}$$

From Eq. (8.7), we then obtain

$$\Omega(E, N, V) = \int_{0<H(\mathbf{X}^\alpha)<E_\alpha} d\mathbf{X}^\alpha \Omega_\beta(E - E_\alpha) = \int_0^E dE_\alpha \Sigma_\alpha(E_\alpha)\Omega_\beta(E - E_\alpha), \tag{9.18}$$

where $\Omega_\beta(E - E_\alpha)$ is the volume occupied by system β when it has energy $E - E_\alpha$. We can let the upper limit go to infinity in Eq. (9.18) since $\Omega_\beta(E - E_\alpha) = 0$ for $E_\alpha > E$. If we then take the derivative with respect to E, we find

$$\Sigma(E, N, V) = \int_0^\infty dE_\alpha \Sigma_\beta(E - E_\alpha)\Sigma_\alpha(E_\alpha). \tag{9.19}$$

Eq. (9.19) gives the law of composition of structure functions.

Let us now divide the phase space into energy shells of width ΔE (very small) and assume the energy is roughly constant in each shell but changes from shell to shell. We can then replace the integration by a summation in Eq. (9.18) and obtain

$$\Omega(E, N, V) = \sum_{i=1}^{E/\Delta E} \Omega_\alpha(E_{\alpha_i})\Omega_\beta(E - E_{\alpha_i}). \tag{9.20}$$

The sum is over all shells of width ΔE. There will be one term in the sum with energy $\bar{E}_\alpha$ which is larger than all others. Thus, we can write

$$\Omega_\alpha(\bar{E}_\alpha)\Omega_\beta(E - \bar{E}_\alpha) \leqslant \Omega(E, N, V) \leqslant \frac{E}{\Delta} \Omega_\alpha(\bar{E}_\alpha)\Omega_\beta(E - \bar{E}_\alpha). \tag{9.21}$$

If we next divide by h^N and take the logarithm, we obtain

$$S(\bar{E}_\alpha) + S(\bar{E}_\beta) \leqslant S(E, N, V) \leqslant S(\bar{E}_\alpha) + S(\bar{E}_\beta) - k_B \ln\left(\frac{E}{\Delta}\right). \tag{9.22}$$

We now note that the logarithm of the volume of phase space grows as N as we increase the number of degrees of freedom, while $\ln(E/\Delta)$ grows as $\ln(N)$. In the limit $N \to \infty$, we can neglect the term depending on $\ln(E/\Delta)$ and obtain

$$S(E, N, V) = S(\bar{E}_\alpha) + S(\bar{E}_\beta). \tag{9.23}$$

Thus, in the thermodynamic limit, the entropy given by Eq. (9.23) is additive.

In practice, when computing the thermodynamic properties of a system of energy E, it is easier to work with the total volume of phase space $\Omega(E, N, V)$, enclosed by the surface of energy E, than with the volume, $\Omega_{\Delta E}(E, N, V)$, of the energy shell $E \to E + \Delta E$. In the thermodynamic limit, both quantities give the same entropy. To see this let us divide the volume enclosed by the surface of energy E into $E/\Delta E$ shells of width ΔE. These shells will range in energy from 0 to E. Shells with low energy will have a smaller volume than shells with higher energy. The total volume can be written as a sum of the volumes of all the shells:

$$\Omega(E, N, V) = \sum_{i=1}^{E/\Delta E} \Omega_{\Delta E}(E_i, N, V). \tag{9.24}$$

The shell with largest volume will be that with energy $E \to E + \Delta E$. Thus, we can write

$$\Omega_{\Delta E}(E, N, V) \leqslant \Omega(E, N, V) \leqslant \frac{E}{\Delta V} \Omega_{\Delta E}(E, N, V), \tag{9.25}$$

and by the same argument used to obtain Eq. (9.23), we can write *in the thermodynamic limit*

$$S(E, N, V) = k_B \ln \frac{\Omega(E, N, V)}{C^N} = k_B \ln \frac{\Omega_{\Delta E}(E, N, V)}{C^N}. \tag{9.26}$$

Thus, in the thermodynamic limit, we need to find only the total volume of phase space enclosed by the surface $H(X^N) = E$ to obtain thermodynamic properties.

2. Closed Systems: The Canonical Ensemble

A closed system can exchange heat with its surroundings and as a consequence will have a fluctuating total energy. We therefore need to find a probability density for the total energy which corresponds to an extremum of the entropy. This again can be done by the method of Lagrange multipliers except that we now have two constraints. We must require that the probability density be normalized in the whole phase space Γ,

$$\int_\Gamma d\mathbf{X}^N \rho(\mathbf{X}^N) = 1, \tag{9.27}$$

and that the total energy have a fixed average value:

$$\langle E \rangle = \int_\Gamma d\mathbf{X}^N H(\mathbf{X}^N) \rho(\mathbf{X}^N). \tag{9.28}$$

If we introduce a second Lagrange multiplier, α_E, the condition that $\rho(\mathbf{X}^N)$ corresponds to an extremum of the entropy is

$$\delta\left[\int d\mathbf{X}^N(\alpha_0\rho(\mathbf{X}^N) + \alpha_E H(\mathbf{X}^N)\rho(\mathbf{X}^N) - k_B\rho(\mathbf{X}^N)\ln[C^N\rho(\mathbf{X}^N)])\right] = 0, \tag{9.29}$$

where $C^N = h^{3N}$ for distinguishable particles and $C^N = N!\,(h)^{3N}$ for indistinguishable particles (cf. Sec. 9.B.1). This in turn leads to the equation

$$\alpha_0 + \alpha_E H(\mathbf{X}^N) - k_B \ln[C^N\rho(\mathbf{X}^N)] - k_B = 0 \tag{9.30}$$

or

$$\rho(\mathbf{X}^N) = \frac{1}{C^N}\exp\left[\frac{\alpha_0}{k_B} - 1 + \frac{\alpha_E}{k_B}H(\mathbf{X}^N)\right]. \tag{9.31}$$

We will determine the two Lagrange multipliers from Eqs. (9.27) and (9.28). We can first find a relation between α_0 and α_E if we require that $\rho(\mathbf{X}^N)$ be normalized to one. We obtain

$$Z_N(V, \alpha_E) \equiv \exp\left(1 - \frac{\alpha_0}{k_B}\right) = \frac{1}{C^N}\int d\mathbf{X}^N \exp\left[\frac{\alpha_E}{k_B}H(\mathbf{X}^N)\right] \tag{9.32}$$

where $Z_N(V, \alpha_E)$ is called the partition function. We next must determine α_E. To do this, let us multiply Eq. (9.30) by $\rho(\mathbf{X}^N)$ and integrate. We find

$$(\alpha_0 - k_B)\int d\mathbf{X}^N\rho(\mathbf{X}^N) + \alpha_E\int d\mathbf{X}^N H(\mathbf{X}^N)\rho(\mathbf{X}^N) - k_B\int d\mathbf{X}^N\rho(\mathbf{X}^N)\ln[C^N\rho(\mathbf{X}^N)] = 0. \tag{9.33}$$

If we now identify the average energy with the internal energy $U = \langle E\rangle$, Eq. (9.33) becomes

$$-k_B \ln Z_N(V, \alpha_E) + \alpha_E U + S = 0. \tag{9.34}$$

However, we know that the fundamental equation for the Helmholtz free energy can be written $A - U + ST = 0$ (cf. Sec. 2.F). Thus, we make the identification

$$\alpha_E = -\frac{1}{T} \tag{9.35}$$

and

$$A = -k_B T \ln Z_N(T, V), \tag{9.36}$$

where the partition function is now written

$$Z_N(T, V) = \frac{1}{C^N}\int d\mathbf{X}^N\, e^{-\beta H(\mathbf{X}^N)}, \tag{9.37}$$

with $\beta = (k_B T)^{-1}$, and the probability density is written

$$\rho(\mathbf{X}^N) = \frac{1}{C^N Z_N(T, V)}\, e^{-\beta H(\mathbf{X}^N)}. \tag{9.38}$$

An ensemble of identical systems whose probability density is given by Eq. (9.38) is called a canonical ensemble. It is important to note that Eq. (9.30) was obtained by extremizing the Gibbs entropy. Historically, the Gibbs entropy was chosen because it gave this result[1,2] for the probability density of the energy of a system at temperature T. At high temperatures, where equipartition of energy occurs, Eq. (9.38) gives the correct relation between the temperature and the average energy.

Eq. (9.36) is the fundamental equation for a closed system. From it we can obtain all thermodynamic properties. For example, the entropy is given by $S = -(\partial A/\partial T)_{V,N}$, the pressure is given by $P = -(\partial A/\partial V)_{T,N}$, and the chemical potential is given by $\mu = (\partial A/\partial N)_{T,V}$.

For a quantum system, a similar analysis yields, for the density operator,

$$\hat{\rho}^N = e^{\beta[A(T,V,N) - \hat{H}^N]} \tag{9.39}$$

and, for the partition function,

$$Z(T, V, N) = e^{-\beta A(T,V,N)} = \mathrm{Tr}\, e^{-\beta \hat{H}^N}. \tag{9.40}$$

The trace in Eq. (9.40) can be evaluated with respect to any convenient set of orthonormal basis states.

The partition function is particularly simple to evaluate for objects which are distinguishable because it can be written as a product of partition functions for each object. As an example, let us consider a gas of N distinguishable non-interacting particles and assume that each particle can have any one of the discrete energies ϵ_α, where $\alpha = 1, 2, \ldots, m$. If the number of particles with energy ϵ_α is n_α, then the total energy, $E_j\{n_\alpha\}$, of the system for a given set of occupation numbers $\{n_\alpha\} = (n_1, n_2, \ldots, n_m)$ is $E_j\{n_\alpha\} = \sum_{\alpha=1}^m n_\alpha \epsilon_\alpha$, where $N = \sum_{\alpha=1}^m n_\alpha$. Since the particles are distinguishable, the total number of different states of the system with energy $E_j\{n_\alpha\}$ is $N!/n_1!\, n_2! \cdots n_m!$. In evaluating the partition function, we must sum over all different states of the system consistent with the restriction $N = \sum_{\alpha=1}^m n_\alpha$. We then find

$$Z_N(T)_{\mathrm{Dist}} = \sum_{\substack{(j) \\ (\Sigma n_\alpha = N)}} \frac{N!}{n_1!\, n_2! \cdots n_m!} \exp\left\{-\beta \sum_{\alpha=1}^m n_\alpha \epsilon_\alpha\right\} = \left(\sum_{\alpha=1}^m e^{-\beta \epsilon_\alpha}\right)^N,$$

where we have used the multinomial theorem. The quantity $z_1(T) = \sum_{\alpha=1}^m e^{-\beta\epsilon_\alpha}$ is the partition function for a single particle. If we have a gas of indistinguishable particles, then each set $\{n_\alpha\}$ corresponds to *one* state and the partition function becomes

$$Z_N(T)_{\mathrm{Indist}} = \sum_{\substack{(j) \\ (\Sigma n_\alpha = N)}} \exp\left\{-\beta \sum_{\alpha=1}^m n_\alpha \epsilon_\alpha\right\}.$$

For this case the partition function cannot be factorized and it is easier to use the grand partition function to evaluate the properties of the gas. At high temperatures, where the effects of quantum statistics are not important, the partition function can be used to compute the thermodynamic properties of a gas of indistinguishable particles if we note that

$$Z_N(T)_{\mathrm{Indist}} \approx \frac{1}{N!} Z_N(T)_{\mathrm{Dist}}.$$

However, at lower temperatures, where quantum statistics becomes important, this is no longer true and it is easier to use the grand canonical ensemble.

In Eq. (9.38), we have obtained the probability density for the total energy of a system with N degrees of freedom at a fixed temperature, T, and average energy $\langle E\rangle$. As usual, it is important to determine the size of the standard deviation of the distribution so that we can determine the size of energy fluctuations in the system. Let us write the normalization condition in the form

$$\frac{1}{C^N}\int_\Gamma d\mathbf{X}^N\, e^{-\beta(H^N(\mathbf{X}^N)-A(T,V,N))} = 1 \tag{9.41}$$

and take the derivative with respect to β, holding particle number and volume constant. We then find

$$U = \langle E\rangle = \left(\frac{\partial\beta A}{\partial\beta}\right)_{N,V} \tag{9.42}$$

and

$$\langle(E - \langle E\rangle^2)\rangle = k_B T^2 C_V, \tag{9.43}$$

where C_V is the constant volume heat capacity, and U is the internal energy. The heat capacity and the internal energy are proportional to the number of degrees of freedom. That is, $C_V \sim N$. Thus, the fractional deviation

$$\sqrt{\langle(E - \langle E\rangle)^2\rangle}/\langle E\rangle \sim N^{-1/2}$$

and goes to zero as the number of degrees of freedom becomes infinite. This means that the fluctuations in energy become very small relative to the magnitude of the energy itself. In the thermodynamic limit, most microstates will have an energy approximately equal to the average energy $U = \langle E\rangle$ and the canonical ensemble becomes equivalent to the microcanonical ensemble.

3. Open Systems: Grand Canonical Ensemble

An open system can exchange both heat and matter with its surroundings and, therefore, both the energy and the particle number will fluctuate. However, for an open system in equilibrium we can require that both the average energy $\langle E\rangle$ and the average particle number $\langle N\rangle$ be fixed. To obtain the probability density for the energy and particle number, we proceed as before. However, we must introduce a third Lagrange multiplier and must allow for variable particle number. Thus, the normalization takes the form

$$\sum_{N=0}^{\infty}\int d\mathbf{X}^N \rho(\mathbf{X}^N) = 1, \tag{9.44}$$

where we have summed over all possible numbers of particles in the system. The average energy is defined

$$\langle E\rangle = \sum_{N=0}^{\infty}\int d\mathbf{X}^N H(\mathbf{X}^N)\rho(\mathbf{X}^N), \tag{9.45}$$

and the average particle number is defined

$$\langle N\rangle = \sum_{N=0}^{\infty}\int d\mathbf{X}^N N\rho(\mathbf{X}^N). \tag{9.46}$$

To obtain an extremum of the entropy subject to the above constraints, we write

$$\delta\left[\sum_{N=0}^{\infty}\int d\mathbf{X}^N(\alpha_0\rho(\mathbf{X}^N) + \alpha_E H(\mathbf{X}^N)\rho(\mathbf{X}^N) + \alpha_N N\rho(\mathbf{X}^N) - k_B\rho(\mathbf{X}^N)\ln[C\rho(\mathbf{X}^N)])\right] = 0. \tag{9.47}$$

This leads us to the condition

$$\alpha_0 + \alpha_E H(\mathbf{X}^N) + \alpha_N N - k_B - k_B\ln[C^N\rho(\mathbf{X}^N)] = 0. \tag{9.48}$$

We can now use the normalization condition Eq. (9.44) to introduce the grand partition function,

$$\mathscr{Z}(\alpha_E, V, \alpha_N) = \sum_{N=0}^{\infty}\frac{1}{C^N}\int d\mathbf{X}^N \exp(\alpha_E H(\mathbf{X}^N) + \alpha_N N) = e^{(1-\alpha_0/k_B)}. \tag{9.49}$$

To identify α_E and α_N, we multiply Eq. (9.48) by $\rho(\mathbf{X}^N)\,d\mathbf{X}^N$, integrate over phase space coordinates and sum over all possible particle numbers. We then find

$$-k_B\ln\mathscr{Z}(\alpha_E, V, \alpha_N) + \alpha_E\langle E\rangle + S + \alpha_N\langle N\rangle = 0. \tag{9.50}$$

If we compare Eq. (9.50) with the fundamental equation for the grand potential $\Omega = U - TS - \mu N$, we can make the identifications

$$\alpha_E = -\frac{1}{T}, \tag{9.51}$$

$$\alpha_N = \frac{\mu}{T}, \tag{9.52}$$

and

$$\Omega(T, V, \mu) = -k_B T\ln\mathscr{Z}(T, V, \mu). \tag{9.53}$$

Eq. (9.53) is the fundamental equation for an open system. From Eqs. (9.48), (9.51), and (9.52) the probability density can be written

$$\rho(\mathbf{X}^N) = \frac{1}{C^N}\,e^{\beta[\Omega(T,V,\mu) - H(\mathbf{X}^N) + \mu N]}. \tag{9.54}$$

An ensemble of open systems governed by a probability density of the form given in Eq. (9.54) is called a grand canonical ensemble. The entropy is given by

$S = -(\partial\Omega/\partial T)_{V,\mu}$, the pressure is given by $P = -(\partial\Omega/\partial V)_{T,\mu}$, and the average particle number is given by $\langle N\rangle = (\partial\Omega/\partial\mu)_{T,V}$.

For a quantum system, the expressions are analogous. The density operator is given by

$$\hat{\rho} = e^{\beta[\Omega(T,V,\mu)+\mu\hat{N}-\hat{H}]}, \tag{9.55}$$

where $\hat{N}$ is the number operator and $\hat{H}$ is the Hamiltonian. The grand partition function is defined

$$\mathscr{Z}(T, V, \mu) = e^{-\beta\Omega(T,V,\mu)} = \text{Tr}\, e^{-\beta(\hat{H}-\mu\hat{N})}. \tag{9.56}$$

The trace in Eq. (9.56) can be evaluated with respect to any convenient set of basis states.

It is useful to determine how the fluctuations in particle number depend on the average number of particles in the system. If we write the normalization condition in the form

$$\text{Tr}\, e^{-\beta(\hat{H}-\mu\hat{N}-\Omega(T,V,\mu))} = 1 \tag{9.57}$$

and take derivatives with respect to μ holding temperature and volume fixed, we find

$$\langle N\rangle = -\left(\frac{\partial\Omega}{\partial\mu}\right)_{V,T} \tag{9.58}$$

and

$$\langle(N-\langle N\rangle)^2\rangle = -k_BT\left(\frac{\partial^2\Omega}{\partial\mu^2}\right)_{T,V} = k_BT\left(\frac{\partial\langle N\rangle}{\partial\mu}\right)_{T,V}. \tag{9.59}$$

Thus, the fractional deviation behaves as

$$\frac{\sqrt{(N-\langle N\rangle)^2}}{\langle N\rangle} \sim \frac{1}{\sqrt{\langle N\rangle}}.$$

As the average number of particles increases, the size of the fluctuations in particle number becomes small compared to the size of the average particle number. Most microstates will have a particle number approximately equal to the average number and we retrieve the canonical ensemble.

It is useful to write Eq. (9.59) in terms of the isothermal compressibility. We first use Eqs. (2.6), (2.93*c*), and (2.127) to obtain

$$\left(\frac{\partial\langle N\rangle}{\partial\mu}\right)_{T,V} = n\left(\frac{\partial\langle N\rangle}{\partial P}\right)_{T,V} = \frac{\langle N\rangle^2}{V}\kappa_T, \tag{9.60}$$

where we have used the fact that $n = \langle N\rangle/V$, $n = (\partial\langle N\rangle/\partial V)_{T,\mu}$, and $n = (\partial\langle N\rangle/\partial V)_{T,P}$. Thus, the variance can be written

$$\langle(N-\langle N\rangle)^2\rangle = \frac{k_BT\langle N\rangle^2}{V}\kappa_T. \tag{9.61}$$

Near a critical point, the compressibility can become infinite and, therefore, fluctuations in the particle number (or density) become very large.

C. HEAT CAPACITY OF SOLIDS

To first approximation, solids may be viewed as an ordered array of atoms, each one fixed to a lattice site and free to oscillate about the lattice site. Thus, we may treat a solid composed of N atoms as a collection of $3N$ harmonic oscillators. The first attempts to compute the thermodynamic properties of a solid using such a picture were classical and were unable to reproduce the observed heat capacity of solids. Experimentally (cf. Fig. 9.1), one finds that at high temperatures the heat capacity C_V of a typical solid is roughly constant with $C_V \sim 6$ cal/K mole, but as the temperature approaches $T = 0$ K, the heat capacity goes to zero as T^3. Classical theory does not give the correct behavior at low temperatures. Einstein and Debye showed that it is necessary to treat solids quantum mechanically to get the correct behavior for thermodynamic quantities at low temperatures. Before we discuss the theories of Einstein and Debye, it is worthwhile to compute the heat capacity using a classical theory.

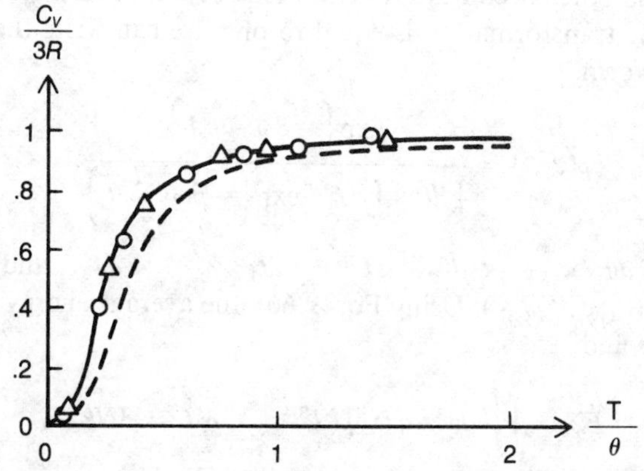

Fig. 9.1. The solid line is the heat capacity curve predicted by Debye theory; the dashed line is the heat capacity curve predicted by Einstein theory; the circles and triangles are experimental values for the heat capacity of Al ($\theta_D = 390$ K) and Cu ($\theta_D = 315$ K), respectively (based on Ref. 9.6).

1. Classical Theory of Solids

Let us assume the solid is composed of $3N$ *independent* distinguishable classical oscillators (each oscillator is at a different lattice site). The Hamiltonian of a single harmonic oscillator is

$$H(p, q) = \frac{p^2}{2m} + \frac{m\omega^2}{2} q^2, \tag{9.62}$$

where ω is the frequency of the oscillator. The probability density for a system of $3N$ independent oscillators in the canonical ensemble is

$$\rho(\mathbf{p}^N, \mathbf{q}^N) = \exp\left\{-\beta \sum_{i=1}^{3N} \left(\frac{p_i^2}{2m} + \frac{m\omega}{2} q_i^2\right)\right\}$$
$$\times \left[\int \cdots \int dp^{3N}\, dq^{3N} \exp\left\{-\beta \sum_{i=1}^{3N} \left(\frac{p_i^2}{2m} + \frac{m\omega}{2} q_i^2\right)\right\}\right]^{-1}. \tag{9.63}$$

The integration in Eq. (9.63) is done most easily in terms of action-angle variables. We shall make the canonical transformations

$$p = -(2m\omega J)^{1/2} \sin\theta \tag{9.64}$$

and

$$q = \left(\frac{2J}{m\omega}\right)^{1/2} \cos\theta. \tag{9.65}$$

The Hamiltonian then becomes a function only of action variables, J. Since the Jacobian of the transformation is equal to one, we can write the probability density in the form

$$\rho(J^{3N}) = \frac{\exp\left\{-\beta\omega \sum_{i=1}^{3N} J_i\right\}}{\int d\theta^{3N} \int dJ^{3N} \exp\left\{-\beta\omega \sum_{i=1}^{3N} J_i\right\}}, \tag{9.66}$$

where $d\theta^{3N} = d\theta_1 \times \cdots \times d\theta_{3N}$, $dJ^{3N} = dJ_1 \times \cdots \times dJ_{3N}$, and J^{3N} denotes the set $J^{3N} = (J_1, \ldots, J_{3N})$. Using Eq. (9.66), the average energy $\langle E\rangle$ is easily computed. We find

$$\langle E\rangle = \iint d\theta^{3N}\, dJ^{3N} \rho(J^{3N}) \sum_{i=1}^{3N} \omega J_i = 3Nk_B T. \tag{9.67}$$

The average energy (internal energy, since $U = \langle E\rangle$) depends on temperature and particle number and is independent of the frequency of the oscillator. The classical heat capacity is

$$C_V = \left(\frac{\partial U}{\partial T}\right)_{V,N} = 3Nk_B = 5.96 \frac{\text{cal}}{\text{K mole}} \tag{9.68}$$

and is independent of temperature. Thus, classical theory gives the correct high-temperature result but not the correct low-temperature result.

2. Einstein Theory of Solids[4,5]

The first theory to give the correct qualitative behavior for solids at all but very low temperatures was due to Einstein. He assumed the solid was composed of

$3N$ independent distinguishable quantum oscillators. The Hamiltonian for a single quantum harmonic oscillator is

$$\hat{H} = \hbar\omega(\hat{N} + \tfrac{1}{2}), \tag{9.69}$$

where $\hat{N}$ is the number operator. The possible energies available to the oscillator are now discrete rather than continuous as in the classical case.

The Helmholtz free energy in the canonical ensemble is given by

$$A(T, V, N) = -k_B T \ln Z_N(T, V), \tag{9.70}$$

where the partition function $Z_N(T, V)$ is written

$$Z_N(T, V) = \mathrm{Tr}\exp\left(-\beta\hbar\omega \sum_{i=1}^{3N} (\hat{N}_i + \tfrac{1}{2})\right). \tag{9.71}$$

We can evaluate the trace using a complete set of orthonormal eigenstates, $\{|n\rangle\}$, of the number operator, $\hat{N}$, where

$$\hat{N}|n\rangle = n|n\rangle \qquad n = 0, 1, 2, \ldots \tag{9.72}$$

and

$$\langle n|n'\rangle = \delta_{n,n'} \tag{9.73}$$

($\delta_{n,n'}$ is the Kronecker delta function). The partition function then becomes

$$Z_N(T, V) = \sum_{n_1=0}^{\infty} \cdots \sum_{n_{3N}=0}^{\infty} \exp\left\{-\beta\hbar\omega \sum_{i=1}^{3N} (n_i + \tfrac{1}{2})\right\} \tag{9.74}$$

or

$$Z_N(T, V) = \left(\frac{e^{-\beta(\hbar\omega/2)}}{1 - e^{-\beta\hbar\omega}}\right)^{3N}. \tag{9.75}$$

In Eq. (9.75), we have used the identity $(1 - x)^{-1} = \sum_{n=0}^{\infty} x^n$. If we take the logarithm of the partition function, we obtain the following expression for the Helmholtz free energy of an Einstein solid:

$$A(T, V, N) = \frac{3N\hbar\omega}{2} + 3Nk_B T \ln(1 - e^{-\beta\hbar\omega}). \tag{9.76}$$

The first term can be interpreted as the zero point free energy of the solid, that is, the energy that remains even at $T = 0$ K due to the uncertainty relations. From Eq. (2.92*a*), we can obtain the entropy of the solid

$$S = -\left(\frac{\partial A}{\partial T}\right)_{V,N} = -3Nk_B \ln(1 - e^{-\beta\hbar\omega}) + \frac{3Nk_B\left(\dfrac{\hbar\omega}{k_B T}\right) e^{-\beta\hbar\omega}}{(1 - e^{-\beta\hbar\omega})}, \tag{9.77}$$

and after some algebra the heat capacity becomes

$$C_V = T\left(\frac{\partial S}{\partial T}\right)_{V,N} = \frac{3Nk_B\left(\frac{\hbar\omega}{k_BT}\right)^2 e^{-\beta\hbar\omega}}{(1 - e^{-\beta\hbar\omega})^2}. \tag{9.78}$$

It is interesting to note that the entropy goes to zero at $T = 0$ K, in accordance with the third law.

It is convenient to introduce a parameter $\theta_E = \hbar\omega/k_B$ called the Einstein temperature. The heat capacity can then be written in the form

$$\frac{C_V}{3Nk_B} = \frac{\left(\frac{\theta_E}{T}\right)^2 e^{\theta_E/T}}{(e^{\theta_E/T} - 1)^2} \equiv F_E\left(\frac{\theta_E}{T}\right). \tag{9.79}$$

The function $F_E(\theta_E/T)$ is called the Einstein function. A plot of $F(\theta_E/T)$ versus T/θ_E is shown in Fig. 9.1. The Einstein function has qualitatively the correct behavior except that in the limit $T \to 0$ K it goes to zero exponentially rather than as T^3 as is observed experimentally. The correct behavior at low temperature was first obtained by Debye.

3. Debye Theory of Solids[5,6]

Debye assumed that the oscillators which composed the solid were coupled. Then the Hamiltonian has the more complicated form

$$H(\mathbf{p}^N, \mathbf{q}^N) = \sum_{i=1}^{3N} \frac{p_i^2}{2m} + \sum_{i,j}^{3N} A_{ij}q_iq_j, \tag{9.80}$$

where m is the mass of an oscillator. The matrix, A_{ij}, contains information about interactions between oscillators. However, a Hamiltonian of the type given in Eq. (9.80) can always be diagonalized. The diagonalized form of the Hamiltonian will look like the sum of Hamiltonians for independent oscillators except that all frequencies will be different:

$$H(\mathbf{P}^N, \mathbf{Q}^N) = \sum_{i=1}^{3N} \frac{P_i^2}{2m} + \sum_{i=1}^{3N} \frac{m\omega_i^2}{2} Q_i^2. \tag{9.81}$$

In this case, the harmonic oscillators which describe the thermal properties of the solid are collective modes, that is, sound waves or phonons. The phonon Hamiltonian can be written in the form

$$\hat{H} = \sum_{i=1}^{3N} \hbar\omega_i(\hat{N}_i + \tfrac{1}{2}) \tag{9.82}$$

where $\hat{N}_i$ is again the number operator. The partition function for the Debye solid takes the form

$$Z_N(T, V) = \sum_{n_1=0}^{\infty} \cdots \sum_{n_{3N}=0}^{\infty} \exp\left\{-\beta\hbar \sum_{i=1}^{3N} (n_i + \tfrac{1}{2})\omega_i\right\}. \tag{9.83}$$

If we take the logarithm of the partition function we can write the Helmholtz free energy as

$$A(T, V, N) = \frac{\hbar}{2}\sum_{i=1}^{3N} \omega_i + k_B T \sum_{i=1}^{3N} \ln(1 - e^{-\beta\hbar\omega_i}). \tag{9.84}$$

Eq. (9.84) is the "fundamental equation" for a harmonic solid.

The Debye theory neglects the discrete structure of the crystal and considers the solid as a continuous isotropic medium. It thus assumes that the dispersion relation for sound waves $\omega = kc$, (where c is the speed of sound, $k = 2\pi/\lambda$ is the wave number, and λ is the wavelength of the sound waves) holds, not just for the long-wavelength sound waves, but for all vibrations in the crystal. These waves, which form standing waves will have allowed wavelengths determined by the size of the crystal, namely, $\lambda = 2L/n$, where $n = (1, 2, \ldots, \infty)$, and are the same as for a string which is pinned at both ends. A given wave will have a dispersion relation of the form

$$\omega_i^2 = c^2[k_{xi}^2 + k_{yi}^2 + k_{zi}^2] = c^2\left[\left(\frac{\pi n_{xi}}{L_x}\right)^2 + \left(\frac{\pi n_{yi}}{L_y}\right)^2 + \left(\frac{\pi n_{zi}}{L_z}\right)^2\right], \tag{9.85}$$

where L_x, L_y, and L_z are the lengths of the crystal in the x, y, and z directions, respectively.

We may plot the allowed values of ω_j as points in a three-dimensional frequency space (cf. Fig. 9.2). The distance between points in the x-direction is $\pi c/L_x$; in the y-direction, $\pi c/L_y$; and in the z-direction, $\pi c/L_z$. The volume per point is therefore $(\pi c)^3/L_xL_yL_z$. The number of points per unit volume is $V/(\pi c)^3$ where $V = L_xL_yL_z$. The total number of allowed values of ω_j less than some value ω is given by $\frac{1}{8}(4\pi\omega^3/3)(V/(\pi c)^3)$ where $\frac{1}{8}(4\pi\omega^3/3)$ is the volume of $\frac{1}{8}$

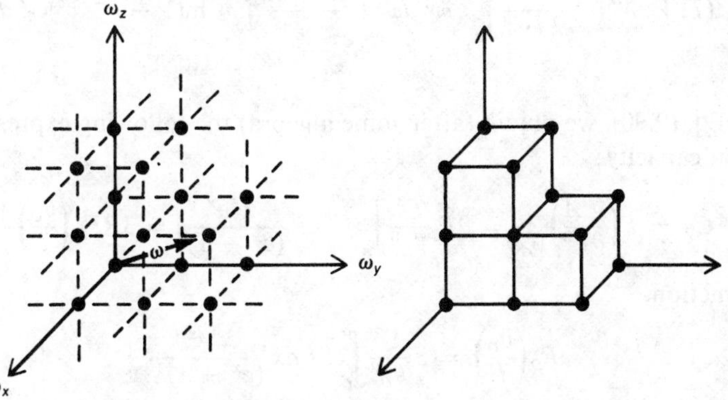

Fig. 9.2. Allowed values of frequency plotted in a three-dimensional frequency space.

of a sphere of radius ω (the frequency must be positive). Thus, the number of points in the range $\omega \to \omega + d\omega$ is given by

$$dn_j = \frac{V}{2\pi^2 c^3}\,\omega^2\,d\omega. \tag{9.86}$$

In general, there will be two transverse sound modes and one longitudinal mode in the solid. The transverse and longitudinal modes will have different velocities. We may take this into account by writing dn_j in the form

$$dn_j = \frac{V}{2\pi^2}\left(\frac{2}{c_t^3} + \frac{1}{c_l^3}\right)\omega^2\,d\omega, \tag{9.87}$$

where c_t is the velocity of the transverse modes and c_l is the velocity of the longitudinal mode.

Since there is a finite number of modes, there will be a maximum possible frequency for the modes. The total number of modes is found by integrating over dn_j. The range of ω goes from 0 to some maximum frequency, ω_D,

$$3N = \int dn_j = \int_0^{\omega_D} \frac{V}{2\pi^2}\left(\frac{2}{c_t^3} + \frac{1}{c_l^3}\right)\omega^2\,d\omega = \frac{V}{2\pi^2}\left(\frac{2}{c_t^3} + \frac{1}{c_l^3}\right)\frac{{\omega_D}^3}{3}. \tag{9.88}$$

If we solve for ${\omega_D}^3$ we find

$${\omega_D}^3 = \frac{18N\pi^2}{V}\left(\frac{2}{c_t^3} + \frac{1}{c_l^3}\right)^{-1} \tag{9.89}$$

(ω_D is called the Debye frequency).

We now have enough information to integrate the expression for the free energy. For large values of N the allowed frequencies will lie close together and we can replace the summations in Eq. (9.84) by integrations. We then find

$$A(T, V, N) = \frac{\hbar}{2}\frac{9N}{{\omega_D}^3}\int_0^{\omega_D} \omega^3\,d\omega + \frac{k_B T 9N}{{\omega_D}^3}\int_0^{\omega_D} \ln(1 - e^{-\beta\hbar\omega})\omega^2\,d\omega. \tag{9.90}$$

From Eq. (9.90), we obtain (after some algebra) the following expression for the heat capacity:

$$C_V = T\left(\frac{\partial^2 A}{\partial T^2}\right)_{V,N} = 9R\frac{T^3}{{\theta_D}^3}\int_0^{\theta_D/T} dx\,\frac{x^4 e^x}{(e^x - 1)^2} = 3RF_D\left(\frac{\theta_D}{T}\right). \tag{9.91}$$

The function,

$$F_D\left(\frac{\theta_D}{T}\right) = 3\frac{T^3}{{\theta_D}^3}\int_0^{\theta_D/T} dx\,\frac{x^4 e^x}{(e^x - 1)^2}, \tag{9.92}$$

is called the Debye function, $x = \beta\hbar\omega$, and θ_D is called the Debye temperature. The Debye temperatures for a number of monatomic crystals are given in

Table 9.1. The Debye theory gives very good agreement with experiment, as can be seen from Fig. 9.1.

	θ_D K		θ_D K
Na	150	Co	385
Ag	215	Ni	375
Cu	315	Pb	88
Be	1000	Al	390
Zn	250	Pt	225
Fe	420	Cd	172

Table 9.1. Values of the Debye temperature for some typical monatomic substances (based on Ref. 9.5).

The low- and high-temperature behavior of the heat capacity are easily obtained from Eqs. (9.91) and (9.92). At very low temperature ($T \to 0$ K) we can replace the upper limit by infinity and do the integral exactly. We find

$$C_V \sim \frac{9RT^3}{\theta_D{}^3} \int_0^\infty dx \frac{x^4 e^x}{(e^x - 1)^2} = \frac{12}{5} \frac{T^3}{\theta_D{}^3} \pi^4 R. \tag{9.93}$$

Thus, at low temperature the specific heat behaves as T^3. In Fig. 9.3, we compare the T^3 law with the experimentally observed behavior of KCl. The agreement is excellent.

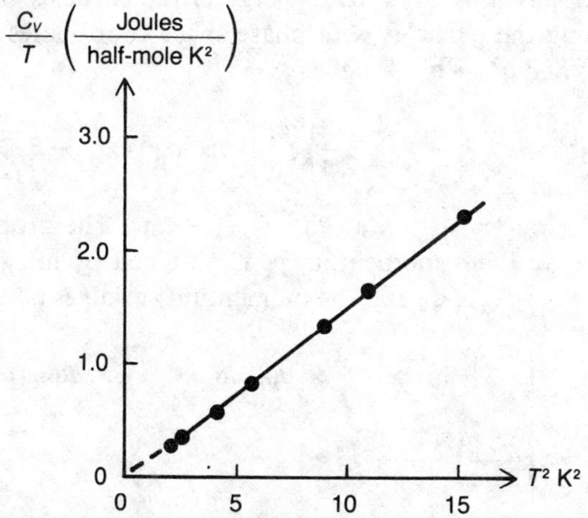

Fig. 9.3. A plot of C_V/T versus T^2 for KCl: the experimental points satisfy the T^3 law predicted by Debye theory (based on Ref. 9.28).

At high temperature the integral can be expanded in powers of x. Then we find

$$C_V \sim 9R \frac{T^3}{\theta_D{}^2} \int_0^{\theta_0/T} x^2\, dx = 3R. \tag{9.94}$$

and we obtain the classical result.

The Debye theory, while giving good agreement with experiment, is not perfect. In real solids the Debye temperature, obtained by comparison of Eq. (9.93) with heat capacity data, varies with temperature. This variation is a direct consequence of the limiting assumptions[7] of the Debye theory.

D. IDEAL GASES

In a gas, identical particles must be regarded as indistinguishable. Therefore, at low temperatures, where the thermal wavelength of particles becomes large and overlapping of wave functions occurs, the statistics of the particles plays a crucial role in determining the thermodynamic behavior of the gas. In this section, we shall first use kinetic theory and the canonical ensemble to obtain the equation of state of a classical gas. Then we will use the grand canonical ensemble to obtain the thermodynamic properties of boson and fermion gases and obtain their classical limit.

1. Classical Ideal Gas: Kinetic Theory

Let us consider a dilute gas of N particles contained in a volume V. We assume that the gas is in equilibrium and that the interaction energy between particles is negligible compared to the kinetic energy of the particles. The probability density of finding the particles with phase space coordinates in the interval $\mathbf{p}^N \to \mathbf{p}^N + d\mathbf{p}^N$ and $\mathbf{q}^N \to \mathbf{q}^N + d\mathbf{q}^N$ is

$$\rho(\mathbf{p}^N, \mathbf{q}^N) = \exp\left\{-\beta \sum_{i=1}^{N} \frac{p_i{}^2}{2m}\right\} \Big/ \iint d\mathbf{p}^N\, d\mathbf{q}^N \exp\left\{-\beta \sum_{i=1}^{N} \frac{p_i{}^2}{2m}\right\} \tag{9.95}$$

where $d\mathbf{p}^N\, d\mathbf{q}^N = d\mathbf{p}_1 \times \cdots \times d\mathbf{p}_N\, d\mathbf{q}_1 \times \cdots \times d\mathbf{q}_N$. The probability density $\rho_1(\mathbf{p}_1)$ that particle 1 has momentum $\mathbf{p}_1$ is obtained by integrating over all position variables $\mathbf{q}_1, \ldots, \mathbf{q}_N$ and the momentum variables $\mathbf{p}_2, \ldots, \mathbf{p}_N$:

$$\begin{aligned} \rho_1(\mathbf{p}_1) &= \int \cdots \int d\mathbf{p}_2 \times \cdots \times d\mathbf{p}_N\, d\mathbf{q}_1 \times \cdots \times d\mathbf{q}_N \rho(\mathbf{p}^N, \mathbf{q}^N) \\ &= \left(\frac{\beta}{2\pi m}\right)^{3/2} e^{-\beta p_1^2/2m}. \end{aligned} \tag{9.96}$$

$\rho_1(\mathbf{p}_1)$ is called the Maxwell-Boltzmann distribution and is normalized to one. Similarly we can write an expression for the normalized probability density

$F(\mathbf{v}_1)$ of finding a particle with velocity $\mathbf{v}_1 \to \mathbf{v}_1 + d\mathbf{v}_1$ where $\mathbf{v}_1 = \mathbf{p}_1/m$. It is given by

$$F(\mathbf{v}_1) = \left(\frac{\beta m}{2\pi}\right)^{3/2} e^{-\beta m v_1^2/2} \tag{9.97}$$

and is normalized to one when we integrate over $\mathbf{v}_1$.

Let us now compute the pressure on the walls due to collisions by particles with the walls. We first remember that pressure is defined as force/area or as momentum transfer/area-time. We will assume that the walls are smooth. Then momentum is transferred in a direction perpendicular to the wall (we shall choose the z-axis to be perpendicular to the wall). The amount of momentum transferred to the wall during each collision is $2mv_z$, where v_z is the z-component of velocity of the particle.

To compute the pressure we will consider a small area element of the wall, dS, and we will choose the origin of coordinates at the center of the element as indicated in Fig. 9.4. Particles having velocity $\mathbf{v} \to \mathbf{v} + d\mathbf{v}$ which collide with dS in time dt are contained in a cylinder of volume $v_z\, dt\, dS$. The number of particles with velocity $\mathbf{v} \to \mathbf{v} + d\mathbf{v}$ which collide with the element dS is

$$dn(\mathbf{v}) = nF(\mathbf{v})\, d\mathbf{v} v_z\, dt\, dS, \tag{9.98}$$

where n is the number density of the gas, $n = N/V$. The pressure exerted on dS due to particles with velocity $\mathbf{v} \to \mathbf{v} + d\mathbf{v}$ is

$$dP(\mathbf{v}) = (2mv_z)v_z nF(\mathbf{v})\, d\mathbf{v}. \tag{9.99}$$

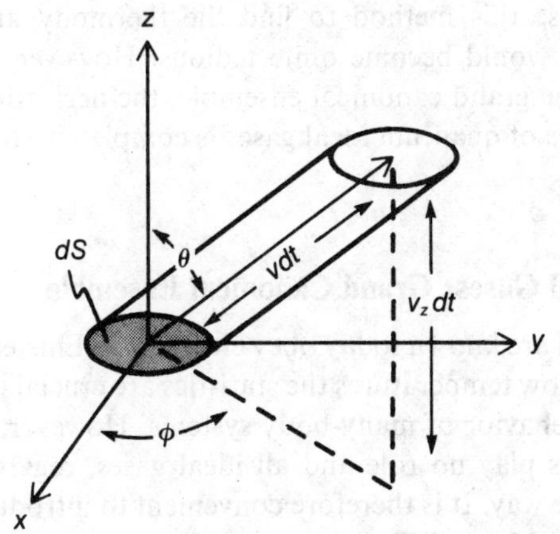

Fig. 9.4. Particles with velocity $\mathbf{v} \to \mathbf{v} + d\mathbf{v}$ collide with dS in time dt.

The pressure due to particles with speed $v \to v + dv$ is found by integrating Eq. (9.99) over angles $\theta = 0 \to \pi/2$ and $\phi = 0 \to 2\pi$. If we remember that $v_z = v \cos\theta$, we obtain

$$dP(v) = \tfrac{1}{3}nmv^2 f(v)\, dv \tag{9.100}$$

where

$$f(v)\, dv = 4\pi\left(\frac{\beta m}{2\pi}\right)^{3/2} v^2\, e^{-\beta m v^2/2}\, dv \tag{9.101}$$

is the probability that a particle has speed in the interval $v \to v + dv$. The total pressure is obtained by integrating over the speed

$$P = \tfrac{1}{3}nm \int_0^\infty dv v^2 f(v) = \tfrac{1}{3}nm\langle v^2\rangle. \tag{9.102}$$

The average kinetic energy is given by

$$\tfrac{1}{2}m\langle v^2\rangle = \frac{1}{2}\int_0^\infty dv m v^2 f(v) = \tfrac{3}{2}k_B T. \tag{9.103}$$

Thus, the canonical ensemble yields the correct equipartition of energy among the degrees of freedom for a classical system. From Eqs. (9.102) and (9.103) we obtain for the ideal gas equation of state:

$$PV = \frac{N}{N_A}(k_B N_A)T = nRT \tag{9.104}$$

where N_A is Avogadro's number and $R = k_B N_A$ is the gas constant. The internal energy is the sum of the average kinetic energies of the particles:

$$U = \tfrac{3}{2}NRT \tag{9.105}$$

as we expect.

If we had to use this method to find the thermodynamic properties of quantum gases, it would become quite tedious. However, as we shall show below, by using the grand canonical ensemble, the derivation of the thermodynamic properties of quantum ideal gases is completely straightforward and relatively simple.

2. Quantum Ideal Gases: Grand Canonical Ensemble

All particles which are known today obey either Bose-Einstein or Fermi-Dirac statistics. At very low temperatures the statistics are crucial in determining the thermodynamic behavior of many-body systems. However, at high temperatures, the statistics play no role and all ideal gases, regardless of statistics, behave in the same way. It is therefore convenient to introduce a third type of statistics, namely Maxwell-Boltzmann statistics, which gives the high-temperature behavior of both Fermi-Dirac and Bose-Einstein systems.

We shall assume that our gas is contained in a cubic box of volume $V = L^3$, where L is the length of a side of the box. When working with quantum systems, it is usually easiest to use the grand canonical ensemble since it then is not necessary to conserve particle number. For an ideal gas, the grand partition function can be written

$$\mathscr{Z}(T, V, \mu) = \mathrm{Tr}\, e^{-\beta(\hat{H}_0 - \mu\hat{N})}, \tag{9.106}$$

where $\hat{H}_0$ is the kinetic energy. It is most convenient to evaluate the trace in the number representation since both $\hat{H}_0$ and $\hat{N}$ are diagonal in that representation.

When evaluating the trace in Eq. (9.106), we must be careful to count each possible state of the system only once. The grand partition function in general will have the form

$$\mathscr{Z}(T, V, \mu) = \sum_{n_0=0}^{\infty} \cdots \sum_{n_j=0}^{\infty} \cdots \sum_{n_\infty=0}^{\infty} S(N, n_1, \ldots, n_\infty) \times \exp\left\{-\beta\left(\sum_{l=0}^{\infty} n_l(\epsilon_l - \mu)\right)\right\}, \tag{9.107}$$

where $\epsilon_l = \hbar^2 k_l^2/2m$ is the kinetic energy of a particle in momentum state $\mathbf{p}_l = \hbar\mathbf{k}_l$. The summation $\sum_{l=0}^{\infty}$ is over all allowed momentum states for the system. The lth momentum state will contain n_l particles. If we know the entire set of occupation numbers $\{n_l\}$ for a given state of the system, then the energy E of that state is determined, since $E = \sum_{l=0}^{\infty} n_l \epsilon_l$. The summation over the occupation numbers in front of the factor $S(N, n_0, \ldots, n_\infty)$ in Eq. (9.107) gives us all different possible configurations of the occupation numbers, that is, all different possible states of the system. The factor $S(N, n_0, \ldots, n_\infty)$ is a statistical factor which ensures that we count each state only once. The difference between Maxwell-Boltzmann, Fermi-Dirac, and Bose-Einstein gases lies in the factor $S(N, n_0, \ldots, n_\infty)$.

To obtain the grand partition function for a Maxwell-Boltzmann gas, it is convenient to find it first for a gas of distinguishable particles. Let us assume that we are given a set of occupation numbers $\{n_l\}$ such that $\sum_{l=0}^{\infty} n_l = N$. Then $S(N, n_0, \ldots, n_\infty)$ is the number of different states which correspond to that set of occupation numbers. For distinguishable particles

$$S_{\mathrm{dis}}(N, n_0, \ldots, n_\infty) = \frac{N!}{n_0! \cdots n_j! \cdots n_\infty!} \tag{9.108}$$

since $S_{\mathrm{dis}}(N, n_0, \ldots, n_\infty)$ is the number of ways that N distinguishable particles can be distributed among the momentum states $l = 0, \ldots, \infty$ such that the set of occupation numbers $\{n_l\}$ remains fixed. Thus, the grand partition function for distinguishable particles becomes

$$\mathscr{Z}_{\mathrm{dis}}(T, V, \mu) = \sum_{n_0=0}^{\infty} \cdots \sum_{n_j=0}^{\infty} \cdots \sum_{n_\infty=0}^{\infty} \frac{N!}{n_0! \cdots n_j! \cdots n_\infty!} \times \exp\left\{-\beta \sum_{l=0}^{\infty} n_l(\epsilon_l - \mu)\right\}. \tag{9.109}$$

Because of the factor $N!$ in the numerator, the evaluation of the grand partition function for distinguishable particles is cumbersome. It is easier to use the canonical ensemble for the case of N-particles on a lattice, as we have seen in Sec. 9.C.

The grand partition function for Maxwell-Boltzmann particles is obtained from $\mathscr{Z}_{\text{dis}}(T, V, \mu)$ by inserting a factor $(N!)^{-1}$ into the expression for $S(N, n_1, \ldots, n_\infty)$. This is called correct Boltzmann counting and gives the proper form of the partition function for indistinguishable particles at high temperature. Thus, for Maxwell-Boltzmann particles we have

$$\mathscr{Z}_{\text{MB}}(T, V, \mu) \equiv \sum_{n_0=0}^{\infty} \cdots \sum_{n_j=0}^{\infty} \cdots \sum_{n_\infty=0}^{\infty} \frac{1}{n_0! \cdots n_j! \cdots n_\infty!} \times \exp\left\{-\beta \sum_{l=0}^{\infty} n_l(\epsilon_l - \mu)\right\}. \tag{9.110}$$

At high temperatures we expect all of the occupation numbers to be of order $n_l \approx 1$ or 0. The grand partition function for Maxwell-Boltzmann particles no longer contains the factor $N!$ and is easy to evaluate.

For a gas of indistinguishable Bose-Einstein particles, each different set of occupation numbers $\{n_l\}$ will correspond to one possible state. Also, there is no restriction on the number of particles that can occupy a given momentum state. Thus, for a Bose-Einstein gas the grand partition function is

$$\mathscr{Z}_{\text{BE}}(T, V, \mu) = \sum_{n_0=0}^{\infty} \cdots \sum_{n_j=0}^{\infty} \cdots \sum_{n_\infty=0}^{\infty} \exp\left\{-\beta \sum_{l=0}^{\infty} n_l(\epsilon_l - \mu)\right\}. \tag{9.111}$$

At high temperature, where all occupation numbers become of order $n_l = 0$ or 1, the Bose-Einstein partition function becomes equivalent to the Maxwell-Boltzmann partition function.

For a gas of indistinguishable Fermi-Dirac particles, again, each different set of occupation numbers corresponds to one possible state and the exclusion principle restricts the occupation number of each momentum state to 0 or 1. Thus, the grand partition function becomes

$$\mathscr{Z}_{\text{FD}}(T, V, \mu) = \sum_{n_0=0}^{1} \cdots \sum_{n_j=0}^{1} \cdots \sum_{n_\infty=0}^{1} \exp\left\{-\beta \sum_{l=0}^{\infty} n_l(\epsilon_l - \mu)\right\}. \tag{9.112}$$

We have not explicitly inserted spin dependence into Eq. (9.112). However, it is a straightforward matter to do so. Again, at high temperatures, this partition function reduces to the Maxwell-Boltzmann case.

These grand partition functions contain all possible thermodynamic information about Maxwell-Boltzmann, Fermi-Dirac, and Bose-Einstein gases. We shall now use them to compute and compare the thermodynamic properties for the three gases.

3. Maxwell-Boltzmann Gas

The grand partition function for a Maxwell-Boltzmann gas (cf. Eq. (9.110)) can be rewritten in the form

$$\mathscr{Z}_{\text{MB}}(T, V, \mu) = \prod_{l=0}^{\infty}\left(\sum_{n_l=0}^{\infty} \frac{1}{n_l!} e^{-\beta n_l(\epsilon_l - \mu)}\right) = \prod_{l=0}^{\infty} (\exp[e^{-\beta(\epsilon_l - \mu)}]), \tag{9.113}$$

where we have used the identity $e^x = \sum_{n=0}^{\infty} (1/n!)x^n$. From Eq. (9.113) we can obtain the grand potential:

$$\Omega_{\text{MB}}(T, V, \mu) = -k_B T \ln \mathscr{Z}_{\text{MB}}(T, V, \mu) = -k_B T \sum_{l=0}^{\infty} e^{-\beta(\epsilon_l - \mu)}. \tag{9.114}$$

If the volume is large enough, the particle energies will be closely spaced and we can change the summation in Eq. (9.114) to an integration. Thus,

$$\sum_l \rightarrow \frac{V}{(2\pi)^3}\int d\mathbf{k}_l = \frac{V}{h^3}\int d\mathbf{p}_l, \tag{9.115}$$

and the grand potential becomes

$$\Omega_{\text{MB}}(T, V, \mu) = \frac{-k_B T V}{h^3}\int d\mathbf{p}\, e^{-\beta((p^2/2m) - \mu)}. \tag{9.116}$$

The integral in Eq. (9.116) is simple. If we note that

$$\int_0^{\infty} x^2 e^{-r^2 x^2}\, dx = \frac{\sqrt{\pi}}{4r^3},$$

we find

$$\Omega_{\text{MB}}(T, V, \mu) = -k_B T V e^{\beta\mu}\left(\frac{m k_B T}{2\pi\hbar^2}\right)^{3/2}. \tag{9.117}$$

Eq. (9.117) is the fundamental equation for a Maxwell-Boltzmann gas.

We can now compute various thermodynamic quantities. For example, the average number of particles is given by

$$\langle N \rangle = -\left(\frac{\partial \Omega_{\text{MB}}}{\partial \mu}\right)_{T,V} = V e^{\beta\mu}\left(\frac{m k_B T}{2\pi\hbar^2}\right)^{3/2} = \frac{V e^{\beta\mu}}{\lambda_T{}^3}, \tag{9.118}$$

where the quantity

$$\lambda_T \equiv \left(\frac{2\pi\hbar^2}{m k_B T}\right)^{1/2} \tag{9.119}$$

is the thermal wavelength and is a measure of the spread of the wave function of the particles. As long as λ_T is much less than the average interparticle spacing, statistics will be unimportant. We can revert Eq. (9.118) and write the chemical potential in terms of the average number of particles $\langle N \rangle$. Thus,

$$\mu = k_B T \ln\left(\frac{\langle N \rangle \lambda_T{}^3}{V}\right). \tag{9.120}$$

The entropy of the Maxwell-Boltzmann gas is given by

$$S = -\left(\frac{\partial \Omega_{\mathrm{MB}}}{\partial T}\right)_{V,\mu} = \left(\frac{5k_B}{2} - \frac{\mu}{T}\right)\frac{V\,e^{\beta\mu}}{\lambda_T{}^3} = \frac{5k_B\langle N \rangle}{2} - k_B\langle N \rangle \ln\left(\frac{\langle N \rangle}{V}\,\lambda_T{}^3\right). \tag{9.121}$$

The last expression for the entropy is called the Sackur-Tetrode equation. The pressure is given by

$$P = -\left(\frac{\partial \Omega_{\mathrm{MB}}}{\partial V}\right)_{T,\mu} = \frac{k_B T\,e^{\beta\mu}}{\lambda_T{}^3} = \frac{\langle N \rangle k_B T}{V}. \tag{9.122}$$

This is just the expected ideal gas result.

It is interesting to compare these equations with the results derived in Sec. 2.I. Having now derived the classical ideal gas equations microscopically, we have a much better idea of their limitations, and we have explicitly evaluated all constants in terms of known quantities. It is important to note that these equations depend on Planck's constant, a purely quantum mechanical quantity. Also, we can now see the origin of the factors of h that appeared in the constant, C^N, in Sec. 9.B.

4. Bose-Einstein Gas[8,9]

Let us now consider an ideal Bose-Einstein gas. The grand partition function in Eq. (9.111) can be written in the form

$$\mathscr{Z}_{\mathrm{BE}}(T, V, \mu) = \prod_{l=0}^{\infty}\left(\sum_{n_l=0}^{\infty} e^{-\beta n_l(\epsilon_l - \mu)}\right) = \prod_{l=0}^{\infty}\left(\frac{1}{1 - e^{-\beta(\epsilon_l - \mu)}}\right). \tag{9.123}$$

From Eq. (9.53), the grand potential can be written

$$\Omega_{\mathrm{BE}}(T, V, \mu) \equiv -k_B T \ln \mathscr{Z}_{\mathrm{BE}}(T, V, \mu) = k_B T \sum_{l=0}^{\infty} \ln(1 - e^{-\beta(\epsilon_l - \mu)}). \tag{9.124}$$

The ideal Bose-Einstein gas is particularly interesting because it can undergo a phase transition. We get our first indication of this by looking at the way particles are distributed in momentum space. The average number of particles is defined

$$\langle N \rangle = -\left(\frac{\partial \Omega_{\mathrm{BE}}}{\partial \mu}\right)_{T,V} = \sum_{l=0}^{\infty}\left(\frac{e^{-\beta(\epsilon_l - \mu)}}{1 - e^{-\beta(\epsilon_l - \mu)}}\right) = \sum_{l=0}^{\infty}\left(\frac{1}{e^{\beta(\epsilon_l - \mu)} - 1}\right) \tag{9.125}$$

Since the particles are independent we may write Eq. (9.125) in the form

$$\langle N \rangle = \sum_{l=0}^{\infty} \langle n_{\mathbf{p}_l} \rangle \tag{9.126}$$

where

$$\langle n_{\mathbf{p}_l} \rangle = \left(\frac{1}{e^{\beta(\epsilon_l - \mu)} - 1} \right) = \left(\frac{1}{e^{\beta\epsilon_l} z^{-1} - 1} \right) \tag{9.127}$$

and $z = e^{\beta\mu}$ is called the fugacity. The quantity $\langle n_{\mathbf{p}_l} \rangle$ is the average number of particles in the momentum state $\mathbf{p}_l$. Note that since the exponential, $e^{\beta\epsilon_l}$, can have a range of values $1 < e^{\beta\epsilon_l} < \infty$, the fugacity can only have values $0 < z < 1$. Otherwise, $\langle n_{\mathbf{p}_l} \rangle$ could become negative. Thus, for an ideal Bose-Einstein gas, the chemical potential μ must be negative. In Eq. (9.127), the state with zero momentum is particularly troublesome because $\langle n_0 \rangle$ can become very large as $z \to 1$,

$$\langle n_0 \rangle = \left(\frac{z}{1 - z} \right) \xrightarrow[z \to 1]{} \infty. \tag{9.128}$$

That is, as $z \to 1$, the zero momentum state can become macroscopically occupied. This is precisely what happens at the phase transition.

To anticipate the onset of a phase transition, we shall isolate all terms belonging to the zero momentum state in the expressions for the thermodynamic quantities. Furthermore, we shall assume that we have a large volume (and large number of particles) so that we can replace the summation over momenta by an integration (cf. Eq. (9.115)). Then the grand potential per unit volume can be written in the form

$$\frac{\Omega_{BE}(T, V, \mu)}{V} = \frac{k_B T}{V} \ln(1 - z) - \frac{k_B T}{\lambda_T{}^3} g_{5/2}(z) \tag{9.129}$$

and the number density can be written

$$\frac{\langle N \rangle}{V} = \frac{\langle n_0 \rangle}{V} + \frac{1}{\lambda_T{}^3} g_{3/2}(z), \tag{9.130}$$

where λ_T is the thermal wavelength. The functions $g_{5/2}(z)$ and $g_{3/2}(z)$ are defined

$$g_{5/2}(z) = -\frac{4}{\sqrt{\pi}} \int_0^{\infty} dx x^2 \ln(1 - z\, e^{-x^2}) = \sum_{\alpha=1}^{\infty} \frac{z^{\alpha}}{\alpha^{5/2}} \tag{9.131}$$

(we have made the change of variables $x = \beta p^2/2m$) and

$$g_{3/2}(z) = z \frac{\partial}{\partial z} g_{5/2}(z) = \sum_{\alpha=1}^{\infty} \frac{z^{\alpha}}{\alpha^{3/2}}. \tag{9.132}$$

For high temperatures and/or low density, the terms $k_B T \ln(1 - z)/V$ and $\langle n_0 \rangle / V$ in Eqs. (9.129) and (9.130), respectively, will be negligible and can be omitted. The series expansion in Eq. (9.131) is obtained by expanding the argument of the integral in powers of z and integrating each term.

Let us now look at Eq. (9.130) for the limiting values of z. The function $g_{3/2}(z)$ is a bounded, positive, monotonically increasing function of z. When $z = 0$ we find

$$\left(\frac{\langle N\rangle}{V}\lambda_T^3\right)_{z=0} = g_{3/2}(0) = 0. \tag{9.133}$$

Thus the fugacity will be zero for zero density or infinite temperature. When $z = 1$ we find

$$\left(\frac{\langle N\rangle}{V}\lambda_T^3\right)_{z=1} = g_{3/2}(1) = \zeta(\tfrac{3}{2}) = 2.612, \tag{9.134}$$

where $\zeta(\frac{3}{2})$ is the Riemann zeta function. For some critical combination of density and temperature, the fugacity will reach its maximum value. If we decrease the temperature or increase the density past the critical values, the function $g_{3/2}(z)$ can no longer change since z cannot become larger than one. Thus, the entire change in $((\langle N\rangle/V)\lambda_T^3)_{z>1}$ must come from the term $(\langle n_0\rangle/V)\lambda_T^3$ and the zero momentum state begins to take on macroscopic values. Thus we must write

$$\frac{\langle N\rangle}{V}\lambda_T^3 = g_{3/2}(z) \qquad \text{for } \frac{\langle N\rangle}{V}\lambda_T^3 < 2.612 \tag{9.135}$$

and

$$\frac{\langle N\rangle}{V}\lambda_T^3 = \frac{\langle n_0\rangle}{V}\lambda_T^3 + g_{3/2}(1) \qquad \text{for } \frac{\langle N\rangle}{V}\lambda_T^3 \geqslant 2.612. \tag{9.136}$$

The macroscopic occupation of the zero momentum state is called Bose-Einstein condensation. It occurs for values of temperature and density such that

$$\lambda_T^3 \geqslant 2.612\frac{V}{\langle N\rangle}. \tag{9.137}$$

The critical temperature (as a function of density) and the critical density (as a function of temperature) can be found from Eq. (9.137). They are

$$T_c = \left(\frac{2\pi\hbar^2}{mk_B}\right)\left(\frac{\langle N\rangle}{2.612V}\right)^{2/3} \tag{9.138}$$

and

$$\left(\frac{\langle N\rangle}{V}\right)_c = 2.612\left(\frac{mk_BT}{2\pi\hbar^2}\right)^{3/2}. \tag{9.139}$$

The fraction of particles in the zero momentum state can be found from Eqs. (9.136), (9.138) and (9.139). It is

$$\frac{\langle n_0\rangle}{\langle N\rangle} = 1 - \frac{2.612}{\lambda_T^3}\frac{V}{\langle N\rangle} = 1 - \left(\frac{T}{T_c}\right)^{3/2}, \tag{9.140}$$

where $V/\langle N\rangle$ is the specific volume. The fraction is plotted as a function of temperature in Fig. 9.5. We see that it is proportional to the average number of particles below the critical point. This is an indication that ODLRO has set in (cf. Sec. 7.F).

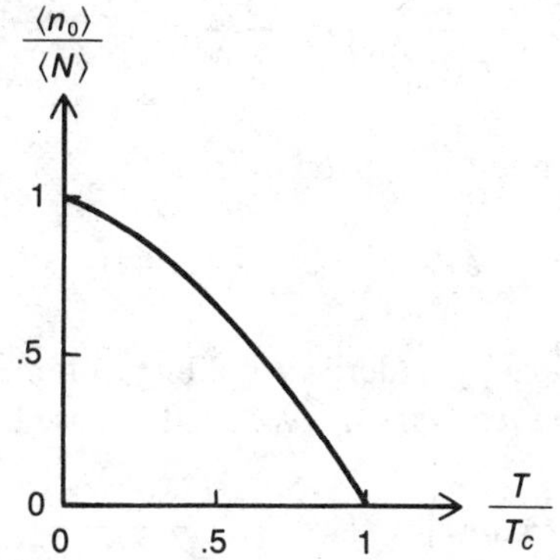

Fig. 9.5. A plot of the fraction of condensed bosons versus the reduced temperature.

Since the pressure is simply related to the grand potential, we can easily write an expression for it, both above and below the critical point. In the thermodynamic limit ($V \to \infty$ and $N \to \infty$ so that N/V = constant) and above the critical point, we find

$$P_> = \lim_{V\to\infty}\left(-\frac{\Omega_{\mathrm{BE}}(T, V, \mu)}{V}\right) = \frac{k_B T}{\lambda_T{}^3}\, g_{5/2}(z), \tag{9.141}$$

while below the critical point, we find

$$P_< = \lim_{V\to\infty}\left(-\frac{\Omega_{\mathrm{BE}}(T, V, \mu)}{V}\right) = \frac{k_B T}{\lambda_T{}^3}\, g_{5/2}(1) = \frac{k_B T}{\lambda_T{}^3}\,(1.342) \tag{9.142}$$

since $\lim_{V\to\infty, z\to 1} V^{-1}\ln(1 - z) = 0$ and $g_{5/2}(1) = 1.342$.

The entropy, in general, can be written in the form

$$\langle S\rangle = -\left(\frac{\partial\Omega_{\mathrm{BE}}}{\partial T}\right)_{V,\mu} = -k_B \sum_{l=0}^{\infty}\left(\langle n_{\mathbf{p}_l}\rangle \ln\langle n_{\mathbf{p}_l}\rangle - (1 + \langle n_{\mathbf{p}_l}\rangle)\ln(1 + \langle n_{\mathbf{p}_l}\rangle)\right). \tag{9.143}$$

For our present purposes it is more interesting to consider the entropy per unit volume. From Eq. (9.141), we obtain (above the critical point)

$$\lim_{V\to\infty}\frac{\langle S\rangle_>}{V} = \lim_{V\to\infty} -\frac{1}{V}\left(\frac{\partial\Omega_{\mathrm{BE}}}{\partial T}\right)_{V,\mu} = \frac{5}{2}\frac{k_B}{\lambda_T{}^3}\, g_{5/2}(z) - \frac{k_B\langle N\rangle}{V}\ln z, \tag{9.144}$$

and from Eq. (9.142), we obtain (below the critical point)

$$\lim_{V\to\infty} \frac{\langle S\rangle_<}{V} = \lim_{V\to\infty} -\frac{1}{V}\left(\frac{\partial \Omega_{\mathrm{BE}}}{\partial T}\right)_{V,\mu} = \tfrac{5}{2}k_B \frac{1}{\lambda_T{}^3}(1.342). \tag{9.145}$$

The heat capacity per unit volume above the critical point is given by

$$c_> = \lim_{V\to\infty}\left(\frac{1}{V}T\left(\frac{\partial S}{\partial T}\right)_{V,\langle N\rangle}\right) = \frac{15}{4}\frac{k_B}{\lambda_T{}^3}g_{5/2}(z) - \tfrac{9}{4}k_B\frac{\langle N\rangle}{V}\frac{g_{3/2}(z)}{g_{1/2}(z)} \tag{9.146}$$

and below the critical point it is given by

$$c_< = \frac{15}{4}k_B\frac{1}{\lambda_T{}^3}(1.342). \tag{9.147}$$

Note that the heat capacity is derived holding volume and particle number fixed. We must consider μ as varying. We can then use Eq. (9.135) to obtain

$$\left(\frac{\partial}{\partial T}(\beta\mu)\right)_{\langle N\rangle,V} = -\frac{3}{2}\frac{\lambda_T{}^3}{V}\frac{1}{T}\frac{\langle N\rangle}{g_{1/2}(z)} \tag{9.148}$$

above the transition point. This equation is useful in deriving Eqs. (9.146) and (9.147). The first thing to notice about Eq. (9.145) is that it obeys the third law. In the limit of zero temperature the entropy approaches zero. In Fig. 9.6, we have plotted isotherms of an ideal Bose-Einstein gas in the P-V plane. The transition from a state in which no particles are condensed to a state in which all particles are condensed corresponds to a first-order phase transition. (It is important to note that the phase transition in liquid He^4 is second order, so there are important differences between an ideal boson fluid and liquid He^4.) As a function of pressure and temperature, the volume and entropy are dis-

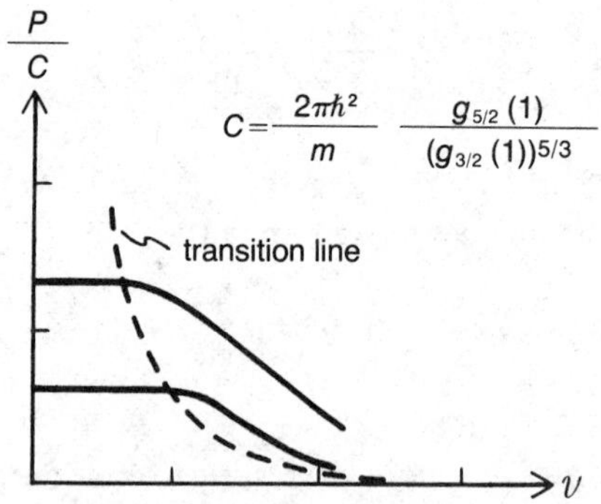

Fig. 9.6. Sketch of some isotherms for a Bose-Einstein gas: the solid lines are isotherms; the dashed line is the line of transition points and obeys the equation $Pv^{5/2} = C$.

continuous. In Fig. 9.7, we have plotted the heat capacity as a function of temperature. At the critical point it exhibits a peak, and in the limit $T \to 0$ K it has a temperature dependence $T^{3/2}$. The phase transition in an ideal Bose-Einstein gas is entirely the result of statistics. As we shall see, an ideal Fermi gas exhibits no such transition.

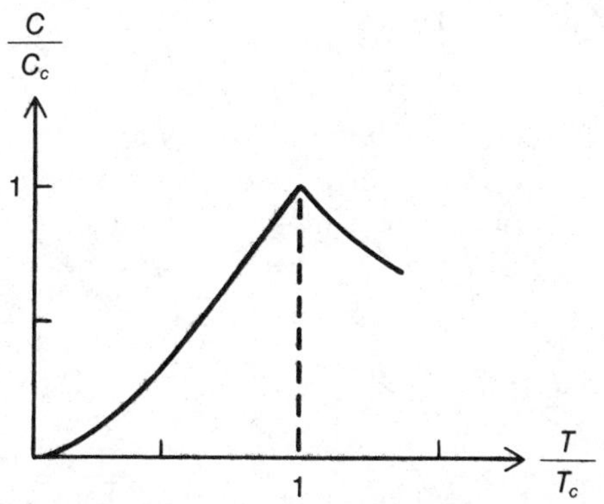

Fig. 9.7. A sketch of the reduced heat capacity per unit volume versus the reduced temperature for a Bose-Einstein gas: c_c is the heat capacity per unit volume at the critical temperature.

The high-temperature behavior of the gas is readily obtained. At high temperature $z \to 0$ and $g_{5/2}(z) \approx g_{3/2}(z) \approx g_{1/2}(z) \approx z$. From Eq. (9.135) we obtain for the density

$$\frac{\langle N \rangle}{V} = \frac{z}{\lambda_T^3}; \tag{9.149}$$

from Eq. (9.141) we obtain for the pressure

$$P = \frac{k_B T}{\lambda_T^3} z = \frac{\langle N \rangle k_B T}{V}; \tag{9.150}$$

and from Eq. (9.146) we obtain for the heat capacity per unit volume

$$c_> = \frac{15k_B}{4\lambda_T^3} z - \tfrac{9}{4} k_B \frac{\langle N \rangle}{V} = \frac{3}{2} \frac{\langle N \rangle k_B}{V}. \tag{9.151}$$

Thus, at high temperature we obtain the same results as for a Maxwell-Boltzmann gas.

5. Fermi-Dirac Gas[8,9]

As a final example of ideal gases, we shall find the thermodynamic properties of an ideal gas of spin $\frac{1}{2}$ Fermi-Dirac particles. The grand partition function for such a system can be written in the form

$$\mathscr{Z}_{\mathrm{FD}}(T, V, \mu) = \sum_{n_{0,\uparrow}=0}^{1} \sum_{n_{0,\downarrow}=0}^{1} \cdots \sum_{n_{j,\uparrow}=0}^{1} \cdots \sum_{n_{\infty,\downarrow}=0}^{1} \times \exp\left\{-\beta \sum_{l=0}^{\infty} \sum_{\lambda_l=\uparrow,\downarrow} n_{l,\lambda_l}(\epsilon_l - \mu)\right\}$$

$$= \prod_{l=0}^{\infty} \left(\sum_{n_{l,\uparrow}=0}^{1} e^{-\beta n_{l,\uparrow}(\epsilon_l - \mu)}\right)\left(\sum_{n_{l,\downarrow}=0}^{1} e^{-\beta n_{l,\downarrow}(\epsilon_l - \mu)}\right), \tag{9.152}$$

where the $n_{l,\uparrow}$ denotes the occupation number for the state with momentum $\mathbf{p}_l$ and spin component $+\frac{1}{2}$ and $n_{l,\downarrow}$ denotes the occupation number of the state with momentum $\mathbf{p}_l$ and spin component $-\frac{1}{2}$. If no magnetic fields are present to distinguish between spin states, then we can write $n_{l,\uparrow} = n_{l,\downarrow} = n_l$ and Eq. (9.152) takes the form

$$\mathscr{Z}_{\mathrm{FD}}(T, V, \mu) = \prod_{l=0}^{\infty} (1 + e^{-\beta(\epsilon_l - \mu)})^g, \tag{9.153}$$

where $g = 2s + 1 = 2$ for spin $s = \frac{1}{2}$ particles, but Eq. (9.153) now applies to particles with spin s. From Eqs. (9.56) and (9.153) we can write the grand potential in the form

$$\Omega_{\mathrm{FD}}(T, V, \mu) = -k_B T g \sum_{l=0}^{\infty} \ln(1 + e^{-\beta(\epsilon_l - \mu)}). \tag{9.154}$$

The factor of g in Eq. (9.154) comes from the degeneracy of the spin states. From Eq. (9.154) we obtain the following expression for the average number of particles:

$$\langle N \rangle = -\left(\frac{\partial \Omega_{\mathrm{FD}}}{\partial \mu}\right)_{V,T} = g \sum_{l=0}^{\infty} \left(\frac{1}{e^{\beta(\epsilon_l - \mu)} + 1}\right) = \sum_{l=0}^{\infty} \langle n_{\mathbf{p}_l} \rangle. \tag{9.155}$$

In Eq. (9.155), $\langle n_{\mathbf{p}_l} \rangle$ is the average number of particles with momentum $\mathbf{p}_l$,

$$\langle n_{\mathbf{p}_l} \rangle = \frac{g}{e^{\beta(\epsilon_l - \mu)} + 1}. \tag{9.156}$$

The entropy can also be obtained from Eq. (9.154) and can be written in the form

$$\langle S \rangle = -\left(\frac{\partial \Omega_{\mathrm{FD}}}{\partial T}\right)_{V,\mu} = -k_B g \sum_{l=0}^{\infty} \left[\frac{\langle n_{\mathbf{p}_l} \rangle}{g} \ln \frac{\langle n_{\mathbf{p}_l} \rangle}{g} + \left(1 - \frac{\langle n_{\mathbf{p}_l} \rangle}{g}\right) \ln\left(1 - \frac{\langle n_{\mathbf{p}_l} \rangle}{g}\right)\right]. \tag{9.157}$$

Inspection of Eq. (9.156) shows that $\langle n_{\mathbf{p}_l} \rangle$ can never become larger than g, corresponding to the g spin states associated with $\mathbf{p}_l$. This is a result of the Pauli exclusion principle. Thus, there can be no macroscopic occupation of any given momentum state for a noninteracting Fermi system and we will have no phase transition analogous to the Bose-Einstein case. However, other interesting things happen.

Let us assume that we have a large volume (and particle number) so that we can replace the summations in Eqs. (9.154) and (9.155) by integrations. We then find

$$\Omega_{FD}(T, V, \mu) = -\frac{gVk_BT}{\lambda_T^3} f_{5/2}(z) \tag{9.158}$$

and

$$\langle N \rangle = \frac{gV}{\lambda_T^3} f_{3/2}(z), \tag{9.159}$$

where

$$f_{5/2}(z) \equiv \frac{4}{\sqrt{\pi}} \int_0^\infty dx x^2 \ln(1 + z\,e^{-x^2}) = \sum_{\alpha=1}^{\infty} \frac{(-1)^{\alpha+1} z^\alpha}{\alpha^{5/2}} \tag{9.160}$$

and

$$f_{3/2}(z) = z \frac{\partial}{\partial z} f_{5/2}(z) = \sum_{\alpha=1}^{\infty} \frac{(-1)^{\alpha+1} z^\alpha}{\alpha^{3/2}}. \tag{9.161}$$

In the above equations, λ_T is the thermal wavelength (cf. Eq. (9.119)), z is the fugacity $z = e^{\beta\mu}$, and we have made the change of variables $x = \beta p^2/2m$.

We can get some idea of the behavior of the chemical potential if we revert[10] Eqs. (9.159) and (9.161) and solve for z. We then find

$$z = \left(\frac{\langle N \rangle \lambda_T^3}{gV}\right) + \frac{1}{2^{3/2}} \left(\frac{\langle N \rangle \lambda_T^3}{gV}\right)^2 + \left(\frac{1}{2^2} - \frac{1}{3^{3/2}}\right)\left(\frac{\langle N \rangle \lambda_T^3}{gV}\right)^3 + \cdots \tag{9.162}$$

The coefficients of various powers of $\langle N \rangle \lambda_T^3/gV$ will always be positive. Thus, for high temperatures $z \sim 0$ and the product $\beta\mu$ will be large and negative. For low temperatures $z \sim \infty$ and the product $\beta\mu$ will be large and positive. At high temperatures the chemical potential will be large and negative and at low temperatures it will be positive although not necessarily large.

In Fig. 9.8 we plot the average occupation number $\langle n_{\mathbf{p}_l} \rangle$ as a function of $\mathbf{p}_l$ at low temperature. For low momentum it has a roughly constant value of g. But for energies of the order of the chemical potential, it starts to fall off rapidly. The dashed line is the plot for zero temperatures. At low temperatures the particles fill the states with lowest momentum and tend to stack up in momentum space. The maximum absolute value of momentum, p_f, of particles at zero temperature is called the Fermi momentum and the region of momentum space with momentum p_f is called the Fermi surface. Particles with momentum less than p_f are said to lie in the Fermi sea.

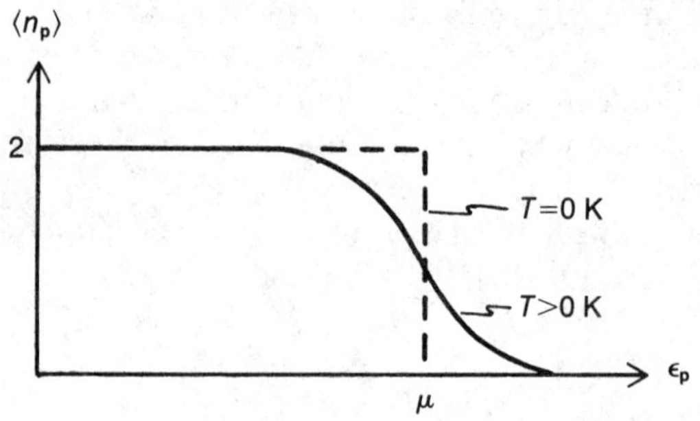

Fig. 9.8. A plot of the average occupation number $\langle n_{\mathbf{p}} \rangle$ of the momentum state **p** for an ideal Fermi gas of spin $\frac{1}{2}$ particles: the dashed line is a plot of $\langle n_{\mathbf{p}} \rangle$ for $T = 0$ K and the solid line is a plot for $T > 0$ K; the quantity μ is the chemical potential and $\epsilon_{\mathbf{p}}$ is the kinetic energy.

It is useful to calculate the thermodynamic properties of the Fermi gas at low temperatures. Let us first consider the integral for $f_{3/2}(z)$. From Eqs. (9.160) and (9.161) we may write $f_{3/2}(z)$ as

$$f_{3/2}(z) = \frac{4}{\sqrt{\pi}} \int_0^\infty dx \frac{x^2}{[e^{x^2-\nu} + 1]} = \frac{2}{\sqrt{\pi}} \int_0^\infty dy \frac{\sqrt{y}}{(e^{y-\nu} + 1)}$$

$$= \frac{4}{3\sqrt{\pi}} \int_0^\infty dy y^{3/2} \frac{e^{y-\nu}}{(1 + e^{y-\nu})^2} \tag{9.163}$$

where $\nu = \beta\mu$. To obtain the last form of the integral, we have integrated by parts. The function $e^{y-\nu}[1 + e^{y-\nu}]^2$ is essentially the derivative of the occupation number and at low temperature is sharply peaked about $y = \nu$ (cf. Fig. 9.9). Thus, to perform the integration in Eq. (9.163), we may expand $y^{3/2}$ in a

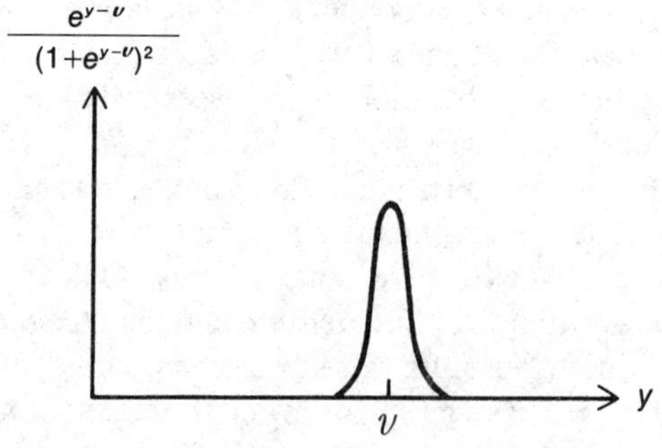

Fig. 9.9. A plot of the function $e^{y-\nu}/(1 + e^{y-\nu})^2$ versus y.

Taylor series about $y = \nu$. If we then let $t = (y - \nu)$, we can write $f_{3/2}(z)$ as

$$f_{3/2}(z) = \frac{4}{3\sqrt{\pi}}\int_{-\nu}^{\infty} dt\, \frac{e^t}{(1 + e^t)^2}\,(\nu^{3/2} + \tfrac{3}{2}\nu^{1/2}t + \tfrac{3}{8}\nu^{-1/2}t^2 + \cdots). \tag{9.164}$$

The contribution from the lower limit in the integral will be of order $e^{-\beta\mu}$. At low temperatures, we can neglect it and extend the lower limit to $-\infty$:

$$f_{3/2}(z) = \frac{4}{3\sqrt{\pi}}\int_{-\infty}^{\infty} dt\, \frac{e^t}{(1 + e^t)^2}\,(\nu^{3/2} + \tfrac{3}{2}\nu^{1/2}t + \tfrac{3}{8}\nu^{-1/2}t^2 + \cdots). \tag{9.165}$$

To evaluate Eq. (9.165), we must evaluate integrals of the form

$$I_n = \int_{-\infty}^{\infty} dt\, \frac{t^n\, e^t}{(1 + e^t)^2}. \tag{9.166}$$

The result is $I_n = 0$ for n odd, $I_0 = 1$, and $I_n = (n - 1)!\,(2n)(1 - 2^{1-n})\zeta(n)$ where $\zeta(n)$ is a Riemann zeta function ($\zeta(2) = \pi^2/6$, $\zeta(4) = \pi^4/90$, $\zeta(6) = \pi^6/95$, etc.). Thus, we finally obtain the low-temperature result:

$$\frac{\langle N\rangle}{V}\frac{\lambda^3}{g} = \frac{4}{3\sqrt{\pi}}\left[(\beta\mu)^{3/2} + \frac{\pi^2}{8}(\beta\mu)^{-1/2} + \cdots\right]. \tag{9.167}$$

If we take the limit $T \to 0$ K in Eq. (9.167) we find the following density-dependent expression for the chemical potential:

$$\mu(T = 0) \equiv \epsilon_f = \frac{\hbar^2}{2m}\left(\frac{6\pi^2\langle N\rangle}{gV}\right)^{2/3}. \tag{9.168}$$

The chemical potential at $T = 0$ K is also called the Fermi energy ϵ_f because at $T = 0$ K it is the maximum energy a particle can have (cf. Fig. 9.8). At very low temperatures, only particles within a distance k_BT of the Fermi surface can participate in physical processes in the gas, because they can change their momentum state. Particles lower down in the Fermi sea have no empty momentum states available for them to jump to and do not contribute to changes in the thermodynamic properties.

Eq. (9.167) may be reverted to find the chemical potential as a function of temperature and density. The result is

$$\mu = \epsilon_f\left[1 - \frac{\pi^2}{12}\left(\frac{k_BT}{\epsilon_f}\right)^2 + \cdots\right]. \tag{9.169}$$

Thus, corrections to the chemical potential are of order T^2.

The internal energy can be computed in a similar manner. At low temperatures it is given by

$$U = \tfrac{3}{5}\langle N\rangle\epsilon_f\left[1 + \tfrac{5}{12}\pi^2\left(\frac{k_BT}{\epsilon_f}\right)^2 + \cdots\right]. \tag{9.170}$$

From Eq. (9.170), we obtain the heat capacity

$$C_V = \left(\frac{\partial U}{\partial T}\right)_{V,\langle N\rangle} = \frac{\langle N\rangle \pi^2}{2}\frac{k_B{}^2 T}{\epsilon_f} + \cdots \tag{9.171}$$

Thus, the heat capacity of an ideal Fermi gas depends linearly on temperature at very low temperatures and goes to zero at $T = 0$ K in accordance with the third law. It is important to note, however, that particles in an ideal Fermi gas can have a large zero point momentum and, therefore, a large pressure and energy even at $T = 0$ K. This is a result of the Pauli exclusion principle.

It is a simple matter to show that at high temperatures all quantities reduce to the corresponding expressions for a Maxwell-Boltzmann gas.

E. MOMENTUM CONDENSATION IN AN INTERACTING FERMI FLUID[11–14]

In the previous section, we found that an ideal Bose gas can condense in momentum space and thereby undergo a phase transition, but an ideal Fermi gas cannot. However, we can obtain momentum condensation of a Fermi gas if we allow a weak attraction to exist between fermions. The model we shall discuss in this section illustrates the basic mechanism for the transition to the condensed phase in a superconductor. To simplify calculations, we shall use the number representation. Those unfamiliar with the number representation may want to skip this section or refer to Appendix B where all quantities are defined.

We shall assume that the fermions have a weak attraction if they have a kinetic energy which lies in an interval $\Delta\epsilon$ on either side of the Fermi surface. Otherwise they behave like an ideal gas. Also, only pairs with equal and opposite momentum and opposite spin components attract one another. With these assumptions, we can write the Hamiltonian of the system in the form

$$\hat{H} = \sum_{\mathbf{k},\lambda} \frac{\hbar^2 k^2}{2m} \hat{a}^{+}_{\mathbf{k},\lambda}\hat{a}_{\mathbf{k},\lambda} + \sum_{\mathbf{k},\mathbf{l}} V_{\mathbf{k},\mathbf{l}}\hat{a}^{+}_{\mathbf{k},\uparrow}\hat{a}^{+}_{-\mathbf{k},\downarrow}\hat{a}_{-\mathbf{l},\downarrow}\hat{a}_{\mathbf{l},\uparrow}, \tag{9.172}$$

where $V_{\mathbf{k},\mathbf{l}} = \langle \mathbf{k},\uparrow; -\mathbf{k},\downarrow|V|\mathbf{l},\uparrow; -\mathbf{l},\downarrow\rangle$ (cf. App. B). Thus, the interaction destroys a pair whose particles have momenta $\mathbf{l}$ and $-\mathbf{l}$ and opposite spin components and creates a pair with momenta $\mathbf{k}$ and $-\mathbf{k}$ and opposite spin components. The potential $V_{\mathbf{k},\mathbf{l}}$ is of the form

$$V_{\mathbf{k},\mathbf{l}} = \begin{cases} -V_0 & \left|\mu - \dfrac{\hbar^2 k^2}{2m}\right| \leqslant \Delta\epsilon \text{ and } \left|\mu - \dfrac{\hbar^2 l^2}{2m}\right| \leqslant \Delta\epsilon, \\ 0 & \text{otherwise} \end{cases} \tag{9.173}$$

where V_0 is a positive constant. To simplify our calculations, we will use a mean field approximation for the interaction and write the Hamiltonian in the form

$$\hat{H} = \sum_{\mathbf{k},\lambda} \frac{\hbar^2 k^2}{2m} \hat{a}^+_{\mathbf{k},\lambda}\hat{a}_{\mathbf{k},\lambda} + \sum_{\mathbf{k},\mathbf{l}} V_{\mathbf{k},\mathbf{l}} X_{\mathbf{k}}{}^+ \hat{a}_{-\mathbf{l},\downarrow}\hat{a}_{\mathbf{l},\uparrow} + \sum_{\mathbf{k},\mathbf{l}} V_{\mathbf{k},\mathbf{l}} X_{\mathbf{l}} \hat{a}^+_{\mathbf{k},\uparrow}\hat{a}^+_{-\mathbf{k},\downarrow}, \tag{9.174}$$

where

$$X_{\mathbf{k}}{}^+ = -\langle \hat{a}^+_{-\mathbf{k}\downarrow}\hat{a}^+_{\mathbf{k}\uparrow}\rangle = -\mathrm{Tr}\, \hat{\rho}\hat{a}^+_{-\mathbf{k}\downarrow}\hat{a}^+_{\mathbf{k}\uparrow} \tag{9.175}$$

and

$$X_{\mathbf{l}} = -\langle \hat{a}_{\mathbf{l}\uparrow}\hat{a}_{-\mathbf{l}\downarrow}\rangle \equiv -\mathrm{Tr}\, \hat{\rho}\hat{a}_{\mathbf{l}\uparrow}\hat{a}_{-\mathbf{l}\downarrow}. \tag{9.176}$$

In Eqs. (9.175) and (9.176), $\hat{\rho}$ is the exact density operator

$$\hat{\rho} = \frac{e^{-\beta(\hat{H}-\mu\hat{N})}}{\mathrm{Tr}\, e^{-\beta(\hat{H}-\mu\hat{N})}} \tag{9.177}$$

with a Hamiltonian given by Eq. (9.174) and $\hat{N}$ is the number operator. The Hamiltonian in Eq. (9.174) does not conserve particle number as does the Hamiltonian in Eq. (9.172). Since we are working with the grand canonical ensemble, this is not a problem. The interaction terms in Eq. (9.174) will only be nonzero when $X_{\mathbf{k}}{}^+$ and $X_{\mathbf{k}}$, which must be determined self-consistently, are nonzero. Thus, at the transition point, gauge symmetry is broken. The quantity $X_{\mathbf{k}}$ may be thought of as a macroscopic wave function for bound pairs with total momentum zero and relative momentum **k** and may be used as an order parameter for the condensed phase. If a macroscopic number of particles become paired, $X_{\mathbf{k}}$ will take on a finite value and the energy of the system will be lowered. The transition to the condensed phase occurs when the thermal energy, which tends to break pairs apart, becomes less important than the interaction energy.

It is useful to rewrite the density operator in the form

$$\hat{\rho} = \frac{e^{-\beta\hat{H}'}}{\mathrm{Tr}\, e^{-\beta\hat{H}'}} \tag{9.178}$$

where the Hamiltonian, $\hat{H}'$, is defined

$$\hat{H}' = \sum_{\mathbf{k}} (\epsilon_{\mathbf{k}}\hat{a}^+_{\mathbf{k}\uparrow}\hat{a}_{\mathbf{k}\uparrow} - \epsilon_{\mathbf{k}}\hat{a}_{-\mathbf{k}\downarrow}\hat{a}^+_{-\mathbf{k}\downarrow}) + \sum_{\mathbf{k},\mathbf{l}} V_{\mathbf{k},\mathbf{l}} X_{\mathbf{k}}{}^+ \hat{a}_{-\mathbf{l},\downarrow}\hat{a}_{\mathbf{l}\uparrow} + \sum_{\mathbf{k},\mathbf{l}} V_{\mathbf{k},\mathbf{l}} X_{\mathbf{l}} \hat{a}^+_{\mathbf{k}\uparrow}\hat{a}^+_{-\mathbf{k}\downarrow} \tag{9.179}$$

with

$$\epsilon_{\mathbf{k}} = \left(\frac{\hbar^2 k^2}{2m} - \mu\right). \tag{9.180}$$

In writing the Hamiltonian in the form given in Eq. (9.179), we have used Fermi anticommutation relations (cf. App. B, Eq. (B.98)) and we now measure the energy from the Fermi surface. Also, we have neglected a numerical factor which cancels out of Eq. (9.178). The Hamiltonian can now be written in the concise form

$$\hat{H}' = \sum_{\mathbf{k}} \hat{\mathbf{A}}_{\mathbf{k}}^{+} \bar{\bar{\mathscr{E}}}_{\mathbf{k}} \hat{\mathbf{A}}_{\mathbf{k}} \tag{9.181}$$

where $\hat{\mathbf{A}}_{\mathbf{k}}$ and $\hat{\mathbf{A}}_{\mathbf{k}}{}^{+}$ are column and row vectors, respectively;

$$\hat{\mathbf{A}}_{\mathbf{k}} = \begin{pmatrix} \hat{a}_{\mathbf{k},\uparrow} \\ \hat{a}^{+}_{-\mathbf{k},\downarrow} \end{pmatrix} \qquad \hat{\mathbf{A}}_{\mathbf{k}}{}^{+} = \widetilde{\hat{a}^{+}_{\mathbf{k},\uparrow},\ \hat{a}_{-\mathbf{k},\downarrow}}; \tag{9.182}$$

and $\bar{\bar{\mathscr{E}}}_{\mathbf{k}}$ is the matrix

$$\bar{\bar{\mathscr{E}}}_{\mathbf{k}} = \begin{pmatrix} \epsilon_{\mathbf{k}} & \Delta_{\mathbf{k}} \\ \Delta_{\mathbf{k}}{}^{+} & -\epsilon_{\mathbf{k}} \end{pmatrix}. \tag{9.183}$$

In Eq. (9.183), $\Delta_{\mathbf{k}}$ is the so-called gap function,

$$\Delta_{\mathbf{k}} \equiv \sum_{\mathbf{l}} V_{\mathbf{k},\mathbf{l}} X_{\mathbf{l}} = -\sum_{\mathbf{l}} V_{\mathbf{k},\mathbf{l}} \langle \hat{a}_{\mathbf{l},\uparrow} \hat{a}_{-\mathbf{l},\downarrow} \rangle, \tag{9.184}$$

and $\Delta_{\mathbf{k}}{}^{+}$ is its Hermitian conjugate. The gap function may also be used as the order parameter for the condensed phase.

As was first shown by Bogoliubov,[12] the Hamiltonian in Eq. (9.181) can be diagonalized by means of a unitary transformation which preserves the fermion anticommutation relations. In so doing, we obtain the Hamiltonian for the collective excitations (called bogolons) of the system. To diagonalize the Hamiltonian, we introduce a 2 × 2 transformation matrix, $\bar{\bar{\mathbf{U}}}_{\mathbf{k}}$, and a new set of column vectors, $\hat{\mathbf{\Gamma}}_{\mathbf{k}}$, such that

$$\hat{\mathbf{A}}_{\mathbf{k}} = \bar{\bar{\mathbf{U}}}_{\mathbf{k}} \hat{\mathbf{\Gamma}}_{\mathbf{k}}. \tag{9.185}$$

The matrix $\bar{\bar{\mathbf{U}}}_{\mathbf{k}}$ is composed of complex functions of **k**,

$$\bar{\bar{\mathbf{U}}}_{\mathbf{k}} = \begin{pmatrix} u_{\mathbf{k}}^{*} & v_{\mathbf{k}} \\ -v_{\mathbf{k}}^{*} & u_{\mathbf{k}} \end{pmatrix}, \tag{9.186}$$

and the column vector $\hat{\mathbf{\Gamma}}_{\mathbf{k}}$ is composed of a new set of creation and annihilation operators, $\hat{\gamma}^{+}_{\mathbf{k},1}$ and $\hat{\gamma}_{\mathbf{k},0}$, respectively:

$$\hat{\mathbf{\Gamma}}_{\mathbf{k}} = \begin{pmatrix} \hat{\gamma}_{\mathbf{k},0} \\ \hat{\gamma}^{+}_{\mathbf{k},1} \end{pmatrix}, \tag{9.187}$$

whose physical significance will become clear shortly. If we require that the functions $u_{\mathbf{k}}$ and $v_{\mathbf{k}}$ satisfy the relation $|u_{\mathbf{k}}|^2 + |v_{\mathbf{k}}|^2 = 1$, then it is easy to show that, if the operators $\hat{a}^{+}_{\mathbf{k},\lambda}$ and $\hat{a}_{\mathbf{k},\lambda}$ obey Fermi anticommutation relations, the operators $\hat{\gamma}_{\mathbf{k},\alpha}$ and $\hat{\gamma}^{+}_{\mathbf{k},\alpha}$ will also obey anticommutation relations:

$$[\hat{\gamma}_{\mathbf{k},\alpha}, \hat{\gamma}^{+}_{\mathbf{k}',\alpha'}]_{+} = \delta_{\mathbf{k},\mathbf{k}'}\,\delta_{\alpha,\alpha'}, \qquad [\hat{\gamma}_{\mathbf{k},\alpha}, \hat{\gamma}_{\mathbf{k}',\alpha'}]_{+} = [\hat{\gamma}^{+}_{\mathbf{k},\alpha}, \hat{\gamma}^{+}_{\mathbf{k}',\alpha'}]_{+} = 0, \tag{9.188}$$

where $\alpha = 0, 1$. Furthermore, the transformation is unitary:

$$\bar{\bar{\mathbf{U}}}_{\mathbf{k}}^{+}\bar{\bar{\mathbf{U}}}_{\mathbf{k}} = \bar{\bar{\mathbf{U}}}_{\mathbf{k}}\bar{\bar{\mathbf{U}}}_{\mathbf{k}}^{+} = \mathbb{1} \tag{9.189}$$

where $\mathbb{1}$ is the 2×2 unit matrix. By reverting Eq. (9.185) we see that $\hat{\gamma}_{\mathbf{k},0}$ decreases the momentum of the system by $\mathbf{k}$ and lowers the spin by $\hbar$ (it destroys a particle with quantum numbers $[\mathbf{k}, \uparrow]$ and creates one with quantum numbers $[-\mathbf{k}, \downarrow]$), whereas $\hat{\gamma}_{\mathbf{k},1}$ increases the momentum by $\mathbf{k}$ and raises the spin by $\hbar$. $\hat{\gamma}_{\mathbf{k},0}$ and $\hat{\gamma}_{\mathbf{k},1}^{+}$ are called the bogolon annihilation and creation operators, respectively.

Let us further require that the transformation diagonalize the Hamiltonian. Thus,

$$\bar{\bar{\mathbf{U}}}_{\mathbf{k}}\bar{\bar{\mathscr{E}}}_{\mathbf{k}}\bar{\bar{\mathbf{U}}}_{\mathbf{k}}^{+} = \mathbb{E}_{\mathbf{k}} \tag{9.190}$$

where $\mathbb{E}_{\mathbf{k}}$ is a 2×2 diagonal matrix containing the exact bogolon excitation energies $E_{\mathbf{k},0}$ and $E_{\mathbf{k},1}$. If we square Eq. (9.190) and use Eqs. (9.183) and (9.189), we find that the excitation energies are independent of the spin variable α and that

$$\mathbb{E}_{\mathbf{k}} = \begin{pmatrix} E_{\mathbf{k}} & 0 \\ 0 & -E_{\mathbf{k}} \end{pmatrix} \tag{9.191}$$

where

$$E_{\mathbf{k}} = (\epsilon_{\mathbf{k}}^{2} + |\Delta_{\mathbf{k}}|^{2})^{1/2}. \tag{9.192}$$

With this transformation, we have succeeded in reducing the interacting Fermi gas of particles to an ideal Fermi gas of bogolons, and the Hamiltonian of the system can be written

$$\hat{H}' = \sum_{\mathbf{k},\alpha} E_{\mathbf{k},\alpha}\hat{\gamma}_{\mathbf{k},\alpha}^{+}\hat{\gamma}_{\mathbf{k},\alpha}. \tag{9.193}$$

The Hamiltonian, written in terms of bogolon operators, looks like that of an ideal Fermi gas. The bogolons are the collective modes of the system and play a role analogous to that of phonons in a Debye solid, although their dispersion relation is quite different.

We now wish to find an equation for the gap function $\Delta_{\mathbf{k}}$. This requires a bit of algebra. Let us first note that

$$\langle\hat{\mathbf{A}}_{\mathbf{k}}\hat{\mathbf{A}}_{\mathbf{k}}^{+}\rangle = \mathrm{Tr}\,\hat{\rho}\hat{\mathbf{A}}_{\mathbf{k}}\hat{\mathbf{A}}_{\mathbf{k}}^{+} \equiv \tfrac{1}{2}(\mathbb{1} + \mathbb{W}_{\mathbf{k}}) \tag{9.194}$$

where

$$\mathbb{W}_{\mathbf{k}} = \begin{pmatrix} 1 - 2N_{\mathbf{k},\uparrow} & -X_{\mathbf{k}} \\ -X_{\mathbf{k}}^{+} & -1 + 2N_{\mathbf{k},\downarrow} \end{pmatrix} \tag{9.195}$$

and

$$N_{\mathbf{k},\sigma} = \langle\hat{a}_{\mathbf{k},\sigma}^{+}\hat{a}_{\mathbf{k},\sigma}\rangle = \mathrm{Tr}\,\rho\hat{a}_{\mathbf{k},\sigma}^{+}\hat{a}_{\mathbf{k},\sigma}. \tag{9.196}$$

The quantity $N_{\mathbf{k},\sigma}$ is the particle occupation number in the interacting system.

It is not the same as the occupation number of an ideal Fermi gas. Let us note that

$$\langle \hat{\Gamma}_{\mathbf{k}} \hat{\Gamma}_{\mathbf{k}}^{+} \rangle \equiv \operatorname{Tr} \hat{\rho} \hat{\Gamma}_{\mathbf{k}} \hat{\Gamma}_{\mathbf{k}}^{+} = \begin{pmatrix} 1 - n_{\mathbf{k},0} & 0 \\ 0 & n_{\mathbf{k},1} \end{pmatrix} \tag{9.197}$$

where $n_{\mathbf{k}\alpha}$ is the bogolon occupation number and is given by

$$\begin{aligned} n_{\mathbf{k},\alpha} &= [\operatorname{Tr} e^{-\beta \hat{H}'}]^{-1} \operatorname{Tr}\left\{ \gamma_{\mathbf{k}\alpha}^{+} \gamma_{\mathbf{k}\alpha} \exp\left[-\beta \sum_{\mathbf{k},\beta} E_{\mathbf{k}\beta} \hat{\gamma}_{\mathbf{k}\beta}^{+} \hat{\gamma}_{\mathbf{k}\beta} \right] \right\} \\ &= (1 + e^{+\beta E_{\mathbf{k},\alpha}})^{-1} = \frac{1}{2}\left(1 - \tanh \frac{\beta E_{\mathbf{k},\alpha}}{2}\right). \end{aligned} \tag{9.198}$$

Since the bogolons are distributed like an ideal Fermi gas, no more than one can occupy each momentum state for a given value of α.

Now that we have expressions for $\langle \hat{\mathbf{A}}_{\mathbf{k}} \hat{\mathbf{A}}_{\mathbf{k}}^{+} \rangle$ and $\langle \hat{\Gamma}_{\mathbf{k}} \hat{\Gamma}_{\mathbf{k}}^{+} \rangle$, we can relate them via the unitary transformation. Thus,

$$\langle \hat{\mathbf{A}}_{\mathbf{k}} \hat{\mathbf{A}}_{\mathbf{k}}^{+} \rangle = \bar{\bar{\mathbf{U}}}_{\mathbf{k}} \langle \hat{\Gamma}_{\mathbf{k}} \hat{\Gamma}_{\mathbf{k}}^{+} \rangle \bar{\bar{\mathbf{U}}}_{\mathbf{k}}^{+}, \tag{9.199}$$

and from Eqs. (9.191), (9.193), (9.197), (9.198), and (9.199) we find

$$\mathbb{W}_{\mathbf{k}} = \tanh\left(\frac{\beta}{2} \bar{\bar{\mathbf{U}}}_{\mathbf{k}} \mathbb{E}_{\mathbf{k}} \bar{\bar{\mathbf{U}}}_{\mathbf{k}}^{+} \right). \tag{9.200}$$

Eq. (9.200) contains a self-consistent equation for the gap function. To extract it, let us now rewrite Eq. (9.200) as follows:

$$\begin{aligned} \mathbb{W}_{\mathbf{k}} &= \tanh\left(\frac{\beta}{2} \bar{\bar{\mathbf{U}}}_{\mathbf{k}} \begin{pmatrix} E_{\mathbf{k}} & 0 \\ 0 & -E_{\mathbf{k}} \end{pmatrix} \bar{\bar{\mathbf{U}}}_{\mathbf{k}}^{+} \right) = \bar{\bar{\mathbf{U}}}_{\mathbf{k}} \begin{pmatrix} \tanh\left(\frac{\beta E_{\mathbf{k}}}{2}\right) & 0 \\ 0 & -\tanh\left(\frac{\beta E_{\mathbf{k}}}{2}\right) \end{pmatrix} \bar{\bar{\mathbf{U}}}_{\mathbf{k}}^{+} \\ &= \bar{\bar{\mathbf{U}}}_{\mathbf{k}} \begin{pmatrix} E_{\mathbf{k}} & 0 \\ 0 & -E_{\mathbf{k}} \end{pmatrix} \bar{\bar{\mathbf{U}}}_{\mathbf{k}}^{+} \frac{1}{E_{\mathbf{k}}} \tanh\left(\frac{\beta E_{\mathbf{k}}}{2}\right) = \bar{\bar{\mathscr{E}}}_{\mathbf{k}} \frac{1}{E_{\mathbf{k}}} \tanh\left(\frac{\beta E_{\mathbf{k}}}{2}\right). \end{aligned} \tag{9.201}$$

If we equate matrix elements, we find

$$1 - 2N_k = \frac{\epsilon_{\mathbf{k}}}{E_{\mathbf{k}}} \tanh\left(\frac{\beta E_{\mathbf{k}}}{2}\right) \tag{9.202}$$

and

$$X_{\mathbf{k}} = -\frac{\Delta_{\mathbf{k}}}{E_{\mathbf{k}}} \tanh\left(\frac{\beta E_{\mathbf{k}}}{2}\right). \tag{9.203}$$

If we next multiply Eq. (9.203) by $V_{\mathbf{k},\mathbf{l}}$ and sum over $\mathbf{k}$, we obtain the famous gap equation

$$\Delta_{\mathbf{l}} = -\sum_{\mathbf{k}} V_{\mathbf{k},\mathbf{l}} \frac{\Delta_{\mathbf{k}}}{E_{\mathbf{k}}} \tanh\left(\frac{\beta E_{\mathbf{k}}}{2}\right), \tag{9.204}$$

which is a self-consistent equation for $\Delta_{\mathbf{l}}$.

We have obtained Eq. (9.204) from the grand canonical ensemble. Therefore, the solutions of the gap equation correspond to extrema of the free energy. However, only one of the solutions at a given temperature corresponds to the stable thermodynamic state (the state which minimizes the free energy) and it must be determined from thermodynamic arguments. We should remember also that the energy, $E_{\mathbf{k}}$, depends on the gap, so Eq. (9.204) is rather complicated. If we now use Eq. (9.173), we find

$$\Delta_{\mathbf{l}} = V_0 \sum_{\mathbf{k}}{}' \frac{\Delta_{\mathbf{k}}}{E_{\mathbf{k}}} \tanh\left(\frac{\beta E_{\mathbf{k}}}{2}\right), \tag{9.205}$$

where the prime on the summation indicates that it is restricted to an energy interval $\Delta\epsilon$ about the Fermi surface. Since the right-hand side does not depend on $\mathbf{l}$, the gap must be a constant for the potential given in Eq. (9.173). Thus,

$$\Delta_{\mathbf{l}} = \begin{cases} \Delta(T) & \text{for } |\epsilon_{\mathbf{l}} - \mu| \leqslant \Delta\epsilon \\ 0 & \text{otherwise} \end{cases} \tag{9.206}$$

(the gap is a function of temperature). If we assume that our system is contained in a large volume, we can change the summation to an integration (cf. Eq. (9.115)) and we can write Eq. (9.205) as

$$1 = V_0 \frac{k_f^2 V}{2\pi^2} \int_{-\Delta\epsilon}^{\Delta\epsilon} \left(\frac{\partial k}{\partial \epsilon_{\mathbf{k}}}\right)_{k_f} d\epsilon_{\mathbf{k}} \frac{\tanh\left[\frac{\beta}{2}(\epsilon_{\mathbf{k}}^2 + \Delta^2(T))^{1/2}\right]}{(\epsilon_{\mathbf{k}}^2 + \Delta^2(T))^{1/2}}. \tag{9.207}$$

In Eq. (9.207), we have made the approximation $k^2\,dk = k_f^2\,dk$ (where k_f is the value of $|\mathbf{k}|$ at the Fermi surface) since the integration is confined to a small region about the Fermi surface. Because the argument is even in $\epsilon_{\mathbf{k}}$, Eq. (9.207) simplifies and we find

$$1 = V_0 N(0) \int_0^{\Delta\epsilon} d\epsilon_{\mathbf{k}} \frac{\tanh\left\{\frac{\beta}{2}(\epsilon_{\mathbf{k}}^2 + \Delta^2(T))^{1/2}\right\}}{(\epsilon_{\mathbf{k}}^2 + \Delta^2(T))^{1/2}}, \tag{9.208}$$

where $N(0) = mVk_f/\pi^2\hbar^2$ is the density of states at the Fermi surface. Eq. (9.208) determines the temperature dependence of the gap $\Delta(T)$ and can be used to find the transition temperature.

The energy of bogolons (measured from the Fermi surface) with momentum $\mathbf{k}$ is $E_{\mathbf{k}} = (\epsilon_{\mathbf{k}}^2 + \Delta^2(T))^{1/2}$. It takes a finite energy to excite them, regardless of their momentum; that is, there is a finite gap in the excitation spectrum. At the critical temperature, T_c, the gap goes to zero and the excitation spectrum reduces to that of an ideal gas. The critical temperature can be obtained from Eq. (9.208). It is the temperature at which the gap becomes zero. Thus,

$$\begin{aligned} 1 &= N(0)V_0 \int_0^{\Delta\epsilon} d\epsilon_{\mathbf{k}} \frac{\tanh\left(\frac{\beta_c \epsilon_{\mathbf{k}}}{2}\right)}{\epsilon_{\mathbf{k}}} = N(0)V_0 \int_0^{\beta_c(\Delta\epsilon/2)} dx \frac{\tanh x}{x} \\ &= N(0)V_0 \ln(1.13\beta_c\,\Delta\epsilon), \end{aligned} \tag{9.209}$$

where $\beta_c = (k_B T_c)^{-1}$, and we have used the fact that

$$\int_0^a \frac{\tanh x}{x}\, dx = \ln(2.26a). \tag{9.210}$$

From Eq. (9.209) we obtain

$$k_B T_c = 1.13\, \Delta\epsilon\, e^{-1/N(0)V_0}. \tag{9.211}$$

Thus, the critical temperature, T_c, varies exponentially as the strength of the attractive interaction varies.

We can also use Eq. (9.208) to find the gap at $T = 0$ K. Since $\tanh \infty = 1$, we can write

$$1 = N(0)V_0 \int_0^{\Delta\epsilon} d\epsilon_{\mathbf{k}} \frac{1}{(\epsilon_{\mathbf{k}}^2 + \Delta^2(0))^{1/2}} = N(0)V_0 \sinh^{-1}\frac{\Delta\epsilon}{\Delta(0)}. \tag{9.212}$$

For weakly coupled systems $N(0)V_0 \ll 1$ and Eq. (9.212) reduces to

$$\Delta(0) = 2\, \Delta\epsilon\, e^{-1/N(0)V_0}. \tag{9.213}$$

Comparing Eqs. (9.211) and (9.213), we obtain the following relation between the critical temperature and the zero temperature gap for weakly coupled systems:

$$\frac{\Delta(0)}{k_B T_c} = 1.764. \tag{9.214}$$

Eq. (9.214) is in good agreement with experimental values of this ratio for superconductors.

Eq. (9.208) may be solved numerically to obtain a plot of the gap as a func-

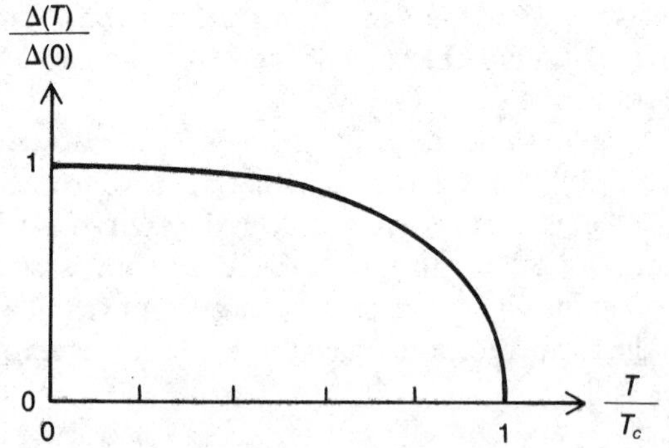

Fig. 9.10. A plot of the energy gap as a function of reduced temperature in the weak coupling limit (based on Ref. 9.14).

tion of temperature (cf. Fig. 9.10). Near the critical temperature, the gap has temperature dependence

$$\frac{\Delta(T)}{\Delta(0)} = 1.74\left(1 - \frac{T}{T_c}\right)^{1/2}. \tag{9.215}$$

Thus, the critical exponent for the gap is $\frac{1}{2}$.

Since the bogolons form an ideal gas, the entropy can be written in the form

$$S = -2k_B \sum_{\mathbf{k}} \{n_{\mathbf{k}} \ln n_{\mathbf{k}} + (1 - n_{\mathbf{k}}) \ln(1 - n_{\mathbf{k}})\} \tag{9.216}$$

(cf. Eq. (9.157)). The heat capacity is easy to find from Eq. (9.216):

$$\begin{aligned} C_V = T\left(\frac{\partial S}{\partial T}\right)_V &= 2\beta k_B \sum_{\mathbf{k}} \frac{\partial n_{\mathbf{k}}}{\partial \beta} \ln\left(\frac{n_{\mathbf{k}}}{1 - n_{\mathbf{k}}}\right) \\ &= 2\beta k_B \sum_{\mathbf{k}} (-1) \frac{\partial n_{\mathbf{k}}}{\partial E_{\mathbf{k}}} \left(E_{\mathbf{k}}^2 + \tfrac{1}{2}\beta \frac{\partial \Delta^2}{\partial \beta}\right). \end{aligned} \tag{9.217}$$

The first term in Eq. (9.217) is continuous at $T = T_c$ but the second term is not since $\partial\Delta^2/\partial\beta$ has a finite value for $T < T_c$ but is zero for $T > T_c$. Near T_c we may let $E_{\mathbf{k}} \to |\epsilon_{\mathbf{k}}|$. Then the heat capacity just below the critical temperature is

$$C_V^{<} = 2\beta_c k_B \sum_{\mathbf{k}} (-1) \frac{\partial n_k}{\partial |\epsilon_{\mathbf{k}}|} \left(\epsilon_{\mathbf{k}}^2 + \tfrac{1}{2}\beta_c \left(\frac{\partial \Delta^2}{\partial \beta}\right)_{T_c}\right) \qquad (T \lessgtr T_c) \tag{9.218}$$

and just above the critical temperature is

$$C_V^{>} = 2\beta_c k_B \sum_{\mathbf{k}} (-1) \frac{\partial n_{\mathbf{k}}}{\partial |\epsilon_{\mathbf{k}}|} \epsilon_{\mathbf{k}}^2 \qquad (T \gtrless T_c). \tag{9.219}$$

The discontinuity in the heat capacity at the critical temperature is

$$\Delta C_V = C_V^{<} - C_V^{>} = -\beta_c^2 k_B \sum_{\mathbf{k}} \left(\frac{\partial \Delta^2}{\partial \beta}\right)_{T_c} \frac{\partial n(|\epsilon_{\mathbf{k}}|)}{\partial |\epsilon_{\mathbf{k}}|} = N(0)\left(-\frac{\partial \Delta^2}{\partial T}\right)_{T_c}, \tag{9.220}$$

where we have expanded the integrand in a Taylor series about the Fermi surface. In Fig. 9.11 we sketch the heat capacity in the transition region. At low temperature, the heat capacity of an ideal or weakly coupled Fermi gas is linear in temperature. The transition to the condensed phase causes a jump in the heat capacity and alters its temperature dependence. The transition to the condensed phase is a continuous transition and the transition point is a λ-point.

The picture we have given here is essentially the same as that given by Bardeen, Schrieffer, and Cooper[11] in their Nobel Prize–winning paper. The choice of interaction energy in Eq. (9.173) comes from the pairing mechanism in superconductors. There the interaction of electrons with lattice phonons creates an attractive interaction between electron pairs of opposite spin which is localized at the Fermi surface.

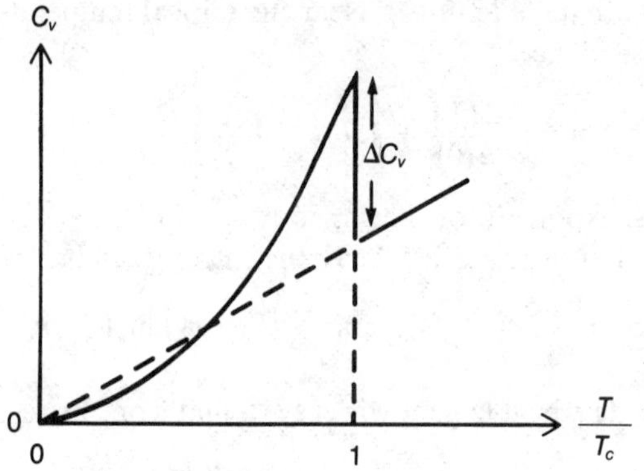

Fig. 9.11. A sketch of the heat capacity curve for a superconductor: the straight dashed line gives the heat capacity in the absence of condensation; the solid line shows the jump in the heat capacity at the transition point and the nonlinear variation below it.

The discussion in this section is based on a method by Balian and Werthamer[13] which was developed to treat more general types of pairing and is especially useful in treating the condensed phases of liquid He^3. For a generalization of this method to the case of systems with spatially varying anisotropic gaps, see Ref. 9.15.

F. ORDER-DISORDER TRANSITIONS

1. General Discussion

Another class of systems which can be studied using methods of equilibrium statistical mechanics are those that exhibit a transition from an ordered to a disordered state. One of the simplest examples of such a system is that of a lattice composed of two different types of objects A and B (cf. Fig. 9.12). We assume that the objects interact but only with their nearest neighbors. Let us call V_{AB} the interaction energy between objects A and B and let us call V_{AA} and

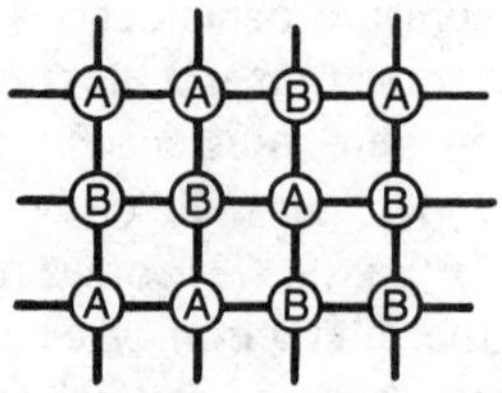

Fig. 9.12. Two-dimensional lattice whose sites carry either A or B values.

V_{BB} the interaction energies between two objects A and two objects B, respectively.

We first consider the system at zero temperature:

(*a*) If $V_{AB} > (V_{AA} + V_{BB})/2$, then a configuration in which objects A all neighbor one another and objects B all neighbor one another would be energetically favored. In this case, the lattice will split into domains, some containing only A and the others containing only B.

(*b*) If $V_{AB} < (V_{AA} + V_{BB})/2$, then the configuration in which A and B neighbor one another will be energetically favored and A and B will alternate on the lattice.

The above possibilities are ordered states. If we now raise the temperature of the system, the thermal energy, k_BT, will tend to randomize the positions of A and B until at some point the system will "melt" and become completely disordered as the temperature is raised further.

The partition function for such a system takes the form

$$Z_N(T) = \sum_{\substack{\text{all lattice}\\ \text{configurations}}} \exp\left(-\beta \sum_{\substack{\text{all near}\\ \text{neighbors}}} V_{ij}\right). \tag{9.221}$$

The mathematical model of such systems is called the Ising model and was originally used by Ising[16] as a model for ferromagnetism. However, the model also describes such systems as lattice gases, binary alloys, "melting" of DNA, etc.

The Ising model can be solved exactly in one and two dimensions, but no one has ever succeeded in solving it analytically in three dimensions. In one dimension it does not exhibit a phase transition, but in two dimensions it does. The Ising model was first solved in two dimensions by Onsager.[9,17–19] It is one of the few exactly soluble models which exhibit a phase transition.

To obtain a more explicit expression for the partition function for the Ising model, let us consider a *d*-dimensional lattice containing N sites and assume that each lattice site is occupied by either an A or a B. Associate with site i a number s_i such that $s_i = +1$ for an A site and $s_i = -1$ for a B site. A particular configuration of the lattice is completely specified when the set of variables $\{s_i\}$ for all lattice sites is known.

If the interaction energy between two nearest-neighbor sites i and j is denoted ϵ_{ij}, and we allow the possibility of an external field F which decreases the energy of an A site and increases the energy of a B site, then the total energy of the lattice for a given configuration $\{s_i\}$ is

$$E\{s_i\} = \sum_{(ij)}^{\gamma N/2} \epsilon_{ij} s_i s_j - F \sum_{i=1}^{N} s_i. \tag{9.222}$$

The summation in the first term is over all different nearest-neighbor pairs. The quantity, γ, is the number of nearest neighbors of a given site. For simplicity, we will always consider the case in which the interaction energy between all nearest-neighbor sites is the same; that is, $\epsilon_{ij} = \epsilon$. Then the partition function of the system may be written

$$Z_N(T, F) = \exp\{-\beta G(T, F, N)\}$$
$$= \sum_{S_1=\pm 1} \cdots \sum_{S_N=\pm 1} \exp\left\{-\beta\left(\epsilon \sum_{(ij)}^{\gamma N/2} S_i S_j - F \sum_{i=1}^{N} S_i\right)\right\} \tag{9.223}$$

where $G(T, F, N)$ is the Gibbs free energy. The relation between the partition function and the thermodynamic potential involves the Gibbs free energy rather than the Helmholtz free energy because the field, F, couples the system mechanically to the outside world.

Let us now introduce the quantity

$$\mathscr{S} = \sum_{i=1}^{N} S_i. \tag{9.224}$$

Then the average $\langle \mathscr{S} \rangle$ is a measure of the relative number of sites occupied by A's and B's. The average $\langle \mathscr{S} \rangle$ is given by

$$\langle \mathscr{S} \rangle = k_B T \frac{\partial}{\partial F} \ln(Z_N(T, F)). \tag{9.225}$$

The number of nearest neighbors, γ, is determined by the form of the lattice (cf. Fig. 9.13).

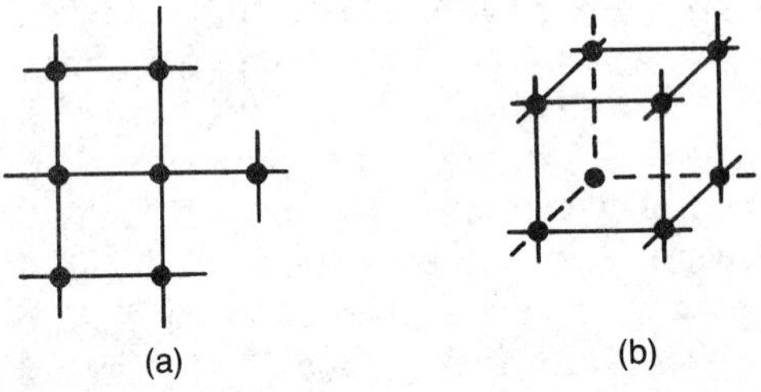

Fig. 9.13. Number of nearest neighbors, γ, for some typical lattices: (*a*) for a two-dimensional square lattice, $\gamma = 4$; (*b*) for a three-dimensional simple cubic lattice, $\gamma = 6$.

The partition function given in Eq. (9.223) is not in a convenient form, because many configurations have the same energy. It is necessary to sort these out and to write the partition function in terms of other variables. We will introduce the following notation:

N_A = the total number of A lattice sites.
$N_B = N - N_A$ = the total number of B lattice sites.

N_{AA} = the total number of A–A nearest-neighbor pairs.
N_{BB} = the total number of B–B nearest-neighbor pairs.
N_{AB} = the total number of A–B nearest-neighbor pairs.

We can find a relation between these five quantities in the following way. Draw the lattice composed of N sites (cf. Fig. 9.14). For each A site draw a line to each of its nearest neighbors. The total number of lines drawn will be γN_A. Then each nearest-neighbor pair A–A will be connected by two lines, each nearest-neighbor pair A–B will be connected by one line, and each nearest-neighbor pair B–B will be connected by no lines. The total number of lines is

$$\gamma N_A = 2N_{AA} + N_{AB}. \tag{9.226}$$

By similar arguments, we can find

$$\gamma N_B = 2N_{BB} + N_{AB} \tag{9.227}$$

and, by definition,

$$N = N_A + N_B. \tag{9.228}$$

Because of Eqs. (9.226)–(9.228), only two of the five parameters N_A, N_B, N_{AA}, N_{BB}, and N_{AB} are independent.

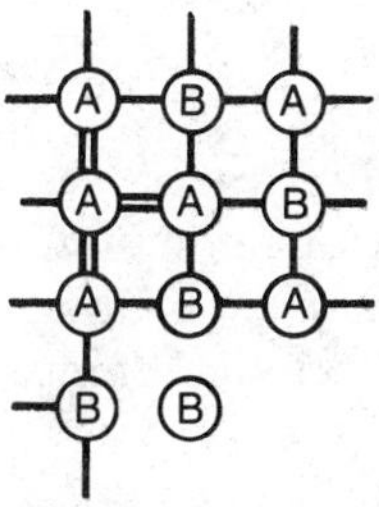

Fig. 9.14. By drawing lines only from A sites to their nearest neighbors, we can determine the number of each type of nearest neighbors.

The summation over lattice sites in the partition function may now be written in terms of the parameters N_A, N_B, N_{AA}, N_{BB}, and N_{AB}. With the help of Eqs. (9.226)–(9.228), we find

$$\sum_{(ij)} s_i s_j = N_{AA} + N_{BB} - N_{AB} = 4N_{AA} - 2\gamma N_A + \frac{\gamma}{2} N \tag{9.229}$$

and

$$\sum_i s_i = N_A - N_B = 2N_A - N, \tag{9.230}$$

and the total energy becomes

$$E(N_A, N_{AA}) = 4\epsilon N_{AA} - 2(\epsilon\gamma + F)N_A + (\tfrac{1}{2}\gamma\epsilon + F)N. \tag{9.231}$$

The partition function may then be written as

$$Z_N(T, F) = e^{-\beta(\frac{1}{2}\gamma\epsilon + F)N} \sum_{N_A=0}^{N} e^{2\beta(\epsilon\gamma + F)N_A} \sum_{N_{AA}} c(N_A, N_{AA})\, e^{-\beta\epsilon 4N_{AA}} \tag{9.232}$$

where $c(N_A, N_{AA})$ is the number of configurations with a given set of values N_A and N_{AA}.

2. Two Applications of the Ising Model

Two interesting applications of the Ising model involve spin systems and lattice gases.[20] The Ising model immediately describes a spin system if we let A represent a spin-up site, $\uparrow$; B represent a spin-down site, $\downarrow$; and F be a magnetic field, $\mathscr{H}$. Then $\langle\mathscr{S}\rangle = \langle M\rangle$, the average magnetization of the system.

The application of the Ising model to a lattice gas requires a little more explanation. A lattice gas consists of a lattice containing N sites, some of which are filled with atoms and some are empty. Here A = filled sites and B = empty sites. The spacing between sites is a. The potential energy between atoms is

$$V = \begin{cases} \infty & r = 0 \\ -\epsilon_0 & r = a \\ 0 & \text{otherwise.} \end{cases} \tag{9.233}$$

The potential therefore contains a hard core at $r = 0$ (no two atoms occupy the same site) and an attractive region (atoms interact with atoms on neighboring sites). If we let N_A denote the number of atoms and N_{AA} denote the number of nearest-neighbor interactions between atoms, then the total energy of the system is

$$E = -\epsilon_0 N_{AA} \tag{9.234}$$

and the partition function for a given value of N_A is

$$Z_{N_A}(T, N) = \sum_{N_{AA}} c(N_A, N_{AA})\, e^{\beta\epsilon_0 N_{AA}} \tag{9.235}$$

where $c(N_A, N_{AA})$ is the number of different configurations with a given value of N_A and N_{AA}. If we assume that each lattice site occupies a unit volume, then the volume of the lattice is $V = N$.

If we assume that we have an infinitely large lattice, then we can define a grand partition function for the lattice gas:

$$\begin{aligned} \mathscr{Z}(T, N, \mu) = e^{-\beta\Omega(T,N,\mu)} &= \sum_{N_A=0}^{\infty} e^{\beta\mu N_A} Z_{N_A}(T, N) \\ &= \sum_{N_A=0}^{\infty} e^{\beta\mu N_A} \sum_{N_{AA}} c(N_A, N_{AA})\, e^{\beta\epsilon_0 N_{AA}}, \end{aligned} \tag{9.236}$$

where $\Omega(T, N, \mu)$ is the grand potential. The maximum value of N_A is actually limited by the value of N. It is easy to see that Eq. (9.236) is mathematically equivalent to Eq. (9.232) if we make the correspondence $G - (\frac{1}{2}\gamma\epsilon + F)N \rightarrow \Omega$, $2(\epsilon\gamma + F) \rightarrow \mu$, and $-4\epsilon \rightarrow +\epsilon_0$.

3. Bragg-Williams Approximation to the Ising Model[9,21]

It is possible to exhibit a phase transition in the Ising model by making a very simple but drastic approximation, due to Bragg and Williams, for the number of nearest-neighbor A–A sites N_{AA}. We shall assume that

$$\frac{N_{AA}}{\frac{1}{2}\gamma N} = \left(\frac{N_A}{N}\right)^2 \tag{9.237}$$

and we shall assume that the interaction between A sites is attractive, $\epsilon = -\epsilon_0$, where ϵ_0 is a positive quantity. With the approximation in Eq. (9.237) the fraction of bonds of the type A–A is equal to the value we would expect if the A sites were placed on the lattice at random. Thus, the only way we can lower the energy of the system is to increase the fraction N_A/N. It does not help to place all A sites next to one another in a group. This is, of course, a very unphysical approximation, but nevertheless we shall find a phase transition of sorts.

With the above changes the partition function in Eq. (9.232) takes the form

$$Z_N(T, F) = \sum_{N_A=0}^{N} c(N_A)\, e^{+\beta(\frac{1}{2}\gamma\epsilon_0 - F)N}\, e^{-2\beta(\epsilon_0\gamma - F)N_A}\, e^{2\beta\epsilon_0\gamma(N_A^2/N)}. \tag{9.238}$$

It is now convenient to make the change of variables,

$$N_A = \frac{N}{2}(L + 1).$$

Then we find

$$Z_N(T, F) = \sum_{L=-1}^{1} c(N_A)\, e^{+\beta N[(\epsilon_0\gamma/2)L^2 + FL]}. \tag{9.239}$$

The factor $c(N_A)$ is easy to determine. It is simply the number of combinations of N things taken N_A at a time,

$$c(N_A) = \frac{N!}{(N - N_A)!\, N_A!}. \tag{9.240}$$

If we use Eq. (9.239), we finally obtain

$$Z_N(T, F) = \sum_{L=-1}^{1} \frac{N!}{\left(\frac{N - NL}{2}\right)!\, \left(\frac{N + NL}{2}\right)!}\, e^{\beta N[(\epsilon_0\gamma/2)L^2 + FL]}. \tag{9.241}$$

We shall evaluate the partition function in the limit $N \to \infty$. We first use Stirling's approximation (cf. Eq. (5.37)) to write Eq. (9.241) as

$$Z_N(T, F) = \sum_{L=-1}^{1} \frac{e^{\beta N((\epsilon_0 \gamma L^2/2) + FL)}}{\sqrt{2\pi N}\left(\frac{1-L}{2}\right)^{N((1-L)/2)+\frac{1}{2}}\left(\frac{1+L}{2}\right)^{N((L+1)/2)+\frac{1}{2}}}. \quad (9.242)$$

The next step is to find the Gibbs free energy per lattice site,

$$g(T, F) = \lim_{N\to\infty}\left[-\frac{k_B T}{N}\ln Z_N(T, F)\right]. \quad (9.243)$$

This is easily done if we note the following property of the logarithm of a sum of terms taken to the power N. Let us assume that we have a sum of terms $(A_1)^N + (A_2)^N + (A_3)^N + \cdots$ where $A_1 > A_2 > A_3 > \cdots$. Then we find

$$\lim_{N\to\infty}\frac{1}{N}\ln\{(A_1)^N + (A_2)^N + (A_3)^N + \cdots\}$$

$$= \lim_{N\to\infty}\frac{1}{N}\ln\left\{A_1{}^N\left[1 + \left(\frac{A_2}{A_1}\right)^N + \left(\frac{A_3}{A_1}\right)^N + \cdots\right]\right\} = A_1. \quad (9.244)$$

Thus, it is easy to see that the Gibbs free energy must have the form

$$g(T, F) = -\left(\frac{\epsilon_0\gamma\bar{L}^2}{2} + F\bar{L}\right) + \frac{1}{\beta}\left(\frac{1-\bar{L}}{2}\right)\ln\left(\frac{1-\bar{L}}{2}\right)$$

$$+ \frac{1}{\beta}\left(\frac{1+\bar{L}}{2}\right)\ln\left(\frac{1+\bar{L}}{2}\right) \quad (9.245)$$

where $\bar{L}$ is the value of L which gives the largest term in the sum in Eq. (9.242). Since

$$\langle \sigma \rangle = \lim_{N\to\infty}\frac{\langle \mathscr{S} \rangle}{N} = -\left(\frac{\partial g}{\partial F}\right)_T = \bar{L}, \quad (9.246)$$

$\bar{L}$ is now the order parameter and is a measure of the relative fraction of A sites. The value of $\bar{L}$ which corresponds to a thermodynamic state must minimize $g(T, F)$. If we determine the value of $\bar{L}$ for which $g(T, F)$ is an extremum, we find

$$2\beta(\epsilon_0\gamma\bar{L} + F) = \ln\left(\frac{1+\bar{L}}{1-\bar{L}}\right) \quad (9.247)$$

or, reverting, we obtain

$$\bar{L} = \tanh(\beta\gamma\epsilon_0\bar{L} + \beta F). \quad (9.248)$$

We must now find the solutions of this equation and determine which solution corresponds to a minimum of the free energy, $g(T, F)$.

Let us assume that the external field is zero, for simplicity. Then we must solve the equation

$$\bar{L} = \tanh(\beta\gamma\epsilon_0\bar{L}). \quad (9.249)$$

This can be done by plotting $f(\bar{L}) = \bar{L}$ versus $f(\bar{L}) = \tanh(\beta\gamma\epsilon_0\bar{L})$ and locating the points of intersection (cf. Fig. 9.15). For $\beta\gamma\epsilon_0 < 1$ there is only one intersection at $\bar{L} = 0$. For $\beta\gamma\epsilon_0 > 1$ there are three intersections at $\bar{L} = L_0, 0, -L_0$, where L_0 is determined graphically:

$$(a)\quad \bar{L} = 0, \qquad \beta\gamma\epsilon_0 < 1. \tag{9.250}$$

$$(b)\quad \bar{L} = \begin{cases} L_0 \\ 0 \\ -L_0 \end{cases} \qquad \beta\gamma\epsilon_0 > 1.$$

The critical temperature is thus given by the condition

$$\gamma\epsilon_0 = k_B T_c. \tag{9.251}$$

We can easily determine which of the three solutions in Eq. (9.250*b*) minimizes the free energy by taking the derivative (since $F = 0$, the free energy of interest is the Helmholtz free energy),

$$\frac{\partial^2 a}{\partial \bar{L}^2} = -\epsilon_0\gamma + \frac{1}{2\beta}\left(\frac{1}{1+\bar{L}} + \frac{1}{1-\bar{L}}\right) = -\epsilon_0\gamma + \frac{1}{\beta}\left(\frac{1}{1-\bar{L}^2}\right). \tag{9.252}$$

Thus, for $\bar{L} = 0$ and $\beta\epsilon_0\gamma > 1$, $\partial^2 a/\partial L^2 < 0$ and we have a maximum. For $\bar{L} = \pm L_0$ we must actually compute values of $\beta\gamma\epsilon_0$ and L_0. But we then find a minimum.

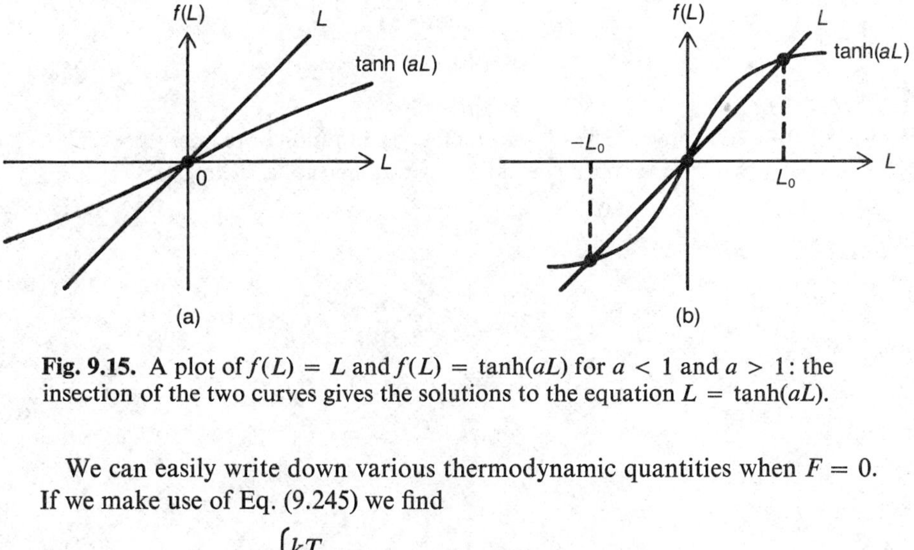

Fig. 9.15. A plot of $f(L) = L$ and $f(L) = \tanh(aL)$ for $a < 1$ and $a > 1$: the insection of the two curves gives the solutions to the equation $L = \tanh(aL)$.

We can easily write down various thermodynamic quantities when $F = 0$. If we make use of Eq. (9.245) we find

$$a(T, L_0) = \begin{cases} \dfrac{kT}{2}\ln\tfrac{1}{4}, & T > T_c \\[2ex] \dfrac{-\gamma\epsilon_0}{2}L_0^2 + \dfrac{1}{2\beta}\ln\left(\dfrac{1-L_0^2}{4}\right), & T < T_c \end{cases} \tag{9.253}$$

Also from Eq. (9.246),

$$\langle \mathscr{s}(T)\rangle = \begin{cases} 0 & T > T_c \\ \pm L_0 & T < T_c \end{cases}. \tag{9.254}$$

We find for the internal energy per site

$$u(T) = a(T) + Ts(T) = \begin{cases} 0 & T > T_c \\ \dfrac{-\gamma\epsilon_0 L_0{}^2}{2} & T < T_c \end{cases}, \tag{9.255}$$

where s is the entropy per lattice site $s = -(\partial a/\partial T)_{L_0}$. Finally, for the specific heat we obtain

$$c(T) = \begin{cases} 0 & T > T_c \\ \dfrac{-\gamma\epsilon_0}{2}\left(\dfrac{dL_0{}^2}{dT^2}\right), & T < T_c \end{cases}. \tag{9.256}$$

Thus, there is a discontinuity in the specific heat.

We can easily find the behavior of L_0 for $T \sim 0$ K and $T \sim T_c$. If we remember that $\tanh x = (e^x - e^{-x})(e^x + e^{-x})^{-1}$, then at $T \sim 0$ K

$$L_0 = 1 - 2\,e^{-2T_c/T}, \tag{9.257}$$

while at $T \lesssim T_c$

$$L_0 = \sqrt{3\left(1 - \frac{T}{T_c}\right)}. \tag{9.258}$$

In Fig. 9.16 we sketch the order parameter as a function of temperature and in Fig. 9.17 we sketch the specific heat for a two-dimensional lattice.

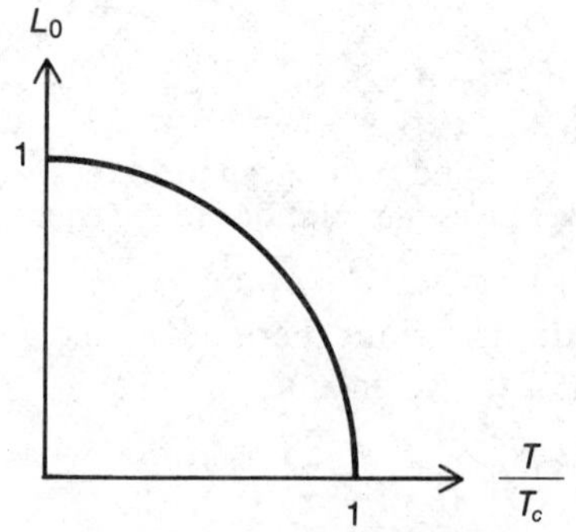

Fig. 9.16. A sketch of the order parameter in the Bragg-Williams approximation to the Ising model.

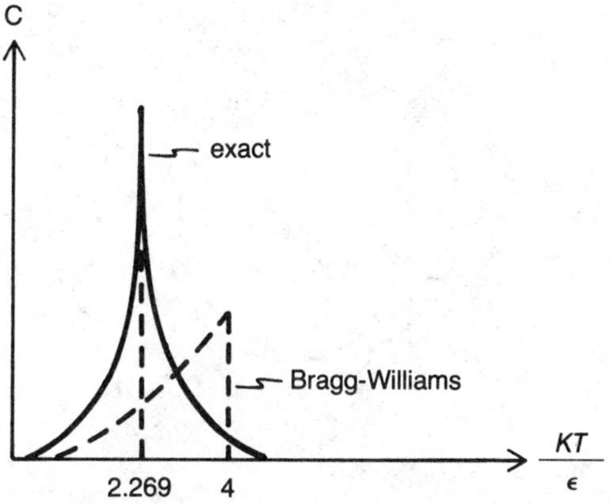

Fig. 9.17. A sketch of the specific heat for the Ising model in two dimensions: the dashed line is the approximate Bragg-Williams result; the solid line is the result of the exact solution of the Ising model.

The Bragg-Williams approximation neglects the effects of short-range order, that is, the possibility that clustering of particles will lower the energy. Above the transition temperature, there is neither short-ranged nor long-ranged order (or even kinetic motion). Thus, the thermodynamic properties are either constant or zero. Below the transition temperature, long-ranged order sets in and the thermodynamic properties become interesting. It is possible to improve on the Bragg-Williams approximation by including energy-lowering effects due to clustering. This has been done by Bethe[22] and gives much better results for the transition region. We will not discuss Bethe's approximation here but refer the reader to the original paper or Ref. 9.9.

4. Exact Solution to the Ising Model

The Ising model has been solved exactly in one and two dimensions. There is a phase transition in two dimensions but not in one dimension. In this section, we shall solve the one-dimensional Ising model exactly and give Onsager's results for the two-dimensional case.

(*a*) *One-Dimensional Ising Model.* Let us assume that the A and B sites are located on a one-dimensional periodic lattice containing N lattice sites (cf. Fig. 9.18). Periodicity is imposed by assuming that $s_{N+1} = s_1$. The total energy for a given configuration $\{s_k\}$ is

$$E\{s_k\} = -\epsilon \sum_{k=1}^{N} s_k s_{k+1} - F \sum_{k=1}^{N} s_k, \tag{9.259}$$

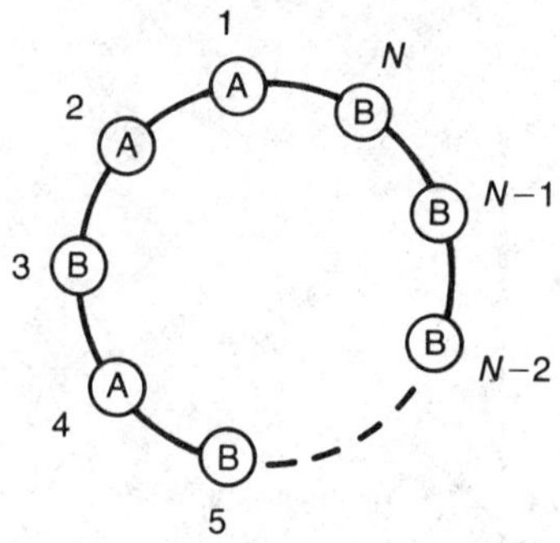

Fig. 9.18. One-dimension Ising model.

where we have assumed that only nearest neighbors interact and we have let $\epsilon_{k,k+1} = -\epsilon$. Thus, the lowest energy state is one in which A sites couple. The partition function is

$$Z_N(T) = \sum_{S_1=\pm 1} \cdots \sum_{S_N=\pm 1} \exp\left[\beta \sum_{k=1}^{N} (\epsilon s_k s_{k+1} + \tfrac{1}{2}F(s_k + s_{k+1}))\right]. \tag{9.260}$$

In Eq. (9.260) we have used the fact that $\sum_k s_k = \frac{1}{2}\sum_k (s_k + s_{k+1})$. It is now convenient to introduce a matrix $\bar{\bar{\mathbf{P}}}$ whose matrix elements are given by

$$\langle s_1|\bar{\bar{\mathbf{P}}}|s_2\rangle = e^{\beta(\epsilon s_1 s_2 + \frac{1}{2}F(s_1+s_2))} \tag{9.261}$$

and which is defined as

$$\bar{\bar{\mathbf{P}}} = \begin{pmatrix} e^{\beta(\epsilon+F)} & e^{-\beta\epsilon} \\ e^{-\beta\epsilon} & e^{\beta(\epsilon-F)} \end{pmatrix}. \tag{9.262}$$

The partition function may then be written

$$\begin{aligned} Z_N(T) &= \sum_{S_1=\pm 1} \cdots \sum_{S_N=\pm 1} \langle s_1|\bar{\bar{\mathbf{P}}}|s_2\rangle\langle s_2|\bar{\bar{\mathbf{P}}}|s_3\rangle\cdots\langle s_N|\bar{\bar{\mathbf{P}}}|s_1\rangle \\ &= \sum_{S_1=\pm 1} \langle s_1|\bar{\bar{\mathbf{P}}}^N|s_1\rangle = \mathrm{Tr}\,\bar{\bar{\mathbf{P}}}^N = \lambda_+{}^N + \lambda_-{}^N, \end{aligned} \tag{9.263}$$

where λ_+ and λ_- are the eigenvalues of the matrix $\bar{\bar{\mathbf{P}}}$, and $\lambda_+ > \lambda_-$. From Eq. (9.262), we obtain the following equation for the eigenvalues of the matrix $\bar{\bar{\mathbf{P}}}$:

$$\lambda_\pm = e^{\beta\epsilon}[\cosh(\beta F) \pm \sqrt{\cosh^2(\beta F) - 2\,e^{-2\beta\epsilon}\sinh(2\beta\epsilon)}]. \tag{9.264}$$

In the limit $N \to \infty$, only the largest eigenvalue, λ_+, contributes to the thermodynamic quantities. This is easily seen if we note that the Gibbs free energy per site is

$$g(T, F) = \lim_{N\to\infty} \frac{1}{N} G_N(T, F) = -k_B T \lim_{N\to\infty} \frac{1}{N} \ln Z_N(T, F) = -k_B T \ln \lambda_+ \tag{9.265}$$

or

$$g(T, F) = -\epsilon - k_B T \ln[\cosh(\beta F) + \sqrt{\cosh^2(\beta F) - 2\, e^{-2\beta\epsilon} \sinh(2\beta\epsilon)}]. \tag{9.266}$$

The order parameter is given by

$$\langle \sigma \rangle = -\left(\frac{\partial g}{\partial F}\right)_T = \frac{\sinh \beta F}{\sqrt{\cosh^2(\beta F) - 2\, e^{-2\beta\epsilon} \sinh(2\beta\epsilon)}}. \tag{9.267}$$

From Eq. (9.267), we see that the one-dimensional Ising model can exhibit no phase transition because when $F \to 0$ the order parameter also goes to zero. Hence, no spontaneous nonzero value of the order parameter is possible.

(*b*) *Two-Dimensional Ising Model.*[9] The two-dimensional Ising model is much more difficult to solve than the one-dimensional case. It was first done by Onsager[17–19] for zero external field and by Yang[23] for a nonzero external field. We shall give only the results here.

For the two-dimensional Ising model, the Helmholtz free energy per site for $F = 0$ is

$$a(T) = -k_B T \ln(2 \cosh 2\beta\epsilon) - \frac{k_B T}{2\pi} \int_0^\pi d\phi \ln \tfrac{1}{2}(1 + \sqrt{1 - \delta^2 \sin^2 \phi}) \tag{9.268}$$

where

$$\delta^2 = \frac{2 \sinh 2\beta\epsilon}{\cosh^2 2\beta\epsilon}. \tag{9.269}$$

The internal energy per site is

$$u(T) = -\epsilon \coth(2\beta\epsilon)\left[1 + \frac{2}{\pi}(2 \tanh^2 2\beta\epsilon - 1) \int_0^{\pi/2} \frac{d\phi}{\sqrt{1 - \delta^2 \sin^2 \phi}}\right]. \tag{9.270}$$

The heat capacity per site is

$$c(T) = \frac{2k_B}{\pi} (\beta\epsilon \coth 2\beta\epsilon)^2 \left\{ 2 \int_0^{\pi/2} \frac{d\phi}{\sqrt{1 - \delta^2 \sin^2 \phi}} - 2 \int_0^{\pi/2} d\phi \sqrt{1 - \delta^2 \sin^2 \phi} \right.$$
$$\left. - (2 - 2 \tanh^2 2\beta\epsilon)\left[\frac{\pi}{2} + (2 \tanh^2 2\beta\epsilon - 1) \int_0^{\pi/2} \frac{d\phi}{\sqrt{1 - \delta^2 \sin^2 \phi}}\right]\right\}. \tag{9.271}$$

The integral

$$\int_0^{\pi/2} \frac{d\phi}{\sqrt{1 - \delta^2 \sin^2 \phi}}$$

has a singularity at $\delta = 1$. This singularity corresponds to the phase transition. From Eq. (9.269), we see that the temperature at which the phase transition occurs is given by

$$\frac{2 \sinh 2\epsilon\beta_c}{\cosh^2 2\epsilon\beta_c} = 1. \tag{9.272}$$

This leads to the relation

$$k_B T_c = (2.269)\epsilon. \tag{9.273}$$

Near $T = T_c$ the specific heat behaves as

$$C(T) = \frac{2k_B}{\pi}(2\epsilon\beta_c)^2\left[-\ln\left|1 - \frac{T}{T_c}\right| + \ln\left(\frac{1}{2\epsilon\beta_c}\right) - \left(1 + \frac{\pi}{4}\right)\right]. \tag{9.274}$$

As $T \to T_c$, the specific heat becomes infinite logarithmically. The spontaneous order parameter per site obtained by Yang is

$$\langle\sigma\rangle = \begin{cases} 0 & T > T_c \\ \dfrac{(1 + e^{-4\beta\epsilon})^{1/4}(1 - 6\, e^{-4\beta\epsilon} + e^{-8\beta\epsilon})^{1/8}}{(1 - e^{-4\beta\epsilon})^{1/2}} & T < T_c \end{cases}. \tag{9.275}$$

The heat capacity for the two-dimensional Ising model is sketched in Fig. 9.16.

G. THE LEE-YANG THEORY OF PHASE TRANSITIONS[9,24]

Lee and Yang have used very simple arguments involving the grand canonical ensemble to arrive at a mechanism by which a phase transition can take place.

Let us consider a classical system of particles interacting via a potential $v(|\mathbf{q}_{ij}|)$ which contains an infinite hard core and a short-range attraction:

$$v(|\mathbf{q}_{ij}|) = \begin{cases} \infty & |\mathbf{q}_{ij}| < a \\ -\epsilon & a \leqslant |\mathbf{q}_{ij}| \leqslant b. \\ 0 & b \leqslant |\mathbf{q}_{ij}| \end{cases} \tag{9.276}$$

Because the particles have an infinite hard core, there is a maximum number, M, which can be fitted into a box of volume V. Therefore, the grand partition function must have the form

$$\begin{aligned} \mathscr{Z}(T, V, \mu) &= \sum_{N=0}^{M} \frac{e^{\beta\mu N}}{N!\, h^{3N}} \iint d\mathbf{p}^N\, d\mathbf{q}^N \exp\left\{\beta \sum_{i=1}^{N} \frac{p_i^2}{2M} + \frac{1}{2} \sum_{i<j=1}^{\frac{1}{2}N(N-1)} V(|\mathbf{q}_{ij}|)\right\} \\ &= \sum_{N=0}^{M} \frac{e^{\beta\mu N}}{\lambda_T^{3N} N!} Q_N(V, T), \end{aligned} \tag{9.277}$$

where λ_T is the thermal wavelength and $Q_N(V, T)$ is called the configuration integral and is defined

$$Q_N(V, T) = \int d\mathbf{q}^N \exp\left\{-\beta \sum_{i<j} V(|\mathbf{q}_{ij}|)\right\}. \tag{9.278}$$

In the last term of Eq. (9.277) we have explicitly performed the momentum integrations. Since we are dealing with classical particles, the phase space coordinates commute and the momentum integrations are trivial. The important information is in the configuration integral. For $N > M$, $Q_N(V, T) = 0$ because of the hard-core region in the potential in Eq. (9.276). Let us now introduce a new variable, $y \equiv e^{\beta\mu}/\lambda_T{}^3$. Then Eq. (9.277) can be written in the form

$$\mathscr{Z}(V, T, y) = \sum_{N=1}^{M} y^N \frac{Q_N(V, T)}{N!}. \tag{9.279}$$

We see that, for finite volume, $\mathscr{Z}(V, T, y)$ is a polynomial of order M in the parameter y. The coefficients of y^N are positive and real for all N.

We can rewrite $\mathscr{Z}(V, T, y)$ in the form

$$\mathscr{Z}(V, T, y) = \prod_{i=1}^{M} \left(1 - \frac{y}{y_i}\right), \tag{9.280}$$

where the quantities y_i are the M roots of the equation:

$$\mathscr{Z}(V, T, y) = 0. \tag{9.281}$$

Because the coefficients, $Q_N/N!$, are all real and positive, none of the roots, y_i, can be real and positive for finite M; otherwise, Eq. (9.281) could never be satisfied. Therefore, for finite M, all roots of y must be either real and negative or complex (cf. Fig. 9.19*a*). As the volume, V, is increased, the number of roots

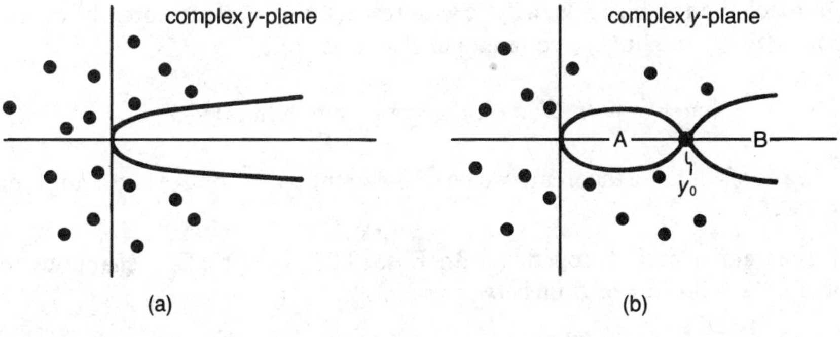

Fig. 9.19. A plot of the roots of $\mathscr{Z}(V, T, y) = 0$: (*a*) for finite V, no roots lie on the positive real axis; (*b*) for infinite volume, roots can lie on the positive real axis and separate regions of different phase; *A* and *B* are regions of different phase.

M will increase and move around in the complex plane. In the limit $V \to \infty$ it can happen that some of the roots will move down (or up) and touch the positive real axis (cf. Fig. 9.19*b*). When this happens, a phase transition occurs because the system can have different behavior for $y < y_0$ and $y > y_0$, where y_0 is the value of the root on the real axis. In general, the pressure $P(y)$ will be continuous across the point y_0, but the density and/or higher derivatives of P will be discontinuous. (We give an explicit example later.)

The pressure and density in the limit of infinite volume are given by

$$\frac{P}{k_B T} = \lim_{V\to\infty}\left[\frac{1}{V}\ln \mathscr{Z}(V, T, y)\right] \tag{9.282}$$

and

$$\frac{1}{v} = \lim_{V\to\infty}\frac{\langle N\rangle}{V} = \lim_{V\to\infty}\left[y\frac{\partial}{\partial y}\frac{1}{V}\ln \mathscr{Z}(V, T, y)\right]. \tag{9.283}$$

In general, the two operations $\lim_{V\to\infty}$ and $y(\partial/\partial y)$ cannot be interchanged freely. However, Lee and Yang proved that for the type of interaction considered in Eq. (9.276) the limit in Eq. (9.282) exists and the operations $\lim_{V\to\infty}$ and $y(\partial/\partial y)$ can be interchanged. The results of Lee and Yang are contained in the following theorems:

Theorem I: For all positive real values of y, $(1/V)\ln \mathscr{Z}(V, T, y)$ approaches, as $V \to \infty$, a limit which is independent of the shape of the volume V. Furthermore, the limit is a continuous monotonically increasing function of y.

Theorem II: If in the complex y-plane a region R containing a segment of the positive real axis is always free of roots, then in this region as $V \to \infty$ all the quantities $(1/V)\ln \mathscr{Z}(V, T, \mu)$ and

$$\left(y\frac{\partial}{\partial y}\right)^n \frac{1}{V}\ln \mathscr{Z}(V, T, \mu) \qquad \text{for } n = (1, 2, \ldots, \infty)$$

approach limits which are analytic with respect to y. Furthermore, the operations $y(\partial/\partial y)$ and $\lim_{V\to\infty}$ commute in R so that

$$\lim_{V\to\infty} y\frac{\partial}{\partial y}\frac{1}{V}\ln \mathscr{Z}(V, T, \mu) = y\frac{\partial}{\partial y}\lim_{V\to\infty}\frac{1}{V}\ln \mathscr{Z}(V, T, \mu).$$

(The proofs of these theorems can be found in the papers of Lee and Yang[14] or Ref. 9.5.)

Theorems I and II, together with Eqs. (9.281) and (9.282), enable us to obtain the following relation between v and P:

$$\frac{1}{v} = y\frac{\partial}{\partial y}\left(\frac{P}{k_B T}\right). \tag{9.284}$$

They tell us that the pressure must be continuous for all y but that the derivatives of the pressure need only be continuous in regions of the positive real y-axis where roots of $\mathscr{Z}(V, T, y)$ do not touch the real axis. At points where roots touch, the derivatives of the pressure can be discontinuous. In general, if $\partial P/\partial y$ is discontinuous, then the system undergoes a first-order phase transition. If $\partial^n P/\partial y^n$ is the first derivative which is discontinuous, then the system undergoes an nth-order phase transition (in the sense of Ehrenfest).

If $1/v = \langle N\rangle/V$ is discontinuous at a point y_0 (where y_0 is a root of $\mathscr{Z} = 0$), then it will increase with y in the direction of increasing y. This can be proved as follows:

$$y\frac{\partial}{\partial y}\left(\frac{1}{v}\right) = \left(y\frac{\partial}{\partial y}\right)^2\left[\frac{1}{V}\ln \mathscr{Z}\right] = \left(y\frac{\partial}{\partial y}\right)\frac{1}{V}\left(\frac{\sum_{N=0}^{\infty} Ny^N Q_N}{\mathscr{Z}}\right)$$
$$= \frac{1}{V}\langle(N - \langle N\rangle)^2\rangle.$$

(Remember that $\langle(N - \langle N\rangle)^2\rangle$ is proportional to N.) The quantity $\langle(N - \langle N\rangle)^2\rangle$ is always greater than zero. Therefore, $1/v$ always increases with y.

In their second paper,[14] Lee and Yang applied this theory of phase transitions to the two-dimensional Ising model. They found that, for the Ising model, the roots of $\mathscr{Z}(V, T, y) = 0$ all lie on the unit circle and close onto the positive real y-axis in the limit of an infinite system. The point at which the roots touch the real axis gives the value of y for which the system undergoes a phase transition.

It is of interest to consider an explicit example.[25] We will consider an expression for $\mathscr{Z}(V, T, y)$ with roots lying on the unit circle:

$$\mathscr{Z}(V, y) = \frac{(1 + y)^V(1 - y)^V}{(1 - y)}, \tag{9.285}$$

where V is an integer. In Eq. (9.285), we have suppressed the temperature dependence. We see that there are no roots for $y = 1$, but $\mathscr{Z}(V, \mu)$ does have V roots $y = -1$ and complex roots of the form $y = e^{i2\pi k/V}$, where $k = 1, 2, \ldots, V - 1$. As the value of V increases, the density of roots on the unit circle increases and the roots get closer to the point $y = 1$ (cf. Fig. 9.20).

The function $(1/V)\ln \mathscr{Z}(V, y)$ has different limits for $y > 1$ and $y < 1$. For $y < 1$, we can write

$$\lim_{V\to\infty}\frac{1}{V}\ln \mathscr{Z}(V, y) = \lim_{V\to\infty}\frac{1}{V}\ln\left(\frac{(1 + y)^V(1 - y)^V}{(1 - y)}\right)$$
$$= \lim_{V\to\infty}\frac{1}{V}[V\ln(1 + y) + \ln(1 - y^V) - \ln(1 - y)]$$
$$= \ln(1 + y),$$

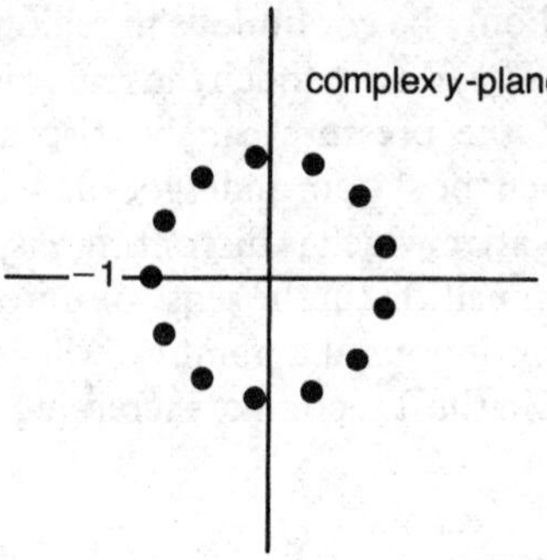

Fig. 9.20. An example where the roots of the grand partition function lie on the unit circle.

and for $y > 1$ we can write

$$\lim_{V\to\infty} \frac{1}{V} \ln \mathscr{Z}(V, y) = \lim_{V\to\infty} \frac{1}{V} \ln \left[\frac{y^V(1+y)^V\left(1 - \frac{1}{y^V}\right)}{(1-y)} \right]$$
$$= \lim_{V\to\infty} \left[\ln y + \ln(1+y) + \frac{1}{V}\ln\left(1 - \frac{1}{y^V}\right) - \frac{1}{V}\ln(1-y) \right]$$
$$= \ln y + \ln(1+y).$$

Therefore, the pressure becomes

$$\frac{P}{k_B T} = \begin{cases} \ln(1+y), & y < 1 \\ \ln(1+y) + \ln y, & y > 1 \end{cases}. \tag{9.286}$$

We see that the pressure is continuous at the point $y = 1$. Using Eq. (9.284) we obtain the following equation for the specific volume:

$$\frac{1}{v} = y \frac{\partial}{\partial y}\left(\frac{P}{k_B T}\right) = \begin{cases} \dfrac{y}{(1+y)}, & y < 1 \\ \dfrac{2y+1}{(1+y)}, & y > 1 \end{cases}. \tag{9.287}$$

Thus, the specific volume has different values at the point $y = 1$ depending on whether it is approached from above or below. We may write the equation of state for the system using Eqs. (9.286) and (9.287):

$$\frac{P}{k_B T} = \begin{cases} \ln \dfrac{v}{v-1}, & v > 2 \\ \ln 2, & 2 > v > \frac{2}{3}. \\ \ln \dfrac{v(1-v)}{(2v-1)^2}, & \frac{2}{3} > v > \frac{1}{2} \end{cases} \tag{9.288}$$

The equation of state exhibits the expected behavior at a first-order phase transition (cf. Fig. 9.21).

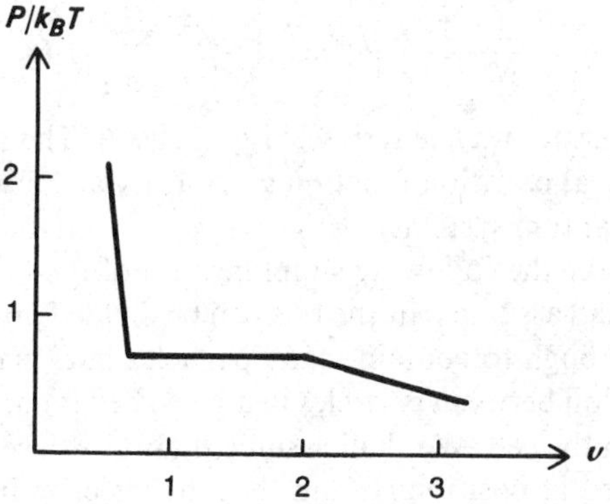

Fig. 9.21. A plot of P/k_BT versus v for a system with grand partition function $\mathscr{Z}(V, y) = (1 + y)^V(1 - y)^V/(1 - y)$.

H. VAN DER WAALS EQUATION

In the chapters on thermodynamics (Chaps. 2–4), we made extensive use of the van der Waals equation. Therefore, it is of interest and it is instructive to derive it from a microscopic theory. In this section, we shall use a model due to Ornstein[26] and stability arguments due to van Kampen[27] to derive the van der Waals equation and the Maxwell construction.

1. Derivation of the van der Waals Equation

Let us consider a box of volume V filled with N particles which obey Maxwell-Boltzmann statistics. We shall assume that the particles have a hard core of radius a and a smooth attractive interaction $-w(|\mathbf{q}_i - \mathbf{q}_j|)$ with a very long range. We shall further assume that the density is high enough and the range of the attractive interaction long enough that many particles interact simultaneously. The partition function for the system can be written

$$Z_N(T, V) = \frac{1}{N!\,h^{3N}} \iint d\mathbf{p}^N\, d\mathbf{q}^N \exp\left[-\beta\left(\sum_{i=1}^{N} \frac{p_i^2}{2m} + \sum_{i<j=1}^{\frac{1}{2}N(N-1)} V(|\mathbf{q}_i - \mathbf{q}_j|)\right)\right], \tag{9.289}$$

where $V(|\mathbf{q}_i - \mathbf{q}_j|)$ is the potential energy of the particles,

$$V(|\mathbf{q}_i - \mathbf{q}_j|) = \begin{cases} +\infty, & |\mathbf{q}_i - \mathbf{q}_j| < a \\ -w(|\mathbf{q}_i - \mathbf{q}_j|), & |\mathbf{q}_i - \mathbf{q}_j| > a \end{cases}. \tag{9.290}$$

For those regions of phase space with $|\mathbf{q}_i - \mathbf{q}_j| < a$, the partition function is identically zero. Because the phase space coordinates commute for classical particles, we can factor the momentum dependence out of the exponential and perform the momentum integrations. The result is

$$Z_N(T, V) = \frac{1}{N!}\frac{1}{\lambda_T^{3N}} \int d\mathbf{q}^N \exp\left[-\beta \sum_{i<j=1}^{\frac{1}{2}N(N-1)} V(\mathbf{q}_i - \mathbf{q}_j)\right], \tag{9.291}$$

where λ_T is the thermal wavelength (cf. Eq. (9.119)). The momentum dependence of the classical partition function gives a trivial contribution (this is no longer true for quantum systems).

We can now make the following simplification of the partition function in Eq. (9.291). We shall assume that the box can be divided into cells of volume Δ which are large enough to contain many particles but small enough that the attractive interaction between particles in a given cell is constant regardless of their separation in the cell. We shall assume that the αth cell contains N_α particles and is located at position $\mathbf{r}_\alpha$, and that the effective interaction between cells α and α' is $W_{\alpha\alpha'} = W(|\mathbf{r}_\alpha - \mathbf{r}_{\alpha'}|)$ (it will be the sum of the interactions between particles in the two cells). We shall denote the phase space volume (for position coordinates) occupied by particles in cell α by

$$\gamma(N_\alpha) = (\Delta - N_\alpha\,\delta)^{N_\alpha}, \tag{9.292}$$

where δ is the volume of the hard core. Then the partition function in Eq. (9.291) can be rewritten in the form

$$\begin{aligned} Z_N(T, V) &= \frac{1}{N!\,\lambda_T^{3N}} \sum_{\{N_\alpha\}}{}' \frac{N!}{\prod_\alpha N_\alpha!}\left[\prod_\alpha \gamma(N_\alpha)\right] \exp\left\{\tfrac{1}{2}\beta \sum_{\alpha,\alpha'} W_{\alpha\alpha'}N_\alpha N_{\alpha'}\right\} \\ &\approx \left(\frac{1}{\lambda_T^{3N}}\right) \sum_{\{N_\alpha\}}{}' e^{\Phi\{N_\alpha\}} \end{aligned} \tag{9.293}$$

where

$$\Phi\{N_\alpha\} = \sum_\alpha \left\{N_\alpha \ln\left(\frac{\Delta - N_\alpha\,\delta}{N_\alpha}\right) + N_\alpha + \tfrac{1}{2}\beta \sum_{\alpha,\alpha'} W_{\alpha\alpha'}N_\alpha N_{\alpha'}\right\}. \tag{9.294}$$

In Eq. (9.293) the product $\prod_\alpha$ is taken over all cells, and $N!/\prod_\alpha N_\alpha!$ is the number of different ways of dividing N distinguishable particles among the cells for a given set of occupation numbers $\{N_\alpha\}$. The prime on the summation in Eq. (9.293) indicates that it is to be taken subject to the condition $\sum_\alpha N_\alpha = N$. In writing Eq. (9.294), we have restricted ourselves to very large systems (the thermodynamic limit) and have neglected terms in the sum which are of relative order $1/N_\alpha$ since for a very large system those terms will give an insignificant contribution. $\Phi\{N_\alpha\}$ is a functional of all the occupation numbers.

In the limit of an infinite volume, the partition function, and therefore the free energy, is given by the largest term in the sum in Eq. (9.293) (cf. Eq. (9.244)). To find this term, we must find the set of occupation numbers which maximizes $\Phi\{N_\alpha\}$ subject to the condition $\sum_\alpha N_\alpha = N$. We can find an extremum of $\Phi\{N_\alpha\}$ subject to the condition $\sum_\alpha N_\alpha = N$ by introducing a Lagrange multiplier χ. We then determine the set $\{N_\alpha\}$ for which

$$\frac{\delta}{\delta N_\alpha}\left[\Phi\{N_{\alpha'}\} - \chi \sum_{\alpha'} N_{\alpha'}\right] = 0. \tag{9.295}$$

This leads to the set of conditions

$$\ln \frac{\Delta - N_\alpha\,\delta}{N_\alpha} - \frac{N_\alpha\,\delta}{\Delta - N_\alpha\,\delta} + \beta \sum_{\alpha'} W_{\alpha\alpha'} N_{\alpha'} = \chi \tag{9.296}$$

for each α. The Lagrange multiplier χ fixes the total number of particles. It can, in principle, be found by solving Eq. (9.296) for N_α in terms of χ. We then perform the summation $\sum_\alpha N_\alpha = N$ and solve for χ as a function of N.

A configuration $\{N_{\alpha'}\}$ which satisfies the set of equations in Eq. (9.296) will correspond to a relative maximum of $\Phi\{N_\alpha\}$ if the determinant

$$\frac{\delta^2\Phi\{N_{\alpha''}\}}{\delta N_\alpha\,\delta N_{\alpha'}} = -\left[\frac{\delta_{\alpha,\alpha'}\,\Delta^2}{N_\alpha(\Delta - N_\alpha)^2} - \beta W_{\alpha\alpha'}\right] \tag{9.297}$$

is negative definite. As shown in Ref. 9.20, a symmetric matrix of the form $a_i\,\delta_{ij} - b_{ij}$, where $b_{ij} = b_{ji} \geqslant 0$, is positive definite if $a_i > \sum_j b_{ij}$ for all i. Thus, the condition that $\Phi\{N_{\alpha''}\}$ be a relative maximum for a given configuration $\{N_{\alpha''}\}$ is

$$\frac{\delta_{\alpha\alpha'}\,\Delta^2}{N_\alpha(\Delta - N_\alpha)^2} > \beta \sum_{\alpha'} W_{\alpha\alpha'} \equiv \frac{\beta W_0}{\Delta} \tag{9.298}$$

for all α. It is useful to note that the left-hand side of Eq. (9.298) as a function of N_α has a minimum value

$$\left(\frac{\Delta^2}{N_\alpha(\Delta - N_\alpha)^2}\right)_{\min} = \frac{27}{4\Delta}. \tag{9.299}$$

Thus, for values of β such that $\beta W_0 < 27/4$, the determinant $\delta^2\Phi\{N_{\alpha''}\}/\delta N_\alpha\,\delta N_{\alpha'}$ will be negative definite for all configurations $\{N_{\alpha''}\}$ and $\Phi\{N_{\alpha''}\}$ will be a concave function of the variables $\{N_{\alpha''}\}$ and will have only one maximum (i.e., no metastable states). Thus, the critical temperature is determined by the condition $\beta_c W_0 = 27/4$.

One possible solution to the set of Eq. (9.296) is obvious. If we set

$$N_\alpha = \frac{N}{V}\Delta \qquad \text{(for all } \alpha\text{)} \tag{9.300}$$

and choose

$$\chi = \ln \frac{V - N\,\delta}{N} - \frac{N\,\delta}{V - N\,\delta} + \beta \frac{N}{V} W_0 \tag{9.301}$$

where

$$W_0 = \sum_{\alpha'} W_{\alpha\alpha'}\,\Delta = \int d\mathbf{r}\, W(\mathbf{r}), \tag{9.302}$$

then we have a solution which corresponds to a uniform density for the fluid. This is what we expect for a fluid in equilibrium unless it has undergone a phase transition and two phases are coexisting. In that case, we expect each phase to have different densities.

Eq. (9.300) yields a relative maximum for $\Phi\{N_\alpha\}$ if

$$-\frac{1}{n(1-n)^2} + \beta W_0 < 0 \tag{9.303}$$

where $n = N/V$ is the density. If the density satisfies Eq. (9.303) when $\beta W_0 < 27/4$, then the solution in Eq. (9.300) will yield the only maximum of $\Phi\{N_\alpha\}$ and the equilibrium state will have only one phase and no metastable state as we would expect above the critical temperature.

If the solution in Eq. (9.300) maximizes the functional $\Phi\{N_\alpha\}$, then it minimizes the free energy density since

$$\begin{aligned} a(n, T) &= -\frac{1}{\beta} \lim_{N,V\to\infty} \frac{1}{V} \ln Z_N(V, T) \equiv -\frac{1}{\beta}\phi(n) \\ &= -\frac{1}{\beta}\left\{n \ln\left(\frac{1-n\,\delta}{n}\right) + n + \tfrac{1}{2}\beta n^2 W_0\right\}. \end{aligned} \tag{9.304}$$

Eq. (9.304) yields the van der Waals equation,

$$P = -\left(\frac{\partial A}{\partial V}\right)_{T,N} = -a(n) - V\left(\frac{\partial a}{\partial V}\right)_{T,N} = \frac{n}{\beta(1-n\,\delta)} - \tfrac{1}{2}n^2 W_0 \tag{9.305}$$

(cf. Eq. (2.12)).

2. Maxwell Construction

To derive the Maxwell construction from stability arguments we must study the behavior of $\Phi\{n\,\Delta\}$ as a function of density. We shall write $\Phi\{n\,\Delta\}$ in the form $\Phi\{n\,\Delta\} = V\phi(n)$, where

$$\phi(n) = n \ln\left[\frac{1-n}{n}\right] + n + \tfrac{1}{2}\beta n^2 W_0 \tag{9.306}$$

and we have chosen the units of δ so that $\delta = 1$. The density now can only take on values in the range $0 < n < 1$. The values of n for which $\phi(n)$ has an extremum are found by solving the equation

$$\phi'(n) = \ln\left[\frac{(1-n)}{n}\right] - \frac{1}{(1-n)} + 1 + \beta W_0 n = \chi \tag{9.307}$$

for n as a function of χ and β; $\phi'(n)$ is the derivative of $\phi(n)$. We can obtain a good idea of the behavior of $\phi'(n)$ if we study its slope $\phi''(n)$, where

$$\phi''(n) = -\frac{1}{n(1-n)^2} + \beta W_0. \tag{9.308}$$

By comparing Eqs. (9.303) and (9.308), we see that values of n for which $\phi''(n)$ is negative correspond to states for which $\Phi\{n\,\Delta\}$ is a relative maximum. By studying Eq. (9.308), we find that, when $\beta W_0 < 27/4$, $\phi''(n)$ will be negative and $\phi'(n)$ will be decreasing and monotonic for all values of n, while for $\beta W_0 > 27/4$, $\phi'(n)$ can have a single region of positive slope. A sketch of $\phi'(n)$ for both cases is given in Fig. 9.22.

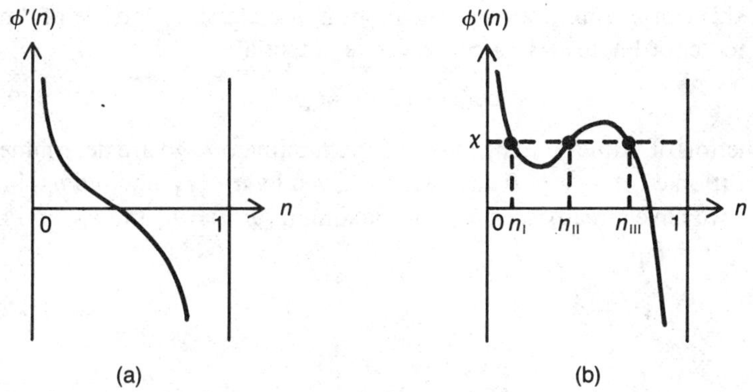

Fig. 9.22. A sketch of $\phi'(n)$: (*a*) the case $\beta W_0 < 27/4$; (*b*) the case $\beta W_0 > 27/4$.

To find the stable states for $\beta W_0 > 27/4$, we must allow for the possibility of a state in which several densities can coexist. Thus we must return to Eq. (9.294) and find its stationary values for a nonuniform density. If we let $N_\alpha = n(\mathbf{r}_\alpha)\,\Delta$, Eq. (9.294) can be written

$$\begin{aligned}\Phi\{n(\mathbf{r})\} &= \int_V n(\mathbf{r})\left\{\ln\left(\frac{1 - n(\mathbf{r})}{n(\mathbf{r})}\right) + 1\right\} d\mathbf{r} + \tfrac{1}{2}\beta \iint_V d\mathbf{r}\, d\mathbf{r}'\omega(\mathbf{r} - \mathbf{r}')n(\mathbf{r})n(\mathbf{r})' \\ &= \int_V \phi(n(\mathbf{r}))\, d\mathbf{r} - \frac{\beta}{4}\iint_V \omega(\mathbf{r} - \mathbf{r}')\{n(\mathbf{r}) - n(\mathbf{r}')\}^2\, d\mathbf{r}\, d\mathbf{r}', \end{aligned} \qquad (9.309)$$

where $\Phi\{n(\mathbf{r})\}$ is a functional of $n(\mathbf{r})$ and $\phi(n(\mathbf{r}))$ is a function of $n(\mathbf{r})$. To find the extrema of $\Phi\{n(\mathbf{r})\}$, we take its variational derivative subject to the condition $\int d\mathbf{r} n(\mathbf{r}) = N$ and set it equal to zero. The result is

$$\frac{\delta\Phi\{n(\mathbf{r})\}}{\delta n(\mathbf{r})} = \phi'(n(\mathbf{r})) - \beta\int \omega(\mathbf{r} - \mathbf{r}')(n(\mathbf{r}) - n(\mathbf{r}'))\, d\mathbf{r}' = \chi. \qquad (9.310)$$

Following van Kampen, we shall limit ourselves to solutions which vary slowly in space (relative to the range of the potential). We then can expand $n(\mathbf{r}) - n(\mathbf{r}')$ in a Taylor series to second order in $(\mathbf{r} - \mathbf{r}')$, and Eq. (9.310) takes the form

$$\phi'(n(\mathbf{r})) + \tfrac{1}{2}\beta\nabla_{\mathbf{r}}^{2}n(\mathbf{r})\int d\mathbf{r}'\mathbf{r}'^{2}\omega(\mathbf{r}') = \chi. \qquad (9.311)$$

If we further assume that $n(\mathbf{r})$ varies only in the x-direction, we find

$$\tfrac{1}{2}\omega_2\beta \frac{d^2n}{dx^2} = -\phi'(n(x)) + \chi \tag{9.312}$$

where

$$\omega_2 = \tfrac{1}{3}\int d\mathbf{r}'\mathbf{r}'^2\omega(\mathbf{r}'). \tag{9.313}$$

Eq. (9.312) can be interpreted as the motion of a classical particle of "mass" $\frac{1}{2}\beta\omega_2$ and "coordinate" n in a conservative potential

$$\psi(n) = \phi(n) - \chi n \tag{9.314}$$

as a function of "time" x. The maxima and minima of $\psi(n)$ are determined by the condition $\phi'(n) - \chi = 0$ and thus are given by $n = n_{\mathrm{I}}$, n_{II}, and n_{III} in Fig. 9.23. As we have already seen, $\psi(n)$ is maximum for $n = n_{\mathrm{I}}$ and n_{III} and minimum for $n = n_{\mathrm{II}}$.

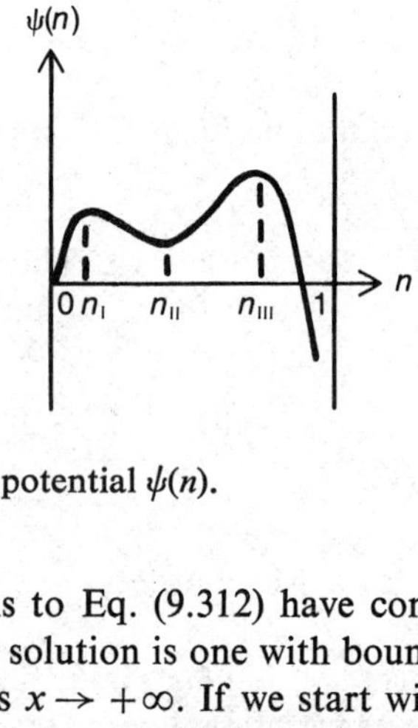

Fig. 9.23. A sketch of the potential $\psi(n)$.

Two possible solutions to Eq. (9.312) have constant density with values $n = n_{\mathrm{I}}$ and n_{III}. Another solution is one with boundary conditions $n \to n_{\mathrm{I}}$ as $x \to -\infty$ and $n \to n_{\mathrm{III}}$ as $x \to +\infty$. If we start with the system in the state $n(-\infty) = n_{\mathrm{I}}$ and we want it to move freely to $n(+\infty) = n_{\mathrm{III}}$, the two maxima in Fig. 9.22 must have the same height (a property of conservative potentials). Thus,

$$\psi(n_{\mathrm{I}}) = \psi(n_{\mathrm{III}}). \tag{9.315}$$

For a given temperature, there will be a unique value of χ for which Eq. (9.315) holds. If we write Eq. (9.315) in the form

$$\int_{n_{\mathrm{I}}}^{n_{\mathrm{III}}} dn\phi'(n) = \chi_0(n_{\mathrm{III}} - n_{\mathrm{I}}), \tag{9.316}$$

we obtain the Maxwell construction (cf. Fig. 9.24) since the line $\chi = \chi_0$ must be

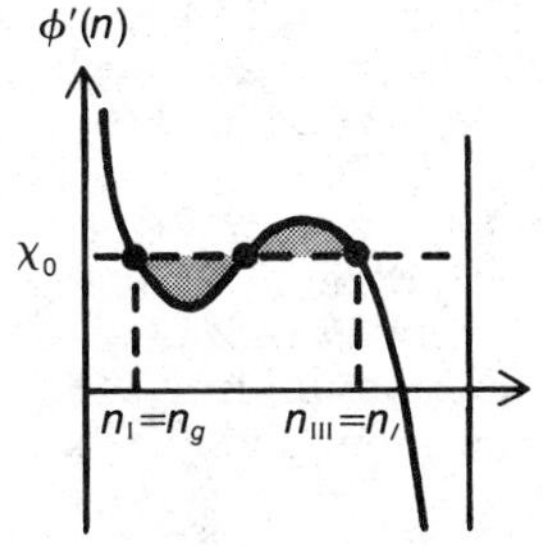

Fig. 9.24. Maxwell construction.

drawn with the shaded areas equal. Thus, for a given temperature, χ_0 is uniquely determined by the equal-area construction. A proof still must be given that this state gives the absolute maximum of $\Phi\{N_\alpha\}$ and therefore is a thermodynamic state. Such a proof has been given in Ref. 9.27.

REFERENCES

1. J. W. Gibbs, *Elementary Principles in Statistical Mechanics* (Dover Publications, New York, 1960).
2. P. and T. Ehrenfest, *The Conceptual Foundations of the Statistical Approach to Mechanics* (Cornell University Press, Ithaca, 1959).
3. A. I. Khinchin, *Mathematical Foundations of Statistical Mechanics* (Dover Publications, New York, 1949).
4. A. Einstein, Ann. Physik *22* 18, 800 (1906); *34* 170 (1911).
5. A. J. Dekker, *Solid State Physics* (Prentice-Hall, Englewood Cliffs, N.J., 1957).
6. P. Debye, Ann. Physik *39* 789 (1912).
7. M. Blackman, Repts. Prog. Phys. *8* 11 (1941).
8. P. Morse, *Thermal Physics* (W. A. Benjamin, New York, 1965).
9. K. Huang, *Statistical Mechanics* (John Wiley & Sons, New York, 1963).
10. Equations for inverting (reverting) series can be found in *Handbook of Chemistry and Physics* (Chemical Rubber Publishing Co., Cleveland, 1962).
11. J. Bardeen, J. R. Schrieffer, and L. N. Cooper, Phys. Rev. *108* 1175 (1957).
12. N. N. Bogoliubov, J.E.T.P. *7* 41, 51 (1958).
13. R. Balian and N. R. Werthamer, Phys. Rev. *131* 1553 (1963).
14. M. Tinkham, *Introduction to Superconductivity* (McGraw-Hill Book Co., New York, 1975).
15. W. K. Wootters, Physica *90A* 317 (1978).
16. E. Ising, Z. Phys. *31* 253 (1925).
17. L. Onsager, Phys. Rev. *65* 117 (1944).
18. B. Kaufman, Phys. Rev. *76* 1232 (1949).
19. B. Kaufman and L. Onsager, Phys. Rev. *76* 1244 (1949).

20. T. D. Lee and C. N. Yang, Phys. Rev. *87* 410 (1952).
21. W. L. Bragg and E. J. Williams, Proc. Roy. Soc. *A145* 699 (1934).
22. H. Bethe, Proc. Roy. Soc. *150* 552 (1935).
23. C. N. Yang, Phys. Rev. *85* 809 (1952).
24. C. N. Yang and T. D. Lee, Phys. Rev. *87* 404, 410 (1952).
25. G. E. Uhlenbeck and G. W. Ford, *Lectures in Statistical Mechanics* (American Mathematical Society, Providence, 1963).
26. L. S. Ornstein, thesis, University of Leiden, 1908.
27. N. G. van Kampen, Phys. Rev. *135* A362 (1964).
28. P. H. Keesom and N. Pearlman, Phys. Rev. *91* 1354 (1953).

PROBLEMS

1. Use the microcanonical ensemble to find the entropy and equation of state of an ideal gas (cf. Prob. 8.4).

2. Use the microcanonical ensemble to find the entropy of an Einstein solid. Find an expression for the internal energy which minimizes the Helmholtz free energy.

3. Consider a lattice with N spin $\frac{1}{2}$ particles with magnetic moment μ. The particles may have their spin components oriented parallel or antiparallel to some direction. Find the total entropy of the system and the configuration which maximizes the entropy.

4. Show that the factor $N!$ due to correct Boltzmann counting resolves the Gibbs paradox.

5. The energy eigenstates for an N-particle system are m-fold degenerate. The orthogonal energy eigenfunctions are denoted $|E, \beta\rangle$ where $\beta = 1, \ldots, m$ and $\hat{H}|E, \beta\rangle = E|E, \beta\rangle$.

   (*a*) Write the density operator $\hat{\rho}$ for the system in the microcanonical ensemble.
   (*b*) Use the definition $S = -k_B \operatorname{Tr} \hat{\rho} \ln \hat{\rho}$ to find an expression for the entropy of the system.

6. Compute the entropy of a Bose-Einstein and a Fermi-Dirac gas, starting from the microcanonical ensemble.

7. Consider a system of noninteracting spin $\frac{1}{2}$ fermions with a magnetic moment, μ, in an external magnetic field. Find a thermodynamic expression for the variance of magnetization fluctuations.

8. A lattice contains N noninteracting atoms at rest, each with spin $\frac{1}{2}$ and magnetic moment μ. If an external magnetic field $\mathscr{H}$ is applied, the atoms each can acquire values of the potential energy $E = -(\alpha/2)s\mathscr{H}$ where $s = \pm 1$. What is the magnetization of the lattice if the lattice has temperature T?

9. A system consists of two particles, each of which has two possible quantum states with energies E_0 and $2E_0$. Write the complete expression for the partition function if:

   (*a*) The particles are distinguishable.
   (*b*) The particles obey Maxwell-Boltzmann statistics.
   (*c*) The particles obey Fermi-Dirac statistics.
   (*d*) The particles obey Bose-Einstein statistics.

10. A box of volume V contains N rigid diatomic molecules. The molecules are composed of atoms of mass M_1 and M_2 held a distance d apart. The molecules can

rotate and therefore absorb energy. On the molecular level the rotation must be treated quantum mechanically. The kinetic energy of rotation is $L^2/2I$, where I is the moment of inertia, and L is the angular momentum. The angular momentum can take on values $L = \hbar\sqrt{l(l + 1)}$, where $l = 0, 1, \ldots, \infty$. For each value of l, the energy is $(2l + 1)$ fold degenerate since L_z can take on values $L_z = -l, \ldots, l - 1, l$, in integer steps. Assume that the energy of rotation and energy of translation are independent. Write the partition function of the system and express it in terms of rotational temperature $\theta_{\text{rot}} = \hbar^2/2Ik_B$. For the hydrogen molecule, assume that $d = 0.75$ Å and calculate θ_{rot}. Show that, below this temperature, the rotational contribution to the thermodynamic properties can be neglected. At very high temperatures, where the angular momentum can be assumed continuous, find the internal energy and the heat capacity of the system.

11. When a closed cubic box of volume V is heated to temperature T, the walls will emit and absorb electromagnetic radiation. When the radiation is in equilibrium with the walls, an equal amount of radiation is emitted and absorbed by the walls, and the radiation forms standing waves in the box. A standing wave of frequency ω and wave vector $\mathbf{k}$ is composed of photons of energy $\hbar\omega$ and momentum $\mathbf{p} = \hbar\mathbf{k}$. Photons are massless bosons. Since photons are continuously emitted and absorbed by the walls, the number of photons in the box continuously changes. Since photons are massless, the chemical potential is zero.

(*a*) Find the average number of photons with momentum p.
(*b*) Find the heat capacity of the photon gas.

12. A box of volume V contains a number of noninteracting particles at temperature T. Each particle has energy states $E_m = mE_0$ where $m = 0, 1, 2, \ldots, \infty$. Find the expression for the entropy per particle if (*a*) the particles are distinguishable or (*b*) the particles are fermions.

13. A system has three distinguishable molecules at rest, each with a quantized magnetic moment which can have z-components $+\frac{1}{2}\mu$ or $-\frac{1}{2}\mu$. Find an expression for the distribution function, f_i (i is the ith configuration), which maximizes the entropy subject to the condition $\sum_i f_i = 1$ and $\sum_i M_{iz} f_i = \gamma\mu$. Compute the entropy and f_i for $\gamma = \frac{1}{2}$.

14. Derive the expression for the Debye heat capacity starting from the internal energy.

15. A one-dimensional Debye solid with N lattice sites coupled harmonically to their nearest neighbors has a Hamiltonian of the form

$$H = \frac{1}{2}\sum_{l=0}^{2N}\frac{p_l^2}{2m} + \tfrac{1}{2}k\sum_{l=0}^{2N}(q_l - q_{l+1})^2.$$

Find the phonon Hamiltonian and thus the frequencies of the phonons, and find the expression for the heat capacity of the lattice.

16. Find the variance for energy fluctuations using the grand canonical ensemble.

17. Compute the magnetization of a noninteracting Fermi gas in the presence of a magnetic field.

18. Show that the entropy for Bose-Einstein and Fermi-Dirac gas (neglecting spin) can be written in the form

$$\langle S\rangle = -k_B\sum_{l=0}^{\infty}(\langle n_{\mathbf{p}_l}\rangle \ln\langle n_{\mathbf{p}_l}\rangle \mp \langle 1 \pm \langle n_{\mathbf{p}_l}\rangle\rangle \ln(1 \pm \langle n_{\mathbf{p}_l}\rangle)),$$

where the top sign holds for Bose-Einstein particles and the bottom sign holds for Fermi-Dirac particles.

19. Derive the entropy and heat capacity of an ideal Bose-Einstein gas (Eqs. (9.144)–(9.147)) starting from Eqs. (9.141) and (9.142) for the pressure.

20. Derive the equations for the internal energy and heat capacity of an ideal Fermi gas which were given in Eqs. (9.170) and (9.171).

21. Show that an ideal Bose-Einstein gas below the transition temperature exhibits off-diagonal, long-range order, while an ideal Fermi-Dirac gas does not.

22. Find the Clausius-Clapeyron equation for an ideal Bose-Einstein gas and sketch the coexistence curve. Show that the line of transition points in the P-v plane obeys the equation

$$Pv^{5/2} = \frac{2\pi\hbar^2}{m} \frac{g_{5/2}(1)}{(g_{3/2}(1))^{5/2}}.$$

23. The virial expansion of the equation of state is given by

$$\frac{PV}{\langle N\rangle k_B T} = \sum_{l=0}^{\infty} a_l(T)\left(\frac{N}{V}\right)^{l-1}.$$

Find the virial coefficient $a_2(T)$ for an ideal Fermi gas. What is $a_2(T)$ for an ideal classical gas?

24. Show that the dispersion relation for bogolons is given by

$$E_k = \left(\left(\frac{\hbar^2 k^2}{2m} - \mu\right)^2 + |\Delta_{\mathbf{k}}|^2\right)^{1/2}$$

and find explicit expressions for the functions $u_{\mathbf{k}}$ and $v_{\mathbf{k}}$ which appear in the unitary transformation matrix in Eq. (9.186).

25. Prove that the bogolon creation and annihilation operators anticommute.

26. Sketch the particle distribution function N_k and the bogolon distribution function n_k as a function of k at zero temperature for the potential considered in Eq. (9.173). (You will probably need the results of Prob. 9.24.)

27. Prove that the Hamiltonian H' for a superconductor (cf. Eq. (9.179)) does not conserve particle number.

28. Write an expression for the free energy of a superconductor and, using the gap as the order parameter, find the coefficients of the first three terms in the Ginzburg-Landau expansion (to third order in the gap).

29. Show that the order parameter, L_0, for the Bragg-Williams approximation to the Ising model has the following temperature dependence:

$$L_0 \approx \begin{cases} 1 - 2e^{-2T_c/T} & T \sim 0\,\text{K} \\ \sqrt{3\left(1 - \dfrac{T}{T_c}\right)} & T \sim T_c \end{cases}.$$

30. Calculate various thermodynamic properties of a lattice gas in the Bragg-Williams approximation. How does the phase transition in the Bragg-Williams approximation differ from a phase transition in a real gas?

10
Equilibrium Fluctuations and Critical Phenomena

A. INTRODUCTORY REMARKS

In Chap. 9, we obtained expressions for the probability distribution for the energy of a closed system and the probability distribution for the energy and particle number of an open system. We found that the size of fluctuations in such systems is partly governed by the response functions. For example, in a closed system the standard deviation of energy fluctuations is proportional to Nc_V, where N is the number of degrees of freedom and c_V is the specific heat. The specific heat is finite and well behaved except near a critical point where it becomes abnormally large and, in some cases, infinite. Thus, near a critical point we expect to encounter abnormally large fluctuations in some thermodynamic quantities. These abnormally large fluctuations are a reflection of the internal adjustments the system is making as it prepares to change to a new phase.

There are many thermodynamic variables which can describe the state of a system and most of them fluctuate about their equilibrium values. In order to get a better idea of the size of the fluctuations, we need a general method of computing variances and higher moments of fluctuations in thermodynamic systems. As Einstein showed, the probability distribution for fluctuations in thermodynamic systems can be expressed in terms of the entropy associated to those fluctuations.

The Einstein method is based on a Taylor series expansion of the entropy about absolute equilibrium, and therefore it breaks down near the critical point where fluctuations can become very large. However, it still gives us valuable insight concerning the behavior of fluctuations near a critical point. In this chapter, we will discuss Einstein fluctuation theory and apply it to fluid systems.

An alternative method for obtaining information about fluctuations in equilibrium systems is linear response theory. Experimentally, we can obtain useful information about equilibrium fluctuations by probing such systems with a weak external field which couples to the thermodynamic quantities of interest. For example, a magnetic field couples to the magnetization and allows us to probe magnetization fluctuations in a system. The way in which the system responds to an external field is determined by the type of fluctuations which occur in it. Indeed, as we shall see, the response function can be expressed directly in terms of the correlation functions for equilibrium fluctuations in a system. In this chapter, we will discuss time-independent linear response theory and obtain a relation between the long-wavelength part of the equilibrium correlation functions and the static response functions. We then will find phenomenological expressions for the spatial variation of correlations in fluid and spin systems. We will find that near critical points fluctuations become correlated over a wide region, indicating that long-range order is setting in.

If we are to describe the thermodynamic behavior of systems as they approach a critical point, we must have a systematic way of treating thermodynamic functions in the neighborhood of the critical point. Such a method exists and is called scaling. We can write the singular part (the part affected by the phase transition) of thermodynamic functions near a critical point in terms of distance from the critical point. Widom was first to point out that, as the distance from the critical point is varied, thermodynamic functions change their scale but not their functional form. The idea of scaling can be expressed mathematically by saying that the thermodynamic functions are *homogeneous functions* of their distance from the critical point. As we shall see, the idea of scaling underlies all theories of critical phenomena and enables us to obtain new equalities between various critical exponents. The scaling behavior of thermodynamic functions near a critical point has been verified experimentally.

Kadanoff was able to apply the idea of scaling in a very clever way to the Ising model and in so doing opened the way for the modern theory of critical phenomena introduced by Wilson. The idea behind Kadanoff scaling is the following. As the correlation length increases, we can rescale (increase) the size of the interacting units on the lattice. That is, instead of describing the lattice in terms of interacting spins, we describe it in terms of interacting *blocks* of spin. We take the average spin of each block and consider it as the basic unit. As the system approaches the critical point, the correlation length gets larger and the block size we use gets larger in such a way that the thermodynamic functions do not change their form, but they do rescale. With this picture, Kadanoff was able to find a relation between the critical exponents associated to the correlation length of fluctuations near the critical point and the critical exponents which were introduced in Sec. 4.L.

Wilson carried Kadanoff's idea a step further and introduced a means of computing critical exponents microscopically. Wilson's approach is based on a systematic rescaling of the effective Hamiltonian which describes a system near the critical point. As the correlation length of a system increases near a critical point, one repeatedly integrates out the effect of shorter ranged correlations and requires that the Hamiltonian retain the same functional form. This leads to nonlinear recursion relations between the effective coupling constants on different length scales. The critical point corresponds to a fixed point of these recursion relations. The eigenvalues of the transformation matrix (linearized about the fixed points) which yields the recursion relations can be expressed in terms of the critical exponents. Therefore, if we can find the eigenvalues, the problem is solved. In this chapter, we will show how Wilson's theory can be applied to some simple models, such as the triangular lattice, the Gaussian model, and the S^4-model. We will leave some examples for homework problems.

B. EINSTEIN FLUCTUATION THEORY[1,2]

1. General Discussion

There is a very simple method, due to Einstein,[2] for obtaining the probability distribution for fluctuations about the equilibrium state. Let us consider a closed adiabatically isolated system with energy in the range $E \to E + \Delta E$. We shall assume that the system is ergodic so that all possible microscopic states of the system are equally probable. Let $\Gamma(E)$ denote the number of microscopic states with energy $E \to E + \Delta E$. Then the entropy of the system is given by

$$S = k_B \ln \Gamma(E). \tag{10.1}$$

Often it is useful to have a more detailed macroscopic description of the system than that given by Eq. (10.1). For example, we might want to subdivide a system into cells and to specify its state according to the thermodynamic behavior of each cell (each cell would be an open system). We can do this if we assume that, in addition to the energy, the macroscopic state of the system is describable in terms of n independent macroscopically measurable parameters (state variables) A_i $(i = 1, \ldots, n)$, such as energy densities, mass densities, etc.

Let $\Gamma(E, A_1, \ldots, A_n)$ denote the number of microstates with energy E and parameters $A_1, \ldots, A_n$. Then the probability that the system is in a macroscopic state described by parameters $E, A_1, \ldots, A_n$ is given by

$$f(E, A_1, \ldots, A_n) = \frac{\Gamma(E, A_1, \ldots, A_n)}{\Gamma(E)}. \tag{10.2}$$

The entropy of the system in a state with parameters $(E, A_1, \ldots, A_n)$ is given by

$$S(E, A_1, \ldots, A_n) = k_B \ln \Gamma(E, A_1, \ldots, A_n). \tag{10.3}$$

Hence,

$$f(E, A_1, \ldots, A_n) = \frac{1}{\Gamma(E)} \exp\left[\frac{1}{k_B} S(E, A_1, \ldots, A_n)\right]. \tag{10.4}$$

The entropy will be a maximum when the system is in an equilibrium state $A_1^0, \ldots, A_n^0$. Any fluctuation about the equilibrium state must cause the entropy to decrease. If we let α_i denote the fluctuations

$$\alpha_i = A_i - A_i^0, \tag{10.5}$$

then we can expand the entropy about its equilibrium value to obtain

$$S(E, A_1, \ldots, A_n) = S(E, A_1{}^0, \ldots, A_n{}^0) - \frac{1}{2}\sum_{i=1}^{n}\sum_{j=1}^{n} g_{ij}\alpha_i\alpha_j + \cdots \tag{10.6}$$

where

$$g_{ij} \equiv -\left[\frac{\partial^2 S}{\partial A_i\, \partial A_j}\right]_{A_i = A_i{}^0;\, A_j = A_j{}^0}. \tag{10.7}$$

The matrix g_{ij} is positive definite since the entropy must decrease and is symmetric since the quantities are state variables. (A matrix is positive definite if every principal minor is positive.) The first-order term in Eq. (10.6) must be identically zero since the fluctuations must cause the entropy to decrease.

We can now substitute Eq. (10.6) into Eq. (10.4) and obtain the following expression for the probability distribution of fluctuations about the equilibrium state:

$$f(\boldsymbol{\alpha}) = C \exp\left\{-\frac{1}{2k_B}\,\bar{\bar{\mathbf{g}}}:\boldsymbol{\alpha\alpha}\right\}. \tag{10.8}$$

The vector $\boldsymbol{\alpha}$ is defined $\boldsymbol{\alpha} = (\alpha_1, \ldots, \alpha_n)$, the (ij)th element of the $n \times n$ matrix $\bar{\bar{\mathbf{g}}}$ is g_{ij}, and $\bar{\bar{\mathbf{g}}}:\boldsymbol{\alpha\alpha} = \sum_{ij} g_{ij}\alpha_i\alpha_j$. We define a normalization constant through the relation

$$C\int_{-\infty}^{\infty} D\boldsymbol{\alpha} \exp\left\{-\frac{1}{2k_B}\,\bar{\bar{\mathbf{g}}}:\boldsymbol{\alpha\alpha}\right\} = C\left(\frac{(2\pi k_B)^n}{|\bar{\bar{\mathbf{g}}}|}\right)^{1/2} = 1 \tag{10.9}$$

where $D\boldsymbol{\alpha} = d\alpha_1 \times \cdots \times d\alpha_n$. Thus,

$$C = \left(\frac{|\bar{\bar{\mathbf{g}}}|}{(2\pi k_B)^n}\right)^{1/2} \tag{10.10}$$

where $|\bar{\bar{\mathbf{g}}}|$ is the determinant of the matrix $\bar{\bar{\mathbf{g}}}$.

We often want to find expectation values of various moments of the fluctuations. To do this, it is convenient to introduce a more general integral,

$$C\int_{-\infty}^{\infty} D\boldsymbol{\alpha} \exp\left\{-\frac{1}{2k_B}\,(\bar{\bar{\mathbf{g}}}:\boldsymbol{\alpha\alpha}) + \mathbf{h}\cdot\boldsymbol{\alpha}\right\} = \exp\{\tfrac{1}{2}k_B\bar{\bar{\mathbf{g}}}^{-1}:\mathbf{hh}\}. \tag{10.11}$$

Then the moment $\langle\alpha_i\alpha_j\rangle$ is defined

$$\langle\alpha_i\alpha_j\rangle = \lim_{\mathbf{h}\to 0}\frac{\partial}{\partial h_i}\frac{\partial}{\partial h_j}\left[C\int_{-\infty}^{\infty} D\boldsymbol{\alpha} \exp\left\{\frac{1}{2k_B}\,(\bar{\bar{\mathbf{g}}}:\boldsymbol{\alpha\alpha}) + \mathbf{h}\cdot\boldsymbol{\alpha}\right\}\right] = k_B(\bar{\bar{\mathbf{g}}}^{-1})_{ij}. \tag{10.12}$$

The average of an odd number of variables, such as $\langle\alpha_i\alpha_j\alpha_k\rangle$, will be zero. Averages of an even number of variables will be nonzero, and because we are averaging with respect to a Gaussian distribution, they will have an especially simple form. As we have seen in the chapter on probability theory, Gaussian distributions are completely specified by their first and second moments. We thus come to the simplest possible form of the famous Wick's theorem, which states: *For a Gaussian distribution, the average of a product of* $2n$ *stochastic variables is equal to the sum of all possible combinations of pairwise averages of the stochastic variables.* For example,

$$\begin{aligned}\langle\alpha_1\alpha_2\alpha_3\alpha_4\rangle &= C\int_{-\infty}^{\infty} D\boldsymbol{\alpha}(\alpha_1\alpha_2\alpha_3\alpha_4)\exp\left\{-\frac{1}{2k_B}(\overline{\mathbf{g}}:\boldsymbol{\alpha}\boldsymbol{\alpha})\right\}\\ &= \lim_{\mathbf{h}\to 0}\frac{\partial}{\partial h_1}\frac{\partial}{\partial h_2}\frac{\partial}{\partial h_3}\frac{\partial}{\partial h_4}\left[C\int_{-\infty}^{\infty} D\boldsymbol{\alpha}\exp\left\{\frac{1}{2k_B}(\overline{\mathbf{g}}:\boldsymbol{\alpha}\boldsymbol{\alpha})+\mathbf{h}\cdot\boldsymbol{\alpha}\right\}\right]\\ &= \langle\alpha_1\alpha_2\rangle\langle\alpha_3\alpha_4\rangle+\langle\alpha_1\alpha_3\rangle\langle\alpha_2\alpha_4\rangle+\langle\alpha_1\alpha_4\rangle\langle\alpha_2\alpha_3\rangle. \qquad (10.13)\end{aligned}$$

As we shall see in Chap. 12, Wick's theorem is of special importance in perturbation expansions since the unperturbed Hamiltonian usually has a generalized Gaussian form.

To develop some intuition for the formalism we have just developed, it is useful to apply it to the case of a fluid.

2. Application to Fluid Systems

Let us consider a closed isolated box with volume V_T, total entropy S_T, total energy E_T, and total number of particles N_T. We shall assume that the box is divided into m cells and that at equilibrium the volume, entropy, internal energy, and number of particles of the ith cell are V_i^0, S_i^0, U_i^0, and N_i^0, respectively (cf. Fig. 10.1). The equilibrium pressure, temperature, and chemical potential of the whole system are denoted P^0, T^0, and μ^0, respectively.

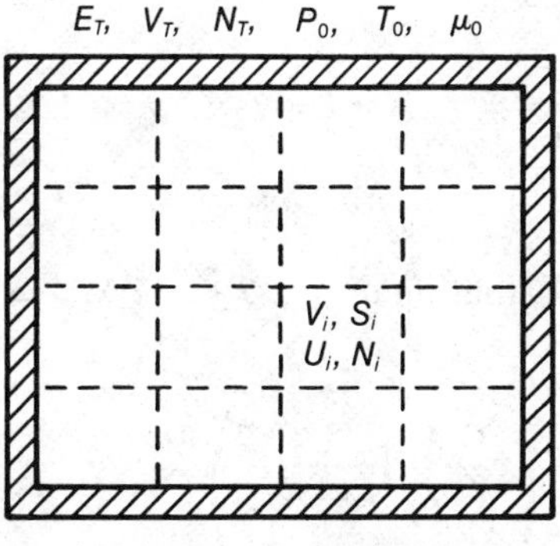

Fig. 10.1. A system with total energy E_T, total volume V_T, total particle number N_T, equilibrium pressure P_0, equilibrium temperature T_0, and equilibrium chemical potential μ_0: to describe local fluctuations about the absolute equilibrium state, we divide it into cells; cell i has volume, entropy, internal energy, and particle number given by V_i, S_i, U_i, and N_i, respectively.

From Eq. (2.153) the entropy change due to deviations in various thermodynamic quantities from their equilibrium state will be

$$\Delta S_T = \frac{1}{2T}\sum_{i=1}^{m}[-\Delta T_i\,\Delta S_i + \Delta P_i\,\Delta V_i - \Delta\mu_i\,\Delta N_i]. \tag{10.14}$$

If we now use the expressions obtained in Sec. 10.B.1, we can write an expression for the probability of a given set of fluctuations. To obtain an explicit expression we must choose a particular set of independent variables. Schematically, we can write the probability as

$$f(\Delta S_T) = \sqrt{\frac{|\bar{\bar{\mathbf{g}}}|}{(2\pi k_B)^{3m}}}\exp\left\{-\frac{1}{2k_BT}\sum_{i=1}^{m}[\Delta T_i\,\Delta S_i - \Delta P_i\,\Delta V_i + \Delta\mu_i\,\Delta N_i]\right\} \tag{10.15}$$

where the matrix $\bar{\bar{\mathbf{g}}}$ is determined once the independent variables are decided upon.

For simplicity, let us assume that the fluctuations take place at constant particle number so that $\Delta N_i = 0$ for all i, and let us choose ΔT_i and ΔV_i to be the independent variables. Then, using the relation $(\partial S/\partial V)_{T,N} = (\partial P/\partial T)_{V,N}$, we can write

$$\Delta T_i\,\Delta S_i - \Delta P_i\,\Delta V_i = \frac{C_V}{T}(\Delta T_i)^2 + \frac{1}{\kappa_T V}(\Delta V_i)^2. \tag{10.16}$$

The vector $\boldsymbol{\alpha}$ can be written $\boldsymbol{\alpha} = (\Delta T_1, \ldots, \Delta T_m; \Delta V_1, \ldots, \Delta V_m)$ and the matrix $\bar{\bar{\mathbf{g}}}$ has elements

$$g_{ij} = \begin{cases} \dfrac{C_V}{T^2} & \text{for } i = j \text{ and } 1 \leqslant i \leqslant m \\[2ex] \dfrac{1}{\kappa_T TV} & \text{for } i = j \text{ and } m+1 \leqslant i \leqslant 2m \\[2ex] 0 & \text{otherwise.} \end{cases} \tag{10.17}$$

The probability of a fluctuation $\boldsymbol{\alpha} = (\Delta T_1, \ldots, \Delta T_m, \Delta V_1, \ldots, \Delta V_m)$ is therefore

$$f(\{\Delta T_i, \Delta V_i\}) = \left(\frac{\left(\dfrac{C_V}{T^2}\right)^m\left(\dfrac{1}{\kappa_T TV}\right)^m}{(2\pi k_B)^{2m}}\right)^{1/2} \times \exp\left[-\frac{1}{2k_BT}\sum_{i=1}^{m}\left[\frac{C_V}{T}(\Delta T_i)^2 + \frac{1}{\kappa_T V}(\Delta V_i)^2\right]\right]. \tag{10.18}$$

We can now obtain expressions for variances and for correlations between fluctuations in various cells. It is easy to see that

$$\langle \Delta V_j \Delta T_j \rangle = \int \cdots \int_{-\infty}^{\infty} d(\Delta V_1) \times \cdots \times d(\Delta V_m)\, d(\Delta T_1) \times \cdots \times d(\Delta T_m) f(\{\Delta T_i, \Delta V_i\})\, \Delta V_j \Delta T_j = 0, \tag{10.19}$$

$$\langle (\Delta V_j)^2 \rangle = TV\kappa_T k_B, \tag{10.20}$$

$$\langle (\Delta T_j)^2 \rangle = \frac{T^2 k_B}{C_V}, \tag{10.21}$$

$$\langle \Delta V_j \rangle = \langle \Delta T_j \rangle = 0, \tag{10.22}$$

and

$$\langle \Delta V_j \Delta V_k \rangle = \langle \Delta T_j \Delta T_k \rangle = 0 \qquad j \neq k. \tag{10.23}$$

Thus, we find that fluctuations in temperature and volume (and therefore density) are statistically independent. This is a consequence of the relations $(\partial S/\partial V)_{T,N} = (\partial P/\partial T)_{V,N}$. One can also show that pressure and entropy fluctuations are statistically independent. However, fluctuations in most other pairs of thermodynamic variables are not independent. The variance of the volume fluctuations is proportional to the isothermal compressibility. Since the compressibility becomes very large near a phase transition, fluctuations in the density can also become very large near a phase transition. If only small fluctuations are probable, there is no difficulty in extending the limits of integration in Eq. (10.6) from $-\infty$ to ∞. However, if we get too close to a critical point, the Gaussian approximation breaks down.

It is important to note that in Eq. (10.23) we found no correlation between various cells. This result was built into the theory because Eq. (10.16) contains no information about coupling between cells. In real systems, there is coupling between cells. This can be included in the theory by expressing ΔS_i and ΔP_i in terms of temperature and volume variations in other cells and not just those of cell i. The more general expression will then contain coupling constants which reflect the strength of the coupling between the cells.

In this section we have analyzed fluid systems by dividing them into discrete cells. This, of course, is a rather artificial way to proceed, but conceptually very simple. It is a simple matter to change the summations over discrete cells to integrations over continuously varying densities, provided the spatial variations have sufficiently long wavelengths (vary slowly enough). We shall leave this as a homework problem.

C. CORRELATION FUNCTIONS AND RESPONSE FUNCTIONS[3,4]

1. General Relations

Let us consider a system in a closed box at a fixed temperature T and volume V. We will assume that the system is described by a set of state variables $\{A_i\}$ and we will apply constant external fields $\{F_j\}$ such that field F_j couples to the quantity A_j and produces a deviation from equilibrium. We will let $a_j(\mathbf{r})$ denote the density of A_j in the volume element $\mathbf{r} \to \mathbf{r} + d\mathbf{r}$. Thus,

$$A_j = \int_V d\mathbf{r} a_j(\mathbf{r}) = \langle a_j \rangle V \tag{10.24}$$

where $\langle a_j \rangle$ is the average value of $a_j(\mathbf{r})$.

The Hamiltonian of the system in the presence of the fields is given by

$$H = H_0 - \sum_j \int d\mathbf{r} a_j(\mathbf{r}) F_j = H_0 - \sum_j A_j F_j. \tag{10.25}$$

The average value of A_j in the presence of the fields is

$$\langle A_j \rangle_{\{F_i\}} = \frac{\operatorname{Tr} e^{-\beta H} A_j}{\operatorname{Tr} e^{-\beta H}} \tag{10.26}$$

and the static isothermal linear response function χ_{ji} is defined

$$\chi_{ji} = \left(\frac{\partial \langle A_j \rangle}{\partial F_i} \right)_{\{F_j = 0\}, T, N} = \beta \left[\frac{\operatorname{Tr} e^{-\beta H_0} A_j A_i}{\operatorname{Tr} e^{-\beta H_0}} - \langle A_i \rangle \langle A_j \rangle \right] \tag{10.27}$$

or

$$\chi_{ji} = \beta \langle (A_j - \langle A_j \rangle)(A_i - \langle A_i \rangle) \rangle. \tag{10.28}$$

χ_{ji} gives the change in A_j due to an applied field F_i. By setting the force $F_j = 0$ in Eq. (10.27) we have neglected any contribution to the response coming from nonlinear effects in the external field. Thus, the equations apply only for very weak external fields.

We can now write the response function in terms of the static correlation function for fluctuations in the medium. From Eqs. (10.24) and (10.28), we write

$$\chi_{ji} = \beta \int d\mathbf{r}_1 \int d\mathbf{r}_2 \langle (a_j(\mathbf{r}_1) - \langle a_j \rangle)(a_i(\mathbf{r}_2) - \langle a_i \rangle) \rangle. \tag{10.29}$$

For a translationally invariant system the response function will depend only on the distance between the fluctuations and not on their absolute position in the medium. Thus, we can write Eq. (10.29) as

$$\chi_{ji} = \beta V \int d\mathbf{r} \langle (a_j(\mathbf{r}) - \langle a_j \rangle)(a_i(0) - \langle a_i \rangle) \rangle = \beta V \int d\mathbf{r} G_{ji}(\mathbf{r}), \tag{10.30}$$

where $\mathbf{r}$ is the distance between fluctuations and

$$C_{ji}(\mathbf{r}) = \langle (a_j(\mathbf{r}) - \langle a_j \rangle)(a_i(0) - \langle a_i \rangle) \rangle \tag{10.31}$$

is the static correlation function for fluctuations a_j and a_i a distance $\mathbf{r}$ apart. If we denote the Fourier transform of $G_{ji}(\mathbf{r})$ by

$$G_{ji}(\mathbf{k}) = \int d\mathbf{r}\, e^{i\mathbf{k}\cdot\mathbf{r}} C_{ji}(\mathbf{r}), \tag{10.32}$$

we find that the static response function and the infinite wavelength component of the correlation function are related by

$$\lim_{V \to \infty} \frac{\chi_{ji}}{V} = \beta G_{ji}(\mathbf{k} = 0). \tag{10.33}$$

Thus, we see that the way in which a system responds to a constant external field is completely determined by the long-wavelength equilibrium fluctuations. In Chap. 4, we found that the static susceptibility becomes infinite as we approach the critical point. Thus, the correlation function will have a large infinite wavelength component indicating that long-range order has set in.

2. Application to Fluid Systems[1]

It is of interest to obtain an explicit expression for the correlation functions $C(\mathbf{r})$ and $G(\mathbf{k})$ in the mean field approximation. In this section we will consider a fluid system and in the next section, a spin system. Let us assume that we have a fluid in a box of volume V and let us divide the fluid into cells of size Δ so that n_α is the density of particles in cell α and $N_\alpha = n_\alpha\Delta$ is the number of particles in cell α. We assume that each cell is large enough (contains enough particles) that the system is locally in equilibrium. That is, each cell is specified by a set of thermodynamic variables, but the values of the thermodynamic variables can change from cell to cell. For a very large system, we can write the partition function in the form (cf. Sec. 9.H)

$$Z_N(T, V) = \sum_{\{n_\alpha\}} e^{-V\phi\{n_\alpha\}}, \tag{10.34}$$

where $\{n_\alpha\}$ is the set of densities of the various cells and specifies one possible configuration for the system and $\phi\{n_\alpha\}$ is a functional of the densities. The summation is taken over all configurations subject to the condition

$$\Delta \sum_{\alpha} n_\alpha = N. \tag{10.35}$$

We may interpret $e^{-V\phi\{n_\alpha\}}$ as the probability for a given configuration. The equilibrium configuration will be the one which maximizes $e^{-V\phi\{n_\alpha\}}$. We expect the equilibrium configuration above the critical point to be the one of uniform density.

Let us now assume that the discrete cells can be replaced by a continuum and that density variations are slow enough that the density is roughly constant in regions of the order of the cell size. We then let

$$\delta n(\mathbf{r}) = n(\mathbf{r}) - n \tag{10.36}$$

denote the deviation of the density from its equilibrium value, n, in a cell located at $\mathbf{r}$. The probability distribution should contain information about the magnitude of fluctuations in the cells and the way in which the fluctuations are distributed from cell to cell. We expect that the most probable configurations are those in which the densities of neighboring cells do not differ much, that is configurations in which the gradients of the density fluctuations are small. Thus, we can write $\phi\{n_\alpha\}$ in the form

$$\phi\{n(\mathbf{r})\} = \beta a(n\beta) + \tfrac{1}{2}C_1(n, \beta)\int d\mathbf{r}\, \delta n^2(\mathbf{r}) + \tfrac{1}{2}C_2(n, \beta)\int d\mathbf{r}(\nabla_{\mathbf{r}}\, \delta n(\mathbf{r}))^2 + \cdots \tag{10.37}$$

where $a(n, \beta)$ is the free energy density. For a translationally invariant system, there can be no terms linear in the gradient since this would impose a preferred direction in the system. Furthermore, since fluctuations must increase the free energy, there can be no terms linear in $\delta n(\mathbf{r})$ because it does not have a definite sign. If we introduce the Fourier transform of the fluctuations,

$$\delta n(\mathbf{r}) = \frac{1}{V}\sum_{\mathbf{k}} e^{-i\mathbf{k}\cdot\mathbf{r}}\, \delta n(\mathbf{k}), \tag{10.38}$$

then Eq. (10.37) takes the form

$$\phi\{\delta n(\mathbf{k})\} = \beta a(n, \beta) + \frac{1}{2V}\sum_{\mathbf{k}} [C_1 + C_2 k^2]\, \delta n(\mathbf{k})\, \delta n(-\mathbf{k}). \tag{10.39}$$

Thus, we can write the probability for a fluctuation $\delta n(\mathbf{k})$ in the form

$$P\{\delta n(\mathbf{k})\} = C\, e^{-\frac{1}{2}[C_1 + C_2 k^2]|\delta n(\mathbf{k})|^2} \tag{10.40}$$

where C is a normalization constant. (Note that $\delta n(-\mathbf{k}) = \delta n^*(\mathbf{k})$, where * denotes complex conjugation.) From Sec. 10.B, the correlation function $\langle\delta n(\mathbf{k})\, \delta n(-\mathbf{k})\rangle$ can immediately be written

$$G(\mathbf{k}) = \langle\delta n(\mathbf{k})\, \delta n(-\mathbf{k})\rangle = \frac{1}{(C_1 + C_2 k^2)}. \tag{10.41}$$

We can Fourier transform Eq. (10.41) to obtain, for large volume V,

$$C(\mathbf{r}) = \frac{1}{V}\sum_{\mathbf{k}} G(\mathbf{k})\, e^{-i\mathbf{k}\cdot\mathbf{r}} = \int \frac{d\mathbf{k}}{(2\pi)^3}\frac{e^{-i\mathbf{k}\cdot\mathbf{r}}}{(C_1 + C_2 k^2)} = \frac{1}{4\pi C_2 r}\, e^{-r\sqrt{C_1/C_2}}. \tag{10.42}$$

The correlation function has a range $\xi = \sqrt{C_2/C_1}$. We can determine how this range changes near a critical point. We first note that volume fluctuations for fixed particle number in Eq. (10.20) are proportional to density fluctuations. Thus, $G(\mathbf{k} = 0) = C_1^{-1} \sim \kappa_T \sim (T - T_c)^{-1}$ (cf. Eq. (4.133)) and, therefore, $\xi \sim (T - T_c)^{-1/2}$. As we approach the critical point, the range of correlations approaches infinity.

3. Application to Spin Systems[5,6]

We can write completely analogous equations for spin systems. The partition function for an Ising system with N lattice sites can be written in the form

$$Z(K) = \sum_{\{S_{\mathbf{n}}\}} \exp\left\{K \sum_{\mathbf{n}} \sum_{\mathbf{e}} S_{\mathbf{n}} S_{\mathbf{n}+\mathbf{e}}\right\}, \tag{10.43}$$

where $K \equiv \beta\epsilon$ is the effective interaction and $\mathbf{e}$ is a vector indicating the positions of various nearest neighbors of site $\mathbf{n}$. The summation is over all possible configurations of the lattice. We can change the summation to an integration if we introduce a weighting factor, $W\{S_{\mathbf{m}}\} = \prod_{\mathbf{m}} \delta(S_{\mathbf{m}}^2 - 1)$. Then Eq. (10.43) can be written

$$Z(K) = \int_{-\infty}^{\infty} \left[\prod_{\mathbf{m}} dS_{\mathbf{m}}\right] W\{S_{\mathbf{m}}\} \exp\left\{K \sum_{\mathbf{n}} \sum_{\mathbf{e}} S_{\mathbf{n}} S_{\mathbf{n}+\mathbf{e}}\right\}. \tag{10.44}$$

Eq. (10.44), for the partition function of the Ising system, is now in a form which allows some interesting generalizations.

Kac and Berlin[7] introduced a variation of the Ising model (called the spherical, or Gaussian, model). They allowed the spins to take on a continuous range of values rather than the discrete values $S = \pm 1$, and they introduced a Gaussian weighting factor:

$$W\{S_{\mathbf{m}}\} = \exp\left[-\frac{b}{2} \sum_{\mathbf{m}} S_{\mathbf{m}}^2\right] \tag{10.45}$$

where b is an arbitrary constant. The Gaussian weighting factor alters the Ising model drastically because it gives the largest weight to spin values $S = 0$. These values do not exist in the Ising model. However, it provides an interesting system in its own right. With the Gaussian weighting factor, the partition function takes the form

$$Z(K) = \int_{-\infty}^{\infty} \left[\prod_{\mathbf{m}} dS_{\mathbf{m}}\right] \exp\left\{K \sum_{\mathbf{n}} \sum_{\mathbf{e}} S_{\mathbf{n}} S_{\mathbf{n}+\mathbf{e}} - \frac{b}{2} \sum_{\mathbf{n}} S_{\mathbf{n}}^2\right\} \tag{10.46}$$

and remains quadratic in the spin factors.

We now introduce a spin field as follows. We let $S_{\mathbf{n}}$ denote the spin at lattice site $\mathbf{n}a$ (we assume an infinitely large cubic lattice with spacing a between lattice sites). Then we introduce the Fourier expansion of $S(\mathbf{k})$,

$$S(\mathbf{k}) = a^d \sum_{\mathbf{n}} S_{\mathbf{n}}\, e^{-i\mathbf{k}\cdot\mathbf{n}a}. \tag{10.47}$$

The components, k_i, of the wave vectors, $\mathbf{k}$, can only take on values in the range $-\pi/a < k_i < \pi/a$ ($i = x, y, z$ for $d = 3$, where d is the dimension of the lattice in configuration space), since wavelengths smaller than the lattice spacing are not allowed. From Eq. (10.47), we can define a continuous spin field,

$$S(\mathbf{x}) = \frac{1}{(2\pi)^d} \int_{-\pi/a}^{\pi/a} \cdots \int_{-\pi/a}^{\pi/a} d\mathbf{k}\, e^{i\mathbf{k}\cdot\mathbf{x}} S(\mathbf{k}), \tag{10.48}$$

where $d\mathbf{k} = dk_1\, dk_2 \times \cdots \times dk_d$. One can easily check that $S(\mathbf{n}a) = S_\mathbf{n}$, but $S(\mathbf{x})$ need not be zero between lattice sites. We can now write the effective spin Hamiltonian as

$$H\{S\} = -K\sum_{\mathbf{n}}\sum_{\mathbf{e}} S_\mathbf{n} S_{\mathbf{n}+\mathbf{e}} + \frac{b}{2}\sum_{\mathbf{n}} S_\mathbf{n}^2$$

$$= \tfrac{1}{2}K\sum_{\mathbf{n}}\left[\sum_{\mathbf{e}}{}' (S_{\mathbf{n}+\mathbf{e}} - S_\mathbf{n})^2 + \left(\frac{b}{K} - 2d\right) S_\mathbf{n}^2\right] \tag{10.49}$$

where the sum $\sum_\mathbf{e}'$ is restricted to positive values of $\mathbf{e}$ and $\sum_\mathbf{e}' = d$ for a cubic lattice. The effective Hamiltonian can be written in terms of wave vectors as

$$H\{S\} = \frac{K}{2}\frac{1}{(2\pi a)^d}\int\cdots\int_{-\pi/a}^{\pi/a} d\mathbf{k} \sum_{\mathbf{e}}{}' |e^{i\mathbf{k}\cdot\mathbf{e}a} - 1|^2 S(\mathbf{k})S(-\mathbf{k})$$

$$+ \frac{K}{2}\frac{1}{(2\pi a)^d}\left(\frac{b}{K} - 2d\right)\int\cdots\int_{-\pi/a}^{\pi/a} d\mathbf{k} S(\mathbf{k})S(-\mathbf{k}). \tag{10.50}$$

If we use the fact that

$$\left(\frac{a}{2\pi}\right)^d \int_{-\pi/a}^{\pi/a} d\mathbf{k}\, e^{-i\mathbf{k}\cdot\mathbf{n}a} = \delta_{\mathbf{n},0}, \tag{10.51}$$

it is easy to show that Eq. (10.50) reduces to Eq. (10.49).

Let us now expand the exponential, keep lowest order terms, and note that $\sum_\mathbf{e}' |\mathbf{k}\cdot\mathbf{e}|^2 = k^2a^2$. We then find

$$H\{S\} = \frac{1}{2}\left(\frac{1}{2\pi}\right)^d \int\cdots\int_{-\pi/a}^{\pi/a} d\mathbf{k}(C_1 + C_2k^2)|S(\mathbf{k})|^2 \tag{10.52}$$

where

$$C_2 = Ka^{-d+2} \tag{10.53}$$

and

$$C_1 = \left(\frac{b}{K} - 2d\right)Ka^{-d}. \tag{10.54}$$

Thus, we again obtain a Gaussian distribution for fluctuations. The correlation functions for the spherical model will have the same behavior and the same critical exponents as for the fluid case.

D. SCALING[3,5]

As we approach the critical point, the distance over which fluctuations are correlated approaches infinity and all effects of the finite lattice spacing are wiped out. There are no natural length scales left. Thus we might expect that in the neighborhood of the critical point, as we change the distance from the critical point, we do not change the form of the free energy but only its *scale*.

The idea of scaling underlies all critical exponent calculations. To understand scaling, we must first introduce the concept of a homogeneous function.

1. Homogeneous Function

A function $F(\lambda x)$ is homogeneous if for all values of λ

$$F(\lambda x) = g(\lambda)F(x). \tag{10.55}$$

The general form of the function $g(\lambda)$ can be found easily. We first note that

$$F(\lambda\mu x) = g(\lambda)g(\mu)F(x) = g(\lambda\mu)F(x) \tag{10.56}$$

so that

$$g(\lambda\mu) = g(\lambda)g(\mu). \tag{10.57}$$

If we take the derivative with respect to μ, we find

$$\frac{\partial}{\partial\mu} g(\lambda\mu) = \lambda g'(\lambda\mu) = g(\lambda)g'(\mu) \tag{10.58}$$

where $g'(\mu) = (\partial/\partial\mu)g(\mu)$. We next set $\mu = 1$ and $g'(1) = p$. Then

$$\lambda g'(\lambda) = pg(\lambda). \tag{10.59}$$

If we integrate from 1 to λ and note that $g(1) = 1$, we find

$$g(\lambda) = \lambda^p. \tag{10.60}$$

Thus,

$$F(\lambda x) = \lambda^p F(x) \tag{10.61}$$

and $F(x)$ is said to be a homogeneous function of degree p. A homogeneous function has a very special form. In Eq. (10.61), if we let $\lambda = x^{-1}$, we obtain

$$F(x) = F(1)x^p. \tag{10.62}$$

Thus, the homogeneous function $F(x)$ has power-law dependence on its arguments.

Let us now consider a generalized homogeneous function of two variables $f(x, y)$. Such a function can always be written in the form

$$f(\lambda^p x, \lambda^q y) = \lambda f(x, y) \tag{10.63}$$

and is characterized by two parameters, p and q. It is convenient to write $f(x, y)$ in another form. We will let $\lambda = y^{-1/q}$. Then

$$f(x, y) = y^{1/q} f\left(\frac{x}{y^{p/q}}, 1\right), \tag{10.64}$$

and we see that the generalized homogeneous function, $f(x, y)$, depends on x and y only through the ratio $x/y^{p/q}$ aside from a multiplicative factor. We can now apply these ideas to themodynamic quantities near the critical point.

2. Widom Scaling[8]

If we assume that the singular part of the thermodynamic potential scales, then we can find a relation between various critical exponents. We will consider magnetic systems since they give a simple picture, and we shall assume that a magnetic induction field, **B**, is present. Let us write the free energy per lattice site in terms of a regular part, $g_r(T, \mathbf{B})$, which does not change as we approach the critical point, and a singular part, $g_s(\epsilon, \mathbf{B})$, which contains the important singular behavior of the system in the neighborhood of the critical point. Then

$$g(T, \mathbf{B}) = g_r(T, \mathbf{B}) + g_s(\epsilon, \mathbf{B}) \tag{10.65}$$

where $\epsilon = (T - T_c)/T_c$. We shall assume that the singular part is a generalized homogeneous function of its parameters,

$$g_s(\lambda^p \epsilon, \lambda^q B) = \lambda g_s(\epsilon, B). \tag{10.66}$$

The critical exponents in Sec. 4.L can all be determined in terms of p and q. Let us first find an expression for β, which is defined (cf. Eq. (4.158))

$$M(\epsilon, B = 0) \approx (-\epsilon)^\beta. \tag{10.67}$$

If we differentiate Eq. (10.66) with respect to B, we obtain

$$\lambda^q M(\lambda^p \epsilon, \lambda^q B) = \lambda M(\epsilon, B). \tag{10.68}$$

If we next let $\lambda = (-\epsilon)^{-1/p}$ and set $B = 0$, we obtain

$$M(\epsilon, 0) = (-\epsilon)^{(1-q)/p} M(-1, 0). \tag{10.69}$$

Thus,

$$\beta = \frac{1-q}{p} \tag{10.70}$$

and we obtain our first relation.

Let us next determine the exponent δ (the degree of the critical isotherm), which is defined (cf. Eq. (4.157))

$$M(0, B) = |B|^{1/\delta} \operatorname{sign} B. \tag{10.71}$$

If we set $\epsilon = 0$ and $\lambda = B^{-1/q}$ in Eq. (10.66), we can differentiate with respect to B and obtain

$$M(0, B) = B^{(1-q)/q} M(0, 1). \tag{10.72}$$

Thus,

$$\delta = \frac{q}{1-q}. \tag{10.73}$$

The magnetic susceptibility is obtained from the thermodynamic relation

$$\chi = \left(\frac{\partial^2 g}{\partial B^2}\right)_{T,V} \sim \begin{cases} (-\epsilon)^{-\gamma'} & T < T_c \\ (\epsilon)^{-\gamma} & T > T_c \end{cases}. \tag{10.74}$$

By differentiating Eq. (10.68), we can write

$$\lambda^{2q}\chi(\lambda^p\epsilon, \lambda^q B) = \lambda\chi(\epsilon, B). \tag{10.75}$$

If we now set $B = 0$, and let $\lambda = (\epsilon)^{-1/p}$, we find

$$\chi(\epsilon, 0) = \epsilon^{(1-2q)/p}\chi(1, 0). \tag{10.76}$$

Thus, the critical exponent for the susceptibility is

$$\gamma = \frac{2q - 1}{p}. \tag{10.77}$$

Similarly, $\gamma' = \gamma$.

The heat capacity at constant B is given by

$$C_B = -\left(\frac{\partial^2 g}{\partial T^2}\right)_B \sim (\epsilon)^{-\alpha} \tag{10.78}$$

(cf. Eq. (4.159)). From Eq. (10.66), we obtain

$$\lambda^{2p}C_B(\lambda^p\epsilon, \lambda^q B) = \lambda C_B(\epsilon, B). \tag{10.79}$$

If we set $B = 0$ and $\lambda = (\epsilon)^{-1/p}$, we find

$$C_B(\epsilon, 0) = \epsilon^{(1-2p)/p}C_B(1, 0) \tag{10.80}$$

and, therefore,

$$\alpha = 2 - \frac{1}{p}. \tag{10.81}$$

Similarly, $\alpha' = \alpha$. Thus, we have obtained the four critical point exponents in terms of the two parameters p and q. If we combine Eqs. (10.70), (10.73), and (10.77), we find

$$\gamma' = \gamma = \beta(\delta - 1). \tag{10.82}$$

From Eqs. (10.70), (10.73), and (10.81), we find

$$\alpha + \beta(\delta + 1) = 2. \tag{10.83}$$

Thus, we obtain exact relations between the critical exponents. It is useful for later reference to express p and q in terms of the critical exponents. We find

$$p = \frac{1}{\beta}\frac{1}{(\delta + 1)} \tag{10.84}$$

and

$$q = \delta\frac{1}{(\delta + 1)}. \tag{10.85}$$

The scaling property for systems near the critical point has been verified experimentally for fluids[9] and magnetic systems.[10]

3. Kadanoff Scaling[11]

Kadanoff has shown how to apply the idea of scaling to the Ising model. Let us consider a d-dimensional Ising system with nearest-neighbor coupling (γ nearest neighbors). The Hamiltonian is

$$H\{S\} = -K \sum_{(ij)}^{\gamma N/2} S_i S_j - B \sum_{i=1}^{N} S_i \tag{10.86}$$

where N is the number of lattice sites. We will divide the lattice into blocks of length La, where a is the distance between sites (cf. Fig. 10.2). The total number

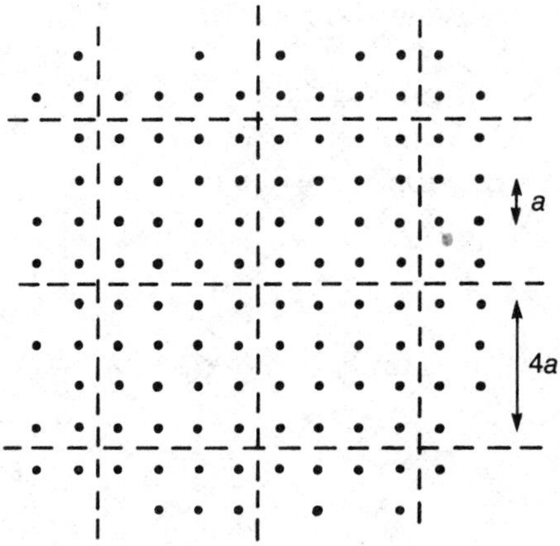

Fig. 10.2. Decomposition of a square lattice into square blocks whose sides have length $La = 4a$.

of spins in each block is L^d. The total number of blocks is NL^{-d}. The total spin in block I is

$$S'_I = \sum_{i \in I} S_i. \tag{10.87}$$

If L is chosen so that $La \ll \xi$ (ξ is the correlation length of spin fluctuations on the lattice (cf. Sec. 10.C)), then the spins in each block will be highly correlated. The block spin can have values ranging from L^d to $-L^d$. It is useful to define a new spin variable, S_I, through the relation

$$S'_I = ZS_I \tag{10.88}$$

where $S_I = \pm 1$ and $Z \approx L^d$. (This assumes that all spins in a block are aligned. It may or may not be true, but it is the assumption that Kadanoff used and the one we will follow here.)

Spins interact with nearest neighbors, so blocks should also interact with nearest neighbors. Thus, the block Hamiltonian will be of the form

$$H\{S_L\} = -K_L \sum_{(IJ)}^{\gamma NL^{-d}/2} S_I S_J - B_L \sum_{I=1}^{NL^{-d}} S_I \tag{10.89}$$

where K_L is the new effective interaction between nearest neighbor *blocks*. The block Hamiltonian looks exactly like the site Hamiltonian except that all quantities are rescaled. Therefore, we expect the free energy per block, $g(\epsilon_L, B_L)$, to have the same functional form as the free energy per site, $g(\epsilon, B)$. Since there are L^d sites per block, we have

$$g(\epsilon_L, B_L) = L^d g(\epsilon, B). \tag{10.90}$$

If we rescale our system and describe it in terms of blocks rather than sites, we reduce the effective correlation length (measured in units of La) and therefore we move farther away from the critical point. Thus, the correlation length will behave as

$$\xi_L(\epsilon_L, B_L) = L^{-1}\xi(\epsilon, B). \tag{10.91}$$

Since rescaling moves us away from the critical point, the temperature ϵ and magnetic field B must also rescale. We assume that

$$\epsilon_L = \epsilon L^x \tag{10.92}$$

where x is positive. Similarly,

$$B \sum_{i=1}^{N} S_i = B \sum_{I=1}^{NL^{-d}} \sum_{i \in I} S_i = B \sum_{I=1}^{NL^{-d}} S_I' = BZ \sum_{I=1}^{NL^{-d}} S_I, \tag{10.93}$$

so that

$$B_L = BZ = L^y B. \tag{10.94}$$

Since $-L^d < Z < L^d$, we must have $y < d$. Eq. (10.90) now becomes

$$g(L^x\epsilon, L^y B) = L^d g(\epsilon, B). \tag{10.95}$$

If we compare Eq. (10.95) with Eq. (10.66), we find $x = pd$ and $y = qd$. Thus,

$$q < 1 \tag{10.96}$$

in agreement with experiment.

The block correlation function is defined

$$C(r_L, \epsilon_L) = \langle S_I S_J \rangle - \langle S_I \rangle \langle S_J \rangle \tag{10.97}$$

where r_L is the distance between blocks I and J in units of La. We can write Eq. (10.97) as

$$\begin{aligned} C(r_L, \epsilon_L) &= Z^{-2}[\langle S_I' S_J' \rangle - \langle S_I' \rangle \langle S_J' \rangle] \\ &= Z^{-2} \sum_{i \in I} \sum_{j \in J} [\langle S_i S_j \rangle - \langle S_i \rangle \langle S_j \rangle] \\ &= Z^{-2}(L^d)^2[\langle S_i S_j \rangle - \langle S_i \rangle \langle S_j \rangle] = Z^{-2}(L^d)^2 C(r, \epsilon) \end{aligned} \tag{10.98}$$

where r is the distance between sites i and j on different blocks. The distances r_L and r are related by the expression

$$r_L = L^{-1}r, \tag{10.99}$$

and we write

$$C(L^{-1}r, \epsilon L^x) = L^{2(d-y)}C(r, \epsilon). \tag{10.100}$$

If we choose $L = r/a$, the correlation function takes the form

$$C(r, \epsilon) = \left(\frac{r}{a}\right)^{2(y-d)} C\left(a, \epsilon\left(\frac{r}{a}\right)^x\right). \tag{10.101}$$

We can now introduce two new exponents for the correlation function. We first define a critical exponent, ν, for the correlation range as

$$\xi = (T - T_c)^{-\nu}. \tag{10.102}$$

For mean field theories $\nu = \frac{1}{2}$. From Eq. (10.42) we see that the correlation function away from the critical point depends on r and ϵ in the combination $r/\xi = r\epsilon^\nu$. In Eq. (10.101) the correlation function depends on r and ϵ in the combination ϵr^x. Thus,

$$x = pd = \nu^{-1}. \tag{10.103}$$

At the critical point, $\epsilon = 0$ and the correlation function varies as

$$C(r, 0) \sim (r)^{2(y-d)}. \tag{10.104}$$

In three dimensions, we expect the correlation function at the critical point to behave as

$$C(r, 0) \sim \left(\frac{1}{r}\right)^{1+\eta} \tag{10.105}$$

where η is another new exponent. For mean field theories $\eta = 0$. In d dimensions, $C(r, 0)$ varies as

$$C(r, 0) = \left(\frac{1}{r}\right)^{(d-2+\eta)} \tag{10.106}$$

and we can make the identification

$$(d - 2 + \eta) = 2(d - y) = 2d(1 - q). \tag{10.107}$$

Thus, the exponents for the correlation function can be written in terms of the exponents for the thermodynamic quantities we have already considered. From Eqs. (10.81) and (10.85) we find

$$\nu = \frac{2 - \alpha}{d} \tag{10.108}$$

and

$$\eta = 2 - d\left(\frac{\delta - 1}{\delta + 1}\right) = 2 - \frac{d\gamma}{2\beta + \gamma}. \tag{10.109}$$

Thus, Kadanoff scaling allows us to obtain new identities between exponents.

E. MICROSCOPIC CALCULATION OF CRITICAL EXPONENTS[5,6,12–15]

The Kadanoff picture of scaling was given firm mathematical foundation by Wilson,[16] who developed a technique for computing the critical exponents microscopically. We shall outline the procedure for the case of spin systems. Let us consider a system described by a partition function:

$$Z(\mathbf{K}, N) = \sum_{\{S_i\}} \exp[-\mathscr{H}(\mathbf{K}, \{S_i\}, N)]. \tag{10.110}$$

The effective Hamiltonian $\mathscr{H}(\mathbf{K}, \{S_i\}, N)$ (which includes temperature), can be written in the form

$$\mathscr{H}(\mathbf{K}, \{S_i\}, N) = K_0 + K_1 \sum_i S_i + K_2 \sum_{(i,j)}^{(1)} S_i S_j + K_3 \sum_{(i,j)}^{(2)} S_i S_j + K_4 \sum_{(i,j,k)}^{(1)} S_i S_j S_k + \cdots \tag{10.111}$$

where $\mathbf{K}$ is an infinite dimensional vector containing all coupling constants, and the summation $\sum^{(i)}$ means that only (ith) nearest neighbors are included. The coupling constants, K_i, contain the temperature. For the Ising model, $K_1 = -\beta B$, $K_2 = -\beta J$ where J is the strength of the coupling between spins, and $K_3 = K_4 = \cdots = 0$.

We can introduce blocks and sum over spins, σ_I, interior to each block. Thus,

$$\begin{aligned} Z(\mathbf{K}, N) &= \sum_{\{S_I, \sigma_I\}} \exp[-\mathscr{H}(\mathbf{K}, \{S_I, \sigma_I\}, N)] \\ &= \sum_{\{S_I\}} \exp[-\mathscr{H}(K_L, \{S_I\}, NL^{-d})] = Z(\mathbf{K}_L, NL^{-d}). \end{aligned} \tag{10.112}$$

Since the new partition function has the same functional form as the old one, we can write the following expression for the free energy density per site:

$$g(\mathbf{K}) = \lim_{N\to\infty} \frac{1}{N} \ln Z(\mathbf{K}, N) = \lim_{N\to\infty} \frac{1}{N} \ln Z(\mathbf{K}_L, NL^{-d}) = L^{-d} g(\mathbf{K}_L). \tag{10.113}$$

The coupling constant vectors of the site spin system and block spin system will be related by a transformation,

$$\mathbf{K}_L = \mathbf{T}(\mathbf{K}), \tag{10.114}$$

where $\mathbf{K}_L$ will be a vector whose elements are nonlinear functions of the original coupling constants. Since our new Hamiltonian is identical in form to the old one, we can repeat the process and transform to even larger blocks nL. After n transformations we find

$$\mathbf{K}_{nL} = \mathbf{T}(\mathbf{K}_{(n-1)L}). \tag{10.115}$$

If the system is not critical, there will be a finite correlation range. Thus, when we transform to larger blocks the effective correlation range appears to shrink and we move away from the critical point. However, when the system is critical, the correlation range is infinite and we reach a fixed point of the transformation. At the fixed point, the transformation **T** can no longer change the vector **K**. Thus, the critical point occurs for values of the coupling constant vectors, $\mathbf{K}^*$, which satisfy the condition

$$\mathbf{K}^* = \mathbf{T}(\mathbf{K}^*). \tag{10.116}$$

The sequence of transformations **T** is called the renormalization group (although **T** only has properties of a semigroup). The motion of the vector **K** in coupling constant space is similar to the motion of the vector describing a state point in phase space, except that the motion of **K** occurs in discrete jumps rather than continuously.

It is useful to illustrate the possible motions of **K** for the case of a two-dimensional vector, $\mathbf{K} = (K_1, K_2)$. The change in block size gives a nonlinear equation for the change in the interaction vector **K**. To locate the critical point we must locate the fixed points of these nonlinear equations. The procedure is identical to that used to find and classify the fixed points of classical dynamical equations (cf. App. D). Let us assume that a fixed point of the transformation $\mathbf{K}_L = \mathbf{T}(\mathbf{K})$ occurs at $\mathbf{K}^* = (K_1^*, K_2^*)$. We will want to know how the vector **K** moves in the neighborhood of the point $\mathbf{K}^*$. As discussed in App. D, we must linearize the transformation about $\mathbf{K}^*$. We will let $\delta\mathbf{K}_L = (\mathbf{K}_L - \mathbf{K}^*)$ and $\delta\mathbf{K} = (\mathbf{K} - \mathbf{K}^*)$. Then for small $\delta\mathbf{K}_L$ and $\delta\mathbf{K}$ we get linearized transformation

$$\delta\mathbf{K}_L = \mathbb{A}\cdot\delta\mathbf{K} \tag{10.117}$$

where

$$\mathbb{A} = \begin{pmatrix} \dfrac{\partial K_{1L}}{\partial K_1} & \dfrac{\partial K_{1L}}{\partial K_2} \\ \dfrac{\partial K_{2L}}{\partial K_1} & \dfrac{\partial K_{2L}}{\partial K_2} \end{pmatrix}_{\substack{K_{1L}=K_1^* \\ K_{2L}=K_2^*}}. \tag{10.118}$$

We next find the eigenvalues and eigenvectors of the matrix $\mathbb{A}$. Since $\mathbb{A}$, in general, will be nonsymmetric, we must use the method of Sec. 6.E. The eigenvectors can be written

$$\delta\mathbf{u}_L = \bar{\bar{\boldsymbol{\Lambda}}}\,\delta\mathbf{u}, \tag{10.119}$$

where $\bar{\bar{\boldsymbol{\Lambda}}}$ is the matrix of eigenvalues,

$$\bar{\bar{\boldsymbol{\Lambda}}} = \begin{pmatrix} \lambda_1 & 0 \\ 0 & \lambda_2 \end{pmatrix}, \tag{10.120}$$

and $\delta\mathbf{u}$ is the right eigenvector,

$$\delta\mathbf{u} = \begin{pmatrix} \delta u_1 \\ \delta u_2 \end{pmatrix}. \tag{10.121}$$

The eigenvalues λ_1 and λ_2 of the matrix $\mathbb{A}$ determine the behavior of trajectories in the neighborhood of the fixed point. In Fig. 10.3 we have drawn the case of a hyperbolic fixed point and its eigencurves. Points along the eigencurves move as

$$\delta u_{nL,1} = (\lambda_1)^n \, \delta u_1 \quad (10.122)$$

$$\delta u_{nL,2} = (\lambda_2)^n \, \delta u_2. \quad (10.123)$$

Thus, for $\lambda > 1$ the point moves away from the fixed point under the transformation, and for $\lambda < 1$ it moves toward the fixed point. The dashed lines represent the trajectories of points which do not lie on the eigencurves. For a hyperbolic fixed point, they will always move away from the fixed point after many transformations. All systems with vectors **K** lying on an eigencurve with eigenvalue $\lambda < 1$ are critical, since with enough transformations they will come arbitrarily close to the fixed point. Such systems are said to exhibit "universality." The behavior of a point along an eigencurve with $\lambda > 1$ is reminiscent of the actual behavior of noncritical systems. As we increase the block size we move away from the critical point. Thus, an eigenvalue $\lambda > 1$ is called relevant and its eigenvector is identified as one of the physical quantities (ϵ or B, for example) which measure the distance of the system from the critical point.

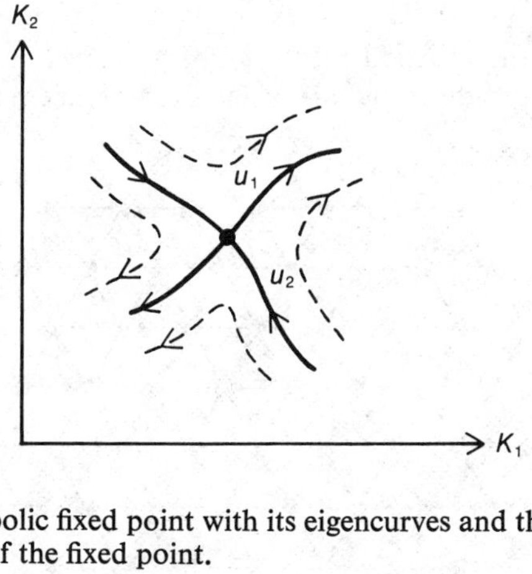

Fig. 10.3. A hyperbolic fixed point with its eigencurves and the flow of points in the neighborhood of the fixed point.

In general we write the singular part of the free energy density in terms of the eigenvectors δu_i and eigenvalues λ_i as follows:

$$g_s(\delta u_1, \delta u_2, \delta u_3, \ldots) = L^{-d} g_s(\lambda_1 \, \delta u_1, \lambda_2 \, \delta u_2, \lambda_3 \, \delta u_3, \ldots) \quad (10.124)$$

(cf. Eq. (10.113)). This looks very much like Widom scaling. Indeed, for the case of an Ising system for which there are two relevant physical parameters which measure the distance of the system from the critical point, we expect that two of the eigenvalues will be relevant, let us say $\lambda_1 > 1$ and $\lambda_2 > 1$. If we compare Eq. (10.124) with Widom scaling of the Ising model,

$$g(\epsilon, B) = \frac{1}{\lambda} g(\lambda^p \epsilon, \lambda^q B), \tag{10.125}$$

we can make the identification $\delta u_1 = \epsilon$ and $\delta u_2 = B$. Thus,

$$\lambda = L^d, \tag{10.126}$$

$$\lambda_1 = (L^d)^p \Rightarrow p = \frac{\ln \lambda_1}{d \ln L}, \tag{10.127}$$

and

$$\lambda_2 = (L^d)^q \Rightarrow q = \frac{\ln \lambda_2}{d \ln L}. \tag{10.128}$$

If we now use Eqs. (10.84) and (10.85) we have expressed the critical exponents in terms of the relevant eigenvalues. In subsequent sections we shall illustrate this procedure for some specific models.

F. CRITICAL EXPONENTS FOR A TRIANGULAR LATTICE[5,6,17]

The calculation of critical exponents is simply illustrated on a two-dimensional lattice of the type considered by Niemeijer and van Leeuven[17] (cf. Fig. 10.4).

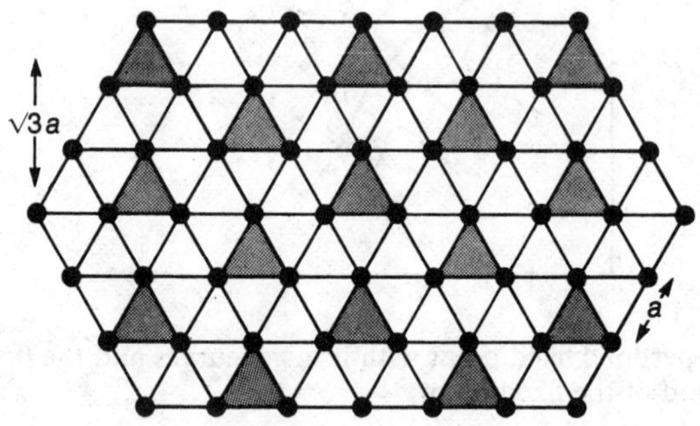

Fig. 10.4. Triangular lattice: for the first block transformation, $L = \sqrt{3}$.

On this lattice, we can form blocks consisting of three spins and in so doing obtain a new triangular lattice in terms of block spins. We assume that the distance between spins is a. Then the distance between the centers of blocks will be $La = \sqrt{3}a$.

The partition function for the spin system is given by

$$Z(\mathbf{K}, B, N) = \sum_{\{S_i\}} \exp[-\mathscr{H}(\mathbf{K}, \{S_i\})] \tag{10.129}$$

where the effective Hamiltonian $\mathscr{H}(\mathbf{K}, \{S_i\})$ is of the form

$$\mathscr{H}(\mathbf{K}, \{S_i\}) = -K \sum_{i \neq j}^{(1)} S_i S_j - B \sum_i S_i. \tag{10.130}$$

The sum $\sum_{ij}^{(1)}$ in Eq. (10.130) is only over nearest neighbors. The coupling constant vector can be written $\mathbf{K} = (-K, -B, 0, \ldots)$. We now transform to the block Hamiltonian. We first define the spin of block I to be

$$S_I = \text{sign}(S_1{}^I + S_2{}^I + S_3{}^I) \tag{10.131}$$

where $S_i{}^I$ is the ith spin in block I. Thus, when two or three spins are up, $S_I = +1$, and when two or three spins are down, $S_I = -1$. For each value of S_I, there are four internal degrees of freedom for a block. We specify these in terms of the variable

$$\sigma_I{}^\alpha = |S_1{}^I + S_2{}^I + S_3{}^I|_\alpha \tag{10.132}$$

where α denotes the four configurations $\downarrow\uparrow\uparrow(\alpha = 1), \uparrow\downarrow\uparrow(\alpha = 2), \uparrow\uparrow\downarrow(\alpha = 3)$, and $\uparrow\uparrow\uparrow(\alpha = 4)$. Thus $\sigma_I{}^1 = 1$, $\sigma_I{}^2 = 1$, $\sigma_I{}^3 = 1$, and $\sigma_I{}^4 = 3$. A given configuration of the lattice is completely specified once the set $\{S_I, \sigma_I\}$ is specified. The partition function can be written

$$Z(\mathbf{K}, N) = \sum_{\{S_I, \sigma_I\}} \exp[-\mathscr{H}\{S_I, \sigma_I\}] \tag{10.133}$$

and the spin block Hamiltonian $\mathscr{H}(K_L, \{S_I\})$ is determined by the condition

$$\exp[-\mathscr{H}(\mathbf{K}_L, \{S_I\})] = \sum_{\{\sigma_I\}} \exp[-\mathscr{H}(\mathbf{K}, \{S_I, \sigma_I\})]. \tag{10.134}$$

Since various blocks are coupled, we must use perturbation theory to find an approximation for the block spin Hamiltonian. This can be done as follows. We first divide the Hamiltonian into two parts,

$$\mathscr{H}(\mathbf{K}, \{S_I, \sigma_I\}) = \mathscr{H}_0(\mathbf{K}, \{S_I, \sigma_I\}) + V(\mathbf{K}, \{S_I, \sigma_I\}), \tag{10.135}$$

where $\mathscr{H}_0$ does not include interaction between blocks,

$$\mathscr{H}_0(\mathbf{K}, \{S_I, \sigma_I\}) = -K \sum_I \sum_{i \in I} \sum_{j \in J} S_i S_j, \tag{10.136}$$

and V contains the effect of interaction between blocks and a contribution from the external field

$$V = -K \sum_{I \neq J} \sum_{i \in I} \sum_{j \in J} S_i S_j - B \sum_I \sum_{i \in I} S_i. \tag{10.137}$$

We then introduce the following expectation value:

$$\langle A\{S_I\}\rangle = \frac{\sum_{\{\sigma_I\}} e^{-\mathscr{H}_0(\mathbf{K}, \{S_I, \sigma_I\})} A(\{S_I, \sigma_I\})}{\sum_{\{\sigma_I\}} e^{-\mathscr{H}_0(\mathbf{K}, \{S_I, \sigma_I\})}}. \tag{10.138}$$

Then, Eq. (10.134) can be written

$$\exp[-\mathscr{H}(\mathbf{K}_L,\{S_I\})] = \left[\sum_{\{\sigma_I\}} e^{-\mathscr{H}_0(\mathbf{K},\{S_I,\sigma_I\})}\right]\langle e^V\rangle. \tag{10.139}$$

The sum $\sum_{\{\sigma_I\}}$ is over all internal spin configurations for all the blocks. Thus,

$$\sum_{\{\sigma_I\}} e^{-\mathscr{H}_0(\mathbf{K},\{S_I,\sigma_I\})} = [Z_0(\mathbf{K})]^M \tag{10.140}$$

where M is the number of blocks and

$$Z_0(K) = \sum_{\sigma_I} e^{K(S_1^I S_2^I + S_1^I S_3^I + S_2^I S_3^I)} = [3\,e^{-K} + e^{3K}]. \tag{10.141}$$

The expectation value, $\langle e^V\rangle$, can be written in a cumulant expansion (cf. Eq. (5.19)):

$$\langle e^V\rangle = e^{\langle V\rangle + \frac{1}{2}[\langle V^2\rangle - \langle V\rangle^2] + \cdots} \tag{10.142}$$

We can obtain an approximation to the block spin Hamiltonian by terminating the cumulant expansion after a few terms. Let us consider the simplest approximation and retain only $\langle V\rangle$. We first write the interaction energy due to coupling between blocks (cf. Fig. 10.5) in the form

$$V_{IJ} = -KS_3{}^J(S_1{}^I + S_2{}^I) = -2KS_3{}^J S_1{}^I. \tag{10.143}$$

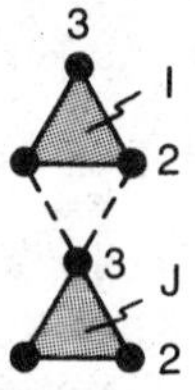

Fig. 10.5. Interaction between two block spins.

Thus,

$$\langle V_{IJ}\rangle = -2K\langle S_3{}^J\rangle\langle S_1{}^I\rangle. \tag{10.144}$$

We now can easily compute the average value $\langle S^J\rangle$,

$$\langle S^J\rangle = Z_0^{-1}(K)\sum_{\sigma_J} S_3{}^J\, e^{K(S_1^J S_2^J + S_2^J S_3^J + S_3^J S_1)}$$

$$= Z_0^{-1}(K) S_J(e^{3K} + e^{-K}). \tag{10.145}$$

Thus,

$$\langle V_{IJ}\rangle = -2K\left(\frac{e^{3K} + e^{-K}}{e^{3K} + 3e^{-K}}\right)^2 S_I S_J \tag{10.146}$$

and

$$\langle V\rangle = \sum_{I\neq J} -2K\left(\frac{e^{3K}+e^{-K}}{e^{3K}+3\,e^{-K}}\right)^2 S_I S_J - \sum_I 3\left(\frac{e^{3K}+e^{-K}}{e^{3K}+3\,e^{-K}}\right) S_I B. \tag{10.147}$$

From Eqs. (10.139), (10.140), (10.142) the block spin Hamiltonian can be written

$$\mathscr{H}\{S_I\} = M\ln Z_0(K) + \langle V\rangle + \tfrac{1}{2}[\langle V^2\rangle - \langle V\rangle^2] + \cdots \tag{10.148}$$

Thus, to lowest order in $\langle V\rangle$ we find

$$\mathscr{H}(S_I, K_L, B_L) = M\ln Z_0(K) - \sum_{I\neq J} 2K\left(\frac{e^{-K}+e^{3K}}{3\,e^{-K}+e^{3K}}\right)^2 S_I S_J - \sum_I 3\left(\frac{e^{3K}+e^{-K}}{e^{3K}+3\,e^{-K}}\right) BS_I. \tag{10.149}$$

The first term is a contribution from spins inside each block and does not contribute to the singular part of the free energy. We therefore can neglect it. The block spin coupling constant and magnetic field can now be written

$$K_L = 2K\left(\frac{e^{3K}+e^{-K}}{e^{3K}+3\,e^{-K}}\right)^2 \tag{10.150}$$

and

$$B_L = 3\left(\frac{e^{3K}+e^{-K}}{e^{3K}+3\,e^{-K}}\right) B. \tag{10.151}$$

Thus, we see that the transformation is highly nonlinear. Fixed points occur when $B^* = 0$ and when $K^* = 0$ or K^* is a solution to the equation

$$\frac{1}{2} = \left(\frac{e^{3K^*}+e^{-K^*}}{e^{3K^*}+3\,e^{-K^*}}\right)^2. \tag{10.152}$$

Thus a second point occurs when

$$K^* = \tfrac{1}{4}\ln(1+2\sqrt{2}) = 0.34. \tag{10.153}$$

The fixed point $B^* = 0$, $K^* = 0$ has eigenvalues $\lambda_K = 0.5$ and $\lambda_B = 1.5$ and corresponds to $T_c = \infty$. For zero field there is no relevant eigenvalue and the fixed point is unphysical. The fixed point $B^* = 0$, $K^* = 0.34$ turns out to be interesting. Let us linearize the transformation about $B^* = 0$, $K^* = 0.34$. We let $\delta K = K - K^*$ and $\delta B = B - B^*$ and obtain the following linearized equation in the neighborhood of the fixed point:

$$\begin{pmatrix}\delta K_L\\ \delta B_L\end{pmatrix} = \begin{pmatrix}\dfrac{\partial K_L}{\partial K} & \dfrac{\partial K_L}{\partial B}\\[2ex] \dfrac{\partial B_L}{\partial K} & \dfrac{\partial B_L}{\partial B}\end{pmatrix}_{\substack{B=B^*=0\\ K=K^*=0.34}} \begin{pmatrix}\delta K\\ \delta B\end{pmatrix} = \begin{pmatrix}1.62 & 0\\ 0 & 2.12\end{pmatrix}\begin{pmatrix}\delta K\\ \delta B\end{pmatrix}. \tag{10.154}$$

The transformation is already diagonal and the eigenvalues are readily obtained. We find

$$\lambda_K = 1.62 \quad \text{and} \quad \lambda_B = 2.12. \tag{10.155}$$

From Eqs. (10.127) and (10.128), we immediately obtain p and q. Thus,

$$p = \frac{\ln 1.62}{2 \ln 1.73} = 0.44 \tag{10.156}$$

and

$$q = \frac{\ln 2.12}{2 \ln 1.73} = 0.68 \tag{10.157}$$

and the critical exponents are $\alpha = -0.27$ and $\delta = 2.1$. An exact solution of this two-dimensional Ising model yields $(\lambda_k)_{\text{exact}} = 1.73$, $(\lambda_B)_{\text{exact}} = 2.80$, $\alpha_{\text{exact}} = 0$, and $\delta_{\text{exact}} = 15$. Thus, we are close for λ_k but our calculation of λ_B is not very good. It is possible to carry the calculations to higher orders in $\langle V^n \rangle$. In so doing, more elements of the vector, **K**, are introduced and better agreement with the exact results is obtained.

G. CRITICAL EXPONENTS FOR THE GAUSSIAN MODEL[5,6,12]

Let us consider the effective Hamiltonian for the Gaussian model in the presence of a magnetic field (cf. Eq. (10.52)):

$$H(r, B', \{S\}) = \frac{1}{2}\left(\frac{1}{2\pi}\right)^d \int_{-\pi/a}^{\pi/a} \cdots \int_{-\pi/a}^{\pi/a} d\mathbf{k}(k^2 + r)S'(\mathbf{k})S'(-\mathbf{k}) + B'S'(\mathbf{k} = 0), \tag{10.158}$$

where we have let $S'(\mathbf{k}) = S(\mathbf{k})(Ka^{2-d})^{1/2}$, $B' = B(Ka^{2-d})^{-1/2}$, and $r = (b/K - 2d)a^{-2}$. The partition function then becomes

$$Z(r, B', \{S\}) = \int DS' \exp\left[-\frac{1}{2}\int \cdots \int_{-\pi/a}^{\pi/a} d\mathbf{k}(k^2 + r)S'(\mathbf{k})S'(-\mathbf{k}) - B'S'(\mathbf{k} = 0)\right] \tag{10.159}$$

where DS' now denotes a functional integral over values of the spin variables $S'(k)$. When expressed in terms of $S'(k)$, the spins form a continuous field rather than a discrete set. The discrete product of integrations in Eq. (10.44) becomes a continuous product of integrations or, in other words, a functional integral.

To obtain a block Hamiltonian, we will introduce a scaling parameter $L > 1$ and divide the momentum intervals $0 < k_i < \pi/a$ into short-wavelength intervals $\pi/La < k_i < \pi/a$ and long-wavelength intervals $0 < k_i < \pi/La$. We then separate the spin fields into their long-wavelength and short-wavelength parts:

$$S'(\mathbf{k}) = \mathscr{S}_L(\mathbf{k}) \quad \text{for} \quad 0 < k_i < \frac{\pi}{La} \qquad (i = 1, \ldots, d) \tag{10.160}$$

and

$$S'(\mathbf{k}) = \sigma_L(\mathbf{k}) \quad \text{for} \quad \frac{\pi}{La} < k_i < \frac{\pi}{a} \qquad (i = 1, \ldots, d). \tag{10.161}$$

From Eq. (10.48), we introduce the corresponding spin fields $\mathscr{S}_L(\mathbf{x})$ and $\sigma_L(\mathbf{x})$. The spin field $\mathscr{S}_L(\mathbf{x})$ is defined

$$\mathscr{S}_L(\mathbf{x}) = \left(\frac{1}{2\pi}\right)^d \int \cdots \int_{-\pi/La}^{\pi/La} d\mathbf{k} \mathscr{S}_L(\mathbf{k})\, e^{i\mathbf{k}\cdot\mathbf{x}} \tag{10.162}$$

and is slowly varying over a region the size of the block, while the spin field $\sigma_L(\mathbf{x})$ is defined

$$\sigma_L(\mathbf{x}) = \left(\frac{1}{2\pi}\right)^d \int \cdots \int_{\pi/La}^{\pi/a} d\mathbf{k} \sigma_L(\mathbf{k})\, e^{i\mathbf{k}\cdot\mathbf{x}} \tag{10.163}$$

where $\int_a^b d\mathbf{k} \equiv \int_a^b d\mathbf{k} + \int_{-b}^{-a} d\mathbf{k}$ and $\sigma_L(\mathbf{x})$ is rapidly varying inside a block.

We can now decompose the Hamiltonian into a long-wavelength and a short-wavelength contribution:

$$\begin{aligned}\mathscr{H}(r, B', \{\mathscr{S}_L, \sigma_L\}) = {} & \frac{1}{2}\left(\frac{1}{2\pi}\right)^d \int \cdots \int_0^{\pi/La} d\mathbf{k}(k^2 + r)\mathscr{S}_L(\mathbf{k})\mathscr{S}_L(-\mathbf{k}) \\ & + B'\mathscr{S}_L(\mathbf{k} = 0) \\ & + \frac{1}{2}\left(\frac{1}{2\pi}\right)^d \int \cdots \int_{\pi/La}^{\pi/a} d\mathbf{k}(k^2 + r)\sigma_L(\mathbf{k})\sigma_L(-\mathbf{k}).\end{aligned} \tag{10.164}$$

The partition function takes the form

$$\begin{aligned} & Z(r, B', \{\mathscr{S}_L, \sigma_L\}) \\ & = \left\{\int \cdots \int D\mathscr{S}_L \exp\left[-\frac{1}{2}\left(\frac{1}{2\pi}\right)^d \int \cdots \int_0^{\pi/La} d\mathbf{k}(k^2 + r)\mathscr{S}_L(\mathbf{k})\mathscr{S}_L(-\mathbf{k}) - B'\mathscr{S}_L(\mathbf{k} = 0)\right]\right\} \\ & \times \left\{\int \cdots \int D\sigma_L \exp\left[-\frac{1}{2}\left(\frac{1}{2\pi}\right)^d \int \cdots \int_{\pi/La}^{\pi/a} d\mathbf{k}(k^2 + r)\sigma_L(\mathbf{k})\sigma_L(-\mathbf{k})\right]\right\},\end{aligned} \tag{10.165}$$

where $D\mathscr{S}_L$ and $D\sigma_L$ again denote functional integrals. The integration $D\sigma_L$ yields a contribution to the nonsingular part of the free energy since it comes from short-wavelength variations in the spin field. Therefore, that term can be neglected and the singular part of the partition function is simply

$$Z(r_L, B'_L, \{\mathscr{S}_L\}) = \int \cdots \int D\mathscr{S}_L \exp\left[-\frac{1}{2}\left(\frac{1}{2\pi}\right)^d \int \cdots \int_0^{\pi/La} d\mathbf{k}(k^2 + r)\mathscr{S}_L(\mathbf{k})\mathscr{S}_L(-\mathbf{k}) - B'\mathscr{S}_L(\mathbf{k} = 0)\right]. \tag{10.166}$$

To return to the form of the partition function in Eq. (10.159), we make a change of scale,

$$k_L = Lk \tag{10.167}$$

and

$$\mathscr{S}_L(\mathbf{k}_L) = Z^{-1}\mathscr{S}_L(\mathbf{k}). \tag{10.168}$$

Then Eq. (10.166) becomes

$$Z(r_L, B'_L, \{\mathscr{S}_L\}) = \int D\mathscr{S}_L \exp\left[-\frac{1}{2}\left(\frac{1}{2\pi}\right)^d Z^2L^{-d} \int \cdots \int_0^{\pi/a} d\mathbf{k}_L\left(\frac{k_L{}^2}{L^2} + r\right)\mathscr{S}_L(\mathbf{k}_L)\mathscr{S}_L(-\mathbf{k}_L) - B'Z\mathscr{S}_L(\mathbf{k}_L = 0)\right]. \tag{10.169}$$

We now choose Z, so the coefficient of $k_L{}^2$ is simply $\frac{1}{2}$. Thus,

$$Z = L^{1+d/2}. \tag{10.170}$$

Then the Hamiltonian takes the form

$$H(r_L, B_L, \{\mathscr{S}_L\}) = \frac{1}{2}\left(\frac{1}{2\pi}\right)^d \int_0^{\pi/a} d\mathbf{k}_L(k_L{}^2 + r_L)\mathscr{S}_L(\mathbf{k}_L)\mathscr{S}_L(-\mathbf{k}_L) + B_L\mathscr{S}_L(\mathbf{k}_L = 0), \tag{10.171}$$

where

$$r_L = L^2 r \tag{10.172}$$

and

$$B_L = L^{1+d/2}B'. \tag{10.173}$$

Eqs. (10.172) and (10.173) define the renormalization group transformation. The fixed point of Eqs. (10.172) and (10.173) occurs for $B^* = 0$ and $r^* = 0$. The eigenvalues are

$$\lambda_r = L^2 \tag{10.174}$$

and

$$\lambda_B = L^{1+d/2}. \tag{10.175}$$

Since they are greater than one they are both relevant; λ_r is associated with temperature variable ϵ and λ_B with the magnetic field. Thus,

$$p = \frac{\ln L^2}{d \ln L} = \frac{2}{d} \tag{10.176}$$

and

$$q = \frac{\ln L^{1+d/2}}{d \ln L} = \frac{2+d}{2d} \tag{10.177}$$

and the critical exponents α, ν, and δ become

$$\alpha = 2 - \frac{d}{2}, \tag{10.178}$$

$$\nu = \tfrac{1}{2}, \tag{10.179}$$

and

$$\delta = \frac{d+2}{d-2}. \tag{10.180}$$

Thus, for $d = 2$, $\alpha = 1$ and $\delta = \infty$. We see that the Gaussian model has little in common with the Ising model in two dimensions.

H. THE S^4-MODEL[5,6,12]

The Gaussian model is not a very realistic approximation to the Ising model. A more interesting case is the S^4-model. Let us again consider Eq. (10.44), but now use as a weighting factor

$$W(S_{\mathbf{m}}) = e^{-(b/2)S_{\mathbf{m}}^2 - uS_{\mathbf{m}}^4}. \tag{10.181}$$

If we choose $b = -4u$, then $W(S_{\mathbf{m}})$ takes the simple form (to within a constant factor)

$$W(S_{\mathbf{m}}) = e^{-u(S_{\mathbf{m}}^2 - 1)^2} \tag{10.182}$$

and gives a fairly good approximation to the Ising model (cf. Fig. 10.6), since

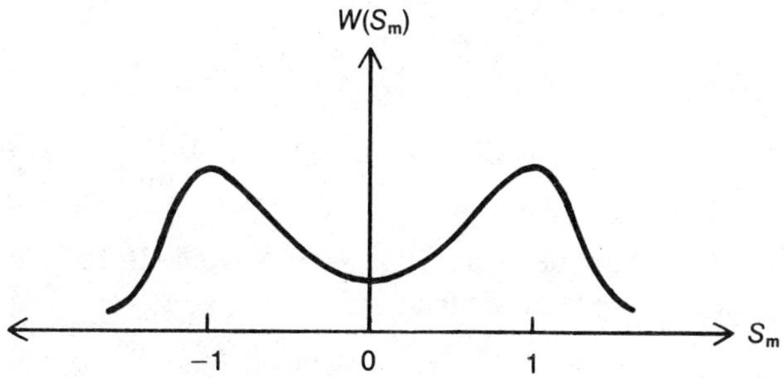

Fig. 10.6. Weighting factor for the S^4-model.

it is peaked at $S_{\mathbf{m}} = \pm 1$ as for the Ising case. We shall compute critical exponents using the weighting factor in Eq. (10.181) and, for simplicity, we will assume there is no magnetic field present. The partition function takes the form

$$Z(K, b, u, \{S_{\mathbf{m}}\}) = \int_{-\infty}^{\infty} \left[\prod_{\mathbf{m}} dS_{\mathbf{m}}\right] \exp\left[K \sum_{\mathbf{n},\mathbf{e}} S_{\mathbf{n}} S_{\mathbf{n}+\mathbf{e}} - \sum_{\mathbf{n}} \left(\frac{b}{2} S_{\mathbf{n}}^{2} + u S_{\mathbf{n}}^{4}\right)\right], \tag{10.183}$$

and the effective Hamiltonian is given by

$$H(r, b, u, \{S\}) = -K \sum_{\mathbf{n},\mathbf{e}} S_{\mathbf{n}} S_{\mathbf{n}+\mathbf{e}} + \sum_{\mathbf{n}} \left(\frac{b}{2} S_{\mathbf{n}}^{2} + u S_{\mathbf{n}}^{4}\right) \tag{10.184}$$

and now contains terms quartic in the spin variables. We can write the effective Hamiltonian in the form

$$\begin{aligned} H(r, u, \{S'\}) = {} & \frac{1}{2}\left(\frac{1}{2\pi}\right)^{d} \int_{-\pi/a}^{\pi/a} d\mathbf{k}(k^{2} + r) S'(\mathbf{k}) S'(-\mathbf{k}) \\ & + u'\left(\frac{1}{2\pi}\right)^{4d} \int \cdots \int_{-\pi/a}^{\pi/a} d\mathbf{k}_1\, d\mathbf{k}_2\, d\mathbf{k}_3\, d\mathbf{k}_4 \\ & \times S'(\mathbf{k}_1) S'(\mathbf{k}_2) S'(\mathbf{k}_3) S'(\mathbf{k}_4)\, \delta(\mathbf{k}_1 + \mathbf{k}_2 + \mathbf{k}_3 + \mathbf{k}_4), \end{aligned} \tag{10.185}$$

where r and $S'(\mathbf{k})$ are defined in Sec. 10.G and $u' = u(Ka^{2-d})^{-2}$. In order to find a scaled Hamiltonian for the system, we follow a procedure similar to that for the triangular lattice (cf. Sec. 10.F). We must first decompose the quadratic part of the effective Hamiltonian into its long-wavelength and short-wavelength parts, as we did in Sec. 10.G, and define the quantities:

$$\mathscr{H}_0(\{\mathscr{S}_L\}, r) = \frac{1}{2}\left(\frac{1}{2\pi}\right)^{d} \int_{-\pi/La}^{\pi/La} d\mathbf{k}(k^{2} + r) \mathscr{S}_L(\mathbf{k}) \mathscr{S}_L(-\mathbf{k}), \tag{10.186}$$

$$\mathscr{H}_0(\{\sigma_L\}, r) = \frac{1}{2}\left(\frac{1}{2\pi}\right)^{d} \int_{\pi/La}^{\pi/a} d\mathbf{k}(k^{2} + r) \sigma_L(\mathbf{k}) \sigma_L(-\mathbf{k}), \tag{10.187}$$

and

$$\begin{aligned} V(\{\mathscr{S}_L, \sigma_L\}, u') = u'\left(\frac{1}{2\pi}\right)^{4d} & \int \cdots \int_{-\pi/a}^{\pi/a} d\mathbf{k}_1\, d\mathbf{k}_2\, d\mathbf{k}_3\, d\mathbf{k}_4 \\ & \times S'(\mathbf{k}_1) S'(\mathbf{k}_2) S'(\mathbf{k}_3) S'(\mathbf{k}_4)\, \delta(\mathbf{k}_1 + \mathbf{k}_2 + \mathbf{k}_3 + \mathbf{k}_4) \end{aligned} \tag{10.188}$$

where $\mathscr{S}_L(\mathbf{k})$ and $\sigma_L(\mathbf{k})$ are defined in Eqs. (10.162) and (10.163). We now can introduce an expectation value defined

$$\langle A(\{\mathscr{S}_L\})\rangle = \frac{\int D\sigma_L\, e^{-\mathscr{H}_0(\{\sigma_L\})} A(\{\sigma_L, \mathscr{S}_L\})}{\int D\sigma_L e^{-\mathscr{H}_0(\{\sigma_L\})}} \tag{10.189}$$

and write the partition function in the form

$$Z(r, u') = \left[\int D\sigma_L\, e^{-\mathscr{H}_0(\{\sigma_L\})}\right]\int D\mathscr{S}_L\, e^{-\mathscr{H}_0(\{\mathscr{S}_L\})}\langle e^{-V(\{\mathscr{S}_L,\sigma_L\})}\rangle$$

$$= \left[\int D\sigma_L\, e^{-\mathscr{H}_0(\{\sigma_L\})}\right]\int D\mathscr{S}_L\, e^{-\mathscr{H}_0(\{\mathscr{S}_L\})-\langle V\rangle+\frac{1}{2}[\langle V^2\rangle-\langle V\rangle^2]} + \cdots \tag{10.190}$$

where we have introduced a cumulant expansion as before. We are only concerned with those parts of the partition function which contribute to the singular part of the free energy. Thus, the term in brackets, which comes from short-wavelength variations, can be neglected and the effective Hamiltonian takes the form

$$\mathscr{H}(\{\mathscr{S}_L\}, r_L, u_L') = \frac{1}{2}\left(\frac{1}{2\pi}\right)^d\int_{-\pi/La}^{\pi/La} d\mathbf{k}(k^2 + r)\mathscr{S}_L(\mathbf{k})\mathscr{S}_L(-\mathbf{k}) + \langle V\rangle - \tfrac{1}{2}[\langle V^2\rangle - \langle V\rangle^2] + \cdots \tag{10.191}$$

After a scale transformation, we seek a new Hamiltonian of the form

$$\mathscr{H}_L(\{\mathscr{S}_L\}, r_L, u_L) = \frac{1}{2}\left(\frac{1}{2\pi}\right)^d\int_{-\pi/a}^{\pi/a} d\mathbf{k}_L({k_L}^2 + r_L)\mathscr{S}_L(\mathbf{k}_L)\mathscr{S}_L(-\mathbf{k}_L) + u_L\left(\frac{1}{2\pi}\right)^{4d}\int\cdots\int_{-\pi/a}^{\pi/a} d\mathbf{k}_{1L}\, d\mathbf{k}_{2L}\, d\mathbf{k}_{3L}\, d\mathbf{k}_{4L} \times \mathscr{S}_L(\mathbf{k}_{1L})\mathscr{S}_L(\mathbf{k}_{2L})\mathscr{S}_L(\mathbf{k}_{3L})\mathscr{S}_L(\mathbf{k}_{4L}). \tag{10.192}$$

If we can write Eq. (10.191) in the same form as Eq. (10.192), then we can obtain the new coupling constant, u_L. We first compute the expectation value, $\langle V\rangle$, and obtain

$$\begin{aligned}\langle V\rangle = &\left[\int D\sigma_L\, e^{-\mathscr{H}_0(\{\sigma_L\})}\right]^{-1}\int D\sigma_L\, e^{-\mathscr{H}_0(\{\sigma_L\})}\left(\frac{1}{2\pi}\right)^{4d} \\ &\times\left\{u'\int\cdots\int_{-\pi/La}^{\pi/La} d\mathbf{k}_1\, d\mathbf{k}_2\, d\mathbf{k}_3\, d\mathbf{k}_4\right. \\ &\times \mathscr{S}_L(\mathbf{k}_1)\mathscr{S}_L(\mathbf{k}_2)\mathscr{S}_L(\mathbf{k}_3)\mathscr{S}_L(\mathbf{k}_4)\,\delta(\mathbf{k}_1+\mathbf{k}_2+\mathbf{k}_3+\mathbf{k}_4) \\ &+ 6u'\int_{-\pi/La}^{\pi/La} d\mathbf{k}_1\int_{-\pi/La}^{\pi/La} d\mathbf{k}_2\left[\int_{\pi/La}^{\pi/a} d\mathbf{k}_3\int_{\pi/La}^{\pi/a} d\mathbf{k}_4\mathscr{S}_L(\mathbf{k}_1)\mathscr{S}_L(\mathbf{k}_2)\right. \\ &\left.\times\,\sigma_L(\mathbf{k}_3)\sigma_L(\mathbf{k}_4)\,\delta(\mathbf{k}_1+\mathbf{k}_2+\mathbf{k}_3+\mathbf{k}_4)\right] \\ &+ u'\int\cdots\int_{\pi/La}^{\pi/a} d\mathbf{k}_1\, d\mathbf{k}_2\, d\mathbf{k}_3\, d\mathbf{k}_4\sigma_L(\mathbf{k}_1)\sigma_L(\mathbf{k}_2)\sigma_L(\mathbf{k}_3)\sigma_L(\mathbf{k}_4) \\ &\left.\times\,\delta(\mathbf{k}_1+\mathbf{k}_2+\mathbf{k}_3+\mathbf{k}_4)\right\}.\end{aligned} \tag{10.193}$$

The factor 6 enters because there are six terms with two factors of $\mathscr{S}_L(\mathbf{k})$ and two factors of $\sigma_L(\mathbf{k})$. They each give identical contributions. The last term can be neglected since it contains only short-wavelength contributions and contributes to the regular part of the free energy. To evaluate the second term, we use Wick's theorem and Eq. (10.12) to obtain

$$\langle \sigma_L(\mathbf{k}_3)\sigma_L(\mathbf{k}_4)\rangle = \frac{(2\pi)^d}{k_3{}^2 + r}\,\delta(\mathbf{k}_3 + \mathbf{k}_4). \tag{10.194}$$

Then the singular part of $\langle V\rangle$ takes the form

$$\begin{aligned}\langle V\rangle_S = u'\left(\frac{1}{2\pi}\right)^{4d} \int\cdots\int_{-\pi/La}^{\pi/La} d\mathbf{k}_1\cdots d\mathbf{k}_4 \\ \times\ \mathscr{S}_L(\mathbf{k}_1)\mathscr{S}_L(\mathbf{k}_2)\mathscr{S}_L(\mathbf{k}_3)\mathscr{S}_L(\mathbf{k}_4)\,\delta(\mathbf{k}_1 + \mathbf{k}_2 + \mathbf{k}_3 + \mathbf{k}_4) \\ +\ 6u'\left(\frac{1}{2\pi}\right)^{3d} \int\cdots\int_{-\pi/La}^{\pi/La} d\mathbf{k}_1\, d\mathbf{k}_2 \mathscr{S}_L(\mathbf{k}_1)\mathscr{S}_L(\mathbf{k}_2) \int_{\pi/La}^{\pi/a} d\mathbf{k}\,\frac{1}{k^2 + r}.\end{aligned} \tag{10.195}$$

We see that $\langle V\rangle$ has a structure similar to the original Hamiltonian, but it yields no correction to the quartic part. The correction to the quartic part can be obtained from the second-order terms in the cumulant expansion.

Many terms contribute to the second-order cumulant, $\frac{1}{2}[\langle V^2\rangle - \langle V\rangle^2]$. To obtain explicit expressions for them, we must use Wick's theorem. We will not attempt to work out all the details here but will summarize the results. We will leave some of the details for a homework problem. As Wilson and Kogut[12] have shown, the only term of interest in $\frac{1}{2}[\langle V^2\rangle - \langle V\rangle^2]$ is the correction to the quartic term in the effective Hamiltonian. They have shown that corrections to the quadratic terms can be neglected. Furthermore, contributions to the second-order cumulant coming from pairwise averages of spins which are both located on the same potential, V, give no contribution since for those terms $\langle V^2\rangle = \langle V\rangle^2$ and they cancel out. Thus, we only retain terms which have at least one pairwise average between spins on different factors, V. The only term which involves four spins, $\mathscr{S}_L(\mathbf{k})$, and which contributes to the second-order cumulant is

$$\begin{aligned}\tfrac{1}{2}[\langle V^2\rangle]_{\text{quartic}} = \frac{1}{2}\left[\int D\sigma_L\, e^{-\mathscr{H}_0(\{\sigma_L\})}\right]^{-1} \int D\sigma_L\, e^{-\mathscr{H}_0(\{\sigma_L\})}\left(\frac{1}{2\pi}\right)^{8d} \\ \times \Bigg\{36u'^2 \int\cdots\int_{-\pi/La}^{\pi/La} d\mathbf{k}_1\, d\mathbf{k}_2\, d\mathbf{k}_7\, d\mathbf{k}_8 \int\cdots\int_{\pi/La}^{\pi/a} d\mathbf{k}_3\, d\mathbf{k}_4\, d\mathbf{k}_5\, d\mathbf{k}_6 \\ \times\ \mathscr{S}_L(\mathbf{k}_1)\mathscr{S}_L(\mathbf{k}_2)\sigma_L(\mathbf{k}_3)\sigma_L(\mathbf{k}_4)\sigma_L(\mathbf{k}_5)\sigma_L(\mathbf{k}_6)\mathscr{S}_L(\mathbf{k}_7)\mathscr{S}_L(\mathbf{k}_8) \\ +\cdots\Bigg\}.\end{aligned} \tag{10.196}$$

The origin of the numerical factor, 36, can be understood if we note that there is a factor, $\frac{1}{2}$, in front of $\langle V^2\rangle_{\text{quartic}}$, there are six terms in each factor V with two

$\mathcal{S}_L$'s and two σ_L's, and there are two different ways of taking pairwise averages for each term in $V^2(\frac{1}{2} \times 6 \times 6 \times 2 = 36)$. If we apply Wick's theorem to Eq. (10.196), we obtain

$$\begin{aligned}\tfrac{1}{2}\langle V^2\rangle_{\text{quartic}} = {} & \left(\frac{1}{2\pi}\right)^{5d} 36u'^2 \int\cdots\int_{-\pi/La}^{\pi/La} d\mathbf{k}_1\cdots d\mathbf{k}_4 \\ & \times \mathcal{S}_L(\mathbf{k}_1)\mathcal{S}_L(\mathbf{k}_2)\mathcal{S}_L(\mathbf{k}_3)\mathcal{S}_L(\mathbf{k}_4)\,\delta(\mathbf{k}_1+\mathbf{k}_2+\mathbf{k}_3+\mathbf{k}_4) \\ & \times \int_{\pi/La}^{\pi/a} d\mathbf{k}\left(\frac{1}{k^2+r}\right)\left(\frac{1}{(\mathbf{k}_3+\mathbf{k}_4-\mathbf{k})^2+r}\right) + \text{other terms.}\end{aligned} \tag{10.197}$$

If we combine contributions from Eqs. (10.195) and (10.197), the Hamiltonian $\mathcal{H}(\{\mathcal{S}_L\}, r_L, u_L)$ in Eq. (10.191) takes the following form:

$$\begin{aligned}\mathcal{H}(\{\mathcal{S}_L\}, r_L, u_L) = {} & \frac{1}{2}\left(\frac{1}{2\pi}\right)^{d}\int_{-\pi/La}^{\pi/La} d\mathbf{k}\,\mathcal{S}_L(\mathbf{k})\mathcal{S}_L(-\mathbf{k}) \\ & \times \left[(k^2+r) + \frac{12u'}{(2\pi)^d}\int_{\pi/La}^{\pi/a} d\mathbf{k}\,\frac{1}{k^2+r}\right] \\ & + \left(\frac{1}{2\pi}\right)^{4d}\int\cdots\int_{-\pi/La}^{\pi/La} d\mathbf{k}_1\cdots d\mathbf{k}_4 \\ & \times \mathcal{S}_L(\mathbf{k}_1)\mathcal{S}_L(\mathbf{k}_2)\mathcal{S}_L(\mathbf{k}_3)\mathcal{S}_L(\mathbf{k}_4)\,\delta(\mathbf{k}_1+\mathbf{k}_2+\mathbf{k}_3+\mathbf{k}_4) \\ & \times \left[u' - \frac{36u'^2}{(2\pi)^d}\int_{\pi/La}^{\pi/a} d\mathbf{k}\left(\frac{1}{k^2+r}\right)\left(\frac{1}{(\mathbf{k}_3+\mathbf{k}_4-\mathbf{k})^2+r}\right)\right] \\ & + \cdots\end{aligned} \tag{10.198}$$

where we have neglected contributions (terms not containing block spins), some second-order quadratic terms, and terms containing six or eight block spins. Also, we neglected all contributions from higher order cumulants. To obtain Eq. (10.198) in a form similar to the original effective Hamiltonian, we must make a change of scale. Using Eqs. (10.167) and (10.168), we find

$$\begin{aligned}\mathcal{H}_L(\{\mathcal{S}_L\}, r_L, u_L) = {} & \frac{1}{2}\frac{Z^2L^{-d}}{(2\pi)^d}\int_{-\pi/a}^{\pi/a} d\mathbf{k}_L\,\mathcal{S}_L(\mathbf{k}_L)\mathcal{S}_L(-\mathbf{k}_L) \\ & \times \left[\frac{k_L{}^2}{L^2} + r + \frac{12u'}{(2\pi)^d}\int_{\pi/La}^{\pi/a} d\mathbf{k}\,\frac{1}{k^2+r}\right] \\ & + \left(\frac{1}{2\pi}\right)^{4d}\frac{Z^4}{(L^d)^3}\int\cdots\int_{-\pi/a}^{\pi/a} d\mathbf{k}_{1L}\times\cdots\times d\mathbf{k}_{4L} \\ & \times \mathcal{S}_L(\mathbf{k}_{1L})\mathcal{S}_L(\mathbf{k}_{2L})\mathcal{S}_L(\mathbf{k}_{3L})\mathcal{S}_L(\mathbf{k}_{4L})\,\delta(\mathbf{k}_{1L}+\mathbf{k}_{2L}+\mathbf{k}_{3L}+\mathbf{k}_{4L}) \\ & \times \left[u' - \frac{36u'^2}{(2\pi)^d}\int_{\pi/La}^{\pi/a} d\mathbf{k}\left(\frac{1}{k^2+r}\right)\left(\frac{1}{(\mathbf{k}_3+\mathbf{k}_4-\mathbf{k})^2+r}\right)\right].\end{aligned} \tag{10.199}$$

Eq. (10.199) is the transformed Hamiltonian that we have been looking for. If we let $Z = L^{1+d/2}$ (cf. Sec. 10.G), then we obtain the following expression for r_L:

$$r_L = L^2\left[r + \frac{12u'}{(2\pi)^d}\int_{\pi/La}^{\pi/a} d\mathbf{k}\,\frac{1}{k^2 + r}\right]. \tag{10.200}$$

The expression for u_L requires more discussion. To obtain an expression for u_L which is independent of the **k**-integrations in Eq. (10.199), we must make an approximation. If we let $\mathbf{k}_3 + \mathbf{k}_4 = 0$, we obtain

$$u_L = L^{\mathscr{E}}\left[u' - \frac{36u'^2}{(2\pi)^d}\int_{\pi/La}^{\pi/a} d\mathbf{k}\left(\frac{1}{k^2 + r}\right)^2\right] \tag{10.201}$$

where $\mathscr{E} = 4 - d$. This approximation is made every time we change the scale of the Hamiltonian. Eqs. (10.200) and (10.201) are the recursion relations we have been looking for. They are correct as long as $|\mathscr{E}| \ll 1$ (this will become clearer below).

To find the fixed points of Eqs. (10.200) and (10.201), we will turn them into differential equations. We first write the recursion relation for a transformation from block L to block sL,

$$r_{sL} = s^2\left[r_L + \frac{12u_L}{(2\pi)^d}\int\cdots\int_{\pi/sa}^{\pi/a} d\mathbf{k}\,\frac{1}{k^2 + r_L}\right] \tag{10.202}$$

and

$$u_{sL} = s^{\mathscr{E}}\left[u_L - \frac{36u_L{}^2}{(2\pi)^d}\int\cdots\int_{\pi/sa}^{\pi/a} d\mathbf{k}\left(\frac{1}{k^2 + r_L}\right)^2\right], \tag{10.203}$$

where s is now considered to be a continuous parameter. Let $s = 1 + h$ for $h \ll 1$. Then, Eqs. (10.202) and (10.203) can be approximated by

$$r_{(1+h)L} = (1 + 2h)\left[r_L + \frac{12u_L}{(2\pi)^d}\int_{(\pi/a)(1-h)}^{\pi/a} d\mathbf{k}\left(\frac{1}{k^2 + r_L}\right)\right] \tag{10.204}$$

and

$$u_{(1+h)L} = (1 + \mathscr{E}h)\left[u_L - \frac{36u_L{}^2}{(2\pi)^d}\int_{(\pi/a)(1-h)}^{\pi/a} d\mathbf{k}\left(\frac{1}{k^2 + r_L}\right)^2\right]. \tag{10.205}$$

We next evaluate the integrals. We can write for small h

$$\frac{1}{(2\pi)^d}\int_{(\pi/a)(1-h)}^{\pi/a} d\mathbf{k}\left(\frac{1}{k^2 + r_L}\right)^n = \frac{2}{(2\pi)^d}\left(\frac{1}{\left(\frac{\pi}{a}\right)^2 + r_L}\right)^n\int_{(\pi/a)(1-h)}^{\pi/a} d\mathbf{k}$$

$$\equiv \left(\frac{1}{\left(\frac{\pi}{a}\right)^2 + r_L}\right)^n\frac{Ah}{12}$$

where A is a constant. Thus, Eqs. (10.204) and (10.205) become (to first order in h)

$$r_{(1+h)L} - r_L = 2hr_L + \frac{u_L Ah}{\left(\left(\frac{\pi}{a}\right)^2 + r_L\right)} \tag{10.206}$$

and

$$u_{(1+h)L} - u_L = \mathscr{E}hu_L - \frac{3u_L{}^2 Ah}{\left(\left(\frac{\pi}{a}\right)^2 + r_L\right)^2}. \tag{10.207}$$

If we next divide by hL and take the limit $h \to 0$, we find

$$L\frac{dr_L}{dL} = 2r_L + \frac{Au_L}{\left(\left(\frac{\pi}{a}\right)^2 + r_L\right)} \tag{10.208}$$

and

$$L\frac{du_L}{dL} = \mathscr{E}u_L - \frac{3Au_L{}^2}{\left(\left(\frac{\pi}{a}\right)^2 + r_L\right)^2}. \tag{10.209}$$

Now let $t \equiv \ln L$ and choose the units of a such that $\pi/a = 1$. Then

$$\frac{dr}{dt} = 2r + \frac{Au}{1+r} \tag{10.210}$$

and

$$\frac{du}{dt} = \mathscr{E}u - \frac{3Au^2}{(1+r)^2}. \tag{10.211}$$

Higher order terms can be neglected as long as $|\mathscr{E}| \ll 1$. The fixed points of Eqs. (10.210) and (10.211) occur when

$$\frac{dr^*}{dt} = 2r^* + \frac{Au^*}{1+r^*} = 0 \tag{10.212}$$

and

$$\frac{du^*}{dt} = \mathscr{E}u^* - \frac{3Au^{*2}}{(1+r^*)^2} = 0. \tag{10.213}$$

There are two sets of fixed points, ($u^* = 0, r^* = 0$) and ($u^* = \mathscr{E}/3A, r^* = -\mathscr{E}/6$). If we let $\delta u = u - u^*$ and $\delta r = r - r^*$, then the linearized transformation becomes (to first order in n^* and r^*)

$$\begin{pmatrix} \dfrac{d\delta r}{dt} \\ \dfrac{d\delta u}{dt} \end{pmatrix} = \begin{pmatrix} 2 - Au^* & A(1 - r^*) \\ 0 & \mathscr{E} - 6Au^* \end{pmatrix} \begin{pmatrix} \delta r \\ \delta u \end{pmatrix} \tag{10.214}$$

and we obtain the eigenvalues $\lambda_r = 2 - Au^*$ and $\lambda_u = \mathscr{E} - 6Au^*$ for small u^* and r^*. Thus, λ_r is a relevant eigenvalue. For the fixed point ($u^* = 0$ and $r^* = 0$) the eigenvalues are $\lambda_r = 2$ and $\lambda_u = \mathscr{E}$ (this is just the Gaussian fixed point), and for the fixed point ($u^* = \mathscr{E}/3A$ and $r^* = -\mathscr{E}/6$) the eigenvalues are $\lambda_r = 2 - \mathscr{E}/3$ and $\lambda_u = -\mathscr{E}$. For $d > 4$, $\mathscr{E} < 0$ and the second fixed point is repulsive and therefore unphysical since both eigencurves are directed away from the fixed point and it can never be reached. Thus, for $d > 4$, the physics is governed by the Gaussian fixed point. For $d < 4$, the Gaussian fixed point is repulsive and for this case the fixed point ($u^* = \mathscr{E}/3A$, $r^* = -\mathscr{E}/6$) is the physical one. For $d = 4$, the fixed points coalesce. In Fig. 10.7, we sketch the trajectories in the neighborhood of the two fixed points.

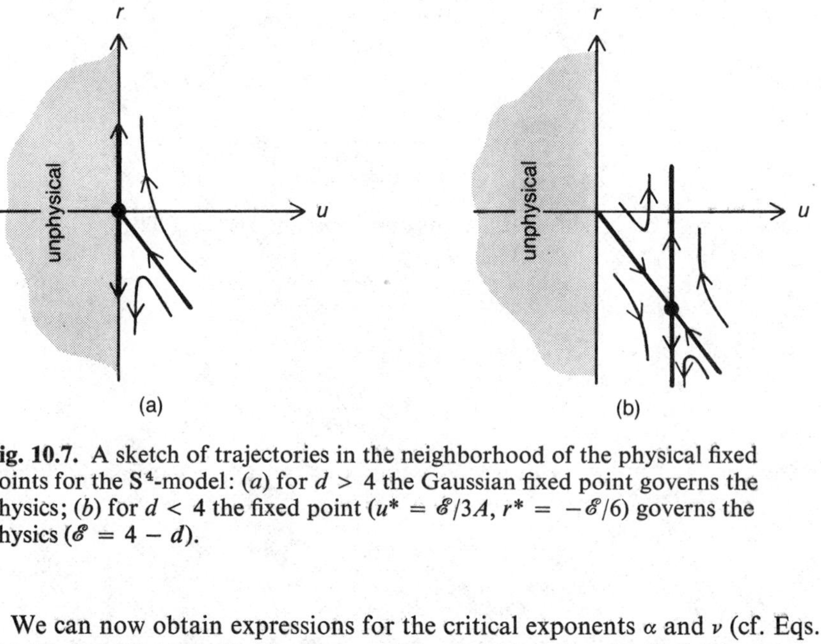

Fig. 10.7. A sketch of trajectories in the neighborhood of the physical fixed points for the S^4-model: (*a*) for $d > 4$ the Gaussian fixed point governs the physics; (*b*) for $d < 4$ the fixed point ($u^* = \mathscr{E}/3A$, $r^* = -\mathscr{E}/6$) governs the physics ($\mathscr{E} = 4 - d$).

We can now obtain expressions for the critical exponents α and ν (cf. Eqs. (10.81) and (10.108)), but it requires a bit of algebra. To obtain the equations for the eigencurves of the S^4-model in the form given in Eq. (10.122), we must rewrite the differential equations in Eq. (10.214) in terms of L. We first write Eq. (10.214) in terms of its eigenvectors and eigenvalues. (Note that the left and right eigenvectors will be different since the matrix on the right-hand side is not symmetric [cf. Sec. 6.E].) For our purposes, it is not necessary to find explicit expressions for the eigenvectors. If we let $\delta u_1(t)$ denote the relevant eigenvector, we can write

$$\frac{d\delta u_1(t)}{dt} = (2 - Au^*)\, \delta u_1(t). \tag{10.215}$$

The solution to Eq. (10.215) can be written

$$\delta u_1(t) = e^{(2-Au^*)t}\,\delta u_1(0). \tag{10.216}$$

If we now remember that $t = \ln L$, we find

$$\delta u_{1,L} = e^{(2-Au^*)\ln L}\,\delta u_1 \tag{10.217}$$

so that $\lambda_1 = \exp[(2 - Au^*)\ln L]$. We then make use of Eq. (10.127) to obtain

$$p = \left(\frac{2 - Au^*}{d}\right), \tag{10.218}$$

and from Eqs. (10.81) and (10.108) we obtain

$$\alpha = 2 - \frac{d}{2 - Au^*} \tag{10.219}$$

and

$$\nu = \frac{1}{2 - Au^*}. \tag{10.220}$$

Since the fixed points which characterize the system are different for $d > 4$ and $d < 4$, the critical exponents will also differ for those two cases.

Let us first consider the case $d > 4$. Then $u^* = 0$ and we find

$$\alpha = 2 - \frac{d}{2} = \frac{\mathscr{E}}{2}, \tag{10.221}$$

where we have used the fact that $\mathscr{E} = 4 - d$, and

$$\nu = \tfrac{1}{2}. \tag{10.222}$$

Thus, we recover the results of the Gaussian model. The exponent, δ, can also be computed in a manner similar to the Gaussian model and the result is

$$\delta = \frac{d + 2}{d - 2} \approx 3 + \mathscr{E} \tag{10.223}$$

as for the Gaussian model.

When $d < 4$, $u^* = \mathscr{E}/3A$, and we find

$$\alpha \approx \frac{\mathscr{E}}{6} \tag{10.224}$$

and

$$\nu \approx \frac{1}{2} + \frac{\mathscr{E}}{12}. \tag{10.225}$$

The magnetic critical exponent can also be computed and the result is

$$\delta \approx 3 + \mathscr{E} \tag{10.226}$$

as for the case $d > 4$. The other exponents can be obtained from the identity in Eqs. (10.82), (10.83), and (10.109). In Table 10.1, we compare the results of the S^4-model with the experimental results, with mean field results, and with the results due to exact calculations for the Ising model. The first thing to note is that, for $d = 4$, the mean field theories and the S^4-model give the same results. For $d = 3$, the S^4-model gives very good agreement with the experimental and Ising values of the exponents (the exact results for the three-dimensional Ising are obtained from a numerical calculation). However, there is really no reason why it should. We have retained only the lowest order approximation for the exponents in the $\mathscr{E}$-expansion. If we take $\mathscr{E} = 1$ (as it is for $d = 3$), then the expansion need not converge. In fact, higher order terms in the series give large contributions and ruin the agreement. For this reason the $\mathscr{E}$-expansion is thought to be an asymptotic expansion, that is, one for which the first few terms give a result close to the result obtained by summing the entire series.

Critical Exponent	Experimental Value	Exact Ising $d = 2$	Exact Ising $d = 3$	Mean Field Theory	S^4-model $d > 4$	S^4-model $d = 4$	S^4-model $d < 4$	S^4-model $d = 3$
α	0–0.2	0	0.12	0	$\mathscr{E}/2$	0	$\mathscr{E}/6$	0.17
β	0.3–0.4	0.125	0.31	$\frac{1}{2}$	$1/2-\mathscr{E}/4$	$\frac{1}{2}$	$1/2-\mathscr{E}/6$	0.33
δ	4–5	15	5.2	3	$3 + \mathscr{E}$	3	$3 + \mathscr{E}$	4
γ	1.2–1.4	1.75	1.25	1	1	1	$1 + \mathscr{E}/6$	1.17
ν	0.6–0.7	1.0	0.64	$\frac{1}{2}$	$\frac{1}{2}$	$\frac{1}{2}$	$1/2 + \mathscr{E}/12$	0.58
η	0.1	0.25	0.056	0	0	0	0	0

Table 10.1. Values of the critical exponents from experiment and from various theories (based on Refs. 10.5 and 10.12).

One may well ask the meaning of an expansion about $d = 4$, and in terms of noninteger d, when real physical systems correspond to $d = 1, 2$, or 3. The answer is that it provides us with a theoretical tool for understanding the role of the dimension of space in physical phenomena, and it has led to a major advance in our understanding of critical phenomena.

REFERENCES

1. L. D. Landau and E. M. Lifshitz, *Statistical Physics* (Pergamon Press, Oxford, 1958).

2. A. Einstein, *Investigations on the Theory of Brownian Movement* (Methuen and Co., London, 1926).
3. H. E. Stanley, *Introduction to Phase Transitions and Critical Phenomena* (Oxford University Press, Oxford, 1971).
4. D. Forster, *Hydrodynamic Fluctuations, Broken Symmetry, and Correlation Functions* (W. A. Benjamin, New York, 1975).
5. F. Ravendal, "Scaling and Renormalization Groups," lecture notes 1975–1976, Nordita, Blegdamsvieg 17 DK-2100, Copenhagen, Denmark. (An exceptionally clear presentation of critical phenomena and the renormalization group which we have followed closely in this chapter.)
6. S. Ma, *Modern Theory of Critical Phenomena* (W. A. Benjamin, New York, 1976).
7. M. Kac and T. H. Berlin, Phys. Rev. *86* 82 (1952).
8. B. Widom, J. Chem. Phys. *43* 3898 (1965).
9. M. S. Green, M. Viscentini-Missoni, and J. M. H. Levelt Sengers, Phys. Rev. Let. *18* 1113 (1967).
10. J. T. Ho and J. D. Litster, Phys. Rev. *22* 603 (1969).
11. L. Kadanoff, Physics *2* 263 (1966).
12. K. Wilson and J. Kogut, Phys. Rep. *12C* 75 (1974).
13. M. E. Fischer, Rev. Mod. Phys. *46* 597 (1974).
14. M. N. Barber, Phys. Rep. *29* 1 (1977).
15. K. Wilson, Rev. Mod. Phys. *47* 773 (1975).
16. K. Wilson, Phys. Rev. *B4* 3174, 3184 (1971).
17. Th. Niemeijer and J. M. J. van Leeuven, Phys. Rev. *31* 1411 (1973).

PROBLEMS

1. If $\bar{\bar{g}}$ is a symmetric $n \times n$ matrix, prove that

$$\int_{-\infty}^{-\infty} D\alpha \exp\left\{-\frac{1}{2k_B}(\bar{\bar{g}}:\alpha\alpha) + \mathbf{h}\cdot\alpha\right\} = \left(\frac{(2\pi k_B)^n}{|\bar{\bar{g}}|}\right)^{1/2} \exp\{\tfrac{1}{2}k_B\bar{\bar{g}}^{-1}:\mathbf{hh}\}$$

where $\boldsymbol{\alpha}$ and $\mathbf{h}$ are vectors of dimension n.

2. Reformulate the Einstein theory for the case in which fluctuations vary slowly and continuously in space.

3. For the system considered in Sec. 10.B.2, find the variance of the enthalpy $\langle(\Delta H_i)^2\rangle$. (*Hint:* use P and S as independent variables.)

4. For the system considered in Sec. 10.B.2, find the variances $\langle(\Delta T_i)^2\rangle$ and $\langle(\Delta P_i)^2\rangle$ for temperature and pressure fluctuations and find the correlations $\langle\Delta P_i\,\Delta T_i\rangle$ and $\langle\Delta P_i\,\Delta T_j\rangle$ $(i \neq j)$. Discuss the meaning of your results.

5. For the triangular lattice, compute the contribution to the block Hamiltonian coming from the term $\frac{1}{2}[\langle V^2\rangle - \langle V\rangle^2]$. Note that the result includes second and third nearest-neighbor coupling between blocks.

6. Using the same method as for the triangular lattice, find the critical exponents for nine spin square blocks on the two-dimensional square lattice for the nearest-neighbor Ising model (cf. Fig. 10.2). Retain terms in the block Hamiltonian to order $\langle V\rangle$.

7. Using the same method as for the triangular lattice, find the critical exponents for five spin blocks on a square lattice for the two-dimensional nearest-neighbor Ising model. Retain terms to order $\langle V \rangle$ (cf. Fig. 10.8).

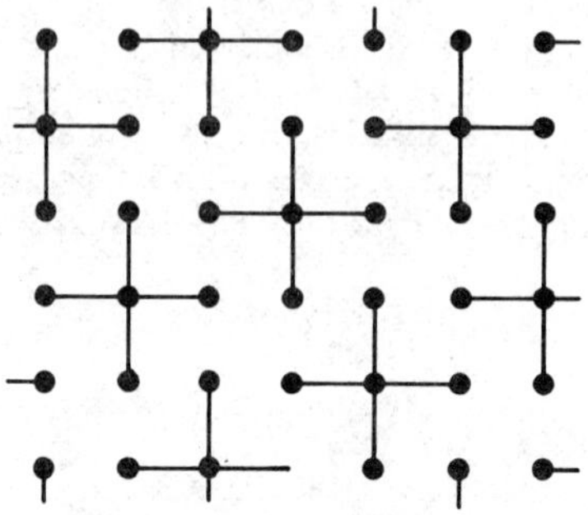

Fig. 10.8. Five spin blocks on a two-dimensional square lattice.

8. Show that $\mathcal{S}_L(\mathbf{x})$ is slowly varying and $\sigma_L(x)$ is rapidly varying inside a block of size La (cf. Sec. 10.G).

9. Write out the entire contribution to the second-order cumulant in the S^4-model.

10. Find the left and right eigenvectors for the linearized transformation matrix (cf. Eq. (10.214)) for the S^4-model.

11
Classical Fluids

A. INTRODUCTORY REMARKS

One of the great successes of equilibrium statistical mechanics has been the theory of interacting classical fluids. For real fluids, the interparticle potential has a hard core and a weak attractive region; consequently, the usual perturbation theories do not work. For dilute or moderately dense fluids, the only small parameter is the density and, therefore, various thermodynamic quantities must be formulated in terms of a density expansion.

In this chapter, we shall restrict ourselves primarily to neutral fluids with spherically symmetric potentials. The quantity of central importance in describing such systems is the radial distribution function, which is a measure of the probability of finding a particle at some distance, q, from any particle in the fluid. The radial distribution function can be measured directly by X-ray scattering experiments or it can be obtained from molecular dynamics experiments. We shall begin this chapter by deriving expressions for various thermodynamic quantities in terms of the radial distribution function using both the canonical and grand canonical ensembles.

In the chapters on thermodynamics, we have already been introduced to the virial expansion of the equation of state. We shall now derive an expression for it, and for the virial coefficients, using microscopic theory. The method involves writing the grand partition function in terms of a cumulant expansion and then expressing the grand potential and pressure in terms of the cumulants or cluster functions. The cluster expansion for the pressure is an expansion in terms of the fugacity. However, with the help of the fugacity expansion for the density, it can be expressed as a power series in the density. The techniques for doing this were first developed by Ursell, whose work was later extended by Mayer. The analysis is greatly facilitated with the use of graph theory which enables one to give a simple physical interpretation to various terms in the virial expansion.

The first few virial coefficients have been computed for a variety of interparticle potentials. We shall discuss the results for the hard-core potential, the square-well potential, and the Lennard-Jones 6–12 potential and compare them with experiment.

The virial expansion gives good results only at low densities. For higher density it is necessary to sum parts of the virial expansion. This is most easily done in terms of the radial distribution function. We shall derive the virial expansion for the radial distribution function and discuss four approximation schemes which allow one to obtain differential or integro-differential equations for the radial distribution function (the hypernetted chain equation, the Percus-Yevick equation, the BGY equation, and the Kirkwood equation). We shall then compare these results with experiment.

Although perturbation expansions in terms of the interparticle potential do not work for strongly coupled fluids, a perturbation scheme due to Zwanzig has been developed in which the effects of the attractive part of the potential are treated as a perturbation on the hard-core system. This method appears to give good results and will be discussed briefly.

For very light molecules at low temperatures, quantum effects introduce significant deviation from classical behavior. We shall conclude this chapter by constructing the virial expansion for degenerate quantum fluids and we will obtain the second virial coefficient for a dilute hard-core boson fluid.

B. THERMODYNAMICS AND THE RADIAL DISTRIBUTION FUNCTION

The thermodynamic properties of a classical fluid can all be determined in terms of the one- and two-body reduced probability densities in configuration space. Before we introduce approximation schemes for studying these functions, we will obtain some exact relations between various thermodynamic quantities and the reduced probability densities.

1. Reduced Probability Densities

It is useful to introduce explicit general expressions for the n-body reduced probability densities in both the canonical and the grand canonical ensemble. We first introduce a phase function for the N-body system which is a sum of n-body functions in configuration space and is of the form

$$O_n^N(\mathbf{q}^N) = \sum_{\alpha=1}^{N!/n!(N-n)!} O_\alpha(\mathbf{q}^\alpha), \tag{11.1}$$

where α denotes a particular n-body grouping of particles and the summation is over all possible combinations of N-particles taken n at a time. The quantity, $\mathbf{q}^\alpha$, denotes the set of configuration space coordinates for all the particles in the group α. In practice, we are usually concerned only with the case $n = 1$ and $n = 2$. For the case $n = 2$, Eq. (11.1) reduces to

$$O_2^N(\mathbf{q}^N) = \sum_{(ij)}^{\frac{1}{2}N(N-1)} O(\mathbf{q}_{ij}), \tag{11.2}$$

where $\mathbf{q}_{ij}$ is the relative displacement between particles i and j.

If we take the expectation value of $O_n{}^N$ in the canonical ensemble, we can write

$$\langle O_n{}^N(V, T)\rangle \equiv \frac{1}{n!}\int\cdots\int d\mathbf{q}_1\cdots d\mathbf{q}_n O_n(\mathbf{q}_1, \ldots, \mathbf{q}_n) n_n{}^N(\mathbf{q}_1, \ldots, \mathbf{q}_n; V, T)$$

$$= Z_N^{-1}(V, T)\frac{1}{h^{3N}N!}\iint d\mathbf{q}^N\, d\mathbf{p}^N \sum_{\alpha=1}^{N!/n!(N-n)!} O_\alpha(\mathbf{q}^\alpha)\, e^{-\beta H^N(\mathbf{p}^N, \mathbf{q}^N)}, \tag{11.3}$$

where $n_n{}^N(\mathbf{q}_1, \ldots, \mathbf{q}_n; V, T)$ is the n-body reduced probability density and

$$H^N(\mathbf{p}^N, \mathbf{q}^N) = \sum_{i=1}^{N}\frac{p_i{}^2}{2m} + \sum_{(ij)=1}^{\frac{1}{2}N(N-1)} v(\mathbf{q}_{ij}) \tag{11.4}$$

is the N-body Hamiltonian (cf. Eq. (7.38)). In this chapter, we will be interested primarily in potentials for spherically symmetric particles. For that case, $v(\mathbf{q}_{ij}) \equiv v(q_{ij})$. For classical particles, the momentum and position coordinates commute and we can write $\exp(-\beta H^N) = \exp(-\beta T^N)\exp(-\beta v^N)$, where T^N is the N-particle kinetic energy and v^N is the N-body potential energy. The momentum integrations in Eq. (11.3) are simple and cancel out of the expression for $\langle O_n{}^N(V, T)\rangle$. Furthermore, because the N-particle potential is a symmetric function of coordinates, each term in the summation $\alpha = 1, 2, \ldots, N!/n!\,(N-n)!$ gives an identical contribution. This can be shown by interchanging dummy variables. Thus, we can write Eq. (11.3) in the form

$$\langle O_n{}^N(V, T)\rangle = Q_N^{-1}(V, T)\frac{N!}{n!\,(N-n)!}\int d\mathbf{q}^N O_n(\mathbf{q}_1, \ldots, \mathbf{q}_n)\, e^{-\beta v^N(\mathbf{q}^N)}, \tag{11.5}$$

where

$$Q_N(V, T) \equiv \int d\mathbf{q}^N\, e^{-\beta v^N(\mathbf{q}^N)} \tag{11.6}$$

is called the configuration integral. Comparison of Eqs. (11.5) and (11.3) yields the following expression for the n-body reduced probability density in the canonical ensemble:

$$n_n{}^N(\mathbf{q}_1, \ldots, \mathbf{q}_n; V, T) = Q_N^{-1}(V, T)\frac{N!}{(N-n)!}\int\cdots\int d\mathbf{q}_{n+1}\cdots d\mathbf{q}_N\, e^{-\beta v^N(\mathbf{q}^N)}. \tag{11.7}$$

The quantity, $n_n{}^N(\mathbf{q}_1, \ldots, \mathbf{q}_n; V, T)$, has the normalization

$$\int\cdots\int d\mathbf{q}_1\cdots d\mathbf{q}_n n_n{}^N(\mathbf{q}_1, \ldots, \mathbf{q}_n; V, T) = \frac{N!}{(N-n)!} \tag{11.8}$$

and is the classical analog of the reduced density matrices introduced in Sec. 7.F.

In the grand canonical ensemble, the expectation value of the function, O_n, can be written

$$\langle O_n^\mu(V, T)\rangle \equiv \frac{1}{n!}\int\cdots\int d\mathbf{q}_1\cdots d\mathbf{q}_n O_n(\mathbf{q}_1, \ldots, \mathbf{q}_n) n_n^\mu(\mathbf{q}_1, \ldots, \mathbf{q}_n; V, T)$$

$$= \mathscr{Z}^{-1}(\mu, V, T)\sum_{N=n}^{\infty}\frac{1}{N!\,h^{3N}}\iint d\mathbf{q}^N\,d\mathbf{p}^N O_n{}^N(\mathbf{q}^N)\,e^{-\beta(H^N(\mathbf{p}^N,\mathbf{q}^N)-\mu N)}, \tag{11.9}$$

where the superscript, μ, is used to denote the fact that the particle number is allowed to fluctuate and the chemical potential is held fixed. The quantity, $\mathscr{Z}(\mu, V, T)$, is the grand partition function (cf. Eq. (9.56)). By arguments similar to those for the canonical ensemble, we can write the n-body reduced probability density in the grand canonical ensemble in the form

$$n_n^\mu(\mathbf{q}_1, \ldots, \mathbf{q}_n; V, T) = \mathscr{Z}^{-1}(\mu, V, T)\sum_{N=n}^{\infty}\frac{1}{(N-n)!\,h^{3N}}\int d\mathbf{p}^N \times\int d\mathbf{q}_{n+1}\cdots d\mathbf{q}_N\,e^{-\beta(H^N(\mathbf{p}^N,\mathbf{q}^N)-\mu N)}. \tag{11.10}$$

It is a simple matter to show that $n_n^\mu(\mathbf{q}_1, \ldots, \mathbf{q}_n; V, T)$ has the normalization

$$\int\cdots\int d\mathbf{q}_1\cdots d\mathbf{q}_n n_n^\mu(\mathbf{q}_1, \ldots, \mathbf{q}_n; V, T) = \left\langle\frac{N!}{(N-n)!}\right\rangle. \tag{11.11}$$

Note that $n_n^N(\mathbf{q}_1, \ldots, \mathbf{q}_n; V, T)$ and $n_n^\mu(\mathbf{q}_1, \ldots, \mathbf{q}_n; V, T)$ both are proportional to the probability density for finding any n-particles at the points $\mathbf{q}_1, \ldots, \mathbf{q}_n$, and not a specific set of particles. If the particles in the gas were distributed completely at random, the reduced distribution functions would take the form

$$n_n^N(\mathbf{q}_1, \ldots, \mathbf{q}_n; V, T) \approx \left(\frac{N}{V}\right)^n \tag{11.12}$$

and

$$n_n^\mu(\mathbf{q}_1, \ldots, \mathbf{q}_n; V, T) \approx \left\langle\frac{N}{V}\right\rangle^n, \tag{11.13}$$

(for large volume and particle number) indicating that there is no spatial correlation between particles.

It is convenient to introduce new distribution functions, $g_n^{N(\mu)}(\mathbf{q}_1, \ldots, \mathbf{q}_n; V, T)$, such that

$$n_n^N(\mathbf{q}_1, \ldots, \mathbf{q}_n; V, T) = \left(\frac{N}{V}\right)^n g_n{}^N(\mathbf{q}_1, \ldots, \mathbf{q}_n; V, T) \tag{11.14}$$

and

$$n_n^{\mu}(\mathbf{q}_1, \ldots, \mathbf{q}_n; V, T) = \left(\frac{\langle N\rangle}{V}\right)^n g_n^{\mu}(\mathbf{q}_1, \ldots, \mathbf{q}_n; V, T). \qquad (11.15)$$

The distribution functions, $g_n^{\mu(N)}(\mathbf{q}_1, \ldots, \mathbf{q}_n; V, T)$, will be useful in subsequent sections.

2. Thermodynamics and Reduced Probability Densities[1-3]

Let us now obtain relations between thermodynamic variables and the reduced two-body probability density in configuration space. The simplest quantity to consider is the internal energy, $U(V, T, N)$. The internal energy is just the expectation value of the Hamiltonian. In the canonical ensemble we can write,

$$U(T, V, N) = Z_N^{-1}(T, V)\frac{1}{N!\,h^{3N}}\iint d\mathbf{p}^N\, d\mathbf{q}^N H^N(\mathbf{p}^N, \mathbf{q}^N)\, e^{-\beta H^N(\mathbf{p}^N, \mathbf{q}^N)}$$

$$= \tfrac{3}{2}Nk_BT + \frac{1}{2}\iint d\mathbf{q}_1\, d\mathbf{q}_2 v(\mathbf{q}_{12}) n_2^N(\mathbf{q}_1, \mathbf{q}_2; V, T). \qquad (11.16)$$

The first term on the right-hand side is the kinetic contribution to the internal energy, and the second term is the contribution due to the interaction of the particles. For spherically symmetric potentials, $v(\mathbf{q}_{ij}) = v(q_{ij})$ where $q_{ij} = |\mathbf{q}_i - \mathbf{q}_j|$, and the two-particle reduced probability density takes the form $n_2^N(\mathbf{q}_1, \mathbf{q}_2; V, T) = n_2^N(\mathbf{q}_{12}; V, T)$. If we change integration variables to relative and center of mass coordinates and integrate over the center of mass, we find

$$U(V, T, N) = \tfrac{3}{2}k_BNT + \frac{1}{2}\frac{N^2}{V}\int dq 4\pi q^2 v(q) g_2^N(q; V, T), \qquad (11.17)$$

where $q \equiv q_{12}$ and $g_2^N(q; V, T)$ is the radial distribution function. The radial distribution function is extremely important because it completely characterizes the behavior of a classical fluid of spherically symmetric particles. It also has a direct physical interpretation. The quantity $(N/V)g(q)4\pi q^2\, dq$ is the average number of particles in a spherical shell of width $q \to q + dq$ at a distance q from any particle in the fluid.

The radial distribution function has been measured experimentally by X-ray scattering techniques and by means of molecular dynamics calculations. In Fig. 11.1 we show the form of the radial distribution function obtained from X-ray scattering experiments on argon. The dashed line in Fig. 11.2 shows a typical interparticle potential (Lennard-Jones 6–12). The parameter, σ, is a measure of the size of the hard core. The radial distribution function goes to zero at about the hard-core radius, indicating that no particles can penetrate the hard core. It has a maximum at about the distance of the minimum of the attractive well, and for larger distances it oscillates to one.

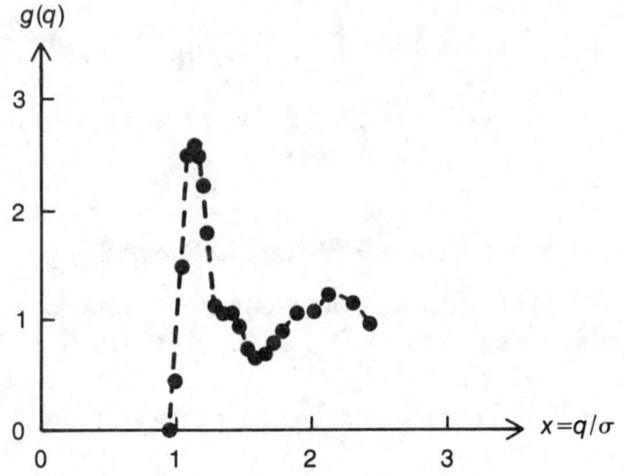

Fig. 11.1. A plot of the radial distribution function of argon at a temperature of 91.8 K and a pressure of 1.8 atm.: the points were obtained from X-ray scattering experiments; σ is a measure of the hard-core radius (based on Ref. 11.29).

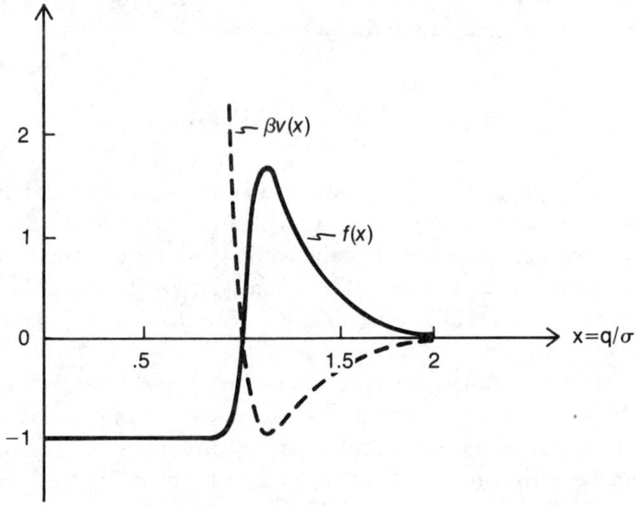

Fig. 11.2. The dashed line is a plot of a typical interparticle potential (Lennard-Jones 6–12) $v(x)$; the solid line is a plot of $f(x) = (\exp(-\beta v(x)) - 1)$; σ is a measure of the hard-core radius, $v(1) = 0$; the plots are given for a fixed temperature.

It is possible to express other thermodynamic quantities in terms of the radial distribution function. We will first compute the pressure by using a rather simple trick. Let us consider a box of volume $V = L^3$ (L is the length of a side) which contains N-particles. We will assume the box is large enough that we can neglect the effects of the walls (in the thermodynamic limit this is rigorously true). From Eqs. (2.92*b*) and (9.36), we can write

$$P = k_BT\left(\frac{\partial}{\partial V}\ln Z_N(V, T)\right)_{T,N} = k_BT\left(\frac{\partial}{\partial V}\ln Q_N(V, T)\right)_{T,N}$$

$$= \frac{k_BTL}{3V}\left(\frac{\partial \ln Q_N(V, T)}{\partial L}\right)_{T,N}. \tag{11.18}$$

The configuration integral can be written

$$Q^N(V, T) = \int_0^L dq_1 \cdots \int_0^L dq_{3N}\, e^{-\beta v^N(\mathbf{q}^N)} = V^N \int_0^1 \cdots \int_0^1 dx_1 \cdots dx_{3N}\, e^{-\beta v^N(L\mathbf{x}^N)}, \tag{11.19}$$

where we have made the change of variables $q_i = Lx_i$. The derivative of $Q^N(V, T)$ with respect to L gives

$$\left(\frac{\partial Q^N(V, T)}{\partial L}\right)_{T,N} = \frac{3N}{L} Q^N(V, T) - \beta V^N$$

$$\times \int_0^1 \cdots \int_0^1 dx_1 \cdots dx_{3N}\left(\frac{\partial v^N(L\mathbf{x}^N)}{\partial L}\right)_{N,T} e^{-\beta v^N(L\mathbf{x}^N)}. \tag{11.20}$$

If we now note

$$\frac{\partial v^N(L\mathbf{x}^N)}{\partial L} = \frac{1}{2}\sum_{(ij)} \frac{\partial \mathbf{q}_{ij}}{\partial L}\cdot\frac{\partial v(\mathbf{q}_{ij})}{\partial \mathbf{q}_{ij}} = \frac{1}{2L}\sum_{(ij)} \mathbf{q}_{ij}\cdot\frac{\partial v(\mathbf{q}_{ij})}{\partial \mathbf{q}_{ij}}, \tag{11.21}$$

we can combine Eqs. (11.18), (11.20), and (11.21) to give

$$\frac{P}{k_BT} = \frac{N}{V} - \frac{\beta}{6V}\iint d\mathbf{q}_1\, d\mathbf{q}_2 \mathbf{q}_{12}\cdot\frac{\partial v(\mathbf{q}_{12})}{\partial \mathbf{q}_{12}}\, n_2{}^N(\mathbf{q}_1, \mathbf{q}_2; V, T). \tag{11.22}$$

For spherically symmetric potentials, Eq. (11.22) takes the form

$$\frac{P}{k_BT} = \frac{N}{V} - \frac{\beta}{6}\left(\frac{N}{V}\right)^2 \int_0^\infty dq 4\pi q^2 q \frac{\partial v(q)}{\partial q}\, g_2{}^N(q; V, T) \tag{11.23}$$

and, again, we find that the pressure depends only on the radial distribution function. Eq. (11.23) is called the pressure equation.

We can find an expression for the chemical potential by using a slightly different trick. If the number of particles is very large, the chemical potential can be written

$$\mu = \left(\frac{\partial A}{\partial N}\right)_{V,T} \approx A(N, V, T) - A(N-1, V, T)$$

$$= -k_BT \ln\left(\frac{Q_N(V, T)}{Q_{N-1}(V, T)}\right) + k_BT \ln \frac{N!}{(N-1)!} + 3k_BT \ln \lambda_T, \tag{11.24}$$

where λ_T is the thermal wavelength. Let us now write the interaction potential in the form

$$v^N(\mathbf{q}^N, \gamma) = \sum_{j=2}^{N} \gamma v(\mathbf{q}_{1j}) + \sum_{2\le i<j\le N} v(\mathbf{q}_{ij}). \tag{11.25}$$

Then the configuration integral, defined in terms of $v^N(\mathbf{q}^N, \gamma)$, has the property that $Q_N(V, T; \gamma)_{\gamma=1} = Q_N(V, T)$ and $Q_N(V, T; \gamma)_{\gamma=0} = VQ_{N-1}(V, T)$. Therefore,

$$\ln \frac{Q_N(V, T)}{Q_{N-1}(V, T)} = \int_0^1 d\gamma \frac{d}{d\gamma} \ln Q_N(V, T) + \ln V \tag{11.26}$$

and

$$\mu = k_BT \ln\left(\frac{N\lambda_T{}^3}{V}\right) - k_BT \int_0^1 d\gamma \frac{d}{d\gamma} \ln Q_N(V, T; \gamma). \tag{11.27}$$

With a little algebra, we can show that

$$\frac{\mu}{k_BT} = \ln \frac{N\lambda_T{}^3}{V} + \frac{1}{Nk_BT}\int_0^1 d\gamma \iint d\mathbf{q}_1\, d\mathbf{q}_2 v(\mathbf{q}_{12}) n_2{}^N(\mathbf{q}_1, \mathbf{q}_2; \gamma, V, T). \tag{11.28}$$

For spherically symmetric potentials, we obtain the following expression for the chemical potential:

$$\frac{\mu}{k_BT} = \ln\left(\frac{N\lambda_T{}^3}{V}\right) + \frac{N}{V}\frac{1}{k_BT}\int_0^1 d\gamma \int_0^\infty dq 4\pi q^2 v(q) g_2{}^N(q; \gamma, V, T). \tag{11.29}$$

Thus, we obtain an expression for the chemical potential in terms of the radial distribution function.

As a final example, we will obtain an expression for the compressibility in the grand canonical ensemble. Let us first note that, from Eq. (11.11), we can write

$$\langle N\rangle = \int d\mathbf{q}_1 n_1{}^\mu(\mathbf{q}_1; V, T) \tag{11.30}$$

and

$$\langle N^2\rangle - \langle N\rangle = \iint d\mathbf{q}_1\, d\mathbf{q}_2 n_2{}^\mu(\mathbf{q}_1, \mathbf{q}_2; V, T). \tag{11.31}$$

If we combine Eqs. (11.30) and (11.31) with Eq. (9.61), we find

$$\frac{1}{\langle N\rangle}\iint d\mathbf{q}_1\, d\mathbf{q}_2[n_2{}^{\mu}(\mathbf{q}_1, \mathbf{q}_2; VT) - n_1{}^{\mu}(\mathbf{q}_1; V, T)n_1{}^{\mu}(\mathbf{q}_2; V, T)]$$

$$= \frac{\langle N\rangle k_B T\kappa_T}{V} - 1, \tag{11.32}$$

where κ_T is the isothermal compressibility. For a system with a spherically symmetric potential, Eq. (11.32) can be written in the form

$$k_B T\left(\frac{\partial n}{\partial P}\right)_{T,\langle N\rangle} = \left[1 + \langle n\rangle \int_0^{\infty} dq 4\pi q^2 [g_2{}^{\mu}(q; VT) - 1]\right] \tag{11.33}$$

where $\langle n\rangle = \langle N\rangle/V$. Eq. (11.33) is called the compressibility equation.

C. VIRIAL EXPANSION OF THE EQUATION OF STATE[1–5]

In order to obtain tractable microscopic expressions for the thermodynamic properties of fluids, we must find some way to approximate the many-body theory; that is, we must find a small parameter that we can use as an expansion parameter. For dilute or moderately dense gases, such a parameter is the density. In this section, we shall first obtain a microscopic expression for the virial expansion of the equation of state (density expansion) for a classical fluid, and then we shall compare the predictions of the microscopic theory to experimental results.

1. Virial Expansion and Cluster Functions

Let us consider a Maxwell-Boltzmann gas of identical particles of mass, m, which interact via two-body short-ranged forces. We shall assume that the potential has a large repulsive core and short-ranged attraction (cf. Fig. 11.2). For such a system the grand partition function can be written in the form

$$\mathscr{Z}(T, V, \mu) = \sum_{N=0}^{\infty} \frac{1}{N!}\frac{1}{\lambda_T^{3N}} e^{\beta N\mu} Q_N(V, T), \tag{11.34}$$

where the momentum integrations have been performed and the configuration integral, $Q_N(V, T)$, is defined in Eq. (11.6).

If the interparticle potential, v_{ij}, is short ranged, then for large values of separation between particles, $v_{ij} = 0$ and $\exp(-\beta v_{ij}) = 1$. It is convenient to introduce a function, f_{ij}, such that

$$f_{ij} = e^{-\beta v_{ij}} - 1. \tag{11.35}$$

The function f_{ij} becomes zero outside the range of the interaction. Furthermore, in the region of the hard core, where $v_{ij} = \infty$, the function $f_{ij} = -1$.

Thus f_{ij} is a much better expansion parameter than v_{ij} (cf. Fig. 11.2). In terms of the function f_{ij}, the configuration integral can be written as

$$Q_N(V, T) = \int \cdots \int d\mathbf{q}_1 \cdots d\mathbf{q}_N \prod_{(ij)}^{\frac{1}{2}N(N-1)} (1 + f_{ij}). \tag{11.36}$$

In Eq. (11.36), the product is taken over all pairs of particles (ij). There are $N(N - 1)/2$ such pairs.

It was first shown by Ursell[6] that the partition function can be written in terms of a cumulant expansion. We first write the configuration integral in the form

$$Q_N(V, T) = \int \cdots \int d\mathbf{q}_1 \cdots d\mathbf{q}_N W_N(\mathbf{q}_1, \ldots, \mathbf{q}_N), \tag{11.37}$$

so that

$$\mathscr{Z}(V, T, \mu) = \sum_{N=0}^{\infty} \frac{1}{N!} \frac{1}{\lambda_T^{3N}} e^{\beta N \mu} \int \cdots \int d\mathbf{q}_1 \cdots d\mathbf{q}_N W_N(\mathbf{q}_1, \ldots, \mathbf{q}_N). \tag{11.38}$$

We now define the following cumulant expansion for the grand partition function:

$$\mathscr{Z}(V, T, \mu) = \exp\left[\sum_{l=0}^{\infty} \frac{1}{l!} \frac{1}{\lambda_T^{3l}} e^{\beta l \mu} \int \cdots \int d\mathbf{q}_1 \cdots d\mathbf{q}_l U_l(\mathbf{q}_1, \ldots, \mathbf{q}_l)\right], \tag{11.39}$$

where $U_l(\mathbf{q}_1, \ldots, \mathbf{q}_l)$ is called a cluster function, or Ursell function. In terms of the cluster functions, the grand potential takes a simple form:

$$\begin{aligned} \Omega(V, T, \mu) &= -k_B T \ln \mathscr{Z}(V, T, \mu) \\ &= -k_B T \sum_{l=1}^{\infty} \frac{1}{l!} \frac{1}{\lambda_T^{3l}} e^{\beta l \mu} \int \cdots \int d\mathbf{q}_1 \cdots d\mathbf{q}_l U_l(\mathbf{q}_1, \ldots, \mathbf{q}_l). \end{aligned} \tag{11.40}$$

If we know the function $W_N(\mathbf{q}_1, \ldots, \mathbf{q}_N)$, then we can easily find the cluster functions $U_N(\mathbf{q}_1, \ldots, \mathbf{q}_N)$. We simply expand Eqs. (11.38) and (11.39) in powers of $(\lambda_T^{-3} \exp(\beta\mu))$ and equate coefficients. We then obtain the following hierarchy of equations:

$$U_1(\mathbf{q}_1) = W_1(\mathbf{q}_1), \tag{11.41a}$$

$$U_2(\mathbf{q}_1, \mathbf{q}_2) = W_2(\mathbf{q}_1, \mathbf{q}_2) - W_1(\mathbf{q}_1)W_1(\mathbf{q}_2), \tag{11.41b}$$

$$\begin{aligned} U_3(\mathbf{q}_1, \mathbf{q}_2, \mathbf{q}_3) = {} & W_3(\mathbf{q}_1, \mathbf{q}_2, \mathbf{q}_3) - W_1(\mathbf{q}_1)W_2(\mathbf{q}_2, \mathbf{q}_3) \\ & - W_1(\mathbf{q}_2)W_2(\mathbf{q}_1, \mathbf{q}_3) - W_1(\mathbf{q}_3)W_2(\mathbf{q}_1, \mathbf{q}_2) \\ & + 2W_1(\mathbf{q}_1)W_1(\mathbf{q}_2)W_1(\mathbf{q}_3), \end{aligned} \tag{11.41c}$$

etc. We can revert Eqs. (11.41) and find

$$W_1(\mathbf{q}_1) = U_1(\mathbf{q}_1), \tag{11.42a}$$

$$W_2(\mathbf{q}_1, \mathbf{q}_2) = U_2(\mathbf{q}_1, \mathbf{q}_2) + U_1(\mathbf{q}_1)U_1(\mathbf{q}_2), \tag{11.42b}$$

$$\begin{aligned} W_3(\mathbf{q}_1, \mathbf{q}_2, \mathbf{q}_3) = {} & U_3(\mathbf{q}_1, \mathbf{q}_2, \mathbf{q}_3) + U_1(\mathbf{q}_1)U_2(\mathbf{q}_2, \mathbf{q}_3) + U_1(\mathbf{q}_2)U_2(\mathbf{q}_1, \mathbf{q}_3) \\ & + U_1(\mathbf{q}_3)U_2(\mathbf{q}_1, \mathbf{q}_2) + U_1(\mathbf{q}_1)U_1(\mathbf{q}_2)U_1(\mathbf{q}_3), \end{aligned} \tag{11.42c}$$

etc. The cumulant expansion we have just introduced is independent of the form of $W_N(\mathbf{q}_1, \ldots, \mathbf{q}_N)$.

For the systems we are interested in, in this chapter, we can write

$$W_N(\mathbf{q}_1, \ldots, \mathbf{q}_N) = \prod_{(ij)}^{\frac{1}{2}N(N-1)} (1 + f_{ij}). \tag{11.43}$$

Thus, the first few terms are

$$W_1(\mathbf{q}_1) = 1, \tag{11.44a}$$

$$W_2(\mathbf{q}_1, \mathbf{q}_2) = (1 + f_{12}), \tag{11.44b}$$

$$W_3(\mathbf{q}_1, \mathbf{q}_2, \mathbf{q}_3) = (1 + f_{12})(1 + f_{13})(1 + f_{23}), \tag{11.44c}$$

$$W_4(\mathbf{q}_1, \mathbf{q}_2, \mathbf{q}_3, \mathbf{q}_4) = (1 + f_{12})(1 + f_{13})(1 + f_{14})(1 + f_{23})(1 + f_{24})(1 + f_{34}), \tag{11.44d}$$

etc. If we expand out the expressions for $W_N(\mathbf{q}_1, \ldots, \mathbf{q}_N)$, we see that the case $N = 1$ contains one term, the case $N = 2$ contains two terms, the case $N = 3$ contains eight terms, and the case $N = 4$ contains sixty-four terms. From Eqs. (11.41) we can find the first few cluster functions. They are

$$U_1(\mathbf{q}_1) = 1, \tag{11.45a}$$

$$U_2(\mathbf{q}_1, \mathbf{q}_2) = f_{12}, \tag{11.45b}$$

and

$$U_3(\mathbf{q}_1, \mathbf{q}_2, \mathbf{q}_3) = f_{12}f_{13} + f_{12}f_{23} + f_{13}f_{23} + f_{12}f_{13}f_{23}. \tag{11.45c}$$

The expression for $U_4(\mathbf{q}_1, \mathbf{q}_2, \mathbf{q}_3, \mathbf{q}_4)$ is too long to write down here, so we will leave it as an exercise. It contains thirty-eight terms. It is easy to see that the functions $W_N(\mathbf{q}_1, \ldots, \mathbf{q}_N)$ and $U_N(\mathbf{q}_1, \ldots, \mathbf{q}_N)$ rapidly become very complicated as N increases.

Since the grand partition function contains terms with arbitrarily large values of N, it is necessary to find a systematic procedure for categorizing and sorting various terms in the expressions for $W_N(\mathbf{q}_1, \ldots, \mathbf{q}_N)$ and $U_N(\mathbf{q}_1, \ldots, \mathbf{q}_N)$. Such a procedure is called graph theory. We can represent $W_N(\mathbf{q}_1, \ldots, \mathbf{q}_N)$ in terms of graphs in the following way:

$$W_N(\mathbf{q}_1, \ldots, \mathbf{q}_N) = \sum (\text{all different } N\text{-particle graphs}). \tag{11.46}$$

An N-particle graph is obtained by drawing N-numbered circles and connecting them together with any number of lines (including zero) subject only to the condition that any two circles cannot be connected by more than one line. Two N-particle graphs are different if the numbered circles are connected differently. An algebraic expression can be associated with an N-particle graph by assigning a factor f_{ij} to each pair of numbered circles that are connected by a line. For example, the 6-particle graph,

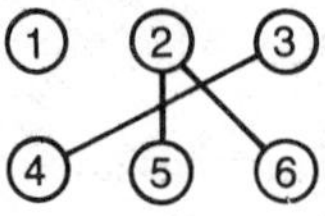

$$= f_{34}f_{25}f_{26} \tag{11.47}$$

and the 5-particle graph,

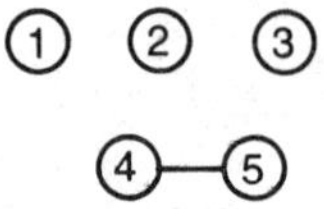

$$= f_{45}. \tag{11.48}$$

The graphical expression for $W_3(\mathbf{q}_1, \mathbf{q}_2, \mathbf{q}_3)$ can be written

$W_3(\mathbf{q}_1, \mathbf{q}_2, \mathbf{q}_3) =$  $+ 3$  $+ 3$  $+$  , (11.49)

where we have grouped together all graphs with the same topological structure before labels are attached.

From Eqs. (11.41), it is not difficult to see that the cluster functions can also be represented by graphs, but of a special form. We find that

$$U_l(\mathbf{q}_1, \ldots, \mathbf{q}_l) = \sum \text{(all different } l\text{-clusters)}. \tag{11.50}$$

An l-cluster is an l-particle graph in which all numbered circles are interconnected. Thus, in an l-cluster it must be possible to reach all numbered circles by continuously following the lines connecting the circles. Graphical expressions for the first few cluster functions are given by

$$U_1(\mathbf{q}_1) = \bullet, \tag{11.51a}$$

$$U_2(\mathbf{q}_1, \mathbf{q}_2) = \bullet\!\!-\!\!\bullet, \tag{11.51b}$$

$U_3(\mathbf{q}_1, \mathbf{q}_2, \mathbf{q}_3) = 3$  $+$  , (11.51c)

and

$U_4(\mathbf{q}_1, \mathbf{q}_2, \mathbf{q}_3, \mathbf{q}_4) = 12$  $+ 4$  $+ 12$ 

$+ 3$  $+ 6$ 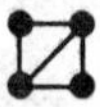 $+$ 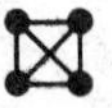 . (11.51d)

Again, we have grouped together all graphs which have the same topological structure before labels are attached.

We see from graph theory that the cluster functions have a special meaning.

An N-particle cluster function $U_N(\mathbf{q}_1, \ldots, \mathbf{q}_N)$ is nonzero only if all N-particles interact. Thus, they contain information about the way in which clusters occur among interacting particles in the gas. By writing the partition function in terms of a cumulant expansion, we have isolated the quantities which have physical content.

There is also another reason for writing $\mathscr{Z}(V, T, \mu)$ as a cumulant expansion, which does not become apparent until we try to integrate over the functions $W_N(\mathbf{q}_1, \ldots, \mathbf{q}_N)$. Let us consider some examples and restrict ourselves to 4-particle graphs.

(*a*) Each unconnected dot gives rise to a factor V. For example,

$$\int\cdots\int d\mathbf{q}_1\cdots d\mathbf{q}_4$$

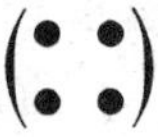

$$= \int\cdots\int d\mathbf{q}_1\cdots d\mathbf{q}_4 1 = V^4$$

and

$$\int\cdots\int d\mathbf{q}_1\cdots d\mathbf{q}_4$$

$$= \int\cdots\int d\mathbf{q}_1\cdots d\mathbf{q}_4 f_{24}f_{34}$$

$$= V\iiint d\mathbf{q}_2\, d\mathbf{q}_3\, d\mathbf{q}_4 f_{24}f_{34}.$$

(*b*) If two parts of a graph are not connected by a line, the corresponding algebraic expression factors into a product. For example,

$$\int\cdots\int d\mathbf{q}_1\cdots d\mathbf{q}_4$$

$$= \int\cdots\int d\mathbf{q}_1\cdots d\mathbf{q}_4 f_{12}f_{34}$$

$$= \left(\iint d\mathbf{q}_1\, d\mathbf{q}_2 f_{12}\right)\left(\iint d\mathbf{q}_3\, d\mathbf{q}_4 f_{34}\right).$$

(*c*) Each completely connected part of a graph (cluster) is proportional to the volume (this comes from integrating over the center of mass of a cluster). For example,

$$\int\cdots\int d\mathbf{q}_1\cdots d\mathbf{q}_4$$

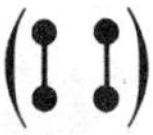

$$= \left(V\int d\mathbf{q}_{12} f_{12}\right)\left(V\int d\mathbf{q}_{34} f_{34}\right)$$

and

$$\int\cdots\int d\mathbf{q}_1\cdots d\mathbf{q}_4$$

$$= V\iiint d\mathbf{q}_1\, d\mathbf{q}_2\, d\mathbf{q}_3 f_{12}f_{23}f_{13}$$

$$= V^2\iint d\mathbf{q}_{31}\, d\mathbf{q}_{21} f(|\mathbf{q}_{31}|)f(|\mathbf{q}_{21}|)f(|\mathbf{q}_{31} - \mathbf{q}_{21}|),$$

where $\mathbf{q}_{31} = \mathbf{q}_3 - \mathbf{q}_1$ and $\mathbf{q}_{21} = \mathbf{q}_2 - \mathbf{q}_1$. In the second example, we have integrated over the position of particle 1 in the cluster. This gives a factor V since the cluster can be anywhere in the box. The positions of particles 2 and 3 are then specified relative to particle 1 (it is useful to draw a picture). Thus, the configuration integral $Q_N(V, T)$ contains terms which have volume dependence of the form V^n, where n can range anywhere from $n = 1$ to $n = N$. The integrals over cluster functions, however, are always proportional to the volume. We can write

$$b_l(V, T) = \frac{1}{V l!}\int \cdots \int d\mathbf{q}_1 \cdots d\mathbf{q}_l U_l(\mathbf{q}_1, \ldots, \mathbf{q}_l), \tag{11.52}$$

where $b_l(V, T)$ is called a cluster integral and, at most, depends on the density. Thus, the expression for the grand potential, $\Omega(V, T, \mu)$, in Eq. (11.40) is proportional to the volume. By rewriting the grand partition function as a cumulant expansion, we have rewritten it in terms of a single function which is proportional to the volume.

We are now in a position to obtain the virial expansion of the equation of state. From Eqs. (11.40) and (11.52), we may write the grand potential as

$$\Omega(V, T, \mu) = -Vk_BT \sum_{l=1}^{\infty} \frac{b_l(T, V)\, e^{\beta l \mu}}{\lambda_T^{3l}}. \tag{11.53}$$

The pressure then becomes

$$P = -\frac{\Omega(V, T, \mu)}{V} = k_BT \sum_{l=1}^{\infty} \frac{b_l(T, V)\, e^{\beta l \mu}}{\lambda_T^{3l}} \tag{11.54}$$

and the particle density is given by

$$\frac{\langle N\rangle}{V} = -\frac{1}{V}\left(\frac{\partial \Omega}{\partial \mu}\right)_{V,T} = \sum_{l=1}^{\infty} \frac{l b_l(V, T)\, e^{\beta l \mu}}{\lambda_T^{3l}}. \tag{11.55}$$

The virial expansion of the equation of state is an expansion in powers of the density,

$$\frac{PV}{\langle N\rangle k_BT} = \sum_{l=1}^{\infty} B_l(T)\left(\frac{\langle N\rangle}{V}\right)^{l-1}. \tag{11.56}$$

To determine the lth virial coefficient, B_l, let us for simplicity consider all equations in the thermodynamic limit ($V \to \infty$, $\langle N\rangle \to \infty$, such that $\langle N\rangle/V =$ constant). Then $b_l(T, V) \xrightarrow[V\to\infty]{} \bar{b}_l(T)$. If we combine Eqs. (11.54), (11.55), and (11.56), we obtain

$$\left(\sum_{l=1}^{\infty} \frac{\bar{b}_l(T)\, e^{\beta l \mu}}{\lambda_T^{3l}}\right)\left(\sum_{n=1}^{\infty} \frac{n\bar{b}_n(T)\, e^{\beta n \mu}}{\lambda_T^{3n}}\right)^{-1} = \sum_{l'=1}^{\infty} B_{l'}\left(\sum_{n'=1}^{\infty} \frac{n'\bar{b}_{n'}(T)\, e^{\beta n' \mu}}{\lambda_T^{3n'}}\right)^{l'-1}. \tag{11.57}$$

If we now expand both sides of Eq. (11.57) and equate coefficients of equal powers of $\lambda_T^{-3} \exp(\beta\mu)$, we obtain the following expressions for the first four virial coefficients:

$$B_1(T) = \bar{b}_1(T) = 1, \tag{11.58}$$

$$B_2(T) = -\bar{b}_2(T), \tag{11.59}$$

$$B_3(T) = 4\bar{b}_2^{\,2}(T) - 2\bar{b}_3(T), \tag{11.60}$$

$$B_4(T) = -20b_2^{\,3}(T) + 18\bar{b}_2(T)\bar{b}_3(T) - 3\bar{b}_4(T), \tag{11.61}$$

etc. The higher order terms in the virial expansion are determined by larger and larger clusters of particles.

The functions f_{ij} depend only on relative coordinates. Consequently, we can show that

$$\frac{1}{V}\int\int\int dq_1\, dq_2\, dq_3 \left(\text{⋀}\right) = \left[\int dq_1 \left(\text{•—•}\right)\right]^2.$$

Similarly,

$$\frac{1}{V}\int\cdots\int dq_1\cdots dq_4 \left(\text{⊓⊔}\right) = \frac{1}{V}\int\cdots\int dq_1\cdots dq_4 \left(\text{▱}\right) = \left[\int dq_1 \left(\text{•—•}\right)\right]^3$$

and

$$\frac{1}{V}\int\cdots\int dq_1\cdots dq_4 \left(\text{▱}\right) = \left[\int dq_1 \left(\text{•—•}\right)\right]\left[\int\cdots\int dq_2\, dq_3\, \text{⋀}\right].$$

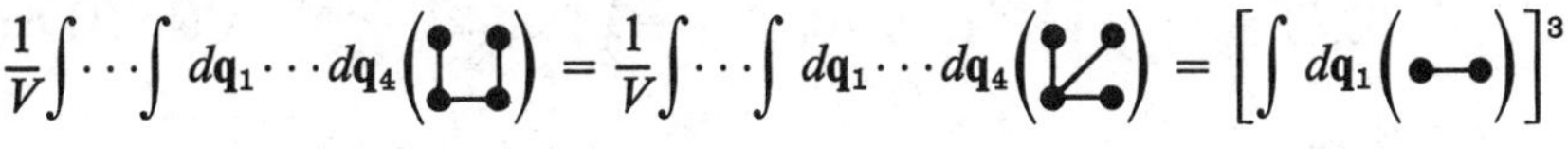

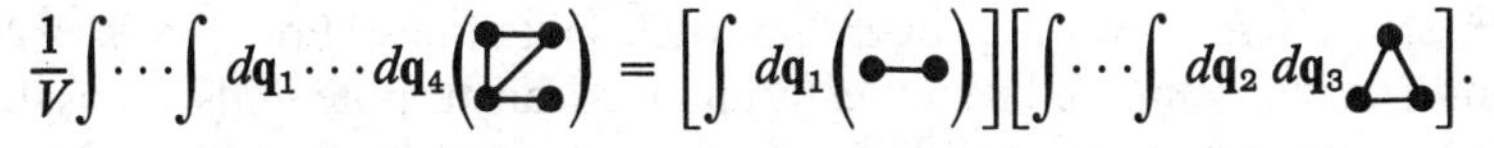

These identities enable us to simplify the expressions for the virial coefficients and write them as

$$B_n(V, T) = -\frac{1}{n}\frac{1}{(n-2)!}\int dq_1\cdots dq_n V_n(q_1, \ldots, q_n), \qquad (11.62)$$

where $V_n(q_1, \ldots, q_n)$ is called a star function and

$$V_n(q_1, \ldots, q_n) = \sum (\text{all different } n\text{-particle star graphs}). \qquad (11.63)$$

An n-particle star graph is a connected graph which is so tightly connected that if we remove any point and all lines that connect to it the remaining graph will still be connected.

In terms of graphs, the first few star functions are given by

$$V_2(q_1, q_2) = \text{•—•}, \qquad (11.64a)$$

$$V_3(q_1, q_2, q_3) = \text{⋀}, \qquad (11.64b)$$

and

$$V_4(q_1, q_2, q_3, q_4) = 3\,\text{⊡} + 6\,\text{▨} + \text{⊠}. \qquad (11.64c)$$

The numerical factors in front of each graph in Eq. (11.64c) indicate that there are that many ways to assign the coordinates to the points and obtain topologically different graphs. Thus, there are that many different graphs. Some higher order star functions are given in Ref. 11.5. The nth-order virial coefficient is composed only of tightly bound connected n-particle graphs.

We are now in a position to compare the predictions of this theory to experimental observations.

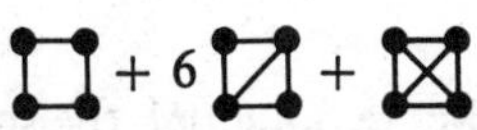

2. Second Virial Coefficient[1,7]

The second virial coefficient gives the correction to the ideal gas equation of state due to two-body clustering. For very dilute gases, two-body clusters give by far the dominant contribution to interaction effects in the fluid and it is sufficient to terminate the virial expansion at second order.

From Eqs. (11.35), (11.52), and (11.59), the second virial coefficient can be written

$$B_2(T) = -\frac{1}{2V}\iint d\mathbf{q}_1\, d\mathbf{q}_2 f(\mathbf{q}_{12}) = -\frac{1}{2}\int d\mathbf{q}_{12}(e^{-\beta v(\mathbf{q}_{12})} - 1), \quad (11.65)$$

where we have changed to center of mass and relative coordinates and have integrated over the center of mass. The behavior of the second virial coefficient has been studied for a variety of interparticle potentials. For very simple potentials it can be computed analytically and for realistic potentials it must be computed numerically. We shall focus on three potentials which historically have been important in understanding the behavior of the virial coefficients.

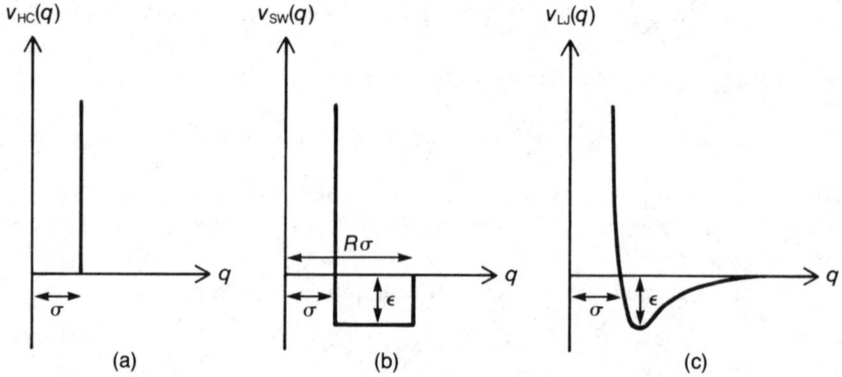

Fig. 11.3. Sketch of various interparticle potentials: (*a*) hard-core potential, (*b*) square-well potential, (*c*) Lennard-Jones 6–12 potential.

They are the hard-core potential, the square-well potential, and the Lennard-Jones 6–12 potential (cf. Fig. 11.3).

(*a*) *Hard-Core Potential.* The hard-core potential is shown in Fig. 11.3*a* and has the form

$$v_{HC}(q) = \begin{cases} \infty & q < \sigma \\ 0 & q > \sigma \end{cases} \quad (11.66)$$

where σ is the radius of the hard core. The second virial coefficient for this potential is simple to compute. We find

$$B_2(T)_{\rm HC} = -\frac{1}{2}\int_0^\sigma dq 4\pi q^2(-1) = \frac{2\pi\sigma^3}{3} \equiv b_0 \tag{11.67}$$

and thus $B_2(T)_{\rm HC}$ is independent of temperature. If we terminate the virial expansion at second order, the equation of state of a gas of hard spheres of radius, σ, takes the form

$$PV = \langle N\rangle k_B T\left[1 + \frac{2\pi\sigma^3}{3}\frac{\langle N\rangle}{V}\right]. \tag{11.68}$$

Thus, the hard core serves to increase the pressure for a given density, as we would expect.

(*b*) *Square-Well Potential.* The square-well potential is shown in Fig. 11.3*b* and has the form

$$v_{\rm SW}(q) = \begin{cases} \infty & 0 < q < \sigma \\ -\epsilon & \sigma < q < R\sigma \\ 0 & R\sigma < q. \end{cases} \tag{11.69}$$

The square-well potential has a hard core of radius, σ, and a square attractive region of depth ϵ and width $(R - 1)\sigma$. The second virial coefficient can be computed analytically and has the form

$$B_2(T)_{\rm SW} = \frac{2\pi\sigma^2}{3}\,[1 - (R^3 - 1)(e^{\beta\epsilon} - 1)]. \tag{11.70}$$

Note that $B_2(T)_{\rm SW}$ differs from $B_2(T)_{\rm HC}$ by a temperature-dependent term. At low temperatures it is negative and at high temperatures it becomes positive. At low temperatures the attractive interaction energy due to the square well can compete with the thermal energy, k_BT, and causes a lowering of the pressure relative to the ideal gas value. At high temperature the hard core becomes dominant and the pressure increases relative to that of an ideal gas.

We can write $B_2(T)_{\rm SW}$ in a reduced form if we let $B^*(T)_{\rm SW} = B(T)_{\rm SW}/b_0$ ($b_0 = 2\pi\sigma^3/3$) and $T^* = k_BT/\epsilon$. Then we find

$$B_2^*(T)_{\rm SW} = [1 - (R^3 - 1)(e^{1/T^*} - 1)]. \tag{11.71}$$

Eq. (11.71) will be useful when we compare the square-well results to experiment.

(*c*) *Lennard-Jones 6–12 Potential.* A potential which gives a very good approximation to the interaction between atoms is the Lennard-Jones 6–12 potential,

$$v_{\rm LJ}(q) = 4\epsilon\left[\left(\frac{\sigma}{q}\right)^{12} - \left(\frac{\sigma}{q}\right)^6\right] \tag{11.72}$$

(cf. Fig. 11.3*c*). The Lennard-Jones potential has a gradually sloping hard core, which takes account of the fact that particles with high energy can, to some extent, penetrate the hard core. When $q = \sigma$, $v_{LJ}(\sigma) = 0$. Thus, $q = \sigma$ is the radius of the hard core when the potential changes from positive to negative. The minimum of the Lennard-Jones potential occurs when $q = (2)^{1/6}\sigma$ and the value of the potential at the minimum is $v_{LJ}((2)^{1/6}\sigma) = -\epsilon$. Thus, ϵ is the depth of the Lennard-Jones potential.

The second virial coefficient for the Lennard-Jones potential can be found analytically in the form of a series expansion. If we integrate Eq. (11.65) by parts and introduce the notation $x = q/\sigma$, $T^* = k_B T/\epsilon$, and $B_2^*(T)_{LJ} = B_2(T)_{LJ}/b_0$, we find

$$B_2^*(T)_{LJ} = \frac{4}{T^*}\int_0^\infty dx\, x^2\left[\frac{12}{x^{12}} - \frac{6}{x^6}\right]\exp\left\{-\frac{4}{T^*}\left[\left(\frac{1}{x}\right)^{12} - \left(\frac{1}{x}\right)^{6}\right]\right\}. \tag{11.73}$$

If we expand $\exp[(4/T^*)(1/x)^6]$ in an infinite series, each term of the series can be computed analytically and we obtain the following expansion for $B_2^*(T)_{LJ}$:

$$B_2^*(T)_{LJ} = \sum_{n=0}^{\infty} \alpha_n \left(\frac{1}{T^*}\right)^{((2n+1)/4)} \tag{11.74}$$

where the coefficients α are defined in terms of gamma functions as

$$\alpha_{(n)} = -\frac{2}{4n!}\Gamma\left(\frac{2n-1}{4}\right). \tag{11.75}$$

Values of the coefficients, α_n, for $n = 0, \ldots, 40$ are given in Ref. 11.1 p. 1119. In Table 11.1, we give all values for $n = 0, \ldots, 12$. The expansion for $B_2^*(T)_{LJ}$

n	α_n	n	α_n
0	+1.7330010	7	−0.0228901
1	−2.5636934	8	−0.0099286
2	−0.8665005	9	−0.0041329
3	−0.4272822	10	−0.0016547
4	−0.2166251	11	−0.0006387
5	−0.1068205	12	−0.0002381
6	−0.0505458		

Table 11.1. Coefficients for the expansion of the second virial coefficient for Lennard-Jones 6–12 potential (based on Ref. 11.1).

converges rapidly for $T^* > 4$, but more slowly for lower values of T^*. Values of $B_2^*(T)_{LJ}$ for T^* ranging from 0.3 to 400 are given in Table 11.2.

T^*	B_2^*	T^*	B_2^*
0.30	−27.8806	4.00	+0.1154
0.40	−13.7988	4.50	+0.1876
0.50	−8.7202	5.00	+0.2433
0.70	−4.7100	10.00	+0.4609
1.00	−2.5381	20.00	+0.5254
1.50	−1.2009	30.00	+0.5269
2.00	−0.6276	40.00	+0.5186
2.50	−0.3126	50.00	+0.5084
3.00	−0.1152	100.00	+0.4641
3.50	+0.0190	400.00	+0.3583

Table 11.2. Values of the reduced second virial coefficient versus the reduced temperature for the Lennard-Jones potential (based on Ref. 11.1).

In Fig. 11.4, we plot $B_2^*(T)$ versus T^* for both the square-well potential and the Lennard-Jones potential. We also give experimental values of $B_2^*(T)$ for a variety of substances. The Lennard-Jones potential gives values of $B_2^*(T)$ in

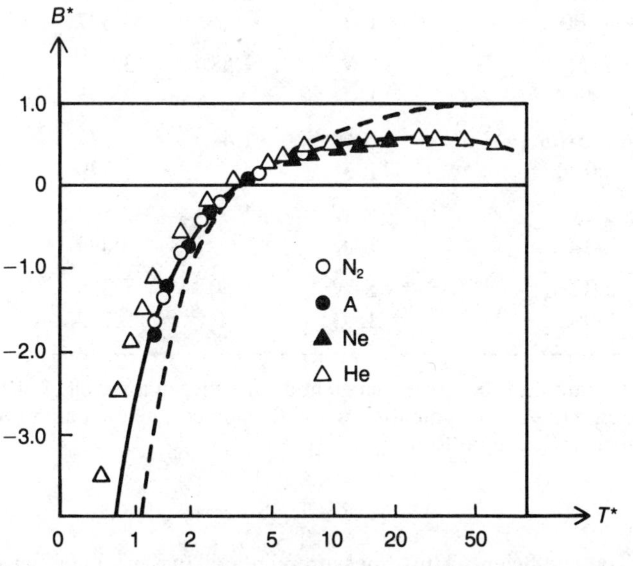

Fig. 11.4. The reduced second virial coefficient: the solid line is the calculated curve for the Lennard-Jones 6–12 potential; the dashed line is the calculated curve for the square-well potential (for $R = 1.6$); the points are experimental values for a variety of gases (based on Ref. 11.1).

good agreement with experimental results. At high temperatures $B_2^*(T)_{LJ}$ and the experimental points for He gas exhibit a maximum, while $B_2^*(T)_{SW}$ does not. The maximum occurs because at high temperatures real particles can penetrate into the hard core and lower the amount of excluded volume. The square-well potential has a hard core with infinite slope and cannot account for this effect, while the Lennard-Jones potential, with a sloping hard core, can account for it. The data points for He deviate from the classical results at low temperature. These deviations are due to quantum effects and will be discussed in Sec. 11.I. The second virial coefficients for all classical gases, when plotted in terms of reduced quantities, are identical. This is an example of the law of corresponding states.

In Table 11.3, we have given a list of values for the parameters ϵ/k_B and σ for various substances, assuming they interact via square-well or Lennard-Jones potentials. These parameters are obtained from experimental values for the

	Potential	R	σ (Å)	ϵ/k_B (K)
Argon (A)	S-W	1.70	3.067	93.3
190°C → 600°C	L-J		3.504	117.7
Krypton (Kr)	S-W	1.68	3.278	136.5
160°C → 600°C	L-J		3.827	164.0
Nitrogen (N_2)	S-W	1.58	3.277	95.2
150°C → 400°C	L-J		3.745	95.2
Carbon dioxide (CO_2)	S-W	1.44	3.571	283.6
0°C → 600°C	L-J		4.328	198.2
Xenon (Xe)	S-W	1.64	3.593	198.5
0°C → 700°C	L-J		4.099	222.3
Methane (CH_4)	S-W	1.60	3.355	142.5
0°C → 350°C	L-J		3.783	148.9

Table 11.3. Values of the parameters σ and ϵ for the square-well (S-W) and Lennard-Jones (L-J) 6–12 potentials taken from experimental data on the second virial coefficient (based on Ref. 11.8).

second virial coefficient. Thus, measurements of the virial coefficients of real gases provide an extremely important way of determining the form of the effective interparticle potential for various molecules. The value of ϵ can be found if we take experimental values of $B_2(T)$ for two different temperatures, say T_1 and T_2, equate the two ratios

$$\left(\frac{B_2(T_2)}{B_1(T_1)}\right)_{\text{exper}} = \left(\frac{B_2^*(\epsilon, T_2)_{\text{theor}}}{B_2^*(\epsilon, T_1)_{\text{theor}}}\right), \tag{11.76}$$

and solve for ϵ. In this ratio, all dependence on σ is eliminated. Once we find ϵ, we can find σ if we note that

$$B_2(T)_{\text{exper}} = \frac{2\pi\sigma^3}{3} B_2^*(\epsilon, T)_{\text{theor}}. \tag{11.77}$$

The only unknown in Eq. (11.77) is σ.

The Lennard-Jones 6–12 potential is perhaps the most widely used interparticle potential, but there are many other forms of potential that may be used to compute the virial coefficients, and some of them give better agreement with experimental results over a wider range of temperature than does the Lennard-Jones potential.

3. Third Virial Coefficient[1,3,7,8]

The third virial coefficient contains the contribution from three-body clusters in the gas and may be written in the form

$$B_3(T) = -\frac{1}{3V}\iiint d\mathbf{q}_1\, d\mathbf{q}_2\, d\mathbf{q}_3 U_3(\mathbf{q}_1, \mathbf{q}_2, \mathbf{q}_3) + \frac{1}{V^2}\left(\iint d\mathbf{q}_1\, d\mathbf{q}_2 U_2(\mathbf{q}_1, \mathbf{q}_2)\right)^2. \tag{11.78}$$

Eq. (11.78) is completely general and does not depend on the form of the interparticle potential. In deriving microscopic expressions for the cluster functions, we have assumed that the N-particle potential was additive, that is, that it could be written as the sum of strictly two-body interactions:

$$v^N(\mathbf{q}^N) = \sum_{(ij)}^{\frac{1}{2}N(N-1)} v_{ij}(\mathbf{q}_{ij}). \tag{11.79}$$

In fact, in a real gas this is not true. For example, if three bodies in a gas interact simultaneously, they become polarized and an additional three-body polarization interaction occurs. This polarization interaction has a significant effect on the third virial coefficients at low temperatures and must be included.[9] Let us write the total three-body interaction in the form

$$v^3(\mathbf{q}^3) = v(\mathbf{q}_{12}) + v(\mathbf{q}_{13}) + v(\mathbf{q}_{23}) + \Delta v_{123}(\mathbf{q}_1, \mathbf{q}_2, \mathbf{q}_3) \tag{11.80}$$

where Δv_{123} is the three-body polarization interaction. The polarization interaction has been computed by several methods[10,11] and has the form

$$\Delta v_{123} = \frac{\nu(1 + 3\cos\gamma_1 \cos\gamma_2 \cos\gamma_3)}{q_{12}q_{13}q_{23}}, \tag{11.81}$$

where q_{ij} are the lengths of the sides of the atomic triangle, γ_i are the internal angles, and ν is a parameter that depends on the polarizability and properties of the two-body interaction. Thus, the polarization interaction is repulsive.

If we include the correction due to polarization effects, we can write the third virial coefficient as

$$B_3(T) = B_3(T)^{\text{add}} + \Delta B_3(T), \tag{11.82}$$

where

$$B_3(T)^{\text{add}} = -\frac{1}{3V}\iiint d\mathbf{q}_1\, d\mathbf{q}_2\, d\mathbf{q}_3 f(\mathbf{q}_{12})f(\mathbf{q}_{13})f(\mathbf{q}_{23}) \tag{11.83}$$

and

$$\Delta B_3(T) = -\frac{1}{3V}\iiint d\mathbf{q}_1\, d\mathbf{q}_2\, d\mathbf{q}_3(e^{-\beta\Delta v_{123}} - 1)\, e^{-\beta(v_{12}+v_{23}+v_{13})}. \tag{11.84}$$

The third virial coefficients for additive potentials and the corrections due to nonadditive polarization effects have been obtained for a number of potentials. We shall again consider the three potentials discussed in Sec. 11.C.2.

(*a*) *Hard-Core Potential.* The third virial coefficient for hard-core potentials is fairly easy to compute. We shall use a very elegant method due to Katsura[12] because it can be generalized to the case of square-well potentials and higher order virial coefficients. It is also possible to compute the third virial coefficient by means of a geometrical analysis.

Let us first integrate Eq. (11.83) over the position of particle 1 and write it in the form

$$B_3(T)^{\text{add}} = -\frac{1}{3}\iint d\mathbf{q}_{31}\, d\mathbf{q}_{21} f(|\mathbf{q}_{31}|)f(|\mathbf{q}_{21}|)f(|\mathbf{q}_{31} - \mathbf{q}_{21}|). \tag{11.85}$$

The Katsura method involves a Fourier transformation of the quantity $f(q)$. If we assume that the interaction potential depends only on the magnitude of the distance between particles (as it does for the hard core), then we can introduce the transform

$$\gamma(t) = \left(\frac{1}{2\pi}\right)^{3/2}\int d\mathbf{q}\, e^{-i\mathbf{t}\cdot\mathbf{q}} f(q) = \left(\frac{2}{\pi}\right)^{1/2}\int_0^\infty dq\, \frac{q}{t}\, f(q)\sin(tq), \tag{11.86}$$

where t is the magnitude of the vector $\mathbf{t}$, q is the magnitude of $\mathbf{q}$, and $\mathbf{t}\cdot\mathbf{q} = tq\cos\theta$, where θ is the angle between $\mathbf{t}$ and $\mathbf{q}$. The inverse transform is

$$f(q) = \left(\frac{1}{2\pi}\right)^{3/2}\int d\mathbf{t}\, e^{i\mathbf{t}\cdot\mathbf{q}}\gamma(t) = \left(\frac{2}{\pi}\right)^{1/2}\int_0^\infty dt\, \frac{t}{q}\, \gamma(t)\sin(tq). \tag{11.87}$$

For hard-core particles $f(q) = -1$ for $0 < q < \sigma$ and $f(q) = 0$ for $\sigma < q$. Thus,

$$\gamma(t) = -\left(\frac{2}{\pi}\right)^{1/2} \int_0^\sigma dq \frac{q}{t} \sin(tq) = -\sigma^3 \frac{J_{3/2}(\sigma t)}{(\sigma t)^{3/2}}, \tag{11.88}$$

where $J_{3/2}(\sigma t)$ is a Bessel function. If we note that

$$f(|\mathbf{q}_{31} - \mathbf{q}_{21}|) = \left(\frac{1}{2\pi}\right)^{3/2} \int dt\, e^{it\cdot(\mathbf{q}_3 - \mathbf{q}_2)} \gamma(t), \tag{11.89}$$

we can write, after some integration,

$$\begin{aligned} B_3(T)_{\text{HC}}^{\text{add}} &= -\frac{(2\pi)^{3/2}}{3} \int dt \gamma^3(t) \\ &= \frac{4\pi(2\pi)^{3/2}\sigma^6}{3} \int_0^\infty dx [J_{3/2}(x)]^3 x^{-5/2}. \end{aligned} \tag{11.90}$$

The integral in Eq. (11.90) can be done[12] and yields

$$B_3(T)_{\text{HC}}^{\text{add}} = \tfrac{5}{8} b_0^2 \tag{11.91}$$

for the additive contribution to the hard-core third virial coefficient. Thus, $B_3(T)_{\text{HC}}^{\text{add}}$ is temperature independent and positive and increases the pressure.

(b) Square-Well Potential. An expression for the additive third virial coefficient for the square-well potential was first obtained analytically by Kihara.[1,7] For $R \leqslant 2$, it has the form

$$\begin{aligned} B_3(T)_{\text{SW}}^{\text{add}} = \tfrac{1}{8} b_0^2 [5 &- (R^6 - 18R^4 + 32R^3 - 15)x \\ &- (2R^6 - 36R^4 + 32R^3 + 18R^2 - 16)x^2 \\ &- (6R^6 - 18R^4 + 18R^2 - 6)x^3], \end{aligned} \tag{11.92a}$$

while for $R \geqslant 2$, it has the form

$$\begin{aligned} B_3(T)_{\text{SW}}^{\text{add}} = \tfrac{1}{8} b_0^2 [5 &- 17x - (-32R^3 + 18R^2 + 48)x^2 \\ &- (5R^6 - 32R^3 + 18R^2 + 26)x^3] \end{aligned} \tag{11.92b}$$

where $x = [e^{\beta\epsilon} - 1]$. In computing the third virial coefficient, the values of R, ϵ, and σ obtained from the second virial coefficient are generally used. Sherwood and Prausnitz have plotted values of $B_3(T)_{\text{SW}}^{\text{add}}$ and the correction $\Delta B_3(T)_{\text{SW}}$ for a variety of values of R. Their results are shown in Fig. 11.5. At low temperature, the contribution from polarization effects can have a large effect. The corrected values give better agreement with experimental results than does the additive part alone. In Fig. 11.6, we compare the additive and corrected results with experimental values of $B_3(T)$ for argon.

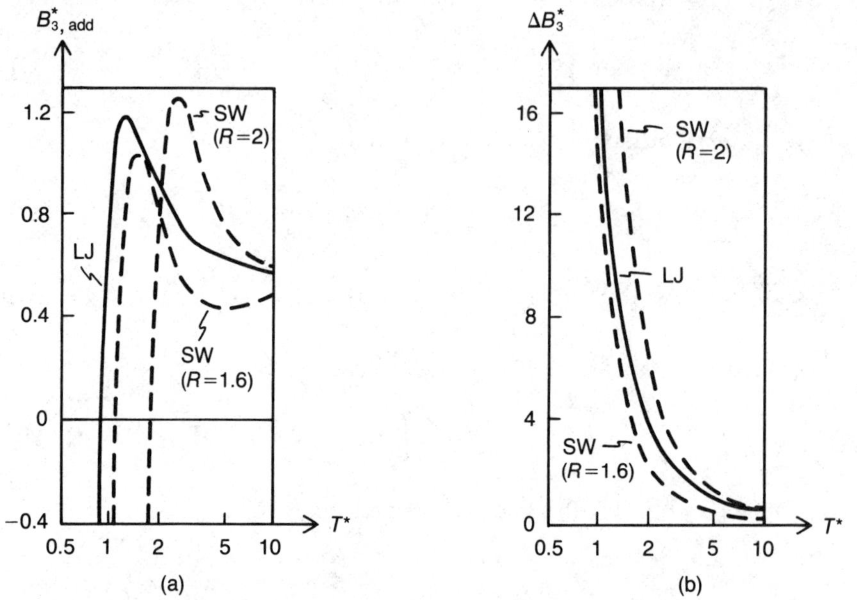

Fig. 11.5. The third virial coefficient: (*a*) the additive contribution as calculated for the Lennard-Jones 6–12 potential and the square-well potential for $R = 1.6$ and $R = 2$; (*b*) the polarization contribution for the potentials in (*a*) (based on Ref. 11.8).

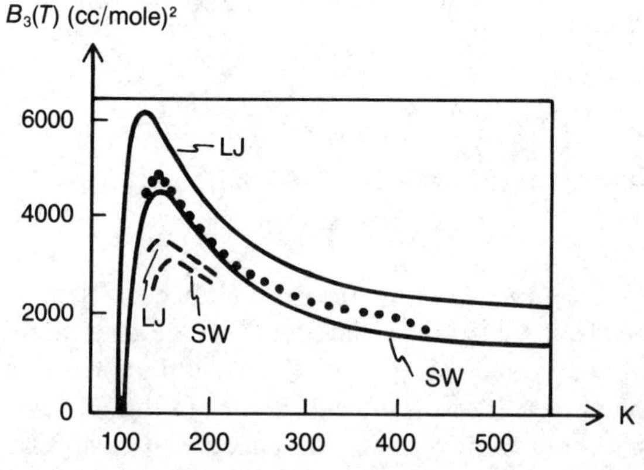

Fig. 11.6. Comparison of observed and calculated third virial coefficient for argon: solid lines include the polarization effects; the dashed lines give only the additive contribution (based on Ref. 11.8).

(*c*) *Lennard-Jones Potential.* The additive third virial coefficient for the Lennard-Jones potential can be obtained analytically in the form of a series expansion in a manner similar to that used for the second virial coefficient. However, it is much more complicated. The series expansion takes the form

$$B_3(T)_{\mathrm{LJ}} = b_0^2 \sum_{n=0}^{\infty} \beta_n \left(\frac{1}{T^*}\right)^{-(n+1)/2}. \tag{11.93}$$

A table of values for the coefficients β_n may be found in Ref. 11.1; we will not give them here. Curves for $B_3(T)_{\mathrm{LJ}}^{\mathrm{add}}$ and $\Delta B_3(T)_{\mathrm{LJ}}$ are shown in Fig. 11.5. They are very similar to the curves for the square-well potential. The Lennard-Jones curves for argon are compared with experimental results in Fig. 11.6. The Lennard-Jones potential does not give extremely good agreement with the experimental values for argon, probably because the potential well has insufficient curvature.[8]

4. Higher Order Virial Coefficients[13]

As we increase the density or decrease the temperature of a gas, at some point it will undergo a transition to a liquid or a solid phase. Past this point, and well before we reach it, the virial expansion ceases to give good agreement with the observed equation of state. It is possible to obtain some idea of the limits of the virial expansion for a hard-sphere gas. Although hard-sphere gases do not exist in nature, they can be constructed on a computer by means of molecular dynamics calculations. In these calculations, the motion of a finite number of particles is followed by solving the equations of motion of the system. The phase space coordinates of all the particles can be obtained at any given time, and the phase functions of interest can be computed. The thermodynamic properties are obtained by time averaging of the phase functions of interest.

The fourth, fifth, and sixth virial coefficients for systems of hard spheres have been computed[13] and the results are

$$B_4(T) = 0.28695 b_0^3, \tag{11.94a}$$

$$B_5(T) = (0.1103 \pm 0.0003) b_0^4, \tag{11.94b}$$

and

$$B_6(T) = (0.0386 \pm 0.0004) b_0^5. \tag{11.94c}$$

The virial expansion for the equation of state is plotted in Fig. 11.7*a* and compared to "experimental" results from molecular dynamics calculations. In the figure, a curve is given for the virial expansion up to and including $B_5(T)$ and the virial expansion up to and including $B_6(T)$. The agreement gets better as we add more terms, but it is not extremely good at higher densities. Better agreement occurs if we construct the so-called Padé approximate to the virial expansion. The Padé approximate is obtained as follows. If we are given a finite expansion for some function $F(x)$,

$$F(x) = a_0 + a_1 x + \cdots + a_n x^n, \tag{11.95}$$

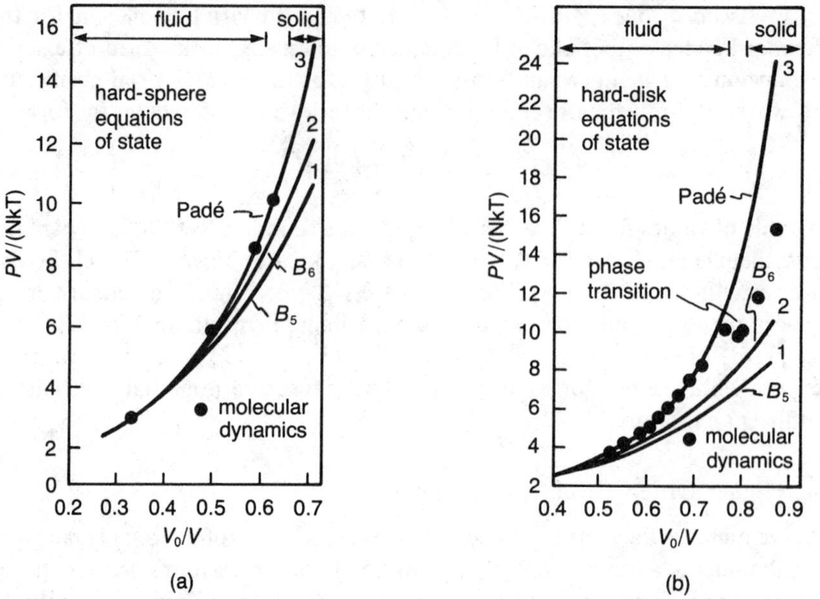

Fig. 11.7. Equation of state for hard-sphere and hard-disk fluid: (*a*) hard-spheres—line 1 includes all virial coefficients up to B_5; line 2 includes all virial coefficients up to B_6; line 3 gives the prediction of the Padé approximate; molecular dynamics results are indicated by the dots; (*b*) hard disks; all quantities have meaning similar to part (*a*) (based on Ref. 11.13).

we can construct the Padé approximate to $F(x)$ by writing it in the form

$$F(x) = \frac{b_0 + b_1 x + b_2 x^2 + \cdots + b_m x^m}{c_0 + c_1 x + \cdots + c_l x^l}. \tag{11.96}$$

If we expand Eq. (11.96) in a power series, we can find the coefficients b_i and c_j from the coefficients a_n. If we know a finite number of coefficients, a_n, then we can find a finite number of coefficients b_i and c_j. By doing this, we replace a finite series expansion by an infinite series expansion.

Ree and Hoover[13] have constructed the Padé approximate to the equation of state for a hard-sphere gas. They obtain

$$\frac{PV}{Nk_BT} = 1 + b_0 \frac{N}{V} \frac{\left(1 + 0.063507 b_0 \frac{N}{V} + 0.017329 b_0{}^2 \left(\frac{N}{V}\right)^2\right)}{\left(1 - 0.561493 b_0 \frac{N}{V} + 0.081313 b_0{}^2 \left(\frac{N}{V}\right)^2\right)}. \tag{11.97}$$

The Padé approximate to the equation of state is plotted in Fig. 11.7*a*. It gives much better results than the simple virial expansion. It is worth noting that the hard-sphere gas exhibits a phase transition from a fluid state to a crystalline state as the density increases. While we will not give the details here, Fig. 11.7*b* shows the results of an analogous calculation for a system of hard disks. The drop in the pressure at about $V_0/V = 0.75$ indicates a transition to a crystalline state. The Padé approximate gives poor agreement for the crystal phase.

Expressions for higher virial coefficients for square-well potentials have been obtained, but, as we can see from the results for hard spheres, a simple virial expansion does not give very good agreement for dense systems. For dense systems, other techniques must be developed, and we shall consider some of them in later sections.

D. VIRIAL EXPANSION OF THE REDUCED PROBABILITY DENSITIES[3,5,14]

Microscopic theories of the liquid state generally focus on the reduced probability densities and, in particular, on the radial distribution function. Therefore, it is useful to find a general expression for the virial expansion of the reduced distribution functions. This can be done quite easily if we introduce a generating function $\mathscr{L}(V, T, \mu; \{\theta\})$ for the n-particle reduced probability density:

$$\mathscr{L}(V, T, \mu; \{\theta\}) \equiv \mathscr{Z}^{-1}(V, T, \mu) \sum_{N=0}^{\infty} \frac{1}{N!} x^N \int d\mathbf{q}_1 \cdots d\mathbf{q}_N W_N(\mathbf{q}_1, \ldots, \mathbf{q}_N; \{\theta\}) \tag{11.98}$$

where

$$W_N(\mathbf{q}_1, \ldots, \mathbf{q}_N; \{\theta\}) \equiv \theta_1\theta_2 \times \cdots \times \theta_N\, e^{-\beta V^N(\mathbf{q}^N)}, \tag{11.99}$$

$\theta_i \equiv \theta(\mathbf{q}_i)$ is some function of $\mathbf{q}_i$, and we have let $x = \lambda_T^{-3} e^{\beta\mu}$; $\mathscr{L}(V, T, \mu; \{\theta\})$ is a functional of the quantities θ_i. If we take the variational derivative of $\mathscr{L}(V, T, \mu; \{\theta\})$ with respect to θ_i and at the end set $\theta_i \to 1$ and $\delta\theta_i/\delta\theta_1 = \delta(\mathbf{q}_i - \mathbf{q}_1)$ ($\delta(\mathbf{q}_i - \mathbf{q}_1)$ is the Dirac delta function), the reduced probability density may be written in the form

$$n_1^{\mu}(\mathbf{q}_1; V, T) = \lim_{\{\theta\}\to 1} \frac{\delta\mathscr{L}}{\delta\theta_1}, \tag{11.100a}$$

$$n_2^{\mu}(\mathbf{q}_1, \mathbf{q}_2; V, T) = \lim_{\{\theta\}\to 1} \frac{\delta^2\mathscr{L}}{\delta\theta_1\, \delta\theta_2}, \tag{11.100b}$$

and, in general,

$$n_n^{\mu}(\mathbf{q}_1, \ldots, \mathbf{q}_n; V, T) = \lim_{\{\theta\}\to 1} \frac{\delta^n\mathscr{L}}{\delta\theta_n \times \cdots \times \delta\theta_2\, \delta\theta_1}. \tag{11.100c}$$

The order in which we take the variational derivative does not matter.

We can now write the generating function in terms of a cumulant expansion:

$$\mathscr{L}(V, T, \mu; \{\theta\}) = \mathscr{Z}^{-1}(V, T, \mu) \exp[\mathscr{K}(V, T, \mu; \{\theta\})] \tag{11.101}$$

where

$$\mathcal{K}(V, T, \mu; \{\theta\}) = \sum_{l=1}^{\infty} \frac{1}{l!} x^l \int dq_1 \cdots dq_l U_l(q_1, \ldots, q_l; \{\theta\}). \quad (11.102)$$

As before, the cluster function $U_l(q_1, \ldots, q_l; \{\theta\})$ can be determined from Eqs. (11.41) and (11.99). In the limit $\{\theta\} \to 1$, we have

$$\lim_{\{\theta\} \to 1} \exp\{\mathcal{K}(V, T, \mu; \{\theta\})\} = \mathcal{L}(V, T, \mu) \quad (11.103)$$

and

$$\lim_{\{\theta\} \to 1} \mathcal{L}(V, T, \mu; \{\theta\}) = 1. \quad (11.104)$$

If we now substitute Eq. (11.101) into Eqs. (11.100) and use Eq. (11.103), we find the following cluster expansion for the reduced probability densities:

$$n_1{}^\mu(q_1; V, T) = \lim_{\{\theta\} \to 1} \frac{\delta \mathcal{K}}{\delta \theta_1} \quad (11.105)$$

and

$$n_2{}^\mu(q_1, q_2; V, T) = \lim_{\{\theta\} \to 1} \left[\frac{\delta^2 \mathcal{K}}{\delta \theta_2 \, \delta \theta_1} + \frac{\delta \mathcal{K}}{\delta \theta_2} \frac{\delta \mathcal{K}}{\delta \theta_1} \right], \quad (11.106)$$

etc. We will not need the higher order reduced probability densities, so we will not write them down here, but they are easily found.

Let us now note that the cluster expansion of $\mathcal{K}(V, T, \mu; \{\theta\})$ has the same structure as that of the grand potential, and the cluster functions $U_l(q_1, \ldots, q_l; \{\theta\})$ have the same form as the Ursell functions and can be expressed in terms of the same graphs. The only difference is that, now, for each point i of a graph, we multiply by a factor θ_i in addition to multiplying by a function f_{ij} for each line connecting points i and j. The graphical expression for $\mathcal{K}(V, T, \mu; \{\theta\})$, to fourth order in x, has the form

$$\mathcal{K}(V, T, \mu; \{\theta\}) = x \int dq_1 \left(\bullet \right) + x^2 \iint dq_1 \, dq_2 \left(\frac{1}{2} \, \bullet\!\!-\!\!\bullet \right)$$

$$+ x^3 \int \cdots \int dq_1 \, dq_2 \, dq_3 \left(\frac{1}{2} \bigwedge + \frac{1}{6} \triangle \right)$$

$$+ x^4 \int \cdots \int dq_1 \cdots dq_4$$

$$\times \left(\frac{1}{2} \sqcup\!\!\!\cdot + \frac{1}{6} \diagup\!\!\!\diagdown + \frac{1}{2} \diagup\!\!\!Z + \frac{1}{8} \square + \frac{1}{4} \boxtimes + \frac{1}{24} \boxtimes \right)$$

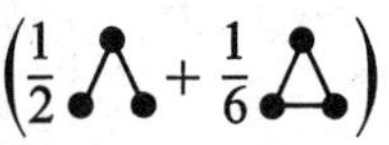

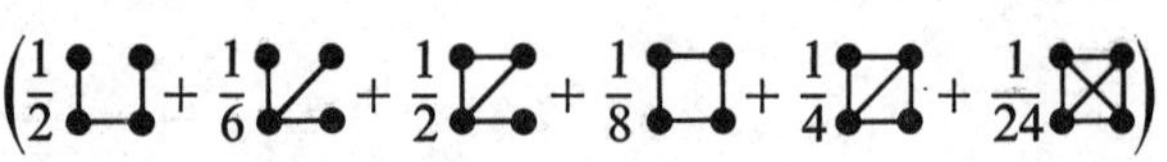

$$+ \cdots \quad (11.107)$$

The numerical factor in front of each graph is the inverse of its symmetry number. It prevents us from counting the contribution from a particular graph more than once when the integrations are performed. The symmetry number is equal to the number of ways of relabeling the graph and leaving it topologically unchanged. When we take a variational derivative of $\mathscr{K}(V, T, \mu; \{\theta\})$ and take the limit $\{\theta\} \to 1$, we in effect remove one point and one integration variable from each of the graphs in all possible ways. We replace the removed point by an open circle. If we then regroup all graphs of identical topological structure, we obtain the following graphical expression for the first variational derivative:

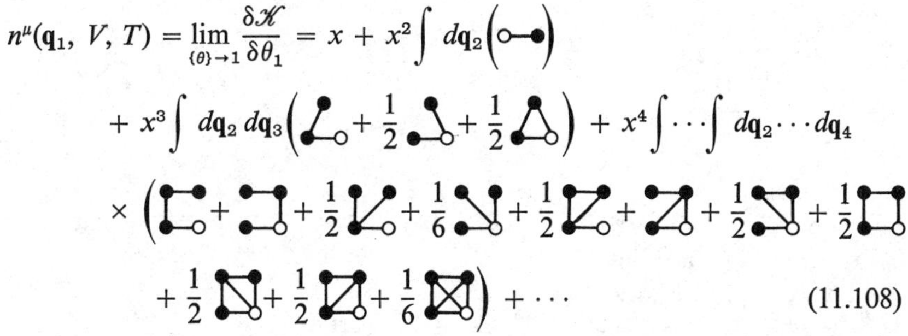

$$n^\mu(\mathbf{q}_1, V, T) = \lim_{\{\theta\} \to 1} \frac{\delta \mathscr{K}}{\delta \theta_1} = x + x^2 \int d\mathbf{q}_2 \left(\circ\!\!-\!\!\bullet \right)$$

$$+ x^3 \int d\mathbf{q}_2\, d\mathbf{q}_3 \left(\mathbf{L}_\circ + \frac{1}{2} \mathbf{\Lambda} + \frac{1}{2} \mathbf{\triangle} \right) + x^4 \int \cdots \int d\mathbf{q}_2 \cdots d\mathbf{q}_4$$

$$\times \left(\mathbf{\Gamma}_\circ + \mathbf{\square} + \frac{1}{2} \mathbf{Z}_\circ + \frac{1}{6} \mathbf{N} + \frac{1}{2} \mathbf{Z} + \mathbf{\triangle} + \frac{1}{2} \mathbf{N} + \frac{1}{2} \mathbf{\square} \right.$$

$$\left. + \frac{1}{2} \mathbf{N} + \frac{1}{2} \mathbf{Z} + \frac{1}{6} \mathbf{\boxtimes} \right) + \cdots \tag{11.108}$$

The open circles are called root, or base, points and the points that are integrated over are called field points.

Because the limit $\{\theta\} \to 1$ is taken, $n_1{}^\mu(\mathbf{q}_1, V, T)$ does not depend on the functions θ_i any longer. Furthermore, each open circle is labeled by the coordinate q_1. We can assign an algebraic expression to each graph by assigning a factor, f_{ij}, to each line connecting points i and j, including the lines attaching to point 1. Since the function f_{ij} depends only on the relative position of points i and j, the one-body reduced probability density, $n_1{}^\mu(\mathbf{q}_1; V, T)$, is independent of $\mathbf{q}_1$ and reduces to Eq. (11.55), as it should. Thus, $n_1{}^\mu(\mathbf{q}_1; V, T) = N/V$. If the Hamiltonian contained a space-dependent external field, this would no longer be true.

The second variational derivative yields

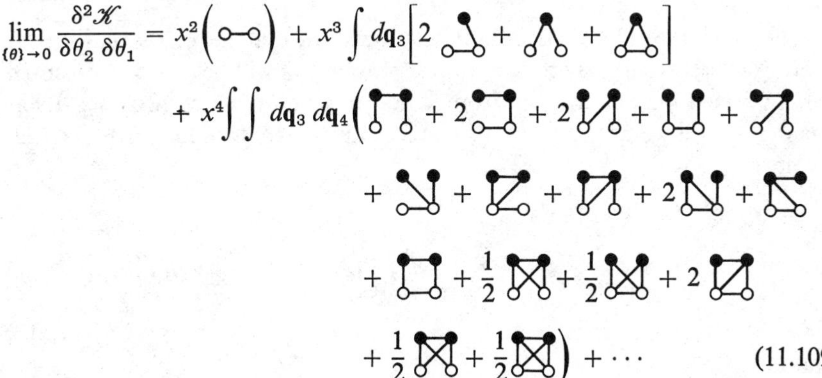

$$\lim_{\{\theta\} \to 0} \frac{\delta^2 \mathscr{K}}{\delta \theta_2\, \delta \theta_1} = x^2 \left(\circ\!\!-\!\!\circ \right) + x^3 \int d\mathbf{q}_3 \left[2 \mathbf{\Lambda} + \mathbf{\triangle} + \mathbf{\triangle} \right]$$

$$+ x^4 \int \int d\mathbf{q}_3\, d\mathbf{q}_4 \left(\mathbf{\Pi} + 2\mathbf{\square} + 2\mathbf{N} + \mathbf{\square} + \mathbf{Z} \right.$$

$$+ \mathbf{N} + \mathbf{Z} + \mathbf{Z} + 2\mathbf{N} + \mathbf{N}$$

$$+ \mathbf{\square} + \frac{1}{2} \mathbf{N} + \frac{1}{2} \mathbf{N} + 2 \mathbf{Z}$$

$$\left. + \frac{1}{2} \mathbf{N} + \frac{1}{2} \mathbf{\boxtimes} \right) + \cdots \tag{11.109}$$

where, for example, $2 \, \circ\!\!-\!\!\bullet$ denotes the two graphs $_1\circ\!\!-\!\!\bullet_2{}^{\bullet 3}$ and $_1\circ\!\!-\!\!\circ_2{}^{\bullet 3}$.

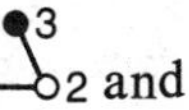

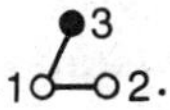

We can combine Eqs. (11.106), (11.108), and (11.109); then, after some algebra, we obtain the following expression for the two-body reduced probability density:

$$n_2^{\mu}(\mathbf{q}_1, \mathbf{q}_2; V, T) = e^{-\beta v(q_{12})}\Big[x^2 + x^3 \int d\mathbf{q}_3 \Big[$$

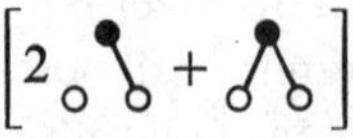

$$\Big]\Big] + x^4 \iint d\mathbf{q}_3\, d\mathbf{q}_4 \Big($$

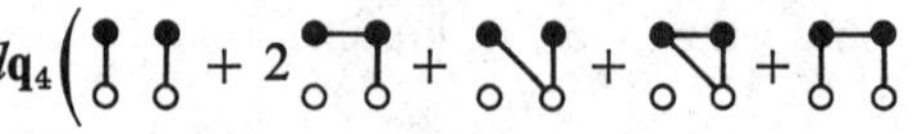

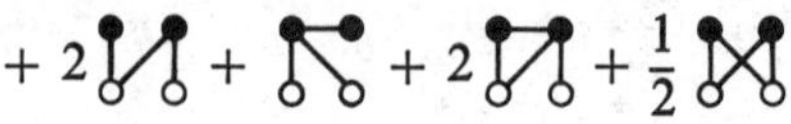

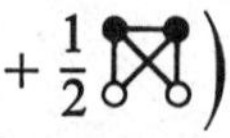

$$\Big) + \cdots \qquad (11.110)$$

Note that we have added a number of graphs together and have factored out the dependence of $v(q_{12})$; that is,

$$\exp(-\beta v(\mathbf{q}_{12})) = \Big(1 +$$

$$\Big).$$

It is now easy to write Eq. (11.110) in the form of a density expansion. If we revert Eq. (11.53), we can write

$$x = \langle n\rangle - 2b_2\langle n\rangle^2 + 8b_2{}^2\langle n\rangle^3 + 3b_3\langle n\rangle^4 + \cdots \qquad (11.111)$$

where b_n is defined in Eq. (11.52) and $\langle n\rangle$ is the average particle density. Substituting Eq. (11.111) into Eq. (11.110), we find, after much cancellation,

$$n_2^{\mu}(\mathbf{q}_1, \mathbf{q}_2; V, T) \equiv \langle n\rangle^2 g(q_{12}) = e^{-\beta v(q_{12})}\Big[\langle n\rangle^2 + \langle n\rangle^3 \int d\mathbf{q}_3 \Big($$

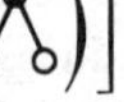

$$\Big)\Big] + \langle n\rangle^4 \iint d\mathbf{q}_3\, d\mathbf{q}_4 \Big($$

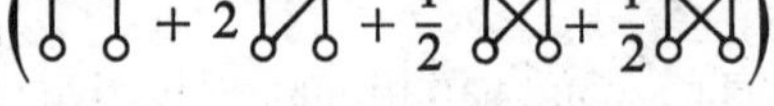

$$\Big) + \cdots \qquad (11.112)$$

where $g(q_{12})$ is the radial distribution function. Note that the graphs that appear in Eq. (11.112) are closely related to star graphs and will be called two-rooted stars (a mixed metaphor). An l-rooted star is a graph which becomes a star when $(l - 1)$ lines are inserted to connect all the roots. Thus, Eq. (11.112) becomes

$$n_2^{\mu}(\mathbf{q}_1, \mathbf{q}_2; V, T) \equiv \langle n\rangle^2 g(q_{12})$$

$$= e^{-\beta v(q_{12})} \sum_{l=2}^{\infty} \frac{\langle n\rangle^l}{(l-2)!} \iint d\mathbf{q}_3 \cdots d\mathbf{q}_l V_l(\mathbf{q}_1\mathbf{q}_2; \mathbf{q}_3, \ldots, \mathbf{q}_l) \qquad (11.113)$$

where

$$V_l(\mathbf{q}_1, \mathbf{q}_2; \mathbf{q}_3 \cdots \mathbf{q}_l) = \sum (\text{all different } l\text{-particle two-rooted stars}).$$

As has been shown in Ref. 11.5, the virial expansion of $n_n^{\mu}(\mathbf{q}_1, \ldots, \mathbf{q}_n; V, T)$ can be written

$$n_n^{\mu}(\mathbf{q}_1, \ldots, \mathbf{q}_n; V, T)$$
$$= W_n(\mathbf{q}_1, \ldots, \mathbf{q}_n) \sum_{l=n}^{\infty} \frac{\langle n \rangle^l}{(l-n)!} \int d\mathbf{q}_{n+1} \cdots d\mathbf{q}_l V_l(\mathbf{q}_1, \ldots, \mathbf{q}_n; \mathbf{q}_{n+1} \cdots \mathbf{q}_l) \tag{11.114}$$

where $V_l(\mathbf{q}_1, \ldots, \mathbf{q}_n; \mathbf{q}_{n+1}, \ldots, \mathbf{q}_l)$ is the sum of all different l-particle, n-rooted stars.

We can now use the results of this section to find approximate equations for the radial distribution function for dense fluids.

E. ORNSTEIN-ZERNIKE EQUATION AND APPROXIMATION SCHEMES[3,14]

The compressibility equation (11.33) can be written in terms of the structure function, $h(q)$, in the form

$$k_B T \left(\frac{\partial \langle n \rangle}{\partial P} \right)_T = 1 + \langle n \rangle \int h(q)\, d\mathbf{q} \tag{11.115}$$

where $h(q) = g(q) - 1$. The structure function contains all possible information about correlations between particles a distance q apart. Near a phase transition, where $(\partial \langle n \rangle / \partial P)_T \to \infty$, it must become very long ranged so the integral will diverge. Ornstein and Zernike[15] introduced the idea of decomposing the structure function into a short-ranged part, $c(q)$, and a long-ranged part. They wrote the structure function in the form

$$h(q_{12}) = c(q_{12}) + \langle n \rangle \int d\mathbf{q}_3 c(q_{13}) h(q_{32}). \tag{11.116}$$

Eq. (11.116) is called the Ornstein-Zernike equation. The first term, $c(q_{12})$ (called the direct correlation function), contains the effects of short-ranged correlations and the second contains the effect of long-ranged correlations and allows for interactions between the particles 1 and 2 which first propagate through the medium. If we introduce the Fourier transform

$$\bar{h}(\mathbf{k}) = \left(\frac{1}{2\pi} \right)^{3/2} \int d\mathbf{q}_{12}\, e^{i\mathbf{k}\cdot(\mathbf{q}_1 - \mathbf{q}_2)} h(q_{12}) \tag{11.117}$$

and a similar one for $c(q_{12})$, we can write Eq. (11.116) in the form

$$\bar{h}(\mathbf{k}) = \bar{c}(\mathbf{k}) + \langle n \rangle \bar{c}(\mathbf{k}) \bar{h}(\mathbf{k}), \tag{11.118}$$

where $\bar{c}(\mathbf{k})$ is the Fourier transform of the direct correlation function $c(q_{12})$.

In terms of the quantity, $\bar{h}(\mathbf{k})$, the compressibility equation takes the form

$$k_B T \left(\frac{\partial \langle n \rangle}{\partial P} \right)_T = 1 + \langle n \rangle \bar{h}(\mathbf{k} = 0). \tag{11.119}$$

If we take the reciprocal of Eq. (11.115), we find

$$\frac{1}{k_B T}\left(\frac{\partial P}{\partial\langle n\rangle}\right)_T = \frac{1}{1 + \langle n\rangle\bar{h}(\mathbf{k}=0)} = 1 + \langle n\rangle\bar{c}(\mathbf{k}=0) = 1 + \langle n\rangle\int d\mathbf{q}\, c(q). \tag{11.120}$$

Since $(\partial P/\partial\langle n\rangle)_T \to 0$ near the critical point, $c(q)$ must remain short ranged.

We can obtain a graphical expression for $c(q)$ as follows. We first rewrite Eq. (11.116) to obtain

$$c(q_{12}) = h(q_{12}) - \langle n\rangle \int d\mathbf{q}_3 c(q_{13}) h(q_{32}) \tag{11.121}$$

and iterate it to obtain a series expansion for $c(q_{12})$ in terms of products of structure functions, $h(q_{ij})$. We then substitute the graphical expression for $h(q_{ij})$ (cf. Eqs. (11.112) and (11.115)) into Eq. (11.121). There will be many cancellations and we find

$c(q_{12}) =$  $+ \langle n\rangle \int d\mathbf{q}_3 \Big($  $\Big)$

$+ \langle n\rangle^2 \iint d\mathbf{q}_3\, d\mathbf{q}_4 \Big($ 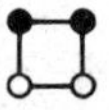 $+ \frac{1}{2}$ 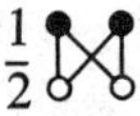 $+ \frac{1}{2}$ 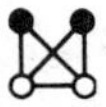 $+ 2$ 

$+ \frac{1}{2}$  $+ \frac{1}{2}$ 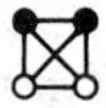 $\Big) + \cdots$ (11.122)

It is easy to see that the graphs which contribute to $c(q_{12})$ consist of all l-particle two-rooted stars with no nodes. A node is a single field point on a graph which, if removed, allows the graph to fall into two parts. Thus,

$$c(q_{12}) = \sum (\text{all two-rooted stars with no nodes}). \tag{11.123}$$

The nodes make $h(q_{12})$ long ranged because they allow the graph to open and stretch out. In the graphs for $h(q_{12})$, the bond between root points 1 and 2 is given by $\exp(-\beta v(q_{12}))$ which goes to one as $q_{12} \to \infty$, whereas for most graphs in $c(q_{12})$, the bond between root points 1 and 2 is represented by $f(q_{12})$ which goes to zero as $q_{12} \to \infty$.

The diagrams which contribute to $h(q) = g(q) - 1$ are of three types. There are chains, bundles, and elementary clusters. *Chains* are rooted stars with nodes. *Bundles* are rooted stars with parallel links between roots and with no nodes. If the root points of a bundle are removed, it will fall into two or more pieces. *Elementary clusters* are rooted stars which are neither chains nor bundles. Some examples of each type of graph are given in Fig. 11.8. We shall introduce the following notation: we will let

$$\mathscr{C}(q_{12}) = \sum_{l=3}^{\infty} \int\cdots\int d\mathbf{q}_3 \cdots d\mathbf{q}_l \text{ (all different two-rooted stars with nodes [chains])} \tag{11.124a}$$

$$\mathscr{B}(q_{12}) = \sum_{l=3}^{\infty} \int \cdots \int d\mathbf{q}_3 \cdots d\mathbf{q}_l \text{ (all different two-rooted stars}$$
$$\text{which form bundles)} \qquad (11.124b)$$

$$\mathscr{E}(q_{12}) = \sum_{l=3}^{\infty} \int \cdots \int d\mathbf{q}_3 \cdots d\mathbf{q}_l \text{ (all different elementary clusters).}$$

$$(11.124c)$$

With the help of this notation we can now find a general integral equation for the radial distribution function, $g(q)$.

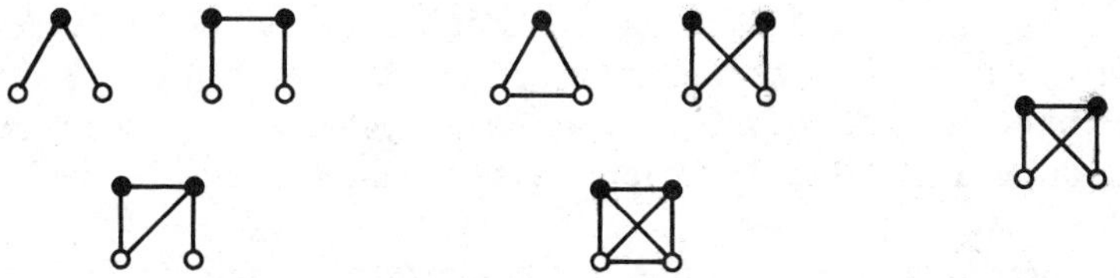

(a)   (b)   (c)

Fig. 11.8. Different types of rooted stars: (a) chains—the lower graph is called a netted chain; (b) bundles; (c) elementary clusters.

A two-rooted bundle has the special property that it can be written as products of chains and elementary clusters, each depending only on q_{12}. For example,

$$\iint d\mathbf{q}_3\, d\mathbf{q}_4 \left(\tfrac{1}{2} \, \rlap{\raise2pt\hbox{$\bowtie$}} \right) \equiv \frac{1}{2} \left(\int d\mathbf{q}_3 \, \wedge \right)^2.$$

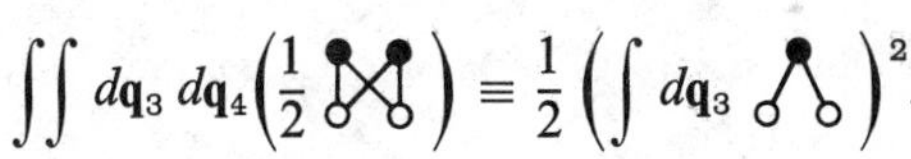

Thus, we can write

$$g(q_{12}) = e^{-\beta v(q_{12})} \exp[\mathscr{C}(q_{12}) + \mathscr{E}(q_{12})]. \qquad (11.125)$$

If we expand the second exponential in a power series, we reproduce all two-rooted stars. If we take the logarithm of Eq. (11.125), we find

$$\ln g(q_{12}) = -\beta v(q_{12}) + \mathscr{C}(q_{12}) + \mathscr{E}(q_{12}). \qquad (11.126)$$

From the definition of $h(q_{12})$ and $c(q_{12})$, we can also write

$$h(q_{12}) = \mathscr{C}(q_{12}) + \mathscr{B}(q_{12}) + \mathscr{E}(q_{12}) \qquad (11.127)$$

and

$$c(q_{12}) = \mathscr{B}(q_{12}) + \mathscr{E}(q_{12}). \qquad (11.128)$$

If we now introduce one last identity, we can obtain a completely general integral equation for $g(q_{12})$. Let us note that the chains can be written in the form

$$\mathscr{C}(q_{12}) = \langle n \rangle \int [\mathscr{B}(q_{13}) + \mathscr{E}(q_{13})] h(q_{32})\, d\mathbf{q}_3. \qquad (11.129)$$

The terms $[\mathscr{B}(q_{13}) + \mathscr{E}(q_{13})]$ represent all possible links that can occur between root point 1 and the first node, and $h(q_{32})$ represents all possible links between the first node and final root point 2. We can now eliminate $\mathscr{B}(q_{12})$ and $\mathscr{C}(q_{12})$ from Eqs. (11.126)–(11.129) and obtain the following exact integral equation for $g(q)$:

$$\ln g(q_{12}) + \beta v(q_{12}) - \mathscr{E}(q_{12}) = n \int d\mathbf{q}_3 (g(q_{13}) - 1 - \ln g(q_{13}) - \beta v(q_{13}) + \mathscr{E}(q_{13}))(g(q_{32}) - 1). \tag{11.130}$$

Eq. (11.130) cannot be solved exactly because $\mathscr{E}(q)$ cannot be determined exactly. However, two very important approximate equations for $g(q_{12})$ do emerge from it which give fairly good agreement with the observed properties of dense fluids. They are the hypernetted chain equation and the Percus-Yevick equation.

The hypernetted chain equation[14] is obtained by neglecting $\mathscr{E}(q)$ in Eq. (11.130). Thus, we obtain the following closed equation for $g_{\mathrm{HNC}}(q_{12})$:

$$\ln(g_{\mathrm{HNC}}(q_{12})) + \beta v(q_{12}) = n \int d\mathbf{q}_3 (g_{\mathrm{HNC}}(q_{13}) - 1 - \ln g_{\mathrm{HNC}}(q_{13}) - \beta v(q_{13}))(g_{\mathrm{HNC}}(q_{32}) - 1) \tag{11.131}$$

or, in terms of the direct correlation function,

$$c_{\mathrm{HNC}}(q_{12}) = g_{\mathrm{HNC}}(q_{12}) - 1 - \ln g_{\mathrm{HNC}}(q_{12}) - \beta v(q_{12}). \tag{11.132}$$

Eq. (11.131) is called the hypernetted chain equation. Thus, the hypernetted chain equation involves the approximation of $c(q_{12})$ by bundles and excludes all contributions from elementary clusters. The bundles may or may not have direct links between the roots.

The Percus-Yevick equation[16] is obtained by omitting all elementary clusters and all bundles which do not have direct links between the roots. Thus, the direct correlation function in the Percus-Yevick approximation has very short range. It can be written in the form

$$c_{\mathrm{PY}}(q_{12}) = f(q_{12})\, e^{\beta v(q_{12})} g_{\mathrm{PY}}(q_{12}). \tag{11.133}$$

By substituting into the Ornstein-Zernike equation, we find

$$g_{\mathrm{PY}}(q_{12}) = e^{-\beta v(q_{12})} + e^{-\beta v(q_{12})} \frac{\langle N \rangle}{V} \int d\mathbf{q}_3 f(q_{13})\, e^{\beta v(q_{13})} g_{\mathrm{PY}}(q_{13}) h_{\mathrm{PY}}(q_{32}). \tag{11.134}$$

Eq. (11.134) is called the Percus-Yevick equation.

Amazingly enough, the Percus-Yevick equation for a hard-sphere system can be solved analytically. Theile[17] showed that the solution is a piecewise analytic function. Either the pressure equation (11.23) or the compressibility equation (11.33) can be used to obtain the equation of state from the exact solution to the Percus-Yevick equation. Since the radial distribution function obtained from the Percus-Yevick equation is approximate, we do not expect the pressure and compressibility equations to give the same results for the equation of state. The equation of state obtained from the pressure equation is

$$P = \frac{N}{V} k_B T \frac{(1 + 2x + 3x^2)}{(1 - x)^2} \tag{11.135}$$

and that obtained from the compressibility equation is

$$P = \frac{\langle N \rangle}{V} k_B T \frac{(1 + x + x^2)}{(1 - x)^3}, \tag{11.136}$$

where $x = \langle N \rangle b_0 / 4V$ is the ratio of the volume of the N-particles to the total volume. In both equations, the pressure becomes infinite at $x = 1$, which is a density greater than the closely packed density $x_{cp} = \pi\sqrt{2}/6$. Thus, the equations give unphysical results for very dense systems and they do not predict the phase transition to a crystalline structure which is observed in molecular dynamics experiments. However, as we shall see in Sec. 11.G, at lower densities they give good agreement with the results of molecular dynamics experiments.

Both the Percus-Yevick and the hypernetted chain equations are closed equations for the radial distribution function and can be solved numerically for arbitrary potentials. We shall compare the predictions of these two equations to experimental results in Sec. 11.G. First, however, it is useful to introduce a completely different method for obtaining closed equations for the radial distribution. This method involves closure of the hierarchy of equations for the reduced probability densities.

F. SUPERPOSITION PRINCIPLE[3,18]

It is possible to obtain a hierarchy of integro-differential equations for the reduced probability densities which is in some respects similar to the BBGKY hierarchy obtained in Chap. 7. Let us write the equation for the reduced probability density in the form

$$n_n^{\mu}(\mathbf{q}_1, \ldots, \mathbf{q}_n; V, T) = \frac{1}{\mathscr{Z}(V, T, \mu)} \sum_{N=n} \frac{x^N}{(N - n)!} \int d\mathbf{q}_{n+1} \cdots d\mathbf{q}_N \, e^{-\beta V^N} \tag{11.137}$$

and take the derivative with respect to $\mathbf{q}_1$. We find

$$\frac{\partial}{\partial \mathbf{q}_1} n_n^{\mu}(\mathbf{q}_1, \ldots, \mathbf{q}_n; V, T) = -\beta \sum_{j=2}^{n} \left(\frac{\partial v_{1j}}{\partial \mathbf{q}_1}\right) n_n^{\mu}(\mathbf{q}_1, \ldots, \mathbf{q}_n; V, T)$$
$$- \beta \int d\mathbf{q}_{n+1} \left(\frac{\partial v_{1,n+1}}{\partial \mathbf{q}_1}\right) n_{n+1}^{\mu}(\mathbf{q}_1, \ldots, \mathbf{q}_{n+1}; V, T) \tag{11.138}$$

where $v_{ij} \equiv v(q_{ij})$. Eq. (11.138) is called the BGY hierarchy (Bogoliubov,[19] Born,[20] Green,[20] and Yvon[21]) for the n-body reduced probability density. From it we can obtain the equation for the radial distribution function $g(q_{12})$ in the form

$$\beta \frac{\partial}{\partial \mathbf{q}_1} g^{\mu}(q_{12}) + \left(\frac{\partial v_{12}}{\partial \mathbf{q}_1}\right) g^{\mu}(q_{12}) = -\frac{\langle N \rangle}{V} \int d\mathbf{q}_3 \left(\frac{\partial v_{13}}{\partial \mathbf{q}_1}\right) g_3^{\mu}(\mathbf{q}_1, \mathbf{q}_2, \mathbf{q}_3). \tag{11.139}$$

Thus, the equation for $g^{\mu}(q_{12})$ depends on the three-body distribution $g_3^{\mu}(\mathbf{q}_1, \mathbf{q}_2, \mathbf{q}_3)$. As we can see from Eq. (11.138), the equation for $g_3^{\mu}(\mathbf{q}_1, \mathbf{q}_2, \mathbf{q}_3)$ depends on $g_4^{\mu}(\mathbf{q}_1, \mathbf{q}_2, \mathbf{q}_3, \mathbf{q}_4)$, etc. In order to solve Eq. (11.139), we must find some way to terminate the hierarchy of Eq. (11.138). The simplest and historically one of the most important methods is called the superposition approximation and was originally introduced by Kirkwood.[22] The superposition approximation involves the replacement of the three-body distribution, $g_3^{\mu}(\mathbf{q}_1, \mathbf{q}_2, \mathbf{q}_3)$, by a product of two-body distributions,

$$g_3^{\mu}(\mathbf{q}_1, \mathbf{q}_2, \mathbf{q}_3) = g^{\mu}(q_{12}) g^{\mu}(q_{13}) g^{\mu}(q_{23}). \tag{11.140}$$

Thus, the probability of finding three particles in a given configuration is assumed to be equal to the probability that each pair will be found there independent of the third particle. One expects such an approximation to be good only at low densities, which is indeed the case, as we shall see. If the superposition approximation is made on Eq. (11.139), it can be written in the form

$$\beta \frac{\partial}{\partial \mathbf{q}_1} [\ln g_{\mathrm{BGY}}^{\mu}(q_{12})] = -\frac{\partial v_{12}}{\partial \mathbf{q}_1} - \langle n \rangle \int d\mathbf{q}_3 \left(\frac{\partial v_{13}}{\partial \mathbf{q}_1}\right) g_{\mathrm{BGY}}^{\mu}(q_{13}) g_{\mathrm{BGY}}^{\mu}(q_{23}). \tag{11.141}$$

Eq. (11.141) is called the BGY equation and is a closed integro-differential equation for the radial distribution function.

Another equation for the radial distribution was first obtained by Kirkwood and Boggs,[23] starting from the canonical ensemble. We proceed in a manner similar to that used in Sec. 11.B to obtain the expression for the chemical potential. Let us write the n-body reduced distribution function in terms of the N-body potential given in Eq. (11.26). Then,

$$n_n^N(\mathbf{q}_1, \ldots, \mathbf{q}_N; V, T, \gamma) = \frac{1}{Q_N(V, T, \gamma)} \frac{N!}{(N-n)!} \times \int \cdots \int d\mathbf{q}_{n+1} \cdots d\mathbf{q}_N \, e^{-\beta v^N(\mathbf{q}^N; \gamma)} \tag{11.142}$$

where

$$Q_N(V, T, \gamma) = \int \cdots \int d\mathbf{q}_1 \cdots d\mathbf{q}_N \, e^{-\beta v^N(\mathbf{q}^N; \gamma)} \tag{11.143}$$

is the configuration integral defined in terms of the potential in Eq. (11.25). If we take the derivative of Eq. (11.142) with respect to γ, we can write it in the form

$$\frac{1}{\beta} \frac{\partial}{\partial \gamma} \ln n_n^N(\mathbf{q}_1, \ldots, \mathbf{q}_n; V, T, \gamma) = -\sum_{j=2}^{n} v_{1j} + \frac{1}{N} \iint d\mathbf{q}_1 \, d\mathbf{q}_2 v_{12} n_2^N(\mathbf{q}_1 \mathbf{q}_2; V, T, \gamma) - \int d\mathbf{q}_{n+1} v_{1,n+1} n_{n+1}^N(\mathbf{q}_1, \ldots, \mathbf{q}_{n+1}; V, T, \gamma) \times [n_n^N(\mathbf{q}_1, \ldots, \mathbf{q}_n; V, T, \gamma)]^{-1}. \tag{11.144}$$

Let us next note that

$$n_n^N(\mathbf{q}_1, \ldots, \mathbf{q}_n; V, T; \gamma = 0) = \frac{N}{V} n_{n-1}^{N-1}(\mathbf{q}_1, \ldots, \mathbf{q}_n; V, T, \gamma) \tag{11.145}$$

and integrate Eq. (11.144) over γ from 0 to γ'. Then we obtain

$$\begin{aligned} \frac{1}{\beta} \ln n_n^N(\mathbf{q}_1, \ldots, \mathbf{q}_n; V, T, \gamma') &= \frac{1}{\beta} \ln \frac{N}{V} + \frac{1}{\beta} \ln n_{n-1}^{N-1}(\mathbf{q}_2, \ldots, \mathbf{q}_n; VT) \\ &\quad - \gamma' \sum_{j=2}^{N} v_{1j} + \frac{1}{N} \int_0^{\gamma'} d\gamma \iint d\mathbf{q}_1 \, d\mathbf{q}_2 v_{12} n_2^N(\mathbf{q}_1, \mathbf{q}_2; V, T, \gamma) \\ &\quad - \int_0^{\gamma'} d\gamma \int d\mathbf{q}_{n+1} v_{1,n+1} n_{n+1}^N(\mathbf{q}_1, \ldots, \mathbf{q}_{n+1}; V, T, \gamma) \\ &\quad \times [n_n^N(\mathbf{q}_1, \ldots, \mathbf{q}_n; V, T, \gamma]^{-1}. \end{aligned} \tag{11.146}$$

Eq. (11.146) yields the following equation for the two-body reduced probability density:

$$\frac{1}{\beta}\ln n_2^N(\mathbf{q}_1, \mathbf{q}_2; V, T, \gamma')$$

$$= \frac{2}{\beta}\ln\frac{N}{V} - \gamma' v_{12} + \frac{1}{N}\int_0^{\gamma'} d\gamma \iint d\mathbf{q}_1\, d\mathbf{q}_2 v_{12} n_2^N(\mathbf{q}_1, \mathbf{q}_2; V, T, \gamma)$$

$$- \int_0^{\gamma'} d\gamma \int d\mathbf{q}_3 v_{13} n_3^N(\mathbf{q}_1, \mathbf{q}_2, \mathbf{q}_3; V, T, \gamma)[n_2^N(\mathbf{q}_1, \mathbf{q}_2; V, T, \gamma)]^{-1}. \tag{11.147}$$

The equation for the radial distribution function is given by

$$\frac{1}{\beta}\ln g^N(q_{12}) = -v_{12} + \frac{N}{V}\int_0^{\gamma'} d\gamma \int d\mathbf{q}_{12} v_{12} g^N(q_{12})$$

$$- \frac{N}{V}\int_0^{\gamma'} d\gamma \int d\mathbf{q}_3 v_{13} g_3^N(\mathbf{q}_1, \mathbf{q}_2, \mathbf{q}_3; V, T, \gamma)[g^N(q_{12})]^{-1}. \tag{11.148}$$

If we now introduce the superposition approximation for $g_3(\mathbf{q}_1, \mathbf{q}_2, \mathbf{q}_3)$, we find

$$\frac{1}{\beta}\ln g_K^N(q_{12}, \gamma) = -v(q_{12}) - \frac{N}{V}\int_0^{\gamma'} d\gamma \int d\mathbf{q}_3 v(q_{13}) g_K^N(q_{13}, \gamma)$$

$$\times [g_K^N(q_{23}, \gamma) - 1]. \tag{11.149}$$

Eq. (11.149) is called the Kirkwood equation for the radial distribution function. It, too, is an integral equation which must be solved numerically.

In this and the previous section, we have obtained four equations for the radial distribution function: the hypernetted chain equation, the Percus-Yevick equation, the BGY equation, and the Kirkwood equation. In the next section we shall compare the predictions of these equations with experimental results.

G. EXPERIMENTAL RESULTS FOR DENSE FLUIDS

In Sec. 11.B, we derived two equations for the equation of state of a classical fluid: the pressure equation (11.23) and the compressibility equation (11.33). Both equations involve the radial distribution function, $g(q)$, and give the same expressions for the equation of state in the thermodynamic limit if we use the exact expression for $g(q)$. If we use an approximate expression for $g(q)$, the two equations give different results. One criterion for deciding which approximation scheme gives the best description of a dense fluid is to determine which gives the most consistent results when the two equations of state are used.

For low densities, one way of comparing various approximations for the radial distribution function is to compute the virial coefficients using both the pressure equation and the compressibility equation. This has been done by a number of authors for the hypernetted chain equation, the Percus-Yevick equation, the BGY equation, and the Kirkwood equation. All four approximations give exact results for the second and third virial coefficients, B_2 and B_3, but deviate from the exact results for higher order virial coefficients. The results for B_4, B_5, and B_6 are given in Table 11.4 for a hard-sphere gas. We see that the pressure and compressibility equations always give different results. However, among the four approximation schemes, the Percus-Yevick equation gives the most consistent and accurate results for the virial coefficients.

	B_4/b_0^3	B_5/b_0^4	B_6/b_0^5
Exact	0.2869	0.1103 ± 0.0005	0.0386 ± 0.0004
BGY (P)	0.2252	0.0475	
(C)	0.3424	0.1335	
K (P)	0.1400		
(C)	0.4418		
HNC (P)	0.4453	0.1447	0.0382
(C)	0.2092	0.0493	0.0281
PY (P)	0.2500	0.0859	0.0273
(C)	0.2969	0.1211	0.0449

Table 11.4. The fourth, fifth, and sixth virial coefficients for a hard-sphere fluid using the BGY equation, the Kirkwood equation (K), the hypernetted chain equation (HNC), and the Percus-Yevick equation (PY). The values using both the pressure equation and compressibility equation are given (based on Ref. 11.30).

Approximate expressions for the radial distribution functions $g_{HNC}(q)$, $g_{PY}(q)$, $g_{BGY}(q)$, and $g_K(q)$ can be obtained numerically ($g_{PY}(q)$ can be obtained exactly) for a hard-sphere gas over a wide range of densities, and the equation of state has been computed from both the pressure equation and the compressibility equation. The results of these calculations are given in Fig. 11.9 and are compared with the results of molecular dynamics calculations. We see again that, of the four approximation schemes, the Percus-Yevick equation gives the best results and that, when the Percus-Yevick approximation is used, the compressibility equation gives better results than does the pressure equation. In Fig. 11.10, we compare the radial distribution function for a hard-sphere gas obtained from molecular dynamics calculations and from the Percus-Yevick equation. Again, the results are good.

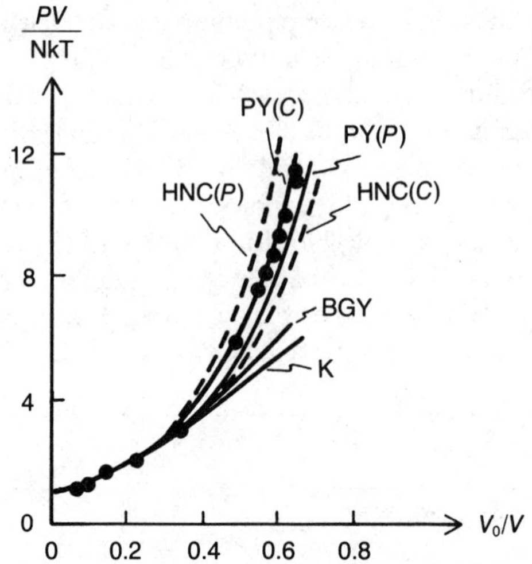

Fig. 11.9. A comparison of the results of the four approximation schemes with the results of molecular dynamics calculations (based on Ref. 11.30).

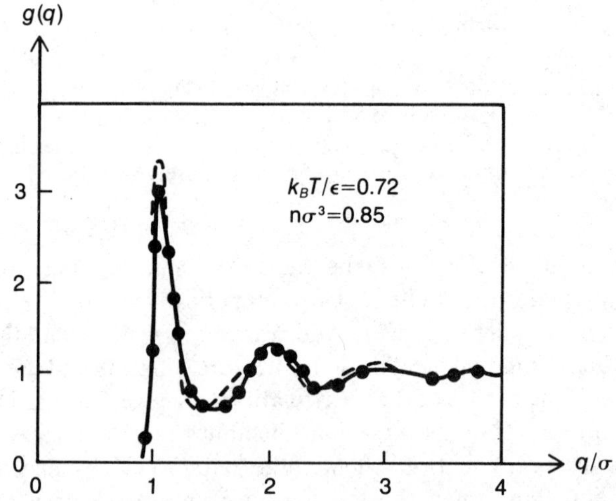

Fig. 11.10. The radial distribution function: the dashed line is obtained from the Percus-Yevick equation, the solid line from Barker-Henderson perturbation theory; the dots are the results of molecular dynamics experiments (based on Ref. 11.25).

Extensive numerical studies have also been done for the Lennard-Jones 6–12 potential. Watts has computed the equation of state for a number of isotherms using the Percus-Yevick equation and finds that the system undergoes a phase transition. His results are shown in Fig. 11.11. There is a large region where the Percus-Yevick equation has no solutions. This region separates different branches of each isotherm of the equation of state and corresponds to the coexistence region.

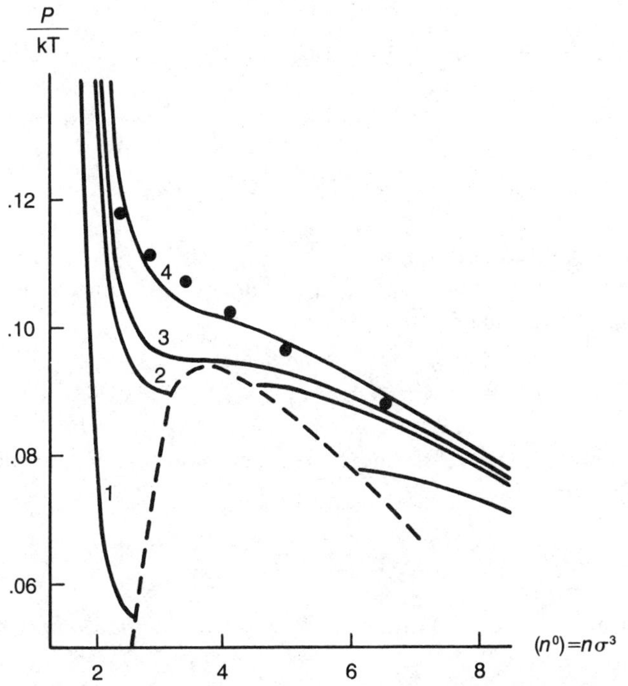

Fig. 11.11. Isotherms of the equation of state for a Lennard-Jones fluid obtained by solving the Percus-Yevick equation: *(1)* $T^* = 1.2$; *(2)* $T^* = 1.263$; *(3)* $T^* = 1.275$; *(4)* $T^* = 1.3$; the dots are experimental points for argon at $T^* = 1.3$; for the region enclosed by the dashed line, the Percus-Yevick equation has no solutions (based on Ref. 11.31).

H. PERTURBATION THEORIES[3,24–26]

There is still another method for treating dense fluids which appears to give better results than any of those discussed so far. This method is due to Zwanzig,[24] who noted that the qualitative behavior of dense fluids is determined by the hard core, while the attractive part of the interparticle potential adds only small corrections to the hard-core behavior. Thus, it is possible to treat the attraction between molecules as a perturbation on the hard-core system.

The perturbation theory is easy to work out. For classical systems, the kinetic contributions to the free energy factor from the configuration contributions. The interesting behavior of the system is contained in the configuration integral. Let us assume that the interaction, v^N, may be divided into two parts, $V_0{}^N$ and $V_1{}^N$, so that

$$v^N = V_0{}^N + V_1{}^N, \tag{11.150}$$

where $V_0{}^N$ is the hard-core contribution and $V_1{}^N$ is the contribution from the attractive part of the interaction. The configuration free energy, $\bar{A}^N$, may be written

$$e^{-\beta \bar{A}^N} = \frac{Q_N}{N!} = \frac{1}{N!}\int\cdots\int d\mathbf{q}_1\cdots d\mathbf{q}_N\, e^{-\beta V^N}, \tag{11.151}$$

where Q_N is the configuration integral.

Let us now introduce the configuration integral, $Q_N^{(0)}$, and configuration free energy, $A_0{}^N$, for the hard-core system:

$$e^{-\beta A_0^N} = \frac{Q_N^{(0)}}{N!} = \frac{1}{N!}\int\cdots\int d\mathbf{q}_1\cdots d\mathbf{q}_N\, e^{-\beta V_0^N}. \tag{11.152}$$

The probability density for finding the hard-core system in configuration $\mathbf{q}_1, \ldots, \mathbf{q}_N$ will be

$$\rho_0^N(\mathbf{q}_1, \ldots, \mathbf{q}_N) = \frac{e^{-\beta V_0^N}}{Q_N^{(0)}} \tag{11.153}$$

and is normalized to one. We now can write the configuration free energy in the form

$$\begin{aligned} e^{-\beta \bar{A}^N} &= e^{-\beta A_0^N}\int\cdots\int d\mathbf{q}_1\cdots d\mathbf{q}_N \rho_0^N(\mathbf{q}_1, \ldots, \mathbf{q}_N)\, e^{-\beta V_1^N} \\ &= e^{-\beta A_0^N}\langle e^{-\beta V_1^N}\rangle. \end{aligned} \tag{11.154}$$

If we rewrite Eq. (11.154) in terms of a cumulant expansion, we obtain the following expression for the configuration free energy:

$$\bar{A}^N = A_0{}^N + \langle V_1{}^N\rangle - \frac{\beta}{2}[\langle (V_1{}^N)^2\rangle - \langle V_1{}^N\rangle^2] + \cdots \tag{11.155}$$

where the expectation value is defined relative to $\rho_0^N(\mathbf{q}_1, \ldots, \mathbf{q}_N)$. Thus,

$$\langle V_1{}^N\rangle = \int\cdots\int d\mathbf{q}_1\cdots d\mathbf{q}_N V_1{}^N \rho_0^N(\mathbf{q}_1, \ldots, \mathbf{q}_N). \tag{11.156}$$

If we assume that the attractive part of the potential can be written as a sum of two-body interactions; that is, $V_1{}^N = \sum_{(ij)} v_{ij}^{(1)}$; we obtain

$$\langle V_1{}^N \rangle = \frac{1}{2}\frac{N^2}{V}\int d\mathbf{q}_{12} v^1(q_{12}) g^{(0)}(q_{12}) \tag{11.157}$$

where $g^0(q_{12})$ is the radial distribution function of the hard-core system. Similar expressions can be obtained for higher order terms in the cumulant expansion, but they involve higher order reduced probability densities for the hard-core system.

This type of theory has been used with great success by a number of authors to describe the behavior of interacting systems of particles with hard cores and weak attractions. We shall not discuss the various theories in detail here. However, in Fig. 11.10 we give some results of Henderson and Barker on the radial distribution function of a Lennard-Jones system using the perturbation theory approach. The dots are molecular dynamics results. The dashed line is the prediction of the Percus-Yevick equation and the solid line is the result of perturbation theory. We see that the agreement is excellent.

I. QUANTUM CORRECTIONS TO THE VIRIAL COEFFICIENTS[1,27]

For fluids composed of molecules of small mass, such as He, the classical expressions for the virial coefficients do not give very good results at lower temperatures (cf. Fig. 11.4). For such particles, the thermal wavelength $\lambda_T = (2\pi\hbar^2/mk_BT)^{1/2}$ will be relatively large and quantum corrections must be taken into account. There are two kinds of quantum effects which must be considered: *diffraction effects*, which are important when the thermal wavelength is the size of the radius of the molecules, and the *effects of statistics*, which are important when the thermal wavelength is the size of the average distance between particles.

To find general expressions for the virial coefficients, we proceed along lines similar to those for a classical fluid except that, for the quantum case, momentum variables no longer commute with position variables and cannot be eliminated immediately. The grand partition function for a quantum system can be written in the form

$$\mathscr{Z}(V, T, \mu) = \sum_{N=0}^{\infty} \frac{e^{\beta\mu N}}{N!\,\lambda_T{}^{3N}} \operatorname{Tr}_N \hat{W}_N(\beta) \tag{11.158}$$

where λ_T is the thermal wavelength,

$$\hat{W}_N(\beta) = \lambda_T^{3N} N!\, e^{-\beta \hat{H}^N}, \tag{11.159}$$

and $\hat{H}^N$ is the N-body Hamiltonian; $\hat{W}_N(\beta)$ depends on the phase space coordinate operators for N-particles. As before, we can expand the grand partition function in a cumulant expansion,

$$\mathscr{Z}(V, T, \mu) = \exp\left[\sum_{l=0}^{\infty} \frac{e^{\beta\mu l}}{l!\,\lambda_T^{3l}} \operatorname{Tr}_l \hat{U}_l(\beta)\right] \tag{11.160}$$

where $\hat{U}_l(\beta)$ depends on the phase space coordinate operators for l-particles. If we equate coefficients of equal powers of the parameter $e^{\beta\mu}/\lambda_T{}^3$ in Eqs. (11.159) and (11.160), we obtain

$$\operatorname{Tr}_1 \hat{U}_1(\beta) = \operatorname{Tr}_1 \hat{W}_1(\beta), \tag{11.161a}$$

$$\operatorname{Tr}_2 \hat{U}_2(\beta) = \operatorname{Tr}_2 \hat{W}_2(\beta) - (\operatorname{Tr}_1 \hat{U}_1(\beta))^2, \tag{11.161b}$$

$$\operatorname{Tr}_3 \hat{U}_3(\beta) = \operatorname{Tr}_3 \hat{W}_3(\beta) - 3(\operatorname{Tr}_1 \hat{U}_1(\beta))(\operatorname{Tr}_2 \hat{U}_2(\beta)) + 2(\operatorname{Tr}_1 \hat{U}_1(\beta))^3, \tag{11.161c}$$

etc. From Eq. (11.160) we can write the grand potential in the form

$$\Omega(V, T, \mu) = -k_B T \sum_{l=1}^{\infty} \frac{e^{\beta\mu l}}{l!\,\lambda_T^{3l}} \operatorname{Tr}_l \hat{U}_l(\beta), \tag{11.162}$$

and the average particle density takes the form

$$\langle n\rangle = -\frac{1}{V}\left(\frac{\partial\Omega}{\partial\mu}\right)_{V,T} = \sum_{l=1}^{\infty} \frac{l\,e^{\beta\mu l}}{\lambda_T^{3l}} b_l(V, T), \tag{11.163}$$

where

$$b_l(V, T) \equiv \frac{\lambda_T^{3l}}{l!\,V} \operatorname{Tr}_l \hat{U}_l(\beta). \tag{11.164}$$

The virial expansion for the equation of state may be written

$$\frac{P}{\langle n\rangle k_B T} = \sum_{l=1}^{\infty} \mathscr{B}_l(T)\langle n\rangle^{l-1} \tag{11.165}$$

where the virial coefficients, $\mathscr{B}_l(T)$, are related to the quantities $b_l(V, T)$ through Eqs. (11.56)–(11.59). Thus, $\mathscr{B}_1 = b_1$, $\mathscr{B}_2 = -b_2$, $\mathscr{B}_3 = 4b_2{}^2 - 2b_3$, etc.

We can now obtain some virial coefficients for quantum systems. We shall first consider the case of ideal Bose-Einstein and Fermi-Dirac gases and obtain the second virial coefficient for them. Then we shall consider interacting bosons and fermions and find general expressions for the second virial coefficients for those systems.

1. Ideal Quantum Gases

We shall find the second virial coefficient for ideal boson and fermion gases and leave the higher order virial coefficients as a homework problem. We shall assume that the boson gas is at high enough temperature and low enough density that momentum condensation has not set in. For an ideal gas, the second virial coefficient takes the form

$$\mathscr{B}_2{}^0(T) = -\frac{1}{2!\,V}\mathrm{Tr}_2\,\hat{U}_2^0(\beta)$$

$$= -\frac{1}{2!\,V}\mathrm{Tr}_2\,\hat{W}_2^0(\beta) + \frac{1}{2!\,V}(\mathrm{Tr}_1\,\hat{W}_1^0(\beta))^2, \tag{11.166}$$

where

$$\hat{W}_2{}^0(\beta) = \lambda_T{}^6 2!\,e^{-\beta(\hat{T}_1+\hat{T}_2)}, \tag{11.167}$$

$$\hat{W}_1(\beta) = \lambda_T{}^3\,e^{-\beta\hat{T}_1}, \tag{11.168}$$

and $\hat{T}_i$ is the kinetic energy operator for particle i: $\hat{T}_i = \hat{p}_i{}^2/2m$ (cf. App. B, Eq. (B.22)). To evaluate the trace we will use the conventions of App. B. We can use any convenient set of symmetric or antisymmetric states to evaluate the trace. In the position representation, Eq. (11.166) takes the form

$$\mathscr{B}_2^{(0)}(T) = -\frac{\lambda_T{}^6}{2!\,V}\iint d\mathbf{r}_1\,d\mathbf{r}_2\langle\mathbf{r}_1,\mathbf{r}_2|e^{-\beta(\hat{T}_1+\hat{T}_2)}|\mathbf{r}_1,\mathbf{r}_2\rangle^{(p)}$$

$$+\frac{\lambda_T{}^6}{2!\,V}\left(\int d\mathbf{r}_1\langle\mathbf{r}_1|e^{-\beta\hat{T}_1}|\mathbf{r}_1\rangle\right)^2 \tag{11.169}$$

where we now let $|\mathbf{r}_1,\mathbf{r}_2\rangle^{(p)} = |\mathbf{r}_1,\mathbf{r}_2\rangle + \epsilon|\mathbf{r}_2,\mathbf{r}_1\rangle$ with $\epsilon = +1$ for bosons and $\epsilon = -1$ for fermions. The origin of the numerical factors becomes clear if we note that for a general N-body operator $\hat{O}^N$

$$\mathrm{Tr}_N\,\hat{O}^N = \frac{1}{N!}\int d\mathbf{r}_1\cdots\int d\mathbf{r}_N{}^{(S)(A)}\langle\mathbf{r}_1,\ldots,\mathbf{r}_N|\hat{O}^N|\mathbf{r}_1,\ldots,\mathbf{r}_N\rangle^{(S)(A)}$$

$$= \frac{1}{N!}\int d\mathbf{r}_1\cdots\int d\mathbf{r}_N\langle\mathbf{r}_1,\ldots,\mathbf{r}_N|\hat{O}^N|\mathbf{r}_1,\ldots,\mathbf{r}_N\rangle^{(p)} \tag{11.170}$$

where $|\mathbf{r}_1,\ldots,\mathbf{r}_N\rangle^{(p)} = \sum_p (\epsilon)^p|\mathbf{r}_1,\ldots,\mathbf{r}_N\rangle$ (cf. App. B, Eqs. (B.43) and (B.53)). Eq. (11.169) can be simplified to obtain

$$\mathscr{B}_2{}^0(T) = -\epsilon\frac{\lambda_T{}^6}{2!\,V}\iint d\mathbf{r}_1\,d\mathbf{r}_2\langle\mathbf{r}_1|e^{-\beta\hat{T}_1}|\mathbf{r}_2\rangle\langle\mathbf{r}_2|e^{-\beta\hat{T}_2}|\mathbf{r}_1\rangle. \tag{11.171}$$

If statistics are not important, $\epsilon = 0$ and $\mathscr{B}_2{}^0(T) = 0$, as it should. If we use Eqs. (B.7) and (B.10) in App. B, we can perform the integrations to obtain

$$\mathscr{B}_2{}^0(T) = -\epsilon\frac{\lambda_T{}^3}{(2)^{5/2}}. \tag{11.172}$$

Thus, to second order in density, the equation of state takes the form

$$\frac{P}{\langle n\rangle k_B T} = 1 \mp \frac{\lambda_T{}^3}{(2)^{5/2}}\langle n\rangle, \tag{11.173}$$

where the upper sign applies to a Bose-Einstein gas and the lower sign applies to a Fermi-Dirac gas. For bosons, the pressure is lowered relative to that of a classical ideal gas, while for fermions it is increased. Higher order virial coefficients are easy to obtain for this system.

2. Interacting Quantum Gas

Let us now consider the case of an interacting gas. For this case the virial coefficient can be written

$$\mathscr{B}_2(T) = -\frac{1}{2!\,V}\operatorname{Tr}_2 \hat{W}_2(\beta) + \frac{1}{2!\,V}(\operatorname{Tr}_1 \hat{W}_1(\beta))^2 \qquad (11.174)$$

where

$$\hat{W}_2(\beta) = 2!\,\lambda_T{}^6\, e^{-\beta\hat{H}^2} \qquad (11.175)$$

and $\hat{H}^2 = \hat{T}_1 + \hat{T}_2 + v_{12}$. It is easiest to consider the difference,

$$\mathscr{B}_2(T) - \mathscr{B}_2{}^0(T) = -\frac{\lambda_T{}^6}{V}\operatorname{Tr}_2(e^{-\beta\hat{H}^2} - e^{-\beta\hat{H}_0^2}), \qquad (11.176)$$

where $\mathscr{B}_2{}^0(T)$ is the second virial coefficient of the ideal boson or fermion gas and $\hat{H}_0{}^2 = \hat{T}_1 + \hat{T}_2$. If the interaction between particles depends only on relative coordinates $v_{12} = v(\mathbf{q}_{12})$, the center of mass motion separates from the relative motion and we find

$$\mathscr{B}_2(T) - \mathscr{B}_2{}^0(T) = -\frac{\lambda_T{}^6}{V}\operatorname{Tr}_2[e^{-\beta\hat{P}^2/4m}(e^{-\beta\hat{H}_2^{\mathrm{rel}}} - e^{-\beta\hat{H}_{2,0}^{\mathrm{rel}}})]$$

$$= -\frac{\lambda_T{}^6}{V}\sum_{K}\langle\mathbf{K}|e^{-\beta\hbar^2K^2/4m}|\mathbf{K}\rangle\sum_{n}\langle n|e^{-\beta\hat{H}_2^{\mathrm{rel}}} - e^{-\beta\hat{H}_{2,0}^{\mathrm{rel}}}|n\rangle, \qquad (11.178)$$

where $\hat{\mathbf{P}}$ is the center of mass momentum $\hat{\mathbf{P}} = \hat{\mathbf{p}}_1 + \hat{\mathbf{p}}_2$ and where $\hat{H}_2^{\mathrm{rel}}$ and $\hat{H}_{2,0}^{\mathrm{rel}}$ are the Hamiltonians for relative motion in the interacting and non-interacting systems, respectively. The summation over n is over all energy eigenstates for the relative motion. We can integrate out the center of mass dependence if we let $\sum_{\mathbf{K}} \to (V/(2\pi)^3)\int d\mathbf{K}$. Then we find

$$\mathscr{B}_2(T) - \mathscr{B}_2{}^0(T) = -2^{3/2}\lambda_T{}^3\sum_{n(\mathrm{bound})} e^{-\beta\epsilon_n}$$

$$- 2^{3/2}\lambda_T{}^3\sum_{n(\mathrm{cont})}(e^{-\beta\epsilon_n} - e^{-\beta\epsilon_{n,0}}) \qquad (11.178)$$

where the first term contains contributions from the bound states and the second term contains contributions from the eigenvalues in the continuum.

The evaluation of the second term in Eq. (11.178) is easy if we introduce a trick. The eigenfunctions of the relative Hamiltonian in the position representation, $\langle \mathbf{r}|n\rangle = \psi_n(\mathbf{r})$, satisfy the Schrödinger equation,[28]

$$-\frac{\hbar^2}{m}\nabla_{\mathbf{r}}^2\psi_n(\mathbf{r}) + v(\mathbf{r})\psi_n(\mathbf{r}) = \epsilon_n\psi_n(\mathbf{r}), \tag{11.179}$$

and can be written in the form

$$\psi_n(\mathbf{r}) = \sum_{l=0}^{\infty} A_l(2l+1)P_l(\cos\theta)R_l(kr) \tag{11.180}$$

where $P_l(\cos\theta)$ is a Legendre polynomial of lth order and $R_l(kr)$ satisfies the equation

$$\frac{1}{r^2}\frac{d}{dr^2}\left(r^2\frac{dR_l}{dr}\right) + \left(k^2 - U(r) - \frac{l(l+1)}{r^2}\right)R_l(kr) = 0. \tag{11.181}$$

In Eq. (11.181), we have let $\epsilon_n = \hbar^2k^2/2m$ and $U(r) = mv(r)/\hbar^2$, and, in Eq. (11.180), θ is the angle between the direction of $\mathbf{k}$ and $\mathbf{r}$. For bosons, the wave function must be symmetric under interchange of particles, $\psi(\mathbf{r}) = \psi(-\mathbf{r})$, and therefore only even values of l can contribute to Eq. (11.180). For fermions, the wave functions must be antisymmetric, $\psi(\mathbf{r}) = -\psi(-\mathbf{r})$, and only odd values of l can contribute. The solutions, $R_l(kr)$, can be expressed in terms of spherical Bessel functions, $j_l(kr)$, and spherical Neumann functions, $n_l(kr)$. For large values of r (that is, outside the range of the interaction), the most general form of the solution to Eq. (11.181) is

$$R_l(kr) = A_l[\cos\delta_l(k)j_l(kr) - \sin\delta_l(k)n_l(kr)], \tag{11.182}$$

where $\delta_l(k)$ is the phase shift of the wave function relative to its value for noninteracting systems. For noninteracting systems $\delta_l(k) = 0$.

Let us assume that the system is contained in a large spherical container of radius R. Eventually, we will let $R \to \infty$. Let us require that on the surface of the spherical container $R_l(kR) \equiv 0$ for both the interacting and the noninteracting systems. The asymptotic form of Eq. (11.159) for large r is

$$R_l(kr) \xrightarrow[kr\to\infty]{} \frac{A_l}{kr}\sin(kr - \tfrac{1}{2}\pi l + \delta_l(k)). \tag{11.183}$$

If $R_l(kr) \equiv 0$ for $r = R$, then the argument of the sine function must satisfy the condition

$$kR - \tfrac{1}{2}\pi l + \delta_l(k) = m_l\pi \tag{11.184}$$

for the interacting system and

$$kR - \tfrac{1}{2}\pi l = m_l^0\pi \tag{11.185}$$

for the noninteracting system, where m_l and m_l^0 are integers which label the allowed eigenfunctions in the spherical container. Eqs. (11.184) and (11.185) give us a relation between m_l and m_l^0 for the interacting and noninteracting systems. If we take the derivative of Eqs. (11.184) and (11.185), we find

$$\frac{dm_l}{dk} = \frac{R}{\pi} + \frac{1}{\pi}\frac{d\delta_l(k)}{dk} \tag{11.186}$$

and

$$\frac{dm_l^0}{dk} = \frac{R}{\pi}. \tag{11.187}$$

Each of the energy eigenstates for this system is $(2l + 1)$-fold degenerate. Thus, the total density of states for a given value of l and k is

$$\frac{dn_l}{dk} = (2l + 1)\frac{dm_l}{dk}. \tag{11.188}$$

With this expression, we can rewrite the last term in Eq. (11.178) in the form

$$\begin{aligned}\sum_{n(\text{cont})} (e^{-\beta\epsilon_n} - e^{-\beta\epsilon_n^0}) &= \sum_{l=0}^{\infty} (2l + 1)\int_0^{\infty} dk\left(\frac{dm_l}{dk} - \frac{dm_l^0}{dk}\right) e^{-\beta\hbar^2k^2/m} \\ &= \frac{1}{\pi}\sum_{l=0}^{\infty} (2l + 1)\int_0^{\infty} dk\,\frac{d\delta_l(k)}{dk}\,e^{-\beta\hbar^2k^2/m}.\end{aligned} \tag{11.189}$$

The second virial coefficient then takes the form

$$\begin{aligned}\mathscr{B}_2(T) = \mathscr{B}_2^0(T) &- 2^{3/2}\lambda_T^3 \sum_{n(\text{bound})} e^{-\beta\epsilon_n} \\ &- \frac{2^{3/2}\lambda_T^3}{\pi}\sum_{l=0}^{\infty}{}' (2l + 1)\int_0^{\infty} dk\,\frac{d\delta_l(k)}{dk}\,e^{-\beta\hbar^2k^2/m},\end{aligned} \tag{11.190}$$

where the summation over l is restricted to even values of l for bosons and odd values for fermions.

It is worthwhile to compute $\mathscr{B}_2(T)$ for a boson gas of hard spheres of radius σ. Such a system has no bound states, and only the first and third terms in Eq. (11.190) contribute. For $r < \sigma$, $v(r) = \infty$ and we expect that $R_l(kr) \equiv 0$ since the particles cannot penetrate. For $r > \sigma$, the form of $R_l(kr)$ given in Eq. (11.182) can be used since the particles do not interact for those values of r. At $r = \sigma$, the wave functions for $r > \sigma$ and $r < \sigma$ must be equal. Thus,

$$R_l(k\sigma) = A_l[\cos\delta_l(k)j_l(k\sigma) - \sin\delta_l(k)n_l(k\sigma)] = 0 \tag{11.191}$$

or

$$\tan\delta_l(k) = \frac{j_l(k\sigma)}{n_l(k\sigma)}. \tag{11.192}$$

For $(k\sigma) \ll 1$, we can write Eq. (11.192) in the form

$$\tan\delta_l(k) = -\frac{(k\sigma)^{2l+1}}{(2l + 1)[(2l - 1)!!]} \tag{11.193}$$

where $(2l - 1)!! = (2l - 1)(2l - 3)(2l - 5) \times \cdots \times 1$.

For a dilute boson gas, the momentum of the particles will be low and therefore the lowest order partial wave will give good approximation to the virial coefficients. For bosons, only even values of l can contribute. If we let $\delta_0 = -k\sigma$, we obtain the following approximate expression for $\mathscr{B}_2(T)$:

$$\mathscr{B}_2(T) = -\frac{\lambda_T^3}{2^{5/2}} + 2\sigma\lambda_T^2 + \cdots \tag{11.194}$$

Thus, while the statistics tends to lower the pressure of the boson gas, the hard core tends to increase it. For the reduced second virial coefficient, $\mathscr{B}_2^*(T)$, we have

$$\mathscr{B}_2^*(T) = \frac{\mathscr{B}_2(T)}{b_0} = -\frac{3\sqrt{2}}{16\pi}\left(\frac{\lambda_T}{\sigma}\right)^3 + \frac{3}{\pi}\left(\frac{\lambda_T}{\sigma}\right)^2 + \cdots \tag{11.195}$$

When $\lambda_T > \sigma$ the term coming from the statistics begins to dominate.

Until now, there has been very little success in treating degenerate quantum fluids along lines analogous to those used for classical fluids. If we go beyond the second virial coefficient, new complications occur which do not exist in classical fluids. We shall gain a better idea of the nature of these complications in the next chapter when we discuss approximation schemes in quantum fluids.

REFERENCES

1. J. O. Hirschfelder, C. F. Curtiss, and R. B. Bird, *Molecular Theory of Gases and Liquids* (John Wiley & Sons, New York, 1954).
2. T. L. Hill, *Statistical Mechanics* (McGraw-Hill Book Co., New York, 1956).
3. D. A. McQuarrie, *Statistical Mechanics* (Harper and Row, New York, 1976).
4. J. E. Mayer and M. G. Mayer, *Statistical Mechanics* (John Wiley & Sons, New York, 1941).
5. G. E. Uhlenbeck and G. W. Ford in *Studies in Statistical Mechanics*, Vol. 1, ed. J. de Boer and G. E. Uhlenbeck (North-Holland Publishing Co., Amsterdam, 1962).
6. H. D. Ursell, Proc. Cambridge Phil. Soc. *23* 685 (1927).
7. T. Kihara, Rev. Mod. Phys. *25* 831 (1953); *27* 412 (1955).
8. A. E. Sherwood and J. M. Prausnitz, J. Chem. Phys. *41* 413, 429 (1964).
9. H. W. Graben and R. D. Present, Phys. Rev. Let. *9* 247 (1962).
10. B. M. Axilrod, J. Chem. Phys. *19* 719 (1951).
11. Y. Midzuno and T. Kihara, J. Phys. Soc. (Japan) *11* 1045 (1956).
12. S. Katsura, Phys. Rev. *115* 1417 (1959).
13. F. H. Ree and W. G. Hoover, J. Chem. Phys. *40* 939 (1964).
14. J. S. Rowlinson, Repts. Prog. Phys. *28* 169 (1965).
15. Reprinted in *The Equilibrium Theory of Classical Fluids*, ed. H. L. Frisch and J. L. Lebowitz (W. A. Benjamin, New York, 1964).
16. J. K. Percus and G. J. Yevick, Phys. Rev. *110* 1 (1958).
17. E. Thiele, J. Chem. Phys. *39* 474 (1963).
18. G. H. A. Cole, Repts. Prog. Phys. *31* 419 (1968).
19. N. N. Bogoliubov in *Studies in Statistical Mechanics*, Vol. 1, ed. J. de Boer and G. E. Uhlenbeck (North-Holland Publishing Co., Amsterdam, 1962).

20. M. Born and H. S. Green, Proc. Roy. Soc. *A188* 10 (1946).
21. J. Yvon, *Actualites scientifiques et industrielles*, no. 203 (Paris: Hermann and Cie, 1935).
22. J. G. Kirkwood, J. Chem. Phys. *3* 300 (1935).
23. J. G. Kirkwood and E. M. Boggs, J. Chem. Phys. *10* 394 (1942).
24. R. Zwanzig, J. Chem. Phys. *22* 1420 (1954).
25. J. A. Barker and D. Henderson, Ann. Rev. Phys. Chem. *23* 439 (1972).
26. J. D. Weeks, D. Chandler, and H. C. Andersen, J. Chem. Phys. *54* 5237 (1971).
27. K. Huang, *Statistical Mechanics* (John Wiley & Sons, New York, 1963).
28. L. Schiff, *Quantum Mechanics* (McGraw-Hill Book Co., New York, 1968).
29. J. G. Kirkwood, *Theory of Liquids* (Gordon and Breach, New York, 1968).
30. D. Henderson, Ann. Rev. Phys. Chem. *15* 31 (1964).
31. R. O. Watts, J. Chem. Phys. *48* 50 (1968).

PROBLEMS

1. Compute the third virial coefficient for a hard-sphere gas using a geometrical approach in evaluating the integral.

2. Compute the second and third virial coefficients for the Gaussian model, $f(q_{ij}) = \exp(-aq_{ij}^2)$. Sketch the interparticle potential for the Gaussian model.

3. Compute the second and third virial coefficients for a hard-disk (two-dimensional) gas and a hard-rod (one-dimensional) gas.

4. Compute the second virial coefficient for a square-well potential.

5. Compute the second virial coefficient for the Sutherland potential: $v(q) = \infty$ for $q < \sigma$ and $v(q) = -cq^{-\gamma}$ for $q > \sigma$. Sketch the potential.

6. Derive Eqs. (11.74) and (11.75) for the second virial coefficient of a system of particles which interact via a Lennard-Jones 6–12 potential.

7. Compute the van der Waals constants a and b for nitrogen, assuming the molecules interact via a square-well potential.

8. For a system of nonspherically symmetric molecules which interact via the potential $v(\mathbf{q}_1, \mathbf{q}_2) = \phi_1(q_{12}) + \phi_2(q_{12})[\cos^2\theta_1 + \cos^2\theta_2]$, derive the second virial coefficient in terms of an integral over q_{12}. Discuss the shape of the molecules which leads this potential.

9. Find a general expression for the second virial coefficient for a mixture of two types of molecules.

10. Obtain the graphical expression for $g_3(\mathbf{q}_1, \mathbf{q}_2, \mathbf{q}_3)$ to third order in the fugacity and to third order in the density. What graphs are neglected in making the superposition approximation?

11. Draw all graphs which are retained in the expression for the direct correlation function to order $\langle n\rangle^2$ in the hypernetted chain approximation and the Percus-Yevick approximation.

12. For a hard-sphere gas compute the radial distribution function to first order in the density and sketch and discuss its behavior.

13. Compute the constant pressure heat capacity in terms of the second and third virial coefficients.

14. Obtain an expression for the Joule-Thompson coefficient in terms of the second and third virial coefficients.

15. Obtain the virial expansion for the internal energy and entropy of a classical fluid.

16. Obtain an expression for b_l for an ideal fermion and an ideal boson gas. Find the second and third virial coefficients.

17. Compute the second virial coefficient for a dilute hard-sphere Fermi gas. Include the contributions from the first two odd partial waves.

18. Compute the radial distribution function for an ideal boson and an ideal fermion gas. Discuss its behavior.

12 Quantum Fluids

A. INTRODUCTORY REMARKS

If we lower the temperature or increase the density of a fluid sufficiently, at some point the thermal wavelength becomes larger than the interparticle spacing and the effects of quantum statistics begin to play an important role in shaping the thermodynamic behavior of the system. We have already seen in Chap. 9 that ideal fermion or boson systems exhibit similar thermodynamic behavior at high temperature, but radically different behavior at low temperature. In Chap. 11, we found that statistics can have a significant effect on the virial coefficients for hard-sphere systems at low temperature.

In this chapter, we will study the behavior of weakly coupled highly degenerate quantum fluids. For such systems the phase space variables do not commute and a new type of behavior sets in which requires new techniques. We will study quantum fluids using a perturbation expansion rather than the binary expansions used in Chap. 11. Until now, attempts to develop a microscopic theory of moderately dense or dense quantum fluids based directly on a binary expansion have not been as successful as for classical fluids. Most theories of quantum fluids use perturbation expansions (cf. Ref. 12.1 for a discussion).

Perturbation expansions can take us a long way in our understanding of the behavior of quantum fluids. Historically, there have been two approaches to the subject. One approach very widely used in the early days was to study the ground state of the fluid (the thermodynamic state at $T = 0$ K). The ground state can be described in terms of a single wave function. However, this approach is limited because it only describes interacting systems whose state can be obtained from the noninteracting ground state by adiabatically turning on the interaction between the particles. It tells us nothing about the finite temperature behavior of the system. The other approach, and the one we will follow here, is to study the thermodynamic properties directly from the grand partition function, thus giving a fully temperature-dependent theory.

The first step in obtaining a theory of quantum fluids is formal. We will obtain an expression for the grand potential of normal (not superfluid) boson and fermion systems in terms of a perturbation expansion. We will work in the number representation and use Wick's theorem to evaluate various terms in the expansion for the grand potential. As for the case of classical fluids, the grand potential can be expressed in terms of fully connected diagrams, although the diagrams for quantum fluids have quite a different structure from the classical diagrams and reflect the fact that the phase space coordinates do not commute.

To gain some familiarity with the theory, we will apply it to the case of an electron gas in a uniform positive-charge background. In this system, collective effects play a dominant role in determining the thermodynamic behavior of the system. The electrons can polarize the medium by repelling surrounding electrons, thus creating a positive charge in their immediate neighborhood. This polarization of the medium is a collective effect which drastically alters the thermodynamic behavior of the system and can be taken into account only by summing infinite numbers of diagrams. We will consider a high density, weakly coupled electron gas and we will obtain expressions for the internal energy at zero temperature and the equation of state (the Debye-Hückel equation) at high temperature. We will find that at zero temperature the density expansion is not given by a powers series but has terms logarithmic in the density parameter. This is a common type of behavior in systems where collective effects play a dominant role. We shall see it again in later chapters.

In classical fluids, it is easiest to work with the reduced probability densities. In quantum fluids the corresponding object is called a propagator. The propagator is a temperature-dependent correlation function of creation and annihilation operators. It is more general and contains more information than a reduced density matrix, but in a certain limit the reduced density matrices can be obtained from it. The poles of the Fourier transformed propagator give the spectrum of excitations in the interacting system. We will obtain a diagrammatic expression for the one-body propagator and then find a general equation for it (Dyson's equation) in terms of self-energy effects.

To illustrate the use of the propagator, we will obtain the dispersion relations for the excitations in a weakly coupled Fermi fluid and a weakly coupled condensed boson fluid. These excitations also occur in strongly coupled quantum fluids and are important for understanding the thermodynamic behavior of the helium liquids and other quantum systems.

B. GRAND POTENTIAL FOR NORMAL BOSON AND FERMI FLUIDS

When the thermal wavelength of the particles in a fluid becomes of the order of the interparticle spacing, the effects of statistics must be included in any microscopic theory and lead to the appearance of new phenomena. In this section, we will obtain a cumulant expansion for the grand potential for normal boson and fermion systems, that is, systems which have not undergone a transition to a superfluid state. We will consider the case of superfluid boson systems in a later section.

We will derive a microscopic theory of quantum fluids, using perturbation expansions rather than binary expansions as was done for the case of classical fluids. Since we will be working with perturbation expansions throughout this chapter, we will work exclusively in the number representation (although we can derive the same results using any other representation) because it simplifies many calculations.

1. Cumulant Expansion for the Grand Potential

The grand partition function of a quantum fluid can be written in the form

$$\mathscr{Z}(T, V, \mu) = \operatorname{Tr} e^{-\beta\hat{K}}, \tag{12.1}$$

where

$$\hat{K} = \hat{K}_0 + \lambda\hat{V} \tag{12.2}$$

and

$$\hat{K}_0 = \hat{H}_0 - \mu\hat{N}. \tag{12.3}$$

Here, $\hat{H}_0$ is the kinetic energy operator, μ is the chemical potential, $\hat{V}$ is the potential energy operator, and λ is a coupling constant that will be used later. It is useful to introduce an additional operator,

$$\hat{U}(\tau, \tau') \equiv e^{\tau\hat{K}_0} e^{-(\tau-\tau')\hat{K}} e^{-\tau'\hat{K}_0}, \tag{12.4}$$

which obeys the differential equation

$$\frac{\partial}{\partial\tau}\hat{U}(\tau, \tau') = -\lambda\hat{V}(\tau)_0\hat{U}(\tau, \tau') \tag{12.5}$$

where

$$\hat{V}(\tau)_0 = e^{\tau\hat{K}_0}\hat{V}e^{-\tau\hat{K}_0}. \tag{12.6}$$

The grand partition function then takes the form

$$\mathscr{Z}(V, T, \mu) = \operatorname{Tr} e^{-\beta\hat{K}_0}\hat{U}(\beta, 0) \tag{12.7}$$

($\hat{U}(\beta, 0)$ is obtained from $\hat{U}(\tau, \tau')$ by setting $\tau = \beta$ and $\tau' = 0$).

We can integrate Eq. (12.5) to obtain

$$\hat{U}(\tau, \tau') = 1 - \lambda\int_{\tau'}^{\tau} d\tau_1 \hat{V}(\tau_1)_0\hat{U}(\tau_1, \tau') \tag{12.8}$$

(note that $\hat{U}(\tau, \tau) = 1$). The perturbation expansion for $\hat{U}(\tau, \tau')$ is then found by iterating Eq. (12.8):

$$\hat{U}(\tau, \tau') = \sum_{n=0}^{\infty}(-\lambda)^n\int_{\tau'}^{\tau} d\tau_1\int_{\tau'}^{\tau_1} d\tau_2 \times \cdots \times\int_{\tau'}^{\tau_{n-1}} d\tau_n\hat{V}(\tau_1)_0\hat{V}(\tau_2)_0 \times \cdots \times \hat{V}(\tau_n)_0. \tag{12.9}$$

If we now combine Eqs. (12.7) and (12.9), we obtain the following perturbation expansion for the grand partition function:

$$\mathscr{Z}(V, T, \mu) = \sum_{n=0}^{\infty} (-\lambda)^n \int_0^{\beta} d\tau_1 \int_0^{\tau_1} d\tau_2 \times \cdots \times \int_0^{\tau_{n-1}} d\tau_n$$
$$\times \operatorname{Tr}[e^{-\beta\hat{K}_0} \hat{V}(\tau_1)_0 \times \cdots \times \hat{V}(\tau_n)_0]. \qquad (12.10)$$

Note that μ is the *exact* chemical potential and depends on λ. Thus, Eq. (12.10) is not a simple expansion in powers of λ.

In the number representation the operator, $\hat{K}_0$, takes the form

$$\hat{K}_0 = \sum_k \left(\frac{\hbar^2|\mathbf{k}|^2}{2m} - \mu\right) \hat{a}_k{}^+ \hat{a}_k \qquad (12.11)$$

and the potential energy operator takes the form

$$\hat{V} = \frac{1}{4} \sum_{k_1,\ldots,k_4} \langle k_1, k_2|V|k_3, k_4\rangle^{(\pm)} \hat{a}_{k_1}^+ \hat{a}_{k_2}^+ \hat{a}_{k_4} \hat{a}_{k_3} \qquad (12.12)$$

where k_i denotes both momentum and spin variables, $k_i = (\mathbf{k}_i, \sigma_i)$. Eqs. (12.11) and (12.12) may be used for either boson or fermion statistics. The quantity, $\langle k_1, k_2|V|k_3, k_4\rangle^{(\pm)}$, denotes either the symmetrized or antisymmetrized matrix elements and is defined

$$\langle k_1, k_2|V|k_3, k_4\rangle^{(\pm)} = \langle k_1, k_2|V|k_3, k_4\rangle \pm \langle k_1, k_2|V|k_4, k_3\rangle. \qquad (12.13)$$

The top sign applies for bosons and the bottom sign for fermions.

It is useful to write Eq. (12.9) in a slightly different form. Let us first consider an example. If we introduce the Heaviside function, $\theta(x)$, where $\theta(x) = 0$ for $x < 0$ and $\theta(x) = 1$ for $x > 0$, the second-order term in Eq. (12.10) may be rewritten

$$(-1)^2 \int_0^{\beta} d\tau_1 \int_0^{\tau_1} d\tau_2 \hat{V}(\tau_1)_0 \hat{V}(\tau_2)_0$$
$$= (-1)^2 \int_0^{\beta} d\tau_1 \int_0^{\beta} d\tau_2 \hat{V}(\tau_1)_0 \hat{V}(\tau_2)_0 \theta(\tau_1 - \tau_2)$$
$$= \frac{1}{2!} \int_0^{\beta} d\tau_1 \int_0^{\beta} d\tau_2 [\hat{V}(\tau_1)_0 \hat{V}(\tau_2)_0 \theta(\tau_1 - \tau_2) + \epsilon^P \hat{V}(\tau_2) \hat{V}(\tau_1) \theta(\tau_2 - \tau_1)]$$
$$\equiv \frac{1}{2!} \int_0^{\beta} d\tau_1 \int_0^{\beta} d\tau_2 T[\hat{V}(\tau_1)_0 \hat{V}(\tau_2)_0]$$

where T denotes a temperature-ordered product and is defined as the sum of all different temperature orderings of the product of operators $\hat{V}(\tau_1)\hat{V}(\tau_2)$. The factor $\epsilon = +1$ for bosons and $\epsilon = -1$ for fermions. P is the number of permutations of creation and annihilation operators needed to make the reordering $\hat{V}(\tau_1)\hat{V}(\tau_2) \rightarrow \hat{V}(\tau_2)\hat{V}(\tau_1)$. The nth-order term in Eq. (12.9) can be rewritten in a similar manner and the grand partition function takes the form

$$\mathscr{Z}(V, T, \mu) = \{\mathrm{Tr}\, e^{-\beta\hat{K}_0}\} \sum_{n=0}^{\infty} \lambda^n W_n(V, T, \mu) \tag{12.14}$$

where

$$W_n(V, T, \mu) = \{\mathrm{Tr}\, e^{-\beta\hat{K}_0}\}^{-1} \frac{(-1)^n}{n!} \int_0^\beta d\tau_1 \times \cdots \times \int_0^\beta d\tau_n \times \mathrm{Tr}\{e^{-\beta\hat{K}_0} T[\hat{V}(\tau_1)_0 \times \cdots \times \hat{V}(\tau_n)_0]\}. \tag{12.15}$$

The temperature-ordered product $T[\hat{V}(\tau_1)_0 \times \cdots \times \hat{V}(\tau_n)_0]$ contains the sum of all $n!$ different permutations of the n potential operators $\hat{V}(\tau)_0$ multiplied by the appropriate Heaviside functions. For fermions, each term in the temperature-ordered product which requires an odd number of permutations of creation and annihilation operators to reorder the potential energy operators must be multiplied by a minus sign. (Note that, if the potential energy operators contain an even number of creation and annihilation operators, no minus signs are ever needed.)

We can now rewrite the grand partition function in the form of a cumulant expansion:

$$\mathscr{Z}(V, T, \mu) = \{\mathrm{Tr}\, e^{-\beta\hat{K}_0}\}\left\{\exp\left[\sum_{l=1}^{\infty} \lambda^l U_l(V, T, \mu)\right]\right\}. \tag{12.16}$$

The cluster functions, $U_n(V, T, \mu)$, can be defined in terms of the functions $W_n(V, T, \mu)$ by expanding Eqs. (12.14) and (12.16) in powers of λ and equating terms which depend on the same power in λ. Thus,

$$U_1 = W_1, \tag{12.17a}$$

$$U_2 = W_2 - \tfrac{1}{2}W_1^2, \tag{12.17b}$$

$$U_3 = W_3 - W_1W_2 + \tfrac{1}{3}W_1^3, \tag{12.17c}$$

etc. From Eqs. (9.56) and (12.16) the grand potential can be written in the form

$$\Omega(V, T, \mu) = -\frac{1}{\beta}\ln \mathrm{Tr}\, e^{-\beta\hat{K}_0} - \frac{1}{\beta}\sum_{l=1}^{\infty} \lambda^l U_l(V, T, \mu). \tag{12.18}$$

Thus, we obtain a cumulant expansion for the grand potential.

We now must find explicit expressions for the cluster functions $U_n(V, T, \mu)$. This is relatively simple because we can use Wick's theorem (cf. Sec. 10.B) to evaluate the trace in Eq. (12.15).

2. Wick's Theorem[2-5]

Since the operator, $\hat{K}_0$, depends only on pairs of creation and annihilation operators, the probability density $\hat{\rho}_0 = e^{-\beta\hat{K}_0}/\text{Tr}\, e^{-\beta\hat{K}_0}$ is the field theoretic analog of a generalized Gaussian distribution. Thus, Wick's theorem can be used to evaluate the trace in Eq. (12.15). Wick's theorem states that *the average of a product of operators is equal to the sum of all possible ways of averaging the product in pairs.*

Before we use Wick's theorem, we must establish a few properties of the creation and annihilation operators. Let us first note that

$$\frac{\partial}{\partial\tau}\hat{a}_k(\tau)_0 = [\hat{K}_0, \hat{a}_k(\tau)_0]_- = -\epsilon_k{}^0\hat{a}_k(\tau)_0 \tag{12.19}$$

and

$$\frac{\partial}{\partial\tau}\hat{a}_k{}^+(\tau)_0 = [\hat{K}_0, \hat{a}_k{}^+(\tau)_0]_- = \epsilon_k{}^0\hat{a}_k{}^+(\tau)_0 \tag{12.20}$$

where $\hat{a}_k(\tau)_0 = e^{\tau\hat{K}_0}\hat{a}_k\, e^{-\tau\hat{K}_0}$ and $\epsilon_k{}^0 = (\hbar^2|\mathbf{k}|^2/2m) - \mu$. Thus, the temperature dependence of the creation and annihilation operators is

$$\hat{a}_k{}^+(\tau)_0 = e^{+\epsilon_k^0\tau}\hat{a}_k{}^+ \tag{12.21}$$

and

$$\hat{a}_k(\tau)_0 = e^{-\epsilon_k^0\tau}\hat{a}_k, \tag{12.22}$$

respectively, for both bosons and fermions. Let us further note that

$$\hat{a}_k\, e^{-\beta\hat{K}_0} = e^{-\beta\hat{K}_0}\hat{a}_k\, e^{-\epsilon_k^0\beta} \tag{12.23}$$

and

$$\hat{a}_k{}^+\, e^{-\beta\hat{K}_0} = e^{-\beta\hat{K}_0}\hat{a}_k{}^+\, e^{\epsilon_k^0\beta}. \tag{12.24}$$

With these results the expectation value of a product of creation and annihilation operators can be written

$$\begin{aligned}\langle\hat{a}_k{}^+\hat{a}_{k'}\rangle_0 &= \text{Tr}\,\hat{\rho}_0\hat{a}_k{}^+\hat{a}_{k'} = \mp\,\delta_{k,k'} \pm \text{Tr}\,\hat{\rho}_0\hat{a}_{k'}\hat{a}_k^+ \\ &= \mp\,\delta_{k,k'} \pm e^{+\beta\epsilon_k^0}\langle\hat{a}_k{}^+\hat{a}_{k'}\rangle_0\end{aligned} \tag{12.25}$$

where we have made use of the commutation relations for boson and fermion operators and Eq. (12.24) (cf. App. B). The top sign applies to bosons and the bottom sign applies to fermions. Rearranging terms, we obtain

$$\langle\hat{a}_k^+\hat{a}_{k'}\rangle_0 = \delta_{k,k'}n_0(k) = \frac{\delta_{k,k'}}{(e^{\beta\epsilon_k^0} \mp 1)}. \tag{12.26}$$

The quantity, $n_0(k)$, is the momentum distribution function for free bosons $(-)$ or fermions $(+)$. Similarly,

$$\langle\hat{a}_k\hat{a}_{k'}^+\rangle_0 = \delta_{k,k'}(1 \pm n_0(k)) \tag{12.27}$$

(the top sign is for bosons and the bottom sign is for fermions) and

$$\langle \hat{a}_k \hat{a}_{k'} \rangle_0 = \langle \hat{a}_k{}^+ \hat{a}_{k'}^+ \rangle_0 = 0. \tag{12.28}$$

If we consider a product of four operators, we evaluate the trace as follows. By successively applying boson or fermion commutation relations, we can write

$$\begin{aligned}\mathrm{Tr}\{\hat{\rho}^0 \hat{a}_{k_1}^+ \hat{a}_{k_2}^+ \hat{a}_{k_4} \hat{a}_{k_3}\} = &-\delta_{k_1,k_4}\, \mathrm{Tr}\{\hat{\rho}^0 \hat{a}_{k_2}^+ \hat{a}_{k_3}\} \mp \delta_{k_1,k_3}\, \mathrm{Tr}\{\hat{\rho}^0 \hat{a}_{k_2}^+ \hat{a}_{k_4}\} \\ &\pm e^{-\beta \epsilon_{k_1}^0}\, \mathrm{Tr}\{\hat{\rho}^0 \hat{a}_{k_1}^+ \hat{a}_{k_2}^+ \hat{a}_{k_4} \hat{a}_{k_3}\}.\end{aligned} \tag{12.29}$$

After rearranging terms, we finally obtain

$$\mathrm{Tr}\{\hat{\rho}^0 \hat{a}_{k_1}^+ \hat{a}_{k_2}^+ \hat{a}_{k_4} \hat{a}_{k_3}\} = [\langle \hat{a}_{k_1}^+ \hat{a}_{k_3} \rangle \langle \hat{a}_{k_2}^+ \hat{a}_{k_4} \rangle \pm \langle \hat{a}_{k_1}^+ \hat{a}_{k_4} \rangle \langle \hat{a}_{k_2}^+ \hat{a}_{k_3} \rangle]. \tag{12.30}$$

The sign in front of a given term in the sum of products of pairwise averages is always positive for bosons but will be negative for fermions if it requires an odd number of permutations to rearrange the product of operators into the appropriate pairs. If we apply the above procedure to the expectation value $\mathrm{Tr}\{\hat{\rho}_0 \hat{a}_{k_1}^+ \hat{a}_{k_2}^+ \hat{a}_{k_4} \hat{a}_{k_3} \hat{a}_{k_5}^+ \hat{a}_{k_6}^+ \hat{a}_{k_8} \hat{a}_{k_7}\}$ we obtain a sum of twenty-four terms (some identical), each containing a product of four pairwise averages.

The expression for $W_n(V, T, \mu)$ (cf. Eq. (12.15)) contains temperature-ordered products of operators. Therefore, we must generalize Wick's theorem to the case of temperature-ordered products. It will be useful to introduce the idea of a contraction: *a contraction is the average of a temperature-ordered pair of operators.* Let us consider the temperature-ordered product

$$T[\hat{a}_k(\tau_1)_0 \hat{a}_{k'}^+(\tau_2)_0] \equiv \hat{a}_k(\tau_1)_0 \hat{a}_{k'}^+(\tau_2)_0 \theta(\tau_1 - \tau_2) \pm \hat{a}_{k'}^+(\tau_2) \hat{a}_k(\tau_1) \theta(\tau_2 - \tau_1). \tag{12.31}$$

The *contraction* of $\hat{a}_k(\tau_1)_0$ and $\hat{a}_{k'}^+(\tau_2)_0$ is denoted $G_0(k, k'; \tau_1 - \tau_2)$ and is defined

$$\begin{aligned}G_0(k, k'; \tau_1 - \tau_2) &= -\langle T[\hat{a}_k(\tau_1)_0 \hat{a}_{k'}^+(\tau_2)_0] \rangle_0 = \mp \langle T[a_{k'}^+(\tau_2) a_k(\tau_1)] \rangle_0 \\ &= -\delta_{k,k'}[(1 \pm n_0(k))\theta(\tau_1 - \tau_2) \pm n_0(k)\theta(\tau_2 - \tau_1)]\, e^{-(\tau_1 - \tau_2)\epsilon_k^0}.\end{aligned} \tag{12.32}$$

If we write

$$G_0(k, k'; \tau_1 - \tau_2) = \delta_{k,k'} G_0(k; \tau_1 - \tau_2), \tag{12.33}$$

we obtain

$$G_0(k; \tau_1 - \tau_2) = -[(1 \pm n_0(k))\theta(\tau_1 - \tau_2) \pm n_0(k)\theta(\tau_2 - \tau_1)]\, e^{-(\tau_1 - \tau_2)\epsilon_k^0}. \tag{12.34}$$

The quantities $G_0(k, k'; \tau_1 - \tau_2)$ and $G_0(k; \tau_1 - \tau_2)$ are also called free propagators.

Wick's theorem for a temperature-ordered product of operators can now be stated as follows: *the average of a temperature-ordered product of creation and annihilation operators is equal to the sum of all possible ways of contracting the product in pairs*. The sign in front of a given term in the sum is always positive for bosons but is negative for fermions if it requires an odd number of permutations to rearrange the product of operators into the appropriate pairs.

It is useful to illustrate Wick's theorem for temperature-ordered products with some examples. Let us first note that the temperature-ordered product gives all possible orderings of the different potential operators, $\hat{V}(\tau)$. Within a given potential operator $\hat{V}(\tau)$ the creation and annihilation operators all have the same temperature and are always ordered with the annihilation operators to the right. We therefore can consider the temperature-ordered product as a temperature ordering of creation and annihilation operators in which all possible orderings occur. If we now use the identities

$$\theta(\tau_1 - \tau_2)\theta(\tau_1 - \tau_3) = \theta(\tau_1 - \tau_2)\theta(\tau_2 - \tau_3) + \theta(\tau_1 - \tau_3)\theta(\tau_3 - \tau_2) \tag{12.35}$$

and

$$\theta(\tau_1 - \tau_2)\theta(\tau_2 - \tau_1) = 0, \tag{12.36}$$

we obtain

$$\begin{aligned}\langle T[\hat{a}_{k_1}^+(\tau_1)_0\hat{a}_{k_2}^+(\tau_2)_0\hat{a}_{k_3}(\tau_3)_0\hat{a}_{k_4}(\tau_4)_0]\rangle_0 &= \delta_{k_2,k_3}\,\delta_{k_1,k_4}G_0(k_1;\tau_1 - \tau_4)G_0(k_2;\tau_2 - \tau_3) \\ &\pm \delta_{k_1,k_3}\,\delta_{k_2,k_4}G_0(k_1;\tau_1 - \tau_3)G_0(k_2;\tau_2 - \tau_4).\end{aligned} \tag{12.37}$$

A product of three creation and three annihilation operators will yield a sum of six terms, each involving a product of three contractions.

3. Diagrams[5–7]

With the help of Wick's theorem, we can obtain algebraic expressions for the functions, $W_n(V, T, \mu)$. From Eqs. (12.12) and (12.15), we can write

$$\begin{aligned}W_1(V, T, \mu) = \frac{(-1)}{4}\int_0^\beta d\tau_1 \sum_{k_1\ldots k_4} \langle k_1, k_2|V|k_3, k_4\rangle^{(\pm)} \\ \times \mathrm{Tr}[e^{-\beta\hat{K}_0}T\{\hat{a}_{k_1}^+(\tau_1)\hat{a}_{k_2}^+(\tau_1)\hat{a}_{k_4}(\tau_1)\hat{a}_{k_3}(\tau_1)\}].\end{aligned} \tag{12.38}$$

Thus, for the case $n = 1$, all operators have the same temperature dependence. If we use Wick's theorem, we obtain

$$W_1(V, T, \mu) = \frac{(-1)}{2}\beta \sum_{k_1,k_2} \langle k_1, k_2|V|k_1, k_2\rangle^{(\pm)}G_0(k_1; 0)G_0(k_2; 0) \tag{12.39}$$

where

$$G_0(k; 0) = n_0(k). \tag{12.40}$$

For $n = 2$, we find

$$\begin{aligned}
W_2(V, T, \mu) = {} & \frac{1}{8} \sum_{k_1,k_4} \int_0^\beta d\tau_1 \int_0^\beta d\tau_2 \langle k_1, k_2 | V | k_3, k_4 \rangle^{(\pm)} \langle k_3, k_4 | V | k_1, k_2 \rangle^{(\pm)} \\
& \times G_0(k_1; \tau_1 - \tau_2) G_0(k_2; \tau_1 - \tau_2) \\
& \times G_0(k_3; \tau_2 - \tau_1) G_0(k_4; \tau_2 - \tau_1) \\
& \pm \frac{1}{2} \sum_{k_1 \ldots k_4} \int_0^\beta d\tau_1 \int_0^\beta d\tau_2 \langle k_1, k_2 | V | k_3, k_4 \rangle^{(\pm)} \langle k_3, k_4 | V | k_1, k_2 \rangle^{(\pm)} \\
& \times G_0(k_1; \tau_1 - \tau_2) G_0(k_3; \tau_2 - \tau_1) \\
& \times G_0(k_2; 0) G_0(k_4; 0) \\
& + \frac{1}{8} \left[(-1) \sum_{k_1 k_2} \int_0^\beta d\tau_1 \langle k_1, k_2 | V | k_1, k_2 \rangle^{(\pm)} G_0(k_1; 0) G_0(k_2; 0) \right]^2 .
\end{aligned} \tag{12.41}$$

Higher order terms rapidly become more complicated and it is useful to introduce diagrammatic expressions for $W_n(V, T, \mu)$ (for classical fluids we call them graphs and for quantum fluids we call them diagrams). The diagrams that represent $W_n(V, T, \mu)$ are different from the classical graphs. For quantum fluids, we represent interactions by dots and particles by lines, the reverse of what we do for classical fluids.

We may represent $W_n(V, T, \mu)$ as follows:

$$W_n(T, V, \mu) = \sum (\text{all different } n\text{th-order 0-diagrams}). \tag{12.42}$$

An nth-order 0-diagram is a collection of n vertices and $2n$ directed lines such that each vertex has two lines entering and two lines leaving. No directed lines can be broken. Two 0-diagrams are different if they have different topological structure. An algebraic expression can be associated to a 0-diagram according to the rules in Table 12.1. We may represent $W_1(V, T, \mu)$, $W_2(V, T, \mu)$, and $W_3(V, T, \mu)$ as follows:

$W_1(V, T, \mu) =$  (12.43*a*)

$W_2(V, T, \mu) =$ 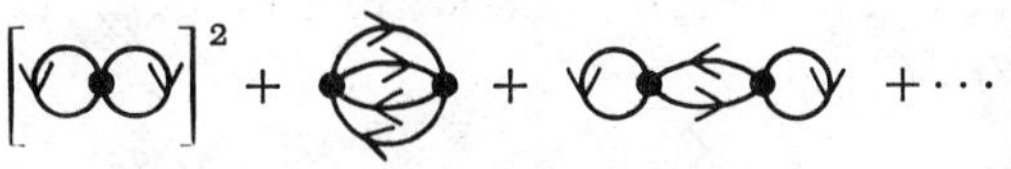 (12.43*b*)

$W_3(V, T, \mu) =$ 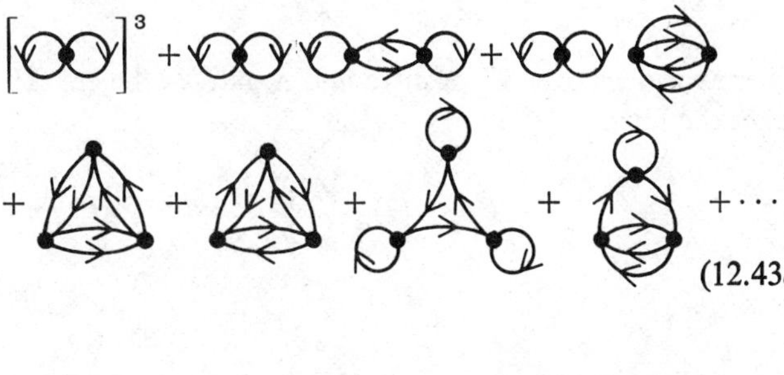

(12.43*c*)

If we substitute the expressions for $W_n(T, V, \mu)$ into Eqs. (12.17), we obtain expressions for the cluster functions $U_n(V, T, \mu)$. We find that the diagrams which contribute to $U_n(V, T, \mu)$ have a special structure. They consist only of connected nth-order 0-diagrams. Thus,

$$U_n(V, T, \mu) = \sum (\text{all different connected } n\text{th-order 0-diagrams}). \tag{12.44}$$

Connected 0-diagrams are 0-diagrams for which all the vertices can be reached by continuously following directed lines. As a result, the cluster functions $U_n(V, T, \mu)$ are each proportional to the volume, whereas the functions $W_n(V, T, \mu)$ can contain terms with volume dependence ranging from V to V^n. The first few cluster functions can be represented by the following diagrams:

$U_1(V, T, \mu) =$  (12.45a)

$U_2(V, T, \mu) =$ 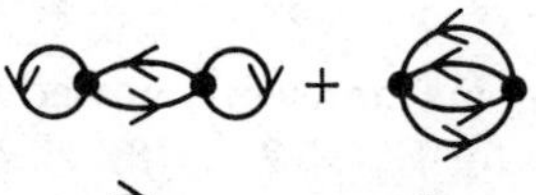 $+\cdots$ (12.45b)

$U_3(V, T, \mu) =$ 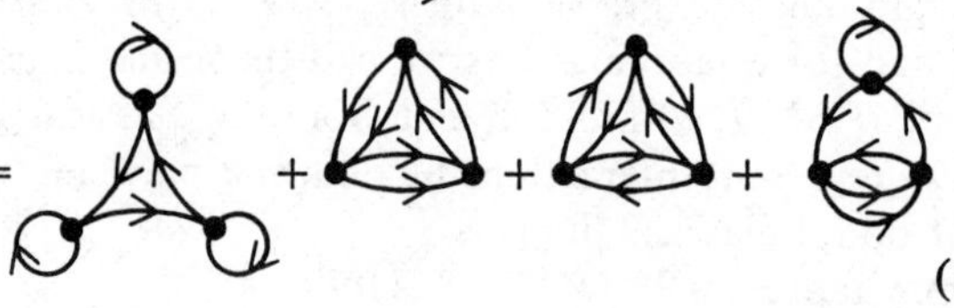 $+\cdots$ (12.45c)

etc. The rules for associating algebraic expressions to the connected 0-diagrams are given in Table 12.1.

The rules for evaluating the 0-diagrams have been given in terms of momentum and temperature variables. In practice, it is useful to Fourier transform the diagrams and write them in terms of momenta and frequencies. We first note that

$$\int_0^\beta d\tau\, e^{-in\pi\tau/\beta} = \beta\, \delta_{n,0} \tag{12.46}$$

for n even, and we introduce the Fourier transform $\bar{G}_0(k, \omega_n)$ from the equation

$$G_0(k, \tau) = \frac{1}{\beta} \sum_n e^{-i\omega_n\tau} \bar{G}_0(k, \omega_n) \tag{12.47}$$

where n ranges over all positive and negative integers. The inverse transform takes the form

$$\bar{G}_0(k, \omega_n) = \frac{1}{2} \int_{-\beta}^{\beta} d\tau\, e^{i\omega_n\tau} G_0(k, \tau). \tag{12.48}$$

Table 12.1. Rules for temperature-dependent diagrams

A. Rules for Temperature-Dependent 0-*Diagrams*

(i) Label the lines from 1 to l, where l is the number of lines and assign to the ith line a momentum and spin index k_i.

(ii) Label each vertex from 1 to v, where v is the number of vertices, and assign to the ith vertex a temperature τ_i.

(iii) Assign to each vertex a factor,

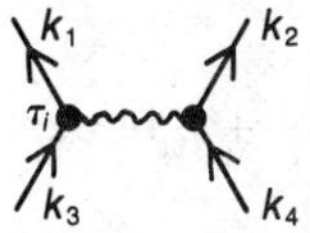

$$= (-1)\langle k_1, k_2|V|k_3, k_4\rangle^{(\pm)}.$$

We have drawn the vertex as a small spring to keep track of the way lines enter. This is necessary for fermions. For bosons, there is no sign ambiguity and it could be drawn as a dot

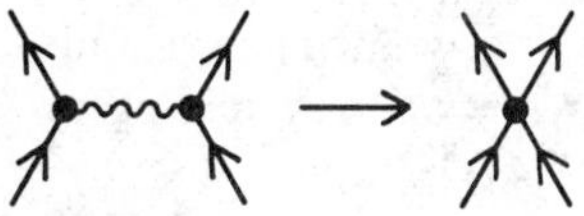

(iv) Assign to each line a factor

$$k_1 = G_0(k_1; \tau_i - \tau_j).$$

(v) For fermions, assign a factor $(-1)^F$, where F is the number of fermion loops. (A fermion loop is a solid line that can pass through many vertices but closes on itself when the vertex is drawn as a small spring.) Assign to each diagram a factor S^{-1}, where S is the symmetry number of the diagram and is defined as the number of permutations of labels which leave the diagram topologically unchanged (for this counting, all vertices should be considered as dots).

(vi) Sum over all momenta and spins and integrate over all temperatures from 0 to β.

B. Rules for Temperature-Dependent 1-*Diagrams*

(i) Label the internal lines from 1 to l, where l is the number of internal lines, and assign a momentum/spin index $k_i = (\mathbf{k}_i, \lambda_i)$ to the ith line. Label the outgoing external line by $k = (\mathbf{k}, \lambda)$ and the incoming external line by $k' = (\mathbf{k}', \lambda')$. Assign temperatures τ and τ' to the ends of the outgoing and incoming lines, respectively.

(ii) Assign factors as in Rules (A.ii)–(A.v).

(iii) Sum over momenta and spins on internal lines and integrate over all internal temperatures from 0 to β.

If we examine Eq. (12.48), we find that the possible values of n are restricted to even values for bosons and odd values for fermions. To show this, let us expand Eq. (12.48) as follows:

$$\begin{aligned}\bar{G}_0(k, \omega_n) &= \frac{1}{2}\int_{-\beta}^{0} d\tau\, e^{i\omega_n\tau} G_0(k, \tau) + \frac{1}{2}\int_{0}^{\beta} d\tau\, e^{i\omega_n\tau} G_0(k, \tau) \\ &= \pm\frac{1}{2}\int_{-\beta}^{0} d\tau\, e^{i\omega_n\tau} G_0(k, \tau + \beta) + \frac{1}{2}\int_{0}^{\beta} d\tau\, e^{i\omega_n\tau} G_0(k, \tau)\end{aligned} \tag{12.49}$$

If we next make the change of variables $\tau \to \tau' + \beta$ in the first term, we can write Eq. (12.49) as

$$\bar{G}_0(k, \omega_n) = \tfrac{1}{2}(1 \pm e^{-in\pi}) \int_0^{\beta} d\tau\, e^{i\omega_n\tau} G_0(k, \tau). \tag{12.50}$$

We now note that $\frac{1}{2}(1 + e^{-in\pi}) = 1$ for n even and $\frac{1}{2}(1 + e^{-in\pi}) = 0$ for n odd. Thus, for bosons we must restrict n to even integers. For fermions, we restrict n to odd integers and Eq. (12.49) becomes

$$\bar{G}_0(k, \omega_n) = \int_0^{\beta} d\tau\, e^{i\omega_n\tau} G_0(k, \tau) \tag{12.51}$$

where

$$\omega_n = \begin{cases} \dfrac{2n\pi}{\beta} & \text{bosons} \\[2ex] \dfrac{(2n + 1)\pi}{\beta} & \text{fermions} \end{cases} \tag{12.52}$$

(n now ranges over all integers).

From Eqs. (12.33) and (12.51) we can write an explicit expression for $G_0(k, \omega_n)$. For bosons we find

$$G_0(k, \omega_n) = -\int_0^{\beta} d\tau\, e^{i\omega_n\tau}[(1 + n_0(k))]\, e^{-\tau\epsilon_k^0} = \frac{1}{(\omega_n i - \epsilon_k{}^0)} \tag{12.53}$$

where $\omega_n = 2n\pi/\beta$, and for fermions we find

$$G_0(k, \omega_n) = -\int_0^{\beta} d\tau\, e^{i\omega_n\tau}(1 - n_0(k))\, e^{-\tau\epsilon_k^0} = \frac{1}{(\omega_n i - \epsilon_k{}^0)} \tag{12.54}$$

where $\omega_n = (2n + 1)\pi/\beta$.

If we substitute Eq. (12.47) into the temperature-dependent 0-diagrams, then to each line we can associate an integer n (or "frequency"). If we then integrate over the temperature, we obtain for each vertex a delta function (cf. Eq. (12.46)) which causes the "frequencies" of lines entering and leaving the vertex to be conserved. We can write the 0-diagrams in terms of "frequencies" rather than temperature according to the rules in Table 12.2.

In terms of "frequencies," $U_1(V, T, \mu)$ and $U_2(V, T, \mu)$ can be written

$$U_1(V, T, \mu) = \lim_{\delta \to 0} \frac{(-1)}{2} \sum_{k_1, k_2} \sum_{n_1, n_2} \langle 12|V|12\rangle^{(\pm)} \bar{G}_0(1)\bar{G}_0(2)\, e^{i\omega_1 \delta}\, e^{i\omega_2 \delta} \tag{12.55}$$

Table 12.2. Rules for "frequency"-dependent diagrams

A. Rules for "Frequency"-Dependent 0-*Diagrams*

(i) Label each line from 1 to l, where l is the number of lines, and associate to the ith line an integer n_i and momentum/spin index k_i.

(ii) Assign to each vertex a factor

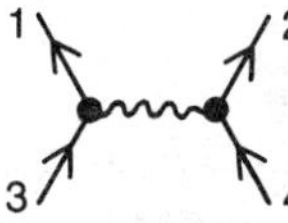

$$= -\beta\, \delta_{n_1+n_2, n_3+n_4} \langle k_1, k_2|V|k_3, k_4\rangle^{(\pm)} \equiv (-1)\langle 12|V|34\rangle^{(\pm)}$$

(cf. Table 12.1, Rule (A.iii)).

(iii) Assign to each line a factor

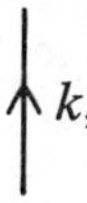

$$k, n = \frac{1}{\beta}\, \bar{G}_0(k, \omega_n) = \frac{1}{\beta}\left(\frac{1}{i\omega_n - \epsilon_k^0}\right)$$

where $\omega_n = 2n\pi/\beta$ for bosons and $\omega_n = (2n + 1)\pi/\beta$ for fermions.

(iv) For fermions, assign a factor $(-1)^F$, where F is the number of fermion loops; and a factor S^{-1} (cf. Table 12.1, Rule (A.v)).

(v) For each loop that passes through only one vertex, assign a convergence factor $e^{i\omega_n \delta}$, where n is the integer associated with the line and δ is positive and has values such that $\beta > \delta > 0$.

(vi) Sum over all integers and over all momenta and spins.

B. Rules for "Frequency"-Dependent 1-*Diagrams*

(i) Label the internal lines from 1 to l, where l is the number of internal lines, and associate to the ith line an integer n_i and momentum/spin label k_i. Assign labels (k, n) and (k', n') to the outgoing and incoming lines, respectively.

(ii) Assign factors as in Rules (A.ii)–(A.v).

(iii) Sum over integers n_i and momentum and spin indices on internal lines. Multiply by a factor β^2.

---

and

$$U_2(V, T, \mu) = \frac{1}{8} \sum_{k_1,\ldots,k_4} \sum_{n_1,\ldots,n_4} \langle 12|V|34\rangle^{(\pm)} \langle 34|V|12\rangle^{(\pm)}$$
$$\times \bar{G}_0(1)\bar{G}_0(2)\bar{G}_0(3)\bar{G}_0(4)$$
$$\pm \lim_{\delta\to 0} \frac{1}{2} \sum_{k_1,\ldots,k_4} \sum_{n_1,\ldots,n_4} \langle 12|V|32\rangle^{(\pm)} \langle 34|V|14\rangle^{(\pm)}$$
$$\times \bar{G}_0(1)\bar{G}_0(2)\bar{G}_0(3)\bar{G}_0(4)\, e^{i\omega_2\delta}\, e^{i\omega_4\delta} \tag{12.56}$$

where we have let $\bar{G}_0(i) = \bar{G}_0(k_i, \omega_i)$. The convergence factor, $e^{i\omega\delta}$, has been included for each line in a temperature-dependent diagram which has a factor $G_0(k_i, 0)$ associated with it. It enables us to locate mathematically those lines which only depend on $n_0(k)$. It is helpful to illustrate the use of the convergence factor with a specific example. Let us evaluate the sum

$$\lim_{\delta\to 0} \sum_{n=-\infty}^{\infty} e^{-i(2n\pi\delta/\beta)} \left(\frac{1}{\dfrac{2in\pi}{\beta} - \epsilon_k{}^0} \right).$$

It is easiest to do this in terms of a contour integration.[8] The function $\beta(e^{\beta z} - 1)^{-1}$ has poles at $z = 2\pi in/\beta$, each with residue 1. Thus, we can write the sum in the form

$$\lim_{\delta\to 0} \frac{1}{\beta} \sum_{n=-\infty}^{\infty} e^{-(2n\pi i/\beta)\delta} \left(\frac{1}{\dfrac{2i\pi n}{\beta} - \epsilon_k{}^0} \right) = \lim_{\delta\to 0} \frac{1}{2\pi i} \int_C dz \frac{1}{(e^{\beta z} - 1)} \frac{e^{\delta z}}{(z - \epsilon_k{}^0)} \tag{12.57}$$

where the poles of the function $(e^{\beta z} - 1)^{-1}(z - \epsilon_k{}^0)^{-1}$ and the contour C are shown in Fig. 12.1. Let us now consider the contour Γ shown in Fig. 12.2.

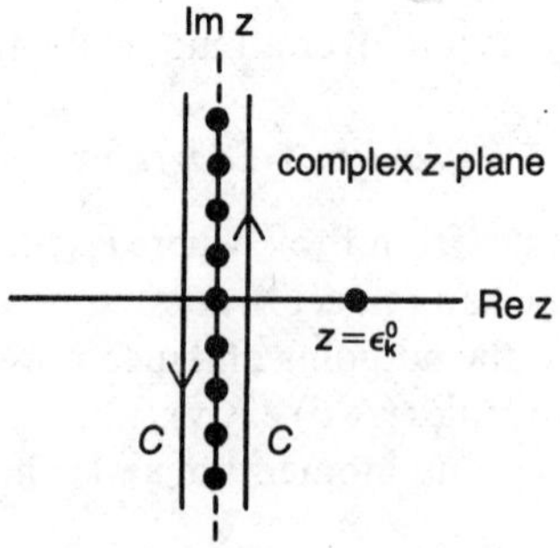

Fig. 12.1. Poles of the function $(e^{\beta z} - 1)^{-1}(z - \epsilon_k{}^0)^{-1}$ and the contour C.

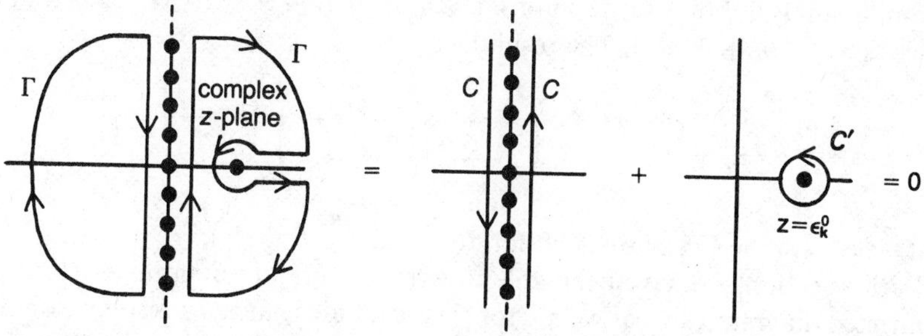

Fig. 12.2. The contour Γ can be used to change the contour integration over the poles of $(e^{\beta z} - 1)^{-1}$ to one over the pole of $(z - \epsilon_k{}^0)^{-1}$.

Since Γ encloses no poles, integration over it will be zero. However, we obtain a relation between contributions from different parts of the contour Γ. Since

$$\lim_{\mathrm{Re}\, z \to +\infty} \frac{e^{\delta z}}{(e^{\beta z} - 1)} \to e^{-z(\beta - \delta)} \to 0$$

and

$$\lim_{\mathrm{Re}\, z \to -\infty} \frac{e^{\delta z}}{(e^{\beta z} - 1)} \to e^{\delta z} \to 0,$$

there will be no contribution from the semicircle at infinity. Similarly, the contributions along the positive real axis cancel. Thus, we find

$$\begin{aligned} 0 &= \frac{1}{2\pi i} \oint_{\Gamma} dz \frac{e^{\delta z}}{(e^{\beta z} - 1)(z - \epsilon_k{}^0)} \\ &= \frac{1}{2\pi i} \oint_{C} dz \frac{e^{\delta z}}{(e^{\beta z} - 1)(z - \epsilon_k{}^0)} + \frac{1}{2\pi i} \oint_{C'} dz \frac{e^{\delta z}}{(e^{\beta z} - 1)(z - \epsilon_k{}^0)} \end{aligned}$$

or

$$\frac{1}{2\pi i} \oint_{C} dz \frac{e^{\delta z}}{(e^{\beta z} - 1)(z - \epsilon_k{}^0)} = \frac{-1}{2\pi i} \oint_{C'} dz \frac{e^{\delta z}}{(e^{\beta z} - 1)(z - \epsilon_k{}^0)}. \tag{12.58}$$

We now change the integration over contour C to one over contour C'. Since the contour C' encloses a simple pole, we immediately obtain

$$\lim_{\delta \to 0} \frac{1}{\beta} \sum_{u=-\infty}^{\infty} e^{i(2\pi n\delta/\beta)} \left(\frac{1}{\dfrac{2in\pi}{\beta} - \epsilon_k{}^0} \right) = \frac{-1}{(e^{\beta\epsilon_k^0} - 1)}. \tag{12.59}$$

A similar analysis holds for fermions if we note that $-\beta(e^{\beta z}+1)^{-1}$ has simple poles at $z = (2n+1)\pi i/\beta$. The result is

$$\lim_{\delta \to 0} \frac{1}{\beta} \sum_{n=-\infty}^{\infty} e^{i((2n+1)\pi\delta/\beta)} \left(\frac{1}{i\dfrac{(2n+1)\pi}{\beta} - \epsilon_k{}^0} \right) = \frac{1}{(e^{\beta\epsilon_k^0}+1)} \tag{12.60}$$

and we obtain the Fermi distribution function.

In this section, we have been able to write the grand potential as a sum of connected diagrams, just as we did for classical fluids, and as a result we obtain an expression for the grand potential which is proportional to the volume, as we should.

C. DIRECT AND EXCHANGE INTERACTIONS

It is useful, at this point, to review some properties of the symmetrized interactions

$$\begin{aligned}\langle k_1, k_2|V|k_3, k_4\rangle^{(\pm)} &= \langle \mathbf{k}_1\lambda_1; \mathbf{k}_2, \lambda_2|V|\mathbf{k}_3\lambda_3; \mathbf{k}_4\lambda_4\rangle \\ &\pm \langle \mathbf{k}_1, \lambda_1, \mathbf{k}_2, \lambda_2|V|\mathbf{k}_4, \lambda_4; \mathbf{k}_3, \lambda_3\rangle\end{aligned} \tag{12.61}$$

which appear in our expression for the grand potential. As discussed in App. B, the kets $|\mathbf{k}_i\lambda_i; \mathbf{k}_j, \lambda_j\rangle$ denote a direct product of spin and momentum eigenfunctions. The first term in Eq. (12.61) is called the direct term and the second is called the exchange term. The exchange term is purely quantum mechanical in origin. In the classical limit, where $h \to 0$ (h is Planck's constant), the exchange contribution goes to zero. If we have a spin-dependent potential, $V(\hat{\mathbf{q}}_{ij}; \hat{\mathbf{s}}_i, \hat{\mathbf{s}}_j)$, the matrix elements may be written

$$\begin{aligned}&\langle \mathbf{k}_1\lambda_1; \mathbf{k}_2\lambda_2|V|\mathbf{k}_3\lambda_3; \mathbf{k}_4\lambda_4\rangle \\ &\quad = \delta_{\mathbf{k}_1+\mathbf{k}_2,\mathbf{k}_3+\mathbf{k}_4} \frac{1}{V}\int d\mathbf{r}\, e^{i(\mathbf{k}_1-\mathbf{k}_3)\cdot\mathbf{r}} \langle \lambda_1\lambda_2|V(\mathbf{r}; \hat{\mathbf{s}}_1, \hat{\mathbf{s}}_2)|\lambda_3\lambda_4\rangle.\end{aligned} \tag{12.62}$$

For spin-independent potential, we find

$$\langle \mathbf{k}_1\lambda_1; \mathbf{k}_2\lambda_2|V|\mathbf{k}_3\lambda_3; \mathbf{k}_4\lambda_4\rangle = \delta_{\lambda_1,\lambda_3}\,\delta_{\lambda_2,\lambda_4}\,\delta_{\mathbf{k}_1+\mathbf{k}_2,\mathbf{k}_3+\mathbf{k}_4}\bar{V}(\mathbf{k}_1-\mathbf{k}_3) \tag{12.63}$$

where

$$\bar{V}(\mathbf{k}_1-\mathbf{k}_3) = \frac{1}{V}\int d\mathbf{r}\, e^{-i(\mathbf{k}_1-\mathbf{k}_3)\cdot\mathbf{r}}V(\mathbf{r}) \tag{12.64}$$

is the Fourier transform of the potential and $\mathbf{k}_1 - \mathbf{k}_3$ is the momentum transfer during the interaction.

If we introduce center of mass and relative coordinates $\mathbf{K}_{12} = \mathbf{k}_1 + \mathbf{k}_2$, $\mathbf{K}_{34} = \mathbf{k}_3 + \mathbf{k}_4$, $\mathbf{k}_{12} = \frac{1}{2}(\mathbf{k}_1 - \mathbf{k}_2)$, and $\mathbf{k}_{34} = \frac{1}{2}(\mathbf{k}_3 - \mathbf{k}_4)$, and use the fact that $\mathbf{k}_1 - \mathbf{k}_3 = \mathbf{k}_4 - \mathbf{k}_2$, we can write $(\mathbf{k}_1 - \mathbf{k}_3) = \mathbf{k}_{12} - \mathbf{k}_{34}$. Combining the above results, we obtain

$$\langle k_1, k_2|V|k_3, k_4\rangle$$
$$= \delta_{\mathbf{K}_{12},\mathbf{K}_{34}}[\delta_{\lambda_1,\lambda_3}\,\delta_{\lambda_2,\lambda_4}\bar{V}(\mathbf{k}_{12} - \mathbf{k}_{34}) \pm \delta_{\lambda_1,\lambda_4}\,\delta_{\lambda_2,\lambda_3}\bar{V}(\mathbf{k}_{12} + \mathbf{k}_{34})]. \tag{12.65}$$

It is possible to expand the potentials $V(\mathbf{k}_{12} - \mathbf{k}_{34})$ and $V(\mathbf{k}_{12} + \mathbf{k}_{34})$ in a partial wave expansion. We then obtain

$$V(\mathbf{k}_{12} - \mathbf{k}_{34}) = 4\pi \sum_{l=0}^{\infty} (2l + 1)P_l(\boldsymbol{\varkappa}_{12}\cdot\boldsymbol{\varkappa}_{34})V_l(|\mathbf{k}_{12}|, |\mathbf{k}_{34}|), \tag{12.66}$$

and

$$V(\mathbf{k}_{12} + \mathbf{k}_{34}) = 4\pi \sum_{l=0}^{\infty} (2l + 1)(-1)^l P_l(\boldsymbol{\varkappa}_{12}, \boldsymbol{\varkappa}_{34})V_l(|\mathbf{k}_{12}|, |\mathbf{k}_{34}|) \tag{12.67}$$

where $\boldsymbol{\varkappa}_{ij}$ is the unit vector, $\boldsymbol{\varkappa}_{ij} = \mathbf{k}_{ij}/|\mathbf{k}_{ij}|$, and

$$V_l(|\mathbf{k}_{12}|, |\mathbf{k}_{34}|) = \int dr\, r^2 V(r) j_l(|\mathbf{k}_{12}|r) j_l(|\mathbf{k}_{34}|r) \tag{12.68}$$

($j_l(kr)$ is a spherical Bessel function and $P_l(\boldsymbol{\varkappa}_{12}\cdot\boldsymbol{\varkappa}_{34})$ is a Legendre polynomial). If we combine Eqs. (12.65)–(12.67), we obtain

$$\langle k_1, k_2|V|k_3, k_4\rangle^{(\pm)} = \delta_{\mathbf{K}_{12},\mathbf{K}_{34}}(4\pi) \sum_{l=0}^{\infty} (2l + 1)[\delta_{\lambda_1,\lambda_3}\,\delta_{\lambda_2,\lambda_4} \pm (-1)^l\,\delta_{\lambda_1,\lambda_4}\,\delta_{\lambda_2,\lambda_3}]$$
$$\times P_l(\boldsymbol{\varkappa}_{12}, \boldsymbol{\varkappa}_{34})V_l(|\mathbf{k}_{12}|, |\mathbf{k}_{34}|). \tag{12.69}$$

For bosons, we set the spin delta functions equal to one and find there is no interaction between bosons for odd l. Similarly, for spin $\frac{1}{2}$ fermions, there is no interaction between parallel spin fermions for even l or antiparallel spin fermions for odd l. This is a consequence of the symmetry and antisymmetry, respectively, of the boson and fermion wave functions.

For spin $\frac{1}{2}$ fermions, it is useful to write Eq. (12.69) in still another form. If we consider a spherically symmetric spin-independent potential, the total angular momentum and the total spin of the interacting particles will be separately conserved. We can describe the interaction in terms of the total spin of the interacting particles, rather than in terms of the spins of the individual particles. The total spin is $\hat{\mathbf{S}} = \frac{1}{2}(\hat{\boldsymbol{\sigma}}_1 + \hat{\boldsymbol{\sigma}}_2)$, where $\hat{\boldsymbol{\sigma}}_1$ and $\hat{\boldsymbol{\sigma}}_2$ are the Pauli spin operators for the individual particles. It is convenient to introduce the spin exchange operator $\frac{1}{2}(1 + \hat{\boldsymbol{\sigma}}_1\cdot\hat{\boldsymbol{\sigma}}_2)$, which has the effect

$$\tfrac{1}{2}(1 + \hat{\boldsymbol{\sigma}}_1\cdot\hat{\boldsymbol{\sigma}}_2)|\lambda_1, \lambda_2\rangle = |\lambda_2, \lambda_1\rangle. \tag{12.70}$$

We can now write Eq. (12.69) in the form

$$\langle k_1, k_2|V|k_3, k_4\rangle = \delta_{\mathbf{K}_{12},\mathbf{K}_{34}}[\langle\lambda_1\lambda_2|\lambda_3\lambda_4\rangle\overline{V}_s(\mathbf{k}_{12}, \mathbf{k}_{34}) + \langle\lambda_1\lambda_2|\hat{\boldsymbol{\sigma}}_1\cdot\hat{\boldsymbol{\sigma}}_2|\lambda_3\lambda_4\rangle\overline{V}_a(\mathbf{k}_{12}, \mathbf{k}_{34})] \tag{12.71}$$

where

$$\overline{V}_s(\mathbf{k}_{12}, \mathbf{k}_{34}) = \overline{V}(\mathbf{k}_{12} - \mathbf{k}_{34}) - \tfrac{1}{2}\overline{V}(\mathbf{k}_{12} + \mathbf{k}_{34}) \tag{12.72}$$

and

$$\overline{V}_a(\mathbf{k}_{12}, \mathbf{k}_{34}) = -\tfrac{1}{2}\overline{V}(\mathbf{k}_{12} + \mathbf{k}_{34}) \tag{12.73}$$

are called the symmetric and antisymmetric (under spin exchange) parts of the interaction. If we evaluate Eq. (12.71) for different spin values, the only nonzero values are

$$\begin{aligned}\langle\mathbf{k}_1\uparrow; \mathbf{k}_2\uparrow|V|\mathbf{k}_3, \uparrow; \mathbf{k}_4\uparrow\rangle^{(-)} &= \langle\mathbf{k}_1\downarrow; \mathbf{k}_2\downarrow|V|\mathbf{k}_3\downarrow; \mathbf{k}_4\downarrow\rangle^{(-)}\\ &= \delta_{\mathbf{K}_{12},\mathbf{K}_{34}}[\overline{V}_s(\mathbf{k}_{12}, \mathbf{k}_{34}) - \overline{V}_a(\mathbf{k}_{12}, \mathbf{k}_{34})],\end{aligned} \tag{12.74}$$

$$\begin{aligned}\langle\mathbf{k}_1\uparrow; \mathbf{k}_2\downarrow|V|\mathbf{k}_3\uparrow; \mathbf{k}_4\downarrow\rangle^{(-)} &= \langle\mathbf{k}_1\downarrow, \mathbf{k}_2\uparrow|V|\mathbf{k}_3\downarrow; \mathbf{k}_4\uparrow\rangle^{(-)}\\ &= \delta_{\mathbf{K}_{12},\mathbf{K}_{34}}[\overline{V}_s(\mathbf{k}_{12}, \mathbf{k}_{34}) + \overline{V}_a(\mathbf{k}_{12}, \mathbf{k}_{34})],\end{aligned} \tag{12.75}$$

and

$$\begin{aligned}\langle\mathbf{k}_1\uparrow; \mathbf{k}_2\downarrow|V|\mathbf{k}_3\downarrow; \mathbf{k}_4\uparrow\rangle^{(-)} &= \langle\mathbf{k}_1\downarrow; \mathbf{k}_2\uparrow|V|\mathbf{k}_3\uparrow; \mathbf{k}_4\downarrow\rangle^{(-)}\\ &= \delta_{\mathbf{K}_{12},\mathbf{K}_{34}}2\overline{V}_a(\mathbf{k}_{12}, \mathbf{k}_{34}).\end{aligned} \tag{12.76}$$

These results will be useful for one of the homework problems.

D. ELECTRON GAS[5,8–10]

1. Effective Hamiltonian

Let us consider a gas of electrons which interact via the Coulomb force and are free to move in the presence of a uniform positive background charge distribution. We will assume the system is confined to a spherical container of radius R and volume $V = 4\pi R^3/3$. We will always be interested in the thermodynamic limit $R \to \infty$, $N \to \infty$, such that $n = N/V =$ constant. For a system of N electrons we can write the Hamiltonian as

$$\hat{H}^N = \hat{H}^N_{\text{el}} + \hat{V}_{\text{el-B}} + V_{\text{B}} \tag{12.77}$$

where $\hat{H}_{\text{el}}^N$ is the Hamiltonian of the electron system,

$$\hat{H}_{\text{el}}^N = \sum_{i=1}^{N} \frac{|\mathbf{p}_i|^2}{2m} + \frac{1}{2} \sum_{i<j=1}^{N} \frac{e^2\, e^{-\alpha|\hat{\mathbf{q}}_i - \hat{\mathbf{q}}_j|}}{|\hat{\mathbf{q}}_i - \hat{\mathbf{q}}_j|}; \tag{12.78}$$

$\hat{V}_{\text{el-B}}^N$ is the electron-background interaction

$$\hat{V}_{\text{el-B}} = -e^2 \sum_{i=1}^{N} \int d\mathbf{r}\, \frac{n_+(\mathbf{r})\, e^{-\alpha|\hat{\mathbf{q}}_i - \mathbf{r}|}}{|\hat{\mathbf{q}}_i - \mathbf{r}|} \tag{12.79}$$

($en_+(\mathbf{r})$ is the charge density of the background); and the self-interaction of the background is

$$V_{\text{B}} = +\frac{e^2}{2} \iint d\mathbf{r}_1\, d\mathbf{r}_2\, \frac{e^{-\alpha|\mathbf{r}_1 - \mathbf{r}_2|}}{|\mathbf{r}_1 - \mathbf{r}_2|}\, n_+(\mathbf{r}_1) n_+(\mathbf{r}_2). \tag{12.80}$$

The factor $\exp(-\alpha|\hat{\mathbf{q}}_i - \hat{\mathbf{q}}_j|)$ is a convergence factor which allows us to consider each term in Eq. (12.77) separately. Without it each term would be infinite in the thermodynamic limit because of the long range of the Coulomb potential. However, $\hat{H}^N$ itself is well behaved since the system is neutral. We will do all calculations with $\alpha \neq 0$ and set $\alpha = 0$ after the thermodynamic limit is taken.

Let us now consider various terms in Eq. (12.77). We assume that the density of the positive-charge background is equal to the average density of the electrons. This ensures neutrality of the system. Thus, $n_+(\mathbf{r}) = \langle N \rangle / V$. The integral in Eq. (12.80) can be done easily. If we neglect contributions which go to zero when we take the limit $R \to \infty$ and then $\alpha \to 0$, we find

$$V_{\text{B}} = \tfrac{1}{2} e^2 \frac{\langle N \rangle^2}{V} \frac{4\pi}{\alpha^2}. \tag{12.81}$$

As $\alpha \to 0$, $V_{\text{B}} \to \infty$.

It is useful to write the Hamiltonian in the number representation. From App. B, we obtain

$$\begin{aligned} \hat{H} = {} & \tfrac{1}{2} e^2 \frac{\langle N \rangle^2}{V} \frac{4\pi}{\alpha^2} + \sum_{\mathbf{k},\lambda} \frac{\hbar^2 |\mathbf{k}|^2}{2m} \hat{a}_{\mathbf{k},\lambda}^{+} \hat{a}_{\mathbf{k},\lambda} \\ & - e^2 \int d\mathbf{r}\, \frac{\langle N \rangle}{V} \sum_{\mathbf{k},\lambda} \langle \mathbf{k} | \frac{e^{-\alpha|\hat{\mathbf{q}} - \mathbf{r}|}}{|\hat{\mathbf{q}} - \mathbf{r}|} | \mathbf{k} \rangle \hat{a}_{\mathbf{k},\lambda}^{+} \hat{a}_{\mathbf{k},\lambda} \\ & + \frac{e^2}{2} \sum_{k_1 \ldots k_4} \langle k_1, k_2 | V | k_3, k_4 \rangle \hat{a}_{k_1}^{+} \hat{a}_{k_2}^{+} \hat{a}_{k_4} \hat{a}_{k_3}. \end{aligned} \tag{12.82}$$

The third term in Eq. (12.82) can be evaluated to obtain

$$-\frac{\langle N \rangle}{V} e^2 \int d\mathbf{r} \sum_{\mathbf{k},\lambda} \langle \mathbf{k} | \frac{e^{-\alpha|\hat{\mathbf{q}} - \mathbf{r}|}}{|\hat{\mathbf{q}} - \mathbf{r}|} | \mathbf{k} \rangle \hat{a}_{\mathbf{k},\lambda}^{+} \hat{a}_{\mathbf{k},\lambda} = -e^2 \frac{\langle N \rangle}{V} \frac{4\pi}{\alpha^2} \hat{N} \tag{12.83}$$

where $\hat{N}$ is the number operator and we have used the definition $\hat{N} = \sum_{\mathbf{k},\lambda} \hat{a}_{\mathbf{k},\lambda}^{+} \hat{a}_{\mathbf{k},\lambda}$. The fourth term in Eq. (12.82) can be separated into a part with no momentum transfer and a part with finite momentum transfer. The part with no momentum transfer can be written

$$\frac{e^2}{2}\sum_{k_1k_2} \bar{V}(0)a_{k_1}^+a_{k_2}^+a_{k_1}a_{k_2} = \frac{e^2 4\pi}{2V\alpha^2}(\hat{N}^2 - \hat{N}) \tag{12.84}$$

by commuting $\hat{a}_{k_1}$ to the left of $\hat{a}_{k_2}^+$. Thus, the total Hamiltonian takes the form

$$\hat{H} = \frac{1}{2}\frac{e^2 4\pi}{V\alpha^2}(\langle N\rangle^2 - 2\langle N\rangle\hat{N} + \hat{N}^2 - \hat{N}) + \sum_{\mathbf{k},\lambda}\frac{\hbar^2|\mathbf{k}|^2}{2m}\hat{a}_{\mathbf{k},\lambda}^+\hat{a}_{\mathbf{k},\lambda}$$
$$+ \frac{e^2}{2}\sum_{k_1\ldots k_4}{}' \langle k_1, k_2|V|k_3, k_4\rangle a_{k_1}^+a_{k_2}^+a_{k_3}a_{k_4}. \tag{12.85}$$

The prime on the last summation rules out any term with $\mathbf{k}_1 = \mathbf{k}_3$. In the thermodynamic limit, $\langle\hat{N}\rangle^n \approx \langle\hat{N}^n\rangle$. The contributions proportional to $\langle\hat{N}\rangle^2$ cancel. In the thermodynamic limit, the total energy becomes infinite while the energy per particle remains finite. Thus, the term proportional to $\hat{N}/V$ is negligible in that limit, and we obtain the following expression for the effective Hamiltonian of the electron system:

$$\hat{H} = \sum_{\mathbf{k},\lambda}\frac{\hbar^2|\mathbf{k}|^2}{2m}\hat{a}_{\mathbf{k},\lambda}^+\hat{a}_{\mathbf{k},\lambda} + \frac{e^2}{4}\sum_{k_1\ldots k_4}{}' \langle k_1, k_2|V|k_3, k_4\rangle^{(-)}\hat{a}_{k_1}^+\hat{a}_{k_2}^+\hat{a}_{k_4}\hat{a}_{k_3}. \tag{12.86}$$

In Eq. (12.86) we have antisymmetrized the second term, and the prime means that both direct and exchange terms with $\mathbf{k}_1 - \mathbf{k}_3 = 0$ are excluded. An explicit expression for the matrix element $\langle k_1, k_2|V|k_3, k_4\rangle$ can be obtained using the screened Coulomb potential, $V(q) = e^2q^{-1}\exp(-\alpha q)$. It can be written

$$\langle k_1, k_2|V|k_3, k_4\rangle = \delta_{\mathbf{k}_1+\mathbf{k}_2,\mathbf{k}_3+\mathbf{k}_4}\,\delta_{\lambda_1,\lambda_3}\,\delta_{\lambda_2,\lambda_4}\frac{1}{V}\frac{4\pi\, e^2}{|\mathbf{k}_1 - \mathbf{k}_3|^2 + \alpha^2} \tag{12.87}$$

if we neglect terms that go to zero when $R \to \infty$.

It is instructive to obtain an expression for the internal energy to order e^2 at $T = 0$ K. We first find the grand potential. To lowest order

$$\Omega_0(V, T, \mu) = -\frac{1}{\beta}\operatorname{Tr} e^{-\beta(\hat{H}^0 - \mu\hat{N})} = \frac{4}{3}\frac{V}{4\pi^2}\left(\frac{2m}{\hbar^2}\right)^{3/2}\int_0^\infty d\epsilon\,\frac{\epsilon^{3/2}}{e^{\beta(\epsilon-\mu)} + 1} \tag{12.88}$$

(we have integrated Eq. (9.158) by parts). If we now make a change of variables, $x = \beta(\epsilon - \mu)$ and $y = \beta\mu$, we obtain

$$\Omega_0(V, T, \mu) = \frac{4}{3}\frac{V}{4\pi^2}\left(\frac{2m}{\hbar^2}\right)^{3/2}\left(\frac{1}{\beta}\right)^{5/2}\int_{-y}^\infty dx\,\frac{(x + y)^{3/2}}{e^x + 1}. \tag{12.89}$$

For $T \to 0$ K and therefore $y \to \infty$, the integral yields

$$\int_{-y}^{\infty} dx \frac{(x+y)^{3/2}}{e^x + 1} = \tfrac{2}{5}y^{5/2} + \frac{\pi^2}{4} y^{1/2} + \cdots \tag{12.90}$$

and the grand potential takes the form

$$\Omega_0(V, T = 0, \mu(0)) = \frac{2V}{\pi^2}\left(\frac{2m}{\hbar^2}\right)^{3/2} \frac{[\mu(0)]^{5/2}}{15}. \tag{12.91}$$

Note that $\mu(0)$ is the *exact* chemical potential at $T = 0$ K and therefore depends on e^2.

The first-order 0-diagram gives

$$\Omega_1(V, T, \mu) =$$

$$= -\frac{1}{2}\sum_{k_1, k_2}{}' \langle k_1, k_2|V|k_1, k_2\rangle^{(-)} n_0(k_1) n_0(k_2). \tag{12.92}$$

In Eq. (12.92), the direct contribution only involves zero momentum transfer and has already been removed. Therefore, only the exchange term contributes and we find

$$\Omega_1(V, T, \mu) = -\frac{4\pi e^2}{2V}\sum_{k_1 k_2}{}' \delta_{\lambda_1,\lambda_2} \frac{n_0(\mathbf{k}_1) n_0(\mathbf{k}_2)}{|\mathbf{k}_1 - \mathbf{k}_2|^2 + \alpha^2}. \tag{12.93}$$

Eq. (12.93) is easy to evaluate at $T = 0$ K. We first note that $n_0(\mathbf{k}) = \theta(\mu(0) - (\hbar^2|\mathbf{k}|^2/2m))$ at $T = 0$ K. If we let $\gamma \equiv ((2m/\hbar^2)\mu(0))^{1/2}$, we can write

$$\Omega_1(V, T = 0, \mu(0)) = -\frac{4\pi e^2 V}{(2\pi)^6}\iint d\mathbf{k}_1\, d\mathbf{k}_2 \frac{1}{|\mathbf{k}_1 - \mathbf{k}_2|^2}\theta(\gamma - |\mathbf{k}_1|)\theta(\gamma - |\mathbf{k}_2|) \tag{12.94}$$

where the spin summation has been performed. We next make the change of variables $\mathbf{Q} = \mathbf{k}_1 - \mathbf{k}_2$ and $\mathbf{k}' = \mathbf{k}_1 + \frac{1}{2}\mathbf{Q}$, and Eq. (12.94) takes the form

$$\Omega_1(V, T = 0, \mu(0)) = -\frac{4\pi e^2 V}{(2\pi)^6}\int d\mathbf{Q}\int d\mathbf{k}' \frac{1}{Q^2} \times \theta(\gamma - |\mathbf{k}' + \tfrac{1}{2}\mathbf{Q}|)\theta(\gamma - |\mathbf{k}' - \tfrac{1}{2}\mathbf{Q}|). \tag{12.95}$$

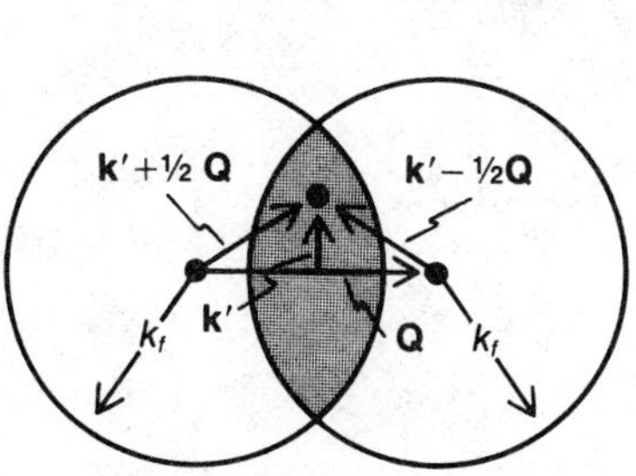

Fig. 12.3. Shaded area is the region of integration for the integral $\int d\mathbf{k}'\theta(\gamma - |\mathbf{k}' + \frac{1}{2}\mathbf{Q}|)\theta(\gamma - |\mathbf{k}' - \frac{1}{2}\mathbf{Q}|)$.

The region of integration over $\mathbf{k}'$ is shown in Fig. 12.3. The integration over $\mathbf{k}'$ may be done geometrically and yields

$$\int d\mathbf{k}'\theta(\gamma - |\mathbf{k}' + \tfrac{1}{2}\mathbf{Q}|)\theta(\gamma - |\mathbf{k}' - \tfrac{1}{2}\mathbf{Q}|) = \frac{4\pi}{3}\gamma^3(1 - \tfrac{3}{2}x + \tfrac{1}{2}x^3)\theta(1 - x) \qquad (12.96)$$

where $x = Q/2\gamma$. The final expression for $\Omega_1(V, T = 0, \mu(0))$ is

$$\Omega_1(V, T = 0, \mu(0)) = -\frac{e^2 V m^2 \mu^2(0)}{\pi^3 \hbar^4}. \qquad (12.97)$$

Note that $\mu(0)$ is the chemical potential of the interacting system.

We can now compute the internal energy to order e^2 at $T = 0$ K for fixed V and $\langle N \rangle$. The internal energy at $T = 0$ K is defined

$$U(V, T = 0, \langle N \rangle) = \Omega(V, T = 0, \mu(0)) + \mu(0)\langle N \rangle. \qquad (12.98)$$

We must first expand the chemical potential in powers of e^2. Thus, we write

$$\mu(0) = \epsilon_f + \mu_1 e^2 + \mu_2 e^4 + \cdots \qquad (12.99)$$

where μ_1 and μ_2 are independent of e^2 and

$$\epsilon_f = \frac{\hbar^2}{2m}\left(\frac{3\pi^2\langle N \rangle}{V}\right)^{2/3} \equiv \frac{\hbar^2 k_f^2}{2m} \qquad (12.100)$$

is the Fermi energy for an ideal gas of spin $\frac{1}{2}$ particles (including the spin degeneracy factor) and k_f is the Fermi momentum. Next we expand the average particle number in powers of e^2 by means of Taylor expansion about $\mu(0) = \epsilon_f$. Thus,

$$\langle N \rangle = -\left(\frac{\partial \Omega}{\partial \mu}\right)_{V,T=0} = -\left[\left(\frac{\partial \Omega_0}{\partial \mu}\right)_{V,T}\right]_{\epsilon_f} - e^2\left[\frac{1}{e^2}\left(\frac{\partial \Omega_1}{\partial \mu}\right)_{V,T}\right]_{\epsilon_f} - \left[\left(\frac{\partial^2 \Omega_0}{\partial \mu^2}\right)_{V,T}\right]_{\epsilon_f} \mu_1 e^2 + \cdots \qquad (12.101)$$

and we find

$$\langle N \rangle = -\left[\left(\frac{\partial \Omega_0}{\partial \mu}\right)_{V,T}\right]_{\epsilon_f} \qquad (12.102)$$

and

$$\mu_1 = -\left[\frac{1}{e^2}\left(\frac{\partial \Omega_1}{\partial \mu}\right)_{V,T}\right]_{\epsilon_f} \Bigg/ \left[\left(\frac{\partial^2 \Omega_0}{\partial \mu^2}\right)_{V,T}\right]_{\epsilon_f}. \qquad (12.103)$$

We now expand the internal energy in a Taylor series

$$U(V, T = 0, \langle N \rangle) = \Omega_0(V, T = 0, \epsilon_f) + \Omega_1(V, T = 0, \epsilon_f) + \left[\left(\frac{\partial \Omega_0}{\partial \mu}\right)_{V,T}\right]_{\epsilon_f} \mu_1 e^2 + \cdots + \epsilon_f\langle N \rangle + e^2\mu_1\langle N \rangle + \cdots \qquad (12.104)$$

Using Eq. (12.102), this reduces to

$$U(V, T = 0, \langle N \rangle) = \Omega_0(V, T = 0, \epsilon_f) + \Omega_1(V, T = 0, \epsilon_f) + \epsilon_f\langle N \rangle + \cdots \tag{12.105}$$

to order e^2 in the coupling constant. We can combine Eqs. (12.91), (12.97), and (12.105) to obtain

$$U(V, T = 0, \langle N \rangle) = \tfrac{3}{5}\langle N \rangle \epsilon_f - \frac{e^2 V m^2 \epsilon_f^{\,2}}{\pi^3 \hbar^4} + \cdots \tag{12.106}$$

Eq. (12.106) is often written in a slightly different form. If we introduce the Bohr radius, $a_0 = \hbar^2/me^2$, write the volume as $V = \frac{4}{3}\pi r_0^{\,3}\langle N \rangle$ ($r_0^{\,3}$ is a measure of the interparticle spacing), and introduce a parameter $r_s = r_0/a_0$, which is a measure of the density of the system, we find

$$U(V, T = 0, \langle N \rangle) = \frac{e^2\langle N \rangle}{2a_0}\left[\frac{2.21}{r_s^{\,2}} - \frac{0.916}{r_s}\right] + \cdots \tag{12.107}$$

Eq. (12.107) gives the high-density limit ($r_s \to 0$) of an electron gas. In that limit, the contribution from kinetic energy dominates. In Eq. (12.107), the exchange energy gives a negative contribution to the internal energy. At high density, the system forms a gas. However, as the density of the system decreases, the internal energy becomes negative and goes through a minimum, thus causing the system to become bound.

If we try to extend the expansion in powers of $r_s^{\,1}$ to higher orders, we find that there is one type of diagram in each order, the polarization diagram, which gives an infinite contribution. Thus, we cannot obtain higher order terms as a simple perturbation expansion but must sum an infinite series of terms to obtain the next correction.

2. Polarization Diagrams

The polarization diagrams give a divergent contribution because, in the limit in which the momentum transfer, Q, goes to zero, they become infinite. To find corrections to the internal energy from polarization effects we must retain the polarization diagrams from each order in the perturbation expansion and sum them. Let us write the grand potential in the form

$$\Omega(V, T, \mu) = \Omega_0(V, T, \mu) + \Omega_1(V, T, \mu) + \Delta\Omega(V, T, \mu) + \cdots \tag{12.108}$$

where $\Delta\Omega(V, T, \mu)$ is the sum of all polarization diagrams. $\Delta\Omega(V, T, \mu)$ can be expressed diagrammatically as

$\Delta\Omega(V, T, \mu) =$

$+ \cdots$ (12.109)

and algebraically as

$$\Delta\Omega(V, T, \mu) = \sum_{l=2}^{\infty} \frac{-1}{2n\beta} \sum_{k_1 \ldots k_{2l}} \sum_{n_1 \ldots n_{2l}} \langle 1, 4|V|2, 3\rangle^{(-)} \langle 3, 6|V|4, 5\rangle^{(-)} \times \cdots$$
$$\times \langle 2l-3, 2l|V|2l-2, 2l-1\rangle^{(-)} \langle 2l-1, 2|V|2l, 1\rangle^{(-)}$$
$$\times \bar{G}_0(1)\bar{G}_0(2) \times \cdots \times \bar{G}_0(2l)$$
$$+ \frac{1}{8\beta} \sum_{k_1 \ldots k_4} \sum_{n_1 \ldots n_4} \langle 1, 4|V|2, 3\rangle^{(-)} \langle 3, 2|V|4, 1\rangle^{(-)}$$
$$\times \bar{G}_0(1) \times \cdots \times \bar{G}_0(4). \qquad (12.110)$$

The extra term is included because the symmetry number of the second-order diagram is $\frac{1}{8}$. The second-order term can be summed back in later.

Because of momentum conservation, the contribution from the direct part of the interaction at each vertex of a polarization diagram depends on the same momentum transfer $\mathbf{Q} = \mathbf{k}_1 - \mathbf{k}_2 = \mathbf{k}_3 - \mathbf{k}_4 = \mathbf{k}_5 - \mathbf{k}_6 = \cdots = \mathbf{k}_{2l-1} - \mathbf{k}_{2l}$. Thus, if we retain only the direct part of the interaction in the lth-order term, we get a contribution of the form $(4\pi e^2/(Q^2 + \alpha^2))^l$, which in the limit $\alpha \to 0$ will be highly divergent for $Q \to 0$. Contributions from the exchange part of the interaction will not be divergent since they have the form $4\pi e^2/(\mathbf{k}_i - \mathbf{k}_j + \mathbf{Q})$ and do not diverge for $\alpha \to 0$, $Q \to 0$. Thus, the dominant contribution to the grand potential comes from the direct terms in Eq. (12.110). The second-order term in Eq. (12.110) is a special case. The product of exchange terms must be included since they are equal to the product of direct terms. With this approximation we can write Eq. (12.110) in the form

$$\Delta\Omega(V, T, \mu) = \frac{1}{2} \sum_{l=2}^{\infty} \frac{-1}{l\beta} \sum_{k_1 \ldots k_{2l}} \sum_{n_1 \ldots n_{2l}} \langle 1, 4|V|2, 3\rangle \langle 3, 6|V|4, 5\rangle$$
$$\times \cdots \times \langle 2l-1, 2|V|2l, 1\rangle \bar{G}_0(1)\bar{G}_0(2) \times \cdots \times \bar{G}_0(2l) \qquad (12.111)$$

where $\Delta\Omega(V, T, \mu)$ now depends only on the direct interaction.

It is convenient to introduce a "frequency" transfer $\nu = n_1 - n_2 = n_3 - n_4 = \cdots = n_{2l-1} - n_{2l}$ (note that ν can only assume even values). We then integrate over "frequency" and momentum delta functions to obtain

$$\Delta\Omega(V, T, \mu) = \frac{1}{\beta} \sum_{l=2}^{\infty} \frac{-1}{2l\beta} \sum_{\mathbf{Q}} \sum_{\nu} \left(\frac{4\pi e^2}{Q^2 + \alpha^2}\right)^l \chi_0{}^l(\mathbf{Q}, \nu)$$
$$= \frac{1}{2\beta} \sum_{\mathbf{Q},\nu} \left\{ \ln\left[1 - \frac{4\pi e^2}{Q^2 + \alpha^2} \chi_0(\mathbf{Q}, \nu)\right] + \frac{4\pi e^2}{Q^2 + \alpha^2} \chi_0(\mathbf{Q}, \nu) \right\} \qquad (12.112)$$

where

$$\chi_0(\mathbf{Q}, z_\nu) = \frac{2}{\beta V} \sum_{\mathbf{k}} \sum_{n} \left(\frac{1}{\omega_n i - \epsilon_{\mathbf{k}}{}^0}\right) \left(\frac{1}{(\omega_n + z_\nu)i - \epsilon^0_{\mathbf{k}+\mathbf{Q}}}\right), \qquad (12.113)$$

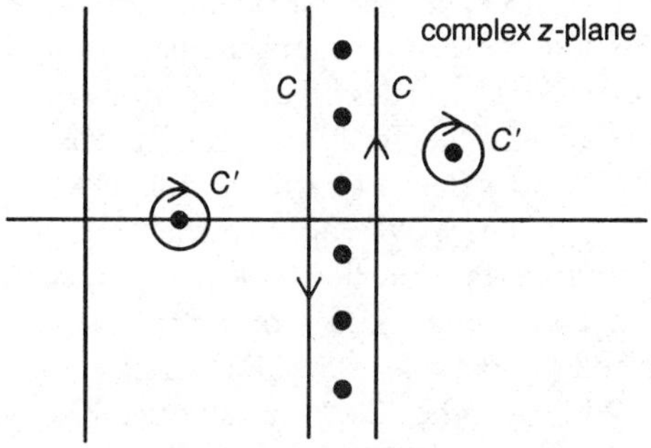

Fig. 12.4. Integration contours for the particle-hole propagator.

$z_\nu = 2\pi\nu/\beta$, and $\omega_n = (2n + 1)\pi/\beta$. Eqs. (12.112) and (12.113) give the leading contribution to the grand potential of an electron gas in a uniform positive-charge background. We can simplify Eq. (12.113). We first note that the Fermi distribution $n^0(z) = [\exp(\beta(z - \mu)) + 1]^{-1}$ has poles at $z_n = (2n + 1)\pi i/\beta + \mu$, where n ranges over all integers. The residue of $n^0(z)$ at z_n is $-1/\beta$. Thus, we can write $\chi_0(\mathbf{Q}, \nu)$ as

$$\chi_0(\mathbf{Q}, z_\nu) = -\frac{2}{V}\oint_C \frac{dz}{(2\pi i)} n^0(z) \sum_{\mathbf{k}} \left(\frac{1}{z - \epsilon_{\mathbf{k}}^0}\right)\left(\frac{1}{z + z_\nu i - \epsilon_{\mathbf{k}+\mathbf{Q}}^0}\right) \quad (12.114)$$

where the poles of $n^0(z)$ and the contour C are shown in Fig. 12.4. The integration over contour C can be replaced by one over contour C' if we note that the contribution from any contour at $|z| = \infty$ is zero. Thus, we evaluate Eq. (12.114) at the poles of the Fermi propagator and not the distribution function, $n^0(z)$. The result is

$$\chi_0(\mathbf{Q}, z_\nu) = \frac{2}{V}\sum_{\mathbf{k}} \left(\frac{n^0(\mathbf{k}) - n^0(\mathbf{k} + \mathbf{Q})}{z_\nu i + \epsilon_{\mathbf{k}}^0 - \epsilon_{\mathbf{k}+\mathbf{Q}}^0}\right). \quad (12.115)$$

Notice that $\chi_0(\mathbf{Q}, z_\nu)$ describes the propagation of bosons since it depends on frequencies $z_\nu = 2\nu\pi i/\beta$. It is useful to rewrite the expression for $\chi_0(\mathbf{Q}, z_\nu)$. We first add and subtract the term $n^0(\mathbf{k})n^0(\mathbf{k} + \mathbf{Q})$ in the numerator

$$\chi_0(\mathbf{Q}, z_\nu) = \frac{-2}{V}\sum_{\mathbf{k}} \frac{n^0(\mathbf{k} + \mathbf{Q})(1 - n^0(\mathbf{k})) - n^0(\mathbf{k})(1 - n^0(\mathbf{k} + \mathbf{Q}))}{z_\nu i - \epsilon_{\mathbf{k}}^0 + \epsilon_{\mathbf{k}+\mathbf{Q}}^0}. \quad (12.116)$$

Next, we let $\mathbf{k} + \mathbf{Q} \rightarrow -\mathbf{k}$ in part of Eq. (12.116). Then $\chi_0(\mathbf{Q}, z_\nu)$ takes the form

$$\chi_0(\mathbf{Q}, z_\nu) = -\frac{4}{V}\sum_{\mathbf{k}} n^0(\mathbf{k})(1 - n^0(\mathbf{k} + \mathbf{Q}))\frac{\epsilon_{\mathbf{k}}^0 - \epsilon_{\mathbf{k}+\mathbf{Q}}^0}{(z_\nu i)^2 - (\epsilon_{\mathbf{k}}^0 - \epsilon_{\mathbf{k}+\mathbf{Q}}^0)^2}. \quad (12.117)$$

The function $\chi_0(\mathbf{Q}, z_\nu)$ is often called a particle-hole propagator and describes the "propagation" of a particle-hole pair at the Fermi surface rather than just a single particle as does $\bar{G}(i)$ ($\bar{G}(i)$ describes the "propagation" of a particle in terms of its energy relative to the Fermi surface). The distribution $(1 - n^0(\mathbf{k}))$ gives the distribution of holes (lack of particles) and $n^0(\mathbf{k})$ is the distribution of particles in terms of their energies relative to the Fermi surface; $\chi_0(\mathbf{Q}, z_\nu)$ is determined by the distribution of particle-hole pairs. By summing the polarization diagrams we have included a coherent contribution from these objects. They form a boson collective mode in the system and result from polarization effects. An electron will repel other electrons in its neighborhood, thus creating an attractive screening cloud or holes around it. This is a highly coherent effect involving all particles in the medium. The polarization diagrams describe this screening effect.

We will now use the polarization diagrams to find the thermodynamic properties of a high-temperature electron gas and corrections to the density expansion for a zero-temperature electron gas.

3. Classical Electron Gas

At high temperature, $e^{\beta\mu} \ll 1$, and the distribution function takes the form

$$n(\mathbf{k}) \approx e^{-\beta\epsilon_{\mathbf{k}}^0} \tag{12.118}$$

while the particle-hole propagator for $\nu \neq 0$ becomes

$$\begin{aligned}\chi_0(\mathbf{Q}, \nu) &\xrightarrow[T\to\infty]{} \frac{4}{V}\left(\frac{1}{2\nu\pi k_B T}\right)^2 \sum_{\mathbf{k}} n(\mathbf{k})(1 - n(\mathbf{k} + \mathbf{Q}))(\epsilon_{\mathbf{k}}^0 - \epsilon_{\mathbf{k}+\mathbf{Q}}^0)\\ &= \frac{4}{V}\left(\frac{1}{2\nu\pi k_B T}\right)^2 \sum_{\mathbf{k}} n^0(\mathbf{k})(\epsilon_{\mathbf{k}}^0 - \epsilon_{\mathbf{k}+\mathbf{Q}}^0) = -\frac{\hbar^2 Q^2}{m(2\pi\nu k_B T)^2}.\end{aligned} \tag{12.119}$$

Thus, for $\nu \neq 0$, $\chi_0(\mathbf{Q}, \nu) \sim Q^2$ and the product $\lim_{\alpha\to 0}(Q^2 + \alpha^2)^{-1}\chi_0(\mathbf{Q}, z_\nu)$ is independent of Q. At high temperature, the divergent contribution in each order comes from the term $\nu = 0$. Let us then consider the propagator for $\nu = 0$:

$$\chi_0(\mathbf{Q}, 0) = \frac{2}{V}\sum_{\mathbf{k}}\left(\frac{n^0(\mathbf{k}) - n^0(\mathbf{k} + \mathbf{Q})}{\epsilon_{\mathbf{k}}^0 - \epsilon_{\mathbf{k}+\mathbf{Q}}^0}\right). \tag{12.120}$$

Both numerator and denominator vanish at $\mathbf{Q} = 0$, and the sum as a whole is well behaved. The integration can be performed. If we let $\sum_{\mathbf{k}} \to V/(2\pi)^3 \int d\mathbf{k}$, and note that $\mathbf{k}\cdot\mathbf{Q} = kQ\cos\theta$, we find

$$\begin{aligned}\chi_0(\mathbf{Q}, 0) &= \frac{4}{(2\pi)^3}\int d\mathbf{k}\, \frac{n^0(\mathbf{k})}{\dfrac{\hbar^2 kQ\cos\theta}{m} - \dfrac{\hbar^2 Q^2}{2m}}\\ &= \frac{m}{Q(\pi\hbar)^2}\int_0^\infty k\, dk \ln\left|\frac{2k - Q}{2k + Q}\right| e^{\beta\mu}\, e^{-\beta k^2\hbar^2/2m}.\end{aligned} \tag{12.121}$$

If we make the change of variables $Q = x/\lambda_T$ and $k = y/\lambda_T$ (λ_T is the thermal wavelength $\lambda_T = (2\pi\beta\hbar^2/m)^{1/2}$), Eq. (12.121) takes the form

$$\chi_0(\mathbf{Q}, 0) = -\frac{2\beta\, e^{\beta\mu}}{\lambda_T^{\,3}}\, \phi(\lambda_T Q) \tag{12.122}$$

where

$$\phi(x) = \frac{1}{x\pi}\int_0^\infty dy y\, e^{-y^2/4\pi} \ln\left|\frac{2y + x}{2y - x}\right|. \tag{12.123}$$

One can easily check that $\phi(0) = 1$ and $\phi(x) \sim 8\pi/x^2$ for $x \gg 1$. We now write an expression for the leading contribution to the grand potential for a classical electron gas:

$$\Delta\Omega(T, V, \mu) = \frac{1}{2\beta}\sum_{\mathbf{Q}}\left\{\ln\left[1 + \frac{8\pi\, e^2\beta\, e^{\beta\mu}}{Q^2\lambda_T^{\,3}}\, \phi(\lambda_T Q)\right] - \frac{8\pi\, e^2\beta\, e^{\beta\mu}}{Q^2\lambda_T^{\,3}}\, \phi(\lambda_T Q)\right\}. \tag{12.124}$$

If we let $\sum_{\mathbf{Q}} \to (V/(2\pi)^3)\int d\mathbf{Q}$ and make a change of variables $\lambda_T^{\,2}Q^2 = e^2A^2z^2$ where $A^2 = 8\pi\beta\lambda_T^{-1}\, e^{\beta\mu}$, we can write

$$\Delta\Omega(T, V, \mu) = \frac{2V\, e^3\sqrt{\beta}}{\sqrt{\pi}}\left(\frac{2e^{\beta\mu}}{\lambda_T^{\,3}}\right)^{3/2}\int_0^\infty dz z^2\{\ln[1 + z^{-2}\phi(eAz)] - z^{-2}\phi(eAz)\}. \tag{12.125}$$

To find the leading contribution to $\Delta\Omega(T, \mu, V)$ for small e^2, we can set $e = 0$ in the integrand since neglected terms are of order e^5 or higher. Then, since $\phi(0) = 1$, we find

$$\begin{aligned}\Delta\Omega(T, V, \mu) &= \frac{2V\, e^3\sqrt{\beta}}{\sqrt{\pi}}\left(\frac{2\, e^{\beta\mu}}{\lambda_T^{\,3}}\right)^{3/2}\int_0^\infty dz z^2\{\ln[1 + z^{-2}] - z^{-2}\}\\ &= -\tfrac{2}{3}V\sqrt{\beta}\, e^3\sqrt{\pi}\left(\frac{2\, e^{\beta\mu}}{\lambda_T^{\,3}}\right)^{3/2}.\end{aligned} \tag{12.126}$$

Eq. (12.126) gives the leading contribution from the polarization diagrams for small coupling constant at high temperature. It is proportional to e^3 even though all terms we have kept are proportional to powers of e^2 and the lowest order term is proportional to e^4. The terms we have neglected in writing Eq. (12.126) are of order e^4.

We now write the complete expression for the grand potential of a classical electron gas for small coupling constant. If we note that the contribution from the exchange term, $\Omega_1(V, T, \mu)$, goes to zero as $h \to 0$ (classical limit) and that for an ideal classical gas $\Omega_0(V, T, \mu) = -2V\, e^{\beta\mu}/\beta\lambda_T^{\,3}$ (cf. Eq. (9.117)), we obtain

$$\Omega(V, T, \mu) = -\frac{2}{\lambda_T^{\,3}\beta}\, e^{\beta\mu}\left[1 + \tfrac{2}{3}(2\pi)^{1/2}\left(\frac{e^2\beta}{\lambda_T}\right)^{3/2} e^{\beta\mu/2}\right]. \tag{12.127}$$

The average particle density is

$$\frac{\langle N\rangle}{V} = -\frac{1}{V}\left(\frac{\partial\Omega}{\partial\mu}\right)_{T,V} = \frac{2}{\lambda_T{}^3}\, e^{\beta\mu}\left[1 + \sqrt{2\pi}\left(\frac{e^2\beta}{\lambda_T}\right)^{3/2} e^{\beta\mu/2}\right]. \quad (12.128)$$

Eq. (12.128) can be reverted to obtain an expression for μ. To lowest order in e^2, we find

$$\beta\mu = \ln\left[\frac{\langle N\rangle}{V}\frac{\lambda_T{}^3}{2}\left(1 - \sqrt{\pi}(e^2\beta)^{3/2}\left(\frac{N}{V}\right)^{1/2}\right)\right]. \quad (12.129)$$

The pressure is given by

$$P = -\frac{\Omega}{V} = \frac{\langle N\rangle}{V}\frac{1}{\beta}\left[1 - \frac{\sqrt{\pi}}{3}(e^2\beta)^{3/2}\left(\frac{N}{V}\right)^{1/2}\right]. \quad (12.130)$$

Eq. (12.130) is the Debye-Hückel equation[11] for the equation of state of a classical ionized gas. The correction is of order e^3 and thus cannot be obtained by a simple perturbation expansion. By summing the polarization diagrams, we have included coherent screening effects which change the form of the equation of state in a nontrivial way.

4. Zero Temperature Limit

At very low temperature, we can change the summation over z_ν to an integration since discrete changes in ν cause very small changes in z_ν. Thus, the correction to the grand potential due to polarization diagrams takes the form

$$\Delta\Omega(V, T, \mu) = \frac{V}{4\pi(2\pi)^3}\int d\mathbf{Q}\int_{-\infty}^{\infty} dz \times\left\{\ln\left[1 - \frac{4\pi e^2}{Q^2}\chi_0(\mathbf{Q}, z_\nu)\right] + \frac{4\pi e^2}{Q^2}\chi_0(\mathbf{Q}, z_\nu)\right\}. \quad (12.131)$$

The dominant contribution comes from $Q \approx 0$. In that limit we can write

$$\chi_0(\mathbf{Q}, z) = -2\int\frac{d\mathbf{k}}{(2\pi)^3}\frac{n^0(\mathbf{k}+\mathbf{Q}) - n^0(\mathbf{k})}{zi - \epsilon^0_{\mathbf{k}+\mathbf{Q}} + \epsilon_{\mathbf{k}}{}^0} \underset{Q\to 0}{\approx} -2\int\frac{d\mathbf{k}}{(2\pi)^3}\frac{\mathbf{Q}\cdot\nabla_{\mathbf{k}} n^0(\mathbf{k})}{zi - \dfrac{\hbar^2}{m}\mathbf{k}\cdot\mathbf{Q}} \quad (12.132)$$

where we have expanded the numerator of Eq. (12.132) in a Taylor series about $Q = 0$. At zero temperature, the Fermi distribution is a step function and its derivative is a delta function. Thus, $\nabla_{\mathbf{k}} n(\mathbf{k}) = -\boldsymbol{\varkappa}\,\delta(k - \gamma)$ where $\gamma = \sqrt{2m\mu(0)}$ and $\boldsymbol{\varkappa} = \mathbf{k}/|\mathbf{k}|$. The integrations in Eq. (12.132) can be performed to give

$$\chi_0(\mathbf{Q}, z) \underset{Q\to 0}{\approx} -\frac{\gamma m}{\pi^2\hbar^2} R\left(\frac{mz}{\hbar^2 Q\gamma}\right) \quad (12.133)$$

where

$$R(x) = \int_0^1 \frac{dyy^2}{y^2 + x^2} = 1 - x \arctan\left(\frac{1}{x}\right). \tag{12.134}$$

Eq. (12.133) can be used to evaluate Eq. (12.131) for $Q \approx 0$. It is useful to introduce a cutoff, Q_0, on the Q integration in Eq. (12.131). Then for $Q < Q_0$ we can use Eq. (12.133), while for $Q > Q_0$ we must use the full expression for $\chi_0(\mathbf{Q}, z)$. However, if we are interested in an expansion in powers of e^2, we can expand the integrand for $Q > Q_0$ in powers of e^2 and keep the lowest order term. If we do this and make the change of variables $y = Q/\gamma$ and $z = \hbar^2 Q\gamma x/m$, we find

$$\begin{aligned} \Delta\Omega(V, T, \mu) = {} & \frac{V\hbar^2\gamma^5}{8\pi^3 m} \int_{-\infty}^{\infty} dx \int_0^{y_0} dyy^3 \\ & \times \left\{ \ln\left[1 + \frac{4m}{\gamma\hbar^2\pi}\frac{e^2}{y^2} R(x)\right] - \frac{4me^2}{\gamma\hbar^2\pi y^2} R(x) \right\} \\ & + \Delta\Omega(V, T, \mu; y > y_0), \end{aligned} \tag{12.135}$$

where

$$\Delta\Omega(V, T, \mu; y > y_0) = -\frac{1}{2}\frac{V\hbar^2\gamma^5}{8\pi^3 m} \int_{-\infty}^{\infty} dx \int_{y_0}^{\infty} y^3\, dy \left[\frac{4\pi e^2}{\gamma^2 y^2} \chi_0\left(\gamma y; \frac{\hbar^2\gamma^2 yx}{m}\right)\right]^2. \tag{12.136}$$

In Eq. (12.135), the integral over y can be done

$$\begin{aligned} & \int_0^{y_0} dyy^3 \left[\ln\left[1 + \frac{A(x)\, e^2}{y^2}\right] - \frac{A(x)\, e^2}{y^2} \right] \\ & = \tfrac{1}{4} A^2(x)\, e^2 \left\{ \frac{y_0^4}{A^2(x)\, e^4} \left[\ln\left(1 + \frac{A(x)\, e^2}{y_0^2}\right) - \frac{A(x)\, e^2}{y_0^2} \right] - \ln\left(1 + \frac{y_0^2}{A(x) e^2}\right) \right\} \\ & \approx \tfrac{1}{4} A^2(x)\, e^4 [\ln (A(x)\, e^2) - \tfrac{1}{2} - 2 \ln y_0] + O(e^6), \end{aligned} \tag{12.137}$$

where

$$A(x) = \frac{4m}{\gamma\hbar^2\pi} R(x). \tag{12.138}$$

If we note that

$$\int_{-\infty}^{\infty} dxR^2(x) = \tfrac{2}{3}\pi(1 - \ln 2) \tag{12.139}$$

and

$$\int_{-\infty}^{\infty} dxR^2(x) \ln R(x) = -(0.551)\tfrac{2}{3}\pi(1 - \ln 2) \tag{12.140}$$

(Eq. (12.140) must be integrated numerically), Eq. (12.135) can be written

$$\Delta\Omega(V, T = 0, \mu) = \lim_{y_0 \to 0} \frac{V\gamma^3 e^4 m}{2\pi^5 \hbar^2} \tfrac{2}{3}\pi(1 - \ln 2)\left[\ln \frac{4m e^2}{\gamma\hbar^2 \pi} - 1.051 - 2 \ln y_0\right]$$
$$+ \Delta\Omega(V, T, \mu; y > y_0). \quad (12.141)$$

In the limit $y_0 \to 0$ both $\ln y_0$ and $\Delta\Omega(V, T, \mu; y > y_0)$ are divergent. However, when added together, they give a finite result. (We shall leave it to an exercise to show this.)

We can write Eq. (12.141) in terms of the density factor r_s. It then takes the form

$$\Delta\Omega(V, T = 0, \mu(0)) = \frac{e^2\langle N\rangle}{2a_0} [(0.0622) \ln r_s - 1.077] + \delta \quad (12.142)$$

where

$$\delta = -\frac{e^2\langle N\rangle}{2a_0} (1.244) \ln y_0 + \Delta\Omega(V, T = 0, \mu(0); y > y_0) \quad (12.143)$$

is independent of r_s. The contributions from the second-order exchange diagrams will also be independent of r_s. There will be higher order terms with dependence $r_s \ln(r_s)$ coming from higher order contributions from $\Delta\Omega(V, T = 0, \mu(0))$. If all exchange effects are included, the internal energy takes the form[5]

$$U(V, T = 0, \langle N\rangle)$$
$$= \frac{e^2\langle N\rangle}{2a_0}\left[\frac{2.21}{r_s^2} - \frac{0.916}{r_s} - 0.094 + 0.0622 \ln r_s + O(r_s \ln r_s)\right]. \quad (12.144)$$

Thus, the density expansion for a high-density electron gas is a nonanalytic function of the density. This type of result is common to many-body systems in which collective behavior plays an important role.

At very low densities, the electron system can form a crystalline state called a Wigner solid. The procedure for obtaining a low-density expansion for the internal energy is discussed in Refs. 12.12 and 12.13.

E. PROPAGATORS FOR NORMAL BOSON AND FERMI FLUIDS[5–7]

1. Physical Interpretation

The quickest way to obtain an understanding of the microscopic processes leading to the observed thermodynamic behavior of a quantum system is through a study of the propagator

$$G(k, k'; \tau - \tau') = -\langle T(\hat{a}_k(\tau)\hat{a}_{k'}^+(\tau'))\rangle \tag{12.145}$$

where the average is taken with respect to the exact density operator

$$\langle O\rangle \equiv \frac{\text{Tr}\, e^{-\beta\hat{K}}\hat{O}}{\text{Tr}\, e^{-\beta\hat{K}}} \tag{12.146}$$

and the temperature dependence of the creation and annihilation operators is given by

$$\hat{a}_k(\tau) = e^{\tau\hat{K}}\hat{a}_k\, e^{-\tau\hat{K}} \tag{12.147}$$

and

$$\hat{a}_k{}^+(\tau) = e^{\tau\hat{K}}\hat{a}_k{}^+\, e^{-\tau\hat{K}}. \tag{12.148}$$

The propagators, $G(k, k'; \tau - \tau')$, are the analog in quantum fluids of the one-body reduced probability densities that we introduced for classical fluids in Sec. 11.B, except that $G(k, k'; \tau - \tau')$ is more general and contains more information than a reduced probability density.

The propagator, $G(k, k'; \tau - \tau')$, contains all possible information about expectation values of one-body operators since the one-body reduced density matrix can be obtained from it through the relation

$$\langle k|\hat{\rho}_1|k'\rangle = \lim_{\tau'\to\tau^+} G(k, k', \tau - \tau') \tag{12.149}$$

(cf. Eq. (7.93)). The particle momentum distribution function for the interacting system is given by

$$n(k) = \lim_{\tau'\to\tau^+} G(k, k; \tau - \tau'). \tag{12.150}$$

The limit $\tau' \to \tau^+$ indicates that we let τ' approach τ and finally equal τ in such a way that $\tau' > \tau$.

Once we have the propagator, we can compute all thermodynamic properties of the system. Let us derive an expression for the internal energy and grand potential in terms of $G(k, k'; \tau - \tau')$. We first consider the internal energy $U(V, T, \langle N\rangle) = \langle\hat{H}\rangle$. To obtain an expression for it we must derive an equation of motion for $\hat{a}_k(\tau)$ in the interacting system. It is fairly easy to show that

$$\begin{aligned}\frac{\partial}{\partial\tau}\hat{a}_{k_1}(\tau) &= [\hat{K}, \hat{a}_{k_1}(\tau)]_- \\ &= -\sum_{k_2,k_3,k_4} \langle k_1, k_2|V|k_3, k_4\rangle\hat{a}_{k_2}^+(\tau)\hat{a}_{k_4}(\tau)\hat{a}_{k_3}(\tau)\end{aligned} \tag{12.151}$$

for both boson and fermion systems. If we take the temperature derivative of $G(k, k'; \tau - \tau')$ and use Eq. (12.151), we find

$$\langle \hat{V} \rangle \equiv \frac{1}{2} \sum_{k_1, \ldots, k_4} \langle k_1, k_2 | V | k_3, k_4 \rangle \langle a_{k_1}^+ a_{k_2}^+ a_{k_4} a_{k_3} \rangle$$
$$= \mp \frac{1}{2} \sum_k \lim_{k \to k'} \lim_{\tau' \to \tau +} \left[-\frac{\partial}{\partial \tau} - \epsilon_k{}^0 \right] G(k, k'; \tau - \tau'). \quad (12.152)$$

The internal energy can now be written

$$U(V, T, \langle N \rangle) = \langle \hat{T} \rangle + \langle \hat{V} \rangle$$
$$= \mp \frac{1}{2} \sum_k \lim_{k \to k'} \lim_{\tau' \to \tau +} \left[-\frac{\partial}{\partial \tau} + \epsilon_k{}^0 + 2\mu \right] G(k, k'; \tau - \tau') \quad (12.153)$$

where $\hat{T}$ is the kinetic energy operator.

From Eq. (12.153), we can derive an expression for the grand potential. Let us write $K(\lambda) = \hat{K}_0 + \lambda \hat{V}$ and for the grand partition function write

$$\mathscr{Z}(\lambda) = e^{-\beta \Omega(\lambda)} = \operatorname{Tr} e^{-\beta \hat{K}(\lambda)} \quad (12.154)$$

where $\Omega(\lambda)$ is the grand potential. If we expand the exponential under the trace and take the derivative of $\mathscr{Z}(\lambda)$ with respect to λ, we find

$$\frac{\partial \mathscr{Z}}{\partial \lambda} = \sum_{l=1}^{\infty} \frac{1}{l!} (-\beta)^l \frac{\partial}{\partial \lambda} \operatorname{Tr}(\hat{K}_0 + \lambda \hat{V})^l$$
$$= -\beta \operatorname{Tr}(e^{-\beta \hat{K}} \hat{V}) = -\frac{\beta}{\lambda} e^{-\beta \Omega} \langle \lambda \hat{V} \rangle. \quad (12.155)$$

From the derivative of $\mathscr{Z}(\lambda)$, we can find the derivative of $\Omega(\lambda)$:

$$\frac{\partial \Omega}{\partial \lambda} = \frac{1}{\lambda} \langle \lambda \hat{V} \rangle. \quad (12.156)$$

If we now integrate Eq. (12.156) from $\lambda = 0$ to $\lambda = 1$, we obtain

$$\Omega(V, T, \mu) = \Omega_0(V, T, \mu) + \int_0^1 d\lambda \frac{1}{\lambda} \langle \lambda \hat{V} \rangle$$
$$= \Omega_0(V, T, \mu) \mp \int_0^1 d\lambda \frac{1}{\lambda} \sum_k \lim_{k' \to k} \lim_{\tau' \to \tau +} \frac{1}{2} \left[-\frac{\partial}{\partial \tau} - \epsilon_k^0 \right] G(k, k'; \tau - \tau'). \quad (12.157)$$

Thus, we can derive all thermodynamic quantities from the propagator.

The propagator has one important property that the one-body reduced density matrix does not have. When we Fourier transform the temperature dependence of the propagator, the poles of the Fourier transformed propagator give the spectrum of the excitations in the system. This is most easily seen if we evaluate $G(k, k'; \tau - \tau')$ in terms of eigenstates of $\hat{K}$, which we shall denote $\{|\Psi_\alpha^\mu\rangle\}$. We will let $|\Psi_0^\mu\rangle$ denote the eigenvector with the smallest eigenvalue. At $T = 0$ K, the entropy $S = 0$, and the grand potential will be given by

$$\Omega(T = 0, V, \mu) = \langle\Psi_0^\mu|\hat{K}|\Psi_0^\mu\rangle = \kappa_0 = \langle\hat{H}\rangle - \mu\langle\hat{N}\rangle \quad (12.158)$$

where κ_0 is the eigenvalue of $\hat{K}$ in the state $|\Psi_0^\mu\rangle$ and μ, the chemical potential, is fixed once the average particle number is fixed:

$$\langle N\rangle = -\left(\frac{\partial\Omega}{\partial\mu}\right)_{T,V}. \quad (12.159)$$

The eigenvalues of $\hat{K}$ correspond to the allowed values of the free energy in the ground state and for excited states when $T = 0$.

It is useful to determine how the full propagator depends on the excitation spectrum. Let us first note that $G(k, k'; \tau - \tau')$ is periodic in τ since for $\tau > 0$

$$\begin{aligned} G(k, k'; \tau) &= -\mathrm{Tr}\, e^{-\beta\hat{K}}\hat{a}_k(\tau)\hat{a}_{k'}^+(0) = -\mathrm{Tr}\, \hat{a}_k(\tau)\, e^{-\beta\hat{K}}\, e^{\beta\hat{K}}\hat{a}_{k'}^+(0)\, e^{-\beta\hat{K}} \\ &= -\mathrm{Tr}\, e^{-\beta\hat{K}}\hat{a}_{k'}^+(\beta)\hat{a}_k(\tau) = \pm G(k, k'; \tau - \beta) \end{aligned} \quad (12.160)$$

and for $\tau < 0$

$$G(k, k'; \tau) = \pm G(k, k'; \tau + \beta). \quad (12.161)$$

Thus,

$$G(k, k'; \tau) = G(k, k'; \tau \pm 2\beta) \quad (12.162)$$

and the propagator is periodic in τ with period 2β.

It is possible to write the propagator in terms of the exact excitation spectrum if we use the complete set of states $\{|\Psi_\alpha^\mu\rangle\}$ to evaluate the trace in Eq. (12.145). Then,

$$\begin{aligned} G(k, k'; \tau - \tau') = \sum_{\alpha,\alpha'} [&e^{-\beta\kappa_\alpha}\langle\Psi_\alpha^\mu|\hat{a}_k|\Psi_{\alpha'}^\mu\rangle\langle\Psi_{\alpha'}^\mu|a_{k'}^+|\Psi_\alpha^\mu\rangle \\ &\times e^{-(\tau-\tau')(\kappa_{\alpha'}-\kappa_\alpha)}\theta(\tau - \tau') \\ &\pm e^{-\beta\kappa_\alpha}\langle\Psi_{\alpha'}^\mu|\hat{a}_{k'}^+|\Psi_\alpha^\mu\rangle\langle\Psi_\alpha^\mu|\hat{a}_k|\Psi_{\alpha'}^\mu\rangle \\ &\times e^{-(\tau-\tau')(\kappa_{\alpha'}-\kappa_\alpha)}\theta(\tau' - \tau)]. \end{aligned} \quad (12.163)$$

The Fourier transform of $G(k, k'; \tau - \tau')$ is defined

$$\bar{G}(k, k'; \omega_n) = \int_0^\beta d\tau\, e^{i\omega_n\tau}G_0(k, k'; \tau) \quad (12.164)$$

where $\omega_n = 2\pi n/\beta$ for bosons and $\omega_n = (2n + 1)\pi/\beta$ for fermions as before. From Eqs. (12.163) and (12.164), we obtain

$$\bar{G}(k, k'; \omega_n) = -\sum_{\alpha\alpha'} \frac{\langle\Psi_\alpha^\mu|\hat{a}_k|\Psi_{\alpha'}^\mu\rangle\langle\Psi_{\alpha'}^\mu|\hat{a}_{k'}^+|\Psi_\alpha^\mu\rangle(e^{-\beta\kappa_{\alpha'}} - e^{-\beta\kappa_\alpha})}{i\omega_n - (\kappa_{\alpha'} - \kappa_\alpha)}. \quad (12.165)$$

Thus, we have the important result that the poles of $\bar{G}(k, k'; \omega_n)$ give the exact excitation free energies for the system considered. The numerator gives a weighting factor for each type of excitation. We might interpret the numerator as follows. We have the system in state $|\Psi_\alpha{}^\mu\rangle$. We then add a particle of momentum/spin k'. This causes the system to change to a superposition of states $|\Psi^\mu_{\alpha'}\rangle$. The probability amplitude for going from state $|\Psi_\alpha{}^\mu\rangle$ to $|\Psi^\mu_{\alpha'}\rangle$ is $\langle\Psi^\mu_{\alpha'}|\hat{a}^+_{k'}|\Psi_\alpha{}^\mu\rangle$. We then remove a particle of momentum/spin k; $\langle\Psi_\alpha{}^\mu|\hat{a}_k|\Psi^\mu_{\alpha'}\rangle$ is the probability amplitude of going back to state $|\Psi_\alpha{}^\mu\rangle$.

2. Diagrammatic Expansion

For many-body systems, we generally cannot find the eigenvalues of $\hat{K}$. We therefore must use some approximation scheme for determining the excitation spectrum, and we are forced to return to perturbation theory. However, the results are quite beautiful. We can write the full propagator in the following form:

$$G(k, k'; \tau - \tau') = [\mathscr{Z}(T, V, \mu)]^{-1} \operatorname{Tr} e^{-\beta\hat{K}_0} \times [\hat{U}(\beta, \tau)\hat{a}_k(\tau)_0\hat{U}(\tau, \tau')a^+_{k'}(\tau')_0\hat{U}(\tau', 0)\theta(\tau - \tau') \pm \hat{U}(\beta, \tau')\hat{a}^+_{k'}(\tau')_0\hat{U}(\tau', \tau)\hat{a}_k(\tau)_0\hat{U}(\tau, 0)\theta(\tau' - \tau)] \tag{12.166}$$

where $\hat{U}(\tau, \tau')$ is defined in Eq. (12.9). In order to simplify Eq. (12.166), we note the following result for an arbitrary operator $\hat{O}(\tau)_0$:

$$\hat{U}(\beta, \tau)\hat{O}(\tau)_0\hat{U}(\tau, 0) = \sum_{n=0}^{\infty}\sum_{m=0}^{\infty}(-1)^{m+n}\frac{1}{n!}\frac{1}{m!}\int_\tau^\beta d\tau_1 \times \cdots \times \int_\tau^\beta d\tau_n \int_0^\tau d\tau_1' \times \cdots \times \int_0^\tau d\tau_m' \times [T[\hat{V}(\tau_1)_0 \times \cdots \times \hat{V}(\tau_n)_0]]\hat{O}(\tau)_0[T[\hat{V}(\tau_1')_0 \times \cdots \times \hat{V}(\tau_m')_0]]$$
$$= \sum_{l=0}^{\infty}(-1)^l\frac{1}{l!}\int_0^\beta d\tau_1 \times \cdots \times \int_0^\beta d\tau_l T[\hat{V}(\tau_1)_0 \times \cdots \times \hat{V}(\tau_l)_0\hat{O}(\tau)_0]. \tag{12.167}$$

Eq. (12.167) may be understood as follows. Consider the lth-order contribution from the right-most term which has n factors $V(\tau_i)$ to the left and m factors $V(\tau_i)$ to the right of $\hat{O}(\tau)_0$ (where $l = m + n$). There are $l!/m!\,n!$ ways of permuting the factors, $V(\tau_i)$, so temperature variables to the left and right of $\hat{O}(\tau)_0$ change. The sum of all these terms yields the middle expression in Eq. (12.167). A similar analysis of Eq. (12.166) yields for the propagator

$$G(k, k'; \tau - \tau') = -[\mathscr{Z}(T, V, \mu)]^{-1}\sum_{n=0}^{\infty}(-\lambda)^n\frac{1}{n!}\int_0^\beta d\tau_1 \times \cdots \times \int_0^\beta d\tau_n \operatorname{Tr}[e^{-\beta\hat{K}_0}T[\hat{V}(\tau_1)_0 \times \cdots \times \hat{V}(\tau_n)_0\hat{a}_k(\tau)\hat{a}^+_{k'}(\tau')]]. \tag{12.168}$$

Since the average is taken with respect to the state $e^{-\beta\hat{K}_0}$, which is quadratic in field operators, we can make use of Wick's theorem to evaluate the trace.

We will now evaluate Eq. (12.168) order by order in λ. Let us write

$$G(k, k'; \tau - \tau') = \sum_{l=0}^{\infty} \lambda^l G_l(k, k'; \tau - \tau') \tag{12.169}$$

and

$$\mathscr{G}_n(k, k'; \tau - \tau') = -(-1)^n \frac{1}{n!} \int_0^\beta d\tau_1 \times \cdots \times \int_0^\beta d\tau_n \operatorname{Tr}\{\hat{\rho}_0 T[\hat{V}(\tau_1)_0 \hat{V}(\tau_2)_0 \cdots \hat{V}(\tau_n)_0 \hat{a}_k(\tau) \hat{a}_{k'}^+(\tau')]\} \tag{12.170}$$

where $\hat{\rho}_0 = e^{-\beta\hat{K}_0}/\operatorname{Tr} e^{-\beta\hat{K}_0}$, and make use of Eq. (12.16). We find for the zeroth-order term

$$\begin{aligned} G_0(k, k'; \tau - \tau') = \mathscr{G}_0(k, k'; \tau - \tau') &= -\operatorname{Tr}[\hat{\rho}_0 \hat{a}_k(\tau) \hat{a}_{k'}^+(\tau')] \\ &= \delta_{k,k'} G_0(k; \tau - \tau'). \end{aligned} \tag{12.171}$$

The first-order contribution is

$$\begin{aligned} G_1(k, k'; \tau - \tau') &= \mathscr{G}_1(k, k'; \tau - \tau') - \mathscr{G}_0(k, k'; \tau - \tau') U_1(V, T, \langle N \rangle) \\ &= \mp \sum_{k_1} \int_0^\beta d\tau_1 \langle k, k_1 | V | k', k_1 \rangle^{(\pm)} \\ &\quad \times G_0(k; \tau - \tau_1) G_0(k'; \tau_1 - \tau') G_0(k_1; 0^+). \end{aligned} \tag{12.172}$$

The factor involving $U_1(V, T, N)$ cancels with one of the terms coming from $\mathscr{G}_1(k, k'; \tau - \tau')$. The second-order contribution is given by

$$\begin{aligned} &G_2(k, k'; \tau - \tau') \\ &\quad = \mathscr{G}_2(k, k'; \tau - \tau') - \mathscr{G}_0(k, k'; \tau - \tau') U_2(V, T, \langle N \rangle) \\ &\quad\quad - \mathscr{G}_1(k, k'; \tau - \tau') U_1(V, T, \langle N \rangle) + \tfrac{1}{2} \mathscr{G}_0(k, k'; \tau - \tau') U_1^2(V, T, \langle N \rangle) \\ &\quad = \sum_{k_1 k_2 k_3} \int_0^\beta \int_0^\beta d\tau_1\, d\tau_2 \langle k_2, k_1 | V | k_3, k_1 \rangle^{(\pm)} \langle k_3, k | V | k', k_2 \rangle^{(\pm)} G_0(k_1; 0) \\ &\quad\quad \times G_0(k_2; \tau_2 - \tau_1) G_0(k_3; \tau_1 - \tau_2) G_0(k; \tau - \tau_2) G_0(k'; \tau_2 - \tau') \\ &\quad\quad + \sum_{k_1 k_2 k_3} \int_0^\beta \int_0^\beta d\tau_1\, d\tau_2 \langle k, k_1 | V | k_2, k_1 \rangle^{(\pm)} \langle k_2, k_3 | V | k', k_3 \rangle^{(\pm)} G_0(k_1; 0) \\ &\quad\quad \times G_0(k_3; 0) G_0(k_2; \tau_2 - \tau_1) G_0(k; \tau - \tau_2) G_0(k'; \tau_2 - \tau') \\ &\quad\quad \pm \frac{1}{2} \sum_{k_1 k_2 k_3} \int_0^\beta \int_0^\beta d\tau_1\, d\tau_2 \langle k, k_3 | V | k_2, k_1 \rangle^{(\pm)} \langle k_2, k_1 | V | k', k_3 \rangle^{(\pm)} \\ &\quad\quad \times G_0(k_2; \tau_2 - \tau_1) G_0(k_1; \tau_2 - \tau_1) G_0(k_3; \tau_1 - \tau_2) \\ &\quad\quad \times G_0(k; \tau - \tau_2) G_0(k'; \tau_1 - \tau'). \end{aligned} \tag{12.173}$$

Again we find that terms involving the cluster functions $U_n(V, T, N)$ cancel with terms coming from $\mathscr{G}_{n'}(k, k'; \tau - \tau')$ and no terms remain which can be written as a product of factors with independent momentum and temperature dependence. This type of cancellation occurs to all orders in λ, and therefore $G(k, k'; \tau - \tau')$ to all orders in λ depends only on fully connected terms. This behavior is similar to that of the one-body reduced probability density of classical fluids which also depends only on fully connected terms.

It is useful to write $G(k, k'; \tau - \tau')$ in terms of diagrams. We obtain the following expression:

$$G(k, k'; \tau - \tau') = \sum (\text{all different connected 1-diagrams}). \quad (12.174)$$

A connected 1-diagram is a collection of vertices entirely interconnected by internal directed lines, with one broken line entering and one broken line leaving. Two connected 1-diagrams are different if they have different topological structure. Temperature-dependent algebraic expressions can be associated with the connected 1-diagrams according to the rules in Table 12.1.B. We can also express the Fourier transform $\bar{G}(k, k'; \omega_n)$ (cf. Eq. (12.48)) in terms of connected 1-diagrams,

$$\bar{G}(k, k'; \omega_n) = (\text{all different connected 1-diagrams}). \quad (12.175)$$

The rules for evaluating $\bar{G}(k, k'; \omega_n)$ are given in Table 12.2.B.

The propagator, $G(k, k'; \omega_n)$ to order λ^2 is given by the following diagrammatic expression:

$\bar{G}(k, k'; \omega_n) =$ 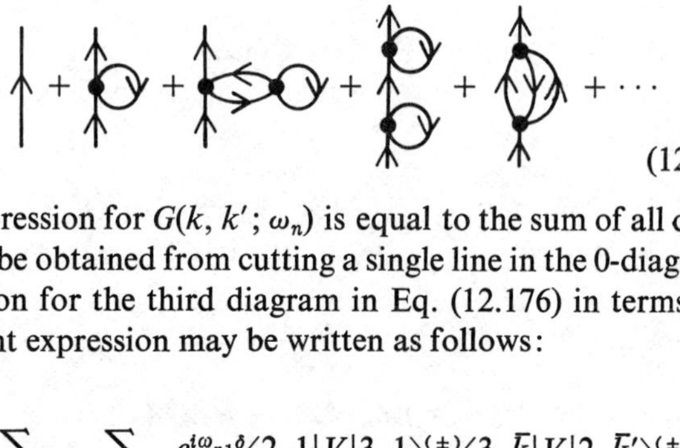 $+ \cdots$

(12.176)

The diagrammatic expression for $G(k, k'; \omega_n)$ is equal to the sum of all different diagrams that can be obtained from cutting a single line in the 0-diagrams. An algebraic expression for the third diagram in Eq. (12.176) in terms of a "frequency"-dependent expression may be written as follows:

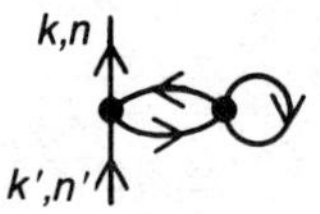

$$= \sum_{k_1, k_2, k_3} \sum_{n_1, n_2, n_3} e^{i\omega_{n1}\delta} \langle 2, 1|V|3, 1\rangle^{(\pm)} \langle 3, \bar{k}|V|2, \bar{k}'\rangle^{(\pm)}$$
$$\times \bar{G}_0(1)\bar{G}_0(2)\bar{G}_0(3)\bar{G}_0(k, \omega_n)\bar{G}(k', \omega_{n'}) \quad (12.177)$$

where we have used the notation

$$\langle 3, \bar{k}|V|2, \bar{k}'\rangle^{(\pm)} = -\beta\, \delta_{k_3+k, k'+k_2}\, \delta_{n_3+n, n'+n_2} \langle k_3, k|V|k_2, k'\rangle^{(\pm)}.$$

It is now possible to find a very concise expression for the exact propagator, $\bar{G}(k, k'; \omega_n)$.

F. DYSON'S EQUATIONS AND SELF-ENERGY STRUCTURES[5–7,14]

In Fig. 12.5, we have displayed some typical connected 1-diagrams which contribute to the exact propagator $G(k, k'; \omega_n)$. Various terms in the expression for

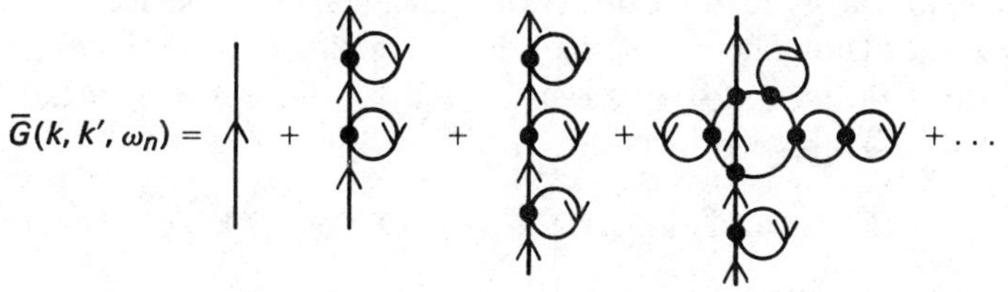

Fig. 12.5. Some typical reducible 1-diagrams.

the propagators look like strings of oddly shaped beads. Indeed, we can rewrite the propagator in the form

$\bar{G}(k, k'; \omega_n) \equiv$ 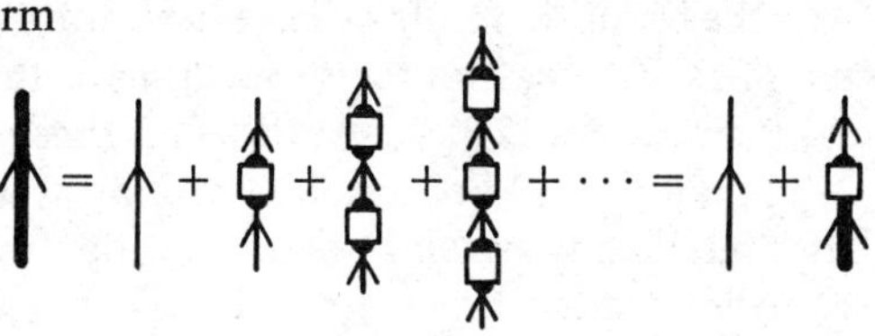 (12.178)

where

 $= \sum$ (all different irreducible connected 1-diagrams). (12.179)

Irreducible 1-diagrams cannot be cut into two parts by cutting one line. We associate algebraic expressions to them in the same way as with the connected 1-diagrams except that no factors are associated with external lines. Some examples of irreducible connected 1-diagrams are given by

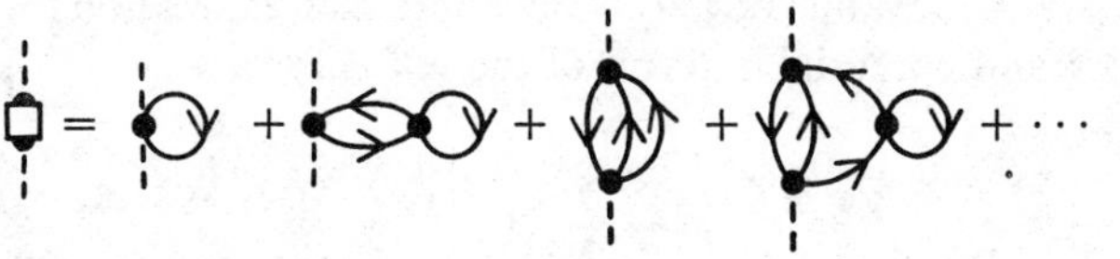 (12.180)

The vertex in Eq. (12.179) is called the self-energy and is denoted

$= \Sigma(k, k'; \omega_n)\, \delta_{n,n'}$. (12.181)

For translationally invariant systems, $\Sigma(k, k'; \omega_n)$ can be written in the form $\Sigma(k, k'; \omega_n) = \delta_{\mathbf{k},\mathbf{k}'}\, \Sigma_{\lambda,\lambda'}(\mathbf{k}, \omega_n)$.

From Eq. (12.178) and the rules in Table 12.2.B we obtain the following integral equation for the propagator:

$$\bar{G}_{\lambda,\lambda'}(\mathbf{k}, \omega_n) = \delta_{\lambda,\lambda'}\, \bar{G}_0(\mathbf{k}, \omega_n) + \sum_{\lambda_1} \bar{G}_0(\mathbf{k}, \omega_n)\, \Sigma_{\lambda,\lambda_1}(\mathbf{k}, \omega_n) \bar{G}_{\lambda_1,\lambda'}(\mathbf{k}, \omega_n). \tag{12.182}$$

For a system in which the full propagator is diagonal in spin space (systems in which the total z-component of spin of the interacting particles is conserved), we can write $\bar{G}(\mathbf{k}, \omega_n)_{\lambda\lambda'} = \bar{G}(\mathbf{k}, \omega_n)\, \delta_{\lambda\lambda'}$ and $\Sigma_{\lambda\lambda'}(\mathbf{k}, \omega_n) = \delta_{\lambda,\lambda'}\, \Sigma(\mathbf{k}, \omega_n)$. Then Eq. (12.182) takes the form

$$\bar{G}(\mathbf{k}, \omega_n) = \bar{G}_0(\mathbf{k}, \omega_n) + \bar{G}_0(\mathbf{k}\omega_n)\, \Sigma(\mathbf{k}\omega_n) \bar{G}(\mathbf{k}, \omega_n). \tag{12.183}$$

Eq. (12.183) is called Dyson's equation. Another form is

$$\bar{G}(\mathbf{k}, \omega_n) = \frac{1}{i\omega_n - \epsilon_{\mathbf{k}}{}^0 - \Sigma(\mathbf{k}, \omega_n)}. \tag{12.184}$$

Eq. (12.184) is the beautiful result we have been after. All we have to do to find the spectrum of excitations and their weighting factors is to find the poles of $\bar{G}(\mathbf{k}, \omega_n)$ and their residue. Thus, we must find those values of ω_n for which $i\omega_n - \epsilon^0{}_{\mathbf{k}} - \Sigma(\mathbf{k}, \omega_n) = 0$. For a noninteracting system, the poles occur for $i\omega_n = \epsilon_{\mathbf{k}}{}^0$. For interacting systems, we must compute the self-energy $\Sigma(\mathbf{k}, \omega_n)$ and then solve the equation $i\omega_n - \epsilon_{\mathbf{k}}{}^0 - \Sigma(\mathbf{k}, \omega_n) = 0$ for ω_n. The excitations in an interacting system can be quite different from those in a noninteracting system. In general we cannot find an exact expression for $\Sigma(\mathbf{k}, \omega_n)$, but we can find very good approximations under suitable conditions.

In performing the resummation which leads to Eq. (12.184), we have vastly improved the convergence of our expansion for $\bar{G}(\mathbf{k}, \omega_n)$. Terms in the expansion for $\bar{G}(\mathbf{k}, \omega_n)$, which look like a chain of beads, have a polynomial dependence on the inverse temperature β. Thus, as $T \to 0$ K and $\beta \to \infty$ these terms converge very slowly. By solving the integral equation for $\bar{G}(\mathbf{k}, \omega_n)$ we have explicitly resummed and removed this polynomial dependence on β.

From Eqs. (12.153) and (12.157) we can obtain an expression for the internal energy and grand potential in terms of the self-energy,

$$U(V, T, \langle N \rangle) = \mp(2s + 1)\frac{1}{\beta}\sum_{\mathbf{k}}\sum_{n} e^{i\omega_n\delta}\left[\epsilon_{\mathbf{k}}{}^0 + \mu + \frac{1}{2}\Sigma(\mathbf{k}, \omega_n)\right]\bar{G}(\mathbf{k}, \omega_n) \tag{12.185}$$

where s is the spin and

$$\Omega(V, T, \mu) = \Omega_0(V, T, \mu) \mp \left(\frac{2s + 1}{2\beta}\right)\sum_{\mathbf{k}}\sum_{n}\int_0^1 \frac{d\lambda}{\lambda}\, e^{i\omega_n\delta}\, \Sigma(\mathbf{k}, \omega_n)\bar{G}(\mathbf{k}, \omega_n). \tag{12.186}$$

Thus, we have reduced the problem of finding the thermodynamic properties and excitation spectrum of a quantum fluid to that of computing the self-energy.

The poles of the exact propagator, $\bar{G}(\mathbf{k}, \omega_n)$, contain information about the exact excitations of the interacting system. Since these are the ones that contribute to physical processes in the interacting system, it is more correct to write all thermodynamic quantities in terms of exact propagators and not free propagators as we have done until this point. In so doing, we also remove the slow convergence due to polynomial dependence on β, and we obtain an expression for thermodynamic quantities in terms of the true physical processes occurring in the fluid. It is easy to write $\Sigma(\mathbf{k}, \omega_n)$ in terms of the exact propagators. We simply add together all "strings of beads" and obtain

$$\Sigma(\mathbf{k}, \omega_n) = \sum (\text{all different irreducible connected skeleton 1-diagrams}). \tag{12.187}$$

An irreducible connected skeleton 1-diagram cannot be cut into two parts by cutting two lines with the same momentum and frequency dependence. Thus, no "strings of beads" can appear in the skeleton diagrams, and all lines are now associated with exact propagators and not free propagators. The irreducible skeleton diagrams are given by

$\Sigma(\mathbf{k}, \omega_n) =$  $+$  $+ \cdots$ (12.188)

We cannot, however, make a simple replacement of free propagators by exact propagators in the diagrams for the grand potential because we overcount terms. A very clear discussion of how to obtain a diagrammatic expression of the grand potential in terms of exact propagators has been given by Luttenger and Ward.[15]

G. EXCITATIONS IN A WEAKLY COUPLED FERMI FLUID[5] AT LOW TEMPERATURE

To illustrate the use of the exact propagator, we shall obtain the excitation spectrum and heat capacity for a low-temperature weakly coupled Fermi fluid of spin $\frac{1}{2}$ particles. For a weakly coupled fluid we can approximate the self-energy by the lowest order diagram in Eq. (12.186) (this is called the Hartree-Fock approximation). For a system with a spherically symmetric spin independent interaction, we can write

$\Sigma_{\mathrm{HF}}(k, k'; \omega_n, T) =$  $= \delta_{\lambda,\lambda'}\, \delta_{\mathbf{k},\mathbf{k}'}\, \Sigma(\mathbf{k}, \omega_{n'}; T),$ (12.189)

where

$$\Sigma_{HF}(\mathbf{k}, \omega_n; T) = \Sigma_{HF}(\mathbf{k}, 0; T)$$

$$= \frac{1}{\beta}\frac{1}{V}\sum_{n'}\sum_{\mathbf{k}'}(2\bar{V}(0) - \bar{V}(\mathbf{k}-\mathbf{k}'))\, e^{i\omega_{n'}\delta}\bar{G}_{HF}(\mathbf{k}', \omega_{n'}; T). \tag{12.190}$$

The factor 2 in front of the direct interaction comes from the spin summation. The self-energy in the Hartree-Fock approximation is independent of the frequency. This makes the calculation of the excitation spectrum and of thermodynamic properties particularly simple. The exact propagator in the Hartree-Fock approximation is

$$\bar{G}_{HF}(\mathbf{k}, \omega_n; T) = \frac{1}{i\omega_n - \epsilon_{\mathbf{k}}^0 - \Sigma_{HF}(\mathbf{k}, 0; T)}. \tag{12.191}$$

Thus, we have a completely self-consistent expression for the self-energy.

The exact propagator, $\bar{G}_{HF}(\mathbf{k}, \omega_n; T)$, depends on temperature through both its dependence on ω_n and the dependence of the self-energy, $\Sigma_{HF}(\mathbf{k}, 0; T)$, on temperature. To obtain a low-temperature expression for the self-energy, we will expand $\bar{G}(\mathbf{k}, \omega_n; T)$ about its zero-temperature value. We first write

$$\bar{G}_{HF}(\mathbf{k}, \omega_n; T)^{-1} = \bar{G}_0(\mathbf{k}, \omega_n)^{-1} - \Sigma_{HF}(\mathbf{k}, 0; T) \tag{12.192}$$

and

$$\bar{G}_{HF}(\mathbf{k}, \omega_n; 0)^{-1} = \bar{G}_0(\mathbf{k}, \omega_n)^{-1} - \Sigma_{HF}(\mathbf{k}, 0; 0). \tag{12.193}$$

If we subtract Eq. (12.193) from Eq. (12.192) and rearrange terms, we obtain, to lowest order in temperature,

$$\bar{G}_{HF}(\mathbf{k}, \omega_n; T) \approx \bar{G}_{HF}(\mathbf{k}, \omega_n; 0) + \bar{G}_{HF}(\mathbf{k}, \omega_n; 0)$$

$$\times [\Sigma_{HF}(\mathbf{k}, 0; T) - \Sigma_{HF}(\mathbf{k}, 0; 0)]\, \bar{G}_{HF}(\mathbf{k}, \omega_n; 0) + \cdots, \tag{12.194}$$

where the temperature dependence of $\bar{G}_{HF}(\mathbf{k}, \omega_n; 0)$ enters only through its dependence on ω_n. Eqs. (12.190) and (12.194) yield the following low-temperature expression for the self-energy:

$$\Sigma_{HF}(\mathbf{k}, 0; T) = \frac{1}{\beta}\frac{1}{V}\sum_{n'}\sum_{\mathbf{k}'}(2\bar{V}(0) - \bar{V}(\mathbf{k}-\mathbf{k}'))\, e^{i\omega_{n'}\delta}\bar{G}_{HF}(\mathbf{k}', \omega_{n'}; 0)$$

$$+ \frac{1}{\beta}\frac{1}{V}\sum_{n'}\sum_{\mathbf{k}'}(2\bar{V}(0) - \bar{V}(\mathbf{k}-\mathbf{k}'))[\bar{G}_{HF}(\mathbf{k}', \omega_{n'}; 0)]^2\, e^{i\omega_{n'}\delta}$$

$$\times [\Sigma_{HF}(\mathbf{k}', 0; T) - \Sigma_{HF}(\mathbf{k}', 0; 0)]. \tag{12.195}$$

Let us next note that

$$\frac{1}{\beta}\sum_n e^{i\omega_n\delta}[\overline{G}_{\rm HF}(\mathbf{k}, \omega_n; 0)]^2 = \frac{\partial}{\partial E_{\mathbf{k}}}\left\{\frac{1}{\beta}\sum_n e^{i\omega_n\delta}[i\omega_n - (E_{\mathbf{k}} - \mu)]^{-1}\right\} = \frac{\partial N_{\mathbf{k}}(T)}{\partial E_{\mathbf{k}}} \tag{12.196}$$

where

$$E_{\mathbf{k}} = \frac{\hbar^2 k^2}{2m} + \Sigma_{\rm HF}(\mathbf{k}, 0; 0) \tag{12.197}$$

and

$$N_{\mathbf{k}}(T) = (e^{\beta(E_{\mathbf{k}}-\mu)} + 1)^{-1}. \tag{12.198}$$

The energy $E_{\mathbf{k}}$ is the exact excitation energy (in the Hartree-Fock approximation) at zero temperature and Eq. (12.197) is its dispersion relation. The self-energy now takes the form

$$\begin{aligned}\Sigma_{\rm HF}(\mathbf{k}, 0; T) &= \frac{1}{V}\sum_{\mathbf{k}'}(2\overline{V}(0) - \overline{V}(\mathbf{k} - \mathbf{k}'))N_{\mathbf{k}'}(T)\\ &\quad + \frac{1}{V}\sum_{\mathbf{k}'}(2\overline{V}(0) - \overline{V}(\mathbf{k} - \mathbf{k}'))\\ &\quad \times [\Sigma_{\rm HF}(\mathbf{k}', 0; T) - \Sigma_{\rm HF}(\mathbf{k}', 0; 0)]\frac{\partial N_{\mathbf{k}'}}{\partial E_{\mathbf{k}'}}.\end{aligned} \tag{12.199}$$

Thus, we have obtained an expression for the self-energy entirely in terms of the excitation energy, $E_{\mathbf{k}}$.

We next obtain an expression for the heat capacity, $C_{V,N}$. We first note that

$$\langle\hat{K}\rangle = \langle\hat{H}\rangle - \mu\langle\hat{N}\rangle = \Omega + TS \tag{12.200}$$

and

$$\left(\frac{\partial\langle K\rangle}{\partial T}\right)_{V,\mu} = \left(\frac{\partial\Omega}{\partial T}\right)_{V,\mu} + S + T\left(\frac{\partial S}{\partial T}\right)_{V,\mu} = T\left(\frac{\partial S}{\partial T}\right)_{V,\mu} \tag{12.201}$$

since $S = -(\partial\Omega/\partial T)_{V,\mu}$. Furthermore, we note that

$$\langle\hat{K}\rangle = \langle\hat{H}\rangle - \mu\langle\hat{N}\rangle = \frac{1}{\beta}\sum_{\mathbf{k}}\sum_n e^{i\omega_n\delta}\left[i\omega_n + \frac{\hbar^2k^2}{2m} - \mu\right]\overline{G}(\mathbf{k}, \omega_n; T). \tag{12.202}$$

The derivation of the heat capacity requires several steps. We first obtain an expression for $T(\partial S/\partial T)_{V,\mu}$. Then we integrate $(\partial S/\partial T)_{V,\mu}$ with respect to T holding V and μ fixed to obtain the entropy $S(T, V, \mu)$. (At most, we neglect a temperature-independent term in doing this. However, since the entropy must be zero at zero temperature, the temperature-independent term must be zero, so we are all right.) Finally, we express the chemical potential in terms of T, V, and $\langle N\rangle$ and find the entropy as a function of T, V, and $\langle N\rangle$. Once we have an expression for $S(V, T, N)$ it is an easy matter to obtain the heat capacity.

The low-temperature expression for $\langle K\rangle$ may be written

$$\langle K\rangle \equiv K(T, V, \mu) = \frac{1}{\beta}\sum_{\mathbf{k}}\sum_{n} e^{i\omega_n\delta}\left(i\omega_n + \frac{\hbar^2k^2}{2m} - \mu\right)\bar{G}_{\mathrm{HF}}(\mathbf{k}, \omega_n; 0)$$
$$+ \frac{1}{\beta}\sum_{\mathbf{k}}\sum_{n} e^{i\omega_n\delta}\left(i\omega_n + \frac{\hbar^2k^2}{2m} - \mu\right)[\bar{G}_{\mathrm{HF}}(\mathbf{k}, \omega_n; 0)]^2$$
$$\times [\Sigma_{\mathrm{HF}}(\mathbf{k}, 0; T) - \Sigma_{\mathrm{HF}}(\mathbf{k}, 0; 0)]. \tag{12.203}$$

However,

$$\frac{1}{\beta}\sum_{n} e^{i\omega_n\delta}\left(i\omega_n + \frac{\hbar^2k^2}{2m} - \mu\right)\bar{G}_{\mathrm{HF}}(\mathbf{k}, \omega_n; 0)$$
$$= [2(E_{\mathbf{k}} - \mu) - \Sigma_{\mathrm{HF}}(\mathbf{k}, 0, 0)]N_{\mathbf{k}}(T) \tag{12.204}$$

and

$$\frac{1}{\beta}\sum_{n} e^{i\omega_n\delta}\left(i\omega_n + \frac{\hbar^2k^2}{2m} - \mu\right)[\bar{G}_{\mathrm{HF}}(\mathbf{k}, \omega_n; 0)]^2$$
$$= N_{\mathbf{k}}(T) + [2(E_{\mathbf{k}} - \mu) - \Sigma_{\mathrm{HF}}(\mathbf{k}, 0; 0)]\frac{\partial N_{\mathbf{k}}(T)}{\partial E_{\mathbf{k}}}. \tag{12.205}$$

Therefore, at low temperature we can write

$$K(T, V, \mu) - K(0, V, \mu)$$
$$= \sum_{\mathbf{k}} [2(E_{\mathbf{k}} - \mu) - \Sigma_{\mathrm{HF}}(\mathbf{k}, 0; 0)]\,(N_{\mathbf{k}}(T) - N_{\mathbf{k}}(0))$$
$$+ \sum_{\mathbf{k}}\left[N_{\mathbf{k}}(0) + [2(E_{\mathbf{k}} - \mu) - \Sigma_{\mathrm{HF}}(\mathbf{k}, 0; 0)]\frac{\partial N_{\mathbf{k}}(0)}{\partial E_{\mathbf{k}}}\right]$$
$$\times [\Sigma_{\mathrm{HF}}(\mathbf{k}, 0; T) - \Sigma_{\mathrm{HF}}(\mathbf{k}, 0, 0)]. \tag{12.206}$$

Eq. (12.206) simplifies considerably if we note that $\partial N_{\mathbf{k}}(0)/\partial E_{\mathbf{k}} = -\delta(E_{\mathbf{k}} - \mu)$ and use the zero temperature limit of Eq. (12.199) and the relation

$$\Sigma_{\mathrm{HF}}(\mathbf{k}, 0; T) - \Sigma_{\mathrm{HF}}(\mathbf{k}, 0; 0)$$
$$= \frac{1}{V}\sum_{\mathbf{k}'} [2\bar{V}(0) - \bar{V}(\mathbf{k} - \mathbf{k}')]$$
$$\times \left\{N_{\mathbf{k}'}(T) - N_{\mathbf{k}'}(0) + \frac{\partial N_{\mathbf{k}'}(0)}{\partial E_{\mathbf{k}'}}[\Sigma_{\mathrm{HF}}(\mathbf{k}', 0; T) - \Sigma_{\mathrm{HF}}(\mathbf{k}', 0; 0)]\right\}. \tag{12.207}$$

Then we obtain

$$K(T, V, \mu) - K(0, V, \mu) = 2\sum_{\mathbf{k}} (E_{\mathbf{k}} - \mu)[N_{\mathbf{k}}(T) - N_{\mathbf{k}}(0)]. \quad (12.208)$$

Thus, the expectation value of $\hat{K}$ takes a very simple form.

We now obtain an expression for the entropy. From Eq. (12.208) we can write

$$\left(\frac{\partial K}{\partial T}\right)_{V,\mu} = T\left(\frac{\partial S}{\partial T}\right)_{V,\mu} = -2\beta \sum_{\mathbf{k}} (E_{\mathbf{k}} - \mu)^2 \frac{e^{\beta(E_{\mathbf{k}}-\mu)}}{(e^{\beta(E_{\mathbf{k}}-\mu)} + 1)^2}. \quad (12.209)$$

The integral in Eq. (12.209) is of the same type that we encountered for an ideal Fermi gas. The argument is sharply peaked about $E_{\mathbf{k}} - \mu = 0$. If we change the summation to an integration ($\sum_{\mathbf{k}} \rightarrow V/(2\pi)^3 \int d\mathbf{k}$) and make the change of variables $t = \beta(E_{\mathbf{k}} - \mu)$, we can integrate (cf. Sec. 9.D) and obtain

$$T\left(\frac{\partial S}{\partial T}\right)_{V,\mu} = \tfrac{1}{3}Vk_B{}^2T\left[k^2 \frac{dk}{dE_{\mathbf{k}}}\right]_{E_{\mathbf{k}}=\mu}. \quad (12.210)$$

Thus, the entropy is given by

$$S(T, V, \mu) = \tfrac{1}{3}Vk_B{}^2T\left[k^2 \frac{dk}{dE_{\mathbf{k}}}\right]_{E_{\mathbf{k}}=\mu}. \quad (12.211)$$

To find the heat capacity, $C_{N,V}$, we must write Eq. (12.211) in terms of the average particle number. We first note that the Fermi momentum, k_f, at $T = 0$ K is defined through the relation

$$\langle N \rangle = 2\frac{V}{(2\pi)^3}\int dk\,\theta(\mu(0) - E_{\mathbf{k}}) = \frac{Vk_f{}^3}{3\pi^2} \quad (12.212)$$

($\theta(\mu(0) - E_{\mathbf{k}})$ is the Heaviside function). At zero temperature the Fermi momentum and the chemical potential are related through the equation

$$\mu(0) = E_{\mathbf{k}}\Big|_{k=k_f} = \frac{\hbar^2 k_f{}^2}{2m} + \Sigma_{\mathrm{HF}}(k_f, 0; 0). \quad (12.213)$$

The derivative of the excitation energy at the Fermi surface is

$$\frac{\partial E_{\mathbf{k}}}{\partial k}\Big|_{k_f} = \frac{\hbar^2 k_f}{m} + \frac{\partial \Sigma_{\mathrm{HF}}(\mathbf{k}, 0; 0)}{\partial k}\Big|_{k=k_f} \equiv \frac{\hbar^2 k_f}{m^*}, \quad (12.214)$$

where m^* is, by definition, the effective mass of excitations at the Fermi surface:

$$\frac{1}{m^*} = \frac{1}{m} + \frac{1}{\hbar^2 k_f}\frac{\partial \Sigma_{\mathrm{HF}}(\mathbf{k}, 0; 0)}{\partial k}\Big|_{k=k_f}. \quad (12.215)$$

The excitation energy near the Fermi surface may be expressed in terms of a Taylor expansion about the Fermi surface and has the form

$$E_{\mathbf{k}} = E_{k_f} + \frac{\hbar^2 k_f}{m^*}(k - k_f) + \cdots \tag{12.216}$$

It is the same as the expansion of the excitation energy of an ideal gas about the Fermi surface except the particle mass is replaced by the effective mass.

If we combine Eqs. (12.211), (12.212), and (12.214), we obtain for the heat capacity

$$C_{V,N} = T\left(\frac{\partial S}{\partial T}\right)_{V,N} = \langle N\rangle k_B{}^2 T \frac{m^* \pi^2}{\hbar^2 k_f{}^2} + \cdots \tag{12.219}$$

Thus, both the excitations and the heat capacity of an interacting Fermi fluid look very much like those of an ideal Fermi gas except that they depend on an effective mass which itself is dependent on the form of the interaction potential.

This result is not restricted to weakly interacting fluids. Galitskii[16] has shown that this behavior persists even for hard-sphere particles. These excitations in a Fermi fluid were first called quasiparticles by Landau,[17] who based his very successful phenomenological theory of Fermi liquids on the existence of this type of excitation. Indeed, the properties of liquid He^3, which is a very dense, strongly coupled system of fermions, are described very well by assuming that the important excitations in the liquid are quasiparticles. Such a picture is even used to describe the superfluid phases of liquid He^3.[18]

H. WEAKLY COUPLED CONDENSED BOSON FLUID AT ZERO TEMPERATURE[5,19,20]

It is possible, using the methods of this chapter, to obtain the zero-temperature excitation spectrum for a weakly coupled boson fluid which has undergone a transition to a superfluid phase (a phase in which the zero-momentum state is macroscopically occupied). The order parameter for the condensed phase is the density of particles, n_0, in the zero-momentum state. The approach we will use is due to Hugenholtz and Pines.[20] We assume that the system is in the condensed phase (assume order parameter n_0 exists), compute various thermodynamic quantities, and then determine if, for nonzero n_0, the resulting expressions lead to a stable thermodynamic state.

1. Exact Propagator

If the zero-momentum state is macroscopically occupied (this means that ODLRO has set in [cf. Sec. 7.F]), the expectation value,

$$N_0 = \langle \hat{a}_0{}^+ \hat{a}_0 \rangle, \tag{12.218}$$

will be proportional to the average number of particles, $\langle N \rangle$, in the system. N_0 is the average occupation number of the zero-momentum state for the interacting system. The expectation values of $\hat{a}_0^+$ and $\hat{a}_0$ will be of order $(N_0 + 1)^{1/2}$ and $(N_0)^{1/2}$, respectively (cf. App. B). If N_0 is a large number, then the difference between $\hat{a}_0^+$ and $\hat{a}_0$ becomes negligible and we can treat them as numbers rather than operators. Thus, we shall make the following replacement. We let

$$\hat{a}_{\mathbf{k}}^+ = \sqrt{N_0} + \hat{\alpha}_{\mathbf{k}}^+ \tag{12.219}$$

and

$$\hat{a}_{\mathbf{k}} = \sqrt{N_0} + \hat{\alpha}_{\mathbf{k}} \tag{12.220}$$

where $\hat{\alpha}_{\mathbf{k}}^+$ and $\hat{\alpha}_{\mathbf{k}}$ are creation and annihilation operators for particles with nonzero momentum. If we limit ourselves to spherically symmetric potentials, the operator $\hat{K}$ takes the form

$$\hat{K} = \hat{K}_0 + \hat{K}_1, \tag{12.221}$$

where

$$\hat{K}_0 = E_0 - \mu N_0 + \hat{K}_0', \tag{12.222}$$

$$E_0 = \tfrac{1}{2} N_0^2 \langle 0, 0 | V | 0, 0 \rangle, \tag{12.223}$$

and

$$\hat{K}_0' = \sum_{\mathbf{k}} \left(\frac{\hbar^2 k^2}{2m} - \mu \right) \hat{\alpha}_{\mathbf{k}}^+ \hat{\alpha}_{\mathbf{k}}. \tag{12.224}$$

The interaction term $\hat{K}_1$ can be written

$$\hat{K}_1 = \sum_{i=1}^{6} \hat{V}_i, \tag{12.225}$$

where

$$\hat{V}_1 = \tfrac{1}{4} N_0 \sum_{\mathbf{k}_1, \mathbf{k}_2} \langle 0, 0 | V | \mathbf{k}_1, \mathbf{k}_2 \rangle^{(+)} \hat{\alpha}_{\mathbf{k}_1} \hat{\alpha}_{\mathbf{k}_2}, \tag{12.226}$$

$$\hat{V}_2 = \tfrac{1}{4} N_0 \sum_{\mathbf{k}_1, \mathbf{k}_2} \langle \mathbf{k}_1, \mathbf{k}_2 | V | 0, 0 \rangle^{(+)} \hat{\alpha}_{\mathbf{k}_1}^+ \hat{\alpha}_{\mathbf{k}_2}^+, \tag{12.227}$$

$$\hat{V}_3 = N_0 \sum_{\mathbf{k}_1, \mathbf{k}_2} \langle \mathbf{k}_1, 0 | V | \mathbf{k}_2, 0 \rangle^{(+)} \hat{\alpha}_{\mathbf{k}_1}^+ \hat{\alpha}_{\mathbf{k}_2}, \tag{12.228}$$

$$\hat{V}_4 = \tfrac{1}{2} \sqrt{N_0} \sum_{\mathbf{k}_1, \mathbf{k}_2, \mathbf{k}_3} \langle \mathbf{k}_1, \mathbf{k}_2 | V | \mathbf{k}_3, 0 \rangle^{(+)} \hat{\alpha}_{\mathbf{k}_1}^+ \hat{\alpha}_{\mathbf{k}_2}^+ \hat{\alpha}_{\mathbf{k}_3}, \tag{12.229}$$

$$\hat{V}_5 = \tfrac{1}{2} \sqrt{N_0} \sum_{\mathbf{k}_1, \mathbf{k}_2, \mathbf{k}_3} \langle 0, \mathbf{k}_1 | V | \mathbf{k}_2, \mathbf{k}_3 \rangle^{(+)} \hat{\alpha}_{\mathbf{k}_1}^+ \hat{\alpha}_{\mathbf{k}_2}^+ \hat{\alpha}_{\mathbf{k}_3}, \tag{12.230}$$

and

$$\hat{V}_6 = \frac{1}{4} \sum_{\mathbf{k}_1, \ldots, \mathbf{k}_4} \langle \mathbf{k}_1, \mathbf{k}_2 | V | \mathbf{k}_3, \mathbf{k}_4 \rangle^{(+)} \hat{\alpha}_{\mathbf{k}_1}^{+} \hat{\alpha}_{\mathbf{k}_2}^{+} \hat{\alpha}_{\mathbf{k}_4} \hat{\alpha}_{\mathbf{k}_3}. \quad (12.231)$$

The terms depending on $\langle 0, 0 | V | \mathbf{k}_1, 0 \rangle^{(+)}$ and $\langle \mathbf{k}_1, 0 | V | 0, 0 \rangle^{(+)}$ are zero because of momentum conservation. The average number of particles in the system is given by

$$\langle N \rangle = N_0 + \sum_{\mathbf{k}} \langle \hat{\alpha}_{\mathbf{k}}{}^{+} \hat{\alpha}_{\mathbf{k}} \rangle. \quad (12.232)$$

The order parameter N_0 is determined by the condition that the grand potential be an extremum with respect to variations in N_0 (the thermodynamically stable value of N_0 yields a minimum for $\Omega(V, T, \mu)$). Thus,

$$\left(\frac{\partial \Omega}{\partial N_0} \right)_{T,\mu,V} = 0. \quad (12.233)$$

Since we wish to find the excitation spectrum, we must study the full propagator, which is defined

$$G(\mathbf{k}, \mathbf{k}', \tau - \tau') = N_0 + G'(\mathbf{k}, \mathbf{k}'; \tau - \tau') \quad (12.234)$$

where

$$G'(\mathbf{k}, \mathbf{k}'; \tau - \tau') = \sum_{l=0}^{\infty} \frac{(-1)^l}{l!} \int_0^{\beta} d\tau_1 \times \cdots$$
$$\times \int_0^{\beta} d\tau_l \, \mathrm{Tr}\{\hat{\rho}_0 T[\hat{K}_1(\tau_1)_0 \times \cdots \times \hat{K}_1(\tau_l)_0 \hat{\alpha}_{\mathbf{k}}(\tau) \hat{\alpha}_{\mathbf{k}'}^{+}(\tau')]\}$$
$$(12.235)$$

and $\hat{\rho}_0 = e^{-\beta \hat{K}_0}/\mathrm{Tr}\, e^{-\beta \hat{K}_0}$. Since $\hat{K}_0$ is quadratic in the operators, $\hat{\alpha}_{\mathbf{k}}$ and $\hat{\alpha}_{\mathbf{k}}{}^{+}$, we can use Wick's theorem to evaluate the trace. We find that $G'(\mathbf{k}, \mathbf{k}'; \tau - \tau')$ depends only on fully connected contractions, just as for the case of non-condensed systems. We can Fourier transform the temperature dependence of the propagator and write it in terms of frequencies, and we can write an expression for the Fourier-transformed propagator in terms of diagrams as follows:

$$\bar{G}'(\mathbf{k}, \mathbf{k}'; \omega_n) = \sum (\text{all different connected } 1'\text{-diagrams}). \quad (12.236)$$

A connected 1′-diagram is a collection of vertices which may have directed dashed line attachments but which are entirely interconnected by directed solid lines. At least two of the lines attached to the vertex must be solid lines. Dashed lines always have one free end (it may be the head or tail). One solid line must enter and one solid line must leave each diagram. Two diagrams are different if they have different topological structures. Algebraic expressions may be

Table 12.3. Rules for diagrams (condensed boson fluid)

Rules for Frequency-Dependent 1′*-Diagrams*

(i) Label the internal solid lines from 1 to l, where l is the number of internal solid lines, and assign a momentum $\mathbf{k}_i$ and integer n_i to the ith line. Label the outgoing external line by $\mathbf{k}$ and the incoming external line by $\mathbf{k}'$.

(ii) To each solid line assign a factor,

$\mathbf{k}, n$  $$= \frac{1}{\beta} G_0(\mathbf{k}, \omega_n) = \frac{1}{\beta} \frac{1}{i\omega_n - \epsilon_{\mathbf{k}}^0}.$$

where $\omega_n = 2\pi n/\beta$.

(iii) To each dashed line assign a factor,

 $$= \sqrt{N_0}.$$

(iv) To the various vertices assign factors,

$$= -\langle 0, 0|V|\mathbf{k}_1, \mathbf{k}_2\rangle^{(+)}\beta\, \delta_{n_1+n_2,0}.$$

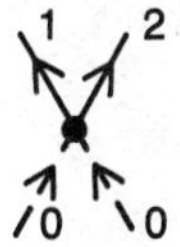

$$= -\langle \mathbf{k}_1, \mathbf{k}_2|V|0, 0\rangle^{(+)}\beta\, \delta_{n_1+n_2,0}.$$

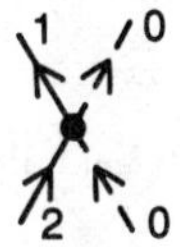

$$= -\langle \mathbf{k}_1, 0|V|\mathbf{k}_2, 0\rangle^{(+)}\beta\, \delta_{n_1,n_2}.$$

$$= -\langle \mathbf{k}_1, 0|V|\mathbf{k}_2, \mathbf{k}_3\rangle^{(+)}\beta\, \delta_{n_1,n_2+n_3}.$$

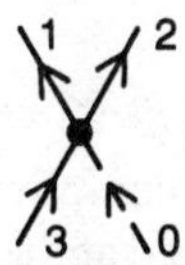

$$= -\langle \mathbf{k}_1, \mathbf{k}_2|V|0, \mathbf{k}_3\rangle^{(+)}\beta\, \delta_{n_1+n_2,n_3}$$

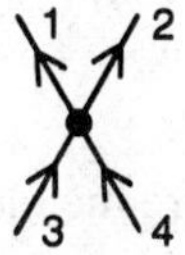

$$= -\langle \mathbf{k}_1, \mathbf{k}_2|V|\mathbf{k}_3, \mathbf{k}_4\rangle^{(+)}\beta\, \delta_{n_1+n_2,n_3+n_4}$$

(v) Assign a factor S^{-1} to the diagram, where S is the symmetry number.

(vi) Sum over all internal momenta and sum over all integers.

---

associated with 1-diagrams according to the rules in Table 12.3. To second order, $G(\mathbf{k}, \mathbf{k}'; \omega_n)$ can be expressed diagrammatically as follows:

$\bar{G}'(\mathbf{k}, \mathbf{k}'; \omega_n) =$ 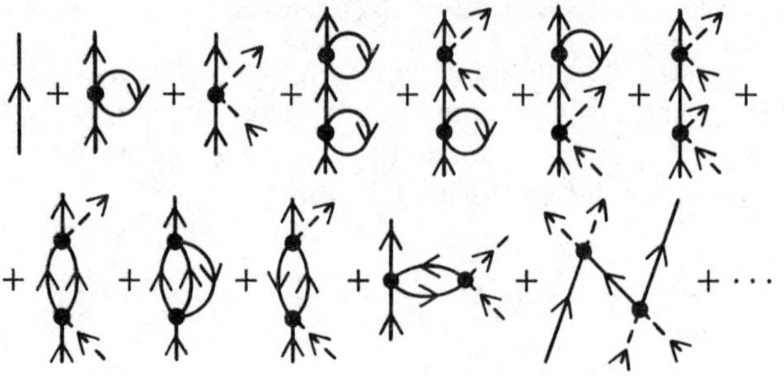

(12.237)

The dashed lines in Eq. (12.237) have zero momentum associated to them. The number of excitations at each vertex is not conserved. Indeed, it is possible to think of the condensate as an external field which can absorb and emit excitations. (Note that any groups of vertices with only one solid line leaving or one solid line entering are not allowed because of momentum conservation.) Algebraic expressions for two of the 1′-diagrams may be written as follows:

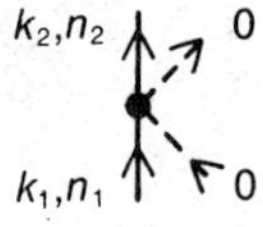

$$= -\beta\, \delta_{n_1,n_2} \langle \mathbf{k}_2, 0|V|\mathbf{k}_1, 0\rangle N_0 \bar{G}_0(\mathbf{k}_1, \omega_{n_1}) \bar{G}_0(\mathbf{k}_2, \omega_{n_2})$$

(12.238)

and

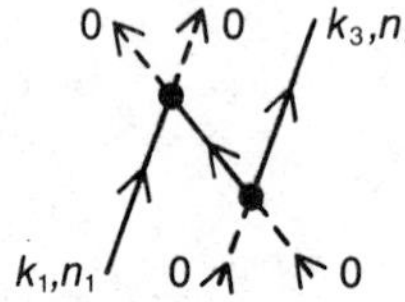

$$= \frac{\beta N_0^2}{4} \sum_{k_2 n_2} \langle 0, 0|V|\mathbf{k}_1, \mathbf{k}_2\rangle \langle \mathbf{k}_2, \mathbf{k}_3|V|0, 0\rangle \times \bar{G}_0(\mathbf{k}_1, \omega_{n_1}) \bar{G}_0(\mathbf{k}_2, \omega_{n_2}) \bar{G}_0(\mathbf{k}_3, \omega_n). \quad (12.239)$$

We can now write Dyson's equations for the system.

2. Dyson's Equations

Dyson's equations for the condensed phase are more complicated than for the noncondensed phase because particles are not conserved at the vertices. For the condensed phase, we must introduce three self-energies:

$\Sigma_{11}(\mathbf{k}, \mathbf{k}'; \omega_n, \omega_{n'}) =$ 

$= \sum$ (all different irreducible connected 1′-diagrams with one incoming and one outgoing solid line), (12.240)

$\Sigma_{21}(\mathbf{k}, \mathbf{k}'; \omega_n, \omega_{n'}) =$ 

$= \sum$ (all different irreducible connected 1′-diagrams with two incoming solid lines), (12.241)

and

$\Sigma_{12}(\mathbf{k}, \mathbf{k}'; \omega_n, \omega_{n'}) =$ 

$= \sum$ (all different irreducible connected 1′-diagrams with two outgoing solid lines). (12.242)

Because of momentum conservation during the interactions, we can write

$$\Sigma_{11}(\mathbf{k}, \mathbf{k}'; \omega_n, \omega_{n'}) = \delta_{\mathbf{k},\mathbf{k}'}\, \delta_{n,n'}\, \Sigma_{11}(\mathbf{k}, \omega_n), \tag{12.243}$$

$$\Sigma_{21}(\mathbf{k}, \mathbf{k}'; \omega_n, \omega_{n'}) = \delta_{\mathbf{k}+\mathbf{k}',0}\, \delta_{n+n',0}\, \Sigma_{21}(\mathbf{k}, \omega_n), \tag{12.244}$$

and

$$\Sigma_{12}(\mathbf{k}, \mathbf{k}'; \omega_n, \omega_{n'}) = \delta_{\mathbf{k}+\mathbf{k}',0}\, \delta_{n+n',0}\, \Sigma_{12}(\mathbf{k}, \omega_n). \tag{12.245}$$

Some examples of 1-diagrams which contribute to these self-energies are

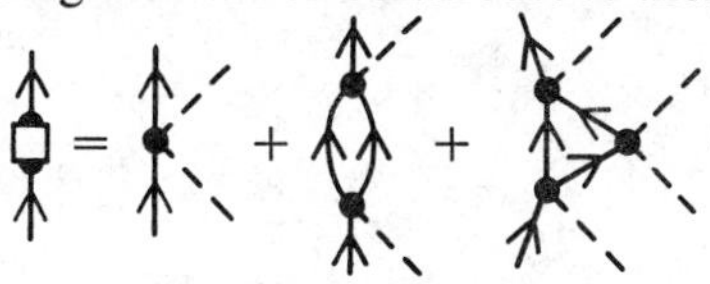
(12.246)

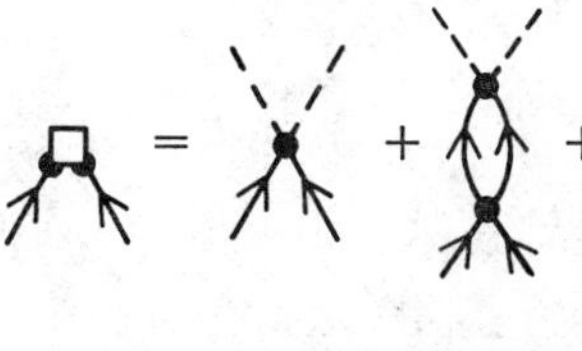
(12.247)

and

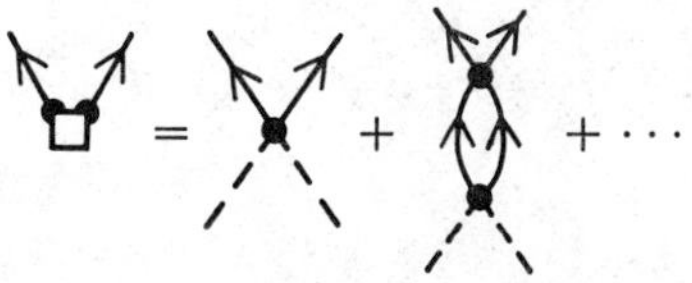
(12.248)

Now that we have introduced new self-energies, we must also introduce new propagators. Thus, we write the full propagator as

$= \bar{G}(\mathbf{k}, \mathbf{k}'; \omega_n, \omega_{n'}) = \delta_{\mathbf{k},\mathbf{k}'}\, \delta_{n,n'} \bar{G}(\mathbf{k}, \omega_n)$ (12.249)

and write two anomalous propagators as

$= \bar{G}_{21}(\mathbf{k}, \mathbf{k}'; \omega_n, \omega_{n'}) = \delta_{\mathbf{k}+\mathbf{k}',0}\, \delta_{n+n',0} \bar{G}_{21}(\mathbf{k}, \omega_n)$

(12.250)

and

$$\mathbf{k}, n$$  $$\mathbf{k}', n = \bar{G}_{12}(\mathbf{k}, \mathbf{k}'; \omega_n, \omega_{n'}) = \delta_{\mathbf{k}+\mathbf{k}',0}\, \delta_{n+n',0} \bar{G}_{12}(\mathbf{k}, \omega_n). \tag{12.251}$$

A careful study of the basic diagrams shows that the full propagators obey the following equations:

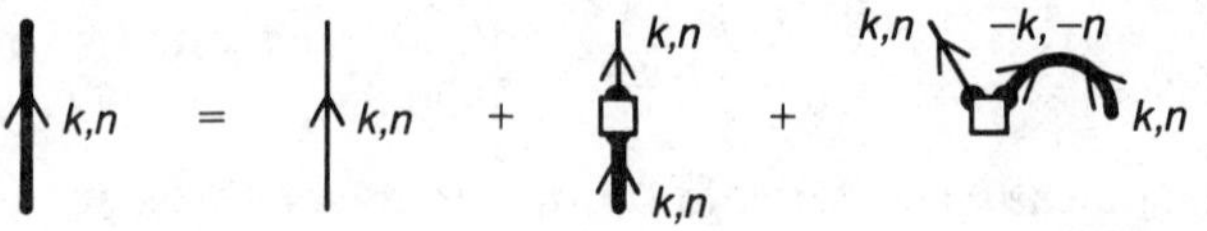

(12.252)

where

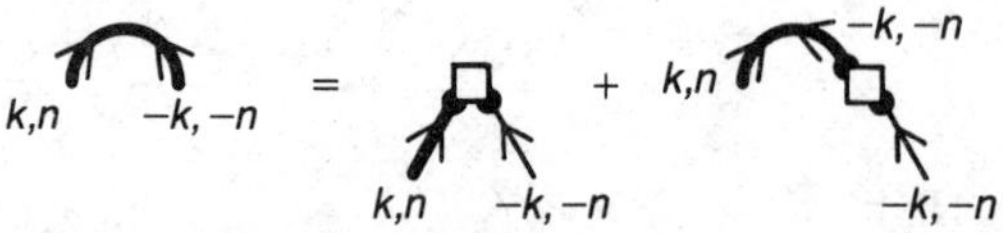

(12.253)

and

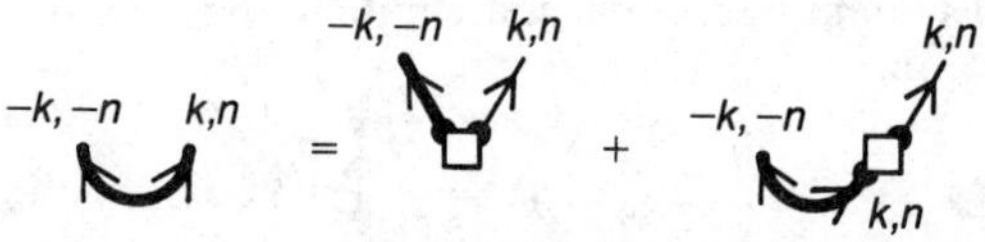

(12.254)

The algebraic expressions associated with Eqs. (12.252)–(12.254) are, respectively,

$$\begin{aligned}\bar{G}(\mathbf{k}, \omega_n) &= \bar{G}_0(\mathbf{k}, \omega_n) + \bar{G}_0(\mathbf{k}, \omega_n)\, \Sigma_{11}\,(\mathbf{k}, \omega_n)\bar{G}(\mathbf{k}, \omega_n) \\ &\quad + \bar{G}_0(\mathbf{k}, \omega_n)\, \Sigma_{12}\,(\mathbf{k}, \omega_n)\bar{G}_{12}(\mathbf{k}, \omega_n),\end{aligned} \tag{12.255}$$

$$\begin{aligned}\bar{G}_{21}(\mathbf{k}, \omega_n) &= \bar{G}_0(-\mathbf{k}, \omega_{-n})\, \Sigma_{21}\,(\mathbf{k}, \omega_n)\bar{G}(\mathbf{k}, \omega_n) \\ &\quad + \bar{G}_0(-\mathbf{k}, \omega_{-n})\, \Sigma_{11}\,(-\mathbf{k}, \omega_{-n})\bar{G}_{21}(\mathbf{k}, \omega_n),\end{aligned} \tag{12.256}$$

and

$$\begin{aligned}\bar{G}_{12}(\mathbf{k}, \omega_n) &= \bar{G}_0(\mathbf{k}, \omega_n)\, \Sigma_{12}\,(\mathbf{k}, \omega_n)\bar{G}(-\mathbf{k}, \omega_{-n}) \\ &\quad + \bar{G}_0(\mathbf{k}, \omega_n)\, \Sigma_{11}\,(\mathbf{k}, \omega_n)\bar{G}_{12}(\mathbf{k}, \omega_n).\end{aligned} \tag{12.257}$$

Eqs. (12.255)–(12.257) may be solved to give

$$\begin{aligned}\bar{G}(\mathbf{k}, \omega_n) = {}& \{[\bar{G}_0(\mathbf{k}, \omega_n)]^{-1} - \Sigma_{11}\,(\mathbf{k}, \omega_n)\} \\ & \times \{[\bar{G}_0(\mathbf{k}, \omega_n)]^{-1}([\bar{G}_0(-\mathbf{k}, \omega_{-n})]^{-1} - \Sigma_{11}\,(-\mathbf{k}, \omega_{-n})) \\ & - ([\bar{G}_0(-\mathbf{k}, \omega_{-n})]^{-1} - \Sigma_{11}\,(-\mathbf{k}, \omega_{-n}))\, \Sigma_{11}\,(\mathbf{k}, \omega_n) \\ & - \Sigma_{12}\,(\mathbf{k}, \omega_n)\, \Sigma_{21}\,(\mathbf{k}, \omega_n)\}^{-1},\end{aligned} \tag{12.258}$$

$$\bar{G}_{21}(\mathbf{k}, \omega_n) = \Sigma_{21}(\mathbf{k}, \omega_n)\bar{G}(\mathbf{k}, \omega_n)\{[\bar{G}_0(-\mathbf{k}, \omega_{-n})]^{-1} - \Sigma_{11}(-\mathbf{k}, \omega_{-n})\}^{-1}, \tag{12.259}$$

and

$$\bar{G}_{12}(\mathbf{k}, \omega_n) = \Sigma_{12}(\mathbf{k}, \omega_n)\bar{G}(-\mathbf{k}, \omega_{-n})\{[\bar{G}_0(\mathbf{k}, \omega_n)]^{-1} - \Sigma_{11}(\mathbf{k}, \omega_n)\}^{-1}. \tag{12.260}$$

Eqs. (12.258)–(12.260) give a completely general expression for the propagators in a condensed boson fluid. Before we obtain the dispersion relation for excitations in the condensed boson fluid, it is necessary to obtain an expression for the chemical potential.

3. Chemical Potential

We shall restrict ourselves to an expression for the chemical potential of a condensed boson fluid at zero temperature. At zero temperature, we can write

$$\Omega(0, V, \mu) = \langle H\rangle_{T=0} - \mu\langle N\rangle_{T=0}. \tag{12.261}$$

Thus, the condition that $\Omega(T, V, \mu)$ be an extremum with respect to variations in N_0 (cf. Eq. (12.233)) leads to the equation

$$\mu(T = 0) = \frac{\partial\langle V\rangle_{T=0}}{\partial N_0}. \tag{12.262}$$

Following the usual methods, we can write the expectation value, $\langle\hat{V}\rangle$, as

$$\langle\hat{V}\rangle = \sum_{m=0}^{\infty}\frac{1}{m!}\int_0^{\beta} d\tau_1 \times \cdots \times \int_0^{\beta} d\tau_m \times \mathrm{Tr}\{e^{-\beta K_0}T[\hat{V}(\tau_1) \times \cdots \times \hat{V}(\tau_m)\hat{V}]_{\mathrm{conn}}\}. \tag{12.263}$$

Eq. (12.263) can be evaluated by inserting the potential

$$\hat{V} = E_0 + \sum_{i=1}^{6}\hat{V}_i \tag{12.264}$$

and then taking all possible contractions. The result is similar in structure to the expressions for the self-energies. Thus, Eq. (12.262) enables us to relate the chemical potential to the self-energy. Hugenholtz and Pines obtained the following result at zero temperature:

$$\mu = \left(\frac{\partial\langle V\rangle}{\partial N_0}\right)_{V,\mu} = \Sigma_{11}(\mathbf{k} = 0, \omega_n)_{T=0\,\mathrm{K}} - \Sigma_{12}(\mathbf{k} = 0, \omega_n)_{T=0\,\mathrm{K}}. \tag{12.265}$$

(It is a worthwhile exercise to derive this result.)

4. Excitations

For a dilute boson fluid, the self-energy $\Sigma_{11}(\mathbf{k}, W_n)$ to lowest order in λ is given by the diagrams

$$\Sigma_{11}(\mathbf{k}, \omega_n) =$$ 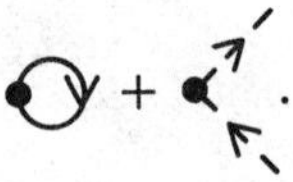 (12.266)

Let us approximate the self-energies by their value at zero temperature. If we note that $n_{\mathbf{k}}^0(T = 0) \equiv 0$, the first diagram is identically zero and we obtain

$$\Sigma_{11}(\mathbf{k}, \omega_n)_{T=0} =$$  $$= N_0[\bar{V}(0) + \bar{V}(\mathbf{k})], \tag{12.267}$$

$$\Sigma_{12}(\mathbf{k}, \omega_n)_{T=0} =$$  $$= N_0\bar{V}(\mathbf{k}), \tag{12.268}$$

and

$$\Sigma_{21}(\mathbf{k}, \omega_n)_{T=0} =$$  $$= N_0\bar{V}(\mathbf{k}). \tag{12.269}$$

From Eq. (12.265), the chemical potential to lowest order in the interaction is

$$\mu = N_0\bar{V}(0). \tag{12.270}$$

It is now a simple matter to compute the full propagators. We find

$$\bar{G}(\mathbf{k}, \omega_n) = \frac{\omega_n + [\epsilon_{\mathbf{k}}^0 + N_0\bar{V}(\mathbf{k})]}{\omega_n^2 - (E_{\mathbf{k}})^2} \tag{12.271}$$

and

$$\bar{G}_{12}(\mathbf{k}, \omega_n)_{T=0\,\mathrm{K}} = \bar{G}_{21}(\mathbf{k}, \omega_n)_{T=0\,\mathrm{K}} = \frac{N_0\bar{V}(\mathbf{k})}{\omega_n^2 - (E_{\mathbf{k}})^2} \tag{12.272}$$

where $E_{\mathbf{k}}$ is the energy of excitations and is given by

$$E_{\mathbf{k}} = \left[\left(\frac{\hbar^2 k^2}{2m}\right)^2 + \frac{\hbar^2 k^2}{m} N_0\bar{V}(\mathbf{k})\right]^{1/2}. \tag{12.273}$$

In the long-wavelength limit, the excitation energy becomes

$$E_{\mathbf{k}} \approx \hbar|\mathbf{k}|\left[\frac{N_0\bar{V}(0)}{m}\right]^{1/2} + \cdots \tag{12.274}$$

Thus, the long-wavelength excitations of a condensed weakly interacting boson fluid are phonons with speed:

$$c = \left[\frac{N_0\bar{V}(0)}{m}\right]^{1/2}. \tag{12.275}$$

While we have derived this result for a weakly coupled boson fluid, it has been shown by Beliaev[19] that a similar dispersion relation also holds for a hard-sphere boson fluid. This type of low momentum dispersion relation has been observed in liquid He^4. In Fig. 12.6, we show the energy of excitations in liquid He^4 as a function of wave vector. These results are obtained from neutron scattering experiments and indicate the existence of phonons in the liquid at very low wave vector.

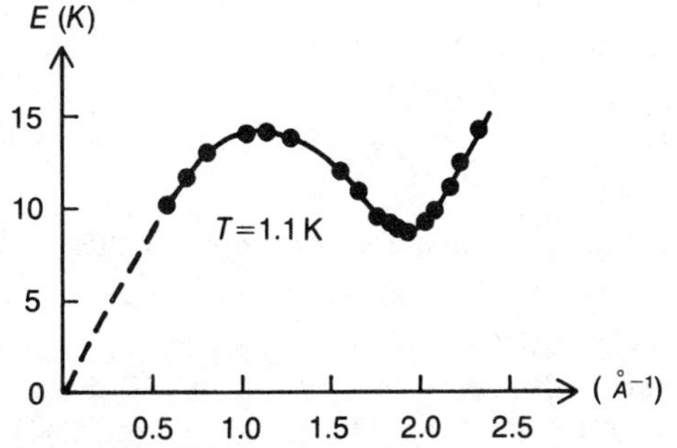

Fig. 12.6. Energy of excitations in liquid He^4 as measured by neutron scattering experiments at $T = 1.1$ K: the slope of extrapolated curve at $K = 0$ Å^{-1} is 239 ± 5 m/sec, in good agreement with direct measurements on the speed of first sound (based on Ref. 12.21).

REFERENCES

1. L. E. Reichl, J. Math. Phys. *17* 2007, 2023, 2034 (1976).
2. G. C. Wick, Phys. Rev. *80* 2681 (1950).
3. T. Matsubara, Prog. Theor. Phys. (Kyoto) *14* 351 (1955).
4. M. Gaudin, Nucl. Phys. *15* 89 (1960).
5. A. L. Fetter and J. D. Walecka, *Quantum Theory of Many-Particle Systems* (McGraw-Hill Book Co., New York, 1971). (This excellent book has been especially useful for this chapter.)
6. A. A. Abrikosov, L. P. Gorkov, and I. E. Dzyaloshinski, *Methods of Quantum Field Theory in Statistical Physics* (Prentice-Hall, Englewood Cliffs, N.J., 1963).
7. R. D. Mattuck, *A Guide to Feynmann Diagrams in the Many-Body Problem* (McGraw-Hill Book Co., New York, 1976).
8. E. W. Montroll and J. C. Ward, Phys. Fluids *1* 55 (1958).
9. M. Gell-Mann and K. A. Brueckner, Phys. Rev. *106* 364 (1957).
10. W. Kohn and J. M. Luttinger, Phys. Rev. *118* 41 (1960).
11. P. Debye and E. Hückel, Z. Physik *24* 185 (1923).
12. E. P. Wigner, Trans. Farad. Soc. *34* 678 (1938).
13. W. J. Carr, Phys. Rev. *122* 1437 (1961).
14. F. J. Dyson, Phys. Rev. *75* 486, 1736 (1949).

15. J. M. Luttenger and J. C. Ward, Phys. Rev. *118* 1417 (1960).
16. V. M. Galitskii, Sov. Phys. J.E.T.P. *7* 104 (1958).
17. L. D. Landau, Soc. Phys. J.E.T.P. *3* 920 (1957); *5* 101 (1957); *8* 70 (1959).
18. D. Rainer and J. W. Serene, Phys. Rev. *B13* 4745 (1976).
19. S. T. Beliaev, Sov. Phys. J.E.T.P. *7* 289, 299 (1958).
20. N. M. Hugenholtz and D. Pines, Phys. Rev. *116* 489 (1959).
21. R. J. Donnelly, *Experimental Superfluidity* (University of Chicago Press, Chicago, 1967).

PROBLEMS

1. Show that $\int d\mathbf{k}\theta(\gamma - |\mathbf{k} + \frac{1}{2}\mathbf{Q}|)\theta(\gamma - |\mathbf{k} - \frac{1}{2}\mathbf{Q}|) = (4\pi/3)\gamma^3(1 - \frac{3}{2}x + \frac{1}{2}x^3) \times \theta(1 - x)$ where $x = Q/2\gamma$ (cf. Eq. (12.96)).

2. For a normal boson or fermion fluid, write down all different diagrams that can contribute to $\bar{G}(k, k'; \omega_n)$ to third order in λ.

3. If a magnetic field, $\mathscr{H}(\mathbf{r})$, is applied to a spin $\frac{1}{2}$ Fermi fluid, a magnetic energy term $\Delta\hat{H} = -\int d\mathbf{r}\mathbf{M}(r)\cdot\mathscr{H}(\mathbf{r})$ must be added to the Hamiltonian, where $\mathbf{M}(\mathbf{r})$ is the magnetization operator $\hat{M}(\mathbf{r}) = (\mu/2)\hat{S}(\mathbf{r})$ (cf. App. B, Eq. (B.102)) and μ is the magnetic moment of each particle. For simplicity, assume that the particles are neutral but interact via a fairly long range interaction. Assume, furthermore, that $\mathscr{H}(\mathbf{r}) = \mathscr{H}_0(\mathbf{r})\hat{\mathbf{z}}$ and that the system is paramagnetic. Obtain an expression for the linear response function, retaining only the contribution from the direct interaction in the polarization diagrams. Include all polarization diagrams. Your expression will include the propagators $\chi_0(\mathbf{Q}, \omega_n)$. What is the physical meaning of Q?

4. It is often useful to treat a neutral Fermi fluid as if the particles interacted via a contact potential, $V(\mathbf{r}) = I\,\delta(\mathbf{r})$, where $\delta(\mathbf{r})$ is the Dirac delta function. This potential is extremely short ranged. When it is used, both the direct and the exchange terms are important and must be retained. This model is often used to treat systems of spin $\frac{1}{2}$ fermions which have enhanced magnetization (such as a system near its Curie point, where spin fluctuations are important). The dominant contribution to the thermodynamic properties in such systems comes from the polarization diagrams. Use this model and the polarization diagrams to compute the self-energy of the system. Assume that the system is paramagnetic, so that the distribution functions and chemical potentials of spin $+\frac{1}{2}$ and spin $-\frac{1}{2}$ particles are the same (even with this approximation the spin contributions must be treated carefully [cf. Eqs. (12.74)–(12.76)]).

5. Construct a generating functional and use it to obtain an expression for the one-body reduced density matrix in terms of connected diagrams (cf. Sec. 11.D). Give rules for obtaining algebraic expressions for your diagrams.

6. For a condensed Fermi fluid with a Hamiltonian

$$\hat{H} = \sum_{k} \frac{\hbar^2 k^2}{2m} a_{\mathbf{k}}{}^{+} a_{\mathbf{k}} + \sum_{\mathbf{k}} \Delta_{\mathbf{k}}{}^{+} \hat{a}_{-\mathbf{k}\downarrow}\hat{a}_{\mathbf{k}\uparrow} + \sum_{\mathbf{k}} \Delta_{\mathbf{k}} \hat{a}^{+}_{\mathbf{k}\uparrow}\hat{a}^{+}_{-\mathbf{k}\downarrow}$$

find an expression for the exact propagator and the anomalous propagators. The quantity $\Delta_{\mathbf{k}}$ is the gap function.

7. Find the temperature dependence of the 0-diagram 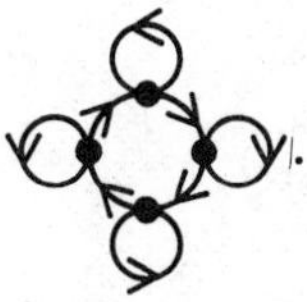.

8. Derive Eq. (12.112) for an electron gas starting from the expression for the self-energy.

9. Compute the effective mass m^* for quasiparticles in a system of neutral spin $\frac{1}{2}$ fermions which interact via a Yukawa potential $V(r) = V_0 \exp(-ar)/r$. Assume the system is close to $T = 0$ K.

10. In Fig. 12.6, the dip in the excitation energy curve at $k = 1.9$ Å^{-1} corresponds to the appearance of a new type of excitation (other than phonons) called rotons. At low temperature, only a few rotons will be excited. Expand the excitation energy $E(k)$ about the minimum at $k = 1.9$ Å^{-1} and find the contribution to the heat capacity due to a noninteracting roton gas. Show that the distribution function for rotons is well approximated by a Maxwell-Boltzmann distribution rather than a Bose-Einstein distribution. Compare the roton and phonon heat capacities and discuss the relative importance of these two types of excitation as a function of temperature.

13 Elementary Transport Theory

A. INTRODUCTORY REMARKS

We now begin the study of systems which are not in equilibrium. In this chapter, we restrict ourselves to dilute gases which are close to equilibrium. We assume that all disturbances are slowly varying in space and have small amplitude. Thus, in any small region, the system will be in equilibrium and its state in that region can be specified by the thermodynamic state variables. However, the values of the thermodynamic variables can vary from one region to another in the fluid.

When a system is disturbed from its equilibrium state, all quantities which are not conserved during collisions decay rapidly to their equilibrium values. After a few collision times, only quantities which are conserved during the collisions remain out of equilibrium. The densities of conserved quantities entirely characterize the nonequilibrium behavior of the fluid after long times. The equations of motion for the densities of the conserved quantities are called the hydrodynamic equations. Examples of conserved quantities are the number of particles, the total momentum of the particles, and the total kinetic energy of the particles (for elastic collisions). If there are inhomogeneities in the densities of these quantities, then particles, momentum, or kinetic energy must be transported from one part of the fluid to another to achieve equilibrium. The rate at which a fluid returns to its equilibrium state is determined by the transport coefficients. They are the constants of proportionality between the gradients in the densities of conserved quantities and the currents that result from those density gradients and lead the system back to equilibrium. The aim of transport theory is to compute the transport coefficients in terms of the microscopic properties of the fluid.

In order to build intuition concerning the nature of transport processes, we will first derive expressions for the coefficients of self-diffusion, shear viscosity, and thermal conductivity using the simplest possible mean free path arguments. The expressions we obtain make it possible to estimate the order of magnitude of the transport coefficients by simple "back of the envelope" calculations.

To obtain a deeper understanding of transport phenomena, we must derive the transport coefficients from a more rigorous microscopic theory. That is, we must derive expressions for them starting from the kinetic equation for the fluid (cf. Sec. 7.C). However, we are then faced with one of the fundamental dilemmas of statistical physics. The exact kinetic equation derived in Chap. 7 is reversible, whereas transport processes are irreversible or dissipative. Boltzmann was the first to derive an irreversible kinetic equation to describe relaxation processes in a fluid. He did it by using probabilistic arguments. Thus, his equation is not based rigorously on the properties of the underlying dynamics of the system. However, the Boltzmann equation works extremely well in providing numerical values for the transport coefficients in dilute fluids and to this day remains one of the great milestones in the history of statistical physics. Much of this chapter is devoted to a study of the Boltzmann equation. We will derive it using arguments similar to Boltzmann's original arguments, and we will indicate its irreversible nature by proving the famous Boltzmann H-theorem. From the H-theorem, Boltzmann was able to obtain a microscopic expression for the entropy of a system near equilibrium which has the proper behavior as the system approaches the equilibrium state.

We will apply Boltzmann's equation to the case of a dilute two-component gas of particles which have different identity but are dynamically the same. For this system, we derive microscopic expressions for the coefficients of self-diffusion, shear viscosity, and thermal conductivity. The method we use to derive the transport coefficients is very simple and elegant and was first introduced by Resibois. We first derive the hydrodynamic equations from the Boltzmann equation, introducing into them the transport coefficients, and we then find the normal mode frequencies of the hydrodynamic equations in terms of the transport coefficients. We next find the hydrodynamic eigenfrequencies of the Boltzmann equation and match these to the normal mode frequencies of the hydrodynamic equations. This gives us the desired microscopic expressions for the transport coefficients.

As a final example, we shall derive the kinetic equation for a weakly coupled quantum fluid using a method which can be generalized to the case of a system with broken symmetry (ODLRO). For quantum fluids, we must obtain the equation of motion, not of reduced probability densities but of a reduced density matrix. The equation we derive is the quantum analog of the Boltzmann equation.

B. ELEMENTARY KINETIC THEORY[1,2]

Before we discuss the full microscopic theory of transport processes based on the Boltzmann equation, it is useful to derive the transport coefficients using very simple mean free path arguments and kinetic theory. These arguments are entirely probabilistic and are independent of the form of the interparticle potential, although they do assume that the interaction is very short ranged compared to the average interparticle spacing. As we will see, the same type of arguments can be used to obtain microscopic expressions for the reaction rates in chemical systems.

1. Mean Free Path

The mean free path, λ, of a particle is the average distance it travels between collisions. We shall assume that collisions occur at random in a gas. We wish to find the probability that a particle can travel a distance r without collision. Because the collisions occur at random, a particle has the same chance of collision in any interval $r \to r + dr$. If the average number of collisions per unit length is $1/\lambda$, then the probability that a collision occurs in an interval dr is dr/λ.

Let $P_0(r)$ denote the probability that no collision occurs in an interval of length r. Then the probability that no collision occurs in interval of length $r + dr$ is

$$P_0(r + dr) = P_0(r)\left(1 - \frac{dr}{\lambda}\right). \tag{13.1}$$

The factor $(1 - (dr/\lambda))$ is the probability that no collision occurs in interval dr. We multiply $P_0(r)$ and $(1 - (dr/\lambda))$ together because the events "collision in length r" and "collision in length dr" are independent. If we expand the left-hand side of Eq. (13.1) in a Taylor series, we obtain

$$\frac{d}{dr} P_0(r) = -P_0(r)\frac{dr}{\lambda} \tag{13.2}$$

and therefore

$$P_0(r) = e^{-r/\lambda}. \tag{13.3}$$

The probability of no collision in length of path r is a Poisson distribution. The probability that a particle suffers its first collision in an interval $r \to r + dr$ is $P_0(r)(dr/\lambda)$. The average distance traveled between collisions is

$$\langle r \rangle = \int_0^\infty rP_0(r)\frac{dr}{\lambda} = \lambda \tag{13.4}$$

which is just our original definition of the mean free path.

2. Collision Frequency

Let us consider a fluid containing several types of particles $A, B, C, \ldots$ etc. Let us assume that particles A have a mass m_A, a diameter d_A, and a number density n_A; particles B have a mass m_B, a diameter d_B, and a number density n_B; etc. Let us further assume that the particles are distributed at random in the fluid and that each type of particle is distributed in the fluid according to the Maxwell-Boltzmann distribution (cf. Eq. (9.97)).

We will find the collision frequency between particles A and B. The radius of the sphere of influence between A and B is

$$d_{AB} = \frac{d_A + d_B}{2} \tag{13.5}$$

(cf. Fig. 13.1). The average relative speed between particles A and B is given by

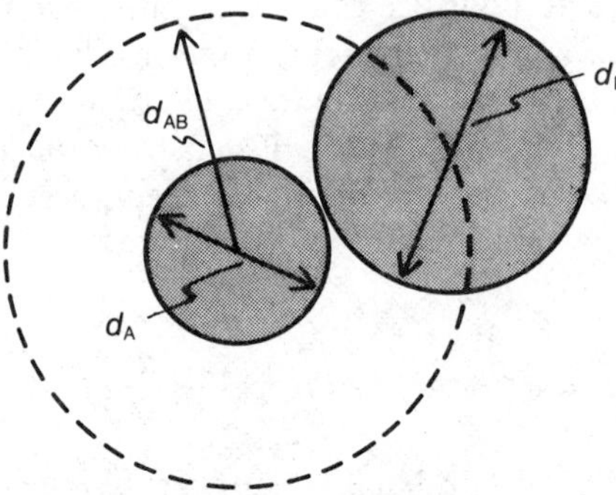

Fig. 13.1. The circle of radius d_{AB} is the sphere of influence for particles A and B.

$$\langle v_r \rangle_{AB} \equiv \int d\mathbf{v}_A \int d\mathbf{v}_B F(\mathbf{v}_A) F(\mathbf{v}_B) |\mathbf{v}_A - \mathbf{v}_B|. \tag{13.6}$$

The center of mass velocity of particles A and B is

$$\mathbf{V}_{cm} = \frac{m_A \mathbf{v}_A + m_B \mathbf{v}_B}{m_A + m_B} \tag{13.7}$$

and the relative velocity is

$$\mathbf{v}_r = \mathbf{v}_A - \mathbf{v}_B. \tag{13.8}$$

The Jacobian of the transformation from coordinates $\mathbf{v}_A$ and $\mathbf{v}_B$ to coordinates $\mathbf{v}_r$ and $\mathbf{V}_{cm}$ is equal to one. Thus, $\langle v_r \rangle_{AB}$ can be rewritten

$$\langle v_r \rangle_{AB} = \left(\frac{\beta M_{AB}}{2\pi}\right)^{3/2} \left(\frac{\beta \mu_{AB}}{2\pi}\right)^{3/2} \iint d\mathbf{v}_r \, d\mathbf{V}_{cm} v_r \, e^{-(\beta/2)(M_{AB} V_{cm}^2 + \mu_{AB} v_r^2)} \tag{13.9}$$

where $M_{AB} = m_A + m_B$ and $\mu_{AB} = m_A m_B/(m_A + m_B)$. If we perform the integrations in Eq. (13.9) we obtain

$$\langle v_r \rangle_{AB} = \left(\frac{8 k_B T}{\pi \mu_{AB}}\right)^{1/2} \tag{13.10}$$

for the average relative speed of particles A and B.

Let us now assume that all particles of mass m_B in the fluid are at rest and a particle of mass m_A moves through the gas with a speed $\langle v_r \rangle_{AB}$. Particle A sweeps out a collision cylinder of radius d_{AB} (radius of the sphere of influence) and volume $\pi d_{AB}^2 \langle v_r \rangle_{AB} t$ in time t. The number of particles B that particle A collides with in time t is $f_{AB} t$, where f_{AB} is the collision frequency

$$f_{AB} = n_B \pi d_{AB}^2 \langle v_r \rangle_{AB}. \tag{13.11}$$

Therefore, the total number of collisions per unit volume per second, ν_{AB}, between particles of type A and of type B is

$$\nu_{AB} = n_A n_B \pi d_{AB}^2 \langle v_r \rangle_{AB} = n_A n_B \pi d_{AB}^2 \left(\frac{8 k_B T}{\pi \mu_{AB}} \right)^{1/2}. \tag{13.12}$$

From Eq. (13.12) we can easily write down the collision frequency ν_{AA} between identical particles,

$$\nu_{AA} = \tfrac{1}{2} n_A{}^2 \pi d_{AA}^2 \langle v_r \rangle_{AB} = \tfrac{1}{2} n_A{}^2 \pi d_{AA}^2 \left(\frac{16 k_B T}{\pi m_A} \right)^{1/2}. \tag{13.13}$$

The extra factor of $\frac{1}{2}$ enters because the particles colliding are identical and it is needed to prevent overcounting.

If we consider a gas of identical particles, the mean free path, λ, the collision frequency for a single particle, f_{AA}, and the average speed, $\langle v \rangle$, are simply related by the equation

$$\lambda = \frac{\langle v \rangle}{f_{AA}} = \langle v \rangle \tau = \frac{1}{\sqrt{2} n_A \pi d_{AA}^2} \tag{13.14}$$

where τ is the collision time (the time between collisions), and the average speed, $\langle v \rangle$, is related to the relative speed by the relation $\langle v_{\mathrm{rel}} \rangle = \sqrt{2} \langle v \rangle$.

3. Self-Diffusion

We can now use the above ideas to obtain an expression for the coefficient of self-diffusion, D. Let us consider a gas of particles which are identical in every way except that some particles have a radioactive tracer attached (or they might have different color or different spin). Self-diffusion involves the transport of tracer particles through a gas of otherwise identical particles. If initially there is an uneven distribution of tracer particles in the gas, the distribution will even out and become uniform through the diffusion process. The rate at which inhomogeneities in the tracer particles become smoothed out is determined by the value of the coefficient of self-diffusion, D.

We shall assume that the density of tracer particles, $n_T(z)$, varies in the z-direction, while the total particle density, n, is held constant. As a first step in obtaining the coefficient of self-diffusion, we draw an imaginary wall in the fluid at $z = 0$ and find the net flux of particles across the wall.

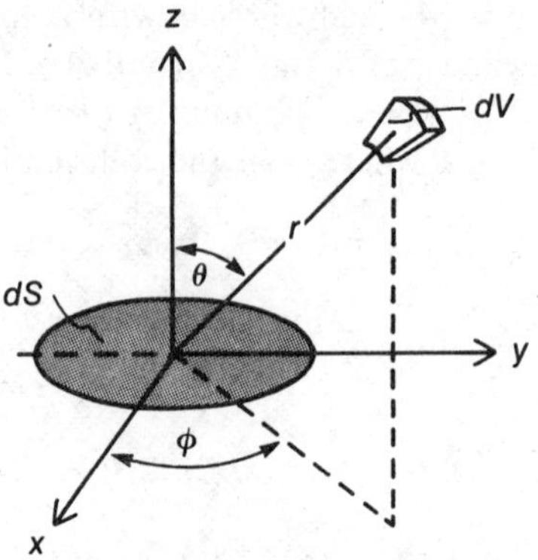

Fig. 13.2. Only a fraction of the particles in volume element, dV, reach the surface area element, dS, without undergoing a collision en route.

Let us first find the number of particles which hit the wall from above. We will look at a segment of the wall, dS, and choose the origin of coordinates to be at the center of dS. We next consider a volume element, dV, of the gas located at position r, θ, ϕ above the wall (cf. Fig. 13.2). The average number of tracer particles undergoing collisions in dV per unit time is $fn_T(z)\,dV = (\langle v\rangle/\lambda)n_T(z)\,dV$, where f is the collision frequency. Particles in dV leave in random directions (any direction is equally likely). The fraction of particles that move toward dS is $d\Omega/4\pi$, where $d\Omega$ is the solid angle subtended by dS and $d\Omega = dS|\cos\theta|/r^2$. Not all particles leaving dV in the direction of dS reach dS. The probability that a tracer particle will reach dS is $e^{-r/\lambda}$ (the probability that it will not undergo a collision). Combining the above results, we obtain the following expression for the number of tracer particles, $dn_T(\mathbf{r})$, which collide in dV, leave directed toward dS, and reach dS without another collision:

$$dn_T(\mathbf{r}) = \left(\frac{\langle v\rangle n_T(z)\,dV}{\lambda}\right)\left(\frac{dS|\cos\theta|}{4\pi r^2}\right)e^{-r/\lambda}. \tag{13.15}$$

The total number hitting a unit area of wall per unit time from above, $\dot{N}_+$, is found by integrating over the entire volume for $z > 0$:

$$\dot{N}_+ = \frac{\langle v\rangle}{4\pi\lambda}\int_0^\infty r^2\,dr\int_0^{\pi/2}\sin\theta\,d\theta\int_0^{2\pi}d\phi\, n(z)\cos\theta\,\frac{e^{-r/\lambda}}{r^2}. \tag{13.16}$$

For the case in which the tracer particles are distributed uniformly throughout the gas, that is, $n(z) =$ constant, Eq. (13.16) reduces to $\dot{N}_+ = n\langle v\rangle/4$.

For small variations in tracer density we may expand $n_T(z)$ in a Taylor series about the origin,

$$n_T(z) = n_T(0) + z\left(\frac{\partial n_T}{\partial z}\right)_0 + \frac{z^2}{2}\left(\frac{\partial^2 n_T}{\partial z^2}\right)_0 + \cdots \tag{13.17}$$

If $n_T(z)$ is a slowly varying function of z, then higher order derivatives $(\partial^2 n_T/\partial z^2)_0$, $(\partial^3 n_T/\partial z^3)_0$, etc. will be small. Because of the factor $e^{-r/\lambda}$ in the integral in Eq. (13.16), only small values of z (values of $z \approx \lambda$) will contribute. Therefore, we can terminate the expansion at $(z^2/2)(\partial^2 n_T/\partial z^2)_0$.

The net transport of tracer particles in the negative z-direction across a unit area of the wall per unit time is given by $(\dot{N}_+ - \dot{N}_-)$, where $\dot{N}_-$ is the number crossing a unit area per unit time in the positive z-direction. The expression for $\dot{N}_-$ is the same as for $\dot{N}_+$ except that θ is integrated from $\theta = (\pi/2) \rightarrow \pi$ and $|\cos\theta|$ is changed to $-\cos\theta$. Therefore,

$$(\dot{N}_+ - \dot{N}_-) = \frac{\langle v \rangle}{4\pi\lambda} \int_0^\infty dr r^2 \int_0^\pi \sin\theta\, d\theta \int_0^{2\pi} d\phi\, n_T(z) \cos\theta\, \frac{e^{-r/\lambda}}{r^2}. \tag{13.18}$$

If we substitute Eq. (13.17) into Eq. (13.18), the first and third terms are identically zero and we obtain

$$(\dot{N}_+ - \dot{N}_-) = \frac{\langle v \rangle \lambda}{3} \left(\frac{\partial n_T}{\partial z}\right)_0. \tag{13.19}$$

If the density increases in the z-direction, then $(\partial n_T/\partial z)_0 > 0$ and $(\dot{N}_+ - \dot{N}_-) > 0$. Therefore, there will be a net transport of particles in the negative z-direction.

If we let $J_D(z)$ denote the number of tracer particles crossing a unit area at z in unit time in the positive z-direction ($J_D(z)$ is the particle flux or current), then

$$J_D(z) = -D \frac{\partial n_T(z)}{dz} \tag{13.20}$$

where

$$D = \frac{\langle v \rangle \lambda}{3} \tag{13.21}$$

is the coefficient of self-diffusion. If the density is a slowly varying function of x, y, and z, then we can write the current as in the more general form

$$\mathbf{J}_D(\mathbf{r}) = -D \nabla_{\mathbf{r}} n_T(\mathbf{r}) \tag{13.22}$$

where $\nabla_{\mathbf{r}}$ denotes the spatial gradient. Eq. (13.22) is called Fick's law.

Let us now consider diffusive flow which is changing in time. We assume that $n_T = n_T(z, t)$ and $J_D = J_D(z, t)$ and consider a region in the fluid bounded by two fixed planes, one at $z = z_0$ and the other at $z = z_0 + dz_0$. The net increase in the number of particles per unit area per unit time between the planes is

$$J_D(z + dz, t) - J_D(z, t) = \left(\frac{\partial J_D}{\partial z}\right) dz = -\frac{\partial}{\partial z}\left(D \frac{\partial n_T}{\partial z}\right) dz. \tag{13.23}$$

If there are no sources or sinks of particles between the planes (particles are conserved), then Eq. (13.23) must equal the rate of increase of number of tracer particles between the planes. Thus,

$$\frac{\partial}{\partial t}(n_T\,dz) = \frac{\partial}{\partial z}\left(D\frac{\partial n_T}{\partial z}\right)$$

and

$$\frac{\partial n_T}{\partial t} = D\frac{\partial^2 n_T}{\partial z^2}. \tag{13.24}$$

We would obtain the same equation if we assumed that the particles performed a random walk through the gas (cf. Chaps. 5 and 6). For variations in three dimensions, Eq. (13.24) can be written

$$\frac{\partial n_T(\mathbf{r}, t)}{\partial t} = D\nabla_{\mathbf{r}}^2 n_T(\mathbf{r}, t). \tag{13.25}$$

Thus, using kinetic theory we have derived again the diffusion equation and we have obtained a mean free path expression for the diffusion coefficient. We can now use these results to find an expression for the coefficients of viscosity and thermal conductivity.

4. Coefficients of Viscosity and Thermal Conductivity

In addition to self-diffusion, other types of transport can occur in a fluid. If a fluid is stirred so that one part moves relative to another part (that is, if the average velocity of particles is a function of position $\langle \mathbf{v} \rangle = \langle \mathbf{v}(\mathbf{r}) \rangle$), then "friction" (viscosity) between the moving parts will bring them to equilibrium and the average velocity of the particles will become uniform throughout the gas. The quantity that is transported is the average velocity of the particles. The rate at which the average velocity is transported is determined by the coefficient of viscosity.

If one part of the fluid is hotter than another part, the particles in the hot region will have a greater average kinetic energy than particles in the cooler region. Transport processes will equalize the average kinetic energy (temperature) in the two regions. The quantity transported is the average kinetic energy and the process which brings the gas to equilibrium is called heat conduction. The rate at which the temperatures in different parts of the gas are equalized is determined by the coefficient of thermal conductivity, K.

It is possible to treat all these transport processes in a unified manner. Let us assume that $A = A(z)$ is the *molecular property* to be transported and that it varies in the z-direction. Let us draw an imaginary plane in the gas at $z = z_0$. When a particle crosses the plane, it transports the value of A it obtained in its last collision and transfers it to another particle in its next collision.

Let $A(z + \Delta z)$ be the value of A transported in the negative z-direction across the plane; $\Delta z = a\lambda$ is the distance above the plane where the particle had

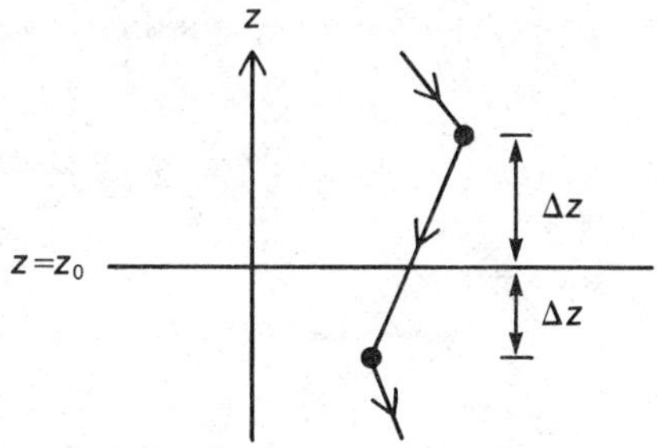

Fig. 13.3. A particle has a collision a distance Δz above the plane and transports property A to another particle a distance Δz below the plane.

its last collision (cf. Fig. 13.3) (λ is the mean free path and a is a proportionality constant). The average number of particles crossing the plane per unit area per unit time is $n\langle v\rangle$. The net amount of A transported in the positive z-direction per unit area per unit time is

$$n\langle v\rangle[A(z_0 - \Delta z) - A(z_0 + \Delta z)] = -2n\langle v\rangle\Delta z\frac{dA}{dz} = -2an\langle v\rangle\lambda\frac{dA}{dz}. \tag{13.26}$$

Let us denote the net amount of A per unit area per unit time (the current) by J_A. Then

$$J_A(z) = -b_A n\langle v\rangle\lambda\frac{dA}{dz} \tag{13.27}$$

where b_A is a proportionality constant. We can determine b_A from our expression for the coefficient of self-diffusion in Eq. (13.21). Let us now apply Eq. (13.27) to the cases of self-diffusion, viscosity, and heat conductivity.

(*a*) *Self-Diffusion.* Let us again consider the system in Sec. 13.B.3. The concentration of tracer particles per particle is $A = n_T(z)/n$. If the density of tracer particles varies in the z-direction, then there will be a concentration gradient $(1/n)(dn_T/dz) = dA/dz$ causing them to diffuse through space. If we let $J_A = J_D$, where J_D is the tracer particle current, then from Eq. (13.27) we obtain

$$J_D(z) = -b_A\langle v\rangle\lambda\frac{dn_T}{dz} = -D\frac{dn_T}{dz} \tag{13.28}$$

and the coefficient of self-diffusion is given by $D = \langle v\rangle\lambda/3$, if we let $b_A = \frac{1}{3}$.

(*b*) *Viscosity.* If a gas is stirred, one part will move relative to another part. Let us assume that the y-component of the average velocity varies in the z-direction. Then $A(z) = m\langle v_y(z)\rangle$ and $J_A = J_{zy}$, where J_{zy} is the net flux of y-

component of momentum per unit area per unit time in the z-direction. From Eq. (13.27) we have

$$J_{zy} = -\tfrac{1}{3}nm\langle v\rangle\lambda \frac{d\langle v_y(z)\rangle}{dz} = -\eta \frac{d\langle v_y(z)\rangle}{dz} \tag{13.29}$$

where

$$\eta = \tfrac{1}{3}nm\langle v\rangle\lambda \tag{13.30}$$

is the coefficient of viscosity. From Eq. (13.14) we know that $\lambda = [\sqrt{2}n\pi d^2]^{-1}$ for hard spheres of diameter d. Therefore,

$$\eta = \frac{m\langle v\rangle}{3\sqrt{2}\pi d^2} \tag{13.31}$$

and the coefficient of viscosity is independent of density—a somewhat surprising result that is verified by experiment.

(*c*) *Heat Conduction.* If the temperature of the gas varies in the z-direction, then

$$A(z) = \tfrac{1}{2}m\langle v^2(z)\rangle = \tfrac{3}{2}k_BT(z) \tag{13.32}$$

and $J_A = J_Q$ is the heat current (the net flux of thermal energy per unit area per unit time). Since $b_A = \frac{1}{3}$, Eq. (13.27) gives

$$J_Q = -K\frac{dT}{dz} \tag{13.33}$$

where

$$K = \frac{n\langle v\rangle\lambda k_B}{2} \tag{13.34}$$

is the coefficient of thermal conductivity.

The coefficient of thermal conductivity can be put in another form. The change in average internal energy per particle as a function of z can be written $\Delta u(z) = mc_VT(z)$, where c_V is the specific heat. Then

$$K = \frac{nm\langle v\rangle\lambda c_V}{3}. \tag{13.35}$$

If we compare Eqs. (13.30) and (13.35), we obtain the following simple relation between the coefficients of viscosity and heat conductivity and the specific heat:

$$\frac{K}{\eta} \approx c_V. \tag{13.36}$$

The above expressions for the transport coefficients have been derived by very simple arguments, but they do describe fairly well the observed qualitative behavior of transport coefficients in dilute gases.

5. Rate of Reaction[3]

In later chapters, we will be interested in transport processes in systems in which chemical reactions can occur. We can use elementary kinetic theory to obtain a qualitative expression for the rate of the reactions. We cannot simply equate the number of collisions between various molecules to the rate at which they undergo chemical reactions. A simple example will illustrate this. Let us consider the reaction $2HI \rightarrow H_2 + I_2$. The radius of the sphere of influence may be obtained from viscosity data on HI gas ($d = 4 \times 10^{-8}$ cm). At a temperature $T = 700$ K, pressure $P = 1$ atm, the collision frequency is easily computed and yields $\nu_{(HI)^2} = 1.3 \times 10^{28}$/sec. If every collision between HI molecules contributed to the reaction, then for a gas containing 10^{23} molecules the reaction would be completed in a fraction of a second. However, experimentally one finds that it takes a much longer time to complete the reaction, the reason being that not all collisions lead to a reaction.

In order to cause a reaction of the type $A + B \rightarrow C + D$, a certain amount of the kinetic energy of A and B must be absorbed during the collision. A and B first form an intermediate state (AB), which then can decay into the products C and D. The intermediate state (AB) is called an activated complex and A and B are called the reactants. The amount of energy which must be absorbed in order to form the activated complex is called the activation energy (cf. Fig. 13.4).

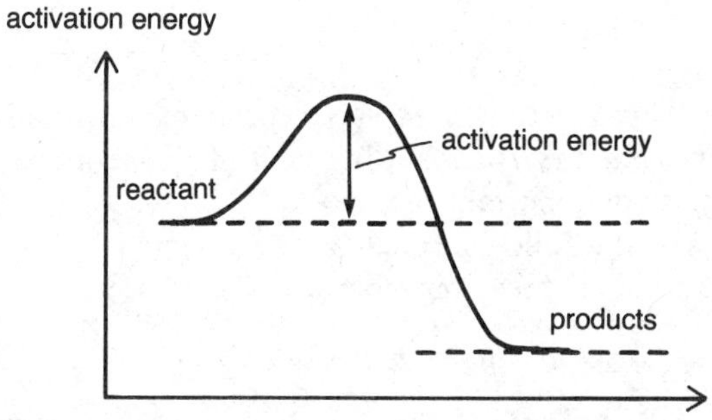

Fig. 13.4. There is an energy barrier that must be surmounted before a chemical reaction can occur.

All of the energy which goes into exciting the activated complex must come from the energy of relative motion of the reactants. Energy in the center of mass motion cannot contribute. If ϵ is the activation energy of the activated complex, then a reaction between A and B can occur only if the relative velocity of A and B is such that

$$\tfrac{1}{2}\mu_{AB}v_r^2 > \epsilon. \tag{13.37}$$

Therefore, to find the rate at which reactions between A and B can take place, we must multiply the collision frequency between A and B by the probability that A and B have a relative velocity greater than $\sqrt{2\epsilon/\mu_{AB}}$.

The probability that the molecules A and B have a center of mass velocity in the range $\mathbf{V}_{cm} \to \mathbf{V}_{cm} + d\mathbf{V}_{cm}$ and a relative velocity in the range $\mathbf{v}_r \to \mathbf{v}_r + d\mathbf{v}_r$ is given by

$$P(\mathbf{V}_{cm}, \mathbf{v}_r)\, d\mathbf{V}_{cm}\, d\mathbf{v}_r = \left(\frac{\beta M_{AB}}{2\pi}\right)^{3/2}\left(\frac{\beta\mu_{AB}}{2\pi}\right)^{3/2} e^{-(\beta/2)(M_{AB}V_{cm}^2 + \mu_{AB}v_r^2)}\, d\mathbf{V}_{cm}\, d\mathbf{v}_r. \tag{13.38}$$

The probability that A and B have a relative velocity in the range $\mathbf{v}_r \to \mathbf{v}_r + d\mathbf{v}_r$ is found by integrating Eq. (13.38) over the center of mass velocity. We then find

$$P(\mathbf{v}_r)\, d\mathbf{v}_r = \left(\frac{\beta\mu_{AB}}{2\pi}\right)^{3/2} e^{-(\beta/2)(\mu_{AB}v_r^2)}\, d\mathbf{v}_r. \tag{13.39}$$

The probability that A and B have a relative velocity $v_r > \sqrt{2\epsilon/\mu_{AB}}$ is found by integrating Eq. (13.39) from $v_r = \sqrt{2\epsilon/\mu_{AB}}$ to $v_r = \infty$ and integrating over all angles. Thus,

$$\begin{aligned}\text{Prob}\left(v_r > \sqrt{\frac{2\epsilon}{\mu_{AB}}}\right) &= \left(\frac{\beta\mu_{AB}}{2\pi}\right)^{3/2}\int_0^{\pi} d\theta \int_0^{2\pi} d\phi \int_{\sqrt{2\epsilon/\mu_{AB}}}^{\infty} dv_r v_r^2 \sin\theta \\ &\quad\times e^{-(\beta/2)(\mu_{AB}v_r^2)} \\ &= f(\beta, \epsilon)\, e^{-\beta\epsilon}.\end{aligned} \tag{13.40}$$

The probability that a reaction takes place depends exponentially on the activation energy. The quantity $f(\beta\epsilon)$ is a function of temperature and activation energy. Its form is not important here.

We can now write the following qualitative expression for the rate of reaction of A and B. The number of reacting molecules, $\dot{n}_R$, per second per unit volume is

$$\dot{n}_R = K(\beta\epsilon)\nu_{AB}\, e^{-\beta\epsilon} \tag{13.41}$$

where ν_{AB} is the collision frequency between molecules A and B and the coefficient $K(\beta\epsilon)$ depends on activation energy and temperature and may also depend on the geometry of the interacting molecules A and B. We may rewrite Eq. (13.41) in the form of a rate equation. If we let N_A denote the number of moles of A per unit volume and use Eq. (13.12), we can write

$$\frac{dN_A}{dt} = -K'(\epsilon, \beta, \mu_{AB})\, d_{AB}\, e^{-\beta\epsilon} N_A N_B \equiv -k_{AB} N_A N_B. \tag{13.42}$$

In Eq. (13.42), $K'(\epsilon\beta\mu_{AB})$ is an arbitrary function of ϵ, β, and μ_{AB} which may depend on the geometry of A and B. The quantity k_{AB} is called the rate constant for the reaction. It depends exponentially on the activation energy. We see that the activation energy is the most important quantity in determining the rate of a chemical reaction, since a small change in ϵ can cause a large change in k_{AB}. Eq. (13.42) gives the rate of decrease of A in the fluid due to the reaction $A + B \rightarrow C + D$. It is a second-order rate equation because it depends on the square of the molar density of species in the fluid. It is the simplest form of rate equation one can have because the rate of change in the concentration of A is determined by only one reaction. There will often be a number of competing reactions taking place in the fluid which determine the rate of change in the concentration of a given type of molecule. Then the rate equation for a given reaction can be a more complicated function of the concentrations.

In general, the expression describing a given chemical reaction is written in the form

$$-\nu_A A - \nu_B B \underset{k_2}{\overset{k_1}{\rightleftharpoons}} \nu_C C + \nu_D D \tag{13.43}$$

where ν_i are the stoichiometric coefficients and tell us how many molecules of each species must combine to enable the reaction to take place. By convention, the stoichiometric coefficients on the left-hand side are negative and those on the right-hand side are positive. The constant k_1 is the rate constant for the forward reaction and k_2 is the rate constant for the backward reaction. The rate of change of A can, in general, be written

$$\frac{dN_A}{dt} = -k_1 N_A^{|\nu_A|} N_B^{|\nu_B|} + k_2 N_C^{\nu_C} N_D^{\nu_D} \tag{13.44}$$

where $|\nu_A|$ denotes the absolute value of ν_A.

When more than one reaction contributes to the rate of change of a given species, the rate equations can become more complicated. For example, the simple process $A \rightarrow B + C$ cannot be explained using collision theory. One must introduce an intermediate in the process. Consider the following two reactions:

$$2A \underset{k_2}{\overset{k_1}{\rightleftharpoons}} A + A^* \tag{13.45}$$

and

$$A^* \overset{k_1}{\rightarrow} B + C. \tag{13.46}$$

Two molecules, A, collide to form one A and an excited A^*. A^* then decays into B and C. The rate equations for this process may be written

$$\frac{dN_A}{dt} = k_1 N_A^2 + k_2 N_A N_{A^*}, \tag{13.47}$$

$$\frac{dN_{A^*}}{dt} = k_1 N_A^2 - k_2 N_A N_{A^*} - k_3 N_A, \tag{13.48}$$

and

$$\frac{dN_B}{dt} = k_2 N_{A^*}. \tag{13.49}$$

We can solve Eqs. (13.47)–(13.49) if we make the assumption that the concentration of A^* is roughly constant, that is, that the rate of formation of A^* is equal to the rate of decay of A^*. Then

$$\frac{dN_{A^*}}{dt} = 0 \Rightarrow N_{A^*} = \frac{k_1 N_A^2}{k_3 + k_2 N_A}. \tag{13.50}$$

With Eq. (13.50), we obtain the following equation for the rate of increase of N_B:

$$\frac{dN_B}{dt} = \frac{k_1 k_3 N_A^2}{k_3 + k_2 N_A}. \tag{13.51}$$

If $k_3 \gg k_2 N_A$ then $dN_B/dt \approx k_1 N_A^2$ and the rate of formation of B looks as if it were governed by the process $2A \xrightarrow{k_1} B + C$. If $k_2 N_A \gg k_3$ then $dN_B/dt = (k_1 k_3/k_2) N_A$ and the rate of formation of B looks as if it were governed by the process $A \xrightarrow{k'} B + C$ where $k' = k_1 k_3/k_2$.

In general, more than one chemical reaction will contribute to the rate of formation or depletion of a given molecule. As we can see from the above example, experimental measurement of the rate equations is not always sufficient for determining the type of chemical processes which are occurring in a fluid.

Now that we have built some intuition about the transport coefficients, we will derive microscopic expressions for them based on a more rigorous microscopic theory, that is, one based on the Boltzmann equation.

C. THE BOLTZMANN EQUATION[4,5]

Until now, we have treated transport processes from a phenomenological point of view and we have obtained very simple expressions for the transport coefficients which are independent of the detailed form of the interaction. Most of the rest of this chapter will be devoted to a derivation of the transport coefficients starting from a kinetic equation (cf. Sec. 7.C).

The kinetic equation given in Sec. 7.C is an exact equation for the one-body reduced probability density, but it cannot be solved. In practice, we try to obtain an approximation to it which contains the important physical features of the system we are considering. In subsequent sections, we shall be interested in one of the simplest types of nonequilibrium systems—a dilute gas of particles which interacts via a short-range, spherically symmetric interaction potential. We shall assume that inhomogeneities in the gas are slowly varying. The form of the kinetic equation which gives the time evolution of the probability density of such a system is called the Boltzmann equation. Although it can be derived directly from Eq. (7.45), the arguments are rather long and tedious. In this section, we shall derive the Boltzmann equation from simple physical arguments.

Let us consider a dilute gas of particles of mass, m, which interact via a spherically symmetric potential $\phi(|\mathbf{q}_i - \mathbf{q}_j|)$. We shall assume that the time a particle spends between collisions is very much longer than the duration of a collision. We shall describe the behavior of the system in terms of a number density $f(\mathbf{p}, \mathbf{q}, t)$ rather than a probability density. The distribution function, $f(\mathbf{p}, \mathbf{q}, t)$, gives the number of particles in the six-dimensional phase space volume element $\mathbf{p} \to \mathbf{p} + d\mathbf{p}, \mathbf{q} \to \mathbf{q} + d\mathbf{q}$. It is related to the probability density $\rho_1(\mathbf{p}, \mathbf{q}, t)$ and the distribution function $F_1(\mathbf{p}, \mathbf{q}, t)$ through the relation

$$f(\mathbf{p}, \mathbf{q}, t) \equiv N\rho_1(\mathbf{p}, \mathbf{q}, t) = \frac{N}{V} F_1(\mathbf{p}, \mathbf{q}, t) \tag{13.52}$$

($\rho_1(\mathbf{p}, \mathbf{q}, t)$ and $F_1(\mathbf{p}, \mathbf{q}, t)$ are defined in Sec. 7.C).

Let us now consider a volume element, $\Delta V_1 = d\mathbf{p}_1 \, \Delta\mathbf{r}$, lying in the region $\mathbf{p}_1 \to \mathbf{p}_1 + d\mathbf{p}_1$, $\mathbf{r} \to \mathbf{r} + \Delta\mathbf{r}$ of the six-dimensional phase space. We shall assume the volume element $\Delta\mathbf{r}$ is large enough to contain many particles and small enough that the distribution function $f(\mathbf{p}, \mathbf{r}, t)$ does not vary appreciably over $\Delta\mathbf{r}$. We wish to find an equation for the rate of change of number of particles in ΔV_1. This change will be due to free streaming of particles into (and out of) ΔV_1 and to scattering of particles into (and out of) ΔV_1 because of collisions. The rate of change of $f(\mathbf{p}_1, \mathbf{r}, t)$ may be written in the form

$$\frac{\partial f_1}{\partial t} = -\dot{\mathbf{q}}_1 \cdot \frac{\partial f_1}{\partial \mathbf{q}_1} + \left.\frac{\partial f_1}{\partial t}\right|_{\text{collisions}}. \tag{13.53}$$

The first term on the right is the contribution due to streaming (flow through the surface of ΔV_1), and the second term is the contribution due to collisions. If an external field were present, then an additional streaming term of the form $-\dot{\mathbf{p}}_1 \cdot \partial f_1/\partial \mathbf{q}_1$ would be present on the right.

We shall assume that the gas is dilute enough that only two-body collisions need be considered. Furthermore, we shall assume that all collisions are elastic. Before we can write an expression for $\partial f/\partial t|_{\text{coll}}$, we must review some aspects of two-body scattering theory.

1. Two-Body Scattering[6]

The two-body Hamiltonian for our system can be written

$$H_2(\mathbf{p}_1, \mathbf{p}_2; \mathbf{q}_1, \mathbf{q}_2) = \frac{p_1^2}{2m} + \frac{p_2^2}{2m} + V(|\mathbf{q}_1 - \mathbf{q}_2|) \tag{13.54}$$

where $V(|\mathbf{q}_1 - \mathbf{q}_2|)$ is the interaction potential and m is the mass of the particles. We will introduce a change of variables to relative coordinates $\mathbf{q} =$

$\mathbf{q}_2 - \mathbf{q}_1$ and $\mathbf{p} = \frac{1}{2}(\mathbf{p}_2 - \mathbf{p}_1)$ and center of mass coordinates $\mathbf{P} = \mathbf{p}_1 + \mathbf{p}_2$ and $\mathbf{Q} = \frac{1}{2}(\mathbf{q}_2 + \mathbf{q}_1)$, respectively. In terms of the relative and center of mass coordinates, the Hamiltonian becomes

$$H_2(\mathbf{P}, \mathbf{p}; \mathbf{Q}, \mathbf{q}) = \frac{P^2}{4m} + \frac{p^2}{m} + V(|\mathbf{q}|). \tag{13.55}$$

We see that the relative and center of mass contributions to the Hamiltonian separate. The center of mass moves through space like a free particle and the interaction reduces to a one-body problem—namely that of a particle of momentum $\mathbf{p}$ and mass $m/2$ in the presence of a fixed central force field.

For elastic collisions, the total kinetic energy is conserved. If we consider the scattering process $(\mathbf{p}_1, \mathbf{p}_2) \rightarrow (\mathbf{p}_1', \mathbf{p}_2')$, then we must have

$$\frac{p_1^2}{2m} + \frac{p_2^2}{2m} = \frac{p_1^{2\prime}}{2m} + \frac{p_2^{2\prime}}{2m} \tag{13.56}$$

where $\mathbf{p}_1$ and $\mathbf{p}_2$ are the momenta of the particles before the collision and $\mathbf{p}_1'$ and $\mathbf{p}_2'$ are the momenta of the particles after the collision. During the collision process the center of mass momentum is conserved. Thus,

$$\mathbf{P} = \mathbf{P}'. \tag{13.57}$$

Furthermore, since we are considering elastic collisions, the magnitude of the relative momentum is conserved,

$$|\mathbf{p}| = |\mathbf{p}'| = \frac{m}{2} g. \tag{13.58}$$

In Eq. (13.58), g is the magnitude of the relative velocity.

The scattering problem may be viewed as that of a beam of particles of momentum $\mathbf{p}$ incident on a fixed central force field. We denote the impact parameter by b and the scattering angle in the laboratory frame by θ_0. The angular momentum of the incident particle is

$$l = \frac{m}{2} gb \tag{13.59}$$

and is conserved. If I is the intensity of the incident beam (number of incident particles/time-area), then the total number of particles $\dot{N}$ scattered into the solid angle $d\Omega = \sin\theta_0 \, d\theta_0 \, d\alpha$ per second is given by (α is the azimuthal angle)

$$\dot{N} = I\sigma_L(g, \theta_0) \sin\theta_0 \, d\theta_0 \, d\alpha = Ib \, db \, d\alpha. \tag{13.60}$$

Eq. (13.60) gives a relation between the impact parameter $b(g, \theta_0)$ and the cross-section, $\sigma(g, \theta_0)$,

$$\sigma_L(g, \theta_0) = \frac{b(g, \theta_0)}{\sin\theta_0} \frac{db}{d\theta_0}. \tag{13.61}$$

The scattering is independent of α because of the symmetry of the potential.

The relation among b, g, and θ_0 is determined by the form of the potential. It may be written in the form

$$\theta_0 = \pi - 2\int_0^{\eta_L} \frac{d\eta}{\left[1 - \eta^2 - \dfrac{4}{mg^2} V\left(\dfrac{b}{\eta}\right)\right]^{1/2}} \tag{13.62}$$

where η_L satisfies the condition $1 - \eta_L{}^2 - (4/mg^2)V(b/\eta_L) = 0$.

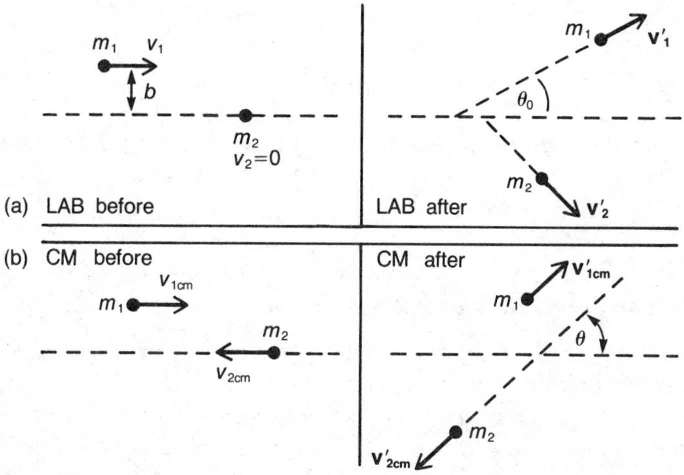

Fig. 13.5. Scattering between two particles: (*a*) laboratory frame; (*b*) center of mass frame.

It is easier to consider the scattering problem in the center of mass frame rather than the laboratory frame. In Fig. 13.5*a* we have drawn the scattering event in the laboratory frame, and in Fig. 13.5*b* we have drawn it in the center of mass frame. The relation between the scattering angles in the laboratory and the center of mass frames is given in Fig. 13.6. Thus,

$$v'_{1cm} \cos\theta + v_{cm} = v'_{1lab} \cos\theta_0 \tag{13.63}$$

and

$$v'_{1cm} \sin\theta = v'_{1lab} \sin\theta_0. \tag{13.64}$$

The azimuthal angle, α, is the same in the two frames. The number of particles per second scattered into the solid angle $d\Omega_0 = \sin\theta_0\, d\theta_0\, d\alpha$ in the laboratory frame must be the same as the number scattered into the solid angle $d\Omega_c = \sin\theta\, d\theta\, d\alpha$. Hence,

$$\sigma_L(g, \theta_0) \sin\theta_0\, d\theta_0\, d\alpha = \sigma_c(g, \theta) \sin\theta\, d\theta\, d\alpha \tag{13.65}$$

where $\sigma_c(g, \theta)$ is the cross-section in the center of mass frame. In the center of mass frame, the scattering event $(\mathbf{p}_1, \mathbf{p}_2) \to (\mathbf{p}_1', \mathbf{p}_2')$ and the reverse scattering event $(\mathbf{p}_1', \mathbf{p}_2') \to (\mathbf{p}_1, \mathbf{p}_2)$ are identical except that the directions of the particles are reversed. Therefore, $\sigma_c(g, \theta)$ is the scattering cross-section for both the forward and the reverse scattering events in the center of mass frame.

We now have enough information to derive the Boltzmann equation.

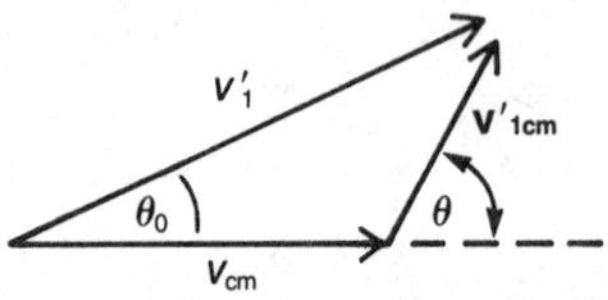

Fig. 13.6. Relation between scattering angles in the laboratory and the center of mass frames.

2. Derivation of the Boltzmann Equation

Let us denote the rate of particles scattered out of ΔV_1 by $\partial f_-/\partial t \, \Delta\mathbf{r} \, d\mathbf{p}_1$. The number of particles in ΔV_1 at time t with coordinates $\mathbf{p}_1 \to \mathbf{p}_1 + d\mathbf{p}_1$ and $\mathbf{r} \to \mathbf{r} + \Delta\mathbf{r}$ is $f(\mathbf{p}_1, \mathbf{r}, t) \, \Delta\mathbf{r} \, d\mathbf{p}_1$. All particles of momentum $\mathbf{p}_2$ lying within a cylinder of volume $d\mathbf{q}_2 = 2\pi b \, db g \, dt$ collide with particles $\mathbf{p}_1$ in time dt. The number of such particles is $f(\mathbf{p}_2, \mathbf{r}, t) 2\pi b \, dbg \, dt \, d\mathbf{p}_2$. The total number of collisions, $N(\mathbf{p}_1\mathbf{p}_2 \to \mathbf{p}_1'\mathbf{p}_2')$, per unit volume in time dt between particles of momentum $\mathbf{p}_1$ and particles of momentum $\mathbf{p}_2$ resulting in new momenta $\mathbf{p}_1'$ and $\mathbf{p}_2'$ is given by

$$N(\mathbf{p}_1\mathbf{p}_2 \to \mathbf{p}_1'\mathbf{p}_2') = 2\pi g b \, db \, dt f(\mathbf{p}_2, \mathbf{r}, t) f(\mathbf{p}_1, \mathbf{r}, t) \, d\mathbf{p}_1 \, d\mathbf{p}_2. \tag{13.66}$$

In Eq. (13.66) we have assumed that the distribution functions do not change appreciably in position space for the volume element ΔV_1 we are considering. Also, we have assumed that the particles $\mathbf{p}_1$ and $\mathbf{p}_2$ are completely uncorrelated. This assumption is called molecular chaos, or "Stosszahl-Ansatz."

In analogy to Eq. (13.66) we may write for the inverse scattering process

$$N(\mathbf{p}_1', \mathbf{p}_2' \to \mathbf{p}_1, \mathbf{p}_2) = 2\pi g b \, db \, dt f(\mathbf{p}_1', \mathbf{r}, t) f(\mathbf{p}_2', r, t) \, d\mathbf{p}_1' \, d\mathbf{p}_2'. \tag{13.67}$$

For elastic collisions, $d\mathbf{p}_1 \, d\mathbf{p}_2 = d\mathbf{p}_1' \, d\mathbf{p}_2'$. (This is easily proved if we note that $d\mathbf{p}_1 \, d\mathbf{p}_2 = d\mathbf{P} \, d\mathbf{p}$ and $d\mathbf{p}_1' \, d\mathbf{p}_2' = d\mathbf{P}' \, d\mathbf{p}'$. Furthermore, for elastic collisions, $d\mathbf{P} = d\mathbf{P}'$ and $p^2 \, dp = p'^2 \, dp'$. Therefore, $d\mathbf{p}_1 \, d\mathbf{p}_2 = d\mathbf{p}_1' \, d\mathbf{p}_2'$.) We may now combine Eqs. (13.60) and (13.65)–(13.67) to obtain the following expression for the net increase, $(\partial f_1/\partial t)_{\text{coll}} \, d\mathbf{p}_1$, in number of particles with momentum

$\mathbf{p}_1 \to \mathbf{p}_1 + d\mathbf{p}_1$ per unit volume per unit time:

$$\left(\frac{\partial f_1}{\partial t}\right)_{\text{coll}} d\mathbf{p}_1 = d\mathbf{p}_1 \int d\mathbf{p}_2 \int d\Omega g\sigma(\theta, g) \times (f(\mathbf{p}_1', r, t)f(\mathbf{p}_2', r, t) - f(\mathbf{p}_1, \mathbf{r}, t)f(\mathbf{p}_2, \mathbf{r}, t)) \tag{13.68}$$

where $d\Omega = \sin\theta \, d\theta \, d\alpha$, and $\sigma(\theta, g)$ is the center of mass collision cross-section. If we now combine Eqs. (13.53) and (13.68), we obtain

$$\frac{\partial f_1}{\partial t} + \dot{\mathbf{q}}_1 \cdot \frac{\partial f_1}{\partial \mathbf{r}} = \int d\mathbf{p}_2 \int d\Omega g\sigma(\theta, g)(f_{1'}f_{2'} - f_1 f_2) \tag{13.69}$$

where $f_i = f(\mathbf{p}_i, \mathbf{r}, t)$. Eq. (13.69) is the Boltzmann equation. Since $f(\mathbf{p}, \mathbf{q}, t) = (N/V)F_1(\mathbf{p}, \mathbf{q}, t)$, the Boltzmann equation has the same form as the exact kinetic equation (7.45) except that the right-hand side is expressed entirely in terms of the distribution function $f(\mathbf{p}_i, \mathbf{r}, t)$. The Boltzmann equation is a nonlinear integro-differential equation for $f(\mathbf{p}_i, \mathbf{r}, t)$.

3. Boltzmann's H-theorem[4]

Boltzmann's equation describes the time evolution of the distribution of particles in six-dimensional phase space for a dilute gas with inhomogeneities which are slowly varying in position space. If no external fields drive the system, it should decay to equilibrium after a long time. Boltzmann's equation describes such behavior. To show this, Boltzmann introduced a function $H(t)$ which he defined

$$H(t) = \iint d\mathbf{q}_1 \, d\mathbf{p}_1 f(\mathbf{p}_1, \mathbf{q}_1, t) \ln f(\mathbf{p}_1, \mathbf{q}_1, t). \tag{13.70}$$

He then showed that, if $f(\mathbf{p}_1, \mathbf{q}_1, t)$ satisfies the Boltzmann equation, $H(t)$ always decreases because of the effect of collisions. The proof goes as follows. We first take the derivative of $H(t)$,

$$\frac{\partial H}{\partial t} = \iint d\mathbf{p}_1 \, d\mathbf{q}_1 \frac{\partial f_1}{\partial t} [\ln f_1 + 1], \tag{13.71}$$

and then substitute Eq. (13.69) into Eq. (13.71). This gives

$$\frac{\partial H}{\partial t} = -\iint d\mathbf{p}_1 \, d\mathbf{q}_1 \left(\dot{\mathbf{q}}_1 \cdot \frac{\partial f_1}{\partial \mathbf{q}_1}\right)[\ln f_1 + 1] + \int d\mathbf{q}_1 \int d\mathbf{p}_1 \int d\mathbf{p}_2 \int d\Omega g\sigma(\theta, g)(f_{1'}f_{2'} - f_1 f_2)[\ln f_1 + 1]. \tag{13.72}$$

We next change the first term on the right-hand side into a surface integral and assume that, for large $\mathbf{p}_1$ and $\mathbf{q}_1$, $f(\mathbf{p}_1, \mathbf{q}_1, t) \to 0$. With this assumption the first term on the right-hand side gives no contribution and we have

$$\frac{\partial H}{\partial t} = \int d\mathbf{q}_1 \int d\mathbf{p}_1 \int d\mathbf{p}_2 \int d\Omega g\sigma(\theta, g)[f_{1'}f_{2'} - f_1 f_2][\ln f_1 + 1]. \tag{13.73}$$

We now can rewrite Eq. (13.73). If we interchange coordinates $\mathbf{p}_1$ and $\mathbf{p}_2$ we obtain

$$\frac{\partial H}{\partial t} = \int d\mathbf{q}_1 \int d\mathbf{p}_1 \int d\mathbf{p}_2 \int d\Omega g\sigma(\theta, g)[f_{1'}f_{2'} - f_1 f_2][\ln f_2 + 1]. \tag{13.74}$$

We next add Eqs. (13.73) and (13.74) and divide by two to obtain

$$\frac{\partial H}{\partial t} = \frac{1}{2}\int d\mathbf{q}_1 \iint d\mathbf{p}_1\, d\mathbf{p}_2 \int d\Omega g\sigma(\theta, g)(f_{1'}f_{2'} - f_1 f_2)[\ln f_1 + \ln f_2 + 2]. \tag{13.75}$$

As a final step, we make a change of dummy variables $\mathbf{p}_1 \leftrightarrow \mathbf{p}_1'$ and $\mathbf{p}_2 \leftrightarrow \mathbf{p}_2'$ in Eq. (13.75), add the result to Eq. (13.75), and divide by two. If we remember that $d\mathbf{p}_1\, d\mathbf{p}_2 = d\mathbf{p}_1'\, d\mathbf{p}_2'$, we find

$$\frac{\partial H}{\partial t} = \frac{1}{4}\int d\mathbf{q}_1 \int d\mathbf{p}_1 \int d\mathbf{p}_2 \int d\Omega g\sigma(\theta, g)(f_{1'}f_{2'} - f_1 f_2)\ln \frac{f_1 f_2}{f_{1'}f_{2'}} \leqslant 0. \tag{13.76}$$

Eq. (13.76) is always less than or equal to zero since a function of the form $(y - x) \ln x/y$ is always less than or equal to zero.

The derivative $\partial H/\partial t$ will be zero only if $f_{1'}f_{2'} = f_1 f_2$ for all collisions. This is the condition of *detailed balance* and is the equilibrium condition for the gas. It can also be written in the form

$$\ln f_1 + \ln f_2 = \ln f_{1'} + \ln f_{2'}. \tag{13.77}$$

At equilibrium, the single particle distribution must be independent of absolute position in the gas. Thus, Eq. (13.77) is a condition which must be satisfied by some function of the momentum of particles before and after collisions. The only microscopic quantities that are conserved in a collision are the total momentum, the kinetic energy, and a constant (particle number). Therefore, in equilibrium $\ln f_1$ must be of the form

$$\ln f_1 = A + B\mathbf{p}_1 + C\frac{p_1^{\,2}}{2m} \tag{13.78}$$

and f_1 reduces to

$$f_1(\mathbf{p}_1) \approx e^{A + \mathbf{B}\cdot\mathbf{p}_1 + C(p_1^2/2m)} \tag{13.79}$$

where A, $\mathbf{B}$, and C are constant. This is the general form of the Maxwell distribution for a gas whose average momentum is not necessarily zero, and is a stationary solution of the Boltzmann equation.

Since $H(t)$ always decreases with time, the negative of $H(t)$ will always increase with time. Boltzmann identified the following quantity as the nonequilibrium entropy $S(t)$:

$$S(t) = -k_B H(t) = -k_B \iint d\mathbf{q}_1\, d\mathbf{p}_1 f(\mathbf{p}_1, \mathbf{q}_1, t) \ln f(\mathbf{p}_1, \mathbf{q}_1, t). \tag{13.80}$$

This entropy differs from the Gibbs entropy in that it depends only on a reduced distribution function and not on the full distribution function as does the Gibbs entropy (cf. Sec. 9.B).

D. LINEARIZED BOLTZMANN EQUATIONS FOR A TWO-COMPONENT SYSTEM

We wish to obtain microscopic expressions for the coefficient of self-diffusion, D, thermal conductivity, K, and shear viscosity, η for a two-component system whose distribution function varies in time and space according to the Boltzmann equation.

Let us consider a dilute gas of identical particles and place a radioactive tracer on half of the particles but not change their properties in any other way. Then the dynamics of the two kinds of particles and the collision cross-sections will be identical. The Boltzmann equation for the normal particles can be written

$$\frac{\partial f_{1,N}}{\partial t} + \dot{\mathbf{p}}_1 \cdot \frac{\partial f_{1,N}}{\partial \mathbf{p}_1} + \dot{\mathbf{q}}_1 \cdot \frac{\partial f_{1,N}}{\partial \mathbf{r}} = \sum_{\alpha,\beta,\gamma} \int d\mathbf{p}_1 \int d\Omega g\sigma(\theta, g)_{\alpha\beta;N\gamma}(f_{1',\alpha} f_{2',\beta} - f_{1,N} f_{2,\gamma}), \tag{13.81}$$

and for the tracer particles it can be written

$$\frac{\partial f_{1,T}}{\partial t} + \dot{\mathbf{p}}_1 \cdot \frac{\partial f_{1,T}}{\partial \mathbf{p}_1} + \dot{\mathbf{q}}_1 \cdot \frac{\partial f_{1,T}}{\partial \mathbf{r}} = \sum_{\alpha,\beta,\gamma} \int d\mathbf{p}_1 \int d\Omega g\sigma(\theta, g)_{\alpha\beta;T\gamma}(f_{1',\alpha} f_{2',\beta} - f_{1,T} f_{2,\gamma}) \tag{13.82}$$

where the summations over α, β, and γ are over normal and tracer particles and we have used the notations $f_{1,N} = f_N(\mathbf{p}_1, \mathbf{r}, t)$ and $f_{1,T} = f_T(\mathbf{p}_1, \mathbf{r}, t)$. The number of tracer and normal particles is conserved during collisions. Thus, only the cross-sections $\sigma_{NN;NN} = \sigma_{NT;NT} = \sigma_{TN;TN} = \sigma_{TT,TT} = \sigma$ are nonzero.

When we compute transport coefficients, it is sufficient to consider systems with small amplitude disturbance from equilibrium. We then expand the normal and tracer particle distributions about absolute equilibrium. If we denote the normal particle equilibrium distribution by $f_N{}^0(\mathbf{p}_1)$ and the tracer particle equilibrium distribution by $f_T{}^0(\mathbf{p}_1)$, we have

$$f^0(\mathbf{p}_1) = f_N{}^0(\mathbf{p}_1) = f_T{}^0(\mathbf{p}_1) = \frac{n_0}{2}\left(\frac{\beta}{2\pi m}\right)^{3/2} e^{-\beta p_1^2/2m}. \tag{13.83}$$

We have assumed that there are $N/2$ tracer and $N/2$ normal particles in the system and have let $n_0 = N/V$.

As a first step in linearizing the Boltzmann equation, we can write

$$f_N(\mathbf{p}_1, \mathbf{r}, t) = f^0(\mathbf{p}_1)[1 + h_N(\mathbf{p}_1, \mathbf{r}, t)] \tag{13.84}$$

and

$$f_T(\mathbf{p}_1, \mathbf{r}, t) = f^0(\mathbf{p}_1)[1 + h_T(\mathbf{p}_1, \mathbf{r}, t)] \tag{13.85}$$

where $h_N(\mathbf{p}_1, \mathbf{r}, t)$ and $h_T(\mathbf{p}_1, \mathbf{r}, t)$ denote small amplitude disturbances in the normal particle and tracer particle distributions, respectively.

If we substitute Eqs. (13.84) and (13.85) into Eqs. (13.81) and (13.82), respectively, and neglect terms of second order or higher in $h_{N(T)}(\mathbf{p}_1, \mathbf{r}, t)$, we obtain

$$\left(\frac{\partial h_{1,N}}{\partial t} + \dot{\mathbf{q}}_1 \cdot \frac{\partial h_{1,N}}{\partial \mathbf{r}}\right) = \sum_{\alpha,\beta,\gamma} \int d\mathbf{p}_2 \int d\Omega g\sigma(\theta, g)_{\alpha\beta N\gamma} f^0(\mathbf{p}_2) \times (h_{1',\alpha} + h_{2',\beta} - h_{1,N} - h_{2,\gamma}) \tag{13.86}$$

and

$$\left(\frac{\partial h_{1,T}}{\partial t} + \dot{\mathbf{q}}_1 \cdot \frac{\partial h_{1,T}}{\partial \mathbf{r}}\right) = \sum_{\alpha,\beta,\gamma} \int d\mathbf{p}_2 \int d\Omega g\sigma(\theta, g)_{\alpha\beta T\gamma} f^0(\mathbf{p}_2) \times (h_{1',\alpha} + h_{2',\beta} - h_{1,T} - h_{2,\gamma}). \tag{13.87}$$

In Eqs. (13.86) and (13.87), we have used kinetic energy conservation to write $f^0(\mathbf{p}_1')f^0(\mathbf{p}_2') = f^0(\mathbf{p}_1)f^0(\mathbf{p}_2)$. Eqs. (13.86) and (13.87) are the linearized Boltzmann equations for the normal and tracer components, respectively.

We can now decouple diffusion effects from viscous and thermal effects. Let us define the total distribution function as

$$h^+(\mathbf{p}_1, \mathbf{r}, t) = h_N(\mathbf{p}_1, \mathbf{r}, t) + h_T(\mathbf{p}_1, \mathbf{r}, t) \tag{13.88}$$

and the distribution function for the difference in normal and tracer distributions as

$$h^-(\mathbf{p}_1, \mathbf{r}, t) = h_N(\mathbf{p}_1, \mathbf{r}, t) - h_T(\mathbf{p}_1, \mathbf{r}, t). \tag{13.89}$$

As long as $h^-(\mathbf{p}_1, \mathbf{r}, t)$ is nonzero, diffusion will occur in the fluid.

The Boltzmann equations for the total distribution, $h^+(\mathbf{p}_1, \mathbf{r}, t)$, and for the difference distribution, $h^-(\mathbf{p}_1, \mathbf{r}, t)$, decouple. If we add Eqs. (13.86) and (13.87) we find

$$\frac{\partial h_1{}^+}{\partial t} + \dot{\mathbf{q}}_1 \cdot \frac{\partial h_1{}^+}{\partial r} = 2\int d\mathbf{p}_2 \int d\Omega g\sigma(\theta, g) f^0(\mathbf{p}_2)(h_{1'}^+ + h_{2'}^+ - h_1{}^+ - h_2{}^+). \tag{13.90}$$

Thus, the total distribution obeys a Boltzmann equation. If we subtract Eq. (13.87) from (13.86), we find

$$\frac{\partial h_1^-}{\partial t} + \dot{\mathbf{q}}_1 \cdot \frac{\partial h_1^-}{\partial \mathbf{r}} = 2\int d\mathbf{p}_2 \int d\Omega g\sigma(\theta, g) f^0(\mathbf{p}_2)(h_{1'}^- - h_1^-). \tag{13.91}$$

Eq. (13.91) is called the Lorentz-Boltzmann equation and is the kinetic equation describing diffusion. It is convenient to introduce the concept of a linearized collision operator. Let us first consider Eq. (13.90). We write it as

$$\frac{\partial h_1^+}{\partial t} + \dot{\mathbf{q}}_1 \cdot \frac{\partial h_1^+}{\partial \mathbf{r}} = \hat{C}_1^+ h_1^+ \tag{13.92}$$

where $\hat{C}_1^+$ is the linearized collision operator which when acting on some arbitrary function $g(\mathbf{p}_1)$ yields

$$\hat{C}_1^+ g(\mathbf{p}_1) = 2\int d\mathbf{p}_2 \int d\Omega g\sigma(\theta, g) f^0(\mathbf{p}_2)(g_{1'} + g_{2'} - g_1 - g_2). \tag{13.93}$$

Let us now introduce a scalar product of two functions $\phi(\mathbf{p}_1)$ and $\chi(\mathbf{p}_1)$,

$$\langle \phi, \chi \rangle \equiv \left(\frac{\beta}{2\pi m}\right)^{3/2} \int d\mathbf{p}_1\, e^{-\beta p_1^2/2m} \phi(\mathbf{p}_1)\chi(\mathbf{p}_1). \tag{13.94}$$

Using Eq. (13.94), it is easy to show that $\hat{C}^+$ is self-adjoint:

$$\langle \Phi, \hat{C}^+\chi \rangle = \langle \hat{C}^+\Phi, \chi \rangle \tag{13.95}$$

(we leave it as a homework problem).

If we write the eigenvalue equation for $\hat{C}^+$ in the form

$$\hat{C}_{\mathbf{p}}^+ \Psi(\mathbf{p}) = \lambda \Psi(\mathbf{p}), \tag{13.96}$$

we can show there are five eigenfunctions of $\hat{C}_{\mathbf{p}}^+$ with eigenvalue zero. They are the five additive constants of the motion: 1, $\mathbf{p}$, and $p^2/2m$. All other eigenvalues of $\hat{C}_{\mathbf{p}}^+$ are negative. We prove this by writing the expectation value of $\hat{C}^+$ in the form

$$\langle \Phi, \hat{C}^+\Phi \rangle = -\frac{N}{V}\left(\frac{\beta}{2\pi m}\right)^3 \iint d\mathbf{p}_1\, d\mathbf{p}_2 \int d\Omega\, e^{-(\beta/2m)(p_1^2+p_2^2)} \times g\sigma(\theta, g)(\Phi_{1'} + \Phi_{2'} - \Phi_1 - \Phi_2)^2. \tag{13.97}$$

Thus, $\langle \Phi, \hat{C}^+\Phi \rangle$ is less than or equal to zero for an arbitrary function $\Phi(\mathbf{p})$. It will be equal to zero only if $(\Phi_{1'} + \Phi_{2'} - \Phi_1 - \Phi_2) = 0$, that is, if $\Phi(\mathbf{p})$ is a linear combination of the five additive constants of motion 1, $\mathbf{p}$, and $p^2/2m$. Therefore, $\hat{C}^+$ is a negative semidefinite operator with a nonpositive spectrum of eigenvalues.

The linearized collision operator $\hat{C}^+$ has one other interesting property. It behaves as a scalar operator with respect to rotations in momentum space.

That is, $\hat{C}_{\mathbf{p}}{}^{+}h(\mathbf{p})$ transforms in the same way under rotation in momentum space as does $h(\mathbf{p})$. Therefore, eigenfunctions of $\hat{C}_{\mathbf{p}}{}^{+}$ have the form

$$\Psi_{r,l,m}(\mathbf{p}) = R_{r,l}(p)\,Y_{l,m}(\mathbf{p}/|\mathbf{p}|) \tag{13.98}$$

where $Y_{l,m}(\mathbf{p}/|\mathbf{p}|)$ are spherical harmonics.

Let us now consider a simple example. Let us solve the linearized Boltzmann equation (13.90) for the case of a spatially homogeneous system. Eq. (13.90) then becomes

$$\frac{\partial h_1{}^{+}}{\partial t} = \hat{C}^{+}h_1{}^{+} = 2\int dp_2 \int d\Omega g\sigma(\theta, g) f_2{}^0[h_{1'} + h_{2'} - h_1 - h_2] \tag{13.99}$$

where $h_1{}^{+} = h^{+}(\mathbf{p}_1, t)$, etc. If we assume that $h_1{}^{+}(\mathbf{p}_1, 0) = \sum_{r,l,m} A_{r,l,m}\Psi_{r,l,m}(\mathbf{p}_1)$, we may write the solution in the form

$$h_1{}^{+}(\mathbf{p}_1, t) = \sum_{r,l,m} e^{\lambda_{r,l}t} A_{r,l,m}\Psi_{r,l,m}(\mathbf{p}_1). \tag{13.100}$$

The fact that the eigenvalues $\lambda_{r,l}$ must all be negative or zero means that $h_1{}^{+}(\mathbf{p}_1, t)$ will decay to a time-independent quantity after long enough time and the system relaxes to equilibrium.

The Lorentz-Boltzmann collision operator, $C_1{}^{-}$, appearing in the equation for the difference distribution has similar properties. If we write

$$\frac{\partial h_1{}^{-}}{\partial t} + \dot{\mathbf{q}}_1 \cdot \frac{\partial h_1{}^{-}}{\partial \mathbf{r}} = C_1{}^{-}h_1{}^{-}, \tag{13.101}$$

the collision operator $\hat{C}_1{}^{-}$, when acting on an arbitrary function $g(\mathbf{p}_1)$ of $\mathbf{p}_1$, yields

$$C_1{}^{-}g(\mathbf{p}_1) = 2\int d\mathbf{p}_2 \int d\Omega g\sigma(\theta, g) f_2{}^0(g_{1'} - g_1). \tag{13.102}$$

The Lorentz-Boltzmann collision operator $C_1{}^{-}$ differs from $C_1{}^{+}$ in that it has only one zero eigenvalue, a constant, while $C_1{}^{+}$ has five.

We can now use the Boltzmann and Lorentz-Boltzmann equations to obtain microscopic expressions for the coefficients of self-diffusion, viscosity, and thermal conductivity.

E. COEFFICIENT OF SELF-DIFFUSION

We shall begin with a derivation of the coefficient of self-diffusion because it is the easiest to obtain. The method we use is due to Resibois[7] and consists of two steps. In the first step, we derive the linearized hydrodynamic equation from the Lorentz-Boltzmann equation and introduce the self-diffusion coefficient into the hydrodynamic equation using Fick's law. We then can find the dispersion relation for the hydrodynamic diffusion mode. In the second step we use Rayleigh-Schrödinger perturbation theory to obtain the hydrodynamic eigenvalues of the Lorentz-Boltzmann equation. We then match the eigenvalue of the hydrodynamic equation to that of the Lorentz-Boltzmann equation and thereby obtain a microscopic expression for the self-diffusion coefficient.

1. Linearized Hydrodynamic Equation

The difference in tracer and normal particle densities at some point $\mathbf{r}$ is given by

$$m(\mathbf{r}, t) = n_N(\mathbf{r}, t) - n_T(\mathbf{r}, t) = \int d\mathbf{p}_1 f^0(\mathbf{p}_1) h^-(\mathbf{p}_1, \mathbf{r}, t) \tag{13.103}$$

(at equilibrium this difference is zero). If we multiply the Lorentz-Boltzmann equation by $f^0(\mathbf{p}_1)$ and integrate over $\mathbf{p}_1$, we obtain

$$\frac{\partial}{\partial t} m(\mathbf{r}, t) + \nabla_{\mathbf{r}} \cdot \mathbf{J}^D(\mathbf{r}, t) = 0 \tag{13.104}$$

where $\mathbf{J}^D(\mathbf{r}, t)$ is the diffusion current and is defined microscopically as

$$\mathbf{J}^D(\mathbf{r}, t) = \int d\mathbf{p}_1 \frac{\mathbf{p}_1}{m} f^0(\mathbf{p}_1) h^-(\mathbf{p}_1, \mathbf{r}, t). \tag{13.105}$$

The contribution from the collision term is identically zero because $C_{\mathbf{p}}^{(-)} 1 = 0$. We now introduce the self-diffusion coefficient using Fick's law,

$$\mathbf{J}^D(\mathbf{r}, t) = -D \nabla_{\mathbf{r}} m(\mathbf{r}, t). \tag{13.106}$$

If we combine Eqs. (13.105) and (13.106), we obtain the following hydrodynamic equation for the self-diffusion process:

$$\frac{\partial}{\partial t} m(\mathbf{r}, t) = D \nabla_{\mathbf{r}}^2 m(\mathbf{r}, t). \tag{13.107}$$

To find the dispersion relation for hydrodynamic modes we define the Fourier transform

$$m(\mathbf{r}, t) = \frac{1}{(2\pi)^4} \int d\mathbf{k} \int d\omega \, e^{i(\mathbf{k} \cdot \mathbf{r} - \omega t)} \bar{m}(\mathbf{k}, \omega) \tag{13.108}$$

which allows us to study each Fourier component of the diffusion equation separately. If we substitute Eq. (13.108) into Eq. (13.107) we obtain

$$-i\omega \bar{m}(\mathbf{k}, \omega) + Dk^2 \bar{m}(\mathbf{k}, \omega) = 0 \tag{13.109}$$

(different Fourier components do not couple, because the hydrodynamic equation is linear). From Eq. (13.109) we obtain the following dispersion relation for the self-diffusion mode:

$$\omega = -iDk^2. \tag{13.110}$$

The diffusion frequency is imaginary, which means that the contribution to the density $m(\mathbf{r}, t)$ with wave vector, k, dies out in a time which depends on the diffusion coefficient and the wave vector, k,

$$m(\mathbf{r}, t) \approx e^{i\mathbf{k}\cdot\mathbf{r}} e^{-Dk^2t}. \tag{13.111}$$

Thus, very long wavelength disturbances take a long time to decay away. This is the characteristic behavior of a hydrodynamic mode. Since the identity of the particles is preserved in each collision, the only way to cause differences in the density of normal and tracer particles to disappear is to physically transport particles from one part of the fluid to another. For very long wavelength disturbances the equalization takes a long time since the particles must be transported over long distances.

2. Eigenfrequencies of the Lorentz-Boltzmann Equation

We can obtain the hydrodynamic eigenfrequencies of the Lorentz-Boltzmann equation in terms of a perturbation expansion in powers of the wave vector, k, by using Rayleigh-Schrödinger perturbation theory. After this has been done, we can equate the eigenfrequency of the hydrodynamic equation to the hydrodynamic eigenfrequency of the Lorentz-Boltzmann equation and thereby obtain a microscopic expression for the coefficient of self-diffusion.

Since Eq. (13.101) is a linear equation for $h^-(\mathbf{p}_1, \mathbf{r}, t)$, we need only consider the equation for one Fourier component and we write

$$h^-(\mathbf{p}, \mathbf{r}, t) = |\Psi_n(\mathbf{p}, \mathbf{k})\rangle_- e^{i\mathbf{k}\cdot\mathbf{r}} e^{-i\omega_n t}; \tag{13.112}$$

we obtain the following eigenvalue equation for $|\Psi_n(\mathbf{p}, \mathbf{k})\rangle$:

$$(C_{\mathbf{p}}^- - ik\boldsymbol{\varkappa}\cdot\mathbf{p})|\Psi_n(\mathbf{p}, \mathbf{k})\rangle = -i\omega_n^-|\Psi_n(\mathbf{p}, \mathbf{k})\rangle \tag{13.113}$$

where $\boldsymbol{\varkappa} = \mathbf{k}/|\mathbf{k}|$ and the eigenfunctions, $|\Psi_n(\mathbf{q}, \mathbf{k})\rangle_-$, are assumed to be orthonormal. For long-wavelength hydrodynamic disturbances, k will be a small parameter. Thus, we can use Rayleigh-Schrödinger perturbation theory to obtain a perturbation expansion for ω_n in powers of k.

Let us assume that both ω_n^- and $|\Psi_n(\mathbf{p}, \mathbf{k})\rangle_-$ can be expanded in powers of k. Then

$$\omega_n = \omega_n^{(0)} + k\omega_n^{(1)} + k^2\omega_n^{(2)} + \cdots \tag{13.114}$$

and

$$|\Psi_n\rangle_- = |\Psi_n^{(0)}\rangle_- + k|\Psi_n^{(1)}\rangle_- + k^2|\Psi_n^{(2)}\rangle_- + \cdots \tag{13.115}$$

If we substitute Eqs. (13.114) and (13.115) into Eq. (13.113), we obtain the following expression for the frequencies $\omega_{n(0)}^-$, $\omega_{n(1)}^-$, and $\omega_{n(2)}^-$:

$$\omega_n^{(0)} = i\,{}_-\langle\Psi_n^{(0)}|\hat{C}_{\mathbf{p}}^-|\Psi_n^{(0)}\rangle_-, \tag{13.116}$$

$$\omega_n^{(1)} = {}_-\langle\Psi_n^{(0)}|\boldsymbol{\varkappa}\cdot\frac{\mathbf{p}}{m}|\Psi_n^{(0)}\rangle_-, \tag{13.117}$$

and

$$\omega_n^{(2)} = {}_-\langle\Psi_n^{\prime(0)}|\left(\varkappa\cdot\frac{\mathbf{p}}{m} - \omega_{n(1)}\right)\frac{1}{-i\hat{C}_{\mathbf{p}}^{-} + \omega_n^{(0)}}\left(\varkappa\cdot\frac{\mathbf{p}}{m} - \omega_{n(1)}\right)|\Psi_n^{\prime(0)}\rangle_- \tag{13.118}$$

where the matrix elements are defined as in Eq. (13.94) and $\langle\Psi_{n'}^{\prime 0}|\Psi_n^{\prime 0}\rangle \equiv \delta_{n,n'}$. At this point we shall restrict our attention to the eigenvalue which reduces to zero when $k \to 0$. This corresponds to the hydrodynamic mode. There will be only one such eigenvalue of $\hat{C}_{\mathbf{p}}^{-}$ since there is only one zero eigenfunction of $\hat{C}_{\mathbf{p}}^{-}$, namely a constant.

We will let $|\Psi_1^{\prime(0)}\rangle_-$ denote the zero eigenfunction of $\hat{C}_{\mathbf{p}}^{-}$ and we will normalize it using the scalar product in Eq. (13.94). Thus,

$$|\Psi_1^{\prime(0)}\rangle_- = 1 \tag{13.119}$$

and we find

$$\omega_1^{(0)} = 0, \tag{13.120}$$

$$\omega_1^{(1)} = \frac{1}{m}\left(\frac{\beta}{2\pi m}\right)^{3/2}\int d\mathbf{p}_1\, e^{-\beta p_1^2/2m}\varkappa\cdot\mathbf{p}_1 \equiv 0, \tag{13.121}$$

and

$$\omega_1^{(2)} = -\frac{1}{m^2}\left(\frac{\beta}{2\pi m}\right)^{3/2}\int d\mathbf{p}_1\, e^{-\beta p_1^2/2m}\varkappa\cdot\mathbf{p}_1\frac{1}{i\hat{C}_{\mathbf{p}}^{-}}\varkappa\cdot\mathbf{p}_1. \tag{13.122}$$

The hydrodynamic eigenfrequency, ω_1, has the correct hydrodynamic behavior. If we now equate Eqs. (13.110) and (13.114), and use Eqs. (13.120)–(13.122), we find

$$D = -\frac{1}{m^2}\left(\frac{\beta}{2\pi m}\right)^{3/2}\int d\mathbf{p}_1\, e^{-\beta p_1^2/2m}\varkappa\cdot\mathbf{p}_1\frac{1}{\hat{C}_{\mathbf{p}}^{(-)}}\varkappa\cdot\mathbf{p}_1. \tag{13.123}$$

Thus, we have obtained a microscopic expression for D. We will discuss how to evaluate the integral in Eq. (13.123) in Sec. 13.G.

F. COEFFICIENTS OF VISCOSITY AND THERMAL CONDUCTIVITY

The Boltzmann collision operator, $\hat{C}_{\mathbf{p}}^{(+)}$, has five zero eigenvalues (it is fivefold degenerate) and therefore we can derive from the linearized Boltzmann equation five linearized hydrodynamic equations: one equation for the total particle density, $n(\mathbf{r}, t) = n_N(\mathbf{r}, t) + n_T(\mathbf{r}, t)$; three equations for the three components of the momentum density, $mn(\mathbf{r}, t)\mathbf{v}(\mathbf{r}, t)$, where $\mathbf{v}(\mathbf{r}, t)$ is the average velocity; and one equation for the average internal energy density. To find the hydrodynamic eigenfrequencies (which will be expressed in terms of the coefficients of viscosity and thermal conductivity), we must find the normal modes of the system of five hydrodynamic equations. We then use the Rayleigh-Schrödinger perturbation theory to find the five *microscopic* hydrodynamic frequencies of the Boltzmann equation. Once they are found we can match them to the five normal mode frequencies of the hydrodynamic equations and thereby obtain microscopic expressions for the coefficients of viscosity and thermal conductivity.

1. Normal Mode Frequencies of the Hydrodynamic Equations[8]

The average particle density is defined by

$$n(\mathbf{r}, t) = \int d\mathbf{p}_1 f^0(\mathbf{p}_1) h^+(\mathbf{p}_1, \mathbf{r}, t). \tag{13.124}$$

If we multiply Eq. (13.90) by $f^0(\mathbf{p}_1)$, integrate over $\mathbf{p}_1$, and make use of the fact that $C_{\mathbf{p}}^{(+)}1 = 0$, we obtain

$$\frac{\partial}{\partial t} n(\mathbf{r}, t) + \nabla_{\mathbf{r}} \cdot \mathbf{J}^n(\mathbf{r}, t) = 0 \tag{13.125}$$

where $\mathbf{J}^n(\mathbf{r}, t)$ is the average particle current and is defined

$$\mathbf{J}^n(\mathbf{r}, t) = \int d\mathbf{p} \frac{\mathbf{p}}{m} f^0(\mathbf{p}) h^+(\mathbf{p}, \mathbf{r}, t). \tag{13.126}$$

It is useful to decompose $\mathbf{J}^n(\mathbf{r}, t)$ into the product $\mathbf{J}^n(\mathbf{r}, t) = n(\mathbf{r}, t)\mathbf{v}(\mathbf{r}, t)$. In the linear regime, $\mathbf{J}^n(\mathbf{r}, t) \approx n_0 \mathbf{v}(\mathbf{r}, t)$, where n_0 is the equilibrium particle density. Eq. (13.125) then becomes

$$\frac{\partial}{\partial t} n(\mathbf{r}, t) + n_0 \nabla_{\mathbf{r}} \cdot \mathbf{v}(\mathbf{r}, t) = 0. \tag{13.127}$$

Eq. (13.127) is the linearized continuity equation and describes the conservation of total particle number (cf. App. A). The continuity equation is entirely reactive. If we reverse time in Eq. (13.127), we do not change the form of the equation. Thus, the continuity equation contains no irreversible effects. These must come from the other hydrodynamic equations.

If we multiply Eq. (13.90) by $\mathbf{p}_1 f^0(\mathbf{p}_1)$, integrate over $\mathbf{p}_1$, and use the fact that $C_{\mathbf{p}}^{(+)}\mathbf{p} \equiv 0$, we obtain

$$m \frac{\partial}{\partial t} \mathbf{J}^n(\mathbf{r}, t) = -\nabla_{\mathbf{r}} \cdot \bar{\bar{P}}(\mathbf{r}, t) \tag{13.128}$$

where $\bar{\bar{P}}(\mathbf{r}, t)$ is the pressure tensor and is defined

$$\bar{\bar{P}}(\mathbf{r}, t) = \frac{1}{m} \int d\mathbf{p} f^0(\mathbf{p}) \mathbf{p}\mathbf{p} h^+(\mathbf{p}, \mathbf{r}, t) \tag{13.129}$$

(we have let $\dot{\mathbf{q}} = \mathbf{p}/m$). The pressure tensor describes the momentum flux, or current, in the system and contains an irreversible part due to viscous effects.

To obtain the equation for the internal energy density, we must take the average of the thermal kinetic energy $(1/2m)(\mathbf{p} - m\mathbf{v}(\mathbf{r}, t))^2$. However, in the linear approximation $\mathbf{v}(\mathbf{r}, t)$ does not contribute. Thus, to find the hydrodynamic equation for the internal energy density, we can multiply Eq. (13.90) by $p_1{}^2/2m$ and integrate over $\mathbf{p}_1$. If we use the fact that $C_{\mathbf{p}}^{(+)}p^2 \equiv 0$, we obtain

$$\frac{\partial}{\partial t} u(\mathbf{r}, t) = -\nabla_{\mathbf{r}} \cdot \mathbf{J}^u(\mathbf{r}, t) \tag{13.130}$$

where $u(r, t)$ is the internal energy per unit volume,

$$u(\mathbf{r}, t) = \int d\mathbf{p} \frac{p^2}{2m} f^0(\mathbf{p}) h^+(\mathbf{p}, \mathbf{r}, t), \tag{13.131}$$

and $\mathbf{J}^u(\mathbf{r}, t)$ is the internal energy current,

$$\mathbf{J}^u(\mathbf{r}, t) = \int d\mathbf{p} \frac{p^2}{2m} \frac{\mathbf{p}}{m} f^0(\mathbf{p}) h^+(\mathbf{p}, \mathbf{r}, t). \tag{13.132}$$

We can write the internal energy density in the form $u(\mathbf{r}, t) = n(\mathbf{r}, t)\, e(\mathbf{r}, t)$, where $e(\mathbf{r}, t)$ is the internal energy per particle. In the linear approximation we find $u(\mathbf{r}, t) = n_0\, e(\mathbf{r}, t) + e_0 n(\mathbf{r}, t)$, and Eq. (13.130) takes the form

$$n_0 \frac{\partial}{\partial t} e(\mathbf{r}, t) + e_0 \frac{\partial}{\partial t} n(\mathbf{r}, t) = -\nabla_{\mathbf{r}} \cdot \mathbf{J}^u(\mathbf{r}, t), \tag{13.133}$$

where e_0 is the equilibrium internal energy per particle. The current $\mathbf{J}^u(\mathbf{r}, t)$ will contain an irreversible part due to thermal conduction.

Eqs. (13.127), (13.128), and (13.133) are the linearized equations of motion for the average particle density, $n(\mathbf{r}, t)$, the momentum density, $m\mathbf{J}^n(\mathbf{r}, t)$, and the internal energy density, $u(\mathbf{r}, t)$. They govern the variation in time and space of these quantities. However, they cannot yet be solved because they contain too many unknowns. As a next step we must find expressions for $\overline{\overline{P}}(\mathbf{r}, t)$ and $\mathbf{J}^u(\mathbf{r}, t)$.

Eq. (13.128) describes the variation in time and space of the average velocity variations $\mathbf{v}(\mathbf{r}, t)$. From Sec. 13.B, we know that, in a real fluid, viscous effects will cause velocity variations to die out. Thus, the pressure tensor must contain the effects of viscous processes in the fluid, that is, effects proportional to gradients in the velocity. Furthermore, we know that a pressure gradient can cause fluid flow and therefore such effects must also be included in $\overline{\overline{P}}(\mathbf{r}, t)$.

For the simple system we are considering, the pressure tensor is symmetric. We can see this immediately from Eq. (13.129). From the discussion in App. C, the pressure tensor breaks into a scalar part and a traceless symmetric part. Since we are dealing with an isotropic fluid, Curie's law (cf. App. C) requires that the scalar part and the traceless symmetric part each have a special form resulting from rotation and inversion symmetry. For the system we are considering, the final expression for the pressure tensor in the linear approximation takes the form

$$P_{ij}(\mathbf{r}, t) = P(\mathbf{r}, t)\,\delta_{ij} - \zeta\,\delta_{ij}(\nabla_{\mathbf{r}}\cdot\mathbf{v}) - \eta\left(\frac{\partial v_i}{\partial x_j} + \frac{\partial v_j}{\partial x_i} - \tfrac{2}{3}\,\delta_{ij}\nabla_{\mathbf{r}}\cdot\mathbf{v}\right) \tag{13.134}$$

where $P(\mathbf{r}, t)$ is the hydrostatic pressure, ζ is the coefficient of bulk viscosity, and η is the coefficient of shear viscosity. If we substitute Eq. (13.134) into Eq. (13.128) we obtain, after some rearrangement,

$$mn_0\frac{\partial}{\partial t}\mathbf{v}(\mathbf{r}, t) = -\nabla_{\mathbf{r}}P(\mathbf{r}, t) + \eta\nabla_{\mathbf{r}}^2\mathbf{v}(\mathbf{r}, t) + (\zeta + \tfrac{1}{3}\eta)\nabla_{\mathbf{r}}(\nabla_{\mathbf{r}}\cdot\mathbf{v}(\mathbf{r}, t)). \tag{13.135}$$

The contribution from the hydrostatic pressure term is reactive: it has the same dependence under time reversal as the left-hand side of Eq. (13.135); whereas, the viscous contribution is dissipative: it changes sign relative to the left-hand side. All terms in Eq. (13.135) are linear in deviations from equilibrium and depend only on first-order gradients. Thus, we have assumed that all spatial variations have very long wavelengths and higher order gradients can be neglected.

Let us now consider the energy equation (13.133). The energy current will have a contribution from the convection of internal energy, a contribution involving the pressure which comes from work done in compression or expansion of regions in the fluid, and a contribution from heat conduction. Thus, in the linear approximation we can write

$$\mathbf{J}^u(\mathbf{r}, t) = n_0\,e_0\mathbf{v}(\mathbf{r}, t) + P_0\mathbf{v}(\mathbf{r}, t) - K\nabla_{\mathbf{r}}T(\mathbf{r}, t) \tag{13.136}$$

where P_0 is the equilibrium pressure, K is the coefficient of thermal conductivity, and $T(\mathbf{r}, t)$ is the temperature. If we substitute Eq. (13.136) into Eq. (13.133) and make use of Eq. (13.127), we find

$$n_0\frac{\partial}{\partial t}e(\mathbf{r}, t) = -P_0\nabla_{\mathbf{r}}\cdot\mathbf{v}(\mathbf{r}, t) + K\nabla_{\mathbf{r}}^2T(\mathbf{r}, t). \tag{13.137}$$

Eq. (13.137) can be simplified if we write it in terms of the local entropy. If we use the thermodynamic relation

$$de = T\,ds + \frac{P}{n^2}\,dn, \tag{13.138}$$

we can write

$$\frac{\partial e}{\partial t} = T\frac{\partial s}{\partial t} + \frac{P}{n^2}\frac{\partial n}{\partial t}. \tag{13.139}$$

If we combine Eqs. (13.137) and (13.139) and again use Eq. (13.127), we obtain

$$n_0 T_0 \frac{\partial}{\partial t} s(\mathbf{r}, t) = K \nabla_{\mathbf{r}}^2 T(\mathbf{r}, t) \tag{13.140}$$

where T_0 is the equilibrium temperature and $s(\mathbf{r}, t)$ is the local entropy per particle.

In Eqs. (13.127), (13.135), and (13.140) we have now expressed the five linearized hydrodynamic equations in terms of the thermodynamic densities. The five equations involve seven variables, but only five variables can be independent. We shall choose the entropy and pressure as the independent variables and expand the rest in terms of them. Thus,

$$n(\mathbf{r}, t) = \left(\frac{\partial n}{\partial P}\right)_s^0 P(\mathbf{r}, t) + \left(\frac{\partial n}{\partial s}\right)_P^0 s(\mathbf{r}, t) \tag{13.141}$$

and

$$T(\mathbf{r}, t) = \left(\frac{\partial T}{\partial P}\right)_s^0 P(\mathbf{r}, t) + \left(\frac{\partial T}{\partial s}\right)_P^0 s(\mathbf{r}, t) \tag{13.142}$$

(we remember that $n(\mathbf{r}, t)$, $T(\mathbf{r}, t)$, $P(\mathbf{r}, t)$, and $s(\mathbf{r}, t)$ go to zero at equilibrium). All partial derivatives refer to the equilibrium state.

If we substitute Eqs. (13.141) and (13.142) into Eqs. (13.127) and (13.140), we obtain the following linearized hydrodynamic equations:

$$\left(\frac{\partial n}{\partial s}\right)_P^0 \frac{\partial s}{\partial t} + \left(\frac{\partial n}{\partial P}\right)_s^0 \frac{\partial P}{\partial t} + n_0 \nabla_{\mathbf{r}} \cdot \mathbf{v} = 0, \tag{13.143}$$

$$mn_0 \frac{\partial}{\partial t} \mathbf{v} = -\nabla_{\mathbf{r}} P + \eta \nabla_{\mathbf{r}}^2 \mathbf{v} + (\zeta + \tfrac{1}{3}\eta) \nabla_{\mathbf{r}} (\nabla_{\mathbf{r}} \cdot \mathbf{v}), \tag{13.144}$$

and

$$T_0 n_0 \frac{\partial s}{\partial t} = K \left(\frac{\partial T}{\partial s}\right)_P^0 \nabla_{\mathbf{r}}^2 s + K \left(\frac{\partial T}{\partial P}\right)_s^0 \nabla_{\mathbf{r}}^2 P. \tag{13.145}$$

If we wish to obtain the hydrodynamic equations for an ideal fluid, we can set the transport coefficients equal to zero in Eqs. (13.143)–(13.145). We can then combine the three equations to obtain the following equation of motion for pressure variations in an ideal fluid:

$$\frac{\partial^2 P}{\partial t^2} - \frac{1}{m} \left(\frac{\partial P}{\partial n}\right)_s^0 \nabla_{\mathbf{r}}^2 P = 0. \tag{13.146}$$

Eq. (13.146) is the equation of motion of sound waves in an ideal fluid where the speed of sound is given by

$$c_0 = \sqrt{\frac{1}{m} \left(\frac{\partial P}{\partial n}\right)_s^0}, \tag{13.147}$$

as it should be.

Eqs. (13.143)–(13.145) are coupled equations for variations in average velocity, entropy, and pressure. We now must find the normal mode solutions. Since the equations are linear, each Fourier component will propagate independently. Thus, we only need to consider a single Fourier component. We will write

$$P(\mathbf{r}, t) = P_{\mathbf{k}}(\omega)\, e^{i(\mathbf{k}\cdot\mathbf{r} - \omega t)}, \tag{13.148}$$

$$s(\mathbf{r}, t) = s_{\mathbf{k}}(\omega)\, e^{i(\mathbf{k}\cdot\mathbf{r} - \omega t)}, \tag{13.149}$$

and

$$\mathbf{v}(\mathbf{r}, t) = \mathbf{v}_{\mathbf{k}}(\omega)\, e^{i(\mathbf{k}\cdot\mathbf{r} - \omega t)}. \tag{13.150}$$

The wave vector, **k**, denotes the direction of propagation of the hydrodynamic wave. If we substitute Eqs. (13.148)–(13.150) into Eqs. (13.143)–(13.145), we obtain

$$\omega\left(\frac{\partial n}{\partial s}\right)_P^0 s_{\mathbf{k}}(\omega) + \omega\left(\frac{\partial n}{\partial P}\right)_s^0 P_{\mathbf{k}}(\omega) - n_0\mathbf{k}\cdot\mathbf{v}_{\mathbf{k}}(\omega) = 0, \tag{13.151}$$

$$mn_0\omega\mathbf{v}_{\mathbf{k}}(\omega) - \mathbf{k}P_{\mathbf{k}}(\omega) = -ik^2\eta\mathbf{v}_{\mathbf{k}}(\omega) - i(\zeta + \tfrac{1}{3}\eta)\mathbf{k}(\mathbf{k}\cdot\mathbf{v}_{\mathbf{k}}(\omega)), \tag{13.152}$$

and

$$T_0n_0\omega s_{\mathbf{k}}(\omega) = -iK\left(\frac{\partial T}{\partial s}\right)_P^0 k^2 s_{\mathbf{k}}(\omega) - ik^2K\left(\frac{\partial T}{\partial P}\right)_s^0 P_{\mathbf{k}}(\omega). \tag{13.153}$$

Eqs. (13.152) and (13.153) may be simplified if we divide the average velocity $\mathbf{v}$ into a longitudinal part $\mathbf{v}^{\parallel}$ and a transverse part $\mathbf{v}^{\perp}$:

$$\mathbf{v} = \mathbf{v}^{\perp} + v^{\parallel}\boldsymbol{\kappa} \tag{13.154}$$

where $\mathbf{v}^{\perp}\cdot\mathbf{k} = 0$. Eqs. (13.151)–(13.153) can then be written in the form

$$(mn_0\omega + ik^2\eta)\mathbf{v}_{\mathbf{k}}^{\perp}(\omega) = 0, \tag{13.155}$$

$$\omega\left(\frac{\partial n}{\partial s}\right)_P^0 s_{\mathbf{k}}(\omega) + \omega\left(\frac{\partial n}{\partial P}\right)_s^0 P_{\mathbf{k}}(\omega) - n_0kv_{\mathbf{k}}^{\parallel}(\omega) = 0, \tag{13.156}$$

$$-kP_{\mathbf{k}}(\omega) + [mn_0\omega + ik^2(\tfrac{4}{3}\eta + \zeta)]v_{\mathbf{k}}^{\parallel}(\omega) = 0, \tag{13.157}$$

and

$$\left[\omega + ik^2\frac{K}{T_0n_0}\left(\frac{\partial T}{\partial s}\right)_P^0\right]s_{\mathbf{k}}(\omega) + ik^2\frac{K}{n_0T_0}\left(\frac{\partial T}{\partial P}\right)_s^0 P_{\mathbf{k}}(\omega) = 0. \tag{13.158}$$

Eq. (13.155) gives the dispersion relation for the transverse velocity waves (shear waves). We see that

$$\omega = -i\frac{\eta k^2}{mn_0}. \tag{13.159}$$

Thus, the shear waves behave very much like the diffusion waves. Any shear disturbance will be damped out.

A solution for the coupled entropy, pressure, and longitudinal velocity modes exists if the determinant of the coefficients of $s_{\mathbf{k}}(\omega)$, $p_{\mathbf{k}}(\omega)$, and $v_{\mathbf{k}}^{\parallel}(\omega)$ is zero, that is, if

$$\det\begin{pmatrix} \omega\left(\dfrac{\partial n}{\partial s}\right)_P^0 & \omega\left(\dfrac{\partial n}{\partial P}\right)_s^0 & -n_0k \\ 0 & -k & [mn_0\omega + ik^2(\tfrac{4}{3}\eta + \zeta)] \\ \left[\omega + ik^2\dfrac{K}{T_0n_0}\left(\dfrac{\partial T}{\partial s}\right)_P^0\right] & ik^2\dfrac{K}{n_0T_0}\left(\dfrac{\partial T}{\partial P}\right)_s^0 & 0 \end{pmatrix} = 0. \tag{13.160}$$

Eq. (13.160) may be rewritten

$$\left[\omega^2 - \frac{k^2}{m}\left(\frac{\partial P}{\partial n}\right)_s^0 + i\frac{\omega k^2}{mn_0}(\tfrac{4}{3}\eta + \zeta)\right]\left[\omega + ik^2\frac{K}{n_0T_0}\left(\frac{\partial T}{\partial s}\right)_P^0\right]$$
$$- i\omega k^2\frac{K}{n_0T_0}\left(\frac{\partial T}{\partial P}\right)_s^0\left(\frac{\partial n}{\partial s}\right)_P^0\left(\frac{\partial P}{\partial n}\right)_s^0\left[\omega + i\frac{k^2}{mn_0}(\tfrac{4}{3}\eta + \zeta)\right] = 0. \tag{13.161}$$

If we use the fact that $c_P = T(\partial s/\partial T)_P^0$ and $c_v = T(\partial s/\partial T)_n^0$ (c_v and c_P are heat capacities per particle), and the identity

$$\left(\frac{\partial T}{\partial P}\right)_s^0\left(\frac{\partial n}{\partial s}\right)_P^0\left(\frac{\partial P}{\partial n}\right)_s^0 = \left(\frac{\partial T}{\partial s}\right)_P^0 - \left(\frac{\partial T}{\partial s}\right)_n^0 = -T\left(\frac{1}{c_v} - \frac{1}{c_P}\right), \tag{13.162}$$

we may rewrite Eq. (13.161) as

$$\left(\omega + ik^2\frac{K}{n_0c_P}\right)\left\{\omega^2 - k^2c_0^2 + i\omega\frac{k^2}{mn_0}\left[\tfrac{4}{3}\eta + \zeta + mK\left(\frac{1}{c_v} - \frac{1}{c_P}\right)\right]\right\}$$
$$- \frac{\omega k^4K}{mn_0^2}\left(\frac{1}{c_v} - \frac{1}{c_P}\right)\left(\tfrac{4}{3}\eta + \zeta - \frac{mK}{c_P}\right) = 0 \tag{13.163}$$

where c_0^2 is the speed of sound in the absence of dissipation. The solutions to Eq. (13.163) may be expanded in powers of k. There are three solutions. To order k^2, they have the following form:

$$\omega_1 = -i\left[\frac{Kk^2}{n_0c_P}\right] \tag{13.164}$$

and

$$\omega_\pm = \pm c_0k - \frac{ik^2}{2mn_0}\left[\tfrac{4}{3}\eta + \zeta + mK\left(\frac{1}{c_v} - \frac{1}{c_P}\right)\right]. \tag{13.165}$$

The first solution (Eq. (13.164)) is purely imaginary and the second two (Eq. (13.165)) are complex. (Small k means long-wavelength disturbances.)

If we substitute ω_1 into Eqs. (13.156)–(13.158) and solve for $s_{\mathbf{k}}(\omega)$, $p_{\mathbf{k}}(\omega)$, and $v_{\mathbf{k}}^{\|}(\omega)$, we find that the lowest order dependence on k for each of these amplitudes must be

$$s_{\mathbf{k}}(\omega) \sim 1, \qquad v_{\mathbf{k}}^{\|}(\omega) \sim k, \quad \text{and} \quad p_{\mathbf{k}} \sim k^2.$$

Therefore, the wave of frequency ω_1 corresponds predominantly to an entropy (or heat) wave. It is damped out exponentially with time.

If we substitute $\omega_\pm$ into Eqs. (13.156)–(13.158), we find that

$$p_{\mathbf{k}} \sim v_{\mathbf{k}}^{\|} \sim 1 \quad \text{and} \quad s_{\mathbf{k}} \sim k.$$

Therefore, the waves of frequency $\omega_\pm$ (waves going in opposite directions) correspond to pressure or longitudinal velocity waves (sound waves). These waves propagate, but in the presence of transport processes they eventually get damped out. Thus, at long wavelengths, there are two shear modes, one heat mode, and two sound modes in the fluid. The frequencies in Eqs. (13.159), (13.164), and (13.165) are the normal mode frequencies of the system. These five frequencies, together with the diffusion frequency, Eq. (13.110), give the six normal mode frequencies of this two-component, dilute gas of dynamically identical particles.

2. Eigenfrequencies of the Boltzmann Equation[7,9]

The second step in deriving microscopic expression for the thermal conductivity and shear viscosity is to obtain the hydrodynamic eigenfrequencies of the linearized Boltzmann equation. Just as in Sec. 13.E, we only need to consider one Fourier component of the linearized Boltzmann equation. If we let

$$h^+(\mathbf{p}, \mathbf{r}, t) = |\Psi_n(\mathbf{p}, \mathbf{k})\rangle_+ \, e^{i\mathbf{k}\cdot\mathbf{r}} \, e^{-i\omega_n t}, \tag{13.166}$$

we obtain the following eigenvalue equation for the eigenvectors:

$$(\hat{C}_{\mathbf{p}}^+ - ik\boldsymbol{\kappa}\cdot\mathbf{p}/m)|\Psi_n(\mathbf{p}, \mathbf{k})\rangle_+ = -i\omega_n|\Psi_n(\mathbf{p}, \mathbf{k})\rangle_+. \tag{13.167}$$

To obtain a perturbation expansion for ω_n in powers of k, we must first look at the eigenvalue problem for the unperturbed operator $\hat{C}_{\mathbf{p}}^+$. We will denote the eigenfunctions of $\hat{C}_{\mathbf{p}}^+$ by $|\phi_n\rangle$. Then

$$\hat{C}_{\mathbf{p}}^+|\phi_n\rangle = -i\omega_n^0|\phi_n\rangle. \tag{13.168}$$

We know that $\hat{C}_{\mathbf{p}}^+$ has five zero eigenvalues and, therefore, five degenerate eigenfunctions. Orthonormalized eigenfunctions of zero eigenvalue are given by

$$|\phi_1\rangle = 1, \tag{13.169a}$$

$$|\phi_2\rangle = \sqrt{\frac{\beta}{m}}\, p_x, \tag{13.169b}$$

$$|\phi_3\rangle = \sqrt{\frac{\beta}{m}}\, p_y, \tag{13.169c}$$

$$|\phi_4\rangle = \sqrt{\frac{\beta}{m}}\, p_z, \tag{13.169d}$$

and

$$|\phi_5\rangle = \sqrt{\frac{2}{3}}\left(-\frac{3}{2} + \frac{\beta}{2m}p^2\right). \tag{13.169e}$$

We shall denote the five eigenfunctions collectively as $|\phi_\alpha\rangle$, where $\alpha = 1, 2, 3, 4,$ and 5. All other eigenfunctions will be denoted collectively as $|\phi_\beta\rangle$ where $\beta = 6, \ldots, \infty$. The eigenfunctions $|\phi_n\rangle$ are assumed orthonormal with respect to the scalar product in Eq. (13.94).

Since the eigenfunctions $|\phi_\alpha\rangle$ are degenerate, we must first find the proper linear combination of them to use for the zero-order approximation to the exact eigenfunctions $|\Psi_\alpha\rangle_+$. That is, we must find some $|\Psi_\alpha^{(0)}\rangle_+$ such that

$$|\Psi_\alpha\rangle_+ = |\Psi_\alpha^{(0)}\rangle_+ + k|\Psi_\alpha^{(1)}\rangle_+ + k^2|\Psi_\alpha^{(2)}\rangle_+ + \cdots \tag{13.170}$$

where

$$|\Psi_\alpha^{(0)}\rangle_+ = \sum_{\alpha'} c_{\alpha\alpha'}|\phi_\alpha\rangle. \tag{13.171}$$

The process of determining the coefficients $c_{\alpha\alpha'}$ will also give us the first-order term, $\omega_\alpha^{(1)}$, in the perturbation expansion of ω_α:

$$\omega_\alpha = \omega_\alpha^{(0)} + k\omega_\alpha^{(1)} + k^2\omega_\alpha^{(2)} + \cdots \tag{13.172}$$

To find the coefficients $c_{\alpha\alpha'}$ and $\omega_\alpha^{(1)}$, let us first insert the expansions in Eqs. (13.170) and (13.172) into Eq. (13.167) and equate coefficients of order k. We then obtain

$$\hat{C}_{\mathbf{p}}{}^{+}|\Psi_\alpha^{(1)}\rangle_+ = i\left(-\omega_\alpha^{(1)} + \frac{\boldsymbol{\varkappa}\cdot\mathbf{p}}{m}\right)|\Psi_\alpha^{(0)}\rangle_+. \tag{13.173}$$

If we multiply Eq. (13.172) by $\langle\phi_{\alpha''}|$ we obtain

$$\omega_\alpha^{(1)}\langle\phi_{\alpha''}, \Psi_\alpha^{(0)}\rangle_+ = \langle\phi_{\alpha''}, (\boldsymbol{\varkappa}\cdot\mathbf{p}/m)\Psi_\alpha^{(0)}\rangle_+ \tag{13.174}$$

(note that since $\hat{C}_{\mathbf{p}}^{(+)}$ is self-adjoint, $\langle\phi_{\alpha''}|$ is the same as its dual vector $|\phi_{\alpha''}\rangle$), or, using Eq. (13.170), we obtain

$$\omega_\alpha^{(1)}c_{\alpha\alpha''} = \sum_{\alpha'} W_{\alpha''\alpha'}c_{\alpha\alpha'} \tag{13.175}$$

where

$$W_{\alpha''\alpha'} = \left\langle \phi_{\alpha''}\, \frac{\boldsymbol{\varkappa}\cdot\mathbf{p}}{m}\, \phi_{\alpha'} \right\rangle$$

$$= \left(\frac{\beta}{2\pi m}\right)^{3/2} \int d\mathbf{p}\; e^{-\beta p^2/2m}\phi_{\alpha''}(\mathbf{p})\, \frac{\boldsymbol{\varkappa}\cdot\mathbf{p}}{m}\, \phi_{\alpha'}(\mathbf{p}). \tag{13.176}$$

If we use the expressions for $|\phi_\alpha\rangle$ ($\alpha = 1, \ldots, 5$) appearing in Eqs. (13.169*a–e*) and assume that **k** lies along the x-axis so that $\mathbf{p}\cdot\boldsymbol{\varkappa} = p_x$, we find by explicit calculation that

$$W_{12} = W_{21} = \left(\frac{1}{m\beta}\right)^{1/2} \tag{13.177}$$

and

$$W_{25} = W_{52} = \left(\frac{2}{3}\right)^{1/2}\left(\frac{1}{m\beta}\right)^{1/2}. \tag{13.178}$$

For all other α'' and α', $W_{\alpha'\alpha''} \equiv 0$.

We may obtain values for $\omega_\alpha^{(1)}$ from Eq. (13.177). We first write it in matrix form and set the determinant of the matrix to zero:

$$\det \overline{\overline{\mathbf{W}}} = 0. \tag{13.179}$$

If we evaluate the above determinant, we obtain the following equation for $\omega_\alpha^{(1)}$:

$$(-\omega^{(1)})^3\left[(\omega^{(1)})^2 - \frac{5}{3}\frac{1}{m\beta}\right] = 0. \tag{13.180}$$

From Eq. (13.177) we obtain the following first-order corrections to ω_α:

$$\omega_1^{(1)} = -\omega_2^{(1)} = \left(\frac{5}{3}\frac{1}{m\beta}\right)^{1/2} = c_0 \tag{13.181}$$

and

$$\omega_3^{(1)} = \omega_4^{(1)} = \omega_5^{(1)} = 0. \tag{13.182}$$

In Eq. (13.181) $c_0 = (\frac{5}{3}(1/m\beta))^{1/2}$ is the velocity of sound of an ideal gas. Notice that we have lifted the degeneracy of only two of the states $|\Psi'_\alpha\rangle_+$. We have to go to higher orders in the perturbation expansion to lift the degeneracy in the rest of the states.

Now that we have expressions for $\omega_\alpha^{(1)}$ and $W_{\alpha\alpha'}$, we can use Eq. (13.175) and the condition that the states $|\Psi_\alpha'^{(0)}\rangle_+$ must be orthogonal to obtain expressions for $c_{\alpha\alpha'}$ and, therefore, for $|\Psi_\alpha'^{(0)}\rangle_+$. Substitution of Eqs. (13.177), (13.178), and (13.179) into Eq. (13.175) gives the following results: $c_{13} = c_{14} = c_{23} = c_{24} = c_{32} = c_{42} = c_{52} = 0$, $c_{15} = (\frac{2}{5})^{1/2}c_{12} = (\frac{2}{3})^{1/2}c_{11}$, $c_{25} = -(\frac{2}{5})^{1/2}c_{22} = (\frac{2}{3})^{1/2}c_{21}$, $c_{35} = -(\frac{3}{2})^{1/2}c_{31}$, $c_{45} = -(\frac{3}{2})^{1/2}c_{41}$, and $c_{51} = -(\frac{2}{3})^{1/2}c_{55}$. The condition of orthonormality of $|\Psi_\alpha'^{(0)}\rangle_+$ gives us $c_{12} = c_{22} = (2)^{-1/2}$, $c_{55} = (\frac{3}{5})^{1/2}$, and $c_{31} = c_{34} = c_{41} = c_{43} = c_{53} = c_{54} = 0$. We therefore obtain the following expressions for the wave functions $|\Psi_\alpha'^{(0)}\rangle_+$:

$$|\Psi_1'^{(0)}\rangle_+ = \frac{1}{\sqrt{2}}\,[(\tfrac{3}{5})^{1/2}|\phi_1\rangle + |\phi_2\rangle + (\tfrac{2}{5})^{1/2}|\phi_5\rangle], \tag{13.183a}$$

$$|\Psi_2'^{(0)}\rangle_+ = \frac{1}{\sqrt{2}}\,[(\tfrac{3}{5})^{1/2}|\phi_1\rangle - |\phi_2\rangle + (\tfrac{2}{5})^{1/2}|\phi_5\rangle], \tag{13.183b}$$

$$|\Psi_3^{(0)}\rangle_+ = |\phi_3\rangle, \tag{13.183c}$$

$$|\Psi_4^{(0)}\rangle_+ = |\phi_4\rangle, \tag{13.183d}$$

and

$$|\Psi_5^{(0)}\rangle_+ = \sqrt{\tfrac{2}{5}}[-|\phi_1\rangle + (\tfrac{3}{2})^{1/2}|\phi_5\rangle]. \tag{13.183e}$$

We will use the states $|\Psi_\alpha^{(0)}\rangle_+$ as the basis states for the perturbation theory. The general expression for $\omega_n^{(2)}$ has been given in Eq. (13.118). We thus find the following expressions for the five hydrodynamic frequencies:

$$\omega_1 = c_0 k + ik^2{}_+\langle\Psi_1^{(0)}|\left(\frac{p_x}{m} - c_0\right)\frac{1}{\hat{C}_{\mathbf{p}}{}^+}\left(\frac{p_x}{m} - c_0\right)|\Psi_1^{(0)}\rangle_+, \tag{13.184a}$$

$$\omega_2 = -c_0 k + ik^2{}_+\langle\Psi_1^{(0)}|\left(\frac{p_x}{m} + c_0\right)\frac{1}{\hat{C}_{\mathbf{p}}{}^+}\left(\frac{p_x}{m} + c_0\right)|\Psi_1^{(0)}\rangle_+, \tag{13.184b}$$

$$\omega_3 = ik^2{}_+\langle\Psi_3^{(0)}|\frac{p_x}{m}\frac{1}{\hat{C}_{\mathbf{p}}{}^+}\frac{p_x}{m}|\Psi_3^{(0)}\rangle_+, \tag{13.184c}$$

$$\omega_4 = ik^2{}_+\langle\Psi_4^{(0)}|\frac{p_x}{m}\frac{1}{\hat{C}_{\mathbf{p}}{}^+}\frac{p_x}{m}|\Psi_4^{(0)}\rangle_+, \tag{13.184d}$$

and

$$\omega_5 = ik^2{}_+\langle\Psi_5^{(0)}|\frac{p_x}{m}\frac{1}{\hat{C}_{\mathbf{p}}{}^+}\frac{p_x}{m}|\Psi_5^{(0)}\rangle_+. \tag{13.184e}$$

We can now make an identification with eigenfrequencies of the linearized hydrodynamic equations. Note that from symmetry arguments $\omega_3 = \omega_4$. The shear modes (cf. Eq. (13.159)) are also doubly degenerate. Thus, we identify ω_3 and ω_4 with the shear modes. The frequencies ω_1 and ω_2 may obviously be identified with the sound modes (cf. Eq. (13.165)) and we identify ω_5 with the entropy (or heat) mode (cf. Eq. (13.164)). If we now equate Eqs. (13.159) and (13.184c), we find for the shear viscosity

$$\eta = -\frac{n_0\beta}{m^2}\left(\frac{\beta}{2\pi m}\right)^{3/2}\int d\mathbf{p}\, e^{-\beta p^2/2m} p_y p_x \frac{1}{\hat{C}_{\mathbf{p}}{}^+} p_y p_x. \tag{13.185}$$

If we equate Eqs. (13.164) and (13.184e), we obtain for the thermal conductivity

$$K = -\frac{2n_0 c_P}{5m^2}\left(\frac{\beta}{2\pi m}\right)^{3/2}\int d\mathbf{p}\, e^{-\beta p^2/2m}\left[\frac{p^2\beta}{2m} - \frac{5}{2}\right] \times p_x \frac{1}{\hat{C}_{\mathbf{p}}{}^+} p_x\left[\frac{p^2\beta}{2m} - \frac{5}{2}\right]. \tag{13.186}$$

Thus, we have obtained the desired expressions for the three transport coefficients D, η, and K. In the next section we discuss how numerical values may be obtained for them.

G. SONINE POLYNOMIALS[10]

The coefficient of self-diffusion, D, the coefficient of shear viscosity, η, and the coefficient of thermal conductivity, K, may be written in the form

$$D = -\frac{1}{m^2}\left(\frac{\beta}{2\pi m}\right)^{3/2}\int d\mathbf{p}\, e^{-\beta p^2/2m} p_x D_x, \tag{13.187}$$

$$\eta = -\frac{n_0\beta}{m^2}\left(\frac{\beta}{2\pi m}\right)^{3/2}\int d\mathbf{p}\, e^{-\beta p^2/2m} p_y p_x B_{xy}, \tag{13.188}$$

and

$$K = -\left(\frac{n_0 k_B}{m^2}\right)\left(\frac{\beta}{2\pi m}\right)^{3/2}\int d\mathbf{p}\, e^{-\beta p^2/2m}\left[\frac{p^2\beta}{2m} - \frac{5}{2}\right] p_x A_x, \tag{13.189}$$

where we have used the ideal gas expression for c_P ($c_P = \frac{5}{2}k_B$). The functions D_x, B_{xy}, and A_x satisfy the equations

$$\hat{C}_{\mathbf{p}}^{-} D_x = p_x, \tag{13.190}$$

$$\hat{C}_{\mathbf{p}}^{+} B_{xy} = p_x p_y, \tag{13.191}$$

and

$$\hat{C}_{\mathbf{p}}^{+} A_x = \left[\frac{p^2\beta}{2m} - \frac{5}{2}\right] p_x. \tag{13.192}$$

(These results are identical to the results obtained using the more traditional Chapman-Enskog procedure.)[8]

It is interesting to note at this point that we can write the expressions for the transport coefficients in an alternative form. Let us consider the viscosity. We can write Eq. (13.188) in the form

$$\eta = n_0\beta\left(\frac{\beta}{2\pi m}\right)^{3/2}\int d\mathbf{p}\, e^{-\beta p^2/2m}\int_0^\infty d\tau J_{xy}\, e^{\hat{C}_{\mathbf{p}}^{+}\tau} J_{xy} \tag{13.193}$$

where

$$J_{xy} \equiv \frac{p_x p_y}{m} = p_x v_y \tag{13.194}$$

is a microscopic current, or flux, of x-component of momentum in the y-direction.

If we denote

$$J_{xy}(\tau) = e^{\hat{C}_{\mathbf{p}}^{+}\tau} J_{xy} \tag{13.195}$$

the viscosity becomes

$$\eta = n_0\beta\int_0^\infty d\tau \langle J_{xy} J_{xy}(\tau)\rangle_{\text{eq}}. \tag{13.196}$$

Eq. (13.196) expresses the viscosity in terms of a current-current correlation function. We shall find in Chap. 15 that this is quite a general result.

In order to obtain numerical values for the transport coefficients, we must expand the unknown functions D_x, A_x, and B_{xy} in terms of a set of orthogonal functions, $S_m{}^n(x)$, called Sonine polynomials. We shall sketch the procedure for the case of the thermal conductivity.

The Sonine polynomials are defined as follows:

$$S_m{}^n(x) = \sum_{p=0}^{n} (-1)^p \frac{\Gamma(m+n+1)x^p}{\Gamma(m+p+1)(n-p)!\,p!} \tag{13.197}$$

where x and m are real numbers and n is an integer; $\Gamma(m + n + 1)$ and $\Gamma(m + p + 1)$ are gamma functions. For the special case $\Gamma(n + \frac{1}{2})$, they are defined

$$\Gamma(n+\tfrac{1}{2}) = \frac{1\cdot 3\cdot 5\cdots(2n-3)(2n-1)\sqrt{\pi}}{2^n}, \tag{13.198}$$

while $\Gamma(n + 1) = n!$. The first two Sonine polynomials are easily found to be

$$S_m{}^0(x) = 1 \tag{13.199}$$

and

$$S_m{}^1(x) = m + 1 - x. \tag{13.200}$$

The Sonine polynomials have the orthogonality property

$$\int_0^\infty dx\, e^{-x}x^m S_m{}^p(x) S_m{}^q(x) = \frac{\Gamma(m+p+1)}{p!}\,\delta_{p,q}. \tag{13.201}$$

We now expand A_x in the following series:

$$A_x = \sum_{r=0}^{\infty} a_r S_{3/2}^r\left(\frac{\beta p^2}{2m}\right) p_x, \tag{13.202}$$

where $S_{3/2}^0\left(\frac{\beta p^2}{2m}\right) = 1$ and

$$S_{3/2}^1\left(\frac{\beta p^2}{2m}\right) = \left(\frac{5}{2} - \frac{p^2\beta}{2m}\right),$$

and substitute it back into Eq. (13.192). We then obtain

$$\hat{C}_p{}^+ \sum_{r=0}^{\infty} a_r S_{3/2}^r\left(\frac{\beta p^2}{2m}\right) p_x = \left(\frac{p^2\beta}{2m} - \frac{5}{2}\right) p_x. \tag{13.203}$$

If we multiply Eq. (13.203) by

$$\left(\frac{\beta}{2\pi m}\right)^{3/2} e^{-\beta p^2/2m} p_x S_{3/2}^q\left(\frac{\beta p^2}{2m}\right)$$

and integrate over $\mathbf{p}$, we obtain an equation of the form

$$\sum_{r=0}^{\infty} M_{qr} a_r = \alpha_q \tag{13.204}$$

where

$$M_{qr} = \left(\frac{\beta}{2\pi m}\right)^{3/2} \int d\mathbf{p}\, e^{-\beta p^2/2m} p_x S^q_{3/2} \hat{C}_{\mathbf{p}}{}^+ S^r_{3/2} p_x \tag{13.205}$$

and

$$\alpha_q = \left(\frac{\beta}{2\pi m}\right)^{3/2} \int d\mathbf{p}\, e^{-\beta p^2/2m} S^q_{3/2} p_x{}^2 \left(\frac{p^2\beta}{2m} - \frac{5}{2}\right) \tag{13.206}$$

Because $\hat{C}_{\mathbf{p}}{}^+ S^0_{3/2} p_x = \hat{C}_{\mathbf{p}}{}^+ p_x = 0$, we see that $M_{0r} = M_{q0} = 0$. We can also simplify α_q. We write α_q in terms of spherical coordinates as

$$\alpha_q = \left(\frac{\beta}{2\pi m}\right)^{3/2} \int_0^{\infty} \int_{-1}^{1} \int_0^{2\pi} p^2\, dp\, d(\cos\theta)\, d\phi\, e^{-\beta p^2/2m} S^q_{3/2}\left(\frac{\beta p^2}{2m}\right)$$
$$\times\, p^2 \cos^2\theta \cos^2\phi \left(\frac{p^2\beta}{2m} - \frac{5}{2}\right). \tag{13.207}$$

The integrations in Eq. (13.207) are easy to perform. We obtain

$$\alpha_q = -\frac{5m}{4\beta}. \tag{13.208}$$

Then Eq. (13.204) can be written

$$\sum_{r=1} M_{qr} a_r = -\frac{5m}{4\beta}\, \delta_{q,1}. \tag{13.209}$$

We now can make approximations to the expansion for A_x given in Eq. (13.202). First we note that the term $r = 0$ does not contribute to the heat conductivity; therefore, we can set $a_0 = 0$. The simplest approximation we can make is to terminate the expansion for A_x after $r = 1$ (that is, set $a_r = 0$ for $r > 1$). Then from Eq. (13.209) we obtain

$$a_1 = -\frac{5m}{4\beta M_{11}} \tag{13.210}$$

and the heat conductivity becomes

$$K = \frac{5 n_0 k_B{}^2 T a_1}{4m}. \tag{13.211}$$

A detailed comparison of the predictions of this theory and experimental data is given in Ref. 13.10 for various forms of the interparticle potential. For dilute gases, the agreement is very good.

H. QUANTUM KINETIC EQUATION

There are a number of methods for deriving the kinetic equation for a quantum fluid. The first derivation was due to Uehling and Uhlenbeck[11] using arguments similar to those we have used in deriving the classical Boltzmann equation. The method we shall use here is due to Peletminskii and Yatsenko[12] and has the advantage that it can be generalized to include the case of systems with broken symmetry.

The basic idea behind the work of Peletminskii and Yatsenko was first formulated by Bogoliubov[13] for classical systems. We know that after a very long time the behavior of a fluid is well described in terms of hydrodynamic equations, and, therefore, the number of degrees of freedom we need is vastly reduced. The idea of Bogoliubov was that a similar reduction occurs on the microscopic level and the exact density matrix varies in time only through its dependence on the one-body reduced density matrix.

1. Basic Model

We shall use the method of Peletminskii and Yatsenko to derive the kinetic equation for a system of fermions which interact via a contact potential, $V(\mathbf{r}) = I\,\delta(\mathbf{r})$, and in which only antiparallel spin particles interact. The total Hamiltonian in the momentum representation can be written in the form

$$\hat{H} = \sum_{\lambda}\sum_{\mathbf{k}} \epsilon_{\mathbf{k}}{}^{0}\hat{a}^{+}_{\mathbf{k},\lambda}\hat{a}_{\mathbf{k},\lambda} + \frac{I}{V}\sum_{\mathbf{k}_1,\ldots,\mathbf{k}_4} \delta_{\mathbf{K}_{12},\mathbf{K}_{34}}\hat{a}^{+}_{\mathbf{k}_1\uparrow}\hat{a}^{+}_{\mathbf{k}_2\downarrow}\hat{a}_{\mathbf{k}_4\downarrow}\hat{a}_{\mathbf{k}_3\uparrow} \tag{13.212}$$

where $\epsilon_{\mathbf{k}}{}^{0} = \hbar^2k^2/2m$ and $\mathbf{K}_{12} = \mathbf{k}_1 + \mathbf{k}_2$. We will let $\Gamma_\lambda(\mathbf{k}_1, \mathbf{k}_2, t)$ denote the one-body reduced density matrix,

$$\Gamma_\lambda(\mathbf{k}_1, \mathbf{k}_2, t) \equiv \langle \mathbf{k}_1\lambda|\hat{\rho}_1(t)|\mathbf{k}_2\lambda\rangle = \mathrm{Tr}\,\hat{\rho}(t)\hat{a}^{+}_{\mathbf{k}_2\lambda}\hat{a}_{\mathbf{k}_1\lambda} \tag{13.213}$$

(cf. Sec. 7.F), where $\hat{\rho}(t)$ is the exact density operator at time t. We will let $\hat{\gamma}_\lambda(\mathbf{k}_1, \mathbf{k}_2)$ denote the operator

$$\hat{\gamma}_\lambda(\mathbf{k}_1, \mathbf{k}_2) = \hat{a}^{+}_{\mathbf{k}_1\lambda}\hat{a}_{\mathbf{k}_2,\lambda}. \tag{13.214}$$

In the following, we will obtain a closed equation of motion for $\Gamma_\lambda(\mathbf{k}_1, \mathbf{k}_2, t)$. It is convenient to introduce an effective free particle Hamiltonian which includes in an average way the interaction of the particles with the medium. We therefore write the total Hamiltonian as

$$\hat{H} = \hat{H}_0 + \hat{V} \tag{13.215}$$

where

$$\hat{H}_0 = \sum_{\lambda}\sum_{\mathbf{k}} \epsilon_{\mathbf{k}}{}^0 \hat{a}_{\mathbf{k}\lambda}^+ \hat{a}_{\mathbf{k}\lambda} + \frac{I}{V}\sum_{\mathbf{k}_1,\ldots,\mathbf{k}_4} \delta_{\mathbf{K}_{12},\mathbf{K}_{34}}[\Gamma_\uparrow(\mathbf{k}_1, \mathbf{k}_3, t)\hat{a}_{\mathbf{k}_2\downarrow}^+ \hat{a}_{\mathbf{k}_4\downarrow} + \Gamma_\downarrow(\mathbf{k}_2, \mathbf{k}_4, t)\hat{a}_{\mathbf{k}_1\uparrow}^+ \hat{a}_{\mathbf{k}_3\uparrow}] \tag{13.216}$$

and

$$\hat{V} = \frac{I}{V}\sum_{\mathbf{k}_1\mathbf{k}_4} \delta_{\mathbf{K}_{12},\mathbf{K}_{34}} \hat{a}_{\mathbf{k}_1\uparrow}^+ \hat{a}_{\mathbf{k}_2\downarrow}^+ \hat{a}_{\mathbf{k}_4\downarrow} \hat{a}_{\mathbf{k}_3\uparrow} - \frac{I}{V}\sum_{\mathbf{k}_1,\ldots,\mathbf{k}_4} \delta_{\mathbf{K}_{12},\mathbf{K}_{34}}[\Gamma_\uparrow(\mathbf{k}_1, \mathbf{k}_3, t)\hat{a}_{\mathbf{k}_2\downarrow}^+ \hat{a}_{\mathbf{k}_4\downarrow} + \Gamma_\downarrow(\mathbf{k}_2, \mathbf{k}_4, t)\hat{a}_{\mathbf{k}_1\uparrow}^+ \hat{a}_{\mathbf{k}_3\uparrow}]. \tag{13.217}$$

Thus, the "free Hamiltonian," $\hat{H}_0$, depends on the one-body reduced density matrix and the interaction term, $\hat{V}$, describes deviations from the mean field behavior. Both $\hat{H}_0$ and $\hat{V}$ depend on time through their dependence on the reduced density matrix.

Let us now consider the density operator $\hat{\rho}(t)$. In general it obeys the equation of motion,

$$\frac{\partial \hat{\rho}(t)}{\partial t} = -\frac{i}{\hbar}[\hat{H}, \hat{\rho}(t)], \tag{13.218}$$

and therefore $\Gamma_\lambda(\mathbf{k}_1, \mathbf{k}_2, t)$ obeys the equation of motion,

$$\frac{\partial \Gamma_\lambda(\mathbf{k}_1, \mathbf{k}_2, t)}{\partial t} = \mathrm{Tr}\,\frac{\partial \hat{\rho}}{\partial t}\hat{\gamma}_\lambda(\mathbf{k}_1, \mathbf{k}_2) = \frac{i}{\hbar}\mathrm{Tr}\,\hat{\rho}(t)][\hat{H}, \hat{\gamma}_\lambda(\mathbf{k}_1, \mathbf{k}_2)] \equiv \mathscr{L}_\lambda(\mathbf{k}_1, \mathbf{k}_2, t), \tag{13.219}$$

where we have cyclically permuted operators under the trace. In order to obtain a closed equation for $\Gamma_\lambda(\mathbf{k}_1, \mathbf{k}_2, t)$, we must find a way to approximate $\mathscr{L}_\lambda(\mathbf{k}_1, \mathbf{k}_2, t)$ in terms of $\Gamma_\lambda(\mathbf{k}_1, \mathbf{k}_2, t)$. This requires a bit of work.

It is useful to find how $\hat{\gamma}_\lambda(\mathbf{k}_1, \mathbf{k}_2)$ evolves under the action of the Hamiltonian $\hat{H}_0$. Let us first write $\hat{H}_0$ in the form

$$\hat{H}_0 = \sum_{\lambda}\sum_{\mathbf{k}_1\mathbf{k}_2} \epsilon_\lambda(\mathbf{k}_1, \mathbf{k}_2)\hat{a}_{\mathbf{k}_1\lambda}^+ \hat{a}_{\mathbf{k}_2\lambda} \tag{13.220}$$

where

$$\epsilon_\uparrow(\mathbf{k}_1, \mathbf{k}_2) = \delta_{\mathbf{k}_1,\mathbf{k}_2}\epsilon_{\mathbf{k}_1}^0 + \frac{I}{V}\sum_{\mathbf{k}_3,\mathbf{k}_4} \delta_{\mathbf{K}_{12},\mathbf{K}_{34}}\Gamma_\downarrow(\mathbf{k}_1, \mathbf{k}_3) \tag{13.221}$$

and

$$\epsilon_\downarrow(\mathbf{k}_1, \mathbf{k}_2) = \delta_{\mathbf{k}_1,\mathbf{k}_2}\epsilon_{\mathbf{k}_1}^0 + \frac{I}{V}\sum_{\mathbf{k}_3,\mathbf{k}_4} \delta_{\mathbf{K}_{12},\mathbf{K}_{34}}\Gamma_\uparrow(\mathbf{k}_1, \mathbf{k}_3). \tag{13.222}$$

The equation of motion for $\hat{\gamma}_\lambda(\mathbf{k}_1, \mathbf{k}_2)$ can then be written

$$\frac{\partial}{\partial s}\hat{\gamma}_\lambda(\mathbf{k}_1, \mathbf{k}_2)_{(s)} = \frac{i}{\hbar}[\hat{H}_0, \hat{\gamma}_\lambda(\mathbf{k}_1, \mathbf{k}_2)_{(s)}] = \frac{i}{\hbar}\hat{E}_\lambda(\mathbf{k}_1, \mathbf{k}_2)\hat{\gamma}_\lambda(\mathbf{k}_1, \mathbf{k}_2)_{(s)} \tag{13.223}$$

where $\hat{E}_\lambda(\mathbf{k}_1, \mathbf{k}_2)$ is an integral operator defined

$$\hat{E}_\lambda(\mathbf{k}_1, \mathbf{k}_2)\hat{\gamma}_\lambda(\mathbf{k}_1, \mathbf{k}_2)_{(s)} = \sum_{\mathbf{k}_3}[\epsilon_\lambda(\mathbf{k}_3, \mathbf{k}_1)\hat{\gamma}_\lambda(\mathbf{k}_3, \mathbf{k}_2)_{(s)} - \epsilon_\lambda(\mathbf{k}_2, \mathbf{k}_3)\hat{\gamma}_\lambda(\mathbf{k}_1, \mathbf{k}_3)_{(s)}]. \tag{13.224}$$

To obtain Eq. (13.224) we have used the Fermi anticommutation relations of App. B. From Eq. (13.223), we find that

$$\hat{\gamma}_\lambda(\mathbf{k}_1, \mathbf{k}_2)_{(s)} = e^{(i/\hbar)\hat{E}(\mathbf{k}_1,\mathbf{k}_2)s}\hat{\gamma}_\lambda(\mathbf{k}_1, \mathbf{k}_2)_{(0)}, \tag{13.225}$$

and, taking the expectation value, we find

$$\Gamma_\lambda(\mathbf{k}_1, \mathbf{k}_2)_{(s)} = e^{(i/\hbar)\hat{E}(\mathbf{k}_1,\mathbf{k}_2)s}\Gamma_\lambda(\mathbf{k}_1, \mathbf{k}_2)_{(0)}. \tag{13.226}$$

We have let s denote time variations due to evolution under H_0. We will let t denote time for evolution under the full Hamiltonian.

2. Bogoliubov Assumption

According to the Bogoliubov assumption, the density operator $\hat{\rho}(t)$, after long times, should have the form

$$\hat{\rho}(t) = \hat{\rho}\{\mathbf{\Gamma}(t)\} + \hat{\rho}'\{\hat{\rho}(0)\}\, e^{-t/\tau_0} \tag{13.227}$$

where $\mathbf{\Gamma}(t)$ denotes a vector containing the function $\Gamma_\lambda(\mathbf{k}_1, \mathbf{k}_2)$ for all values of λ, $\mathbf{k}_1$, $\mathbf{k}_2$. After a characteristic relaxation time, τ_0, all interesting behavior of the system will be contained in $\hat{\rho}\{\mathbf{\Gamma}(t)\}$. The time derivative of $\hat{\rho}\{\mathbf{\Gamma}(t)\}$ can be written

$$\frac{\partial\hat{\rho}\{\mathbf{\Gamma}(t)\}}{\partial t} = \sum_j \frac{\partial\Gamma_j(t)}{\partial t}\frac{\partial\hat{\rho}\{\mathbf{\Gamma}(t)\}}{\partial\Gamma_j(t)} = \sum_j \mathscr{L}_j(t)\frac{\partial\hat{\rho}\{\mathbf{\Gamma}(t)\}}{\partial\Gamma_j(t)}, \tag{13.228}$$

and, thus, from Eq. (3.218) the equation of motion takes the form

$$\sum_j \mathscr{L}_j(t)\frac{\partial\hat{\rho}\{\mathbf{\Gamma}(t)\}}{\partial\Gamma_j(t)} = -\frac{i}{\hbar}[\hat{H}, \hat{\rho}\{\mathbf{\Gamma}(t)\}] \tag{13.229}$$

after a long time. In Eqs. (13.228) and (13.229), we have let j denote the set of variables $(\mathbf{k}_1, \mathbf{k}_2, \lambda)$ and $\sum_j = \sum_\lambda \sum_{\mathbf{k}_1} \sum_{\mathbf{k}_2}$.

We can now use Eq. (13.229) to write a perturbation expansion for $\mathscr{L}_\lambda(\mathbf{k}_1, \mathbf{k}_2, t)$. Let us first write Eq. (13.229) in the form

$$\sum_j \mathscr{L}_j^{(0)}\{\mathbf{\Gamma}(t)\} \frac{\partial \hat{\rho}\{\mathbf{\Gamma}(t)\}}{\partial \Gamma_j(t)} + \frac{i}{\hbar} [\hat{H}_0, \hat{\rho}\{\mathbf{\Gamma}(t)\}]_-$$
$$= -\sum_j (\mathscr{L}_j\{\mathbf{\Gamma}(t)\} - \mathscr{L}_j^0\{\mathbf{\Gamma}(t)\}) \frac{\partial \hat{\rho}\{\mathbf{\Gamma}(t)\}}{\partial \Gamma_j(t)} - \frac{i}{\hbar} [\hat{V}, \hat{\rho}\{\mathbf{\Gamma}(t)\}] = \hat{F}\{\mathbf{\Gamma}(t)\} \tag{13.230}$$

where

$$\mathscr{L}_j^{(0)}\{\mathbf{\Gamma}(t)\} \equiv \frac{i}{\hbar} \operatorname{Tr}\{\hat{\rho}\{\mathbf{\Gamma}(t)\}[\hat{H}_0, \hat{\gamma}_j]\} = \frac{i}{\hbar} \hat{E}_j \Gamma_j(t). \tag{13.231}$$

In Eq. (13.230), $\hat{E}_j$ denotes the integral operator $\hat{E}_\lambda(\mathbf{k}_1, \mathbf{k}_2)$. The arguments leading to Eqs. (13.230) and (13.231) are independent of the parameter s. We therefore can let $\mathbf{\Gamma}(t) \to \mathbf{\Gamma}(t)_s$. From Eqs. (13.231) and (3.226), we obtain the relation

$$\mathscr{L}_j^{(0)}\{\mathbf{\Gamma}(t)_s\} = \frac{\partial}{\partial s} e^{(i/\hbar)E_j s} \Gamma_j(t) = \frac{\partial}{\partial s} \Gamma_j(t)_s. \tag{13.232}$$

Thus, Eq. (13.230) becomes

$$\frac{\partial}{\partial s} \hat{\rho}\{\mathbf{\Gamma}(t)_s\} + \frac{i}{\hbar} [\hat{H}_0, \hat{\rho}\{\mathbf{\Gamma}(t)_s\}] = \hat{F}\{\mathbf{\Gamma}(t)_s\}. \tag{13.233}$$

Let us now write the density operator in the interaction representation:

$$\hat{\rho}_I\{\mathbf{\Gamma}(t)_s\} \equiv e^{i\hat{H}_0 s/\hbar} \hat{\rho}\{\mathbf{\Gamma}(t)_s\}\, e^{-i\hat{H}_0 s/\hbar}. \tag{13.234}$$

Then Eq. (13.233) becomes

$$\frac{\partial}{\partial s} \hat{\rho}_I\{\mathbf{\Gamma}(t)_s\} = e^{i\hat{H}_0 s/\hbar} \hat{F}\{\mathbf{\Gamma}(t)_s\}\, e^{-i\hat{H}_0 s/\hbar}. \tag{13.235}$$

If we integrate Eq. (13.235) from $-\infty$ to s, we find

$$\hat{\rho}_I\{\mathbf{\Gamma}(t)_s\} = \lim_{s \to -\infty} \hat{\rho}_I\{\mathbf{\Gamma}(t)_s\} + \int_{-\infty}^{s} ds'\, e^{i\hat{H}_0 s'/\hbar} \hat{F}\{\mathbf{\Gamma}(t)_s\}\, e^{-i\hat{H}_0 s'/\hbar}. \tag{13.236}$$

Before we can go further, we must introduce boundary conditions for $\hat{\rho}_I\{\mathbf{\Gamma}(t)_s\}$ so that we can evaluate the quantity $\lim_{s\to-\infty} \hat{\rho}_I\{\mathbf{\Gamma}(t)_s\}$.

Following Peletminskii and Yatsenko, let us assume that under free motion (motion governed by $\hat{H}_0$) the density operator phase mixes to a Gibbsian form. That is,

$$\lim_{s \to \infty} e^{-i\hat{H}_0 s/\hbar} \hat{\rho}\{\mathbf{\Gamma}(t)\}\, e^{i\hat{H}_0 s/\hbar} = \hat{\rho}_0\{\mathbf{\Gamma}(t)_s\} \tag{13.237}$$

where

$$\hat{\rho}_0\{\mathbf{\Gamma}(t)\} = \exp\left[-\sum_j X_j\{\mathbf{\Gamma}(t)\}\hat{\gamma}_j\right]. \tag{13.238}$$

Thus, evolution of $\hat{\rho}\{\mathbf{\Gamma}(t)\}$ under the action of $\hat{H}_0$ sifts out any dependence on operators $\hat{a}_k$ and $\hat{a}_k{}^+$ other than through the combination $\hat{\gamma}_j$. This combination remains because it is not changed by $\hat{H}_0$. The function $X_j(\mathbf{\Gamma}(t))$ depends on the reduced density matrix and is determined from the requirement that

$$\Gamma_j(t) = \mathrm{Tr}\ \hat{\rho}_0\{\mathbf{\Gamma}(t)\}\hat{\gamma}_j. \tag{13.239}$$

From the boundary condition in Eq. (13.237) we find the relation

$$\lim_{s\to-\infty} e^{i\hat{H}_0 s/\hbar}\hat{\rho}\{\mathbf{\Gamma}(t)_s\}\, e^{-i\hat{H}_0 s/\hbar} = \hat{\rho}_0\{\mathbf{\Gamma}(t)\}. \tag{13.240}$$

Let us also note that $\hat{\rho}_0\{\mathbf{\Gamma}(t)_s\} = e^{-i\hat{H}_0 s/\hbar}\hat{\rho}_0\{\mathbf{\Gamma}(t)\}e^{i\hat{H}_0 s/\hbar}$. Thus, Eq. (13.236) can be rewritten in the form

$$\hat{\rho}\{\mathbf{\Gamma}(t)\} = \hat{\rho}_0\{\mathbf{\Gamma}(t)\} + \int_{-\infty}^{0} ds'\, e^{i\hat{H}_0 s'/\hbar}F\{\mathbf{\Gamma}(t)_{s'}\}\, e^{-i\hat{H}_0 s'/\hbar} \tag{13.241}$$

and we have obtained an integral equation for the density operator $\hat{\rho}\{\mathbf{\Gamma}(t)\}$.

3. Kinetic Equation

For a weakly coupled system, the kinetic equation, Eq. (13.219), can be terminated at second order in the interaction. Thus,

$$\frac{\partial}{\partial t}\Gamma_\lambda(\mathbf{k}_1, \mathbf{k}_2, t) = \mathscr{L}_\lambda^{(0)}(\mathbf{k}_1, \mathbf{k}_2, t) + \mathscr{L}_\lambda^{(1)}(\mathbf{k}_1, \mathbf{k}_2, t) + \mathscr{L}_\lambda^{(2)}(\mathbf{k}_1, \mathbf{k}_2, t) + \cdots \tag{13.242}$$

The expression for $\mathscr{L}^{(0)}(\mathbf{k}_1, \mathbf{k}_2, t)$ is given in Eq. (13.231): $\mathscr{L}_\lambda^{(1)}(\mathbf{k}_1, \mathbf{k}_2, t)$ is defined

$$\mathscr{L}_j^{(1)} = \frac{i}{\hbar}\mathrm{Tr}[\hat{\rho}_0\{\mathbf{\Gamma}(t)\}[\hat{V}, \hat{\gamma}_j]], \tag{13.243}$$

and $\mathscr{L}_j^{(2)}$ is defined

$$\mathscr{L}_j^{(2)} = \frac{i}{\hbar}\mathrm{Tr}\{\hat{\rho}^{(1)}\{\mathbf{\Gamma}(t)\}[\hat{V}, \hat{\gamma}_j]\}, \tag{13.244}$$

where

$$\hat{\rho}^{(1)}\{\mathbf{\Gamma}(t)\} = \int_{-\infty}^{0} ds'\, e^{i\hat{H}_0 s'/\hbar}F^{(1)}(\{\mathbf{\Gamma}(t)_{(s')}\})\, e^{-i\hat{H}_0 s'/\hbar} \tag{13.245}$$

and

$$F^{(1)}\{\mathbf{\Gamma}(t)_{(s)}\} = -\frac{i}{\hbar}[\hat{V}, \hat{\rho}_0\{\mathbf{\Gamma}(t)_s\}] - \sum_j \mathscr{L}_j^{(1)}\{\Gamma(t)_s\}\frac{\partial\hat{\rho}_0\{\Gamma(t)_s\}}{\partial\Gamma_j(t)_s}. \tag{13.246}$$

Eqs. (13.242) and (13.243) give the quantum kinetic equation for the Fermi system to second order in the interaction, $\hat{V}$. We shall soon see that, because we have included mean field effects in $\hat{H}_0$, $\mathscr{L}_j^{(1)}$ is identically zero. Thus, we will obtain an expression similar to the Boltzmann equation.

We can now make use of a generalized Wick's theorem to simplify the above equations considerably. Let us first note that an expression of the form

$$\sum_j X_j \hat{\gamma}_j = \sum_{\lambda} \sum_{\mathbf{k}_1 \mathbf{k}_2} X_{\mathbf{k}_1 \mathbf{k}_2, \lambda}(\mathbf{\Gamma}(t)) \hat{a}^+_{\mathbf{k}_1 \lambda} \hat{a}_{\mathbf{k}_2 \lambda} \tag{13.247}$$

which appears in $\hat{\rho}\{\mathbf{\Gamma}(t)\}$ can always be diagonalized via a generalized Bogoliubov-Valatin transformation (a generalization of the transformation we used for superconductors). Thus, we can, in principle, transform to a new set of operators in which Eq. (13.247) is diagonal. We can then use Wick's theorem to evaluate the trace in Eqs. (13.243) and (13.244). Because of this, the expectation value of a product of any number of creation and annihilation operators will separate into a sum of all possible combinations of pairwise averages of the products. As a result we find that

$$\mathscr{L}_j^{(1)}\{\mathbf{\Gamma}(t)_s\} \equiv 0 \tag{13.248}$$

because the second and third terms in Eq. (13.217) will cancel the first after the trace has been taken. We therefore can write

$$\hat{\rho}^{(1)}\{\mathbf{\Gamma}(t)\} = -\frac{i}{\hbar}\int_{-\infty}^{0} ds\, e^{i\hat{H}_0 s/\hbar}[\hat{V}, \rho_0\{\mathbf{\Gamma}(t)_s\}]\, e^{-i\hat{H}_0 s/\hbar} \tag{13.249}$$

and

$$\begin{aligned} \mathscr{L}_j^{(2)}\{\mathbf{\Gamma}(t)\} &= -\left(\frac{i}{\hbar}\right)^2 \int_{-\infty}^{0} ds\, \mathrm{Tr}[\hat{V}(s), \hat{\rho}_0\{\mathbf{\Gamma}(t)\}][\hat{V}, \hat{\gamma}_j] \\ &= \left(\frac{i}{\hbar}\right)^2 \int_{-\infty}^{0} ds\, \mathrm{Tr}\{\hat{\rho}_0\{\mathbf{\Gamma}(t)\}[\hat{V}(s), [\hat{V}, \hat{\gamma}_j]]\} \end{aligned} \tag{13.250}$$

where

$$\hat{V}(s) = e^{i\hat{H}_0 s/\hbar}\hat{V}\, e^{-i\hat{H}_0 s/\hbar}. \tag{13.251}$$

The equation of motion for $\Gamma_\lambda(\mathbf{k}_1, \mathbf{k}_2, t)$, to second order in the interaction, takes the form

$$\begin{aligned} \frac{\partial \Gamma_\lambda(\mathbf{k}_1, \mathbf{k}_2, t)}{\partial t} &= \frac{i}{\hbar}\hat{E}_\lambda(\mathbf{k}_1, \mathbf{k}_2)\Gamma_\lambda(\mathbf{k}_1, \mathbf{k}_2, t) \\ &\quad + \left(\frac{i}{\hbar}\right)^2 \int_{-\infty}^{0} ds\, \mathrm{Tr}\{\hat{\rho}_0\{\mathbf{\Gamma}(t)\}[\hat{V}(s), \{\hat{V}, \hat{\gamma}_\lambda(\mathbf{k}_1, \mathbf{k}_2)\}]\} \end{aligned} \tag{13.252}$$

Eq. (13.252) is the kinetic equation for the Fermi system. The first term on the right is the streaming term; the second term is the collision integral.

4. Spatially Homogeneous System

To illustrate the important features of Eq. (13.252) it is easiest to consider the case of a system with no spatial variations. Then $\mathbf{k}_1 = \mathbf{k}_2 = \mathbf{k}$ and Eq. (13.252) takes the form

$$\frac{\partial\Gamma_\lambda(\mathbf{k}, \mathbf{k}, t)}{\partial t} = \left(\frac{i}{\hbar}\right)^2 \int_{-\infty}^{2} ds\, \mathrm{Tr}\{\hat{\rho}_0\{\mathbf{\Gamma}(t)\}[\hat{V}(s), \{\hat{V}, \hat{\gamma}_\lambda(\mathbf{k}, \mathbf{k})\}]\} \tag{13.253}$$

where $\Gamma_\lambda(\mathbf{k}, \mathbf{k}, t)$ is just the number density of fermions with momentum $\mathbf{k}$ and spin λ at time t. The density operator $\hat{\rho}_0\{\mathbf{\Gamma}(t)\}$ takes the form

$$\hat{\rho}_0\{\mathbf{\Gamma}(t)\} = \exp\left[-\sum_\lambda \sum_{\mathbf{k}_1} X_{\mathbf{k}_1\lambda}\{\mathbf{\Gamma}(t)\}\hat{a}^+_{\mathbf{k}_1\lambda}\hat{a}_{\mathbf{k}_1\lambda}\right], \tag{13.254}$$

and the Hamiltonian takes the form

$$\hat{H}_0 = \sum_\lambda \sum_{\mathbf{k}} \epsilon_{\mathbf{k}\lambda}\hat{a}^+_{\mathbf{k}\lambda}a_{\mathbf{k}\lambda} \tag{13.255}$$

where

$$\epsilon_{\mathbf{k}\uparrow} = \epsilon_{\mathbf{k}}{}^0 + In_\downarrow, \tag{13.256}$$

$$\epsilon_{\mathbf{k}\downarrow} = \epsilon_{\mathbf{k}}{}^0 + In_\uparrow, \tag{13.257}$$

and

$$n_\lambda = \frac{1}{V}\sum_{\mathbf{k}} \Gamma_\lambda(\mathbf{k}, \mathbf{k}, t). \tag{13.258}$$

Since the total number of particles with a given spin component is conserved during the interaction, n_λ must be independent of time. For simplicity we shall assume that $n_\uparrow = n_\downarrow = n$. Then $\epsilon_{\mathbf{k}\uparrow} = \epsilon_{\mathbf{k}\downarrow} = \epsilon_{\mathbf{k}}{}^0 + In$.

Let us now find an explicit expression for the collision integral on the right-hand side of Eq. (13.252). We first note that

$$\hat{V}(s) = \sum_{\mathbf{k}_1,\ldots,\mathbf{k}_4} \langle k_1, k_2|V|k_3, k_4\rangle \hat{a}^+_{\mathbf{k}_1\uparrow}(s)\hat{a}^+_{\mathbf{k}_2\downarrow}(s)\hat{a}_{\mathbf{k}_3\downarrow}(s)\hat{a}_{\mathbf{k}_4\uparrow}(s) - gn_\uparrow \sum_{\mathbf{k}_1} \hat{a}^+_{\mathbf{k}_1\downarrow}(s)\hat{a}_{\mathbf{k}_1\downarrow}(s) - gn_\downarrow \sum_{\mathbf{k}_1} \hat{a}^+_{\mathbf{k}_1\uparrow}(s)\hat{a}_{\mathbf{k}_1\uparrow}(s) \tag{13.259}$$

where

$$\hat{a}^+_{\mathbf{k}\lambda}(s) = e^{i\hat{H}_0 s/\hbar}\hat{a}^+_{\mathbf{k}\lambda}\, e^{-i\hat{H}_0 s/\hbar} = e^{i\epsilon_{\mathbf{k}} s/\hbar}\hat{a}^+_{\mathbf{k}\lambda} \tag{13.260}$$

and

$$\hat{a}_{\mathbf{k}\lambda}(s) = e^{i\hat{H}_0 s/\hbar}\hat{a}_{\mathbf{k}\lambda}\, e^{-i\hat{H}_0 s/\hbar} = e^{-i\epsilon_{\mathbf{k}} s/\hbar}\hat{a}_{\mathbf{k}\lambda}. \tag{13.261}$$

Since $\hat{\rho}_0\{\mathbf{\Gamma}(t)\}$ depends only on products of creation and annihilation operators in the form $\hat{a}^+_{\mathbf{k}_1\lambda}\hat{a}_{\mathbf{k}_1\lambda}$, we can use Wick's theorem (cf. Sec. 10.B) to evaluate the trace on the right-hand side of Eq. (13.253). We first note that contributions from contractions between creation and annihilation operators in the same potential function cancel out. If we note that $[\hat{H}_0, \hat{\gamma}_\lambda(\mathbf{k}, \mathbf{k})] = 0$ and $[\hat{\rho}_0, \hat{\gamma}_\lambda(\mathbf{k}, \mathbf{k})] = 0$, we can write $\mathscr{L}_1^{(2)}(\mathbf{k}, \mathbf{k}, t)$ in the form

$$\mathscr{L}_\uparrow^{(2)}(\mathbf{k}, \mathbf{k}, t) = \frac{1}{2}\left(\frac{i}{\hbar}\right)^2 \int_{-\infty}^{\infty} ds\, \mathrm{Tr}\{\hat{\rho}_0\{\mathbf{\Gamma}(t)\}[\hat{V}(s), \{\hat{V}, \hat{\gamma}_\uparrow(\mathbf{k}, \mathbf{k})\}]\}$$

$$= \frac{1}{2}\left(\frac{i}{\hbar}\right)^2 \int_{-\infty}^{\infty} ds \lim_{\delta\to 0} \sum_{\mathbf{k}_1\ldots\mathbf{k}_8} \langle k_1, k_2|V|k_3, k_4\rangle\langle k_5, k_6|V|k_7, k_8\rangle$$
$$\times\, e^{(is/\hbar)(\epsilon_{\mathbf{k}_1} + \epsilon_{\mathbf{k}_2} - \epsilon_{\mathbf{k}_3} - \epsilon_{\mathbf{k}_4})} e^{-\delta|s|}$$
$$\times\, \mathrm{Tr}\{\hat{\rho}_0\{\mathbf{\Gamma}(t)\}[\hat{a}^+_{\mathbf{k}_1\uparrow}\hat{a}^+_{\mathbf{k}_2\downarrow}\hat{a}_{\mathbf{k}_4\downarrow}\hat{a}_{\mathbf{k}_3\uparrow}, (\delta_{\mathbf{k}_5,\mathbf{k}}\hat{a}^+_{\mathbf{k}\uparrow}\hat{a}^+_{\mathbf{k}_6\downarrow}\hat{a}_{\mathbf{k}_8\downarrow}\hat{a}_{\mathbf{k}_7\uparrow}$$
$$- \delta_{\mathbf{k}_7,\mathbf{k}}\hat{a}^+_{\mathbf{k}_5\uparrow}\hat{a}^+_{\mathbf{k}_6\downarrow}a_{\mathbf{k}_8\downarrow}a_{\mathbf{k}\uparrow})]\} \qquad (13.262)$$

where we have inserted a convergence factor $e^{-\delta|s|}$ to ensure that the limits are well defined. If we use Wick's theorem, we find

$$\mathscr{L}_\uparrow^{(2)}(\mathbf{k}, \mathbf{k}, t) = \frac{1}{2}\left(\frac{i}{\hbar}\right)^2 \frac{I^2}{V^2} \lim_{\delta\to 0} \int_{-\infty}^{\infty} ds \sum_{\mathbf{k}_1\mathbf{k}_2\mathbf{k}_3} (e^{(is/\hbar)(\epsilon_{\mathbf{k}_1} + \epsilon_{\mathbf{k}_2} - \epsilon_{\mathbf{k}_3} - \epsilon_{\mathbf{k}})} e^{-\delta|s|})$$
$$\delta(\mathbf{k}_1 + \mathbf{k}_2 - \mathbf{k} - \mathbf{k}_3)$$
$$\times\, [n^\uparrow(\mathbf{k}_1, t)n_\downarrow(\mathbf{k}_2, t)(1 - n_\uparrow(\mathbf{k}, t))(1 - n_\downarrow(\mathbf{k}_3, t))$$
$$- n_\uparrow(\mathbf{k}, t)n_\downarrow(\mathbf{k}_3, t)(1 - n_\uparrow(\mathbf{k}_1, t))(1 - n_\downarrow(\mathbf{k}_2, t))] \qquad (13.263)$$

where

$$n_\lambda(\mathbf{k}, t) = \Gamma_\lambda(\mathbf{k}, \mathbf{k}; t) \qquad (13.264)$$

is the momentum distribution function (cf. Eq. 13.239).

Let us now perform the time integration in Eq. (13.263):

$$I(s) \equiv \lim_{\delta\to 0} \int_{-\infty}^{0} ds(e^{(is/\hbar)(\epsilon_{\mathbf{k}_1} + \epsilon_{\mathbf{k}_2} - \epsilon_{\mathbf{k}_3} - \epsilon_{\mathbf{k}} - i\delta)} + e^{(is/\hbar)(\epsilon_{\mathbf{k}_3} + \epsilon_{\mathbf{k}} - \epsilon_{\mathbf{k}_1} - \epsilon_{\mathbf{k}_2} - i\delta)})$$

$$= \lim_{\delta\to 0} \frac{\hbar}{i}\left[\frac{1}{(\epsilon_{\mathbf{k}_1} + \epsilon_{\mathbf{k}_2} - \epsilon_{\mathbf{k}_3} - \epsilon_{\mathbf{k}} - i\delta)} + \frac{1}{(\epsilon_{\mathbf{k}_3} + \epsilon_{\mathbf{k}} - \epsilon_{\mathbf{k}_1} - \epsilon_{\mathbf{k}_2} - i\delta)}\right]. \qquad (13.265)$$

Eq. (13.265) is easily evaluated if we note that one definition of the Dirac delta function is

$$\delta(x) = \lim_{\delta\to 0} \frac{1}{2\pi i}\left[\left(\frac{1}{x - i\delta}\right) - \left(\frac{1}{x + i\delta}\right)\right]. \qquad (13.266)$$

Thus, Eq. (13.265) becomes

$$I(s) = 2\pi\hbar\, \delta(\epsilon_{\mathbf{k}_1} + \epsilon_{\mathbf{k}_2} - \epsilon_{\mathbf{k}_3} - \epsilon_{\mathbf{k}}). \qquad (13.267)$$

The collision integral $\mathscr{L}_{\lambda}^{(2)}(\mathbf{k}, \mathbf{k}, t)$ will be nonzero only if the argument of the delta function can be zero. For a quantum system in a finite box, this can never happen if the spectrum is nondegenerate, because the energy takes on discrete values and the argument, in general, will not be zero. However, in the thermodynamic limit the energy spectrum becomes continuous and the collision integral $\mathscr{L}_{\lambda}^{(2)}(\mathbf{k}, \mathbf{k}, t)$ will be nonzero. When this happens we have an *H*-theorem and the possibility of dissipation.

One can look at the phenomenon of irreversibility in a slightly different way. In evaluating various quantities with respect to the momentum eigenstates of the unperturbed Hamiltonian H_0, we have essentially decomposed the non-interacting system into its various harmonic motions. The behavior of the system is then very similar to a set of weakly coupled harmonic oscillators. If the natural frequencies of the harmonic oscillators are commensurate (integral multiples of one another), the oscillators can resonate and freely transfer energy among themselves. If the natural frequencies are incommensurate, they will not transfer energy easily. Thus, when the argument of the delta function can take on zero values it means that energy can freely flow between the modes and relaxation of the system to its equilibrium state can occur.

The resonance behavior leading to irreversibility in quantum systems is closely related to the resonance behavior leading to chaotic behavior in the classical anharmonic oscillator systems in Chap. 8. In classical systems, resonance behavior can occur in systems with a finite number of degrees of freedom because classical systems generally have continuous energy spectrum. In quantum systems, the energy spectrum and, therefore, the spectrum of the Liouville operator become continuous only in the thermodynamic limit.

If we make use of the above results, we can write Eq. (13.253) in the form

$$\frac{\partial}{\partial t} n_{\uparrow}(\mathbf{k}_1, t) = -\pi\left(\frac{i}{\hbar}\right)^2 \frac{I^2}{V^2} \sum_{\mathbf{k}_1\mathbf{k}_2\mathbf{k}_3} \delta(\mathbf{k}_1 + \mathbf{k}_2 - \mathbf{k}_3 - \mathbf{k})\, \delta(\epsilon_{\mathbf{k}_1} + \epsilon_{\mathbf{k}_2} - \epsilon_{\mathbf{k}} - \epsilon_{\mathbf{k}_3})$$
$$\times [n_{\uparrow}(\mathbf{k}_1 t) n_{\downarrow}(\mathbf{k}_2 t)(1 - n_{\uparrow}(\mathbf{k} t))(1 - n_{\downarrow}(\mathbf{k}_3 t))$$
$$- n_{\uparrow}(\mathbf{k} t) n_{\downarrow}(\mathbf{k}_3 t)(1 - n_{\uparrow}(\mathbf{k}_1 t))(1 - n_{\downarrow}(\mathbf{k}_2 t))]. \qquad (13.268)$$

A similar equation can be written for $n_{\downarrow}(\mathbf{k}_1, t)$.

Eq. (13.268) is the kinetic equation describing the decay to equilibrium of the momentum distribution for a spatially uniform quantum gas with a Hamiltonian of the form given in Eq. (13.212). It is easy to show that the equilibrium distribution is the ideal Fermi distribution,

$$n_{\lambda}(\mathbf{k}) = [1 + \exp \beta(\epsilon_{\mathbf{k}\lambda} - \mu)]^{-1}. \qquad (13.269)$$

Derivation of the kinetic equation for a spatially varying system is a little more involved and we will not attempt it here. We only note that to carry through the calculation for the spatially varying case we must first transform to a representation in which $\hat{\rho}_0\{\mathbf{\Gamma}(t)\}$ can be written in diagonal form. Then a derivation similar to that given above can be performed, at least formally.

The methods discussed in this section were first developed by Peletminskii and Yatsenko to derive kinetic equations for systems with broken symmetry. Indeed, we can generalize $\hat{\gamma}_j$ to include any operators which transform into themselves under the action of $\hat{H}_0$, that is, any set of operators $\{\gamma_j\}$ which satisfy the condition

$$[\hat{H}_0, \hat{\gamma}_j] = \sum_{\mathbf{k}} a_{jk}\hat{\gamma}_k. \tag{13.270}$$

As shown by Galaiko[14] and by Shumeiko,[15] the method is readily generalized to derive kinetic equations for Fermi superfluids. In that case, the set $\hat{\gamma}_j$ contains operators of the form $\hat{a}^+_{\mathbf{k}_1,\lambda_1}\hat{a}_{\mathbf{k}_2,\lambda_2}$, $\hat{a}^+_{\mathbf{k}_1,\lambda_1}\hat{a}^+_{\mathbf{k}_2,\lambda_2}$, and $\hat{a}_{\mathbf{k}_1,\lambda_1}\hat{a}_{\mathbf{k}_2,\lambda_2}$. The derivation of the microscopic expressions for the transport coefficients in a system of Landau quasiparticles in a Fermi liquid using the methods of this chapter has been given in Ref. 13.16.

REFERENCES

1. R. D. Present, *Kinetic Theory of Gases* (McGraw-Hill Book Co., New York, 1958).
2. E. H. Kennard, *Kinetic Theory of Gases* (McGraw-Hill Book Co., NewYork, 1938).
3. W. J. Moore, *Physical Chemistry* (Prentice-Hall, Englewood Cliffs, N.J., 1963).
4. L. Boltzmann, *Lectures on Gas Theory* (University of California Press, Berkeley, 1964).
5. G. E. Uhlenbeck and G. W. Ford, *Lectures in Statistical Mechanics* (American Mathematical Society, Providence, 1963).
6. H. Goldstein, *Classical Mechanics* (Addison-Wesley Publishing Co., Reading, Mass., 1950).
7. P. Resibois, J. Stat. Phys. *2* 21 (1970).
8. J. D. Fock and G. W. Ford in *Studies in Statistical Mechanics*, Vol. 5, ed. J. de Boer and G. E. Uhlenbeck (North-Holland Publishing Co., Amsterdam, 1970).
9. R. Balescu, *Equilibrium and Non-Equilibrium Statistical Mechanics* (Wiley-Interscience, New York, 1975).
10. S. Chapman and T. G. Cowling, *The Mathematical Theory of Non-uniform Gases* (Cambridge University Press, Cambridge, 1970).
11. E. A. Uehling and G. E. Uhlenbeck, Phys. Rev. *43* 552 (1933).
12. S. Peletminskii and A. Yatsenko, Sov. Phys. J.E.T.P. *26* 773 (1968).
13. N. N. Bogoliubov in *Studies in Statistical Mechanics*, Vol. 3, ed. J. de Boer and G. R. Uhlenbeck (North-Holland Publishing Co., Amsterdam, 1962).
14. V. P. Galaiko, Sov. Phys. J.E.T.P. *34* 203 (1972).
15. V. S. Shumeiko, Sov. Phys. J.E.T.P. *36* 330 (1973).
16. L. E. Reichl, Trans. Th. Stat. Phys. *6* 107, 129 (1977).

PROBLEMS

1. Estimate the value of the coefficient of viscosity of argon gas at 25°C and 1 atm pressure. Compare your estimate with the experimentally observed value of $\eta = 2.27 \times 10^{-4}$ g cm^{-1} sec. Argon has an atomic weight of 39.9 and at low temperature forms a closely packed solid with density 1.65 g/cm^3.

2. The momentum distribution in an ideal gas with a particle density $n = N/V$ is

$$f(\mathbf{p}) = \left(\frac{\beta}{2\pi m}\right)^{3/2} e^{-\beta p^2/2m}(1 + \epsilon \cos \theta)$$

where θ is the angle between $\mathbf{p}$ and the z-axis.

(*a*) Compute the drift velocity of the gas.
(*b*) What is the net number of particles per unit area per unit time traveling in the positive z-direction?

3. Find the number of molecules colliding with a surface per unit area per unit time with normal component of velocity exceeding the value v_0.

4. A furnace wall is constructed in two layers of thickness, $x_1 = 10$ cm and $x_2 = 20$ cm, such that the layers have thermal conductivity $K_1 = 0.002$ cal/sec cm deg°C and $K_2 = 0.004$ cal/sec cm deg°C, respectively. The inner surface is maintained at a temperature of 600°C and the outer surface at a temperature of 460°C. Compute:

(*a*) The heat current per unit area.
(*b*) The temperature at the interface.

5. Two coaxial cylinders are maintained at constant temperature. The inner cylinder has radius R_1 and temperature T_1 and the outer cylinder has temperature T_2 and radius R_2. The cylinders both have length L. If a heat-conducting material with heat-conduction coefficient K lies between the two cylinders, how much heat flows between them each second?

6. Prove that the linearized collision operators $\hat{C}_{\mathbf{p}}^+$ and $\hat{C}_{\mathbf{p}}^-$ are self-adjoint.

7. Prove that $\langle \Phi, \hat{C}^+\Phi \rangle$ and $\langle \Phi, \hat{C}^-\Phi \rangle$ are always less than or equal to zero for arbitrary functions $\Phi = \Phi(\mathbf{p})$.

8. Assume that the time evolution of a dilute gas can be approximated by a simple exponential decay, $e^{\hat{C}^+t} \sim e^{-t/\tau}$ and $e^{\hat{C}^-t} \sim e^{-t/\tau_D}$. Calculate the coefficients of shear viscosity, thermal conductivity, and diffusion. Show that your results are in qualitative agreement with results obtained using mean free path arguments.

9. The linearized Boltzmann collision operator $\hat{C}_{\mathbf{p}}^+$ is a rather complicated object. Sometimes it is useful to introduce approximate collision operators and study the behavior of Boltzmann-like equations. Show that the behavior of the collision operator defined by the equation

$$\hat{J}_{\mathbf{p}}h(\mathbf{p}, t) = -\gamma h(\mathbf{p}, t) + \gamma\left(\frac{\beta}{2\pi m}\right)^{3/2} \int d\mathbf{p}_2\, e^{-\beta p_2^2/2m}$$
$$\times \left[1 + \frac{\beta}{m}\mathbf{p}_1\cdot\mathbf{p}_2 + \frac{2}{3}\left(\frac{\beta p_2{}^2}{2m} - \frac{3}{2}\right)\left(\frac{\beta p_1{}^2}{2m} - \frac{3}{2}\right)\right] h(\mathbf{p}_2, t)$$

has all the important properties of the linearized Boltzmann collision operator. How does it differ from $\hat{C}_{\mathbf{p}}^+$? What is the meaning of γ?

10. The full nonlinear hydrodynamic equations for a simple fluid are given by the equations

$$\frac{\partial n}{\partial t} + \nabla_{\mathbf{r}}\cdot(n\mathbf{v}) = 0, \qquad mn\left(\frac{\partial}{\partial t} + \mathbf{v}\cdot\nabla_{\mathbf{r}}\right)\mathbf{v} = -\nabla_{\mathbf{r}}\cdot\overline{\overline{\mathbf{P}}}$$

and

$$n\left(\frac{\partial}{\partial t} + \mathbf{v}\cdot\nabla_{\mathbf{r}}\right)e = -\nabla_{\mathbf{r}}\cdot\mathbf{J}^{Q} - \frac{1}{2}\sum_{ij} P_{ij}\left(\frac{\partial v_i}{\partial x_j} + \frac{\partial v_j}{\partial x_i}\right),$$

where n is the total particle density, $\mathbf{v} \equiv \langle\mathbf{p}/m\rangle$, $\overline{\overline{\mathbf{P}}} = mn\langle\mathbf{V}\mathbf{V}\rangle$, $\mathbf{J}^Q = \frac{1}{2}mn\langle\mathbf{V}V^2\rangle$, $e = \frac{1}{2}m\langle V^2\rangle$, and $\mathbf{V} = \mathbf{p}/m - \mathbf{v}$. The expectation value $\langle A\rangle$ is defined

$$\langle A\rangle = \int d\mathbf{p} f(\mathbf{p}, \mathbf{r}, t)A(\mathbf{p}) \Big/ \int d\mathbf{p} f(\mathbf{p}, \mathbf{r}, t)$$

where $f(\mathbf{p}, \mathbf{r}, t)$ is the particle distribution function. Starting from the full nonlinear Boltzmann equation for $f(\mathbf{p}, \mathbf{r}, t)$ (cf. Eq. (13.69)), derive the hydrodynamic equations given above. (These are called the Navier-Stokes equations.)

11. Show that the Dirac delta function is defined

$$\delta(x) = \lim_{\delta\to 0}\frac{1}{2\pi i}\left[\left(\frac{1}{x - i\delta}\right) - \left(\frac{1}{x + i\delta}\right)\right].$$

12. Using the procedure of Sec. 13.G for approximating the coefficient of thermal conductivity, find an analogous approximation for the coefficients of shear viscosity and self-diffusion in terms of Sonine polynomials.

13. Consider a Fermi gas which has condensed into a superfluid state so that the expectation values $\langle \hat{a}^{+}_{\mathbf{k}\uparrow}\hat{a}^{+}_{-\mathbf{k}\downarrow}\rangle$ and $\langle \hat{a}_{-\mathbf{k}\downarrow}\hat{a}_{\mathbf{k}\uparrow}\rangle$ are nonzero. Separate the Hamiltonian as in Eqs. (13.216) and (13.217) but include mean field contributions from $\langle \hat{a}^{+}_{\mathbf{k}\uparrow}\hat{a}^{+}_{-\mathbf{k}\downarrow}\rangle$ and $\langle \hat{a}_{-\mathbf{k}\downarrow}\hat{a}_{\mathbf{k}\uparrow}\rangle$. For a system with no spatial variations, find the equation of motion for the reduced density matrix of bogolons (cf. Sec. 9.E and use the Bogoliubov transformation to diagonalize the Hamiltonian).

14. For a single component dissipative isotropic fluid, solve the linearized hydrodynamic equations for $n(\mathbf{k}, t)$, $T(\mathbf{k}, t)$, and $\mathbf{v}(\mathbf{k}, t)$ in terms of their initial values $n(\mathbf{k}, 0)$, $T(\mathbf{k}, 0)$, and $\mathbf{v}(\mathbf{k}, 0)$, respectively.

14
Hydrodynamics and Onsager's Relations

A. INTRODUCTORY REMARKS

Hydrodynamic equations describe the long-wavelength, low-frequency phenomena in a large variety of systems, including dilute gases, liquids, solids, liquid crystals, superfluids, and chemically reacting systems. For complicated systems, transport processes are often coupled. For example, in a multicomponent system, it is possible to have a temperature gradient drive a particle current and a concentration gradient drive a heat current. Some complicated systems can have as many as ten or twenty (or more) transport coefficients to describe the decay to equilibrium from the hydrodynamic regime.

In 1932, Onsager showed that the reversibility of the dynamical laws on the microscopic level requires that certain relations exist between transport coefficients describing coupled transport processes. Onsager's relations are of immense importance because they enable us to link seemingly independent transport processes and thereby reduce the number of experiments that must be performed in order to measure all the transport coefficients.

In this chapter, we will derive Onsager's relations and apply them to transport processes in reacting multicomponent systems and superfluid systems. At the same time we will show how the hydrodynamic equations for such complicated systems can be derived from a knowledge of the thermodynamics and symmetry properties of a system.

B. ONSAGER'S RELATIONS[1–5]

Systems which are out of equilibrium generally return to the equilibrium state through a variety of transport processes which may or may not be coupled to one another. Onsager,[1] without any reference to a particular physical system, was able to derive relations between the transport coefficients for coupled transport processes. These relations are a general consequence of the invariance under time reversal of Hamiltonian dynamics. There are essentially three aspects of the proof: (*a*) it is based on the general theory of time-dependent fluctuations; (*b*) it uses the time-reversal invariance of the mechanical equations of motion; and (*c*) it assumes that fluctuations about the equilibrium state decay, *on the average*, according to the same laws that govern the decay of macroscopic deviations from equilibrium, that is, the hydrodynamic equations. We shall first outline the proof for the case of state variables which do not change sign under time reversal (such as particle density and energy density). We shall denote such variables by α. Then we shall generalize the relations to include the case of fluctuations β which do change sign under time reversal (such as momentum density) and we shall at the same time include the possibility of an external magnetic field.

1. Time-Dependent Correlation Functions and Microscopic Reversibility

The first step in deriving Onsager's relations is to show that the time-reversal invariance of Hamiltonian dynamics requires that the time-dependent correlation functions of macroscopic fluctuations, $\boldsymbol{\alpha} = (\alpha_1, \alpha_2, \ldots, \alpha_n)$, obey the relations

$$\langle \alpha_i \alpha_j(\tau) \rangle = \langle \alpha_i(\tau) \alpha_j \rangle. \tag{14.1}$$

The quantities $\alpha_i = A_i - A_i^0$ and $\alpha_j = A_j - A_j^0$ are the fluctuations about the equilibrium values of the state variables A_i and A_j (cf. Sec. 10.B). Eq. (14.1) tells us that the correlation between a fluctuation α_i at time $t = 0$ and a fluctuation α_j at time $t = \tau$ is the same as that of a fluctuation α_j at time $t = 0$ and a fluctuation α_i at time $t = \tau$. The quantities α_i and α_j can correspond to fluctuations in the same state variables at different points in space. Thus, Eq. (14.1) can also be turned into an equation relating correlations between space- and time-dependent fluctuations.

To establish Eq. (14.1), we note that the correlation matrix $\langle \boldsymbol{\alpha\alpha}(\tau) \rangle$ can be written

$$\langle \boldsymbol{\alpha\alpha}(\tau) \rangle \equiv \iint d\boldsymbol{\alpha}\, d\boldsymbol{\alpha}' \boldsymbol{\alpha\alpha}' f(\boldsymbol{\alpha}) P(\boldsymbol{\alpha}|\boldsymbol{\alpha}', \tau) \tag{14.2}$$

where $P(\boldsymbol{\alpha}|\boldsymbol{\alpha}', \tau)$ is the conditional probability that the fluctuation has value $\boldsymbol{\alpha}'$ at time $t = \tau$, given that it had value $\boldsymbol{\alpha}$ at time $t = 0$ (we are now using the notation of Sec. 6.B).

For a closed isolated system, $f(\boldsymbol{\alpha})$ can be written

$$f(\boldsymbol{\alpha}) = \left(\frac{|\bar{\bar{\mathbf{g}}}|}{(2\pi k_B)^n} \right)^{1/2} e^{-\bar{\bar{\mathbf{g}}}:\boldsymbol{\alpha\alpha}/2k_B} \tag{14.3}$$

(cf. Eq. (10.8)), and the change in entropy which results from these fluctuations is

$$\Delta S = -\tfrac{1}{2}\bar{\bar{\mathbf{g}}}:\boldsymbol{\alpha\alpha}. \tag{14.4}$$

It is useful to introduce a generalized force, **X**, which is defined

$$\mathbf{X} = \bar{\bar{\mathbf{g}}} \cdot \boldsymbol{\alpha} = -\left(\frac{\partial \Delta S}{\partial \boldsymbol{\alpha}}\right), \tag{14.5}$$

and a generalized current, **J**, defined

$$\mathbf{J} = \frac{d\boldsymbol{\alpha}}{dt}. \tag{14.6}$$

Then the time rate of change of the entropy due to fluctuations is

$$\frac{d\,\Delta S}{dt} = -\mathbf{J}\cdot\mathbf{X}. \tag{14.7}$$

For a resistor held at constant temperature, where J is the electric current and X is the applied electric field, Eq. (14.7) is proportional to the rate at which energy is dissipated through Joule heating.

We must now remember that $\boldsymbol{\alpha}$ is a macroscopic variable. Thus, for each value of $\boldsymbol{\alpha}$ there are many possible microscopic states of the system. We can relate the joint probability distribution $f(\boldsymbol{\alpha})P(\boldsymbol{\alpha}|\boldsymbol{\alpha}', \tau)$ for fluctuations $\boldsymbol{\alpha}$ and $\boldsymbol{\alpha}'$ to the microscopic probability density in the following way:

$$f(\boldsymbol{\alpha})P(\boldsymbol{\alpha}|\boldsymbol{\alpha}', \tau) = \frac{1}{\Omega_{\Delta E}(E)} \iint\limits_{\substack{(\boldsymbol{\alpha}\to\boldsymbol{\alpha}+d\boldsymbol{\alpha})\\(E\to E+\Delta E)}} d\mathbf{q}^N\, d\mathbf{p}^N$$

$$\times \iint\limits_{(\boldsymbol{\alpha}'\to\boldsymbol{\alpha}'+d\boldsymbol{\alpha}')} d\mathbf{q}'^N\, d\mathbf{p}'^N P(\mathbf{p}^N, \mathbf{q}^N|\mathbf{p}'^N, \mathbf{q}'^N, \tau). \tag{14.8}$$

In Eq. (14.8) we have used the fact that $\rho(\mathbf{p}^N, \mathbf{q}^N) = \Omega_{\Delta E}(E)^{-1}$ for a closed isolated system ($\Omega_{\Delta E}(E)$ is the volume of the energy shell). The phase space integrations are restricted to the energy shell and to trajectories with values of $\boldsymbol{\alpha}$ and $\boldsymbol{\alpha}'$ appearing in the left-hand side of Eq. (14.8); $P(\mathbf{p}^N, \mathbf{q}^N|\mathbf{p}'^N, \mathbf{q}'^N, \tau)$ is the conditional probability that a system be in a state $(\mathbf{p}'^N, \mathbf{q}'^N)$ at time $t = \tau$, given that it was in the state $(\mathbf{p}^N, \mathbf{q}^N)$ at time $t = 0$. Since classical systems are completely deterministic, we must have

$$P(\mathbf{p}^N, \mathbf{q}^N|\mathbf{p}'^N, \mathbf{q}'^N, \tau) = \delta(\mathbf{q}'^N - \mathbf{q}^N - \Delta\mathbf{q}^N(\mathbf{p}^N, \mathbf{q}^N, \tau))$$

$$\times\ \delta(\mathbf{p}'^N - \mathbf{p}^N - \Delta\mathbf{p}^N(\mathbf{p}^N, \mathbf{q}^N, \tau)) \tag{14.9}$$

where $\Delta\mathbf{q}^N$ and $\Delta\mathbf{p}^N$ are uniquely determined from Hamilton's equations.

Because Hamilton's equations are causal and time-reversal invariant, reversal of all momenta in the system will cause the system to retrace its steps. This implies that

$$P(\mathbf{q}^N, \mathbf{p}^N|\mathbf{q}'^N, \mathbf{p}'^N, \tau) = P(\mathbf{q}^{N\prime}, -\mathbf{p}'^N|\mathbf{q}^N, -\mathbf{p}^N, \tau). \tag{14.10}$$

Since $\boldsymbol{\alpha}$ and $\boldsymbol{\alpha}'$ are even functions of momentum, we can combine Eqs. (14.8) and (14.10) to obtain

$$f(\boldsymbol{\alpha})P(\boldsymbol{\alpha}|\boldsymbol{\alpha}', \tau) = f(\boldsymbol{\alpha}')P(\boldsymbol{\alpha}'|\boldsymbol{\alpha}, \tau). \tag{14.11}$$

From Eq. (14.11), Eq. (14.1) follows easily.

2. Regression of Fluctuations

Eq. (14.1) is important because we can use it to find relations between various transport coefficients. Let us first introduce the conditional average $\langle\boldsymbol{\alpha}(\tau)\rangle_{\boldsymbol{\alpha}_0}$, which is the average value of $\boldsymbol{\alpha}$ at time t, given that the initial value of $\boldsymbol{\alpha}$ was $\boldsymbol{\alpha}_0$. We then can write

$$\langle\boldsymbol{\alpha}(t)\rangle_{\boldsymbol{\alpha}_0} \equiv \int d\boldsymbol{\alpha}\boldsymbol{\alpha}P(\boldsymbol{\alpha}_0|\boldsymbol{\alpha}, \tau) \tag{14.12}$$

for the conditional average.

Onsager assumed that, *on the average*, the fluctuations decay according to the same linear laws (hydrodynamic equations) that govern the decay to equilibrium of systems which are macroscopically out of equilibrium. Thus, the average fluctuation, $\langle\boldsymbol{\alpha}(t)\rangle_{\boldsymbol{\alpha}_0}$, obeys an equation of the form

$$\frac{d}{dt}\langle\boldsymbol{\alpha}(t)\rangle_{\boldsymbol{\alpha}_0} = -\mathbb{M}\cdot\langle\boldsymbol{\alpha}(t)\rangle_{\boldsymbol{\alpha}_0}. \tag{14.13}$$

Eq. (14.13) has the solution

$$\langle\boldsymbol{\alpha}(t)\rangle_{\boldsymbol{\alpha}_0} = e^{-\mathbb{M}t}\cdot\boldsymbol{\alpha}_0. \tag{14.14}$$

The time derivative in Eq. (14.13) must be used with caution. It is defined in the following sense:

$$\frac{d\langle\boldsymbol{\alpha}(t)\rangle_{\boldsymbol{\alpha}_0}}{dt} \equiv \frac{\langle\boldsymbol{\alpha}(t+\tau)\rangle_{\boldsymbol{\alpha}_0} - \langle\boldsymbol{\alpha}(t)\rangle_{\boldsymbol{\alpha}_0}}{\tau} \tag{14.15}$$

where τ is a small time interval whose values are bounded by inequalities,

$$T_0 \ll \tau \ll T. \tag{14.16}$$

T_0 is the time between collisions and T is the time it takes the fluctuation to decay to equilibrium. The limitation in Eq. (14.16) rules out fluctuations which are too small, that is, fluctuations which decay to equilibrium in a few collision times. Similarly, Eq. (14.16) is not valid when the fluctuation has just been created. It takes a few collision times for it to settle down to a hydrodynamic decay.

Eq. (14.1) imposes a condition on the matrix $\mathbb{M}$. If we expand Eq. (14.14) for short times,

$$\langle\boldsymbol{\alpha}(\tau)\rangle_{\boldsymbol{\alpha}_0} = \boldsymbol{\alpha}_0 - t\mathbb{M}\cdot\boldsymbol{\alpha}_0 + O(t^2), \tag{14.17}$$

and substitute it into Eq. (14.1), we obtain

$$\langle \boldsymbol{\alpha}_0 \mathbb{M}\cdot\boldsymbol{\alpha}_0 \rangle = \langle \mathbb{M}\cdot\boldsymbol{\alpha}_0\boldsymbol{\alpha}_0 \rangle. \tag{14.18}$$

If we now use the fact that $\mathbb{M}\cdot\boldsymbol{\alpha} = \boldsymbol{\alpha}^T\cdot\mathbb{M}^T$ and use Eq. (10.12) for the variance in the fluctuations, we obtain

$$\bar{\bar{\mathbf{g}}}^{-1}\cdot\mathbb{M}^T = \mathbb{M}\cdot\bar{\bar{\mathbf{g}}}^{-1} \tag{14.19}$$

where T denotes the transpose. We can define a new tensor

$$\mathbb{L} \equiv \mathbb{M}\cdot\bar{\bar{\mathbf{g}}}^{-1}; \tag{14.20}$$

then Eq. (14.19) becomes

$$\mathbb{L} = \mathbb{L}^T \quad \text{or} \quad L_{ij} = L_{ji}. \tag{14.21}$$

Eqs. (14.21) are called Onsager's relations. If we make use of the generalized force $\mathbf{X}$ (cf. Eq. (14.5)), the time rate of change of the fluctuation can be written

$$\frac{d}{dt}\langle \boldsymbol{\alpha}(t)\rangle_{\boldsymbol{\alpha}_0} = -\mathbb{L}\cdot\langle \mathbf{X}(t)\rangle_{\boldsymbol{\alpha}_0}. \tag{14.22}$$

Eq. (14.22) has the same form as the linearized hydrodynamic equations (cf. Sec. 13.F). Eq. (14.21) is so useful that Onsager received a Nobel prize for deriving it. The matrix $\mathbb{L}$ is a matrix of transport coefficients. Eq. (14.22) tells us that a force resulting from a fluctuation α_i can cause a flux of A_j, and a force arising from a fluctuation α_j can cause a flux of A_i. Eq. (14.21) tells us that the transport coefficients for the two processes are the same. For example, a particle concentration gradient can drive a heat current, and a temperature gradient can drive a particle current. The transport coefficients for the two processes are the same although the processes physically appear to be very different.

C. ONSAGER'S RELATIONS WHEN A MAGNETIC FIELD IS PRESENT[4]

When a magnetic field, $\mathbf{B}$, is present or when we consider a system in a rotating container where Coriolis forces are present, the Onsager relations must be generalized. When we reverse the time and velocities in the equations of motion, we must also reverse the magnetic field, $\mathbf{B}$, in order to keep the equations of motion invariant, since the magnetic force on particles with velocity $\mathbf{v}$ and charge q is $\mathbf{F} = q\mathbf{v} \times \mathbf{B}$. If we have a rotating container we must reverse both the velocity of the particles and the angular velocity $\mathbf{w}$ of the container to keep the Coriolis force $\mathbf{F} = 2m\mathbf{v} \times \mathbf{w}$ from changing sign.

In the presence of a magnetic field and fluctuations $\boldsymbol{\beta}$ (the fluctuations $\boldsymbol{\beta}$

change sign under time reversal), the distribution function must satisfy the relation

$$f(\boldsymbol{\alpha}, \boldsymbol{\beta}, \mathbf{B}) = f(\boldsymbol{\alpha}, -\boldsymbol{\beta}, -\mathbf{B}) \tag{14.23}$$

under time reversal (here $\boldsymbol{\alpha} = (\alpha_1, \alpha_2, \ldots, \alpha_n)$ and $\boldsymbol{\beta} = (\beta_1, \ldots, \beta_m)$). This is easily understood if we write $f(\boldsymbol{\alpha}, \boldsymbol{\beta}, \mathbf{B})$ in terms of microscopic states:

$$f(\boldsymbol{\alpha}, \boldsymbol{\beta}, \mathbf{B}) = \frac{1}{\Omega_{\Delta E}(E)} \underset{\substack{\boldsymbol{\alpha} \to \boldsymbol{\alpha} + d\boldsymbol{\alpha} \\ \boldsymbol{\beta} \to \boldsymbol{\beta} + d\boldsymbol{\beta} \\ E \to E + \Delta E}}{\int \cdots \int} d\mathbf{r}^N \, d\mathbf{p}^N. \tag{14.24}$$

For both classical and quantum systems, energy is an even function of the momentum and the magnetic field. Therefore, if we reverse time and the magnetic field in Eq. (14.24), the integral does not change and Eq. (14.23) follows immediately.

When a magnetic field and fluctuations $\boldsymbol{\beta}$ are present, we can write the distribution function in the more general form

$$f(\boldsymbol{\alpha}, \boldsymbol{\beta}, \mathbf{B}) = C \exp\left(-\frac{1}{2k_B} (\bar{\bar{\mathbf{g}}}:\boldsymbol{\alpha\alpha} + \bar{\bar{\mathbf{m}}}:\boldsymbol{\alpha\beta} + \bar{\bar{\mathbf{n}}}:\boldsymbol{\beta\alpha} + \bar{\bar{\mathbf{h}}}:\boldsymbol{\beta\beta}) \right) \tag{14.25}$$

where C is the normalization constant and the tensors $\bar{\bar{\mathbf{m}}}$ and $\bar{\bar{\mathbf{n}}}$ can be represented by $m \times n$ rectangular matrices. The tensors $\bar{\bar{\mathbf{g}}}$, $\bar{\bar{\mathbf{m}}}$, $\bar{\bar{\mathbf{n}}}$, and $\bar{\bar{\mathbf{h}}}$ may now depend on the magnetic field. In order to satisfy Eq. (14.23), they must have the following behavior when the magnetic field is reversed:

$$\bar{\bar{\mathbf{g}}}(\mathbf{B}) = \bar{\bar{\mathbf{g}}}(-\mathbf{B}), \qquad \bar{\bar{\mathbf{h}}}(\mathbf{B}) = \bar{\bar{\mathbf{h}}}(-\mathbf{B}) \tag{14.26}$$

and

$$\bar{\bar{\mathbf{m}}}(\mathbf{B}) = -\bar{\bar{\mathbf{m}}}(-\mathbf{B}), \qquad \bar{\bar{\mathbf{n}}}(\mathbf{B}) = -\bar{\bar{\mathbf{n}}}(-\mathbf{B}). \tag{14.27}$$

Note that, when $\mathbf{B} = 0$, $\bar{\bar{\mathbf{m}}} = \bar{\bar{\mathbf{n}}} = 0$. The matrix $\bar{\bar{\mathbf{m}}}$ is simply the transpose of $\bar{\bar{\mathbf{n}}}$:

$$\bar{\bar{\mathbf{m}}} = \bar{\bar{\mathbf{n}}}^T. \tag{14.28}$$

The change in the entropy due to fluctuations in the presence of a magnetic field is

$$\Delta S = -\tfrac{1}{2}(\bar{\bar{\mathbf{g}}}:\boldsymbol{\alpha\alpha} + \bar{\bar{\mathbf{m}}}:\boldsymbol{\alpha\beta} + \bar{\bar{\mathbf{n}}}:\boldsymbol{\beta\alpha} + \bar{\bar{\mathbf{h}}}:\boldsymbol{\beta\beta}), \tag{14.29}$$

and we must introduce two generalized forces:

$$\mathbf{X} = -\frac{\partial\, \Delta S}{\partial \boldsymbol{\alpha}} = \bar{\bar{\mathbf{g}}}\cdot\boldsymbol{\alpha} + \tfrac{1}{2}\boldsymbol{\beta}\cdot\bar{\bar{\mathbf{m}}} + \tfrac{1}{2}\bar{\bar{\mathbf{n}}}\cdot\boldsymbol{\beta} = \bar{\bar{\mathbf{g}}}\cdot\boldsymbol{\alpha} + \bar{\bar{\mathbf{n}}}\cdot\boldsymbol{\beta} \tag{14.30}$$

and

$$\mathbf{Y} = -\frac{\partial\, \Delta S}{\partial \boldsymbol{\beta}} = \bar{\bar{\mathbf{h}}}\cdot\boldsymbol{\beta} + \bar{\bar{\mathbf{m}}}\cdot\boldsymbol{\alpha}. \tag{14.31}$$

(Note that in Eqs. (14.30) and (14.31) we have used Eq. (14.28).) The various time-independent correlation functions take the form:

$$\langle \boldsymbol{\alpha\alpha} \rangle = k_B(\bar{\bar{\mathbf{g}}} - \mathbf{n}\cdot\bar{\bar{\mathbf{h}}}^{-1}\cdot\bar{\bar{\mathbf{m}}})^{-1}, \tag{14.32a}$$

$$\langle \boldsymbol{\alpha\beta} \rangle = -k_B(\bar{\bar{\mathbf{g}}} - \bar{\bar{\mathbf{n}}}\cdot\bar{\bar{\mathbf{h}}}^{-1}\cdot\bar{\bar{\mathbf{m}}})^{-1}\cdot\bar{\bar{\mathbf{n}}}\cdot\bar{\bar{\mathbf{h}}}^{-1}, \tag{14.32b}$$

$$\langle \boldsymbol{\beta\alpha} \rangle = -k_B(\bar{\bar{\mathbf{h}}} - \bar{\bar{\mathbf{m}}}\cdot\bar{\bar{\mathbf{g}}}^{-1}\cdot\bar{\bar{\mathbf{n}}})^{-1}\cdot\bar{\bar{\mathbf{m}}}\cdot\bar{\bar{\mathbf{g}}}^{-1}, \tag{14.32c}$$

and

$$\langle \boldsymbol{\beta\beta} \rangle = k_B(\bar{\bar{\mathbf{h}}} - \bar{\bar{\mathbf{m}}}\cdot\bar{\bar{\mathbf{g}}}^{-1}\cdot\bar{\bar{\mathbf{n}}})^{-1}. \tag{14.32d}$$

In the presence of a magnetic field, Eq. (4.10) is generalized to the form

$$P(\mathbf{q}^N, \mathbf{p}^N|\mathbf{q}^{N'}, \mathbf{p}'^N; \tau, \mathbf{B}) = P(\mathbf{q}'^N, -\mathbf{p}'^N|\mathbf{q}^N - \mathbf{p}^N; \tau, -\mathbf{B}), \tag{14.33}$$

and the joint probability density for fluctuations is generalized to the form

$$f(\boldsymbol{\alpha}, \boldsymbol{\beta}, \mathbf{B})P(\boldsymbol{\alpha}, \boldsymbol{\beta}|\boldsymbol{\alpha}', \boldsymbol{\beta}', \tau, \mathbf{B}) = f(\boldsymbol{\alpha}', -\boldsymbol{\beta}', -\mathbf{B})P(\boldsymbol{\alpha}', -\boldsymbol{\beta}'|\boldsymbol{\alpha}, -\boldsymbol{\beta}, \tau, -\mathbf{B}). \tag{14.34}$$

From Eq. (14.33) it follows that the time-dependent correlation functions given by expressions analogous to Eq. (14.2) satisfy the following relations:

$$\langle \boldsymbol{\alpha\alpha}(\tau); \mathbf{B} \rangle = \langle \boldsymbol{\alpha}(\tau)\boldsymbol{\alpha}; -\mathbf{B} \rangle, \tag{14.35a}$$

$$\langle \boldsymbol{\alpha\beta}(\tau); \mathbf{B} \rangle = -\langle \boldsymbol{\alpha}(\tau)\boldsymbol{\beta}; -\mathbf{B} \rangle, \tag{14.35b}$$

$$\langle \boldsymbol{\beta\alpha}(\tau); \mathbf{B} \rangle = -\langle \boldsymbol{\beta}(\tau)\boldsymbol{\alpha}; -\mathbf{B} \rangle, \tag{14.35c}$$

and

$$\langle \boldsymbol{\beta\beta}(\tau); \mathbf{B} \rangle = \langle \boldsymbol{\beta}(\tau)\boldsymbol{\beta}; -\mathbf{B} \rangle \tag{14.35d}$$

under time reversal.

The equations describing the regression of fluctuations can be written as in Eq. (14.13) except that now the matrix $\mathbb{M}$ depends on the magnetic field. Thus,

$$\frac{d}{dt}\begin{pmatrix} \langle \boldsymbol{\alpha}(t) \rangle_{\boldsymbol{\alpha}_0\boldsymbol{\beta}_0} \\ \langle \boldsymbol{\beta}(t) \rangle_{\boldsymbol{\alpha}_0\boldsymbol{\beta}_0} \end{pmatrix} = -\mathbb{M}(\mathbf{B})\cdot\begin{pmatrix} \langle \boldsymbol{\alpha}(t) \rangle_{\boldsymbol{\alpha}_0\boldsymbol{\beta}_0} \\ \langle \boldsymbol{\beta}(t) \rangle_{\boldsymbol{\alpha}_0\boldsymbol{\beta}_0} \end{pmatrix} \tag{14.36}$$

where

$$\mathbb{M}(\mathbf{B}) = \begin{pmatrix} \bar{\bar{\mathbf{M}}}^{\alpha\alpha}(\mathbf{B}) & \bar{\bar{\mathbf{M}}}^{\alpha\beta}(\mathbf{B}) \\ \bar{\bar{\mathbf{M}}}^{\beta\alpha}(\mathbf{B}) & \bar{\bar{\mathbf{M}}}^{\beta\beta}(\mathbf{B}) \end{pmatrix} \tag{14.37}$$

or

$$\frac{d}{dt}\begin{pmatrix} \langle \boldsymbol{\alpha}(t) \rangle_{\boldsymbol{\alpha}_0\boldsymbol{\beta}_0} \\ \langle \boldsymbol{\beta}(t) \rangle_{\boldsymbol{\alpha}_0\boldsymbol{\beta}_0} \end{pmatrix} = -\mathbb{L}(\mathbf{B})\cdot\begin{pmatrix} \langle \mathbf{X}(t) \rangle_{\boldsymbol{\alpha}_0\boldsymbol{\beta}_0} \\ \langle \mathbf{Y}(t) \rangle_{\boldsymbol{\alpha}_0\boldsymbol{\beta}_0} \end{pmatrix}. \tag{14.38}$$

If we use arguments similar to those in Sec. 14.B, Onsager's relations take the form

$$\mathbb{L}^{\alpha\alpha}(-\mathbf{B})^T = \mathbb{L}^{\alpha\alpha}(\mathbf{B}), \tag{14.39a}$$

$$\mathbb{L}^{\beta\alpha}(-\mathbf{B})^T = -\mathbb{L}^{\alpha\beta}(\mathbf{B}), \tag{14.39b}$$

$$\mathbb{L}^{\alpha\beta}(-\mathbf{B})^T = -\mathbb{L}^{\beta\alpha}(\mathbf{B}), \tag{14.39c}$$

and

$$\mathbb{L}^{\beta\beta}(-\mathbf{B})^T = \mathbb{L}^{\beta\beta}(\mathbf{B}). \tag{14.39d}$$

Eqs. (14.39*a–d*) give the general form of Onsager's relations. If we were considering a rotating system, then we would replace the magnetic field, **B**, by the angular velocity, $\boldsymbol{\omega}$.

D. MECHANOCALORIC EFFECT AND THERMOMOLECULAR PRESSURE EFFECT[3–5]

We shall now use Onsager's relations to show how two very different effects in a classical gas can be related. Let us consider an ideal gas which is contained in a closed isolated box divided in half by a partition containing a small hole (we shall say more about the size of the hole later). The total mass and energy of the particles in the box is constant although mass and energy can be transferred from one side to another through the hole. We assume that only one kind of particle is present.

Since mass and energy are conserved, any changes in the mass and energy on the two sides are related. Thus,

$$\Delta M_1 = -\Delta M_2 \equiv \Delta M \tag{14.40}$$

and

$$\Delta U_1 = -\Delta U_2 \equiv \Delta M \tag{14.41}$$

where ΔM_i and ΔU_i are the change in mass and energy, respectively, in compartment i. We shall assume that the two sides are almost in the same equilibrium state and expand about the common equilibrium state. The change in entropy due to flow of energy and matter through the hole is

$$\Delta S = 2\left[\frac{1}{2}\left(\frac{\partial^2 S}{\partial U^2}\right)_M^0 (\Delta U)^2 + \left(\frac{\partial}{\partial U}\left(\frac{\partial S}{\partial M}\right)_U^0\right)_M^0 (\Delta U\,\Delta M) + \frac{1}{2}\left(\frac{\partial^2 S}{\partial M^2}\right)_U^0 (\Delta M)^2\right] \tag{14.42}$$

(cf. Sec. 2.H). The time rate of change of the entropy is given by

$$\frac{d(\Delta S)}{dt} = 2\left[\frac{d(\Delta U)}{dt}\left\{\left(\frac{\partial^2 S}{\partial U^2}\right)_M^0 (\Delta U) + \left(\frac{\partial}{\partial M}\left(\frac{\partial S}{\partial U}\right)_M\right)_U^0 (\Delta M)\right\} + \frac{d(\Delta M)}{dt}\left\{\left(\frac{\partial}{\partial U}\left(\frac{\partial S}{\partial M}\right)_U\right)_M^0 (\Delta U) + \left(\frac{\partial^2 S}{\partial M^2}\right)_U (\Delta M)\right\}\right] \tag{14.43}$$

or, more compactly,

$$\frac{d(\Delta S)}{dt} = 2\left[\frac{d(\Delta U)}{dt}\Delta\left(\frac{\partial S}{\partial U}\right)_M + \frac{d(\Delta M)}{dt}\Delta\left(\frac{\partial S}{\partial M}\right)_U\right]. \tag{14.44}$$

We now define an energy current, J_u, and a mass current, J_m, according to the relations

$$J_u \equiv \frac{d(\Delta U)}{dt} \tag{14.45}$$

and

$$J_m = \frac{d(\Delta M)}{dt}, \tag{14.46}$$

respectively, and we define the corresponding forces, X_u and X_m, according to the relations

$$X_u = -\Delta\left(\frac{\partial S}{\partial U}\right)_m \tag{14.47}$$

and

$$X_m = -\Delta\left(\frac{\partial S}{\partial M}\right)_u, \tag{14.48}$$

respectively (cf. Eqs. (14.5) and (14.6)). Eq. (14.44) then takes the form

$$\frac{d(\Delta S)}{dt} = -2[J_u X_u + J_m X_m]. \tag{14.49}$$

We can find expressions for the forces X_u and X_m from thermodynamics. For a system with constant volume, the first law may be written

$$dS = \frac{1}{T}dU - \frac{\mu}{T}dM \equiv \left(\frac{\partial S}{\partial U}\right)_M dU + \left(\frac{\partial S}{\partial M}\right)_U dM \tag{14.50}$$

where μ is the chemical potential per unit mass. If we compare Eqs. (14.47), (14.48), and (14.50), we obtain

$$X_u = -\Delta\left(\frac{1}{T}\right) \tag{14.51}$$

and

$$X_m = \Delta\left(\frac{\mu}{T}\right). \tag{14.52}$$

The energy force X_u depends on the temperature difference between the two sides, while the mass force X_m depends on a temperature difference and/or a chemical potential difference between the two sides. It is more convenient to write Eq. (14.52) in terms of pressure and temperature differences. We can do this if we make use of the Gibbs-Duheim equation,

$$d\mu = -s\,dT + \frac{1}{\rho}\,dP, \tag{14.53}$$

where s is the entropy per unit mass and $1/\rho$ is the volume per unit mass. Eqs. (14.51) and (14.52) then become

$$X_u = \frac{1}{T^2}\,\Delta T \tag{14.54}$$

and

$$X_m = -\frac{h}{T^2}\,\Delta T + \frac{1}{\rho T}\,\Delta P \tag{14.55}$$

where h is the enthalpy per unit mass,

$$h = u + \frac{1}{\rho}\,P = sT + \mu, \tag{14.56}$$

and, in (14.56), u is the internal energy per unit mass.

Let us now restrict ourselves to the linear regime and assume that the currents J_u and J_m are driven by terms linear in the forces

$$J_u = -L_{uu}X_u - L_{um}X_m \tag{14.57}$$

and

$$J_m = -L_{mu}X_u - L_{mm}X_m \tag{14.58}$$

where

$$L_{um} = L_{mu} \tag{14.59}$$

(Onsager's relations). We can rewrite the energy and mass currents in terms of pressure and temperature differences between the two sides of the box. We find

$$J_u = -\frac{L_{um}}{\rho T}\,\Delta P + \frac{L_{um}h - L_{uu}}{T^2}\,\Delta T \tag{14.60}$$

and

$$J_m = -\frac{L_{mm}}{T\rho}\,\Delta P + \frac{L_{mm}h - L_{mu}}{T^2}\,\Delta T. \tag{14.61}$$

We can now use the Onsager relations to connect two completely different effects in the box. Let us first consider the mechanocaloric effect.

1. Mechanocaloric Effect

Let us assume that the two compartments have the same temperature, $\Delta T = 0$, and a different pressure, $\Delta P \neq 0$. Then there will be a flow of internal energy and mass, both of which are driven by the pressure difference,

$$J_u = -\frac{L_{um}}{\rho T}\Delta P \tag{14.62}$$

and

$$J_m = -\frac{L_{mm}}{\rho T}\Delta P. \tag{14.63}$$

Thus, the mass and energy currents are related according to the relation

$$J_u = \frac{L_{um}}{L_{mm}} J_m. \tag{14.64}$$

The quantity L_{um}/L_{mm} is a proportionality constant which determines the amount of energy per unit mass that is transferred by particles as they move through the hole. It must be computed microscopically. We will consider two cases: (*a*) a Knudsen gas in which the mean free path of particles is much larger than the hole in the partition and (*b*) a Boyle gas in which the hole is much larger than the mean free path.

(*a*) *Knudsen Gas.* In a Knudsen gas, all particles that reach the hole will pass through freely. They do not have to do work against other particles in the hole because the chance of collisions with other particles is very small. We shall assume we are dealing with an ideal gas. We want to find the energy per unit mass carried through the hole. We will first use kinetic theory to find the average energy per particle carried through the hole.

The number of particles with velocity $\mathbf{v} \to \mathbf{v} + d\mathbf{v}$ which reach a surface area element dS in time dt is

$$dN(\mathbf{v}) = nF(\mathbf{v})\, d\mathbf{v} v_z\, dt\, dS \tag{14.65}$$

(cf. Chap. 9 and Fig. 9.4), where

$$F(\mathbf{v}) = \left(\frac{\beta m}{2\pi}\right)^{3/2} e^{-\beta m v^2/2} \tag{14.66}$$

and $n = N/V$ is the number density. The number of particles with speed $v \to v + dv$ which reach the surface per unit area per unit time is

$$d\dot{n}(v) = n\pi\left(\frac{\beta m}{2\pi}\right)^{3/2} e^{-\beta m v^2/2} v^3\, dv. \tag{14.67}$$

The average energy of each particle which reaches the surface is

$$\langle \tfrac{1}{2}mv^2\rangle = \frac{\int \frac{1}{2}mv^2\, dN}{\int dN} = 2kT. \tag{14.68}$$

Therefore, the energy transferred through the hole per particle is $2kT$. The energy per unit mass transferred through the hole is

$$\frac{L_{um}}{L_{mm}} = \frac{N_A}{M_{mw}} 2kT = \frac{2RT}{M_{mw}} \tag{14.69}$$

where N_A is Avogadro's number, M_{mw} is the molecular weight of the particles, and R is the gas constant.

(*b*) *Boyle Gas.* If the hole in the partition is much larger than the mean free path, then the particles have to push their way through because they will continually collide with other particles as they go through. They not only carry internal energy u with them but they also do work as they go. The work per unit mass done by the particles is $(1/\rho)P$. Therefore, we must add this work term to the expression for the energy per unit mass carried by the particles as they pass through the hole. We then obtain

$$\frac{L_{um}}{L_{mm}} = u + \frac{1}{\rho}P = h \tag{14.70}$$

where h is the enthalpy per unit mass. The information we have obtained from the mechanocaloric effect can now be used to study the thermomolecular pressure effect.

2. Thermomolecular Pressure Effect

Let us find the relation between the temperature and pressure of the two sides which must exist if there is no *net* matter flow through the hole, $J_m = 0$, but there is energy flow, $J_u \neq 0$. If there is no net mass transfer through the hole ($J_m = 0$), we obtain

$$\frac{\Delta P}{\Delta T} = \frac{h - \dfrac{L_{mu}}{L_{mm}}}{T\dfrac{1}{\rho}} \tag{14.71}$$

(cf. Eq. (14.61)), which relates the temperature and pressure differences between the two sides. Note that this expression depends on L_{mu} whereas the mechanocaloric effect depended on L_{um}. Without Onsager's relations we could go no further because we could not determine L_{mu}. However, because of microscopic reversibility, we know that $L_{um} = L_{mu}$ and we can obtain an expression for Eq. (14.71) for both a Knudsen gas and a Boyle gas.

(*a*) *Knudsen Gas.* If we combine Eqs. (14.59), (14.69), and (14.71) and note that the enthalpy per unit mass of an ideal gas is $h = 5RT/2M_{mw}$, we find

$$\frac{\Delta P}{\Delta T} = \frac{1}{2}\frac{P}{T}. \tag{14.72}$$

We can then integrate Eq. (14.72) to obtain

$$\frac{P_1}{\sqrt{T_1}} = \frac{P_2}{\sqrt{T_2}}. \tag{14.73}$$

Therefore, for a Knudsen gas there can be both a pressure and a temperature difference between the two sides even though there is no net flow of matter. Note that this has close analogy to the fountain effect in a superfluid (cf. Sec. 3.F).

(*b*) *Boyle Gas.* If we combine Eqs. (14.59), (14.69), and (14.71) and again note that $h = 5RT/2M_{mw}$, we find

$$\frac{\Delta P}{\Delta T} = 0. \tag{14.74}$$

For a Boyle gas, if there is no net flow of matter, there can be no pressure difference although there can be a temperature difference between the two sides.

E. MINIMUM ENTROPY PRODUCTION[3,4]

It is useful to discuss some general properties of the entropy production in a hydrodynamic system. In App. A, we have written the general balance equation for the flow of an arbitrary quantity D whose density was denoted $\mathscr{D}(\mathbf{x}, t)$. Real fluids out of equilibrium are dissipative because of transport processes and chemical reactions which lead to an increase in the entropy of the fluid. If we let $s(\mathbf{x}, t)$ denote the entropy per unit mass of the fluid and $\rho(\mathbf{x}, t)$ denote the mass density, the balance equation for the entropy can be written

$$\frac{\partial \rho s}{\partial t} = -\nabla_{\mathbf{r}} \cdot (\rho s \mathbf{v} + \mathbf{J}_s') + \sigma_s, \tag{14.75}$$

where $\rho s \mathbf{v}$ is the convective entropy current and would exist in the absence of dissipation. $\mathbf{J}_s'$ is an entropy current which arises because of dissipative effects (as we shall see in subsequent sections) and σ_s is an entropy source term which arises because of dissipative effects.

Let us consider the balance equation for a closed isolated system. For simplicity, we shall assume that the system has a constant volume (that is, no work can be done on it). For such a system, no heat or matter can be transported through the walls. Thus, we can integrate Eq. (14.75) over the volume of the system and find the following equation for the total time rate of change of the entropy in the box:

$$\frac{dS}{dt} = \int \frac{\partial \rho s}{\partial t} \, dV = -\int_V \nabla_{\mathbf{r}} \cdot (\rho s \mathbf{v} + \mathbf{J}_s') \, dV + \int_V \sigma_s \, dV. \tag{14.76}$$

The integral over the divergence can be changed to a surface integral. However, since no heat or matter can flow through the walls, there will be no entropy flow either and the surface term will give no contribution. Thus,

$$\frac{dS}{dt} = \int_V \sigma_s \, dV \tag{14.77}$$

is the rate of entropy production in the box due to spontaneous processes in the box. The second law of thermodynamics tells us that for reversible processes the entropy of the box does not change, but for irreversible or spontaneous processes it must increase. Therefore,

$$\frac{dS}{dt} = \int_V \sigma_s \, dV > 0 \tag{14.78}$$

and the local entropy production σ_s (time rate of change of entropy per unit volume) must be positive.

The local entropy production can be written in the form

$$\sigma_s = -\sum_j J_j X_j \geqslant 0 \tag{14.79}$$

where J_j is a local current, or flux, and X_j is a local force (the values of σ_s, J_j, and X_j are given for one cell in the fluid). Near equilibrium the currents J_j may be approximated by the linear relation

$$J_j = -\sum_j L_{ij} X_j \tag{14.80}$$

where L_{ij} as a symmetric matrix

$$L_{ij} = L_{ji} \tag{14.81}$$

(Onsager's relation). If we substitute Eq. (14.80) into Eq. (14.79), we find in the linear regime

$$\sigma_s = \sum_i \sum_j L_{ij} X_i X_j. \tag{14.82}$$

Therefore, for a system in which spontaneous processes take place, the phenomenological coefficients form a symmetric positive definite matrix. (By definition, a symmetric matrix is positive definite if every principal minor is positive. Principal minors are formed by deleting any number of rows and columns and by taking the determinant of the remaining matrix. The rows and columns deleted must be the same; that is, if the ith and jth rows are deleted, then the ith and jth columns must be deleted.)

Let us consider the case of a 2 × 2 matrix:

$$\mathbb{L} = \begin{pmatrix} L_{11} & L_{12} \\ L_{21} & L_{22} \end{pmatrix}.$$

If $\mathbb{L}$ is symmetric and positive definite, then we know that $L_{12} = L_{21}, L_{11} > 0$, $L_{22} > 0$, and $\det \mathbb{L} = L_{11}L_{22} - L_{12}L_{21} > 0$ for a positive entropy production. In general, for an $n \times n$ matrix we have the relations

$$L_{ii} > 0 \qquad \text{(for all } i) \tag{14.83}$$

and

$$L_{ii}L_{kk} > L_{ik}^2 \tag{14.84}$$

in addition to the Onsager relations, $L_{ik} = L_{ki}$.

We can now ask under what conditions the entropy production will be minimum. For simplicity, let us first consider the case with two currents and two forces,

$$\sigma = \sum_{i=1}^{2} \sum_{j=1}^{2} L_{ij}X_iX_j > 0, \tag{14.85}$$

where we have taken the forces to be independent variables. The fluxes are then dependent. If we allow the forces to vary freely, the condition for minimum entropy production is given by the equations

$$\left(\frac{\partial \sigma}{\partial X_1}\right)_{X_2} = 2L_{11}X_1 + 2L_{12}X_2 \equiv -2J_1 = 0 \tag{14.86}$$

and

$$\left(\frac{\partial \sigma}{\partial X_2}\right)_{X_1} = 2L_{22}X_2 + 2L_{21}X_1 = -2J_2 = 0. \tag{14.87}$$

(Note that we have assumed that L_{ij} is constant.) When there are no constraints on the forces, the state of minimum entropy production is the equilibrium state. The entropy production will be minimum (zero) when all forces and fluxes vanish.

Let us now hold one force fixed. That is, we let X_1 = constant (this could be the case of a box with a constant temperature gradient maintained across it). Then the condition for minimum entropy production becomes simply

$$\left(\frac{\partial \sigma_s}{\partial X_2}\right) = 2L_{22}X_2 + 2L_{21}X_1 \equiv -2J_2 = 0. \tag{14.88}$$

If we hold X_1 = constant then J_2 will be zero in a state of minimum entropy production but J_1 will be nonzero and constant. Thus, the state of minimum entropy production will be a stationary state. This fact was first established by Prigogine.[3]

We can generalize this result to an arbitrary number of forces and currents. Let us consider a system with n independent forces X_i ($i = 1, \ldots, n$). The entropy production is

$$\sigma_s = \sum_{i,j=1}^{n} L_{ij}X_iX_j. \tag{14.89}$$

Let us now hold the forces X_i $(i = 1, \ldots, k)$ constant. The condition for minimum entropy production becomes

$$\left(\frac{\partial \sigma_s}{\partial X_i}\right)_{\{X_j \neq X_i\}} = \sum_{j=1}^{n} (L_{ij} + L_{ji}) X_j = 0 \tag{14.90}$$

$(i = k + 1, \ldots, n)$. Using Onsager's relations, Eq. (14.90) becomes

$$2J_i = 0. \tag{14.91}$$

for $(i = k + 1, \ldots, n)$. In other words, if we hold X_i $(i = 1, \ldots, k)$ fixed, then $J_i \equiv 0$ $(i = k + 1, \ldots, n)$ when the system is in a state of minimum entropy production, and the currents J_i $(i = 1, \ldots, k)$ will be constant. Such a state is a kth-order stationary state.

In the linear regime, the stationary states are always stable. That is, fluctuations cannot drive a system away from the stationary state. (This is no longer true in the nonlinear regime, as we shall see in Chap. 17.) Let us consider a system specified by n independent forces. We will hold forces $X_1, \ldots, X_k$ fixed and allow fluctuations to occur in forces $X_{k+1}, \ldots, X_n$. Thus, we consider the entropy production to be a function of the forces $X_{k+1}, \ldots, X_n$. We will let $X_{k+1}^0, \ldots, X_n{}^0$ denote the values of the forces $X_{k+1}, \ldots, X_n$ in the steady state. The entropy production near the steady state can be expanded in a Taylor series of the form

$$\begin{aligned}\sigma_s(X_{k+1}, \ldots, X_n) &= \sigma_s(X_{k+1}^0, \ldots, X_n{}^0) + \sum_{i=k+1}^{n} \left(\frac{\partial \sigma_n}{\partial X_i}\right)^0_{\{X_j \neq X_i\}} \Delta X_i \\ &\quad + \frac{1}{2} \sum_{i,j=k+1}^{n} \left[\frac{\partial}{\partial X_i}\left(\frac{\partial \sigma_n}{\partial X_j}\right)_{\{X_k \neq X_j\}}\right]^0_{\{X_l \neq X_i\}} \Delta X_i \, \Delta X_j + \cdots \\ &= \sigma_s(X_{k+1}^0, \ldots, X_n{}^0) + \frac{1}{2} \sum_{i,j=k+1}^{n} L_{ij} \, \Delta X_i \, \Delta X_j + \cdots\end{aligned} \tag{14.92}$$

The quantity

$$\Delta P = \frac{1}{2} \int dV \sum_{i,j=k+1}^{n} L_{ij} \, \Delta X_i \, \Delta X_j > 0 \tag{14.93}$$

is called the excess entropy production and is always positive since the coefficients L_{ij} form a positive definite symmetric matrix. Since the stationary state is a state of minimum entropy production, we also have

$$\frac{d\Delta P}{dt} = -\int dV \sum_{j=k+1}^{n} \Delta J_j \frac{d\Delta X_j}{dt} = -\int dV \sum_{i,j=k+1}^{n} g_{ji} \frac{d\alpha_i}{dt} \frac{d\alpha_j}{dt} \leqslant 0. \tag{14.94}$$

Thus, ΔP, the excess entropy function, is a Liapounov function (cf. App. D). Eqs. (14.93) and (14.94) guarantee the stability of stationary states in the linear regime. In the nonlinear regime, Eqs. (14.93) and (14.94) need not be satisfied and we have the possibility of forming unstable stationary states as various parameters are changed and of having phase transitions far from equilibrium.

F. SINGLE-COMPONENT NORMAL ISOTROPIC FLUID

Hydrodynamic equations enable us to describe, in terms of a few variables, the time-dependent macroscopic behavior of a fluid. In Chap. 13, we derived hydrodynamic equations from the Boltzmann equation. However, hydrodynamics has a far wider applicability than just dilute gases. Hydrodynamic equations describe the long-wavelength, low-frequency behavior of liquids, liquid crystals, superfluids, spin lattices, and solids. They can be derived from conservation laws, thermodynamics, and the symmetry properties of a system. In the remainder of this chapter we shall derive the hydrodynamic equations for some complicated isotropic systems. In order to illustrate the techniques, however, we shall begin in this section with a simple one-component isotropic fluid.

1. Conservation of Mass: Continuity Equation[6,7]

Let us consider the mass, M, of a volume element $V(t)$ moving with the fluid. Since mass is not created or destroyed, the total mass of the volume element must remain constant. Thus,

$$\frac{dM}{dt} = \frac{d}{dt}\iiint_{V(t)} \rho(\mathbf{r}, t)\, dV_t = \iiint_{V(t)} \left(\frac{d\rho}{dt} + \rho\nabla_{\mathbf{r}}\cdot\mathbf{v}\right) dV_t = 0 \qquad (14.95)$$

where $\rho(\mathbf{r}, t)$ is the mass density, $\mathbf{v}(\mathbf{r}, t)$ is the average velocity of the fluid, and we have used App. A, Eq. (A.17). Since the volume element $V(t)$ is arbitrary, the integrand must be zero and we find

$$\frac{d\rho}{dt} + \rho\nabla_{\mathbf{r}}\cdot\mathbf{v} = 0. \qquad (14.96)$$

Or, using the convective derivative, we find

$$\frac{\partial\rho}{\partial t} + \nabla_{\mathbf{r}}\cdot\mathbf{J} = 0, \qquad (14.97)$$

where $\mathbf{J} = \rho\mathbf{v}$ is the mass current or momentum density and is reactive, $d\rho/dt$ gives the time rate of change of the mass density for an observer moving with the fluid, and $\partial\rho/\partial t$ gives the time rate of change of the mass density at a fixed point in the fluid.

2. Momentum Balance Equation

The total momentum $\mathscr{P}(t)$ of a fluid element obeys Newton's law. The time rate of change of the momentum of the fluid element must equal the sum of the forces acting on the element. Thus,

$$\frac{d\mathscr{P}(t)}{dt} = \frac{d}{dt}\iiint_{V(t)} \mathbf{J}\, dV_t = \iiint_{V(t)} \rho\mathbf{F}\, dV_t + \iint_{S(t)} \mathbf{f}\, dS_t \tag{14.98}$$

where $\mathbf{F}$ is an external force per unit mass acting on the element (such as gravitation or an electric field) and $\mathbf{f}$ is a force per unit area acting on the walls of the volume element due to the surrounding medium; $S(t)$ is the surface of the volume element $V(t)$. In general for a nonideal fluid, $\mathbf{f}$ can be directed differently from the normal to the surface element. If we write the surface element as a vector $d\mathbf{S}_t$ perpendicular to the surface of the volume element, then we can write $\mathbf{f}\, dS_t = d\mathbf{S}_t \cdot \overline{\overline{\mathbf{P}}}$, where $\overline{\overline{\mathbf{P}}}$ is the pressure tensor of the fluid, $\mathbf{f} = \hat{\mathbf{n}} \cdot \overline{\overline{\mathbf{P}}}$, and $\hat{\mathbf{n}}$ is the unit vector perpendicular to dS_t. If we note that

$$\iint_{S(t)} d\mathbf{S}_t \cdot \overline{\overline{\mathbf{P}}} = \iiint_{V(t)} \nabla_{\mathbf{r}} \cdot \overline{\overline{\mathbf{P}}}\, dV_t,$$

then Eq. (14.98) can be written

$$\frac{d\rho\mathbf{v}}{dt} + \rho\mathbf{v}(\nabla_{\mathbf{r}} \cdot \mathbf{v}) = \rho\mathbf{F} + \nabla_{\mathbf{r}} \cdot \overline{\overline{\mathbf{P}}} \tag{14.99}$$

(cf. App. A, Eq. (A.18)). For an ideal fluid, the only force on the walls of $V(t)$ is due to the hydrostatic pressure, which is always perpendicular to the walls and pointed inward. Thus, for an ideal fluid, $\mathbf{f} = -P\hat{\mathbf{n}}$ and $\overline{\overline{\mathbf{P}}} = -P\overline{\overline{\mathbf{U}}}$ ($\overline{\overline{\mathbf{U}}}$ is the unit tensor). For a nonideal fluid, there is also a dissipative contribution $\overline{\overline{\mathbf{\Pi}}}^D$ to the pressure tensor and Eq. (14.99) takes the form

$$\frac{d\rho\mathbf{v}}{dt} + \rho\mathbf{v}(\nabla_{\mathbf{r}} \cdot \mathbf{v}) = \rho\mathbf{F} - \nabla_{\mathbf{r}} P - \nabla_{\mathbf{r}} \cdot \overline{\overline{\mathbf{\Pi}}}^D \tag{14.100}$$

since $\nabla_{\mathbf{r}} \cdot P\overline{\overline{\mathbf{U}}} = \nabla_{\mathbf{r}} P$. The tensor $\overline{\overline{\mathbf{\Pi}}}^D$ is called the stress tensor. We can now use the continuity equation and the definition of the convective derivative to write Eq. (14.98) as

$$\frac{\partial \mathbf{J}}{\partial t} + \nabla_{\mathbf{r}} \cdot (P\overline{\overline{\mathbf{U}}} + \rho\mathbf{v}\mathbf{v} + \overline{\overline{\mathbf{\Pi}}}^D) = \rho\mathbf{F}. \tag{14.101}$$

The term *nonlinear* in the velocity describes the convection of momentum in the fluid.

3. Energy and Entropy Balance Equations

Let us assume that the external force can be represented by a potential so that $\mathbf{F} = -\nabla_{\mathbf{r}}\phi$. Then the total energy per unit volume $\rho\epsilon$ is

$$\rho\epsilon = \rho u + \tfrac{1}{2}\rho v^2 + \rho\phi \tag{14.102}$$

where ϵ is the specific total energy (total energy per unit mass), u is the specific internal energy, $\frac{1}{2}\rho v^2$ is the kinetic energy per unit volume of the fluid, and ϕ is the specific time-independent potential energy of the fluid due to the external field. The balance equation for the total energy per unit volume can be written

$$\frac{\partial \rho\epsilon}{\partial t} + \nabla_{\mathbf{r}}\cdot(\mathbf{J}_\epsilon^R + \mathbf{J}_\epsilon^D) = 0 \tag{14.103}$$

where $\mathbf{J}_\epsilon^R$ is a reactive energy current and $\mathbf{J}_\epsilon^D$ is a dissipative energy current which has yet to be determined.

From Sec. 14.E, the entropy balance equation for the fluid can be written

$$\frac{\partial \rho s}{\partial t} + \nabla_{\mathbf{r}}\cdot(\mathbf{J}_s^R + \mathbf{J}_s^D) = \sigma_s \tag{14.104}$$

where σ_s is an entropy source term due to dissipative processes in the fluid, $\mathbf{J}_s^R$ is a reactive entropy current, and $\mathbf{J}_s^D$ is a dissipative entropy current. We can now use Eqs. (14.97), (14.101), (14.103), and (14.104) to find the hydrodynamic equations for the fluid.

To derive the hydrodynamic equations, we will need some thermodynamic identities. The fundamental equation for the specific internal energy is

$$u = Ts - P\frac{1}{\rho} + \mu \tag{14.105}$$

where μ is the specific chemical potential. The combined first and second laws of thermodynamics can be written

$$du = T\,ds - P\,d\left(\frac{1}{\rho}\right), \tag{14.106}$$

and the Gibbs-Duheim equation can be written

$$d\mu = -s\,dT + \frac{1}{\rho}\,dP. \tag{14.107}$$

We will make much use of these equations.

If we take the time derivative of the total energy in Eq. (14.102), we find

$$\frac{\partial \rho\epsilon}{\partial t} = \frac{\partial \rho u}{\partial t} + \mathbf{v}\cdot\frac{\partial \mathbf{J}}{\partial t} - (\tfrac{1}{2}v^2 - \phi)\frac{\partial \rho}{\partial t}. \tag{14.108}$$

It is useful in the dissipative case to find the local entropy production and, therefore, the balance equation for the entropy. We first combine Eqs. (14.105) and (14.106) to obtain

$$d\rho u = T\,d\rho s + \mu\,d\rho. \tag{14.109}$$

From Eqs. (14.108) and (14.109) we obtain

$$T\frac{\partial \rho s}{\partial t} = \frac{\partial \rho\epsilon}{\partial t} - \mathbf{v}\cdot\frac{\partial \mathbf{J}^R}{\partial t} + (\tfrac{1}{2}v^2 - \phi - \mu)\frac{\partial \rho}{\partial t}. \tag{14.110}$$

From the balance equations, (14.97), (14.101), and (14.103), we find

$$T\frac{\partial \rho s}{\partial t} = -\nabla_{\mathbf{r}}\cdot(\mathbf{J}_E{}^R + \mathbf{J}_E{}^D) + (\mu - \tfrac{1}{2}v^2 + \phi)\nabla_{\mathbf{r}}\cdot\mathbf{J} + \mathbf{v}\cdot\nabla_{\mathbf{r}}P$$
$$+ \mathbf{v}\cdot[\nabla_{\mathbf{r}}\cdot\rho\mathbf{v}\mathbf{v}] + \mathbf{v}\cdot[\nabla_{\mathbf{r}}\cdot\overline{\overline{\mathbf{\Pi}}}{}^D] - \mathbf{v}\cdot\rho\mathbf{F}. \tag{14.111}$$

We now note that

$$\mathbf{v}\cdot(\nabla_{\mathbf{r}}\cdot\rho\mathbf{v}\mathbf{v}) - \tfrac{1}{2}v^2\nabla_{\mathbf{r}}\cdot\rho\mathbf{v} = \nabla_{\mathbf{r}}\cdot(\rho\mathbf{v}\tfrac{1}{2}v^2) \tag{14.112}$$

(this is most easily seen by writing the equation in terms of components), and we use the Gibbs-Duheim equation to write

$$\nabla_{\mathbf{r}}\mu = -s\nabla_{\mathbf{r}}T + \frac{1}{\rho}\nabla_{\mathbf{r}}P. \tag{14.113}$$

If we combine Eqs. (14.111), (14.112), and (14.113) and rearrange terms, we find

$$T\frac{\partial \rho s}{\partial t} = -\nabla_{\mathbf{r}}\cdot(\mathbf{J}_E{}^R + \mathbf{J}_E{}^D - \mu'\mathbf{J} - \overline{\overline{\mathbf{\Pi}}}{}^D\cdot\mathbf{v}) + \rho s\mathbf{v}\cdot\nabla_{\mathbf{r}}T$$
$$- (\overline{\overline{\mathbf{\Pi}}}{}^D:\nabla_{\mathbf{r}}\mathbf{v}) - \mathbf{v}\cdot\rho\mathbf{F} \tag{14.114}$$

where $\mu' = \mu + \frac{1}{2}v^2 + \phi$ (we have assumed that the stress tensor $\overline{\overline{\mathbf{\Pi}}}{}^D$ is symmetric). If we multiply by $1/T$ and rearrange terms once again, we find

$$\frac{\partial \rho s}{\partial t} = -\nabla_{\mathbf{r}}\cdot\left(\frac{\mathbf{J}_E{}^R + \mathbf{J}_E{}^D - \mu'\mathbf{J} - \overline{\overline{\mathbf{\Pi}}}{}^D\cdot\mathbf{v}}{T}\right) + (\mathbf{J}_E{}^R + \mathbf{J}_E{}^D - \mu'\mathbf{J}$$
$$- \overline{\overline{\mathbf{\Pi}}}{}^D\cdot\mathbf{v} - \rho s\mathbf{v}T)\cdot\nabla_{\mathbf{r}}\left(\frac{1}{T}\right) - \frac{1}{T}\overline{\overline{\mathbf{\Pi}}}{}^D:\nabla_{\mathbf{r}}\mathbf{v} - \frac{1}{T}\mathbf{v}\cdot\rho\mathbf{F}. \tag{14.115}$$

For a nondissipative system there can be no entropy production. If we set all dissipative currents to zero, we find

$$\frac{\partial \rho s}{\partial t} = -\nabla_{\mathbf{r}}\cdot\left(\frac{\mathbf{J}_E{}^R - \mu'\mathbf{J}}{T}\right) + (\mathbf{J}_E{}^R - \mu'\mathbf{J} - \rho s\mathbf{v}T)\cdot\nabla_{\mathbf{r}}\left(\frac{1}{T}\right) - \frac{1}{T}\mathbf{v}\cdot\rho\mathbf{F}. \tag{14.116}$$

The condition that the source term be zero in Eq. (14.116) yields

$$\mathbf{J}_E{}^R = \rho s\mathbf{v}T + \mu'\mathbf{J}, \tag{14.117}$$

and Eq. (14.116) reduces to

$$\frac{\partial \rho s}{\partial t} + \nabla_{\mathbf{r}}\cdot(\rho s\mathbf{v}) = -\mathbf{v}\cdot\rho\mathbf{F}, \tag{14.118}$$

as we would expect for a nondissipative fluid. If we now substitute Eq. (14.117) into Eq. (14.115), we find

$$\frac{\partial \rho s}{\partial t} = -\nabla_{\mathbf{r}}\cdot\left(\rho s\mathbf{v} + \frac{\mathbf{J}_E{}^D - \overline{\overline{\mathbf{\Pi}}}{}^D\cdot\mathbf{v}}{T}\right) - \frac{1}{T^2}(\mathbf{J}_E{}^D - \overline{\overline{\mathbf{\Pi}}}{}^D\cdot\mathbf{v})\cdot\nabla_{\mathbf{r}}T$$

$$-\frac{1}{T}\overline{\overline{\mathbf{\Pi}}}{}^D:\nabla_{\mathbf{r}}\mathbf{v} - \frac{1}{T}\mathbf{v}\cdot\rho\mathbf{F}. \tag{14.119}$$

The dissipative entropy current is

$$\mathbf{J}_s{}^D = \frac{\mathbf{J}_E{}^D - \overline{\overline{\mathbf{\Pi}}}{}^D\cdot\mathbf{v}}{T} \tag{14.120}$$

and the local entropy production is

$$\sigma_s = -\frac{1}{T}\mathbf{J}_s{}^D\cdot\nabla_{\mathbf{r}}T - \frac{1}{T}\overline{\overline{\mathbf{\Pi}}}{}^D:\nabla_{\mathbf{r}}\mathbf{v} - \frac{1}{T}\mathbf{v}\cdot\rho\mathbf{F}. \tag{14.121}$$

We have written the entropy production as a product of generalized forces and fluxes. The entropy flux and the momentum flux are $\mathbf{J}_s{}^D$ and $\overline{\overline{\mathbf{\Pi}}}{}^D$, respectively. The forces are $\nabla_{\mathbf{r}}T$ and $\nabla_{\mathbf{r}}\mathbf{v}$. For an isotropic system, Curie's law (cf. App. D) states that a given force cannot drive a current of a different tensor character. The force $\nabla_{\mathbf{r}}\mathbf{v}$ can be decomposed into a scalar part $(\nabla_{\mathbf{r}}\cdot\mathbf{v})\overline{\overline{\mathbf{U}}}$, a symmetric traceless part $(\nabla_{\mathbf{r}}\mathbf{v})^s$, and an antisymmetric traceless part $(\nabla_{\mathbf{r}}\mathbf{v})^a$. However, since $\overline{\overline{\mathbf{\Pi}}}{}^D$ is assumed symmetric and $\mathbf{J}_s{}^D$ is not a pseudo vector, the force $(\nabla_{\mathbf{r}}\mathbf{v})^a$ does not contribute. Thus, we find

$$\mathbf{J}_s{}^D = -\frac{K}{T}\nabla_{\mathbf{r}}T, \tag{14.122}$$

$$(\overline{\overline{\mathbf{\Pi}}}{}^D)^s = -\eta(\nabla_{\mathbf{r}}\mathbf{v})^s, \tag{14.123}$$

and

$$\pi^D = -\zeta(\nabla_{\mathbf{r}}\cdot\mathbf{v}). \tag{14.124}$$

(We note that, under time reversal, these currents always behave oppositely to their reactive counterparts.) In Eq. (14.124) π^D is the scalar part of the tensor $\overline{\overline{\mathbf{\Pi}}}{}^D$ and $(\overline{\overline{\mathbf{\Pi}}}{}^D)^s$ is the traceless symmetric part. The quantity K is the coefficient of thermal conductivity, η is the shear viscosity, and ζ is the bulk viscosity. The entropy production now takes the form

$$\sigma_s = \frac{K}{T^2}(\nabla_{\mathbf{r}}T)^2 + \eta[(\nabla_{\mathbf{r}}\mathbf{v})^s]^2 + \zeta(\nabla_{\mathbf{r}}\cdot\mathbf{v})^2. \tag{14.125}$$

We see that in order for σ_s to be positive definite K, η, and ζ must be positive. Because there are no coupled transport processes for this case we have no use for Onsager's relations.

To summarize, we can now write the hydrodynamic equations as follows:

$$\frac{\partial \rho}{\partial t} + \nabla_{\mathbf{r}} \cdot \rho \mathbf{v} = 0, \tag{14.126a}$$

$$\frac{\partial (\rho \mathbf{v})}{\partial t} + \nabla_{\mathbf{r}} \cdot (\rho \mathbf{v}\mathbf{v}) = \rho \mathbf{F} - \nabla_{\mathbf{r}} P + \eta \nabla_{\mathbf{r}}^2 \mathbf{v} + (\zeta + \tfrac{1}{3}\eta) \nabla_{\mathbf{r}} (\nabla_{\mathbf{r}} \cdot \mathbf{v}), \tag{14.126b}$$

$$\frac{\partial (\rho \epsilon)}{\partial t} + \nabla_{\mathbf{r}} \cdot (\rho s \mathbf{v} T + \rho \mu' \mathbf{v} - K \nabla_{\mathbf{r}} T + \overline{\overline{\Pi}}{}^D \cdot \mathbf{v}) = 0, \tag{14.126c}$$

and

$$\frac{\partial (\rho s)}{\partial t} + \nabla_{\mathbf{r}} \cdot (\rho s \mathbf{v}) = \frac{K}{T} \nabla_{\mathbf{r}}^2 T - \frac{1}{T} \overline{\overline{\Pi}}{}^D : \nabla_{\mathbf{r}} \mathbf{v} - \frac{1}{T} \mathbf{v} \cdot \rho \mathbf{F} \tag{14.126d}$$

These are the nonlinear hydrodynamic equations for an isotropic, single-component fluid. They could also have been derived from the Boltzmann equation (cf. Prob. 13.10); however, they are more general because they can also be applied to a single-component liquid. In the linear regime, they reduce to the linearized equations in Sec. 13.F.

G. MULTICOMPONENT FLUIDS WITH CHEMICAL REACTIONS[3,4]

We shall now generalize the hydrodynamic equations to the case of multicomponent systems in which chemical reactions can take place. Let us consider a system in which there are p constituents which may undergo chemical reactions. Since mass is conserved during a chemical reaction, for the kth reaction we have

$$\sum_j^{(k)} \nu_{jk} M_j = 0 \tag{14.127}$$

where M_j is the molecular weight of the jth constituent, ν_{jk} are the stoichiometric coefficients introduced in Sec. 3.E, and the sum $\sum_j^{(k)}$ is over all constituents involved in the kth reaction. It is convenient to introduce new stoichiometric coefficients $\bar{\nu}_{jk}$ with the units of mass so that

$$\bar{\nu}_{jk} = \nu_{jk} M_j. \tag{14.128}$$

Then Eq. (14.127) takes the form

$$\sum_j^{(k)} \bar{\nu}_{jk} = 0. \tag{14.129}$$

From Eq. (3.58) the change in the total mass of constituent j due to all chemical reactions may be written

$$dm_j = \sum_{k=1}^{r} \bar{\nu}_{jk}\, d\xi_k \tag{14.130}$$

where ξ_k is the degree of reaction. The change in the mass density is

$$d\rho_j = \sum_{k=1}^{r} \bar{\nu}_{jk} \frac{d\xi_k}{V} \tag{14.131}$$

and the time rate of change of the mass density due to chemical reactions may be written

$$\left.\frac{\partial \rho_j}{\partial t}\right|_{\text{reaction}} = \sum_{k=1}^{r} \bar{\nu}_{jk} \mathscr{J}_k \tag{14.132}$$

where

$$\mathscr{J}_k = \frac{1}{V}\frac{\partial \xi_k}{\partial t} \tag{14.133}$$

gives the rate of the kth reaction and has units of moles/m^3 sec. The quantity $\partial \rho_j/\partial t|_{\text{reaction}}$ acts as a source of molecules j due to chemical reactions.

We can now write the hydrodynamic equations for the fluid. The balance equation for the jth constituent can be written

$$\frac{\partial \rho_j}{\partial t} = -\nabla_{\mathbf{r}} \cdot (\rho_j \mathbf{v} + \mathbf{J}_j^D) + \sum_{k=1}^{r} \bar{\nu}_{jk} \mathscr{J}_k \tag{14.134}$$

where $\mathbf{v}$ is the average velocity of the fluid and $\mathbf{J}_j^D$ is a dissipative diffusion current which is nonzero when the average velocity $\mathbf{v}_j$ of various constituents differs from the average velocity $\mathbf{v}$. Mass conservation imposes the condition

$$\sum_{j=1}^{p} \mathbf{J}_j^D = 0. \tag{14.135}$$

If we use Eq. (14.129), we obtain the continuity equation

$$\frac{\partial \rho}{\partial t} = \sum_{j=1}^{p} \frac{\partial \rho_j}{\partial t} = -\nabla_{\mathbf{r}} \cdot \rho \mathbf{v} \tag{14.136}$$

where

$$\rho = \sum_{j=1}^{p} \rho_j \tag{14.137}$$

is the total mass density.

The remaining balance equations can be written in their usual form. The balance equation for the momentum density is

$$\frac{\partial \rho \mathbf{v}}{\partial t} + \nabla_{\mathbf{r}} \cdot (P\overline{\overline{\mathbf{U}}} + \rho \mathbf{v}\mathbf{v} + \overline{\overline{\mathbf{\Pi}}}^D) = 0 \tag{14.138}$$

where

$$\rho \mathbf{v} = \sum_{j=1}^{p} \rho_j \mathbf{v}_j \tag{14.139}$$

is the center of mass momentum density. The balance equation for the total energy density is

$$\frac{\partial \rho \epsilon}{\partial t} + \nabla_{\mathbf{r}} \cdot (\mathbf{J}_E{}^R + \mathbf{J}_E{}^D) = 0, \tag{14.140}$$

and the balance equation for the entropy density is

$$\frac{\partial \rho s}{\partial t} + \nabla_{\mathbf{r}} \cdot (\rho s \mathbf{v} + \mathbf{J}_s{}^D) = \sigma_s. \tag{14.141}$$

We now proceed as in Sec. 14.F. The total energy density is defined

$$\rho \epsilon = \sum_{j=1}^{p} \tfrac{1}{2} \rho_j v_j{}^2 + \rho u = \tfrac{1}{2} \rho v^2 + \rho u + \frac{1}{2} \sum_j \rho_j (\mathbf{v} - \mathbf{v}_j)^2 \tag{14.142}$$

where we have divided the total energy density into a center of mass part and a diffusion part. The fundamental equation may be written

$$u = sT - \frac{1}{\rho} P + \sum_j \mu_j c_j \tag{14.143}$$

where $c_j = \rho_j/\rho$ is the concentration of the jth constituent and μ_j is its chemical potential. The first and second laws of thermodynamics take the form

$$du = T\, ds - P\, d\left(\frac{1}{\rho}\right) + \sum_j \mu_j\, dc_j \tag{14.144}$$

or

$$d\rho u = T\, d\rho s + \sum_j \mu_j\, d\rho_j \tag{14.145}$$

and the Gibbs-Duheim equation becomes

$$s\, dT - \frac{1}{\rho}\, dP + \sum_j c_j\, d\mu_j = 0. \tag{14.146}$$

From Eq. (14.142) the differential of the total energy density may be written

$$d\rho\epsilon = d\rho u + \mathbf{v} \cdot d\rho\mathbf{v} - \tfrac{1}{2} v^2\, d\rho + \frac{1}{2} \sum_j (\mathbf{v}_j - \mathbf{v})^2\, d\rho_j + \sum_j \mathbf{J}_j \cdot d(\mathbf{v}_j - \mathbf{v}) \tag{14.147}$$

where $\mathbf{J}_j$ is a reactive diffusion current,

$$\mathbf{J}_j = \rho_j (\mathbf{v}_j - \mathbf{v}). \tag{14.148}$$

If we combine Eqs. (14.147) and (14.145), we find

$$T\,d\rho s = d\rho\epsilon - \mathbf{v}\cdot d\rho\mathbf{v} + \frac{v^2}{2}\,d\rho - \frac{1}{2}\sum_j(\mathbf{v}_j - \mathbf{v})^2\,d\rho_j - \sum_j \mathbf{J}_j\cdot d(\mathbf{v}_j - \mathbf{v}) - \sum_j \mu_j\,d\rho_j. \tag{14.149}$$

If we now replace the differentials by partial time derivatives and use the balance equations, we can write

$$T\frac{\partial\rho s}{\partial t} = -\nabla\cdot\left(\mathbf{J}_E{}^R + \mathbf{J}_E{}^D - \rho\mathbf{v}\tfrac{1}{2}v^2 - \sum_j \mu_j'\rho_i\mathbf{v}\right) + \mathbf{v}\cdot(\nabla_{\mathbf{r}}\cdot\overline{\overline{\mathbf{\Pi}}}{}^D) + \mathbf{v}\rho s\nabla T + \sum_j \mu_j'\nabla_{\mathbf{r}}\cdot J_j{}^D - \sum_{jk}\mu_j'\bar{\nu}_{jk}\mathscr{J}_k - \sum_j \mathbf{J}_j\cdot\frac{d(\mathbf{v}_j - \mathbf{v})}{dt} \tag{14.150}$$

where we have made use of Eqs. (14.112) and (14.146) and μ_j' is a chemical potential which contains a contribution from diffusional kinetic energy of the jth constituent,

$$\mu_j' = \mu_j + \tfrac{1}{2}(\mathbf{v}_j - \mathbf{v})^2. \tag{14.151}$$

If we divide Eq. (14.150) by T and rearrange terms, we obtain the following balance equation for the entropy:

$$\frac{\partial\rho s}{\partial t} = -\nabla_{\mathbf{r}}\cdot\left(\frac{\mathbf{J}_E{}^R + \mathbf{J}_E{}^D - \rho\mathbf{v}\frac{1}{2}v^2 - \sum_j\mu_j'\rho_j\mathbf{v} - \overline{\overline{\mathbf{\Pi}}}{}^D\cdot\mathbf{v} - \sum_j\mu_j'\mathbf{J}_j{}^D}{T}\right) + \left(\mathbf{J}_E{}^R + \mathbf{J}_E{}^D - \rho\mathbf{v}\tfrac{1}{2}v^2 - \sum_j\mu_j'\rho_j\mathbf{v} - \overline{\overline{\mathbf{\Pi}}}{}^D\cdot\mathbf{v} - \sum_j\mu_j'\mathbf{J}_j{}^D - \rho s\mathbf{v}T\right)\cdot\nabla_{\mathbf{r}}\left(\frac{1}{T}\right) - \frac{\overline{\overline{\mathbf{\Pi}}}{}^D:\nabla_{\mathbf{r}}\mathbf{v}}{T} - \sum_j\frac{1}{T}\mathbf{J}_j{}^D\cdot\nabla_{\mathbf{r}}\mu_j - \frac{1}{T}\sum_{j,k}\mu_j'\bar{\nu}_{jk}\mathscr{J}_k - \frac{1}{T}\sum_j\mathbf{J}_j\cdot\frac{d(\mathbf{v}_j - \mathbf{v})}{dt}. \tag{14.152}$$

We now set the dissipative currents $\mathbf{J}_E{}^D$, $\overline{\overline{\mathbf{\Pi}}}{}^D$, and $\mathbf{J}_j{}^D$, the rate of chemical reaction $\mathscr{J}_j$, and the acceleration $d(\mathbf{v}_j - \mathbf{v})/dt$ to zero, and we obtain the balance equation for the entropy for an ideal multicomponent system. The condition that the entropy source be zero yields

$$\mathbf{J}_E{}^R = \rho s\mathbf{v}T + \rho\mathbf{v}\tfrac{1}{2}v^2 + \sum_j\mu_j'\rho_j\mathbf{v}. \tag{14.153}$$

Thus, Eq. (14.152) for a dissipative system takes the form

$$\frac{\partial\rho s}{\partial t} = -\nabla_{\mathbf{r}}\cdot(\rho s\mathbf{v} + \mathbf{J}_s{}^D) - \frac{1}{T}\mathbf{J}_s{}^D\cdot\nabla_{\mathbf{r}}T - \frac{1}{T}\overline{\overline{\mathbf{\Pi}}}{}^D:\nabla_{\mathbf{r}}\mathbf{v} - \frac{1}{T}\mathbf{J}_j{}^D\cdot\nabla_{\mathbf{r}}\mu_j - \frac{1}{T}\sum_k\bar{\mathscr{A}}_k\mathscr{J}_k - \frac{1}{T}\sum_j\mathbf{J}_j\cdot\frac{d(\mathbf{v}_j - \mathbf{v})}{dt} \tag{14.154}$$

where the dissipative entropy current is given by

$$\mathbf{J}_s{}^D = \mathbf{J}_E{}^D - \overline{\overline{\mathbf{\Pi}}}{}^D \cdot \mathbf{v} - \sum_j \mu_j' \mathbf{J}_j{}^D \tag{14.155}$$

and

$$\bar{\mathscr{A}}_k = \sum_j \bar{\nu}_{jk} \mu_j \tag{14.156}$$

is the affinity (cf. Sec. 3.E).

The total time derivative $d(\mathbf{v}_j - \mathbf{v})/dt$ is an acceleration which will be large initially but can be neglected after the system reaches the hydrodynamic state. There are four generalized currents for this system, namely the entropy current, $\mathbf{J}_s{}^D$; the stress tensor, $\overline{\overline{\mathbf{\Pi}}}{}^D$; the diffusion current, $\mathbf{J}_j{}^D$; and the rate of reaction, $\mathscr{J}_k$. The generalized forces are the temperature gradient, $\nabla_\mathbf{r} T$; the chemical potential gradient, $\nabla_\mathbf{r} \mu'$; the velocity gradient tensor, $\nabla_\mathbf{r} \mathbf{v}$; and the affinity, $\bar{\mathscr{A}}_k$. The linear relations between the currents and the forces take the following form. For the scalar currents we find

$$\mathscr{J}_k = -l_{k0} \nabla_\mathbf{r} \cdot \mathbf{v} - \sum_{k'=1}^{r} l_{kk'} \bar{\mathscr{A}}_{k'} \tag{14.157}$$

$k = 1, \ldots, r$ and

$$\pi^D = -\zeta \nabla_\mathbf{r} \cdot \mathbf{v} - \sum_{k=1}^{r} l_{0k} \bar{\mathscr{A}}_k, \tag{14.158}$$

and for the vector currents we find

$$\mathbf{J}_s{}^D = -\frac{1}{T} K \nabla_\mathbf{r} T - \sum_{j=1}^{P} L_{sj} \nabla_\mathbf{r} \mu_j' \tag{14.159}$$

and

$$\mathbf{J}_j{}^D = -L_{js} \nabla_\mathbf{r} T - \sum_{j'=1}^{P} L_{jj'} \nabla_\mathbf{r} \mu_{j'}'. \tag{14.160}$$

For the tensor current we find

$$(\overline{\overline{\mathbf{\Pi}}})^s = -\eta (\nabla_\mathbf{r} \mathbf{v})^s. \tag{14.161}$$

As before, ζ and η are the coefficients of bulk and shear viscosity, respectively, and K is the coefficient of thermal conductivity. We see that bulk viscosity and chemical reactions are coupled and heat flow and diffusion are coupled. Onsager's relations tell us that

$$l_{k0} = l_{0k} \tag{14.162}$$

for all k and

$$L_{sj} = L_{js} \tag{14.163}$$

for all j. From Eqs. (14.83) and (14.84) we have the conditions

$$l_{00} > 0, \quad K > 0, \quad \eta > 0, \quad \zeta > 0, \quad L_{jj} > 0 \tag{14.164}$$

and

$$l_{00}l_{kk} > l_{0k}^2, \quad \frac{1}{T}KL_{jj} > L_{js}^2. \tag{14.165}$$

Thus, thermodynamics, symmetry properties, and Onsager's relations tell us a great deal about the hydrodynamic behavior of a multicomponent reacting system.

H. SUPERFLUID HYDRODYNAMICS[8–11]

In this section we shall derive the hydrodynamic equations describing He(II) (liquid He^4 below the λ line [cf. Sec. 4.I]). We assume that He(II) is composed of two interpenetrating fluids, one of which behaves normally and has velocity $\mathbf{v}_n$ (normal fluid) in the laboratory frame, and another which carries no entropy and has velocity $\mathbf{v}_s$ (superfluid) in the laboratory frame. We shall first obtain the hydrodynamic equations for the system and then obtain dispersion relations for first and second sound in the superfluid.

1. Hydrodynamic Equations

In the superfluid rest frame, the normal fluid appears to move with velocity $\mathbf{v}_n - \mathbf{v}_s$ and the momentum density of the system is given by

$$\mathbf{J}_0 = \rho_n(\mathbf{v}_n - \mathbf{v}_s) \tag{14.166}$$

where ρ_n is the density of the normal fluid. The total energy in the superfluid rest frame is

$$\rho\epsilon_0 = \rho u + \tfrac{1}{2}\rho_n(\mathbf{v}_n - \mathbf{v}_s)^2 = T\rho s - P + \rho\mu' + \rho_n(\mathbf{v}_n - \mathbf{v}_s)^2 \tag{14.167}$$

where

$$\mu' = \mu - \frac{1}{2}\frac{\rho_n}{\rho}(\mathbf{v}_n - \mathbf{v}_s)^2 \tag{14.168}$$

and we have used the fundamental equation (14.105). For a system with a current $\mathbf{J}_0$ and chemical potential μ' the combined first and second laws yield for the differential of the total entropy

$$T\,d\rho s = d\rho\epsilon_0 - (\mathbf{v}_n - \mathbf{v}_s)\cdot d\mathbf{J}_0 - \mu'\,d\rho. \tag{14.169}$$

Eqs. (14.167) and (14.169) are the basic thermodynamic equations for He(II) in the superfluid rest frame. Using them, we can derive the hydrodynamic behavior of the fluid. If we take the differential of Eq. (14.167) and subtract Eq. (14.169), we find

$$d\mu' = \frac{1}{\rho}\,dP - s\,dT - \frac{1}{2}\frac{\rho_n}{\rho}\,d(\mathbf{v}_n - \mathbf{v}_s)^2. \tag{14.170}$$

Eq. (14.170) is the Gibbs-Duheim equation for the system in the superfluid rest frame.

The reactive pressure tensor of the normal fluid in the superfluid rest frame is

$$\overline{\overline{\mathbf{\Pi}}}_0 = P\overline{\overline{\mathbf{U}}} + \rho_n(\mathbf{v}_n - \mathbf{v}_s)(\mathbf{v}_n - \mathbf{v}_s). \tag{14.171}$$

The second term arises from convection of momentum.

We can now write the same quantities in the laboratory frame by performing a Galilean transformation. The momentum density $\mathbf{J}$ in the laboratory frame becomes

$$\mathbf{J} = \mathbf{J}_0 + \rho\mathbf{v}_s = \rho_n\mathbf{v}_n + \rho_s\mathbf{v}_s = \rho\mathbf{v} \tag{14.172}$$

where

$$\rho = \rho_n + \rho_s \tag{14.173}$$

and ρ_s is the hydrodynamic mass density of superfluid. The total energy in the laboratory frame is

$$\rho\epsilon = \rho\epsilon_0 + \mathbf{J}_0\cdot\mathbf{v}_s + \tfrac{1}{2}\rho{\mathbf{v}_s}^2, \tag{14.174}$$

and, finally, the stress tensor becomes

$$\overline{\overline{\mathbf{\Pi}}} = \overline{\overline{\mathbf{\Pi}}}_0 + \mathbf{J}_0\mathbf{v}_s + \mathbf{v}_s\mathbf{J}_0 + \rho\mathbf{v}_s\mathbf{v}_s = P\overline{\overline{\mathbf{U}}} + \rho_n\mathbf{v}_n\mathbf{v}_n + \rho_s\mathbf{v}_s\mathbf{v}_s. \tag{14.175}$$

From Eqs. (14.167) and (14.174) the differential of the total energy in the laboratory frame becomes

$$d\rho\epsilon = T\,d\rho s + \nu\,d\rho + \mathbf{v}_n\cdot d\mathbf{J} + \boldsymbol{\lambda}\cdot d\mathbf{v}_s \tag{14.176}$$

where

$$\nu = \mu' + \tfrac{1}{2}(\mathbf{v}_n - \mathbf{v}_s)^2 - \tfrac{1}{2}{\mathbf{v}_n}^2 \tag{14.177}$$

and

$$\boldsymbol{\lambda} = \mathbf{J} - \rho\mathbf{v}_n. \tag{14.178}$$

The hydrodynamic equations for the superfluid can be obtained from Sec. 3.F. We know that a gradient in the chemical potential causes a flow of superfluid (fountain effect). Thus, we can postulate the equation

$$\frac{d\mathbf{v}_s}{dt} = -\nabla_{\mathbf{r}}(\mu^R + \mu^D) \tag{14.179}$$

where μ^R is the reactive part of the chemical potential and μ^D is the dissipative part. Eq. (14.179) implies that superfluid flow is essentially potential flow,

$$\frac{d}{dt}(\nabla_{\mathbf{r}} \times \mathbf{v}_s) = 0. \tag{14.180}$$

Since the total mass of the fluid is conserved, we must have a continuity equation,

$$\frac{\partial \rho}{\partial t} + \nabla_{\mathbf{r}} \cdot \rho \mathbf{v} = 0. \tag{14.181}$$

Similarly, the balance equation for the momentum density is

$$\frac{\partial \mathbf{J}}{\partial t} + \nabla_{\mathbf{r}} \cdot (\overline{\overline{\Pi}}^R + \overline{\overline{\Pi}}^D) = 0, \tag{14.182}$$

the balance equation for the energy density is

$$\frac{\partial \rho \epsilon}{\partial t} + \nabla_{\mathbf{r}} \cdot (\mathbf{J}_E{}^R + \mathbf{J}_E{}^D) = 0, \tag{14.183}$$

and the balance equation for the entropy density is

$$\frac{\partial \rho s}{\partial t} + \nabla_{\mathbf{r}} \cdot (\mathbf{J}_s{}^R + \mathbf{J}_s{}^D) = \sigma_s. \tag{14.184}$$

We can now use the above results to find explicit expressions for various currents. From Eqs. (14.176), (14.179), and (14.182)–(14.184), we write the following equation for the entropy density:

$$\begin{aligned} T\frac{\partial \rho s}{\partial t} &= \frac{\partial \rho \epsilon}{\partial t} - \nu \frac{\partial \rho}{\partial t} - \mathbf{v}_n \cdot \frac{\partial \mathbf{J}}{\partial t} - \boldsymbol{\lambda} \cdot \frac{\partial \mathbf{v}_s}{\partial t} \\ &= -\nabla_{\mathbf{r}} \cdot (\mathbf{J}_E{}^R + \mathbf{J}_E{}^D) + \mu \nabla_{\mathbf{r}} \cdot \mathbf{J} + \mathbf{v}_n \cdot (\nabla_n \cdot \overline{\overline{\Pi}}^D) + \mathbf{v}_n \cdot \nabla_{\mathbf{r}} P \\ &\quad + \boldsymbol{\lambda} \cdot \nabla_{\mathbf{r}} (\mu^R + \mu^D) \end{aligned} \tag{14.185}$$

where we have only retained terms to second order in velocity on the right-hand side. If we now use Eq. (14.170) to write

$$\nabla_{\mathbf{r}} P = \rho \nabla_{\mathbf{r}} \mu' + \rho s \nabla_{\mathbf{r}} T + \tfrac{1}{2} \rho_n \nabla_{\mathbf{r}} (\mathbf{v}_n - \mathbf{v}_s)^2; \tag{14.186}$$

then we can combine Eqs. (14.178), (14.185), and (14.186) and obtain

$$\begin{aligned} \frac{\partial \rho s}{\partial t} &= -\nabla_{\mathbf{r}} \cdot \left[\frac{1}{T} (\mathbf{J}_E{}^R + \mathbf{J}_E{}^D - \mu^R \mathbf{J} - \mu^D \boldsymbol{\lambda} - \overline{\overline{\Pi}}^D \cdot \mathbf{v}_n) \right] \\ &\quad + (\mathbf{J}_E{}^R + \mathbf{J}_E{}^D - \mu \mathbf{J} - T \rho s \mathbf{v}_n - \mu^D \boldsymbol{\lambda} - \overline{\overline{\Pi}}^D \cdot \mathbf{v}_n) \cdot \nabla_{\mathbf{r}} \left(\frac{1}{T} \right) \\ &\quad - \frac{1}{T} [\mu^D \nabla_{\mathbf{r}} \cdot \boldsymbol{\lambda} + \overline{\overline{\Pi}}^D : \nabla_{\mathbf{r}} \mathbf{v}_n]. \end{aligned} \tag{14.187}$$

The form of the reactive energy current, $\mathbf{J}_E{}^R$, is easily found by setting the dissipative currents to zero and requiring the entropy production to be zero. We then find

$$\mathbf{J}_E{}^R = \mu\mathbf{J} + T\rho s\mathbf{v}_n. \tag{14.188}$$

We now can use Eq. (14.188) to simplify Eq. (14.187). The result is

$$\frac{\partial \rho s}{\partial t} = -\nabla_{\mathbf{r}}\cdot[\rho s\mathbf{v} + \mathbf{J}_s{}^D] - \frac{1}{T}\mathbf{J}_s{}^D\cdot\nabla_{\mathbf{r}}T - \frac{1}{T}\mu^D\nabla_{\mathbf{r}}\cdot\boldsymbol{\lambda} - \frac{1}{T}\overline{\overline{\Pi}}{}^D:\nabla_{\mathbf{r}}\mathbf{v}_n \tag{14.189}$$

where the dissipative entropy current is given by

$$\mathbf{J}_s{}^D = \frac{1}{T}(\mathbf{J}_E{}^D - \mu^D\boldsymbol{\lambda} - \overline{\overline{\Pi}}{}^D\cdot\mathbf{v}_n). \tag{14.190}$$

From Eq. (14.189), we see that there are four generalized forces, $\nabla_{\mathbf{r}}T$, $\nabla_{\mathbf{r}}\cdot\boldsymbol{\lambda}$, $\nabla_{\mathbf{r}}\cdot\mathbf{v}_n$, and $(\nabla_{\mathbf{r}}\mathbf{v}_n)^s$.

The transport coefficients for the system are defined by the linear relation between the currents and the forces. There are two scalar currents:

$$\pi^D = -\zeta_1\nabla_{\mathbf{r}}\cdot\boldsymbol{\lambda} - \zeta_2\nabla_{\mathbf{r}}\cdot\mathbf{v}_n \tag{14.191}$$

and

$$\mu^D = -\zeta_3\nabla_{\mathbf{r}}\cdot\boldsymbol{\lambda} - \zeta_4\nabla_{\mathbf{r}}\cdot\mathbf{v}_n. \tag{14.192}$$

There is one vector current,

$$\mathbf{J}_s = -\frac{K}{T}\nabla_{\mathbf{r}}T, \tag{14.193}$$

and one tensor current,

$$(\overline{\overline{\Pi}}{}^D)^s = -\eta(\nabla_{\mathbf{r}}\mathbf{v}_n)^s. \tag{14.194}$$

The transport coefficients ζ_1, ζ_2, ζ_3, ζ_4 are called second viscosities; η is called the first viscosity; and K is the thermal conductivity. Onsager's relations require that

$$\zeta_1 = \zeta_4 \tag{14.195}$$

and the positivity of the entropy production requires that

$$\zeta_1{}^2 < \zeta_2\zeta_3. \tag{14.196}$$

These results are well verified experimentally.

We can now obtain dispersion relations for the various types of sound modes in the absence of dissipation. To obtain the sound modes in the nondissipative case, we set all dissipative currents to zero in Eqs. (14.179) and (14.182)–(14.184) and linearize about absolute equilibrium. We shall write $\rho = \rho^0 + \delta\rho$,

$s = s^0 + \delta s$, and $P = P^0 + \delta P$ where ρ^0, s^0, and P^0 denote the equilibrium density, entropy, and pressure, respectively, and we neglect all terms to second order in the velocity. Then the hydrodynamic equations take the form

$$\frac{\partial\,\delta\rho}{\partial t} + \rho_s{}^0\nabla_{\mathbf{r}}\cdot\mathbf{v}_s + \rho_n{}^0\nabla_{\mathbf{r}}\cdot\mathbf{v}_n = 0, \tag{14.197}$$

$$\rho_s{}^0\frac{\partial\mathbf{v}_s}{\partial t} + \rho_n{}^0\frac{\partial\mathbf{v}_n}{\partial t} + \nabla_{\mathbf{r}}\cdot\delta P = 0, \tag{14.198}$$

$$\rho^0\frac{\partial\,\delta s}{\partial t} + s^0\frac{\partial\,\delta\rho}{\partial t} + \rho^0 s^0\nabla_{\mathbf{r}}\cdot\mathbf{v}_n = 0, \tag{14.199}$$

and

$$\frac{\partial\mathbf{v}_s}{\partial t} + \nabla_{\mathbf{r}}\,\delta\mu = 0. \tag{14.200}$$

Eqs. (14.197)–(14.200) enable us to obtain the dispersion relations for the various types of sound that can exist in an ideal superfluid system.

2. First Sound

To obtain the dispersion relation for first sound (density oscillations) in He(II), we assume that the temperature, T, is constant. We then note that for He(II) the thermal expansivity is very small, $\alpha_P = -(1/\rho)(\partial\rho/\partial T)_P \approx 0$. Therefore, the entropy will be constant since

$$\alpha_P \approx -\kappa_T\left(\frac{\partial P}{\partial T}\right)_\rho = -\frac{\kappa_T}{\rho^2}\left(\frac{\partial S}{\partial\rho}\right)_T \approx 0 \tag{14.201}$$

where κ_T is the isothermal compressibility. The hydrodynamic equations now take a simple form if we use the Gibbs-Duheim equation and choose density and pressure as our independent variables. We obtain

$$\frac{\partial\,\delta\rho}{\partial t} + \rho_s{}^0\nabla_{\mathbf{r}}\cdot\mathbf{v}_s + \rho_n{}^0\nabla_{\mathbf{r}}\cdot\mathbf{v}_n = 0, \tag{14.202}$$

$$\rho_s{}^0\frac{\partial\mathbf{v}_s}{\partial t} + \rho_n{}^0\frac{\partial\mathbf{v}_n}{\partial t} + \nabla_{\mathbf{r}}\,\delta P = 0, \tag{14.203}$$

$$\frac{\partial\,\delta\rho}{\partial t} + \rho^0\nabla_{\mathbf{r}}\cdot\mathbf{v}_n = 0, \tag{14.204}$$

and

$$\rho^0\frac{\partial\mathbf{v}_s}{\partial t} + \nabla_{\mathbf{r}}\,\delta P = 0. \tag{14.205}$$

If we subtract Eq. (14.202) from Eq. (14.204), we find

$$\nabla_{\mathbf{r}}\cdot(\mathbf{v}_n - \mathbf{v}_s) = 0, \tag{14.206}$$

and if we subtract Eq. (14.205) from Eq. (14.203), we find

$$\rho_n{}^0 \frac{\partial}{\partial t}(\mathbf{v}_n - \mathbf{v}_s) = 0. \tag{14.207}$$

If initially $\mathbf{v}_n = \mathbf{v}_s = 0$, then $\mathbf{v}_n = \mathbf{v}_s$ for all subsequent times and the two fluids move in phase. If we take the time derivative of Eq. (14.202) and the divergence of Eq. (14.203) and subtract the two equations, we find

$$\frac{\partial^2 \, \delta\rho}{\partial t^2} - \nabla^2 \, \delta P = 0. \tag{14.208}$$

We can now expand δP in terms of temperature and density. However, since temperature is constant, we find

$$\delta P = \left(\frac{\partial P}{\partial \rho}\right)_T \delta\rho = \frac{1}{\rho\kappa_T}\,\delta\rho \tag{14.209}$$

and Eq. (14.208) becomes

$$\frac{\partial^2 \, \delta\rho}{\partial t^2} - \frac{1}{\rho\kappa_T}\nabla^2 \, \delta\rho = 0. \tag{14.210}$$

Thus, density oscillations propagate with a speed $c_1 = \sqrt{1/\rho\kappa_T}$ and have a dispersion relation $\omega = c_1 k$. Such oscillations are called first sound. Note that the speed of first sound depends on the isothermal compressibility, while the speed of sound in ordinary fluids depends on the adiabatic compressibility.

3. Second Sound

To obtain the dispersion relations for second sound (temperature oscillations), we assume that the mass current in the fluid is zero,

$$\mathbf{J} = \rho_s{}^0\mathbf{v}_s + \rho_n{}^0\mathbf{v}_n = 0. \tag{14.211}$$

If we then make use of the Gibbs-Duheim equation and Eq. (14.211), the linear hydrodynamic equations (14.197)–(14.200) reduce to

$$\frac{\partial \rho}{\partial t} = 0, \tag{14.212}$$

$$\delta\nabla_{\mathbf{r}} P = 0, \tag{14.213}$$

$$\frac{\partial \, \delta s}{\partial t} + s^0 \nabla_{\mathbf{r}} \cdot \mathbf{v}_n = 0, \tag{14.214}$$

and

$$\frac{\partial \mathbf{v}_s}{\partial t} - s^0 \nabla_{\mathbf{r}} \, \delta T = 0. \tag{14.215}$$

We now combine Eqs. (14.211), (14.214), and (14.215) and obtain

$$\frac{\partial^2 \,\delta s}{\partial t^2} - (s^0)^2 \frac{\rho_s^{\,0}}{\rho_n^{\,0}} \nabla_{\mathbf{r}}^{\,2}\, \delta T = 0. \tag{14.216}$$

We next expand the entropy in terms of temperature and pressure and use Eq. (14.213). We then find

$$\frac{\partial^2 \,\delta T}{\partial t^2} - \frac{(s^0)^2 T \rho_s^{\,0}}{c_P \rho_n^{\,0}} \nabla_{\mathbf{r}}^{\,2}\, \delta T = 0. \tag{14.217}$$

Eq. (14.217) gives the dispersion relation for second sound. Second sound consists of temperature waves which propagate with a speed

$$c_2 = \sqrt{\frac{(s^0)^2 T \rho_s^{\,0}}{c_P \rho_n^{\,0}}} \tag{14.218}$$

where c_P is the specific heat at constant pressure. Second sound was first measured by Peshkov[12] who used an oscillating heat source to set up standing waves in a tube. A plot of the speed of second sound as a function of temperature is given in Fig. 14.1. We see that the velocity first peaks at 20.36 m/sec at $T = 1.65$ K and decreases slightly as we lower the temperature, but then increases again and reaches a limiting value of about 150 m/sec at $T = 0$ K.

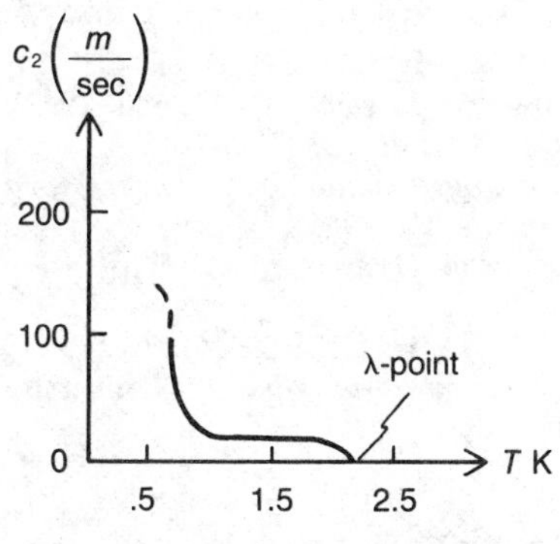

Fig. 14.1. Speed of second sound (based on Ref. 14.13).

Measurements of the speed of second sound enable us to obtain a value of the ratio $\rho_n^{\,0}/\rho^0$. If we remember that $\rho = \rho_n^{\,0} + \rho_s^{\,0}$, then Eq. (14.218) takes the form

$$\frac{\rho_n^{\,0}}{\rho^0} = \frac{(s^0)^2 T^0}{c_P c_2^{\,2} - (s^0)^2 T^0}. \tag{14.219}$$

From Eq. (14.219), measurements of the hydrodynamic density of the normal fluid can be obtained.

Another form of sound occurs when He(II) flows through a porous material. This is called fourth sound. We leave the derivation of its dispersion relation as a problem.

In Eq. (14.179), we have postulated the existence of an additional hydrodynamic equation for He(II). Its origin is not the microscopic conservation laws but broken gauge symmetry in the fluid. We shall return to the hydrodynamics of symmetry broken systems in the next chapter and obtain a deeper understanding of the origin of Eq. (14.179).

REFERENCES

1. L. Onsager, Phys. Rev. *37* 405 (1931); *38* 2265 (1931).
2. H. B. G. Casimir, Rev. Mod. Phys. *17* 343 (1945).
3. I. Prigogine, *Introduction to Thermodynamics of Irreversible Processes* (Wiley-Interscience, New York, 1967).
4. S. R. de Groot and P. Mazer, *Nonequilibrium Thermodynamics* (North-Holland Publishing Co., Amsterdam, 1969).
5. S. R. de Groot, *Thermodynamics of Irreversible Processes* (North-Holland Publishing Co., Amsterdam, 1951).
6. R. Aris, *Tensors and the Basic Equations of Fluid Mechanics* (Prentice-Hall, Englewood Cliffs, N.J., 1962).
7. L. D. Landau and E. M. Lifshitz, *Fluid Mechanics* (Addison-Wesley Publishing Co., Reading, Mass., 1959).
8. P. Mazur and I. Prigogine, Physica *17* 661, 680 (1951).
9. C. P. Enz, Rev. Mod. Phys. *46* 705 (1974).
10. I. M. Khalatnikov, *An Introduction to the Theory of Superfluidity* (W. A. Benjamin, New York, 1965).
11. S. J. Putterman, *Superfluid Hydrodynamics* (North-Holland Publishing Co., Amsterdam, 1974).
12. V. Peshkov, J. Phys. U.S.S.R. *8* 131 (1944); *10* 389 (1946).
13. R. J. Donnelly, *Experimental Superfluidity* (University of Chicago Press, Chicago, 1967).

PROBLEMS

1. The condition for local equilibrium in a fluid is that the stress forces, $\mathbf{f}$ (cf. Eq. (14.98)), add to zero at each point in the fluid. Consider the volume element shown in Fig. 14.2 and prove that the condition for local equilibrium in the fluid requires that $\mathbf{f}$ have the form $\mathbf{f} = \hat{\mathbf{n}} \cdot \overline{\overline{\mathbf{\Pi}}}$, where $\hat{\mathbf{n}}$ is the normal to the surface.

2. Assume that a viscous fluid flows in a square pipe. Choose the z-axis to be the direction of motion of the fluid. Assume that the density of the fluid is constant and that the flow is steady. Starting from the hydrodynamic equations, find the equation relating pressure gradients to velocity gradients in the fluid. Describe what is happening in the pipe. (*Note:* at the walls the velocity is zero.)

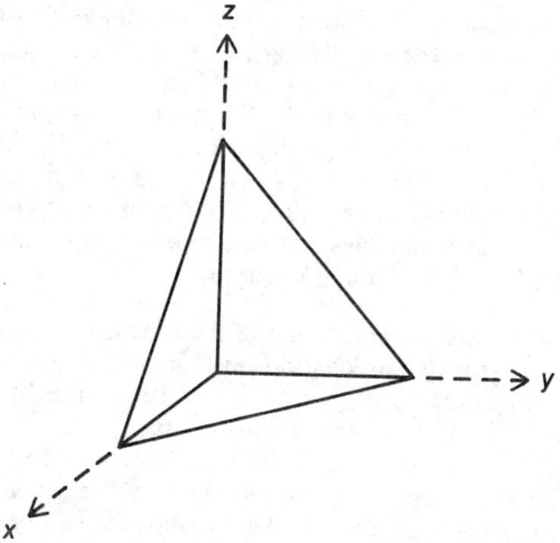

Fig. 14.2. Volume element used to prove that $\mathbf{f} = \hat{\mathbf{n}} \cdot \overline{\overline{\Pi}}$.

3. Consider the flow of He(II) through a porous material. Only the superfluid flows so that $\mathbf{v}_n = 0$. Also, the porous material exchanges momentum with the fluid so that momentum is not conserved and Eq. (14.198) cannot be used. Use Eqs. (14.197), (14.199), and (14.200) and the condition $(\partial P/\partial T)_\rho = 0$ to derive a dispersion relation for density oscillations (fourth sound). Show that fourth sound has a speed $c_4 = \sqrt{(\rho_n/\rho)c_2^2 + (\rho_s/\rho)c_1^2}$, where c_1 and c_2 are the speeds of first and second sound.

4. Find the dispersion relations for second sound in the presence of dissipation.

5. Consider a box with temperature fixed at the boundary so that a constant temperature gradient is maintained across the box. Show that in the linear regime the state of minimum entropy production is a stationary state of the temperature density.

6. An electric field $\mathbf{E}$ and a magnetic field $\mathbf{H}$ are applied to an isotropic binary mixture of charged particles. Assume that the system is unpolarized by the electric or magnetic field so that the electric polarization $\mathbf{P} = 0$ and the magnetization $\mathbf{M} = 0$. Write down all the hydrodynamic equations of the system. Find the local entropy production and the linear relations between the forces and the currents.

7. Consider a volume element $V(t)$ containing extended molecules. The angular momentum per particle of a volume element is $\mathbf{r} \times \mathbf{v}$ where $\mathbf{r}$ refers to arbitrary coordinate system. The intrinsic angular momentum per unit mass of the molecules due to their shape is $\mathbf{l}$ so that the total angular momentum is $\mathbf{L} = \rho(\mathbf{r} \times \mathbf{v}) + \rho\mathbf{l}$. A number of torques can be exerted on $V(t)$. The external field exerts a torque/volume $\rho\mathbf{r} \times \mathbf{F}$ on the fluid element and a torque/volume $\rho\tau$ on the extended molecules. The stress forces, $\mathbf{f}$, exert a torque $\mathbf{r} \times \mathbf{f}$ per unit area and there may be a couple torque/area $\boldsymbol{\tau}$ between molecules at the surface of $V(t)$. Find the equation of motion of $\mathbf{l}$ and show that it is coupled to the antisymmetric part of the stress tensor $(\overline{\overline{\Pi}})^a$. For point particles, show that the antisymmetric part of the stress tensor is zero.

8. Consider a binary system which can undergo a single chemical reaction (for example, $A_2 \rightleftharpoons 2A$). Assume the following: (*a*) the system is enclosed in a box between two walls at $z = d$ and $z = -d$; (*b*) it is in a stationary state; (*c*) a temperature gradient exists in the *z*-direction; (*d*) the phenomenological coefficients are constant; and (*e*) at the walls there are no diffusion currents. If at one wall ($z = -d$) $T = T_0$ and at the other wall ($z = d$) $T = T_0 + T$, find the temperature distribution in the box for small deviations from equilibrium. Assume that viscous effects can be neglected and note that all quantities can be expressed in terms of μ_i/T. You may want to make use of the Gibbs-Duheim equation.

9. A metal is a binary system consisting of a fixed positive ion lattice and mobile electrons. Derive the hydrodynamic equations for this system assuming that it is isotropic and that a constant electric field exists in the metal. Measure the electron flow relative to the lattice and neglect viscous effects.

(*a*) Use Onsager's relations and the positivity of entropy production to obtain all possible information about the phenomenological coefficients for this system.

(*b*) Consider the thermocouple in Fig. 14.3 composed of two different metals A and B. The junctions coupling the two metals are kept at different temperatures, one at T and the other at $T + \Delta T$. Metal A has a capacitor inserted such that the temperature of both plates of the capacitor is T_0. Assume the system is in a stationary state and that no current flows. (Note that the chemical potential at each plate is the same.) Find the potential drop across the capacitor in terms of the phenomenological coefficients of the two metals and ΔT. This is called the thermoelectric effect.

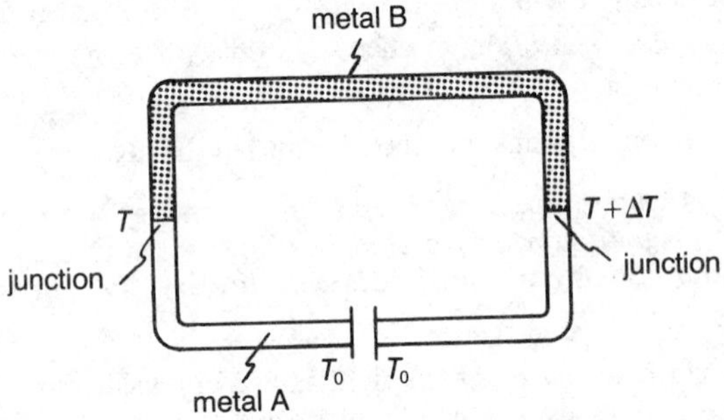

Fig. 14.3. A thermocouple.

(*c*) Consider the junction of the two metals in the thermocouple. Assume that a current flows across the junction but that it is maintained at constant temperature. Find the rate at which heat must be absorbed at the junction to maintain a steady state. This is called the Peltier effect. (*Hint:* express the entropy production in terms of the current rather than the forces.) How are the thermoelectric and Peltier effects related?

10. Consider a two-component, nonreacting system of uncharged particles. Assume that viscous effects do not contribute and write the hydrodynamic equations and entropy production for this system. Next assume that the system is divided into two parts by a membrane permeable to one of the substances.

(*a*) Assume the temperature is maintained constant and is the same on both sides of the membrane and that a concentration gradient is maintained between both sides. For the steady state, express the osmotic pressure in terms of the concentration difference and the phenomenological coefficients.

(*b*) Because of Onsager's relations, osmotic pressure measurements will give information about another quite different effect in the system. What is this other effect?

15 Fluctuation-Dissipation Theorem

A. INTRODUCTORY REMARKS

Fluctuations about the equilibrium state decay *on the average* according to the same linear macroscopic laws (hydrodynamic equations) which describe the decay of the system from a nonequilibrium state to the equilibrium state. If we can probe the equilibrium fluctuations, we have a means of probing the transport processes in the system. The fluctuation-dissipation theorem shows that it is possible to probe the equilibrium fluctuations by applying a weak external field which couples to particles in the medium but yet is too weak to affect the medium. The system will respond to the field and absorb energy from it in a manner which depends entirely on the spectrum of the equilibrium fluctuations. According to the fluctuation-dissipation theorem, the spectrum of equilibrium fluctuations and the rate of absorption of energy from the external field can be expressed in terms of a response matrix which is related to the correlation matrix for equilibrium fluctuations.

In this chapter, we will derive the fluctuation-dissipation theorem and apply it to a variety of systems. The derivation requires several steps. We first introduce the Wiener-Khinchin theorem, which enables us to relate the spectral density matrix for fluctuations to the time-dependent correlation matrix for fluctuations in the system. We then introduce linear response theory and use the assumption of causality to obtain a relation between the real and imaginary parts of the response matrix. Finally, we use linear response theory to obtain a relation between the response matrix and the correlation matrix, and we obtain an expression for energy absorption in terms of the imaginary part of the response matrix.

One of the simplest applications of linear response theory involves a harmonically bound Brownian particle immersed in a medium. If the Brownian particle is pulled away from its equilibrium position, it will decay back to equilibrium and dissipate energy into the fluid. We will obtain an expression for the linear response function of the Brownian particle and in terms of it derive an expression for the correlation function for equilibrium fluctuations in the position for the oscillator.

Another important application of the fluctuation-dissipation theorem involves the scattering of light from a fluid. The electric field of the incident light wave polarizes the particles in a medium, thus allowing the light to couple to the medium. The light will be scattered by density fluctuations, and by measuring the spectrum of the scattered light we can measure the spectrum of the density fluctuations. We will find that the density fluctuations are of two types: thermal density fluctuations due to fluctuations in the local entropy and mechanical density fluctuations due to damped sound waves. For low-frequency and long-wavelength fluctuations, the spectrum of scattered light can be obtained from the linearized hydrodynamic equations and, therefore, light scattering experiments give us a means of measuring transport coefficients in the fluid.

The poles of the spectral density matrix give us information about the spectrum of fluctuations in an equilibrium system. The poles corresponding to very low frequency, long-wavelength processes are due to hydrodynamic modes in the system. By using projection operator techniques, it is possible to introduce projections onto the space of hydrodynamic modes and onto the space orthogonal to it. We can then write the spectral density matrix for hydrodynamic fluctuations in a very general form and show that hydrodynamic modes arise not only from conserved quantities but also from broken symmetries which occur at a continuous phase transition. We shall complete this chapter by a discussion of the origin of hydrodynamic modes and give examples of hydrodynamic modes that arise from a broken symmetry.

B. WIENER-KHINCHIN THEOREM[1–3]

The Wiener-Khinchin theorem enables us to obtain a relation between the correlation matrix for time-dependent fluctuations and the spectral density matrix of fluctuations for ergodic systems with stationary distribution functions. We shall first derive some properties of time-dependent correlation matrices and then derive the Wiener-Khinchin theorem.

1. Properties of Time-Dependent Correlation Matrices

Let us consider the time-dependent correlation matrix, $\langle \boldsymbol{\alpha}(\tau)\boldsymbol{\alpha}(0)\rangle$, for a system governed by a stationary distribution function (such as a system in equilibrium). The correlation matrix has the property

$$\bar{\bar{\mathbf{C}}}_{\alpha\alpha}(\tau) \equiv \langle \boldsymbol{\alpha}(\tau)\boldsymbol{\alpha}\rangle = \langle \boldsymbol{\alpha}(t+\tau)\boldsymbol{\alpha}(t)\rangle = \langle \boldsymbol{\alpha}(-\tau)\boldsymbol{\alpha}\rangle^T = \bar{\bar{\mathbf{C}}}_{\alpha\alpha}(-\tau)^T \tag{15.1}$$

where we have let $t = -\tau$ and T denotes the transpose of the correlation matrix. From the condition of microscopic reversibility, we know that $\langle \boldsymbol{\alpha}(\tau)\boldsymbol{\alpha} \rangle = \langle \boldsymbol{\alpha}\boldsymbol{\alpha}(\tau) \rangle$ (cf. Sec. 14.B) and, therefore,

$$\bar{\bar{\mathbf{C}}}_{\alpha\alpha}(\tau) = \bar{\bar{\mathbf{C}}}_{\alpha\alpha}(\tau)^T. \tag{15.2}$$

Furthermore, from Sec. 10.B, we have

$$\bar{\bar{\mathbf{C}}}_{\alpha\alpha}(0) = \langle \boldsymbol{\alpha}\boldsymbol{\alpha} \rangle = k_B \bar{\bar{\mathbf{g}}}^{-1} \tag{15.3}$$

where $\bar{\bar{\mathbf{g}}}^{-1}$ depends on the thermodynamic response functions. From Eqs. (14.14) and (15.1) the correlation matrix can be written

$$\bar{\bar{\mathbf{C}}}_{\alpha\alpha}(\tau) = \int d\boldsymbol{\alpha}_0 f(\boldsymbol{\alpha}_0)\boldsymbol{\alpha}_0 \langle \boldsymbol{\alpha}(\tau) \rangle_{\boldsymbol{\alpha}_0} = k_B \bar{\bar{\mathbf{g}}}^{-1} \cdot e^{-\bar{\bar{\mathbf{M}}}|\tau|} \tag{15.4}$$

since $\bar{\bar{\mathbf{M}}}$ is a self-adjoint matrix ($|\tau|$ indicates the absolute value of τ).

2. Spectral Density Matrix

We now will introduce the spectral density matrix of fluctuations and show that it is the Fourier transform of the correlation matrix. Let us introduce a slight modification of the state variable $\boldsymbol{\alpha}(t)$ as follows:

$$\boldsymbol{\alpha}(t; T) \equiv \begin{cases} \boldsymbol{\alpha}(t) & |t| < T \\ 0 & |t| > T \end{cases} \tag{15.5}$$

such that

$$\lim_{T \to \infty} \boldsymbol{\alpha}(t; T) = \boldsymbol{\alpha}(t). \tag{15.6}$$

We next introduce the Fourier transform of $\boldsymbol{\alpha}(t; T)$:

$$\boldsymbol{\alpha}(\omega; T) = \int_{-\infty}^{\infty} dt \boldsymbol{\alpha}(t; T)\, e^{i\omega t} = \int_{-T}^{T} dt \boldsymbol{\alpha}(t; T)\, e^{i\omega t}. \tag{15.7}$$

Since the fluctuations, $\boldsymbol{\alpha}(t)$, are real, we find

$$\boldsymbol{\alpha}^*(\omega; T) = \boldsymbol{\alpha}(-\omega; T) \tag{15.8}$$

(* denotes complex conjugation).

The spectral density matrix is defined

$$\bar{\bar{\mathbf{S}}}_{\alpha\alpha}(\omega) = \lim_{T \to \infty} \frac{1}{T} \boldsymbol{\alpha}^*(\omega; T)\boldsymbol{\alpha}(\omega; T). \tag{15.9}$$

Combining Eqs. (15.6) and (15.8), we can write

$$\bar{\bar{\mathbf{S}}}_{\alpha\alpha}(\omega) = \int_{-\infty}^{\infty} d\tau\, e^{i\omega\tau} \lim_{T \to \infty} \frac{1}{T} \int_{-\infty}^{\infty} dt \boldsymbol{\alpha}(t; T)\boldsymbol{\alpha}(t + \tau; T). \tag{15.10}$$

If we now invoke the ergodic theorem (cf. Sec. 8.B), we can equate the time average in Eq. (15.10) to the phase average of the fluctuations:

$$\langle \boldsymbol{\alpha\alpha}(\tau)\rangle = \lim_{T\to\infty} \frac{1}{T}\int_{-T}^{T} \boldsymbol{\alpha}(t)\boldsymbol{\alpha}(t+\tau)\,dt$$

$$= \lim_{T\to\infty} \frac{1}{T}\int_{-\infty}^{\infty} \boldsymbol{\alpha}(t;T)\boldsymbol{\alpha}(t+\tau;T)\,dt. \qquad (15.11)$$

Then Eqs. (15.10) and (15.11) lead to the relation

$$\bar{\bar{\mathbf{S}}}_{\alpha\alpha}(\omega) = \int_{-\infty}^{\infty} d\tau\, e^{i\omega\tau}\langle \boldsymbol{\alpha\alpha}(\tau)\rangle = \int_{-\infty}^{\infty} d\tau\, e^{i\omega\tau}\bar{\bar{\mathbf{C}}}_{\alpha\alpha}(\tau). \qquad (15.12)$$

Thus, the spectral density matrix is the Fourier transform of the correlation matrix. Eq. (15.12) is called the Wiener-Khinchin theorem.

We can now derive some general properties for the spectral density matrix. From Eq. (15.1) we find that the spectral density matrix is Hermitian,

$$\bar{\bar{\mathbf{S}}}_{\alpha\alpha}(\omega) = \int_{-\infty}^{\infty} d\tau\, e^{i\omega\tau}\bar{\bar{\mathbf{C}}}_{\alpha\alpha}(-\tau)^{T} = \bar{\bar{\mathbf{S}}}{}^{*T}_{\alpha\alpha}(\omega) \qquad (15.13)$$

where we have let $\tau \to -\tau$ to obtain the last term. Furthermore, since $\bar{\bar{\mathbf{C}}}_{\alpha\alpha}(\tau)$ is real, we have

$$\bar{\bar{\mathbf{S}}}{}^{*}_{\alpha\alpha}(\omega) = \int_{-\infty}^{\infty} d\tau\, e^{-i\omega\tau}\bar{\bar{\mathbf{C}}}_{\alpha\alpha}(\tau) = \bar{\bar{\mathbf{S}}}_{\alpha\alpha}(-\omega). \qquad (15.14)$$

It is useful to divide $\bar{\bar{\mathbf{S}}}_{\alpha\alpha}(\omega)$ into its real and imaginary parts, $\bar{\bar{\mathbf{R}}}_{\alpha\alpha}(\omega)$ and $\bar{\bar{\mathbf{I}}}_{\alpha\alpha}(\omega)$, respectively:

$$\bar{\bar{\mathbf{S}}}_{\alpha\alpha}(\omega) = \bar{\bar{\mathbf{R}}}_{\alpha\alpha}(\omega) + i\bar{\bar{\mathbf{I}}}_{\alpha\alpha}(\omega). \qquad (15.15)$$

Then from Eqs. (15.13) and (15.14) we find that $\bar{\bar{\mathbf{R}}}_{\alpha\alpha}(\omega)$ is a real, symmetric matrix and an even function of ω ($\bar{\bar{\mathbf{R}}}_{\alpha\alpha}(\omega) = \bar{\bar{\mathbf{R}}}_{\alpha\alpha}(\omega)^{T} = \bar{\bar{\mathbf{R}}}_{\alpha\alpha}(-\omega)$), while $\bar{\bar{\mathbf{I}}}_{\alpha\alpha}(\omega)$ is a real, antisymmetric matrix and an odd function of ω ($\bar{\bar{\mathbf{I}}}_{\alpha\alpha}(\omega) = -\bar{\bar{\mathbf{I}}}_{\alpha\alpha}(\omega)^{T} = -\bar{\bar{\mathbf{I}}}_{\alpha\alpha}(-\omega)$). Furthermore, we find that $\bar{\bar{\mathbf{S}}}_{\alpha\alpha}(\omega) = \bar{\bar{\mathbf{S}}}_{\alpha\alpha}(\omega)^{T}$. Thus, $\bar{\bar{\mathbf{I}}}_{\alpha\alpha}(\omega) = -\bar{\bar{\mathbf{I}}}_{\alpha\alpha}(\omega) = 0$, and the spectral density matrix is a real, symmetric matrix and an even function of ω. We therefore can write the correlation matrix in the form

$$\bar{\bar{\mathbf{C}}}_{\alpha\alpha}(\tau) = \frac{1}{2\pi}\int_{-\infty}^{\infty} d\omega\, e^{-i\omega\tau}\bar{\bar{\mathbf{S}}}_{\alpha\alpha}(\omega) = \frac{1}{\pi}\int_{0}^{\infty} d\omega\bar{\bar{\mathbf{R}}}_{\alpha\alpha}(\omega)\cos(\omega\tau). \qquad (15.16)$$

If a magnetic field is present or the system is rotating, Eq. (15.16) must be generalized.

3. Spectral Density Matrix and Magnetic Fields

Let us now consider the more general case of fluctuations $\boldsymbol{\alpha}$ and $\boldsymbol{\beta}$ in the presence of a magnetic field $\mathbf{B}$ (cf. Sec. 14.C). Then we define the more general spectral density matrix,

$$\overline{\overline{\mathbf{S}}}_{\alpha\beta}(\omega; \mathbf{B}) = \int \overline{\overline{\mathbf{C}}}_{\alpha\beta}(\tau; \mathbf{B})\, e^{i\omega\tau}\, d\tau \tag{15.17}$$

where $\overline{\overline{\mathbf{C}}}_{\alpha\beta}(\tau; \mathbf{B}) = \langle \boldsymbol{\alpha}(\tau)\boldsymbol{\beta}; \mathbf{B}\rangle$. Similar definitions hold for $\overline{\overline{\mathbf{S}}}_{\beta\beta}(\omega; \mathbf{B})$ and $\overline{\overline{\mathbf{S}}}_{\alpha\alpha}(\omega; \mathbf{B})$. We divide the spectral density matrix into its real and imaginary parts as before, for example,

$$\overline{\overline{\mathbf{S}}}_{\alpha\beta}(\omega; \mathbf{B}) = \overline{\overline{\mathbf{R}}}_{\alpha\beta}(\omega, \mathbf{B}) + i\overline{\overline{\mathbf{I}}}_{\alpha\beta}(\omega, \mathbf{B}). \tag{15.18}$$

Using the conditions of stationarity, microscopic reversibility, and reality, we can show that $\mathbf{R}_{\alpha\alpha}(\omega; \mathbf{B})$, $\mathbf{R}_{\beta\beta}(\omega; \mathbf{B})$, and $\mathbf{I}_{\alpha\beta}(\omega; \mathbf{B})$ do not change sign when the magnetic field changes sign, while $\mathbf{R}_{\alpha\beta}(\omega; \mathbf{B})$, $\mathbf{I}_{\alpha\alpha}(\omega; \mathbf{B})$, and $\mathbf{I}_{\beta\beta}(\omega; \mathbf{B})$ do change sign.

C. CAUSALITY AND RESPONSE MATRICES[3]

Let us assume that external forces, $\mathbf{F} = (F_1, F_2, \ldots, F_n)$, are applied to the system and couple to the various state variables $A_1, A_2, \ldots, A_n$ causing a deviation from their equilibrium values. We shall assume that the deviation in variables $A_1, A_2, \ldots, A_n$ is proportional to the applied force (linear response). Then we can write

$$\langle \boldsymbol{\alpha}(t)\rangle_F = \int_{-\infty}^{\infty} dt'\overline{\overline{\mathbf{K}}}(t - t')\cdot\mathbf{F}(t')\, dt' = \int_{-\infty}^{\infty} \overline{\overline{\mathbf{K}}}(\tau)\cdot\mathbf{F}(t - \tau)\, d\tau. \tag{15.19}$$

The matrix $\overline{\overline{\mathbf{K}}}(t - t')$ is real and is called the response matrix. Since the response must be causal (the response cannot precede the force which causes it), $\overline{\overline{\mathbf{K}}}(t - t')$ must satisfy the causality condition,

$$\begin{aligned} \overline{\overline{\mathbf{K}}}(t - t') &= 0 \qquad t - t' < 0, \\ \overline{\overline{\mathbf{K}}}(\tau) &= 0 \qquad \tau < 0. \end{aligned} \tag{15.20}$$

Since Eq. (15.19) is linear in the force, its Fourier transform has a very simple form. If we note that

$$\langle \boldsymbol{\alpha}(t)\rangle_F = \frac{1}{2\pi}\int_{-\infty}^{\infty} \langle \tilde{\boldsymbol{\alpha}}(\omega)\rangle_F\, e^{-i\omega t}\, d\omega \tag{15.21}$$

and use similar expressions relating $\mathbf{F}(\tau)$ to $\tilde{\mathbf{F}}(\omega)$ and $\overline{\overline{\mathbf{K}}}(\tau)$ to $\overline{\overline{\chi}}(\omega)$, we obtain

$$\langle\tilde{\alpha}(\omega)\rangle_F = \overline{\overline{\chi}}(\omega)\cdot\tilde{\mathbf{F}}(\omega). \tag{15.22}$$

We have also used the following definition for the delta function:

$$\delta(t) = \frac{1}{2\pi}\int_{-\infty}^{\infty} d\omega\, e^{-i\omega t}. \tag{15.23}$$

Thus, a force of a given frequency can only excite a response of the same frequency. This will not be true if the response function depends on the force (nonlinear response).

Let us now study the restrictions imposed on the response matrix by causality. From Eq. (15.20), we can write $\overline{\overline{\chi}}(\omega)$ as

$$\overline{\overline{\chi}}(\omega) = \int_{-\infty}^{\infty} \overline{\overline{\mathbf{K}}}(t)\, e^{i\omega t}\, dt. \tag{15.24}$$

Furthermore, we shall assume that the response matrix relaxes fast enough that the integral

$$\int_0^{\infty} \overline{\overline{\mathbf{K}}}(t)\, dt < \infty \tag{15.25}$$

is finite. Physically, this means that a finite force must give rise to a finite response. It is convenient to continue the frequency dependence of $\overline{\overline{\chi}}(\omega)$ into the complex plane by writing

$$\overline{\overline{\chi}}(z) = \int_0^{\infty} dt\, \overline{\overline{\mathbf{K}}}(t)\, e^{izt}, \tag{15.26}$$

where $z = \omega + i\epsilon$ (ϵ positive). If $\overline{\overline{\chi}}(\omega)$ is well behaved (nonsingular), then $\overline{\overline{\chi}}(z)$ will be well behaved since the extra factor $e^{-\epsilon t}$ introduced simply makes the integral fall off faster. The function $\overline{\overline{\chi}}(z)$ is not singular (has no poles) for z in the upper half complex plane, but we can say nothing about its behavior in the lower half complex plane. Note also that $\overline{\overline{\chi}}(z) \to 0$ as $\epsilon \to \infty$.

The matrix $\overline{\overline{\chi}}(\omega)$ is complex ($\overline{\overline{\chi}}^*(\omega) = \overline{\overline{\chi}}(-\omega)$), but causality enables us to obtain a relation between its real and imaginary parts. We do this by introducing the following trick. Define a new matrix

$$\overline{\overline{\mathbf{f}}}(z) = \frac{\overline{\overline{\chi}}(z)}{z - u} \tag{15.27}$$

and integrate it over a contour C' (cf. Fig. 15.1) so that C' encloses no poles of $\overline{\overline{\mathbf{f}}}(z)$. (Note that u is real.) Then

$$\oint_C \overline{\overline{\mathbf{f}}}(z)\, dz = 0 = \oint_{C'} \frac{\overline{\overline{\chi}}(z)}{z - u}\, dz. \tag{15.28}$$

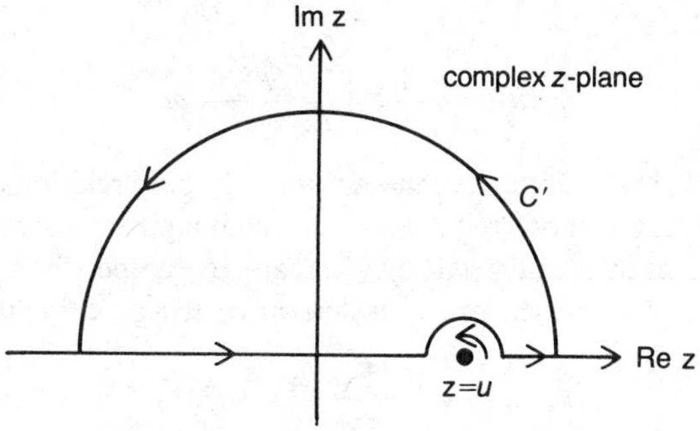

Fig. 15.1. Integration contour used to obtain the Kramers-Kronig relations.

Since $\overline{\overline{\chi}}(z) \to 0$ as $\epsilon \to \infty$ there will be no contribution from the semicircle at infinity. Thus,

$$\oint_{C'} \frac{\overline{\overline{\chi}}(z)}{z-u}\,dz = \int_{-\infty}^{u-r} \frac{\overline{\overline{\chi}}(\omega)}{\omega - u}\,d\omega + \int_{u+r}^{\infty} d\omega\,\frac{\overline{\overline{\chi}}(\omega)}{\omega-u}$$

$$+ \int_{\pi}^{0} d\phi i r\, e^{i\phi}\,\frac{\overline{\overline{\chi}}(u + r\,e^{i\phi})}{u + r\,e^{i\phi} - u} = 0. \tag{15.29}$$

It is useful to introduce the Cauchy principal part,

$$P\int_{-\infty}^{\infty} d\omega\,\frac{\overline{\overline{\chi}}(\omega)}{\omega-u} \equiv \lim_{r\to 0}\left[\int_{-\infty}^{u-r} d\omega\,\frac{\overline{\overline{\chi}}(\omega)}{\omega-u} + \int_{n+r}^{\infty} d\omega\,\frac{\overline{\overline{\chi}}(\omega)}{\omega-u}\right]. \tag{15.30}$$

Eq. (15.29) then gives

$$P\int_{-\infty}^{\infty} d\omega\,\frac{\overline{\overline{\chi}}(\omega)}{\omega-u} = -\lim_{r\to 0}\int_{\pi}^{0} i\,d\phi\overline{\overline{\chi}}(u + r\,e^{i\phi}) = i\pi\overline{\overline{\chi}}(u) \tag{15.31}$$

or

$$\overline{\overline{\chi}}(u) = \frac{1}{\pi i} P\int_{-\infty}^{\infty} \frac{\overline{\overline{\chi}}(\omega)\,d\omega}{\omega-u}. \tag{15.32}$$

Eq. (15.32) is a consequence of causality and allows us to relate the real part, $\overline{\overline{\chi}}'(\omega)$, and the imaginary part, $\overline{\overline{\chi}}''(\omega)$, of the response matrix. Let us write

$$\overline{\overline{\chi}}(\omega) = \overline{\overline{\chi}}'(\omega) + i\overline{\overline{\chi}}''(\omega) \tag{15.33}$$

and make use of Eq. (15.32). We then obtain the following relations between $\overline{\overline{\chi}}'(\omega)$ and $\overline{\overline{\chi}}''(\omega)$:

$$\overline{\overline{\chi}}'(u) = \frac{1}{\pi} P\int_{-\infty}^{\infty} \frac{\overline{\overline{\chi}}''(\omega)}{\omega-u}\,d\omega \tag{15.34}$$

and

$$\bar{\bar{\chi}}''(u) = -\frac{1}{\pi} P \int_{-\infty}^{\infty} \frac{\bar{\bar{\chi}}'(\omega)}{\omega - u} d\omega. \tag{15.35}$$

Eqs. (15.34) and (15.35) are called the Kramers-Kronig relations and enable us to compute the real part of $\bar{\bar{\chi}}(\omega)$ if we know the imaginary part and vice versa. As we shall see, the imaginary part of $\bar{\bar{\chi}}(\omega)$ can be obtained from experiment.

It is useful to generalize the previous derivation to a slightly different matrix:

$$\bar{\bar{\mathbf{f}}}_{+}(z, t) = \frac{e^{izt}\bar{\bar{\chi}}(z)}{z - u} \qquad t > 0 \tag{15.36}$$

and

$$\bar{\bar{\mathbf{f}}}_{-}(z, t) = \frac{e^{-izt}\bar{\bar{\chi}}(z)}{z - u} \qquad t < 0. \tag{15.37}$$

With our choice of sign in the exponential factors, the functions $\bar{\bar{\mathbf{f}}}_{\pm}(z, t)$ will still go to zero as $\epsilon \to \infty$. We then obtain

$$e^{iut}\bar{\bar{\chi}}(u) = \frac{1}{i\pi} P \int_{-\infty}^{\infty} \frac{e^{i\omega t}\bar{\bar{\chi}}(\omega)\, d\omega}{\omega - u} \qquad t > 0 \tag{15.38}$$

and

$$e^{-iut}\bar{\bar{\chi}}(u) = \frac{1}{i\pi} P \int_{-\infty}^{\infty} \frac{e^{-i\omega t}\bar{\bar{\chi}}(\omega)\, d\omega}{\omega - u} \qquad t < 0. \tag{15.39}$$

Eqs. (15.38) and (15.39) will be useful later.

The Kramers-Kronig relations give us a relation between the real and imaginary parts of the response matrix. Therefore, it is only necessary to find one or the other to determine the response $\langle \boldsymbol{\alpha}(t) \rangle$. We can show this explicitly as follows. The Kramers-Kronig relations allow us to rewrite the response matrix, $\bar{\bar{\chi}}(\omega)$, as

$$\bar{\bar{\chi}}(\omega) = \lim_{\eta \to 0} \int \frac{d\omega'}{\pi} \frac{\bar{\bar{\chi}}''(\omega')}{\omega' - \omega - i\eta} \tag{15.40}$$

where we have used the identity

$$\lim_{\eta \to 0} \frac{1}{\omega' - \omega \mp i\eta} = P\left(\frac{1}{\omega' - \omega}\right) \pm i\pi\, \delta(\omega' - \omega). \tag{15.41}$$

The response $\langle \boldsymbol{\alpha}(t) \rangle$ then becomes

$$\begin{aligned} \langle \boldsymbol{\alpha}(t) \rangle &= \frac{1}{2\pi} \int d\omega\, e^{i\omega t} \bar{\bar{\chi}}(\omega) \cdot \tilde{\mathbf{F}}(\omega) \\ &= \lim_{\eta \to 0} \frac{2}{(2\pi)^2} \int_{-\infty}^{\infty} dt' \int_{-\infty}^{\infty} d\omega \int_{-\infty}^{\infty} d\omega' \frac{e^{-i\omega(t - t')}}{\omega' - \omega - i\eta} \bar{\bar{\chi}}''(\omega') \cdot \mathbf{F}(t'). \end{aligned} \tag{15.42}$$

In Eq. (15.42), we have used Eq. (15.40) and have Fourier transformed $\tilde{\mathbf{F}}(\omega)$. If we now make a change of variables $\omega'' = \omega - \omega'$ and introduce the following spectral representation of the Heaviside function,

$$\theta(t - t') = -\int_{-\infty}^{\infty} \frac{d\omega}{(2\pi i)} \frac{e^{-i\omega(t-t')}}{\omega + i\eta}, \tag{15.43}$$

we obtain

$$\langle \boldsymbol{\alpha}(t) \rangle = 2i \int_{-\infty}^{t} dt' \overline{\overline{\mathbf{K}}}''(t - t') \cdot \mathbf{F}(t') \tag{15.44}$$

where $\overline{\overline{\mathbf{K}}}''(t - t')$ is the Fourier transform of $\overline{\overline{\chi}}''(\omega)$. Thus, we have succeeded in writing the response entirely in terms of the imaginary part of the response matrix. The expression for the linear response given in Eq. (15.44) is the one most commonly seen in the literature.

Everything we have done to this point is completely general. Let us now obtain an explicit expression for the response for the case of a constant force which acts for an infinite length of time and then is abruptly shut off. The force we consider has the form

$$\mathbf{F}(t) = \begin{cases} \mathbf{F} & \text{for } t < 0 \\ 0 & \text{for } t > 0 \end{cases}. \tag{15.45}$$

The Fourier transform of the force is

$$\begin{aligned} \tilde{\mathbf{F}}(\omega) &= \mathbf{F} \int_{-\infty}^{0} e^{i\omega t}\, dt = \lim_{\epsilon \to 0} \mathbf{F} \int_{-\infty}^{0} e^{izt}\, dt \\ &= \lim_{\epsilon \to 0} \mathbf{F} \frac{1}{iz} = (-i) \lim_{\epsilon \to 0} \mathbf{F} \left[\frac{\omega}{\omega^2 + \epsilon^2} + \frac{i\epsilon}{\omega^2 + \epsilon^2} \right]. \end{aligned} \tag{15.46}$$

If we now use the following definitions,

$$P\left(\frac{1}{\omega}\right) = \lim_{\epsilon \to 0} \frac{\omega}{\omega^2 + \epsilon^2} \tag{15.47}$$

and

$$\delta(\omega) = \lim_{\epsilon \to 0} \frac{\epsilon}{\omega^2 + \epsilon^2}, \tag{15.48}$$

we obtain

$$\tilde{\mathbf{F}}(\omega) = \mathbf{F} \left[P\left(\frac{1}{i\omega}\right) + \pi\, \delta(\omega) \right]. \tag{15.49}$$

From Eqs. (15.19) and (15.22) the response can be written in the form

$$\langle \boldsymbol{\alpha}(t) \rangle_F = \frac{1}{2\pi} \int_{-\infty}^{\infty} d\omega\, e^{-i\omega t} \overline{\overline{\chi}}(\omega) \cdot \tilde{\mathbf{F}}(\omega). \tag{15.50}$$

Then Eqs. (15.38) and (15.39) yield, for $t < 0$, the expression

$$\overline{\overline{\chi}}(0) = \frac{1}{i\pi} P \int_{-\infty}^{\infty} d\omega \, \frac{e^{-i\omega t}\overline{\overline{\chi}}(\omega)}{\omega}, \tag{15.51}$$

and for $t > 0$, the expression

$$\overline{\overline{\chi}}(0) = \frac{1}{i\pi} P \int_{-\infty}^{\infty} d\omega \, \frac{e^{i\omega t}\overline{\overline{\chi}}(\omega)}{\omega}. \tag{15.52}$$

If we combine Eqs. (15.49), (15.50), and (15.52), the response for $t < 0$ is

$$\langle \boldsymbol{\alpha}(t) \rangle_F = \overline{\overline{\chi}}(0) \cdot \mathbf{F} \tag{15.53}$$

and for $t > 0$

$$\langle \boldsymbol{\alpha}(t) \rangle_F = \frac{1}{i\pi} P \int_{-\infty}^{\infty} d\omega \, \frac{\overline{\overline{\chi}}(\omega) \cdot \mathbf{F}}{\omega} \cos(\omega t). \tag{15.54}$$

Thus, while the force is turned on, the response is constant. When it is turned off, the response becomes time dependent. The variables $A_1, \ldots, A_n$ begin to decay back to their equilibrium values.

D. FLUCTUATION-DISSIPATION THEOREM[3]

The fluctuation-dissipation theorem is extremely important because it relates the response matrix to the correlation matrix for equilibrium fluctuations. As a result, the external field can be used as a probe of equilibrium fluctuations.

To derive the fluctuation-dissipation theorem, let us consider a system to which a constant force is applied from $t = -\infty$ to $t = 0$ and switched off at $t = 0$. We first write the response, $\langle \boldsymbol{\alpha}(t) \rangle_F$, for times $t \geqslant 0$ in terms of the conditional average $\langle \boldsymbol{\alpha}(t) \rangle_{\alpha_0}$,

$$\langle \boldsymbol{\alpha}(t) \rangle_F = \int d\boldsymbol{\alpha}_0 f(\boldsymbol{\alpha}_0, F) \langle \boldsymbol{\alpha}(t) \rangle_{\boldsymbol{\alpha}_0} \qquad \text{for } t \geqslant 0 \tag{15.55}$$

where $f(\boldsymbol{\alpha}_0, \mathbf{F})$ is the distribution function for fluctuations $\boldsymbol{\alpha}_0$ at time $t = 0$ in the presence of a constant external field, $\mathbf{F}$. For times $t > 0$, the field is no longer turned on and we can write

$$\langle \boldsymbol{\alpha}(t) \rangle_{\boldsymbol{\alpha}_0} = e^{-\mathbb{M}t} \cdot \boldsymbol{\alpha}_0 \qquad \text{for } t \geqslant 0 \tag{15.56}$$

(cf. Eq. (14.13)). Combining Eqs. (15.55) and (15.56), we obtain

$$\begin{aligned} \langle \boldsymbol{\alpha}(t) \rangle_F &= e^{-\mathbb{M}t} \cdot \int d\boldsymbol{\alpha}_0 f(\boldsymbol{\alpha}_0, \mathbf{F}) \boldsymbol{\alpha}_0 = e^{-\mathbb{M}t} \cdot \langle \boldsymbol{\alpha}(0) \rangle_F \\ &= e^{-\mathbb{M}t} \cdot \overline{\overline{\chi}}(0) \cdot \mathbf{F} = \frac{1}{i\pi} P \int_{-\infty}^{\infty} d\omega \cos \omega t \, \frac{\overline{\overline{\chi}}(\omega)}{\omega} \cdot \mathbf{F} \end{aligned} \tag{15.57}$$

where we have used Eqs. (15.53) and (15.54). Thus, we find

$$e^{-\mathbb{M}t}\cdot\bar{\bar{\chi}}(0) = \frac{1}{i\pi} P\int_{-\infty}^{\infty} d\omega \cos \omega t \, \frac{\bar{\bar{\chi}}(\omega)}{\omega}. \tag{15.58}$$

If we remember that

$$\bar{\bar{\mathbf{C}}}_{\alpha\alpha}(t) = \langle \boldsymbol{\alpha}(t)\boldsymbol{\alpha} \rangle = k_B\, e^{-\mathbb{M}|t|}\cdot\bar{\bar{\mathbf{g}}}^{-1}, \tag{15.59}$$

we may combine Eqs. (15.58) and (15.59) to obtain

$$\bar{\bar{\mathbf{C}}}_{\alpha\alpha}(t) = \frac{k_B}{i\pi} P\int_{-\infty}^{\infty} d\omega \cos(\omega t)\, \frac{\bar{\bar{\chi}}(\omega)}{\omega}\cdot\bar{\bar{\chi}}^{-1}(0)\cdot\bar{\bar{\mathbf{g}}}^{-1} \tag{15.60}$$

for $t > 0$. Thus, we have obtained a relation between the response matrix, $\bar{\bar{\chi}}(\omega)$, and the equilibrium correlation function for fluctuations.

We can simplify Eq. (15.60) if we can show that $\bar{\bar{\chi}}(0) = (1/T)\bar{\bar{\mathbf{g}}}^{-1}$. The external field, $\mathbf{F}$, does work on the system and increases its internal energy by an amount

$$dU = \mathbf{F}\cdot d\boldsymbol{\alpha}. \tag{15.61}$$

We can expand the differential of the entropy dS and use internal energy and state variables $\boldsymbol{\alpha}$ as independent variables,

$$dS = \left(\frac{\partial S}{\partial U}\right)_{\boldsymbol{\alpha}} dU + \left(\frac{\partial S}{\partial \boldsymbol{\alpha}}\right)_U \cdot d\boldsymbol{\alpha}. \tag{15.62}$$

But $(\partial S/\partial U)_{\boldsymbol{\alpha}} = 1/T$ and $(\partial S/\partial \boldsymbol{\alpha})_U = -\mathbf{X} = -\bar{\bar{\mathbf{g}}}^{-1}\cdot\boldsymbol{\alpha}$. Therefore, we can re-write dS as

$$dS = \left(\frac{\mathbf{F}}{T} - \bar{\bar{\mathbf{g}}}\cdot\boldsymbol{\alpha}\right)\cdot d\boldsymbol{\alpha}. \tag{15.63}$$

For a constant force, $\langle\boldsymbol{\alpha}\rangle = \bar{\bar{\chi}}(0)\cdot\mathbf{F}$ is the expectation value of $\langle\boldsymbol{\alpha}\rangle$ rather than zero, and the entropy will have its maximum for $\boldsymbol{\alpha} = \bar{\bar{\chi}}(0)\cdot\mathbf{F}$. Thus, from Eq. (15.63) we have

$$\left(\frac{\partial S}{\partial \boldsymbol{\alpha}}\right) = \left(\frac{\mathbf{F}}{T} - \bar{\bar{\mathbf{g}}}\cdot\boldsymbol{\alpha}\right), \tag{15.64}$$

and the condition that entropy be maximum at $\boldsymbol{\alpha} = \bar{\bar{\chi}}(0)\cdot\mathbf{F}$ yields

$$\left(\frac{\partial S}{\partial \boldsymbol{\alpha}}\right)_{\boldsymbol{\alpha}=\bar{\bar{\chi}}(0)\cdot\mathbf{F}} = \frac{1}{T}\mathbf{F} - \bar{\bar{\mathbf{g}}}\cdot\bar{\bar{\chi}}(0)\cdot\mathbf{F} = 0 \tag{15.65}$$

or

$$\bar{\bar{\chi}}(0) = \frac{1}{T}\bar{\bar{\mathbf{g}}}^{-1}. \tag{15.66}$$

Combining Eqs. (15.60) and (15.66), we obtain

$$\bar{\bar{\mathbf{C}}}_{\alpha\alpha}(\tau) = \frac{k_B T}{i\pi} P \int_{-\infty}^{\infty} d\omega \, \frac{\bar{\bar{\boldsymbol{\chi}}}(\omega)}{\omega} \cos(\omega t). \tag{15.67}$$

Eq. (15.67) is the famous fluctuation-dissipation theorem. It gives a relation between the linear response function and the correlation function for equilibrium fluctuations.

E. POWER ABSORPTION

The work done by an external force $\mathbf{F}$ to change $\boldsymbol{\alpha}$ by an amount $d\boldsymbol{\alpha}$ is

$$đW = -\mathbf{F} \cdot d\boldsymbol{\alpha} \tag{15.68}$$

(this is work done on the medium). The average rate at which work is done on the medium is just the power $P(t)$ absorbed by the medium:

$$P(t) = \left\langle \frac{đW}{dt} \right\rangle_F = -\mathbf{F}(t) \cdot \langle \dot{\boldsymbol{\alpha}}(t) \rangle_F = -\mathbf{F}(t) \cdot \frac{d}{dt} \int_{-\infty}^{\infty} dt' \bar{\bar{\mathbf{K}}}(t - t') \cdot \mathbf{F}(t'). \tag{15.69}$$

If we write the right-hand side in terms of Fourier transforms $\bar{\bar{\boldsymbol{\chi}}}(\omega)$ and $\tilde{\mathbf{F}}(\omega)$, we obtain

$$P(t) = i\left(\frac{1}{2\pi}\right)^2 \int_{-\infty}^{\infty} d\omega \int_{-\infty}^{\infty} d\omega' \omega' \tilde{\mathbf{F}}(\omega) \cdot \bar{\bar{\boldsymbol{\chi}}}(\omega') \cdot \tilde{\mathbf{F}}(\omega') \, e^{-i(\omega + \omega')t}. \tag{15.70}$$

We can now compute the power absorbed and the total energy absorbed for various types of external forces.

1. Delta Function Force

Let us assume that at time $t = 0$ a delta function force is applied. Then,

$$\mathbf{F}(t) = \mathbf{F} \, \delta(t) \tag{15.71}$$

and

$$\tilde{\mathbf{F}}(\omega) = \mathbf{F}. \tag{15.72}$$

Substituting into Eq. (15.70), we obtain

$$P(t) = i\left(\frac{1}{2\pi}\right)^2 \int_{-\infty}^{\infty} d\omega \int_{-\infty}^{\infty} d\omega' \omega' \bar{\bar{\boldsymbol{\chi}}}(\omega') : \mathbf{FF} \, e^{-i(\omega + \omega')t}. \tag{15.73}$$

(*Note:* $\mathbf{F}\cdot\overline{\overline{\chi}}(\omega)\cdot\mathbf{F} = \overline{\overline{\chi}}(\omega):\mathbf{FF}$.) We can find the total energy absorbed by integrating over all times:

$$W_{\text{abs}} = \int_{-\infty}^{\infty} P(t)\,dt = -\left(\frac{1}{2\pi}\right)\int_{-\infty}^{\infty} d\omega\omega\overline{\overline{\chi}}''(\omega):\mathbf{FF} \qquad (15.74)$$

where $\overline{\overline{\chi}}''(\omega)$ is the imaginary part of the response matrix. Since the total energy absorbed must be a real quantity, only the imaginary part of $\overline{\overline{\chi}}(\omega)$ contributes.

2. Oscillating Force

Now let us consider a monochromatic oscillating force of the form

$$\mathbf{F}(t) = \mathbf{F}\cos\omega_0 t = \tfrac{1}{2}\mathbf{F}(e^{i\omega_0 t} + e^{-i\omega_0 t}). \qquad (15.75)$$

Then

$$\tilde{\mathbf{F}}(\omega) = \pi\mathbf{F}(\delta(\omega + \omega_0) + \delta(\omega - \omega_0)). \qquad (15.76)$$

From Eqs. (15.73) and (15.76), the power absorbed can be written

$$P(t) = -\tfrac{1}{4}[(-i\omega_0)(e^{-i2\omega_0 t} + 1)\overline{\overline{\chi}}(\omega_0) + (i\omega_0)(e^{i2\omega_0 t} + 1)\overline{\overline{\chi}}(-\omega_0)]. \qquad (15.77)$$

As we can see, the instantaneous power absorption oscillates in time. We can find the average power absorbed by taking the time average of Eq. (15.77) over one period of oscillation:

$$\langle\mathbf{P}(t)\rangle = \frac{\omega_0}{\pi}\int_0^{\pi/\omega_0} dtP(t) = \frac{i\omega_0}{4}[\overline{\overline{\chi}}(\omega_0) - \overline{\overline{\chi}}(-\omega_0)]:\mathbf{FF} = \frac{\omega_0}{2}\overline{\overline{\chi}}''(\omega_0):\mathbf{FF} \qquad (15.78)$$

where $\overline{\overline{\chi}}''(\omega_0)$ is the imaginary part of $\overline{\overline{\chi}}(\omega_0)$. Again we see that the average power absorbed depends on the imaginary part of the response matrix. In principle, the average power absorbed can be measured and, therefore, $\overline{\overline{\chi}}''(\omega_0)$ can be measured for all ω_0. The Kramers-Kronig relations allow us to obtain the real part of $\overline{\overline{\chi}}(\omega_0)$ once we know $\overline{\overline{\chi}}''(\omega_0)$. The fluctuation-dissipation theorem relates $\overline{\overline{\chi}}(\omega)$ to the correlation matrix $\overline{\overline{\mathbf{C}}}_{\alpha\alpha}(\tau)$ for equilibrium fluctuations and, therefore, also relates $\overline{\overline{\chi}}(\omega)$ to the spectral density matrix, $\overline{\overline{\mathbf{S}}}_{\alpha\alpha}(\omega)$, of equilibrium fluctuations. Thus, by applying a weak external field to a system, we can probe the equilibrium fluctuations. In the next section, we will apply this theory to a very simple system, namely, that of a harmonically bound Brownian particle in a medium.

F. HARMONICALLY BOUND BROWNIAN PARTICLE[4]

Let us consider a Brownian particle harmonically bound to a wall and immersed in a medium which can exert a force, $F_{\text{med}}(t)$, on it which opposes its

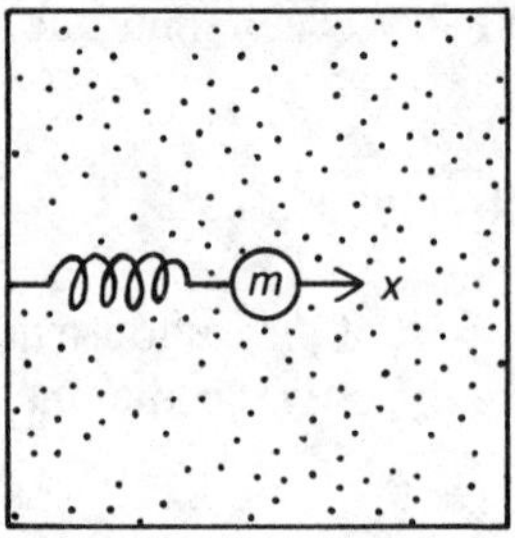

Fig. 15.2. Harmonically bound Brownian particle.

motion (cf. Fig. 15.2). We assume that the mass of the particle is much greater than that of the particles composing the medium. The force, $F_{med}(t)$, can be divided into a completely random part $F_{ran}(t)$ whose average value is zero, $\langle F_{ran}(t)\rangle \equiv 0$, and a frictional part $F_{fric}(t)$ whose average value is proportional to the average velocity of the Brownian particle, $\langle F_{fric}(t)\rangle = -m\gamma\langle\dot{x}(t)\rangle$ (γ is the friction constant). The force, $F_{ran}(t)$, is due to density fluctuations in the medium.

Let us consider one-dimensional motion of the Brownian particle and assume its equilibrium position is at $x = 0$. The equation for fluctuations $x(t)$ about the equilibrium position is

$$m\ddot{x}(t) + m\omega_0{}^2x(t) = F_{med}(t) = F_{fric}(t) + F_{ran}(t), \qquad (15.79)$$

where ω_0 is the natural frequency of the oscillator. In equilibrium, $\langle\ddot{x}(t)\rangle_{eq} = \langle x(t)\rangle_{eq} = \langle F_{med}(t)\rangle = 0$.

We can drive the Brownian particle by applying a time-dependent external force, $F_{ex}(t)$, to it. Eq. (15.79) becomes

$$m\ddot{x}(t) + m\omega_0{}^2x(t) - F_{med}(t) = F_{ex}(t). \qquad (15.80)$$

In the presence of the external force the Brownian particle will be driven away from its equilibrium position and $\langle x(t)\rangle \neq 0$. The equation for the average deviation from equilibrium becomes

$$m\langle\ddot{x}(t)\rangle_F + m\omega_0{}^2\langle x(t)\rangle_F + m\gamma\langle\dot{x}(t)\rangle_F = F_{ex}(t). \qquad (15.81)$$

We will assume that the average response of the Brownian particle to the external force is linear,

$$\langle x(t)\rangle_F = \int_{-\infty}^{\infty} K(t - t')F_{ex}(t')\,dt', \qquad (15.82)$$

and causal so that $K(t - t') \neq 0$ for $t > t'$ but $K(t - t') = 0$ for $t < t'$; $K(t - t')$ is the response function.

The response function is independent of the type of force we apply. Therefore, we can use the simplest type of force to obtain an expression for $\chi(t - t')$. If we apply a delta function force at time $t = 0$,

$$F_{ex}(t) = F\,\delta(t), \tag{15.83}$$

the response is

$$\langle x(t)\rangle_F = K(t)F. \tag{15.84}$$

Combining Eqs. (15.81) and (15.84), we obtain the following equation for $K(t)$:

$$m\ddot{K}(t) + m\omega_0^2K(t) + m\gamma\dot{K}(t) = \delta(t). \tag{15.85}$$

If we take the Fourier transform of Eq. (15.85) we find

$$-m\omega^2\chi(\omega) + m\omega_0^2\chi(\omega) - i\omega\gamma m\chi(\omega) = 1 \tag{15.86}$$

and

$$\chi(\omega) = \frac{1}{m(-\omega^2 + \omega_0^2 - i\gamma\omega)}. \tag{15.87}$$

We can divide $\chi(\omega)$ into real and imaginary parts,

$$\chi(\omega) = \chi'(\omega) + i\chi''(\omega). \tag{15.88}$$

Then we find

$$\chi'(\omega) = \frac{\omega_0^2 - \omega^2}{m((\omega_0^2 - \omega^2)^2 + \gamma^2\omega^2)} \tag{15.89}$$

and

$$\chi''(\omega) = \frac{\gamma\omega}{m((\omega_0^2 - \omega^2)^2 + \gamma^2\omega^2)}. \tag{15.90}$$

Note that, if there is no friction, $\gamma = 0$, there will be no absorption of energy from the particle by the medium because γ will be zero.

From Eq. (15.87), we obtain

$$K(t) = \frac{1}{2\pi}\int_{-\infty}^{\infty} d\omega\chi(\omega)e^{i\omega t} = \frac{1}{2\pi}\int_{-\infty}^{\infty} d\omega\,\frac{e^{i\omega t}}{(-\omega^2 + \omega_0^2 - i\gamma)m}. \tag{15.91}$$

Notice that the poles of $\chi(\omega)$ occur at $\omega = -i\gamma \pm \sqrt{4\omega_0^2 - \gamma^2}$ in the *lower half plane*; $\chi(\omega)$ is well behaved in the upper half plane.

For $t > 0$, we can write Eq. (15.91) as a contour integral, but we must close the contour in the lower half plane so that the integral will be convergent. We then pick up the two poles and get a nonzero result. For $t < 0$ we must close the contour in the upper half plane, but since there are no poles there, we get zero. Therefore, we find

$$K(t) = \frac{\theta(t)\,e^{-(1/2)\gamma t}\sin\{[\omega_0^2 - \frac{1}{4}\gamma^2]^{1/2}t\}}{m[\omega_0^2 - \frac{1}{4}\gamma^2]^{1/2}} \tag{15.92}$$

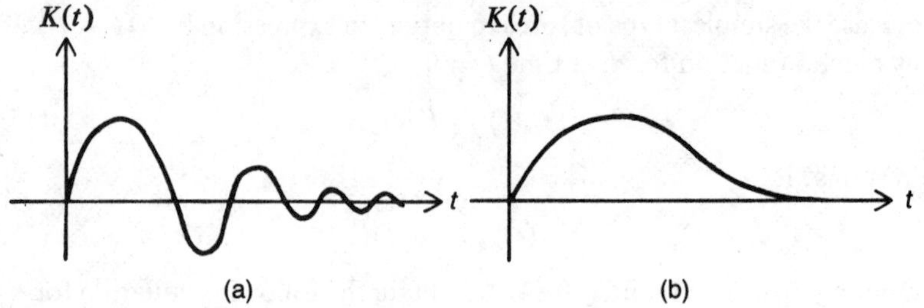

Fig. 15.3. Response function for a harmonically bound Brownian particle: (*a*) underdamped case; (*b*) overdamped case.

where $\theta(t)$ is the Heaviside function ($\theta(t) = 1, t > 0$ and $\theta(t) = 0, t < 0$). We can plot the response for the case $\omega_0 > \gamma/2$ and $\gamma_0 < \gamma/2$ (cf. Fig. 15.3*a* and *b*). If we hit the oscillator with a delta function force, it is displaced from equilibrium and slowly decays back to its equilibrium value. The expression for $K(t)$ obtained in Eq. (15.92) depends only on properties of the medium and not on the type of applied force. We can now substitute Eq. (15.92) back into Eq. (15.82) and find the response to any type of applied force.

From Eq. (15.90) we find immediately that the total energy absorbed by the medium for the case of a delta function force is

$$W_{\text{abs}} = -\frac{1}{(2\pi)}\int_{-\infty}^{\infty} d\omega\, \frac{\gamma F^2\omega^2}{m((\omega_0{}^2 - \omega^2)^2 + \gamma^2\omega^2)}. \tag{15.93}$$

If we invoke the fluctuation-dissipation theorem, we can write down an expression for the correlation function for fluctuations of the Brownian particle about its equilibrium position in the absence of an external field. Combining Eqs. (15.67) and (15.87) we find

$$\langle xx(t)\rangle_{\text{eq}} = \frac{k_B T}{m\pi i} P\int_{-\infty}^{\infty} \frac{\cos \omega t\, d\omega}{\omega(\omega_0{}^2 - \omega^2 - \gamma i\omega)}. \tag{15.94}$$

Thus, we have expressed the correlation function for equilibrium fluctuations in terms of the friction parameter, γ, and the natural frequency, ω_0, of the oscillator.

G. LIGHT SCATTERING[5–10]

The equilibrium fluctuations in a system can be probed by an external field. Since the equilibrium fluctuations in a fluid decay, on the average, according to the laws of hydrodynamics, we should be able to obtain information about transport coefficients simply by scattering light or particles off fluctuations in a fluid at equilibrium. The fact that this can be done was established by Landau and Placzek[5] and by Van Hove.[6]

In 1934, Landau and Placzek showed that it is possible to obtain the time dependence of the density autocorrelation function for a fluid from the hydrodynamic equations. Later, Van Hove, in a very important paper, was able to relate the differential cross-section for scattering of neutrons from a fluid to the density autocorrelation function of the fluid, thus establishing the connection. In this section we shall obtain an expression for the intensity of light scattered from a fluid in terms of the hydrodynamic modes. We will begin with a phenomenological discussion of light scattering and then obtain the phenomenological results from a more rigorous theory which relates the intensity of scattered light to the density correlation function and, therefore, to the hydrodynamic equations.

1. Phenomenology of Light Scattering[9]

A light wave incident on a simple fluid induces oscillating dipoles in the particles of the fluid. These oscillating dipoles re-emit spherical light waves. If the medium is homogeneous, all scattered light waves cancel except those in the forward direction. However, if density fluctuations exist in the fluid, light will be scattered in directions other than the forward direction.

Density fluctuations are both thermal and mechanical in origin:

$$\Delta n = \left(\frac{\partial n}{\partial P}\right)_s \Delta P + \left(\frac{\partial n}{\partial S}\right)_P \Delta S = \left(\frac{\partial n}{\partial P}\right)_T \Delta P + \left(\frac{\partial n}{\partial T}\right)_P \Delta T. \quad (15.95)$$

Thermal density fluctuations result from damped temperature or entropy waves, while mechanical density fluctuations result from sound waves in the fluid. Light which scatters from thermal density fluctuations will be unshifted in frequency, while light which scatters from the sound waves will undergo a frequency shift (Doppler shift).

In general, there will be sound waves with a wide range of wave vectors and frequencies in the fluid. However, for a given scattering angle, θ, only selected sound modes will contribute to the scattering. Light will scatter from the wave fronts of the sound waves and must satisfy the Bragg condition (cf. Fig. 15.4),

$$2\lambda_s \sin\frac{\theta}{2} = \lambda_0 \quad (15.96)$$

where λ_s is the wave length of the sound and λ_0 is the wave length of the incident light. The wave vector of the light wave will be shifted by an amount

$$\Delta k = k - k_0 = 2k_0 \sin\frac{\theta}{2}. \quad (15.97)$$

The frequency of the scattered light will be Doppler shifted by

$$\Omega = \omega - \omega_0 = \pm\frac{2\omega_0 v_s}{c}\sin\frac{\theta}{2} = \pm v_s \Delta k \tag{15.98}$$

where v_s is the speed of sound in the fluid and c is the speed of light.

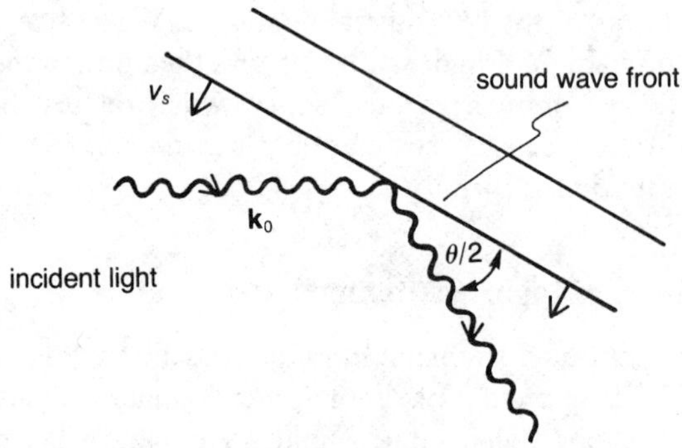

Fig. 15.4. Light scattered from sound wave front.

We can use thermodynamic arguments to find the ratio between the intensity of light scattered by thermal density fluctuations, I_{th}, and that scattered by mechanical density fluctuations, I_{mech}.

We may write

$$\frac{I_{\text{th}}}{I_{\text{mech}}} = \frac{\langle(\Delta n)^2\rangle_{\text{th}}}{\langle(\Delta n)^2\rangle_{\text{mech}}} = \frac{\left(\frac{\partial n}{\partial S}\right)_P^2 \langle(\Delta S)^2\rangle}{\left(\frac{\partial n}{\partial P}\right)_s^2 \langle(\Delta P)^2\rangle} = \frac{\left(\frac{\partial V}{\partial S}\right)_P^2 \langle(\Delta S)^2\rangle}{\left(\frac{\partial V}{\partial P}\right)_s^2 \langle(\Delta P)^2\rangle} \tag{15.99}$$

where $n = N/V$. From thermodynamic fluctuation theory (cf. Sec. 10.B) we obtain the following expressions for the average entropy and pressure fluctuations:

$$\langle(\Delta S)^2\rangle = k_B C_p = k_B T\left(\frac{\partial S}{\partial T}\right)_P \tag{15.100}$$

and

$$\langle(\Delta P)^2\rangle = \frac{k_B T}{V\kappa_s} = -k_B T\left(\frac{\partial P}{\partial V}\right)_s \tag{15.101}$$

where k_B is Boltzmann's constant, C_p is the constant pressure heat capacity, and κ_s is the adiabatic compressibility. If we make use of the chain rule and Maxwell's relations, we may write

$$\frac{I_{\text{th}}}{I_{\text{mech}}} = \left(\frac{\partial P}{\partial S}\right)_V \left(\frac{\partial S}{\partial P}\right)_T = \frac{C_p - C_V}{C_V} = \frac{\kappa_T - \kappa_s}{\kappa_s}, \tag{15.102}$$

where C_V is the constant volume heat capacity and κ_T is the isothermal compressibility.

If we make a plot of the scattered intensity versus shift in frequency, we would expect to find a plot of the form given in Fig. 15.5. There will be one peak in intensity centered at the unshifted frequency. This is called the Rayleigh line. And we will find two peaks with shifted frequency. These are called the Brillouin lines. Such a distribution of intensity is observed experimentally. Since C_p and $\kappa_T \to \infty$ at the critical point, we expect the Rayleigh peak to become very large as we approach the critical point.

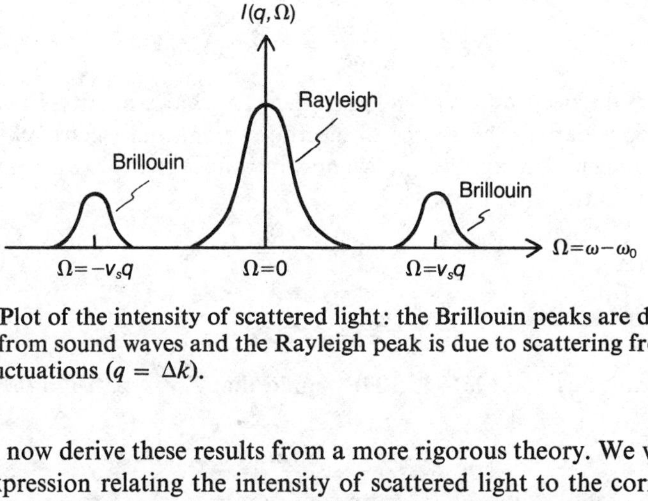

Fig. 15.5. Plot of the intensity of scattered light: the Brillouin peaks are due to scattering from sound waves and the Rayleigh peak is due to scattering from entropy fluctuations ($q = \Delta k$).

We will now derive these results from a more rigorous theory. We will first find an expression relating the intensity of scattered light to the correlation function for density fluctuations in the fluid. Then we will find an expression for the density correlation function from the hydrodynamic equations.

2. Intensity of Scattered Light[7]

We wish to find an expression for the intensity of light scattered from a simple fluid of identical particles. We shall assume that a monochromatic light wave,

$$\mathbf{E}(\mathbf{r}, t) = \mathbf{E}_0\, e^{i(\mathbf{k}_0 \cdot \mathbf{r} - \omega_0 t)}, \tag{15.103}$$

impinges on the fluid, and that the fluid particles have a polarizability α. The induced polarization gives rise to a dipole moment density,

$$\mathbf{P}(\mathbf{r}, t) = \alpha \mathbf{E}(\mathbf{r}, t) \sum_{i=1}^{N} \delta(\mathbf{r} - \mathbf{R}_i(t)) \tag{15.104}$$

where $\mathbf{R}_i(t)$ is the position of the ith particle at time t, which enables the external field to couple to the particles in the fluid. In writing Eq. (15.104) we have neglected multiple scattering effects because we are assuming the polarization is induced entirely by the external field.

The easiest way to obtain the amplitude of the scattered light is to introduce the Hertz potential, $\mathbf{Z}(\mathbf{r}, t)$. This is done in the following way. The electric field $\mathbf{E}(\mathbf{r}, t)$ is defined in terms of the usual scalar potential $\phi(\mathbf{r}, t)$ and vector potential $\mathbf{A}(\mathbf{r}, t)$ as

$$\mathbf{E} = -\nabla_{\mathbf{r}}\phi - \frac{\partial \mathbf{A}}{\partial t}. \tag{15.105}$$

In the Lorentz gauge, the scalar and vector potentials satisfy the equations [8]

$$\nabla_{\mathbf{r}}^2\mathbf{A} - \epsilon\mu\frac{\partial^2\mathbf{A}}{\partial t^2} = -\mu\,\mathbf{J} = -\,\epsilon\mu\frac{\partial \mathbf{P}}{\partial t}, \tag{15.106}$$

where μ is the permeability and ϵ is the permitivity, and

$$\nabla_{\mathbf{r}}^2\phi - \epsilon\mu\frac{\partial^2\phi}{\partial t^2} = -\frac{\rho}{\epsilon} = \nabla_{\mathbf{r}}^2\cdot\mathbf{P}, \tag{15.107}$$

where $\mathbf{J}$ is the electric current and ρ is the electric charge density. For the system we are considering, the source of electric current and electric charge is the dipole moment density, $\mathbf{P}(\mathbf{r}, t)$. We now introduce the Hertz potential $\mathbf{Z}(\mathbf{r}, t)$ by means of the equations

$$\mathbf{A} = \epsilon\mu\frac{\partial \mathbf{Z}}{\partial t} \tag{15.108}$$

and

$$\phi = -\nabla_{\mathbf{r}}\cdot\mathbf{Z}. \tag{15.109}$$

Comparing Eqs. (15.106)–(15.109) we find that $\mathbf{Z}(\mathbf{r}, t)$ satisfies the following equation:

$$\nabla_{\mathbf{r}}^2\mathbf{Z} - \epsilon\mu\frac{\partial^2\mathbf{Z}}{\partial t^2} = -\mathbf{P}. \tag{15.110}$$

Thus, we have an equation of motion for the Hertz potential in which the polarization acts as a source.

Given Eq. (15.110) we now find the Hertz potential for the scattered light wave. The retarded solution to Eq. (15.110) is given by

$$\mathbf{Z}(\mathbf{r}, t) = \frac{1}{4\pi}\int d\mathbf{r}'\int_{-\infty}^{\infty} dt'\,\frac{\mathbf{P}(\mathbf{r}, t')}{|\mathbf{r}' - \mathbf{r}|}\,\delta\!\left(t' - t + \frac{|\mathbf{r}' - \mathbf{r}|}{c}\right) \tag{15.111}$$

(cf. Ref. 15.8). We next introduce the Fourier expansion of the delta function (cf. Eq. (15.23)) and obtain

$$\mathbf{Z}(\mathbf{r}, t) = \frac{1}{2}\left(\frac{1}{2\pi}\right)^2\int d\mathbf{r}'\int_{-\infty}^{\infty} dt'\int_{-\infty}^{\infty} d\omega'\,\frac{\mathbf{P}(\mathbf{r}', t')}{|\mathbf{r}' - \mathbf{r}|}\exp\left\{i\omega'\left(t' - t + \frac{|\mathbf{r}' - \mathbf{r}|}{c}\right)\right\} \tag{15.112}$$

for the outgoing wave. The outgoing electric field is related to $\mathbf{Z}(\mathbf{r}, t)$ through the expression

$$\mathbf{E}(\mathbf{r}, t) = \nabla_{\mathbf{r}}(\nabla_{\mathbf{r}} \cdot \mathbf{Z}) - \epsilon\mu \frac{\partial^2 \mathbf{Z}}{\partial t^2} \tag{15.113}$$

(cf. Eqs. (15.105), (15.107), and (15.108)).

We only need an expression for the scattered light far from the source. If we remember that $\mathbf{r}'$ is restricted to the region of the dipoles and $\mathbf{r}$ is the coordinate of the observer, we can make the following approximation:

$$|\mathbf{r} - \mathbf{r}'| \approx r - \hat{\mathbf{r}} \cdot \mathbf{r}' + \cdots \tag{15.114}$$

where $\hat{\mathbf{r}}$ is the unit vector $\hat{\mathbf{r}} = \mathbf{r}/|\mathbf{r}|$. If we substitute Eqs. (15.104) and (15.114) into Eq. (15.112), we obtain

$$\mathbf{Z}(\mathbf{r}, t) \approx \frac{1}{2}\left(\frac{1}{2\pi}\right)^2 \frac{\alpha \mathbf{E}_0}{r} \int d\mathbf{r}' \int_{-\infty}^{\infty} dt' \int_{-\infty}^{\infty} d\omega'\, e^{i(\mathbf{k}_0 \cdot \mathbf{r}' - \omega_0 t')} \sum_{i=1}^{N} \delta(\mathbf{r}' - \mathbf{R}_i(t'))$$
$$\times \exp\left\{i\omega'\left(t' - t + \frac{r}{c} - \frac{\hat{\mathbf{r}} \cdot \mathbf{r}'}{c}\right)\right\}. \tag{15.115}$$

We next substitute Eq. (15.115) into Eq. (15.113) and neglect terms of order r'/r and smaller. We then obtain the following expression for the scattered electric field:

$$\mathbf{E}'(\mathbf{r}, t) = \frac{1}{2}\left(\frac{1}{2\pi}\right)^2 \frac{\alpha}{c^2 r}(\mathbf{E}_0 - \hat{\mathbf{r}}(\hat{\mathbf{r}} \cdot \mathbf{E}_0)) \int d\mathbf{r}' \int_{-\infty}^{\infty} dt' \int_{-\infty}^{\infty} d\omega' \omega'^2$$
$$\times \exp\{i(\mathbf{k}_0 \cdot \mathbf{r}' - \omega_0 t')\} \exp\left\{i\omega'\left(t' - t + \frac{r}{c} - \frac{1}{c}\hat{\mathbf{r}} \cdot \mathbf{r}'\right)\right\} n(\mathbf{r}', t) \tag{15.116}$$

where

$$n(\mathbf{r}', t) \equiv \sum_{i=1}^{N} \delta(\mathbf{r}' - \mathbf{R}_i(t')) \tag{15.117}$$

is the microscopic density of particles in the medium and we remember that $c^2 = \epsilon_0 \mu_0$ in free space. The spectral intensity of scattered light is defined

$$I(\mathbf{r}, \omega) \equiv \lim_{T \to \infty} \frac{1}{2T} \sqrt{\frac{\epsilon_0}{\mu_0}}\, \mathbf{E}'(\mathbf{r}, \omega; T) \cdot \mathbf{E}'^*(\mathbf{r}, \omega; T) \tag{15.118}$$

where

$$\mathbf{E}'(\mathbf{r}, \omega; T) = \int_{-\infty}^{\infty} dt\, e^{i\omega t} \mathbf{E}'(\mathbf{r}, t) \theta(T - |t|) \tag{15.119}$$

and $\theta(T - |t|)$ is a Heaviside function. If we assume the system is ergodic, we can combine Eqs. (15.118) and (15.119) to obtain

$$I(\mathbf{r}, \omega) = \lim_{T\to\infty} \frac{1}{2T}\sqrt{\frac{\epsilon_0}{\mu_0}} \int_{-\infty}^{\infty} dt \int_{-\infty}^{\infty} dt' \times e^{i\omega(t-t')}\mathbf{E}'(\mathbf{r}, t)\mathbf{E}'^*(\mathbf{r}, t')\theta(T - |t|)\theta(T - |t'|)$$

$$= \frac{1}{2}\sqrt{\frac{\epsilon_0}{\mu_0}} \int_{-\infty}^{\infty} d\tau\, e^{i\omega\tau} \lim_{T\to\infty} \frac{1}{T}\int_{-T}^{T} dt'\mathbf{E}'(\mathbf{r}, t' + \tau)\mathbf{E}'^*(\mathbf{r}, t')$$

$$= \frac{1}{2}\sqrt{\frac{\epsilon_0}{\mu_0}} \int_{-\infty}^{\infty} d\tau\, e^{i\omega\tau}\langle\mathbf{E}'(\mathbf{r}, \tau)\mathbf{E}'^*(\mathbf{r}, 0)\rangle. \qquad (15.120)$$

The average in Eq. (15.120) is taken with respect to a stationary equilibrium ensemble. If we now substitute Eq. (15.116) into Eq. (15.120), we obtain

$$I(\mathbf{r}, \omega) = \frac{1}{8}\sqrt{\frac{\epsilon_0}{\mu_0}} \left(\frac{\alpha}{(2\pi)^2c^2r}\right)^2 E_0^2 \sin^2\phi \int d\mathbf{r}' \int d\mathbf{r}'' \int_{-\infty}^{\infty} dt' \int_{-\infty}^{\infty} dt'' \int_{-\infty}^{\infty} d\omega'$$
$$\times \int_{-\infty}^{\infty} d\omega'' \int_{-\infty}^{\infty} d\tau \omega'^2\omega''^2\, e^{i\omega\tau}\, e^{i(\mathbf{k}_0\cdot\mathbf{r}' - \omega_0 t')}\, e^{-i(\mathbf{k}_0\cdot\mathbf{r}'' - \omega_0 t'')}$$
$$\times \exp\left\{i\omega'\left(t' - \tau + \frac{r}{c} - \frac{1}{c}\hat{\mathbf{r}}\cdot\mathbf{r}'\right)\right\} \exp\left\{-i\omega''\left(t'' + \frac{r}{c} - \frac{1}{c}\hat{\mathbf{r}}\cdot\mathbf{r}''\right)\right\}$$
$$\times \langle n(\mathbf{r}', t')n(\mathbf{r}'', t'')\rangle \qquad (15.121)$$

where we have let $\hat{\mathbf{r}}\cdot\mathbf{E}_0 = E_0\cos\phi$. We next integrate over τ and ω'. This gives us

$$I(\mathbf{r}, \omega) = \frac{1}{16\pi}\sqrt{\frac{\epsilon_0}{\mu_0}} \left(\frac{\alpha}{c^2r2\pi}\right)^2 E_0^2 \sin^2\phi \int d\mathbf{r}' \int d\mathbf{r}'' \int_{-\infty}^{\infty} dt' \int_{-\infty}^{\infty} dt'' \int_{-\infty}^{\infty} d\omega''\omega''^2\omega^2$$
$$\times e^{i(\mathbf{k}_0\cdot\mathbf{r}' - \omega_0 t')}\, e^{-i(\mathbf{k}_0\cdot\mathbf{r}'' - \omega_0 t'')} \exp\left\{i\omega\left(t' + \frac{r}{c} - \frac{1}{c}\hat{\mathbf{r}}\cdot\mathbf{r}'\right)\right\}$$
$$\times \exp\left\{-i\omega''\left(t'' + \frac{r}{c} - \frac{1}{c}\hat{\mathbf{r}}\cdot\mathbf{r}''\right)\right\}\langle n(\mathbf{r}', t')n(\mathbf{r}'', t'')\rangle. \qquad (15.122)$$

We now make the change of variables $\tau = t' - t''$ and note that $\langle n(\mathbf{r}', t')n(\mathbf{r}'', t'')\rangle$ can only depend on the difference $\tau' = t' - t''$. We then can write Eq. (15.122) in the following form:

$$I(\mathbf{r}, \omega) = \frac{1}{8}\sqrt{\frac{\epsilon_0}{\mu_0}} \left(\frac{\alpha}{2\pi rc^2}\right)^2 E_0^2 \sin^2\phi \int d\mathbf{r}' \int d\mathbf{r}'' \int_{-\infty}^{\infty} d\tau\omega^4\, e^{-i(\omega_0 - \omega)\tau}$$
$$\times \exp\left\{i\left(\mathbf{k}_0 - i\frac{\omega}{c}\hat{\mathbf{r}}\right)\cdot(\mathbf{r}' - \mathbf{r}'')\right\}\langle n(\mathbf{r}', \tau)n(\mathbf{r}'')\rangle; \qquad (15.123)$$

or with the change of variables $\boldsymbol{\rho} = \mathbf{r}' - \mathbf{r}''$ we find

$$I(\mathbf{r}, \omega) = \frac{1}{8}\sqrt{\frac{\epsilon_0}{\mu_0}}\left(\frac{\alpha}{2\pi rc^2}\right)^2 E_0^2 \sin^2\phi \int d\boldsymbol{\rho} \int d\mathbf{r}'' \int_{-\infty}^{\infty} d\tau\omega^4$$

$$\times\, e^{-i(\omega_0-\omega)\tau} \exp\left\{i\left(\mathbf{k}_0 - i\frac{\omega}{c}\hat{\mathbf{r}}\right)\cdot\boldsymbol{\rho}\right\}\langle n(\mathbf{r}'' + \boldsymbol{\rho}, \tau)n(\mathbf{r}'')\rangle. \tag{15.124}$$

We next introduce the generalized pair correlation function,

$$C(\boldsymbol{\rho}, \tau) \equiv \frac{1}{N}\int d\mathbf{r}'\langle n(\mathbf{r}' + \boldsymbol{\rho}, \tau)n(\mathbf{r}', 0)\rangle, \tag{15.125}$$

where N is the number of particles in the fluid, and the generalized structure factor (spectral density),

$$S(\mathbf{k}, \Omega) = \int d\boldsymbol{\rho} \int d\tau C(\boldsymbol{\rho}, \tau)\, e^{+i\mathbf{k}\cdot\boldsymbol{\rho}}\, e^{-i\Omega\tau}. \tag{15.126}$$

In terms of these quantities, the intensity may be written

$$I(\mathbf{r}, \omega) = \frac{1}{4} I_0\omega^4\left(\frac{\alpha}{2\pi c^2 r}\right)^2 N \sin^2\phi \int d\boldsymbol{\rho} \int d\tau C(\boldsymbol{\rho}, \tau)$$

$$\times \exp\left\{i\left(\mathbf{k}_0 - \frac{\omega_0}{c}\hat{\mathbf{r}}\right)\cdot\boldsymbol{\rho}\right\} e^{-i(\omega_0-\omega)\tau}$$

$$= \frac{1}{4} I_0\omega^4\left(\frac{\alpha}{2\pi c^2 r}\right)^2 N \sin^2\phi S(\mathbf{k}, \Omega) \tag{15.127}$$

where $I_0 = \frac{1}{2}\sqrt{\frac{\epsilon_0}{\mu_0}}E_0^2$, $\mathbf{k} = \mathbf{k}_0 - (\omega_0/c)\hat{\mathbf{r}}$ and $\Omega = \omega_0 - \omega$.

For a homogeneous system, we may separate the density into an equilibrium part, n_0, and a fluctuation, $\delta n(\mathbf{r}, t)$:

$$n(\mathbf{r}, t) = n_0 + \delta n(\mathbf{r}, t) \tag{15.128}$$

where $\langle\delta n(\mathbf{r}, t)\rangle = 0$; $S(\mathbf{k}, \Omega)$ then becomes

$$S(\mathbf{k}, \Omega) = (2\pi)^2 n_0\, \delta(\mathbf{k})\, \delta(\Omega)$$

$$+ \frac{1}{N}\int d\boldsymbol{\rho} \int d\tau \int d\mathbf{r}'\langle\delta n(\mathbf{r}' + \boldsymbol{\rho}, \tau)\, \delta n(\mathbf{r}', 0)\rangle\, e^{i\mathbf{k}\cdot\boldsymbol{\rho}}\, e^{-i\Omega\tau}. \tag{15.129}$$

The first term in Eq. (15.129) is the contribution due to forward scattering. The second term results from density fluctuations.

3. Hydrodynamic Expressions for Scattered Intensity[10–12]

The long time behavior of the density autocorrelation function $C(\boldsymbol{\rho}, \tau)$ can be obtained from the hydrodynamic equations. The linearized hydrodynamic equations may be written in the form

$$mn_0 \frac{\partial}{\partial t} \mathbf{v}_t(\mathbf{r}, t) = \eta \nabla_{\mathbf{r}}^2 \mathbf{v}_t(\mathbf{r}, t), \tag{15.130}$$

$$\frac{\partial}{\partial t} \delta n(\mathbf{r}, t) + n_0 \nabla_{\mathbf{r}} \cdot \mathbf{v}_l(\mathbf{r}, t) = 0, \tag{15.131}$$

$$mn_0 \frac{\partial}{\partial t} v_l(\mathbf{r}, t) + \frac{mc_0^2}{\gamma} \nabla_{\mathbf{r}}\, \delta n(\mathbf{r}, t) + \frac{mc_0^2 n_0 \alpha_p}{\gamma} \nabla_{\mathbf{r}}\, \delta T(\mathbf{r}, t)$$
$$= (\zeta + \tfrac{4}{3}\eta) \nabla_{\mathbf{r}} (\nabla_{\mathbf{r}} \cdot \mathbf{v}_l(\mathbf{r}, t)), \tag{15.132}$$

and

$$n_0 c_V \frac{\partial}{\partial t} \delta T(\mathbf{r}, t) - \frac{c_V(\gamma - 1)}{\alpha_p} \frac{\partial}{\partial t} \delta n(\mathbf{r}, t) = K \nabla_{\mathbf{r}}^2\, \delta T(\mathbf{r}, t), \tag{15.133}$$

where $\mathbf{v}_t$ is the transverse velocity, $\mathbf{v}_l$ is the longitudinal velocity, m is the mass of the particles,

$$\alpha_p = -\frac{1}{n_0} \left(\frac{\partial n}{\partial T} \right)^0_{P,N}, \qquad \gamma = \frac{c_p}{c_V},$$

c_0 is the speed of sound, and c_V and c_p are the heat capacities per particle at constant volume and pressure, respectively (cf. Sec. 13.F). The transverse part of the average velocity is not coupled to the density. Therefore, it will not contribute to the density correlation function for small fluctuations, and we need only consider Eqs. (15.131)–(15.133). We can eliminate $\mathbf{v}_l(\mathbf{r}, t)$ by taking the divergence of Eq. (15.132) and using Eq. (15.131). We then obtain

$$-m \frac{\partial^2}{\partial t^2} \delta n(\mathbf{r}, t) + \frac{mc_0^2}{\gamma} \nabla_{\mathbf{r}}^2\, \delta n(\mathbf{r}, t) + \frac{(\zeta + \frac{4}{3}\eta)}{n_0} \frac{\partial}{\partial t} \nabla_{\mathbf{r}}^2\, \delta n(\mathbf{r}, t)$$
$$+ \frac{mc_0^2 n_0 \alpha_p}{\gamma} \nabla_{\mathbf{r}}^2\, \delta T(\mathbf{r}, t) = 0 \tag{15.134}$$

and

$$n_0 c_V \frac{\partial}{\partial t} \delta T(\mathbf{r}, t) - \frac{c_V(\gamma - 1)}{\alpha_p} \frac{\partial}{\partial t} \delta n(\mathbf{r}, t) - K \nabla_{\mathbf{r}}^2\, \delta T(\mathbf{r}, t) = 0. \tag{15.135}$$

The function $\langle \delta n(\mathbf{r} + \boldsymbol{\rho}, \tau)\, \delta n(\mathbf{r}, 0) \rangle$ gives the correlation between a density fluctuation at time $t = 0$ and one at time $t = \tau$. To evaluate it we must find $\delta n(\mathbf{r} + \boldsymbol{\rho}, \tau)$, given the value of the fluctuation at time $\tau = 0$.

We first Fourier transform the spatial variables and Laplace transform the time variable according to the equations

$$\delta\tilde{n}(\mathbf{k}, z) = \int d\mathbf{r} \int_0^\infty dt\, e^{-i\mathbf{k}\cdot\mathbf{r}}\, e^{+izt}\, \delta n(\mathbf{r}, t), \tag{15.136}$$

$$-iz\, \delta\tilde{n}(\mathbf{k}, z) = \delta n(\mathbf{k}, 0) + \int_0^\infty dt \left[\frac{\partial}{\partial t}\, \delta n(\mathbf{k}, t)\right] e^{izt}, \tag{15.137}$$

and

$$ik\, \delta n(\mathbf{k}, t) = \int d\mathbf{r}\, e^{-i\mathbf{k}\cdot\mathbf{r}} \nabla_{\mathbf{r}}\, \delta n(\mathbf{r}, t), \tag{15.138}$$

where $z = \omega + i\delta$. Eqs. (15.134) and (15.135) then take the form

$$z^2\, \delta\tilde{n}(\mathbf{k}, z) - \frac{c_0^2}{\gamma} k^2\, \delta\tilde{n}(\mathbf{k}, z) + i\frac{(\zeta + \frac{4}{3}\eta)}{mn_0} k^2 z\, \delta\tilde{n}(\mathbf{k}, z) - \frac{c_0^2 n_0 \alpha_p}{\gamma} k^2\, \delta\tilde{T}(\mathbf{k}, z) = \left(iz - \frac{(\zeta + \frac{4}{3}\eta)}{mn_0} k^2\right) \delta n(\mathbf{k}, 0) \tag{15.139}$$

and

$$\left[\frac{izc_V(\gamma - 1)}{\alpha_p}\right] \delta\tilde{n}(\mathbf{k}, z) + [-in_0 c_V z + Kk]\, \delta\tilde{T}(\mathbf{k}, z) = -\frac{c_V(\gamma - 1)}{\alpha_p}\, \delta n(\mathbf{k}, 0) + n_0 c_V\, \delta T(\mathbf{k}, 0), \tag{15.140}$$

respectively. In Eqs. (15.139) and (15.140), $\delta\tilde{n}$ and $\delta\tilde{T}$ denote Laplace transformed quantities. We have assumed that $(\partial/\partial t)\, \delta n(\mathbf{k}, t = 0) = 0$. This will be true if $\mathbf{v}_l(\mathbf{k}, t = 0) = 0$ initially. We now solve Eqs. (15.139) and (15.140) for $\delta\tilde{n}(\mathbf{k}, z)$ and obtain

$$\delta\tilde{n}(\mathbf{k}, z) = \delta n(\mathbf{k}, 0) \left(\frac{-z^2 - i(a + b)k^2 z + abk^4 + c_0^2\left(1 - \frac{1}{\gamma}\right)k^2}{iz^3 - (a + b)k^2 z - i(c_0^2 k^2 + abk^4)z + \frac{ac_0^2 k^4}{\gamma}}\right). \tag{15.141}$$

We have neglected the term involving $\delta T(\mathbf{k}, 0)$ because in evaluating the density correlation function it gives zero contribution since $\langle \delta n(\mathbf{k}, 0)\, \delta T(\mathbf{k}, 0)\rangle \equiv 0$. Also we have used the notation $a = K/n_0 c_V$ and $b = (\zeta + 4\eta/3)/mn_0$.

We can now find the time dependence of the density fluctuation by taking the inverse transform of Eq. (15.141):

$$\delta n(\mathbf{k}, t) = \frac{1}{2\pi} \int_{-\infty + i\delta}^{\infty + i\delta} dz\, e^{-izt}\, \delta\tilde{n}(\mathbf{k}, z). \tag{15.142}$$

To do this we must find the poles of Eq. (15.141). It is possible to find approximate expressions for the poles by expanding in terms of the quantities ak/c_0 and bk/c_0. For typical light scattering experiments these quantities are of the order 10^{-2} (cf. Ref. 15.8, p. 208). To lowest order in ak^2 and bk^2 the poles of $\delta\tilde{n}(\mathbf{k}, z)$ are given by

$$z = \pm c_0 k - \tfrac{1}{2} i \left[a + b - \frac{a}{\gamma} \right] k^2 = \pm c_0 k - i \Gamma k^2 \tag{15.143}$$

and

$$z = \frac{-iKk^2}{n_0 c_p} = -\frac{iak^2}{\gamma}, \tag{15.144}$$

where in Eq. (15.143) Γ is the sound absorption coefficient

$$\Gamma = \frac{1}{2} \left[\frac{\frac{4}{3}\eta + \zeta}{mn_0} + \frac{K}{n_0} \left(\frac{1}{c_V} - \frac{1}{c_p} \right) \right]. \tag{15.145}$$

If Eq. (15.142) is evaluated, the lowest order terms are

$$\delta n(\mathbf{k}, t) = \delta n(\mathbf{k}, 0) \left[\frac{c_p - c_V}{c_p} e^{-(Kk^2/n_0 c_p)t} + \frac{c_V}{c_p} e^{-\Gamma k^2 t} \cos c_0 k t \right]. \tag{15.146}$$

We can now use this expression to obtain the generalized structure factor.

The part of the generalized structure factor due to nonforward scattering can be written

$$\begin{aligned} S(\mathbf{k}, \Omega) &= \frac{1}{N} \int d\boldsymbol{\rho} \int_{-\infty}^{\infty} dt \int d\mathbf{r}' \langle \delta n(\mathbf{r}' + \boldsymbol{\rho}, t)\, \delta n(\mathbf{r}', t = 0) \rangle\, e^{i\mathbf{k}\cdot\boldsymbol{\rho}}\, e^{-i\Omega t} \\ &= \frac{1}{N} \int_{-\infty}^{\infty} dt\, e^{i\Omega t} \langle \delta n(\mathbf{k}, t)\, \delta n(-\mathbf{k}, 0) \rangle. \end{aligned} \tag{15.147}$$

Note that from Eq. (15.1) the correlation function can depend only on the absolute value of the time. From Eq. (15.146), we can write

$$\begin{aligned} &\langle \delta n(\mathbf{k}, |t|)\, \delta n(-\mathbf{k}, 0) \rangle \\ &\quad = \langle \delta n(\mathbf{k}, 0)\, \delta n(-\mathbf{k}, 0) \rangle \\ &\qquad \times \left\{ \frac{c_p - c_V}{c_p} \exp\left[-\frac{Kk^2 |t|}{n_0 c_p} \right] + \frac{c_V}{c_p} \exp[-\Gamma k^2 |t|] \cos c_0 k t \right\}. \end{aligned} \tag{15.148}$$

If Eq. (15.148) is substituted into (15.147), we obtain

$$S(\mathbf{k}, \Omega) = \frac{1}{N}\langle \delta n(\mathbf{k}, \Omega)\, \delta n(-\mathbf{k}, 0)\rangle \times \left\{ \left(\frac{c_p - c_V}{c_p}\right) \left(\frac{2Kk^2/n_0 c_p}{\left(\frac{Kk^2}{n_0 c_p}\right)^2 + \Omega^2} \right) + \frac{c_V}{c_p}\left[\frac{\Gamma k^2}{(\Gamma k^2)^2 + (\Omega + c_0 k)^2} + \frac{\Gamma k^2}{(\Gamma k^2)^2 + (\Omega - c_0 k)^2} \right] \right\}. \tag{15.149}$$

Away from the critical point, the correlation function $\langle \delta n(\mathbf{k}, 0)\, \delta n(-\mathbf{k}, 0)\rangle$ is independent of $\mathbf{k}$ and proportional to the compressibility. Near the critical point we expect it to behave as $(B + k^2)^{-1}$ (cf. Sec. 10.C and Ref. 15.13), where B is a constant characteristic of the medium.

The frequency-dependent part of $S(\mathbf{k}, \Omega)$ has three terms of the form

$$f(\Omega) = \frac{2\Delta}{\Delta^2 + (\Omega - \Omega')^2} \tag{15.150}$$

where $f(\Omega)$ is a Lorentzian centered at the frequency Ω' with half width at $\frac{1}{2}$ maximum given by Δ. Therefore, the hydrodynamic theory predicts three Lorentzian line shapes, one centered at $\Omega = 0$ and two centered at $\Omega = \pm c_0 k$. The one at $\Omega = 0$ comes from the thermal mode. The two at $\Omega = \pm c_0 k$ come from the sound modes. The Rayleigh line ($\Omega = 0$) has a width $\Delta = Kk^2/n_0 c_p$ and the Brillouin lines $\Omega = \pm c_0 k$ have a width $\Delta = \Gamma k^2$. We see that the measurement of the spectral intensity of light scattered from a fluid can in principle yield a large amount of information about transport processes in the fluid.

H. MICROSCOPIC LINEAR RESPONSE THEORY

As was first shown by Kubo,[14] it is possible to derive the linear response matrix directly from microscopic theory. Let us consider a system to which we apply an external field $F_j(\mathbf{r}, t)$ which couples to microscopic densities $\hat{a}_j(\mathbf{r})$. The total Hamiltonian of the system in the presence of the external field is

$$\hat{\mathscr{H}}_T(t) = \hat{H} + \delta\hat{H}(t) \tag{15.151}$$

where $\hat{H}$ is the Hamiltonian of the system in the absence of the field and

$$\delta\hat{H}(t) = -\int d\mathbf{r}\, \hat{\mathbf{a}}(\mathbf{r}) \cdot \mathbf{F}(\mathbf{r}, t) \tag{15.152}$$

is the perturbed Hamiltonian. We assume that the external field is turned at time $t = -\infty$. The total density matrix in the presence of the field satisfies the equation of motion

$$-i\hbar \frac{\partial \hat{\rho}_T}{\partial t} = [\hat{\mathscr{H}}_T(t), \hat{\rho}_T(t)] \tag{15.153}$$

where $\hat{\rho}_T(t = -\infty) = \hat{\rho}_{eq}$, and $\hat{\rho}_{eq}$ is the equilibrium density matrix in the absence of the field. Let us now expand the total density matrix about its equilibrium value:

$$\hat{\rho}_T(t) = \hat{\rho}_{eq} + \delta\hat{\rho}(t). \tag{15.154}$$

Since $\hat{\rho}_{eq}$ does not depend on time, $[\hat{H}, \hat{\rho}_{eq}] = 0$, and we find

$$-i\hbar \frac{\partial}{\partial t} \delta\hat{\rho}(t) = [\delta\hat{H}(t), \hat{\rho}_{eq}] + [\hat{H}, \delta\hat{\rho}(t)] + [\delta\hat{H}(t), \delta\hat{\rho}(t)]. \tag{15.155}$$

To obtain an expression for $\delta\hat{\rho}(t)$ linear in the applied field, we neglect the nonlinear term $[\delta\hat{H}(t), \delta\hat{\rho}(t)]$ and write

$$-i\hbar \frac{\partial\, \delta\hat{\rho}(t)}{\partial t} = [\delta\hat{H}(t), \hat{\rho}_{eq}] + [\hat{H}, \delta\hat{\rho}(t)]. \tag{15.156}$$

It is now possible to solve Eq. (15.156) for $\delta\hat{\rho}(t)$. We first multiply by dt, then integrate from $t = -\infty$ to t and use the boundary condition $\delta\hat{\rho}(t = -\infty) = 0$. After some reordering of the integration variables, for example,

$$\int_{-\infty}^{t} dt_1 \int_{-\infty}^{t_1} dt_2 \int_{-\infty}^{t_2} dt_3 = \int_{-\infty}^{t} dt_3 \int_{t_3}^{t} dt_1 \int_{t_3}^{t_2} dt_2 = \int_{-\infty}^{t} dt_3 \frac{(t - t_3)^2}{2},$$

we obtain

$$\begin{aligned} \delta\hat{\rho}(t) &= -\frac{i}{\hbar} \int_{-\infty}^{t} dt'\, e^{-i\hat{H}(t-t')/\hbar} [\delta\hat{H}(t), \hat{\rho}_{eq}]\, e^{i\hat{H}(t-t')/\hbar} \\ &= -i \int_{-\infty}^{t} dt'\, e^{-i\hat{L}(t-t')}\, \delta\hat{L}(t')\hat{\rho}_{eq} \end{aligned} \tag{15.157}$$

where $\hat{L}$ and $\delta\hat{L}(t)$ are Liouville operators (cf. Sec. 7.E) and are defined $\hat{L} = \hbar^{-1}[\hat{H},\ ]$ and $\delta\hat{L}(t) = \hbar^{-1}[\delta\hat{H}(t),\ ]$. The response of the system to the external field can be written

$$\begin{aligned} \langle a_i(\mathbf{r}, t)\rangle &= \mathrm{Tr}\, \hat{a}_i(\mathbf{r})\, \delta\hat{\rho}(t) = -i\, \mathrm{Tr} \int_{-\infty}^{t} dt' \hat{a}_i(\mathbf{r})\, e^{-i\hat{L}(t-t')}\, \delta\hat{L}(t')\hat{\rho}_{eq} \\ &= \frac{i}{\hbar} \int_{-\infty}^{t} dt' \int d\mathbf{r}' \langle [\hat{a}_i(\mathbf{r}, t), \hat{a}_j(\mathbf{r}', t')]\rangle_{eq} F_j(\mathbf{r}', t'), \end{aligned} \tag{15.158}$$

where

$$\hat{a}_i(\mathbf{r}, t) = e^{i\hat{H}t/\hbar} \hat{a}_i(\mathbf{r})\, e^{-i\hat{H}t/\hbar} \tag{15.159}$$

and the average $\langle A\rangle_{\text{eq}}$ is defined

$$\langle A\rangle_{\text{eq}} = \text{Tr}\,\hat{\rho}_{\text{eq}}\hat{A}. \tag{15.160}$$

In obtaining Eq. (15.158), we have cyclically permuted operators under the trace and have assumed that $\langle a_i(\mathbf{r})\rangle_{\text{eq}} = 0$. If we compare Eqs. (15.44) and (15.158), we can write the response as

$$\langle a_i(\mathbf{r}, t)\rangle = 2i\int_{-\infty}^{t} dt' \int d\mathbf{r}' K''_{a_i a_j}(\mathbf{r}, \mathbf{r}'; t - t')F_j(\mathbf{r}', t') \tag{15.161}$$

where

$$K''_{a_i a_j}(\mathbf{r}, \mathbf{r}'; t - t') = \frac{1}{2\hbar}\langle[\hat{a}_i(\mathbf{r}, t), \hat{a}_j(\mathbf{r}', t')]\rangle_{\text{eq}}. \tag{15.162}$$

Thus, from microscopic linear response theory we obtain the result that the dissipative part of the response matrix is given by the equilibrium average of the commutator of the densities involved.

Let us now obtain the fluctuation-dissipation theorem from microscopic linear response theory. The position-dependent correlation function in a translationally invariant stationary system can be written

$$C_{a_i a_j}(\mathbf{r}' - \mathbf{r}; \tau) = \langle\hat{a}_i(\mathbf{r}, \tau)\hat{a}_j(\mathbf{r}')\rangle_{\text{eq}}, \tag{15.163}$$

where we have assumed that $\langle a_i(\mathbf{r})\rangle_{\text{eq}} = 0$. The Fourier transform is

$$\begin{aligned} G_{a_i a_j}(\mathbf{k}, \tau) &= \int d(\mathbf{r}' - \mathbf{r})\, e^{-i\mathbf{k}\cdot(\mathbf{r}'-\mathbf{r})}\langle\hat{a}_i(\mathbf{r}, \tau)\hat{a}_j(\mathbf{r}')\rangle_{\text{eq}} \\ &= \frac{1}{V}\iint d\mathbf{r}\, d\mathbf{r}'\, e^{-i\mathbf{k}\cdot(\mathbf{r}'-\mathbf{r})}\langle\hat{a}_i(\mathbf{r}, \tau)\hat{a}_j(\mathbf{r}')\rangle_{\text{eq}} \\ &= \frac{1}{V}\langle\hat{a}_i(-\mathbf{k}, \tau)\hat{a}_j(\mathbf{k})\rangle = \frac{1}{V}\langle\hat{a}_i^*(\mathbf{k}, \tau)\hat{a}_j(\mathbf{k})\rangle. \end{aligned} \tag{15.164}$$

In Eq. (15.164), we have used the definition

$$\hat{a}_i(\mathbf{k}) = \int d\mathbf{r}\, e^{-i\mathbf{k}\cdot\mathbf{r}}\hat{a}_i(\mathbf{r}). \tag{15.165}$$

From Eq. (15.162) the momentum-dependent response matrix can be written

$$K''_{a_i a_j}(\mathbf{k}, \tau) = \frac{1}{2\hbar}\frac{1}{V}\langle[\hat{a}_i^*(\mathbf{k}, \tau), \hat{a}_j(\mathbf{k})]\rangle_{\text{eq}}. \tag{15.166}$$

However,

$$\langle\hat{a}_j\hat{a}_i(\tau)\rangle = \langle\hat{a}_i(\tau - i\hbar\beta)\hat{a}_j\rangle = e^{-i\beta\hbar(\partial/\partial\tau)}\langle\hat{a}_i(\tau)\hat{a}_j\rangle \tag{15.167}$$

where

$$\hat{a}_i(-i\hbar\beta) \equiv e^{\beta\hat{H}}\hat{a}_i\, e^{-\beta\hat{H}} \tag{15.168}$$

and, therefore,

$$K''_{a_ia_j}(\mathbf{k}, \tau) = \frac{1}{2\hbar}(1 - e^{-i\beta\hbar(\partial/\partial\tau)})G_{a_ia_j}(\mathbf{k}, \tau). \tag{15.169}$$

The Fourier transform of Eq. (15.169) takes the form

$$\chi''_{a_ia_j}(\mathbf{k}, \omega) = \frac{1}{2\hbar}(1 - e^{-\hbar\omega\beta})S_{a_ia_j}(\mathbf{k}, \omega) \tag{15.170}$$

where $S_{a_ia_j}(\mathbf{k}, \omega)$ is the spectral density matrix. Eqs. (15.168) and (15.170) were first derived by Kubo[13] and are the form of the fluctuation-dissipation theorem that results from microscopic linear response theory. We see that the imaginary part of the response matrix is expressed directly in terms of the equilibrium correlation function. The full response function can be expressed directly in terms of $S_{a_ia_j}(\mathbf{k}, \omega)$:

$$\chi_{a_ia_j}(\mathbf{k}, \omega) = \lim_{\eta \to 0} \int \frac{d\omega'}{2\pi\hbar} \frac{(1 - e^{-\hbar\omega'\beta})S_{a_ia_j}(\mathbf{k}, \omega')}{\omega' - \omega - i\eta} \tag{15.171}$$

where we have made use of the Kramers-Kronig relations.

From Eq. (15.169) and the properties of the correlation function established earlier, we can obtain some useful properties of $K''_{a_ia_j}(\mathbf{k}, \tau)$ and $\chi''_{a_ia_j}(\mathbf{k}, \omega)$. For example, microscopic reversibility gives

$$\begin{aligned} K''_{a_ia_j}(\tau) &= \frac{1}{2\hbar}(1 - e^{-\beta i\hbar(\partial/\partial\tau)})\langle \hat{a}_i(\tau)\hat{a}_j\rangle \\ &= \frac{1}{2\hbar}(1 - e^{-\beta i\hbar(\partial/\partial\tau)})\langle \hat{a}_j(\tau)\hat{a}_i\rangle \epsilon_j^T \epsilon_i^T \end{aligned} \tag{15.172}$$

where $\epsilon_i^T = +1$ if a_i is even under time reversal and $\epsilon_i^T = -1$ if it is odd under time reversal. From Eqs. (15.166) and (15.172), we then find

$$K''_{a_ia_j}(\mathbf{k}, \tau) = -K''_{a_ia_j}(\mathbf{k}, -\tau)\epsilon_i^T\epsilon_j^T \tag{15.173}$$

and, therefore,

$$\chi''_{a_ia_j}(\mathbf{k}, \omega) = -\epsilon_i^T\epsilon_j^T\chi''_{a_ia_j}(\mathbf{k}, -\omega). \tag{15.174}$$

These identities will be useful later.

I. HYDRODYNAMICS AND LINEAR RESPONSE THEORY[15]

The linear response of a system is independent of the form of the external field. We can use this fact to write the hydrodynamic equations in terms of the linear response function. Let us assume that the field is turned on at time $t = -\infty$ and its strength gradually increases until at time $t = 0$ it has reached its maximum strength. We then abruptly turn it off. We will obtain the response for times $t > 0$. The force can be written

$$F_j(\mathbf{k}, t') = \begin{cases} F_j(\mathbf{k})\, e^{\epsilon t'} & t' \leqslant 0 \\ 0 & t' > 0 \end{cases} \tag{15.175}$$

and the response can be written

$$\delta\langle a_i(\mathbf{k}, t)\rangle = \frac{1}{2\pi}\int_{-\infty}^{\infty} d\omega\, e^{-i\omega t}\chi_{a_i a_j}(\mathbf{k}, \omega)\tilde{F}_j(\mathbf{k}, \omega) \tag{15.176}$$

where $\tilde{F}_j(\mathbf{k}, \omega)$ is the Fourier transform of the force

$$\tilde{F}_j(\mathbf{k}, \omega) = \frac{1}{i(\omega - i\epsilon)} F_j(\mathbf{k}). \tag{15.177}$$

If we substitute Eq. (15.177) into Eq. (15.176) and use Eq. (15.49), we obtain

$$\delta\langle a_i(\mathbf{k}, t)\rangle = \frac{F_j(\mathbf{k})}{2} \chi_{a_i a_j}(\mathbf{k}, \omega = 0) + \frac{F_j(\mathbf{k})}{2\pi i} P\int_{-\infty}^{\infty} d\omega\, \frac{\chi_{a_i a_j}(\mathbf{k}, \omega)\, e^{-i\omega t}}{\omega}, \tag{15.178}$$

and if we use the Kramers-Kronig relation (cf. Eq. (15.39)), we obtain

$$\delta\langle a_i(\mathbf{k}, t = 0)\rangle = \chi_{a_i a_j}(\mathbf{k}, \omega = 0)F_j(\mathbf{k}). \tag{15.179}$$

Let us now write Eq. (15.176) in the form

$$\delta\langle a_i(\mathbf{k}, t)\rangle = \frac{F_j(\mathbf{k})}{2\pi i}\int_{-\infty}^{\infty} d\omega \int_{-\infty}^{\infty} \frac{d\omega'}{\pi} \frac{e^{-i\omega t}}{\omega - i\epsilon} \frac{\chi''_{a_i a_j}(\mathbf{k}, \omega')}{\omega' - \omega - i\eta} \tag{15.180}$$

where we have made use of Eqs. (15.44) and (15.177). In order to perform the integration over ω, we close the contour below and take the contribution from the pole $\omega = \omega' - i\eta$. We get an extra minus sign because of the direction of the contour. The result is

$$\delta\langle a_i(\mathbf{k}, t)\rangle = F_j(\mathbf{k})\int_{-\infty}^{\infty} \frac{d\omega'}{\pi} \frac{e^{-i(\omega' - i\eta)t}}{\omega' - i\epsilon - i\eta} \chi''_{a_i a_j}(\mathbf{k}, \omega). \tag{15.181}$$

Let us now Laplace transform Eq. (15.181). We find

$$\begin{aligned} \delta\langle \tilde{a}_i(\mathbf{k}, z)\rangle &= F_j(\mathbf{k})\int_{0}^{\infty} dt\, e^{izt} \int_{-\infty}^{\infty} \frac{d\omega'}{\pi} \frac{e^{-i(\omega' - i\eta)t}}{\omega' - i\epsilon - i\eta} \chi''_{a_i a_j}(\mathbf{k}, \omega') \\ &= F_j(\mathbf{k})\int_{-\infty}^{\infty} \frac{d\omega'}{\pi i} \frac{\chi''_{a_i a_j}(\mathbf{k}, \omega')}{(\omega'' - \omega' + i\eta' + i\eta)(\omega' - i\epsilon - i\eta)} \end{aligned} \tag{15.182}$$

where $z = \omega'' + i\eta'$. We now let $\epsilon \to 0$ and $\eta \to 0$ and use the relation

$$\frac{1}{\omega'(\omega' - z)} = \frac{1}{(\omega' - z)z} - \frac{1}{z\omega'}. \tag{15.183}$$

Then Eq. (15.182) becomes

$$\delta\langle\tilde{a}_i(\mathbf{k}, z)\rangle = \frac{F_j(\mathbf{k})}{iz}[\chi_{a_i a_j}(\mathbf{k}, z) - \chi_{a_i a_j}(\mathbf{k}, \omega = 0)], \tag{15.184}$$

and from Eq. (15.179) we find

$$\delta\langle\tilde{a}_i(\mathbf{k}, z)\rangle = \frac{1}{iz}[\bar{\bar{\chi}}(\mathbf{k}, z)\bar{\bar{\chi}}^{-1}(\mathbf{k}, \omega = 0) - 1]_{a_i a_j}\,\delta\langle a_j(\mathbf{k}, t = 0)\rangle. \tag{15.185}$$

Thus, we have written the response as an initial value problem. The hydrodynamic poles of $\bar{\bar{\chi}}(\mathbf{k}, z)$ yield the hydrodynamic equations.

J. CORRELATION FUNCTIONS IN TERMS OF PROJECTION OPERATORS[15]

It is useful to introduce projection operators which decompose the system into one space which contains the hydrodynamic densities and one orthogonal to it. This decomposition allows us to generalize the concept of a hydrodynamic mode to include modes which arise not from conserved quantities but from broken symmetries. Projection operators provide a very powerful means of analyzing systems. They were first used in statistical mechanics by Zwanzig,[16] although the idea of projecting onto the macroscopic modes of a system was first used by Mori.[14] They also played an important role in the microscopic theory of irreversible processes developed by Prigogine et al.[18] In this section we shall only be interested in projectors onto the macroscopic modes of system. The discussion in this and subsequent sections follows closely the presentation by Forster in Ref. 15.15.

Let us write a microscopic expression for the correlation function $C_{aa}(\mathbf{r}' - \mathbf{r}; \tau)$:

$$C_{aa}(\mathbf{r}' - \mathbf{r}; \tau) = \langle\hat{a}(\mathbf{r})\hat{a}(\mathbf{r}', -\tau)\rangle_{\text{eq}} = \text{Tr}[e^{-\beta\hat{H}}\hat{a}(\mathbf{r})\, e^{-i\hat{L}\tau}\hat{a}(\mathbf{r}')] \tag{15.186}$$

where $\hat{a}(\mathbf{r})$ is a microscopic density and $\hat{L}$ is the Liouville operator. The Laplace and Fourier transform of $C_{aa}(\mathbf{r}' - \mathbf{r}, \tau)$ is given by the spectral density function:

$$\begin{aligned}\bar{S}_{aa}(\mathbf{k}, z) &= \int d(\mathbf{r}' - \mathbf{r})\, e^{-i\mathbf{k}\cdot(\mathbf{r}'-\mathbf{r})} \int_0^\infty d\tau\, e^{iz\tau} C_{aa}(\mathbf{r}' - \mathbf{r}, \tau) \\ &= \frac{1}{V}\left\langle \hat{a}^*(\mathbf{k}) \frac{i}{z - \hat{L}}\hat{a}(\mathbf{k})\right\rangle_{\text{eq}}.\end{aligned} \tag{15.187}$$

It is useful to consider the spectral density function as a matrix element with respect to the state with the macroscopic mode, $\hat{a}(\mathbf{k})$. Thus, we write it as

$$\bar{S}_{aa}(\mathbf{k}, z) = \langle a(\mathbf{k})| \frac{i}{z - \hat{L}} |a(\mathbf{k})\rangle \tag{15.188}$$

where the bra, $\langle a(\mathbf{k})|$, indicates the complex conjugate function $a^*(\mathbf{k})$. We next introduce a projection operator $P_{\mathbf{k}}$ onto the space of fluctuations $\hat{a}(\mathbf{k})$,

$$P_{\mathbf{k}} = |a(\mathbf{k})\rangle \frac{1}{\langle a(\mathbf{k})|a(\mathbf{k})\rangle} \langle a(\mathbf{k})|, \tag{15.189}$$

and an orthogonal projection,

$$Q_{\mathbf{k}} = 1 - P_{\mathbf{k}}. \tag{15.190}$$

For an isotropic system $\langle a(\mathbf{k})|a(\mathbf{k}')\rangle = 0$ if $\mathbf{k} \neq \mathbf{k}'$ and $P_{\mathbf{k}}P_{\mathbf{k}'} = P_{\mathbf{k}}\,\delta_{\mathbf{k},\mathbf{k}'}$. The projector $Q_{\mathbf{k}}$ projects onto the space orthogonal to the space containing $\hat{a}(\mathbf{k})$. That is, it removes the modes $\hat{a}(\mathbf{k})$. The orthogonal space acts as a bath, or reservoir, which influences the modes $\hat{a}(\mathbf{k})$.

In terms of projection operators, the spectral density function can be written

$$\begin{aligned}\bar{S}_{aa}(\mathbf{k}, z) &= \langle a(\mathbf{k})| \frac{i}{z - \hat{L}P_{\mathbf{k}} - \hat{L}Q_{\mathbf{k}}} |a(\mathbf{k})\rangle \\ &= \langle a(\mathbf{k})| \left(\frac{i}{z - \hat{L}Q_{\mathbf{k}}} + \frac{1}{z - \hat{L}Q_{\mathbf{k}}} \hat{L}P_{\mathbf{k}} \frac{i}{z - \hat{L}}\right) |a(\mathbf{k})\rangle.\end{aligned} \tag{15.191}$$

If we note that

$$\langle a(\mathbf{k})| \frac{i}{z - LQ_k} |a(\mathbf{k})\rangle = \frac{i}{z} \langle a(\mathbf{k})|a(\mathbf{k})\rangle \tag{15.192}$$

and

$$P_{\mathbf{k}} \frac{i}{z - L} |a(\mathbf{k})\rangle = |a(\mathbf{k})\rangle \frac{1}{\langle a(\mathbf{k})|a(\mathbf{k})\rangle} \bar{S}_{aa}(\mathbf{k}, z) \tag{15.193}$$

and solve Eq. (15.191) for $\bar{S}_{aa}(\mathbf{k}, z)$, we find

$$\bar{S}_{aa}(\mathbf{k}, z) = \frac{i\langle a(\mathbf{k})|a(\mathbf{k})\rangle}{z - \dfrac{\langle a(\mathbf{k})|\hat{L}|a(\mathbf{k})\rangle}{\langle a(\mathbf{k})|a(\mathbf{k})\rangle} - \dfrac{\langle a(\mathbf{k})|\hat{L}Q_{\mathbf{k}}[1/(z - \hat{L}Q_{\mathbf{k}})]\hat{L}|a(\mathbf{k})\rangle}{\langle a(\mathbf{k})|a(\mathbf{k})\rangle}}. \tag{15.194}$$

We can now make use of the microscopic equations

$$-i\frac{\partial}{\partial t} |a(\mathbf{k})\rangle = \hat{L}|a(\mathbf{k})\rangle \tag{15.195}$$

and

$$i\frac{\partial}{\partial t}\langle a(\mathbf{k})| = \langle a(\mathbf{k})|\hat{L} \tag{15.196}$$

to obtain

$$\bar{S}_{aa}(\mathbf{k}, z) = \frac{iG_{aa}(\mathbf{k})}{z - i\omega_{aa}(\mathbf{k})G_{aa}^{-1}(\mathbf{k}) + i\sum_{aa}(\mathbf{k}, z)G_{aa}^{-1}(\mathbf{k})}, \tag{15.197}$$

where $G_{aa}(\mathbf{k})$ is the zero time correlation function,

$$G_{aa}(\mathbf{k}) = \langle a(\mathbf{k})|a(\mathbf{k})\rangle; \tag{15.198}$$

$\omega_{aa}(\mathbf{k})$ is a frequency which is usually zero,

$$\omega_{aa}(\mathbf{k}) \equiv \langle \dot{a}(\mathbf{k})|a(\mathbf{k})\rangle; \tag{15.199}$$

and $\sum_{aa}(\mathbf{k}, z)$ is called the memory function,

$$\Sigma_{aa}(\mathbf{k}, z) = \langle \dot{a}(\mathbf{k})|\, Q_{\mathbf{k}} \frac{i}{z - LQ_{\mathbf{k}}} |\dot{a}(\mathbf{k})\rangle. \tag{15.200}$$

As we shall see, if $\bar{S}_{aa}(\mathbf{k}, z)$ has poles which go to zero as $k \to 0$, there will be hydrodynamic modes in the system involving $a(\mathbf{k})$.

It usually happens that there are a number of independent hydrodynamic modes in a system. Then we must consider the correlation matrix

$$C_{a_i a_j}(\mathbf{r} - \mathbf{r}'; \tau) = \langle a_i(\mathbf{r}, \tau)a_j(\mathbf{r}')\rangle_{\text{eq}}. \tag{15.201}$$

The corresponding spectral density matrix can be written

$$\bar{S}_{a_i a_j}(\mathbf{k}, z) = \langle a_i(\mathbf{k})| \frac{i}{z - \hat{L}} |a_j(\mathbf{k})\rangle_{\text{eq}}. \tag{15.202}$$

The projector onto the space of hydrodynamic modes takes the form

$$P_{\mathbf{k}} = \sum_{ij} |a_i(\mathbf{k})\rangle \frac{1}{\langle a_i(\mathbf{k})|a_j(\mathbf{k})\rangle} \langle a_j(\mathbf{k})| = 1 - Q_{\mathbf{k}} \tag{15.203}$$

where $Q_{\mathbf{k}}$ is the orthogonal projector. The equation for the spectral density matrix $\bar{S}_{a_m a_j}(\mathbf{k}, z)$ now takes the form

$$\sum_{l,m}\left[z\,\delta_{ij} - i\omega_{a_i a_l}[G(\mathbf{k})^{-1}]_{a_l a_m} + i\,\Sigma_{a_i a_l}(\mathbf{k}, z)[G(\mathbf{k})^{-1}]_{a_l a_m}\right] \bar{S}_{a_m a_j}(\mathbf{k}, z) = iG_{a_i a_j}(\mathbf{k}). \tag{15.204}$$

We can now use these expressions to show that the hydrodynamic behavior of systems with conserved quantities and that of systems with broken symmetry have quite different origins.

K. GENERAL DEFINITION OF HYDRODYNAMIC EQUATIONS[15]

1. General Form of Hydrodynamic Equations

Hydrodynamic modes, in general, obey linearized equations of motion of the form

$$\frac{\partial}{\partial t}\langle a(\mathbf{r}, t)\rangle - \eta_A \nabla_{\mathbf{r}}^2 \langle a(\mathbf{r}, t)\rangle = 0 \tag{15.205}$$

where η_A is a phenomenological transport coefficient. If we assume that $\langle a(\mathbf{r}, t = 0)\rangle$ is given, we can Fourier and Laplace transform the space and time dependence, respectively, of $\langle a(\mathbf{r}, t)\rangle$ and write

$$\langle \tilde{a}(\mathbf{k}, z)\rangle = \int_0^\infty dt\, e^{izt} \int d\mathbf{r}\, e^{-i\mathbf{k}\cdot\mathbf{r}} \langle a(\mathbf{r}, t)\rangle \tag{15.206}$$

where $z = \omega + i\eta$. Then Eq. (15.205) becomes

$$\langle \tilde{a}(\mathbf{k}, z)\rangle = \frac{i}{z + i\eta_A k^2} \langle a(\mathbf{k}, t = 0)\rangle. \tag{15.207}$$

Eq. (15.207) has a pole at $z = -i\eta_A k^2$. The time dependence of $\langle a(\mathbf{k}, t)\rangle$ is given by

$$\langle a(\mathbf{k}, t)\rangle = e^{-\eta_A k^2 t} \langle a(\mathbf{k}, t = 0)\rangle. \tag{15.208}$$

Eqs. (15.207) and (15.208) are characteristic of hydrodynamic behavior. The mode $\langle a(\mathbf{k}, t)\rangle$ is damped with a lifetime $\tau(\mathbf{k}) = [\eta_A k^2]^{-1}$. The lifetime of a hydrodynamic mode thus depends on the wavelength of the mode. The longer the wavelength the longer the lifetime and, therefore, the longer it takes to smooth out the inhomogeneity. This type of behavior is characteristic of the densities of conserved quantities. They can relax only by being transported physically through the fluid.

The correlation functions will have a similar behavior. We can write the correlation function, $C_{aa}(\mathbf{r}, t)$, as

$$C_{aa}(\mathbf{r}, t) = \langle a(\mathbf{r}, t) | a(0, 0)\rangle_{\text{eq}}. \tag{15.209}$$

From Eqs. (15.205) and (15.208), we obtain

$$\frac{\partial}{\partial t} C_{aa}(\mathbf{r}, t) = \eta_a \nabla_{\mathbf{r}}^2 C_{aa}(\mathbf{r}, t) \tag{15.210}$$

and, therefore, the spectral density function can be written

$$\bar{S}_{aa}(\mathbf{k}, z) = \frac{i}{z + i\eta_a k^2} \langle a(\mathbf{k}) | a(\mathbf{k})\rangle. \tag{15.211}$$

Thus, the spectral density function has a pole at the hydrodynamic frequency.

We can now write the transport coefficient η_a in terms of the spectral density function. Let us first introduce the Fourier transform of $G_{aa}(\mathbf{k}, t)$,

$$S_{aa}(\mathbf{k}, \omega) = \int_{-\infty}^{\infty} dt\, e^{i\omega t} G_{aa}(\mathbf{k}, t). \tag{15.212}$$

The Laplace transform can be written in terms of the Fourier transform if we use the spectral decomposition of the Heaviside function in Eq. (15.43). Then we find

$$\begin{aligned} \bar{S}_{aa}(\mathbf{k}, z) &= \int_0^{\infty} dt\, e^{izt} G_{aa}(\mathbf{k}, t) = -\int_{-\infty}^{\infty} dt \int_{-\infty}^{\infty} \frac{d\omega'}{2\pi i} \frac{e^{i(z-\omega')t}}{\omega' + i\eta} G_{aa}(\mathbf{k}, t) \\ &= \int \frac{d\omega'}{2\pi i} \frac{S_{aa}(\mathbf{k}, \omega')}{\omega' - z}. \end{aligned} \tag{15.213}$$

We can now write $S_{aa}(\mathbf{k}, \omega)$ in terms of $\bar{S}_{aa}(\mathbf{k}, z)$,

$$S_{aa}(\mathbf{k}, \omega) = \lim_{\epsilon \to 0} [\bar{S}_{aa}(\mathbf{k}, \omega + i\epsilon) - \bar{S}_{aa}(\mathbf{k}, \omega - i\epsilon)], \tag{15.214}$$

and from Eq. (15.211) we find

$$S_{aa}(\mathbf{k}, \omega) = \frac{2\eta_a k^2}{\omega^2 + (\eta_a k^2)^2} \langle a(\mathbf{k}) | a(\mathbf{k}) \rangle. \tag{15.215}$$

If we take the limit $k \to 0$ *and then* $\omega \to 0$, we obtain

$$\eta_A = \lim_{\omega \to 0} \left[\lim_{k \to 0} \frac{\omega^2}{k^2} \frac{1}{\langle a(\mathbf{k}) | a(\mathbf{k}) \rangle} S_{aa}(\mathbf{k}, \omega) \right]. \tag{15.216}$$

Thus, we are able to express the transport coefficients in terms of the spectral density function.

Modes with a lifetime which becomes infinite as $\mathbf{k} \to 0$ are not restricted to conserved quantities. They can also occur if a symmetry of the system is broken, and thus broken symmetries can also lead to hydrodynamic equations. We have already seen one example of a hydrodynamic mode due to a broken symmetry for the case of liquid He(II) (cf. Sec. 14.H). The most general definition of a hydrodynamic mode is one whose lifetime becomes infinite as $\mathbf{k} \to 0$, regardless of its origin.

2. Hydrodynamic Modes: Conserved Quantities

Let us now use projection operator techniques to show the origin of hydrodynamic modes coming from conserved quantities. We first note that the frequency $\langle \dot{a}_i(\mathbf{k}) | a_j(\mathbf{k}) \rangle$ is zero if $a_i(\mathbf{k})$ has a well-defined sign under time reversal. This can be seen as follows. From the Kramers-Kronig relations we can write

$$\chi_{a_i a_j}(\mathbf{k}, z) = \int \frac{d\omega'}{\pi} \frac{\chi''_{a_i a_j}(\mathbf{k}, \omega')}{\omega' - z}, \tag{15.217}$$

and from the fluctuation-dissipation theorem we can write

$$\chi_{a_i a_j}(\mathbf{k}, z) = \int \frac{d\omega'}{2\pi\hbar} \frac{(1 - e^{-\hbar\omega'\beta})S_{a_i a_j}(\mathbf{k}, \omega')}{\omega' - z}. \tag{15.218}$$

If $a(\mathbf{r})$ is a conserved quantity, then

$$\lim_{\mathbf{k}\to 0} G_{a_i a_j}(\mathbf{k}, t) = \lim_{k\to 0} \int d\mathbf{r}\, e^{i\mathbf{k}\cdot\mathbf{r}}\langle a_i(\mathbf{r}, t)a_j(0, 0)\rangle$$

$$= \int d\mathbf{r}\langle_i(a\mathbf{r}, t)a_j(0, 0)\rangle = \langle A_i^{\text{tot}}(t)a_j(0, 0)\rangle. \tag{15.219}$$

But if A_i^{tot} is conserved, it is time independent and the spectral density matrix becomes

$$\lim_{\mathbf{k}\to 0} S_{a_i a_j}(\mathbf{k}, \omega) = \langle A_i^{\text{tot}} a_j(0, 0)\rangle \int_{-\infty}^{\infty} dt\, e^{i\omega t} = 2\pi\langle A_i^{\text{tot}}\hat{a}_j(0, 0)\rangle\, \delta(\omega)$$

$$= 2\pi\, \delta(\omega) \lim_{\mathbf{k}\to 0} G_{a_i a_j}(\mathbf{k}, t = 0). \tag{15.220}$$

We now take the static limit of Eq. (15.218) and the limit $\mathbf{k} \to 0$ and we find

$$\lim_{\mathbf{k}\to 0} \chi_{a_i a_j}(\mathbf{k}, 0) = \lim_{\mathbf{k}\to 0} \int \frac{d\omega'}{\hbar}\, \delta(\omega') \frac{(1 - e^{-\hbar\omega'\beta})G_{a_i a_j}(\mathbf{k}, 0)}{\omega'}$$

$$= \beta \lim_{\mathbf{k}\to 0} G_{a_i a_j}(\mathbf{k}, 0). \tag{15.221}$$

Thus, for small k, we can write $G_{a_i a_j}(\mathbf{k})$ in the form

$$G_{a_i a_j}(\mathbf{k}) = \langle a_i(\mathbf{k})|a_j(\mathbf{k})\rangle \equiv \frac{1}{\beta}\chi_{a_i a_j}(\mathbf{k}) = \frac{1}{\beta}\int \frac{d\omega'}{\pi} \frac{\chi''_{a_i a_j}(\mathbf{k}, \omega')}{\omega'}. \tag{15.222}$$

Let us now consider the correlation function $\langle\hat{a}(\mathbf{k})|\hat{a}(\mathbf{k})\rangle$. We can write

$$\langle\dot{a}_i(\mathbf{k})|a_j(\mathbf{k})\rangle = \frac{1}{\beta}\int \frac{d\omega'}{\pi} \frac{\chi''_{\dot{a}_i a_j}(\mathbf{k}, \omega')}{\omega'} = \frac{1}{\beta}\int \frac{d\omega'}{\pi} \chi''_{a_i a_j}(\mathbf{k}, \omega')$$

$$= -\frac{\epsilon_i^T\epsilon_j^T}{\beta}\int \frac{d\omega'}{\pi} \chi''_{a_i a_j}(\mathbf{k}, -\omega') = -\epsilon_i^T\epsilon_j^T\langle\hat{a}_i(\mathbf{k})|\hat{a}_j(\mathbf{k})\rangle \tag{15.223}$$

where we have used Eq. (15.174). Thus, if $\epsilon_i^T = \epsilon_j^T$ we have $\langle\dot{a}_i(\mathbf{k})|a_j(\mathbf{k})\rangle \equiv 0$.

Let us next determine the $\mathbf{k}$ dependence of the memory matrix $\Sigma_{a_i a_j}(\mathbf{k}, z)$. If A_i is a conserved quantity, then for small $\mathbf{k}$ we can write the microscopic balance equation

$$\frac{\partial\hat{a}(\mathbf{k}, t)}{\partial t} = i\mathbf{k}\cdot\hat{\mathbf{J}}^a(\mathbf{k}, t), \tag{15.224}$$

where $\hat{\mathbf{J}}^a(\mathbf{k}, t)$ is the microscopic flux of $\hat{a}(\mathbf{k}, t)$ (cf. Secs. 7.D and 7.H). We then find that $\sum_{aa}(\mathbf{k}, z)$ is proportional to k^2:

$$\Sigma_{aa}(\mathbf{k}, z) = \langle \dot{a}(\mathbf{k}) | Q_{\mathbf{k}} \frac{i}{z - \hat{L}Q_{\mathbf{k}}} Q_{\mathbf{k}} | \dot{a}(\mathbf{k}) \rangle$$

$$= k^2 \langle \boldsymbol{\varkappa} \cdot \hat{\mathbf{J}}^a(\mathbf{k}) | Q_{\mathbf{k}} \frac{i}{z - \hat{L}Q_{\mathbf{k}}} | \boldsymbol{\varkappa} \cdot \hat{\mathbf{J}}^a(\mathbf{k}) \rangle. \tag{15.225}$$

where

$$\boldsymbol{\varkappa} = \mathbf{k}/|\mathbf{k}|.$$

Since $G_{aa}(\mathbf{k})$ tends to a well-behaved thermodynamic function in the limit $\mathbf{k} \to 0$, $\bar{S}_{aa}(\mathbf{k}, z)$ has a hydrodynamic pole. We can show this explicitly if we write $\bar{S}_{aa}(\mathbf{k}, z)$ in the form

$$\bar{S}_{aa}(\mathbf{k}, z) = \frac{iG_{aa}(\mathbf{k})}{z + ik^2 D_{aa}(\mathbf{k}, z)} \tag{15.226}$$

where

$$D(\mathbf{k}, z) = \langle \boldsymbol{\varkappa} \cdot J^a(\mathbf{k}) | Q_{\mathbf{k}} \frac{i}{z - \hat{L}Q_{\mathbf{k}}} Q_{\mathbf{k}} | \boldsymbol{\varkappa} \cdot J^a(\mathbf{k}) \rangle G_{aa}^{-1}(\mathbf{k}). \tag{15.227}$$

$\bar{S}_{aa}(\mathbf{k}, z)$ will have a pole for values of z which satisfy the condition

$$z(\mathbf{k}) + ik^2 D(\mathbf{k}, z(\mathbf{k})) = 0. \tag{15.228}$$

For small k we keep the contribution to second order in k. Since $z(\mathbf{k})$ is at least of second order, we expand $D(\mathbf{k}, z(\mathbf{k}))$ about $\mathbf{k} = 0$ to find

$$z(\mathbf{k}) - ik^2 D(0, 0) = 0 \tag{15.229}$$

where, from Eq. (15.227), we see that

$$\eta_A \equiv D(0, 0) = \lim_{\mathbf{k} \to 0} \sum_{ij} \kappa_i \kappa_j \, d_{ij}(\mathbf{k}, z) \tag{15.230}$$

and

$$d_{ij}(\mathbf{k}, z) = \langle J_i^a(\mathbf{k}) | Q_{\mathbf{k}} \frac{i}{z - \hat{L}Q_{\mathbf{k}}} Q_{\mathbf{k}} | J_j^a(\mathbf{k}) \rangle G_{aa}^{-1}(\mathbf{k}). \tag{15.231}$$

Since $d_{ij}(\mathbf{k}, z)$ becomes independent of $\mathbf{k}$, in the limit $\mathbf{k} \to 0$, the only form possible is

$$\lim_{\mathbf{k} \to 0} d_{ij}(\mathbf{k}, z) = \delta_{ij} \, d(z) \tag{15.232}$$

where

$$d(z) = \frac{1}{3} \lim_{\mathbf{k} \to 0} \langle \mathbf{J}^a(\mathbf{k}) | \cdot Q_{\mathbf{k}} \frac{i}{z - \hat{L}Q_{\mathbf{k}}} Q_{\mathbf{k}} | \mathbf{J}^a(\mathbf{k}) \rangle G_{aa}^{-1}(\mathbf{k}) \tag{15.233}$$

Thus,

$$\eta_A = D(0, 0) = \lim_{\mathbf{k}\to 0, z\to 0} \tfrac{1}{3}\langle \mathbf{J}^a(\mathbf{k})| \cdot Q_{\mathbf{k}} \frac{i}{z - \hat{L}Q_{\mathbf{k}}} Q_{\mathbf{k}}|\mathbf{J}^a(\mathbf{k})\rangle\, G_{aa}^{-1}(\mathbf{k}) \quad (15.234)$$

and we have obtained a general microscopic expression for η_A. We can simplify Eq. (15.234) and at the same time impose an order on the limits $\mathbf{k} \to 0$ and $z \to 0$. Let us note that when $\langle \dot{a}(\mathbf{k})|a(\mathbf{k})\rangle = 0$, $P_{\mathbf{k}}|\dot{a}(\mathbf{k})\rangle = 0$ and $Q_{\mathbf{k}}|\dot{a}(\mathbf{k})\rangle = |\dot{a}(\mathbf{k})\rangle$. Therefore, we can write

$$\langle \dot{a}(\mathbf{k})| \frac{i}{z - L} |\dot{a}(\mathbf{k})\rangle = \langle \dot{a}(\mathbf{k})| Q_{\mathbf{k}} \frac{i}{z - Q_{\mathbf{k}}\hat{L}Q_{\mathbf{k}}} Q_{\mathbf{k}}|\dot{a}(\mathbf{k})\rangle \times \left[1 + \frac{i}{z} \langle \dot{a}(\mathbf{k})| Q_{\mathbf{k}} \frac{i}{z - Q_{\mathbf{k}}\hat{L}Q_{\mathbf{k}}} Q_{\mathbf{k}}|\dot{a}(\mathbf{k})\rangle G^{-1}(\mathbf{k})\right]^{-1}. \quad (15.235)$$

Since

$$\langle \dot{a}(\mathbf{k})| Q_{\mathbf{k}} \frac{i}{z - Q_{\mathbf{k}}\hat{L}Q_{\mathbf{k}}} Q_{\mathbf{k}}|\dot{a}(\mathbf{k})\rangle \approx k^2 \langle \boldsymbol{\varkappa}\cdot \mathbf{J}^a| Q_{\mathbf{k}} \frac{i}{z - Q_{\mathbf{k}}\hat{L}Q_{\mathbf{k}}} Q_{\mathbf{k}}|\boldsymbol{\varkappa}\cdot J^a\rangle,$$

we find

$$\lim_{\mathbf{k}\to 0} \langle \dot{a}(\mathbf{k})| \frac{i}{z - L} |\dot{a}(\mathbf{k})\rangle = \lim_{\mathbf{k}\to 0} \langle \dot{a}(\mathbf{k})| Q_{\mathbf{k}} \frac{i}{z - Q_{\mathbf{k}}\hat{L}Q_{\mathbf{k}}} Q_{\mathbf{k}}|\dot{a}(\mathbf{k})\rangle \quad (15.236)$$

where we must keep $z \neq 0$. Thus, we can write Eq. (15.234) as

$$\eta_A = D(0, 0) = \lim_{z\to 0} \left[\lim_{\mathbf{k}\to 0} \tfrac{1}{3}\langle \mathbf{J}^a(\mathbf{k})| \cdot \frac{i}{z - L} |\mathbf{J}^a(\mathbf{k})\rangle G_{aa}^{-1}(\mathbf{k})\right]$$
$$= \lim_{z\to 0} \lim_{\mathbf{k}\to 0} \frac{1}{3} \int_0^\infty dt \langle \mathbf{J}^a(\mathbf{k})| \cdot e^{-i(z-L)t}|\mathbf{J}^a(\mathbf{k})\rangle G_{aa}^{-1}(\mathbf{k}). \quad (15.237)$$

Eq. (15.237) is the form of the transport coefficients most commonly seen in the literature.

3. Hydrodynamic Modes: Broken Symmetries[15]

In this section we show how hydrodynamic modes appear when a symmetry is broken. Let us first consider the density, $\hat{b}(\mathbf{k}, t)$, of a quantity $\hat{B}$, which is not conserved during collisions, and find the general form of the equation of motion of $\hat{b}(\mathbf{k}, t)$. A typical example of such a quantity might be the local orientation of molecules in the system. The microscopic equation of motion for small **k** will be of the form

$$\frac{\partial}{\partial t}\hat{b}(\mathbf{k}, t) + \mathbf{k}\cdot\hat{\mathbf{J}}^{b}(\mathbf{k}, t) = \hat{\sigma}_{b}(\mathbf{k}, t) \tag{15.238}$$

where $\hat{\sigma}_b(\mathbf{k}, t)$ is a source term and

$$\lim_{\mathbf{k}\to 0} \hat{\sigma}_b(\mathbf{k}, t) = \hat{\sigma}_b(t) = \int d\mathbf{r}\hat{\sigma}(\mathbf{r}, t) \not\equiv 0. \tag{15.239}$$

Thus, the total time rate of change of b will not be zero:

$$\frac{d}{dt}\int d\mathbf{r}\hat{b}(\mathbf{r}, t) = \hat{\sigma}(t) \neq 0. \tag{15.240}$$

We will assume that the zero time correlation function, $G_{bb}(\mathbf{k}) \equiv \langle b(\mathbf{k})|b(\mathbf{k})\rangle$, is well defined in the limit $\mathbf{k} \to 0$. If $\hat{b}(\mathbf{k})$ has a well-defined sign under time reversal, we also have $\langle \dot{b}(\mathbf{k})|b(\mathbf{k})\rangle = 0$. From Eq. (15.194) the spectral density function can be written

$$\bar{S}_{bb}(\mathbf{k}, z) = \frac{iG_{bb}(\mathbf{k})}{z + i\sum_{bb}(\mathbf{k}, z)G_{bb}^{-1}(\mathbf{k})}. \tag{15.241}$$

If $\hat{b}(\mathbf{k}, t)$ is a nonconserved quantity, then to lowest order in **k**

$$\lim_{\mathbf{k}\to 0} \langle b(\mathbf{k})|\, Q_{\mathbf{k}} \frac{i}{z - \hat{L}Q_{\mathbf{k}}} Q_{\mathbf{k}}|b(\mathbf{k})\rangle \to \langle\sigma|\, Q_0 \frac{i}{z - LQ_0} Q_0|\sigma\rangle \tag{15.242}$$

and the memory function does not go to zero as $\mathbf{k} \to 0$. Thus, to lowest order in **k**, the spectral density has a pole independent of **k** and we can write

$$S_{bb}(\mathbf{k}, z) \sim \frac{iG_{bb}(0)}{z + i\dfrac{1}{\tau_b}} \tag{15.243}$$

where τ_b is the lifetime of mode b. The equation of motion of $\langle b(\mathbf{r}, t)\rangle$ will be of the form

$$\frac{\partial}{\partial t}\langle b(\mathbf{r}, t)\rangle = -\frac{1}{\tau_b}\langle \mathbf{b}(\mathbf{r}, t)\rangle \tag{15.244}$$

and the decay will be exponential.

We shall now introduce the idea of a broken symmetry and show how it gives rise to a hydrodynamic mode. We will let $Q(t)$ denote a Hermitian operator which commutes with the Hamiltonian so that

$$\frac{d\hat{Q}(t)}{dt} = \frac{1}{i\hbar}[\hat{Q}, \hat{H}] = 0. \tag{15.245}$$

Thus, $\hat{Q}(t)$ is a constant of the motion (it could be angular momentum, total momentum, particle number, etc.) and generates a continuous symmetry transformation, $U(\phi) = e^{i\phi\hat{Q}/\hbar}$. The Hamiltonian is invariant under this transformation:

$$\hat{U}(\phi)\hat{H}\hat{U}^{+}(\phi) = \hat{H}. \tag{15.246}$$

If we let $\hat{q}(\mathbf{r}, t)$ denote the local density of $\hat{Q}(t)$, we have a microscopic balance equation (for long wavelength):

$$\frac{\partial\hat{q}(\mathbf{r}, t)}{\partial t} + \nabla_{\mathbf{r}}\cdot\hat{\mathbf{J}}^{q}(r, t) = 0. \tag{15.247}$$

Let us now assume that there are two more local (not necessarily conserved) densities, $\hat{a}(\mathbf{r})$ and $\hat{b}(\mathbf{r})$, such that $\hat{Q}$ transforms $\hat{a}(\mathbf{r})$ into $\hat{b}(\mathbf{r})$:

$$\frac{1}{i\hbar}[\hat{Q}, \hat{a}(\mathbf{r})] = \hat{b}(\mathbf{r}). \tag{15.248}$$

If $\hat{Q}$ commutes with the density operator, $\hat{\rho}$, for the system, then we must have $\langle b(\mathbf{r})\rangle \equiv 0$. However, if $[\hat{Q}, \hat{\rho}] \neq 0$, then we can have $\langle b(\mathbf{r})\rangle \neq 0$, since

$$\frac{1}{i\hbar}\,\mathrm{Tr}\,\hat{\rho}[\hat{Q}, \hat{a}(\mathbf{r})] = \frac{1}{i\hbar}\,\mathrm{Tr}[\hat{\rho}, \hat{Q}]\hat{a}(\mathbf{r}) = \mathrm{Tr}\,\hat{\rho}\hat{b}(\mathbf{r}) = \langle b(\mathbf{r})\rangle. \tag{15.249}$$

For arbitrary $\hat{a}(\mathbf{r})$, if $\langle b(\mathbf{r})\rangle \neq 0$ we must have $[\hat{\rho}, \hat{Q}] \neq 0$ and $\hat{Q}$ does not commute with the state of the system. Thus, we say that the symmetry of the state of the system with respect to the continuous symmetry transformation, $\hat{Q}$, has been broken. The quantity $\langle b(\mathbf{r})\rangle$ is the order parameter which arises in a symmetry-breaking phase transition.

The fact that $\langle b(\mathbf{r})\rangle \neq 0$ means that there will be hydrodynamic waves for the density $\langle a(\mathbf{r}, t)\rangle$. These waves are called Goldstone[16] bosons. Let us rewrite Eq. (15.249) as

$$\int d\mathbf{r}\left\langle \frac{1}{i\hbar}[\hat{q}(\mathbf{r}, t), \hat{a}(\mathbf{r}')]\right\rangle = \langle B\rangle \tag{15.250}$$

where $\langle B\rangle$ = constant for a translationally invariant system. Then from Eq. (15.162) we have

$$\int d\mathbf{r}K''_{qa}(\mathbf{r} - \mathbf{r}', t) = K''_{qa}(\mathbf{k} = 0, t) = \frac{i}{2}\langle B\rangle. \tag{15.251}$$

But since $\hat{Q}$ is a conserved quantity, $\hat{Q}(t)$ is constant and $K''_{qa}(\mathbf{k} = 0, t)$ must be independent of t. The Fourier transform of Eq. (15.251) then yields

$$\chi''_{qa}(\mathbf{k} = 0, \omega) = i\pi\langle B\rangle\,\delta(\omega). \tag{15.252}$$

We now note that the $\lim_{\mathbf{k}\to 0}\chi''_{qa}(\mathbf{k}, \omega)$ is uniform (well behaved) if $q(\mathbf{r})$ and $a(\mathbf{r})$ are local densities, that is, depend only on properties of particles in the neighborhood of $\mathbf{r}$. (For systems with short-range interactions this will always be the case, but for long-range interactions it will not.) A function $f(\mathbf{k})$ is uniform at $\mathbf{k} = 0$ (that is, $\lim_{\mathbf{k}\to 0} f(\mathbf{k}) = f(\mathbf{k} = 0) = \int d\mathbf{r} f(\mathbf{r})$) if $f(\mathbf{r})$ falls off rapidly enough so that $\int d\mathbf{r} f(\mathbf{r}) < \infty$. Thus, $\lim_{\mathbf{k}\to 0} \chi''_{qa}(\mathbf{k}, \omega) = \chi''_{qa}(\mathbf{k} = 0, \omega)$ if $\chi''_{qa}(\mathbf{r} - \mathbf{r}', \omega) = (i/\hbar)\langle[\hat{q}(\mathbf{r}), \hat{a}(\mathbf{r}')]\rangle$ vanishes rapidly as $\mathbf{r} - \mathbf{r}' \to \infty$. This will be true if $q(\mathbf{r})$ and $a(\mathbf{r})$ are local densities, that is, depend on field operators in the neighborhood of $\mathbf{r}$ and $\mathbf{r}'$. Then from the commutation relations of field operators $[\hat{q}(\mathbf{r}), \hat{a}(\mathbf{r}')]$ must go to zero for $\mathbf{r} - \mathbf{r}' \to \infty$. However, the fact that $\chi''_{qa}(\mathbf{r}, \omega)$ decays rapidly will not mean that $\chi_{qa}(\mathbf{r}, \omega)$ decays rapidly.

If $\hat{q}(\mathbf{r})$ and $\hat{a}(\mathbf{r})$ are local operators, we can write

$$\lim_{\mathbf{k}\to 0} \chi''_{qa}(\mathbf{k}, \omega) = i\pi\langle B\rangle\, \delta(\omega). \tag{15.253}$$

Thus, there appears to be a mode whose energy goes to zero as $\mathbf{k} \to 0$. Let us assume this is true:

$$\chi''_{qa}(\mathbf{k}, \omega) = i\pi\langle B\rangle\, \delta(\omega - \omega(\mathbf{k})\rangle \tag{15.254}$$

where $\lim_{\mathbf{k}\to 0} \omega(\mathbf{k}) = 0$. From Eq. (15.40) the response function has the form

$$\chi_{qa}(\mathbf{k}, \omega) = \frac{i\langle B\rangle}{\omega(\mathbf{k}) - z}. \tag{15.255}$$

The mode with frequency $\omega(\mathbf{k})$ is the Goldstone boson.

The spectral density function for the modes $\hat{a}(\mathbf{k})$ is given in Eq. (15.197) where we assume that $\langle \dot{a}(\mathbf{k})|a(\mathbf{k})\rangle = 0$ as usual. Since $\hat{a}(\mathbf{k})$ is not a conserved quantity, we will have

$$\lim_{\mathbf{k}\to 0} \langle \dot{a}(\mathbf{k})|\, Q_{\mathbf{k}} \frac{i}{z - \hat{L}Q_{\mathbf{k}}}\, |\dot{a}(\mathbf{k})\rangle \equiv \text{constant} = \sigma_a.$$

The only possibility of obtaining a pole which goes to zero as $\mathbf{k} \to 0$ lies in the quantity $G_{aa}(\mathbf{k})$. Thus, we must determine how $G_{aa}(\mathbf{k})$ depends on $\mathbf{k}$. We first use the Schwartz inequality to write

$$\langle a(\mathbf{k})|a(\mathbf{k})\rangle\langle \dot{q}(\mathbf{k})|\dot{q}(\mathbf{k})\rangle \geqslant |\langle a(\mathbf{k})|\dot{q}(\mathbf{k})\rangle|^2. \tag{15.256}$$

If we assume that $\langle a(\mathbf{k})|a(\mathbf{k})\rangle = \beta^{-1}\chi_{aa}(\mathbf{k})$ (this is true when $\hat{a}(\mathbf{k})$ is a conserved quantity and when $\mathbf{k} = 0$, in general; we assume it is true for $\mathbf{k} \neq 0$), the Schwartz inequality takes the form

$$\chi_{aa}(\mathbf{k})\chi_{\dot{q}\dot{q}}(\mathbf{k}) \geqslant |\chi_{a\dot{q}}(\mathbf{k})|^2 \tag{15.257}$$

or, from Eq. (15.40),

$$\int \frac{d\omega}{\pi} \frac{\chi''_{aa}(\mathbf{k}, \omega)}{\omega} \int \frac{d\omega}{\pi} \frac{\chi''_{\dot{q}\dot{q}}(\mathbf{k}, \omega)}{\omega} \geqslant \left| \int \frac{d\omega}{\pi} \frac{\chi''_{a\dot{q}}(\mathbf{k}, \omega)}{\omega} \right|^2. \tag{15.258}$$

Eq. (15.258) is called the Bogoliubov inequality. Since $\hat{q}(\mathbf{k})$ is a conserved density, we can write $\omega\hat{q} = k\boldsymbol{\varkappa}\cdot\hat{\mathbf{J}}^q$. Thus, from the Bogoliubov inequality and Eq. (15.253), we obtain

$$G_{aa}(\mathbf{k}) = \beta^{-1}\chi_{aa}(\mathbf{k}) = \frac{\langle B\rangle^2}{k^2 R_a} \tag{15.259}$$

where

$$R_a \leqslant \beta \lim_{\mathbf{k}\to 0} \int \frac{d\omega}{\pi} \frac{\chi_{\varkappa\cdot\mathbf{J},\varkappa\cdot\mathbf{J}}(\mathbf{k},\omega)}{\omega} \tag{15.260}$$

is called the stiffness constant. A broken symmetry can give rise to a k^{-2} dependence in $G_{aa}(\mathbf{k})$ and the spectral density function for mode $\hat{a}(\mathbf{k})$ takes the form

$$\tilde{S}_{aa}(\mathbf{k}, z) = \frac{iG_{aa}(\mathbf{k})}{z + ik^2\langle B\rangle^{-2}R_a\sigma_a}. \tag{15.261}$$

Thus, $\hat{a}(\mathbf{k})$ is a hydrodynamic mode even though it is not a conserved quantity. In subsequent sections, we shall apply these ideas to some simple examples.

L. FERROMAGNETIC SYSTEM[15]

One of the simplest examples of a system with broken symmetry is a ferromagnetic solid. Let us consider a rigid lattice with spins localized at the lattice sites, and let us assume that the spins interact via a Hamiltonian of the form

$$\hat{H} = -\frac{1}{2}\sum_{\alpha\neq\beta} J(|\mathbf{r}_\alpha - \mathbf{r}_\beta|)\hat{\mathbf{S}}^\alpha\cdot\hat{\mathbf{S}}^\beta \tag{15.262}$$

where α and β refer to the particular lattice sites. The interaction Hamiltonian depends only on the absolute distance between sites and not on direction. The spin components of a given site satisfy the commutation relation

$$[\hat{S}_i{}^\alpha, \hat{S}_j{}^\alpha] = i\hbar\epsilon_{ijk}\hat{S}_k{}^\alpha. \tag{15.263}$$

Spin components at different sites commute since they are distinguishable. The direction of the spins is defined relative to the lattice. If we rotate all spins at once we do not change the Hamiltonian since it depends only on the relative directions of the spins. Thus, the total spin operator

$$\hat{\mathbf{S}}^{\text{tot}} = \sum_\alpha \hat{\mathbf{S}}^\alpha \tag{15.264}$$

commutes with the total Hamiltonian and is a constant of the motion

$$i\hbar\frac{d\hat{\mathbf{S}}^{\text{tot}}}{dt} = [\hat{\mathbf{S}}^{\text{tot}}, \hat{H}] = 0. \tag{15.265}$$

From Eq. (15.263) the total spin operator satisfies the commutation relation

$$[\hat{S}_i^{\text{tot}}, \hat{S}_j{}^\alpha] = i\hbar\epsilon_{ijk}\hat{S}_k{}^\alpha, \tag{15.266}$$

where $\epsilon_{123} = \epsilon_{312} = \epsilon_{231} = +1, \epsilon_{132} = \epsilon_{321} = \epsilon_{213} = -1$, and $\epsilon_{ijk} = 0$, otherwise, and is the generator of rotations in the system.

Let us now assume that the system has undergone a phase transition and that a spontaneous magnetization appears in the system; that is,

$$\langle \mathbf{S}^{\text{tot}} \rangle = V\mathbf{M}_0 \tag{15.267}$$

where $\mathbf{M}_0$ is the magnetization and V is the volume; $\mathbf{M}_0$ is the order parameter of the ferromagnetic phase. Eq. (15.267) implies that the rotational symmetry of the system has been broken. We can see this as follows. Let us assume that the magnetization is oriented in the z-direction. Thus,

$$\langle S_z^{\text{tot}} \rangle = VM_0 \tag{15.268}$$

but

$$\langle S_x^{\text{tot}} \rangle = \langle S_y^{\text{tot}} \rangle = 0. \tag{15.269}$$

From Eq. (15.266) we can write

$$[\hat{S}_x^{\text{tot}}, \hat{S}_y^{\text{tot}}] = i\hbar \hat{S}_z^{\text{tot}}. \tag{15.270}$$

Thus,

$$\begin{aligned} VM_0 &= \text{Tr}\ \hat{\rho}\hat{S}_z^{\text{tot}} = \frac{1}{i\hbar}\text{Tr}\ \hat{\rho}[\hat{S}_x^{\text{tot}}, \hat{S}_y^{\text{tot}}] \\ &= \frac{1}{i\hbar}\text{Tr}[\hat{\rho}, \hat{S}_x^{\text{tot}}]\hat{S}_y^{\text{tot}} = -\frac{1}{i\hbar}\text{Tr}[\hat{\rho}, \hat{S}_y^{\text{tot}}]\hat{S}_x^{\text{tot}} \end{aligned} \tag{15.271}$$

and, therefore,

$$[\hat{\rho}, \hat{S}_x^{\text{tot}}] \neq 0 \tag{15.272}$$

and

$$[\hat{\rho}, \hat{S}_y^{\text{tot}}] \neq 0. \tag{15.273}$$

By a similar analysis,

$$[\hat{\rho}, \hat{S}_z^{\text{tot}}] = 0. \tag{15.274}$$

Thus, the rotational symmetry of the system about the z-axis is preserved but about the x- and y-axes it is broken. In other words, if we rotate some vector of the system about the z-axis, we do not change the state of the system, because we do not change its angle with respect to $\mathbf{M}_0$. However, if we rotate it about the x- or y-axis we will change its angle with respect to $\mathbf{M}_0$ and therefore may change the state of the system.

As a consequence of this broken symmetry, the state is no longer described by the density matrix $\hat{\rho} = e^{-\beta\hat{H}}$ but by some new density matrix which contains an explicit dependence on $\hat{S}_z^{\text{tot}}$. In fact, we have already seen this in several examples in Chap. 9, for the condensed boson system in Chap. 12, and in the Ginzburg-Landau theory of Chap. 4. If we wish to investigate the possibility of having a phase transition, we can explicitly include the order parameter in the density matrix and then find a temperature-dependent equation for the order parameter. Above the transition temperature the stable solution is the one for which the order parameter is zero. Below the transition temperature the stable solution is one with a nonzero-order parameter. Another means of determining if the symmetry is broken is to break it by applying an external field. In our example of a ferromagnetic system, we would apply a magnetic field $\mathscr{H}_z$ in the z-direction and the state would take the form $\hat{\rho} = \exp(-\beta(\hat{H} - \hat{S}_z^{\text{tot}}\mathscr{H}_z))$. We then compute the magnetization and set the field to zero. If the system is ferromagnetic the magnetization will not vanish when the field is turned off. In this way, we create a preferred direction for the spontaneous magnetization. In both cases, we have broken the symmetry of the density matrix.

Let us now find the hydrodynamic modes in the system. We shall consider long-wavelength processes and therefore neglect any nontranslationally invariant effects due to the lattice. We define the magnetization density by

$$\hat{M}_i(\mathbf{r}) = \sum_\alpha \hat{S}_i^\alpha\, \delta(\mathbf{r} - \mathbf{r}^\alpha) \tag{15.275}$$

and note that

$$\frac{1}{i\hbar}[\hat{S}_y^{\text{tot}}, \hat{M}_x(\mathbf{r}')] = -\hat{M}_z(\mathbf{r}') \tag{15.276}$$

and

$$\frac{1}{i\hbar}[\hat{S}_x^{\text{tot}}, \hat{M}_y(\mathbf{r}')] = \hat{M}_z(\mathbf{r}') \tag{15.277}$$

where

$$\hat{S}_i^{\text{tot}} = \frac{1}{V}\int d\mathbf{r}\hat{M}_i(\mathbf{r}). \tag{15.278}$$

Thus, $\hat{S}_y^{\text{tot}}$ and $\hat{S}_x^{\text{tot}}$ play the role of $\hat{Q}$ and are conserved quantities, $\hat{M}_x(\mathbf{r}')$ and $\hat{M}_y(\mathbf{r}')$ play the role of $\hat{a}(\mathbf{r})$, and $\hat{M}_z(\mathbf{r})$ plays the role of $\hat{b}(\mathbf{r})$. We note that $\hat{M}_i(\mathbf{r})$ is a conserved density and obeys the equation

$$\frac{\partial}{\partial t}\hat{M}_i(\mathbf{r}, t) + \sum_j \nabla_{\mathbf{r}_j}\cdot\hat{J}_{ij}^M(\mathbf{r}, t) = 0. \tag{15.279}$$

In this example, $\hat{q}$, $\hat{a}$, and $\hat{b}$ are all conserved densities.

We now wish to write an equation for the spectral density. We use the Bogoliubov inequality to write

$$\beta G_{M_xM_x}(\mathbf{k}) = \chi_{M_xM_x}(\mathbf{k}) \geqslant \frac{\left|\int \frac{d\omega}{\pi}\frac{1}{\omega}\chi''_{M_x\dot{M}_y}(\mathbf{k}, \omega)\right|^2}{\int \frac{d\omega}{\pi}\frac{1}{\omega}\chi''_{\dot{M}_y\dot{M}_y}(\mathbf{k}, \omega)} \tag{15.280}$$

where $\chi''_{\dot{M}_y\dot{M}_y}(\mathbf{k}, \omega)$ is proportional to the commutator of densities $\dot{M}_y(\mathbf{r})$. Since $\hat{M}_y$ is a conserved density, we have

$$\chi''_{\dot{M}_y\dot{M}_y}(\mathbf{k}, \omega) = \omega^2\chi''_{M_yM_y}(\mathbf{k}, \omega) = k^2\chi''_{\boldsymbol{\kappa}\cdot\mathbf{J}_y^M\boldsymbol{\kappa}\cdot\mathbf{J}_y^M}(\mathbf{k}, \omega). \qquad (15.281)$$

Thus,

$$G_{M_xM_x}(\mathbf{k}) = \beta^{-1}\chi_{M_xM_x}(\mathbf{k}) = \frac{\left|\int \frac{d\omega}{\pi} \chi''_{M_x\dot{M}_y}(\mathbf{k}, \omega)\right|^2}{k^2R_{yy}} \qquad (15.282)$$

where

$$R_{yy} \leqslant \beta \int \frac{d\omega}{\pi} \chi''_{\boldsymbol{\kappa}\cdot\mathbf{J}_y^M\boldsymbol{\kappa}\cdot\mathbf{J}_y^M}(\mathbf{k}, \omega). \qquad (15.283)$$

The numerator of Eq. (15.282) is easily found from Eq. (15.277). We obtain

$$\frac{2}{V}\int d\mathbf{r} \int d\mathbf{r}'\chi''_{M_xM_y}(\mathbf{r}, \mathbf{r}'; t = 0) = \int \frac{d\omega}{\pi} \chi''_{M_xM_y}(\mathbf{k} = 0, \omega) = iM_0. \qquad (15.284)$$

For small **k**,

$$G_{M_xM_x}(\mathbf{k}) = \beta^{-1}\chi_{M_xM_x}(\mathbf{k}) = \frac{{M_0}^2}{k^2R_{yy}}. \qquad (15.285)$$

Similarly,

$$G_{M_yM_y}(\mathbf{k}) = \beta^{-1}\chi_{M_yM_y}(\mathbf{k}) = \frac{{M_0}^2}{k^2R_{xx}}. \qquad (15.286)$$

Thus, if we apply a magnetic field in the x-direction or in the y-direction, it becomes very easy to obtain a large magnetization in the x- or y-direction, respectively, since the response functions $\chi_{M_xM_x}(\mathbf{k})$ and $\chi_{M_yM_y}(\mathbf{k})$ become infinite in the limit $\mathbf{k} \to 0$. This corresponds to the fact that it takes only an infinitesimal external magnetic field to flip the magnetization $\mathbf{M}_0$ from the z-direction to the x- or y-direction.

The spectral density matrix for the system takes the form

$$\bar{\bar{\mathbf{S}}}(\mathbf{k}, z) = \begin{pmatrix} S_{M_xM_x}(\mathbf{k}, z) & S_{M_xM_y}(\mathbf{k}, z) \\ S_{M_yM_x}(\mathbf{k}, z) & S_{M_yM_y}(\mathbf{k}, z) \end{pmatrix}. \qquad (15.287)$$

From Eq. (15.204) the equation for $\bar{\bar{\mathbf{S}}}(\mathbf{k}, z)$ can be written

$$(z\bar{\bar{\mathbf{U}}} - i\bar{\bar{\boldsymbol{\omega}}}(\mathbf{k})\cdot\bar{\bar{\mathbf{G}}}^{-1}(\mathbf{k}) + \bar{\bar{\boldsymbol{\Sigma}}}(\mathbf{k}, z)\cdot\bar{\bar{\mathbf{G}}}^{-1}(\mathbf{k}))\cdot\bar{\bar{\mathbf{S}}}(\mathbf{k}, z) = i\bar{\bar{\mathbf{G}}}(\mathbf{k}) \qquad (15.288)$$

where

$$\bar{\bar{\boldsymbol{\omega}}}(\mathbf{k}) = \begin{pmatrix} G_{\dot{M}_xM_x}(\mathbf{k}) & G_{\dot{M}_xM_y}(\mathbf{k}) \\ G_{\dot{M}_yM_x}(\mathbf{k}) & G_{\dot{M}_yM_y}(\mathbf{k}) \end{pmatrix}. \qquad (15.289)$$

$\overline{\overline{G}}(\mathbf{k})$ has elements $G_{M_iM_j}(\mathbf{k})$ where $i, j = (x, y)$, and the elements of $\overline{\overline{\Sigma}}(\mathbf{k}, z)$ are

$$\Sigma_{M_iM_j}(\mathbf{k}, z) = \langle \dot{M}_i(\mathbf{k})| Q_{\mathbf{k}} \frac{i}{z - Q_{\mathbf{k}}\hat{L}Q_{\mathbf{k}}} Q_{\mathbf{k}}|\dot{M}_j(\mathbf{k})\rangle$$
$$= k^2\langle(\boldsymbol{\varkappa}\cdot\mathbf{J}_i^M)| Q_{\mathbf{k}} \frac{i}{z - Q_{\mathbf{k}}\hat{L}Q_{\mathbf{k}}} Q_{\mathbf{k}}|(\boldsymbol{\varkappa}\cdot\mathbf{J}_j^M)\rangle \equiv k^2 D_{M_iM_j}(\mathbf{k}, z). \tag{15.290}$$

We have used the fact that the magnetization is a conserved quantity.

We now note that $G_{\dot{M}_xM_x}(\mathbf{k}) = G_{\dot{M}_yM_y}(\mathbf{k}) = 0$ because of time-reversal symmetry as usual (cf. Eq. (15.174)). Furthermore, in the absence of a magnetic field $G_{M_xM_y}(\mathbf{k}) = G_{M_yM_x}(\mathbf{k}) = 0$ since M_x and M_y are independent. Thus,

$$\overline{\overline{\omega}}(\mathbf{k}) \approx \begin{pmatrix} 0 & iM_0 \\ -iM_0 & 0 \end{pmatrix}, \tag{15.291}$$

$$\overline{\overline{G}}(\mathbf{k}) = \begin{pmatrix} \dfrac{M_0^2}{k^2R} & 0 \\ 0 & \dfrac{M_0^2}{k^2R} \end{pmatrix}, \tag{15.292}$$

where we have set $R_{xx} = R_{yy} = R$, and

$$\overline{\overline{\omega}}(\mathbf{k})\cdot\overline{\overline{G}}^{-1}(\mathbf{k}) = \begin{pmatrix} 0 & i\dfrac{Rk^2}{M_0} \\ -i\dfrac{Rk^2}{M_0} & 0 \end{pmatrix}. \tag{15.293}$$

Since the magnetization is a conserved quantity, the memory matrix $\overline{\overline{\Sigma}}(\mathbf{k}, z)$ to lowest order in k can be written

$$\overline{\overline{\Sigma}}(\mathbf{k}, z) = \begin{pmatrix} k^2D_{xx} & k^2D_{xy} \\ k^2D_{yx} & k^2D_{yy} \end{pmatrix} \tag{15.294}$$

where $D_{xx} = D_{yy}$ and $D_{xy} = -D_{yx}$.

In the absence of dissipation, Eq. (15.288) takes the form

$$\begin{pmatrix} z & \dfrac{Rk^2}{M_0} \\ -\dfrac{Rk^2}{M_0} & z \end{pmatrix}\begin{pmatrix} \overline{S}_{xx}(\mathbf{k}, z) & \overline{S}_{xy}(\mathbf{k}, z) \\ \overline{S}_{yx}(\mathbf{k}, z) & \overline{S}_{yy}(\mathbf{k}, z) \end{pmatrix} = i\begin{pmatrix} \dfrac{M_0^2}{k^2R} & 0 \\ 0 & \dfrac{M_0^2}{k^2R} \end{pmatrix}. \tag{15.295}$$

The poles of the spectral density function occur when

$$\det\begin{pmatrix} z & \dfrac{Rk^2}{M_0} \\ -\dfrac{Rk^2}{M_0} & z \end{pmatrix} = 0 \tag{15.296}$$

or when $z = \pm i(Rk^2/M_0)$. Thus, in the absence of dissipation, there are two propagating spin waves (undamped). But they differ from sound waves (phonons) in that their dispersion relation depends on k^2 rather than k. The dispersion relations in the presence of dissipation can be found by setting the determinant of

$$\det(z\bar{\bar{\mathbf{U}}} - i\bar{\bar{\boldsymbol{\omega}}}(\mathbf{k})\cdot\bar{\bar{\mathbf{G}}}^{-1}(\mathbf{k}) + i\bar{\bar{\boldsymbol{\Sigma}}}(k, z)\bar{\bar{\mathbf{G}}}^{-1}(k)) = 0$$

and solving for z. Because the spin density is a conserved quantity, the damping terms will be of order k^4.

M. BROKEN SYMMETRY IN SUPERFLUIDS

We would like to say a few words about the symmetry that is broken in superfluids. As we have seen in Sec. 12.H, the condensed phase of He^4 is described by a nonzero-order parameter $\langle \hat{a}_0 \rangle \neq 0$ or equivalently $\langle \hat{\Psi}(r) \rangle \neq 0$. At the λ-transition in He^4 gauge symmetry is broken. The generator of a gauge transformation is the number operator,

$$\hat{N} = \int d\mathbf{r} \hat{\Psi}^+(\mathbf{r})\hat{\Psi}(\mathbf{r}). \tag{15.297}$$

If particles are conserved, the Hamiltonian generally commutes with the number operator,

$$[\hat{H}, \hat{N}] = 0. \tag{15.298}$$

The field operators, however, do not. From the commutation relations in App. B, we find

$$[\hat{N}, \hat{\Psi}(\mathbf{r})] = -\hat{\Psi}(\mathbf{r}) \tag{15.299}$$

and

$$[\hat{N}, \hat{\Psi}^+(\mathbf{r})] = \hat{\Psi}^+(\mathbf{r}). \tag{15.300}$$

The number operator generates a transformation which changes the phase of the field operators. Thus, from Eqs. (15.299) and (15.300), we have

$$e^{i\alpha\hat{N}}\hat{\Psi}(\mathbf{r})\, e^{-i\alpha\hat{N}} = \hat{\Psi}(\mathbf{r})\, e^{-i\alpha} \tag{15.301}$$

and

$$e^{i\alpha\hat{N}}\hat{\Psi}^+(\mathbf{r})\, e^{-i\alpha\hat{N}} = \hat{\Psi}^+(\mathbf{r})\, e^{i\alpha}. \tag{15.302}$$

If the system undergoes a transition to a state in which the field operators have nonzero expectation value, then the gauge symmetry of the state must be broken since

$$\langle \hat{\Psi}(\mathbf{r}) \rangle = \mathrm{Tr}\, \hat{\rho}\hat{\Psi}(\mathbf{r}) = -\mathrm{Tr}\, \hat{\rho}[\hat{N}, \hat{\Psi}(\mathbf{r})] = \mathrm{Tr}[\hat{\rho}, \hat{N}]\hat{\Psi}(\mathbf{r}) \tag{15.303}$$

and, therefore,

$$[\hat{\rho}, \hat{N}] \neq 0. \tag{15.304}$$

Thus, to describe such a system we must use a density operator with broken gauge symmetry built in. This is exactly what we did in Sec. 12.H to describe condensed boson systems and in Sec. 9.E to describe condensed Fermi systems (which are also examples of systems with broken gauge symmetry). The derivation of the superfluid hydrodynamic equations using the methods discussed in this chapter, and other interesting examples, is discussed in the monograph by Forster.[15]

REFERENCES

1. N. Wiener, Acta Math. *55* 117 (1930).
2. A. Khintchine, Math. Annalen *109* 604 (1934).
3. S. R. de Groot and P. Mazur, *Nonequilibrium Thermodynamics* (North-Holland Publishing Co., Amsterdam, 1969).
4. P. C. Martin in *Many Body Physics*, ed. C. de Witt and R. Balian, Les Houches Summer School 1967 (Gordon and Breach, New York, 1968).
5. L. Landau and G. Placzek, Physik Z. Sowjet Union *5* 172 (1934).
6. L. Van Hove, Phys. Rev. *95* 249 (1954).
7. K. I. Komarov and I. Z. Fisher, Sov. Phys. J.E.T.P. *16* 1358 (1963).
8. J. D. Jackson, *Classical Electrodynamics* (John Wiley & Sons, New York, 1962).
9. J. Frenkel, *Kinetic Theory of Liquids* (Oxford University Press, Oxford, 1946).
10. R. D. Mountain, Rev. Mod. Phys. *38* 205 (1966).
11. H. E. Stanley, *Introduction to Phase Transitions and Critical Phenomena* (Oxford University Press, Oxford, 1971).
12. L. Kadanoff and P. C. Martin, Ann. Phys. *24* 419 (1963).
13. Reprinted in *The Equilibrium Theory of Classical Fluids*, ed. H. L. Frisch and J. L. Lebowitz (W. A. Benjamin, New York, 1964).
14. R. Kubo, J. Phys. Soc. *12* 570 (1957).
15. D. Forster, *Hydrodynamic Fluctuations, Broken Symmetry and Correlations Functions* (W. A. Benjamin, New York, 1975).
16. R. Zwanzig in *Lectures in Theoretical Physics*, Vol. 3 (Wiley-Interscience, New York, 1961).
17. H. Mori, Prog. Theor. Phys. (Kyoto) *34* 423 (1965).
18. I. Prigogine, Cl. George, F. Henin, and L. Rosenfeld, Chemica Scripta *4* 5 (1973).
19. J. Goldstone, Nuovo Cimento *19* 154 (1961).
20. S. Ma and G. F. Mazenko, Phys. Rev. B*11* 4077 (1975).

PROBLEMS

1. A paramagnetic system sits in a constant external field $\mathscr{H}_0\hat{z}$ and therefore has constant magnetization $M_0\hat{z}$. Assume this is the equilibrium configuration of the system and that a small uniform time-dependent magnetic field $\mathscr{H}_1(t) = \mathscr{H}_1\hat{x}\cos\omega_0 t$

is applied to the system. The magnetization then becomes time dependent. Its equation of motion can be written

$$\frac{d\langle \mathbf{M} \rangle}{\partial t} = -\gamma \langle \mathbf{M} \rangle \times \mathscr{H}(t) - \mathbf{D}(\langle \mathbf{M} \rangle)$$

(Bloch equations) where $\langle \mathbf{M} \rangle = \operatorname{Tr} \hat{\rho}\mathbf{M}$, $\mathscr{H}(t) = \mathscr{H}_0\hat{z} + \mathscr{H}_1\hat{x}$, and $D(\langle \mathbf{M} \rangle)$ is a damping term due to interactions between particles in the medium. The constant $\gamma = g\mu/\hbar$, where g is the Landé g-factor and μ is the magnetic moment of particles in the medium. Assume that $\langle \mathbf{M}(t) \rangle$ can be written $\langle \mathbf{M}(t) \rangle = M_x(t)\hat{x} + M_y(t)\hat{y} + (M_0 - M_z(t))\hat{z}$. Write the equations of motion for $M_x(t)$, $M_y(t)$, and $M_z(t)$, assuming $D_x = M_x/T_2$, $D_y = M_y/T_2$, and $D_z = M_z/T_1$. The equations you derive will be nonlinear functions of $\mathscr{H}_1$, M_x, M_y, and M_z. We assume that these are small, so the equations can be linearized. If we define the response matrix $\overline{\overline{\mathbf{K}}}(\tau)$ from the equation $\langle \mathbf{M}(t) \rangle = \int_{-\infty}^{\infty} dt' \overline{\overline{\mathbf{K}}}(t - t') \cdot \mathscr{H}(t')$, find $\chi_{xx}(\omega)$ and $K_{xx}(\tau)$. Compute the average power absorbed during the period $T = 2\pi/\omega_0$. Write the expression for $\chi_{xx}(\omega)$ in the limit $1/T_1 \to 0$ and $1/T_2 \to 0$. (Note that $\hat{x}$, $\hat{y}$, and $\hat{z}$ are unit vectors in the x, y, and z directions, respectively.)

2. Hydrodynamic equations describe the evolution of slowly varying modes in a system. We can account for the effects of rapidly fluctuating modes on hydrodynamic evolution by adding a random noise term to the hydrodynamic equations. In general we can write

$$\frac{d\alpha_i(t)}{dt} = -L_i\alpha_i(t) + f_i(t),$$

where L_i is an Onsager coefficient and is related to the inverse damping time of the mode and $f_i(t)$ is a delta correlated noise source, $\langle f_i(t) f_j(t') \rangle = 2\Gamma_i \delta_{ij}\, \delta(t - t')$ and Γ_i is the strength of the noise. Assume that $\langle \alpha_i(t) f_i(t) \rangle = 0$ ($\alpha_i(t)$ varies slowly compared to $f_i(t)$) and show that $\Gamma_i = L_i \langle \alpha_i^2 \rangle_{\text{eq}}$. Show that the power spectral density is given by $S_{ij}(\omega) = 2\Gamma_i \delta_{ij} [\omega^2 + L_i^2]^{-1}$ (Hint: see problem 6.10).

3. Using the microscopic definition of the imaginary part of the response function $K''_{a_i a_j}(\mathbf{r}t, \mathbf{r}'t')$ given in Eq. (15.106), prove that

$$\begin{aligned} K''_{a_i a_j}(\mathbf{r}, t; \mathbf{r}', t') &= -K''_{a_i a_j}(\mathbf{r}', t'; \mathbf{r}, t) = -[K''_{a_i a_j}(\mathbf{r}, t; \mathbf{r}', t')]^* \\ &= -\epsilon_i{}^T \epsilon_j{}^T K''_{a_i a_j}(\mathbf{r}, -t; \mathbf{r}', -t') \end{aligned}$$

and that

$$\begin{aligned} \chi''_{a_i a_j}(\mathbf{r}, \mathbf{r}'; \omega) &= -\chi''_{a_i a_j}(\mathbf{r}', \mathbf{r}; -\omega) = -[\chi''_{a_i a_j}(\mathbf{r}, \mathbf{r}'; -\omega)]^* \\ &= -\epsilon_i{}^T \epsilon_j{}^T \chi''_{a_i a_j}(\mathbf{r}, \mathbf{r}', -\omega) \end{aligned}$$

where * denotes complex conjugation.

4. Prove that

$$\lim_{k \to 0} \chi_{nn}(\mathbf{k}) = n\left(\frac{\partial n}{\partial P}\right)_T, \qquad \lim_{k \to 0} \chi_{ss}(\mathbf{k}) = mnc_pT, \quad \text{and} \quad \lim_{k \to 0} \chi_{ns}(\mathbf{k}) = T\left(\frac{\partial n}{\partial T}\right)_P.$$

5. The response matrix decomposes into two independent parts,

$$\chi_{v_i v_j}(\mathbf{k}, z) = \kappa_i \kappa_j \chi_l(\mathbf{k}, z) + (\delta_{ij} - \kappa_i \kappa_j)\chi_t(\mathbf{k}, z),$$

where $\chi_l(\mathbf{k}, z)$ and $\chi_t(\mathbf{k}, z)$ are the longitudinal and transverse parts and $\boldsymbol{\varkappa}$ is the unit vector $\boldsymbol{\varkappa} = \mathbf{k}/|\mathbf{k}|$. Prove that $\lim_{\mathbf{k}\to 0} \chi_l(\mathbf{k}) = n/m$, where n is the particle density and m is the mass. (*Hint:* use the microscopic response function and the microscopic definition of various densities given in Chap. 7; also use the microscopic continuity equation.) For a classical fluid show that $\lim_{\mathbf{k}\to 0} \chi_t(\mathbf{k}) = n/m$. (This is no longer true for a superfluid.)

6. For a classical fluid, prove that

$$\eta = \lim_{\omega\to 0} \lim_{\mathbf{k}\to 0} \frac{\omega m^2}{k^2} \chi_t''(\mathbf{k}, \omega), \qquad (\tfrac{4}{3}\eta + \zeta) = \lim_{\omega\to 0} \lim_{\mathbf{k}\to 0} \frac{\omega}{k^2} \chi_l''(\mathbf{k}, \omega),$$

and

$$KT = \lim_{\omega\to 0} \lim_{\mathbf{k}\to 0} \left[\frac{\omega}{k^2} \chi_{ss}''(\mathbf{k}, \omega)\right]$$

and, therefore, that

$$\eta(\delta_{ij} + \tfrac{1}{3}\kappa_i\kappa_j) + \zeta\kappa_i\kappa_j = \lim_{\omega\to 0} \lim_{\mathbf{k}\to 0} m^2 \chi_{v_i v_j}(\mathbf{k}, \omega).$$

Then use these results to show that

$$KT = \frac{1}{6Vk_BT} \int_0^\infty dt\, e^{-\epsilon t} \langle \mathbf{J}^s_{\text{tot}}(t) \cdot \mathbf{J}^s_{\text{tot}}(0) \rangle$$

and

$$\eta(\delta_{ij} + \tfrac{1}{3}\kappa_i\kappa_j) + \zeta\kappa_i\kappa_j = \lim_{\mathbf{k}\to 0} \frac{1}{2Vk_BT} \int_0^\infty dt\, e^{-\epsilon t} \langle \boldsymbol{\varkappa} \cdot \overline{\overline{\mathbf{P}}}_{\text{tot}}(t) \overline{\overline{\mathbf{P}}}_{\text{tot}}(0) \cdot \boldsymbol{\varkappa} \rangle$$

where $\mathbf{J}^s_{\text{tot}}$ is the microscopic entropy current and $\overline{\overline{\mathbf{P}}}_{\text{tot}}$ is the total microscopic momentum current.

7. Find the poles for small k of the spectral density function of the ferromagnetic system described in Section 15.L in the presence of dissipation.

8. In a ferromagnetic system, the spin density obeys the linearized hydrodynamic equation

$$\frac{dS(\mathbf{k}, t)}{dt} = -Dk^2 S(\mathbf{k}, t) + f(\mathbf{k}, t),$$

where D is the diffusion coefficient and the noise source, $f(\mathbf{k}, t)$ is delta correlated $\langle f(\mathbf{k}, t) f(\mathbf{k}', t')\rangle = 2\Gamma k^2\, \delta(\mathbf{k} - \mathbf{k}')\, \delta(t - t')$ (cf. problem 15.2). Show that $\Gamma = D\langle S^2(\mathbf{k})\rangle_{\text{eq}}$. Near the critical point $\langle S^2(\mathbf{k})\rangle_{\text{eq}} \approx (k^2 + \xi^2)^{-1}$, where ξ^2 is the correlation range ($\xi \to \infty$ as $T \to T_C$, where T_C is the critical temperature). (cf. Chap. 10.C.3)). Assume that Γ remains well behaved at the critical point. Find the power spectral density and show that as $T \to T_C$, $\omega(\mathbf{k}) \sim k^4$ whereas away from the critical point $\omega(\mathbf{k}) \sim k^2$ for small k. This is called *critical slowing down.* (Note: in this problem we have used a mean field model and as usual it does not quite agree with reality. In fact, Γ does diverge at the critical point (cf. Ref. 15.20).

16
Long Time Tails

A. INTRODUCTORY REMARKS

In recent years, computer experiments and a more careful study of the correlation functions have shown that the picture of transport phenomena given by Boltzmann is not completely correct. It is now realized that many-body processes play a very important role in determining the long time behavior of the correlation functions and, at best, Boltzmann's equation can only describe short time processes. The Boltzmann equation predicts exponential decay of the correlation functions, but simple hydrodynamic arguments give a decay of the form $t^{-d/2}$ (long time tail), where d is the dimension of the system. These long time tails have been observed in computer experiments and are now well established theoretically.

The many-body processes which give rise to the long time tails have forced us to change our whole approach to linear hydrodynamics. It was long thought that corrections to the Navier-Stokes equations could be obtained by expanding the macroscopic currents in terms of higher order gradients, thereby introducing new transport coefficients to account for shorter wavelength effects. However, this is now known to be incorrect. Microscopic calculations have shown that the transport coefficients in the so-called Burnett order (coefficients of the fourth-order gradients) and higher order are infinite; therefore, a simple power series expansion of the hydrodynamic frequencies in terms of the wave vector will not work. The hydrodynamic frequencies are nonanalytic functions of the wave vectors. In this chapter, we will derive the long time tails from both macroscopic and a microscopic point of view and show that their origin is the hydrodynamic modes in the fluid.

B. HYDRODYNAMIC ORIGIN OF THE LONG TIME TAILS

Transport coefficients can be computed by integrating the appropriate time-dependent current correlation functions over an infinite time interval (from

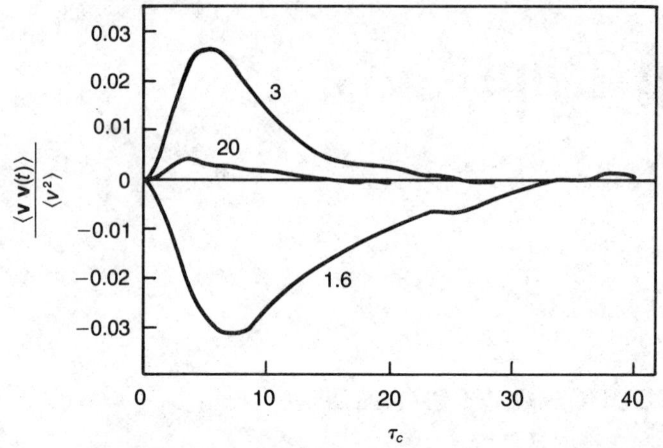

Fig. 16.1. A plot of difference between the velocity autocorrelation function obtained from molecular dynamics calculations for hard spheres and that given by Boltzmann theory: the correlation function is plotted as a function of mean collision times for 108 particles at $V/V_0 = 1.6$, 3, and 20 (based on Ref. 16.1).

$t = 0$ to $t = \infty$) (cf. Chap. 15). Therefore, if the transport coefficients are to be finite, the correlation functions must decay sufficiently rapidly in time. It was long thought that the correlation functions would decay exponentially since this is what is predicted by the Boltzmann equation. However, in 1967, Alder and Wainwright,[1] in a computer calculation of the velocity autocorrelation function for a gas of hard spheres, found that the decay of the autocorrelation function was exponential only for a few mean free times and that after a longer time the decay went as $t^{-3/2}$ (in three dimensions). In other words, it had a long time tail. In Fig. 16.1, we give their results for various densities. The figure gives the difference between the exact velocity autocorrelation function for a hard-sphere gas and that computed using the Boltzmann equation. Both are plotted as a function of the number of collisions that have occurred. The different curves correspond to different particle densities (V_0 is the close-packed volume). We see that after many collisions have occurred the correlation functions decay away very slowly—as $t^{-3/2}$.

The long time tails are now known to be due to the long-lived hydrodynamic modes in the fluid. Derivation of the long time tails based on hydrodynamic arguments has been given by a number of authors. The approach we shall follow is that of Ernst, Hauge, and van Leeuven.[2] In later sections, we shall discuss the tails from a microscopic point of view.

Let us restrict ourselves to consideration of the coefficient of shear viscosity, which can be written

$$\eta = \lim_{\delta \to 0} \beta \int_0^\infty dt\, e^{-\delta t} C_\eta(t) \tag{16.1}$$

where the correlation function $C_\eta(t)$ is defined

$$C_\eta(t) = \frac{1}{V}\langle J_{xy}J_{xy}(t)\rangle_{\text{eq}} = \frac{1}{V}\text{Tr}\{\rho_{\text{eq}}J_{xy}J_{xy}(t)\}. \tag{16.2}$$

The trace indicates an integration over all phase space coordinates, and the kinetic part of the microscopic momentum current is defined

$$J_{xy} = \sum_{\alpha=1}^{N} \frac{p_{\alpha x}p_{\alpha y}}{m} \tag{16.3}$$

(cf. Sec. 7.D). The equilibrium probability density is given by

$$\rho_{\text{eq}}(\mathbf{p}^N, \mathbf{q}^N) = e^{-\beta H^N(\mathbf{p}^N, \mathbf{q}^N)}. \tag{16.4}$$

where the N-body Hamiltonian is defined

$$H^N(\mathbf{p}^N, \mathbf{q}^N) = \sum_{i=1}^{N} \frac{p_i^2}{2m} + \frac{1}{2}\sum_{i\neq j=1}^{N} V(|\mathbf{q}_i - \mathbf{q}_j|). \tag{16.5}$$

We can write Eq. (16.2) in the form

$$C_\eta(t) = \frac{N}{V}\langle j_{xy}(\mathbf{p}_1)J_{xy}(t)\rangle_{\text{eq}} \tag{16.6}$$

where

$$j_{xy}(\mathbf{p}_1) = \frac{p_{1x}p_{1y}}{m} \tag{16.7}$$

by making appropriate changes in dummy variables in Eq. (16.2).

It is now convenient to rewrite Eq. (16.6) in a form in which the currents contain spatial dependence. We shall assume that at time $t = 0$ particle 1 has velocity $\mathbf{p}_1(0)/m = \mathbf{v}_0$ and lies in a neighborhood of the point $\mathbf{r}_0$ with a probability density given by $W(\mathbf{q}_1(0) - \mathbf{r}_0)$. The correlation function $C_\eta(t)$ can be written

$$C_\eta(t) = \frac{N}{V}\int d\mathbf{v}_0 \int d\mathbf{r}_0 j_{xy}(m\mathbf{v}_0)\left\langle \delta\left(\frac{\mathbf{p}_1(0)}{m} - \mathbf{v}_0\right) W(\mathbf{q}_1(0) - \mathbf{r}_0)J_{xy}(t)\right\rangle_{\text{eq}} \tag{16.8}$$

where

$$\int d\mathbf{q}_1 W(\mathbf{q}_1(0) - \mathbf{r}_0) \equiv 1. \tag{16.9}$$

Because the interaction potential, $V(|\mathbf{q}_1 - \mathbf{q}_2|)$, depends only on relative positions of the particles, the probability density, $\rho_{\text{eq}}(\mathbf{p}^N, \mathbf{q}^N)$, is independent of $\mathbf{q}_1$ and the integration over $\mathbf{q}_1$ can be done trivially and gives a factor V (the volume). Inclusion of the distribution $W(\mathbf{q}_1 - \mathbf{r}_0)$ in Eq. (16.8) is a device which will be helpful later in determining the initial conditions. We next rewrite $J_{xy}(t)$ in terms of a momentum flux density by noting

$$J_{xy}(t) = \sum_{i=1}^{N} \frac{p_{ix}(t)p_{iy}(t)}{m} = \int d\mathbf{r} \sum_{i=1}^{N} \frac{p_{ix}(t)p_{iy}(t)}{m}\, \delta(\mathbf{q}_i(t) - \mathbf{r}). \tag{16.10}$$

Eq. (16.8) then becomes

$$C_\eta(t) = \frac{N}{V}\int d\mathbf{v}_0 \int d\mathbf{r}_0 j_{xy}(m\mathbf{v}_0) \int d\mathbf{r} \left\langle \delta\left(\frac{\mathbf{p}_1(0)}{m} - \mathbf{v}_0\right) W(\mathbf{q}_1(0) - \mathbf{r}_0) J_{xy}(\mathbf{r}, t) \right\rangle_{\text{eq}} \tag{16.11}$$

where

$$J_{xy}(\mathbf{r}, t) \equiv \sum_{i=1}^{N} \frac{p_{ix}(t)p_{iy}(t)}{m}\, \delta(\mathbf{q}_i(t) - \mathbf{r}) \tag{16.12}$$

and is a function of the initial phase space variables.

It is now possible to write Eq. (16.8) in terms of a conditional average value, $\langle J_{xy}(\mathbf{r}, t)\rangle_{\text{noneq}}$, through the definition

$$\begin{aligned} &\left\langle \delta\left(\frac{\mathbf{p}_1}{m} - \mathbf{v}_0\right) W(\mathbf{q}_1 - \mathbf{r}_0) J_{xy}(\mathbf{r}, t) \right\rangle_{\text{eq}} \\ &\quad \equiv \left\langle \delta\left(\frac{\mathbf{p}_1}{m} - \mathbf{v}_0\right) W(\mathbf{q}_1 - \mathbf{r}_0) \right\rangle_{\text{eq}} \langle J_{xy}(\mathbf{r}, t)\rangle_{\text{noneq}} \\ &\quad = \frac{1}{V}\left(\frac{\beta}{2\pi m}\right)^{3/2} e^{-\beta m v_0^2/2} \langle J_{xy}(\mathbf{r}, t)\rangle_{\text{noneq}} \end{aligned} \tag{16.13}$$

where we can show by direct integration that

$$\left\langle \delta\left(\frac{\mathbf{p}_1}{m} - \mathbf{v}_0\right) W(\mathbf{q}_1 - \mathbf{r}_0) \right\rangle_{\text{eq}} \equiv \frac{1}{V}\left(\frac{\beta}{2\pi m}\right)^{3/2} e^{-\beta m v_0^2/2}. \tag{16.14}$$

An important assumption we must now make is that the nonequilibrium distribution function approaches a local equilibrium distribution faster than the asymptotic decay of the correlation function. Then, after sufficiently long time, the one-particle distribution function can be written as a local equilibrium distribution function of the form

$$f_1(\mathbf{p}, \mathbf{r}, t) = n(\mathbf{r}, t)\left[\frac{1}{2\pi m k_B T(\mathbf{r}, t)}\right]^{3/2} \exp\left[-\frac{1}{2}\frac{(\mathbf{p} - m\mathbf{v}(\mathbf{r}, t))^2}{m k_B T(\mathbf{r}, t)}\right]. \tag{16.15}$$

The local equilibrium distribution function describes a situation where each small volume element of the fluid is in equilibrium, but the state variables change in value from one volume element to another (the volume elements contain enough particles that they can reach equilibrium locally). We can use Eq. (16.15) to compute the conditional average $\langle J_{xy}(\mathbf{r}, t)\rangle_{\text{noneq}}$, and we obtain

$$\langle J_{xy}(\mathbf{r}, t)\rangle_{\text{noneq}} = \int d\mathbf{p}\, \frac{p_x p_y}{m} f_1(\mathbf{p}, \mathbf{r}, t) = mn(\mathbf{r}, t)v_x(\mathbf{r}, t)v_y(\mathbf{r}, t). \tag{16.16}$$

For small deviations from equilibrium the conditional average, $\langle J_{xy}(\mathbf{r}, t)\rangle_{\text{noneq}}$, can be approximated by

$$\langle J_{xy}(\mathbf{r}, t)\rangle_{\text{noneq}} = mn_0 v_x(\mathbf{r}, t)v_y(\mathbf{r}, t) \tag{16.17}$$

where n_0 is the equilibrium density.

If we combine Eqs. (16.11), (16.13), and (16.17) we obtain the following expression for the correlation function:

$$C_\eta(t) = m^2 n_0^2 \left(\frac{\beta}{2\pi m}\right)^{3/2} \int d\mathbf{v}_0\, e^{-\beta m v_0^2/2} v_{0x} v_{0y} \int d\mathbf{r} v_x(\mathbf{r}, t)v_y(\mathbf{r}, t). \tag{16.18}$$

The time and space dependence of the average velocity, $\mathbf{v}(\mathbf{r}, t)$, can be obtained from the hydrodynamic equations. We will only consider the contribution from the transverse component of velocity (cf. Sec. 13.F). If we Fourier transform the average velocities $v_x(\mathbf{r}, t)$ and $v_y(\mathbf{r}, t)$ through the equation

$$v_x(\mathbf{r}, t) = \frac{1}{(2\pi)^3} \int d\mathbf{k}\, e^{i\mathbf{k}\cdot\mathbf{r}} v_x(\mathbf{k}, t), \tag{16.19}$$

Eq. (16.18) takes the form

$$C_\eta(t) = m^2 n_0^2 \left(\frac{\beta}{2\pi m}\right)^{3/2} \left(\frac{1}{2\pi}\right)^3 \int d\mathbf{v}_0\, e^{-\beta m v_0^2/2} v_{0x} v_{0y} \int d\mathbf{k} v_x(\mathbf{k}, t)v_y(-\mathbf{k}, t). \tag{16.20}$$

From Eq. (13.159) we know that the transverse velocity has a time dependence

$$v_\perp(\mathbf{k}, t) = v_\perp(\mathbf{k}, 0)\, e^{-\eta k^2 t/mn_0}. \tag{16.21}$$

However, before we can evaluate $C_\eta(t)$, we must find an expression for the initial value $\mathbf{v}_\perp(\mathbf{k}, 0)$. This is a nontrivial problem because the hydrodynamic equations hold only after long times when the system has settled down to local equilibrium. The initial values physically have no meaning. Following Ernst, Hauge, and van Leeuven, we may use Eq. (16.13) to compute the *effective* initial value, $v_\perp(\mathbf{k}, 0)$. We proceed as follows. The average velocity $\mathbf{v}(\mathbf{r}, 0)$ is given by

$$\mathbf{v}(\mathbf{r}, 0) = \frac{1}{n_0}\left\langle \sum_i \frac{\mathbf{p}_i(0)}{m}\, \delta(\mathbf{q}_i(0) - \mathbf{r}) \right\rangle_{\text{noneq}}$$

$$= \frac{\left\langle \delta\left(\frac{\mathbf{p}_1(0)}{m} - \mathbf{v}_0\right) W(\mathbf{q}_1(0) - \mathbf{r}_0) \sum_i \frac{\mathbf{p}_i(0)}{m}\, \delta(\mathbf{q}_i(0) - \mathbf{r}) \right\rangle_{\text{eq}}}{n_0 \left\langle \delta\left(\frac{\mathbf{p}_1(0)}{m} - \mathbf{v}_0\right) W(\mathbf{q}_1(0) - \mathbf{r}_0) \right\rangle_{\text{eq}}}. \tag{16.22}$$

If we assume that $W(\mathbf{r} - \mathbf{r}_0)$ varies slowly in space, it will be essentially constant over the region where $\delta(\mathbf{q}_i(0) - \mathbf{r})$ varies. Therefore, $W(\mathbf{r} - \mathbf{r}_0)$ can be taken outside the integral and we obtain

$$\mathbf{v}(\mathbf{r}, 0) = \frac{\mathbf{v}_0}{n_0} W(\mathbf{r} - \mathbf{r}_0). \tag{16.23}$$

Thus, the effective initial velocity is expressed in terms of the distribution $W(\mathbf{r} - \mathbf{r}_0)$.

We can now use the above results to obtain an explicit expression for the transverse velocity autocorrelation function after long times. We find

$$\begin{aligned} C_\eta(t) &= m^2 \left(\frac{\beta m}{2\pi}\right)^{3/2} \int d\mathbf{v}_0\, e^{-\beta m v_0^2/2} v_{0x} v_{0y} \\ &\quad \times \left(\frac{1}{2\pi}\right)^3 \int d\mathbf{k} v_{0x\perp} v_{0y\perp} |W_\mathbf{k}|^2 e^{-2(\eta/mn_0)k^2 t} \\ &= A \frac{1}{\left(\dfrac{\eta}{mn_0} t\right)^{3/2}}, \end{aligned} \tag{16.24}$$

where we have used the fact that $\lim_{\mathbf{k}\to 0} W_\mathbf{k} = 1$ ($W_\mathbf{k}$ is the Fourier transform of the distribution $W(\mathbf{r} - \mathbf{r}_0)$) and A is a constant. We shall leave the computation of the longitudinal part as a homework problem.

From purely hydrodynamic arguments we obtain long time tails in the correlation function, although the Boltzmann equation predicts exponential decay. This indicates that the Boltzmann equation gives too simple a picture of processes occurring in a fluid. As we shall see from the microscopic theory in subsequent sections, this is indeed the case.

C. VIRIAL EXPANSION OF THE VELOCITY AUTOCORRELATION FUNCTION

Expressions for transport coefficients obtained using the Boltzmann equation are limited to systems with very low density. If we wish to obtain expressions for the transport coefficients at higher densities, we must begin by writing them in terms of a virial expansion (expansion in powers of the density). The first attempt in this direction was made by Bogoliubov,[3] who gave a prescription for obtaining a virial expansion for the transport coefficients, but the first actual calculation was done by Choh and Uhlenbeck.[4] The Boltzmann equation includes only effects of two-body collisions. Choh and Uhlenbeck computed the contribution to the transport coefficients due to three-body processes. Their result was well behaved. Somewhat later, Kawasaki and Oppenheim[15] attempted to extend calculations of the transport coefficients to higher order in density. In so doing, they found that all higher order terms are infinite and, therefore, the expansion of the transport coefficients in powers of the density is divergent and ill defined. The divergences in the virial expansion come from secular effects which are mathematically similar to the low temperature divergences we found in quantum gases in Chap. 12. Resummation of the divergent terms leads to well-defined expressions for the transport coefficients but introduces a nonanalytic density expansion for them. As we shall see, the divergences are related to the appearance of long time tails in the correlation functions. One of the simplest correlation functions we can compute, and the one measured in the molecular dynamics calculations of Alder and Wainwright, is the velocity autocorrelation function. In the remainder of this chapter we will follow closely the approach of Dorfmann and Cohen in obtaining a microscopic expression for the velocity autocorrelation function for a classical gas at moderate densities. In so doing, we will understand better the microscopic origin of the long time tails.

If we consider a classical system of N-particles which interact via a spherically symmetric potential, the Hamiltonian has the form

$$H^N(\mathbf{X}^N) = \sum_{i=1}^{N} \frac{p_i^2}{2m} + \sum_{i<j} V(|\mathbf{q}_i - \mathbf{q}_j|) \tag{16.25}$$

and the Liouville operator has the form

$$\hat{\mathscr{H}}^N(\mathbf{X}^N) = \hat{\mathscr{H}}_0^N + \frac{1}{2}\sum_{i \neq j}^{N}\sum^{N} \hat{\Theta}_{ij} \tag{16.26}$$

where

$$\hat{\mathscr{H}}_0^N = \sum_{i=1}^{N} \frac{\mathbf{p}_i}{m}\cdot\frac{\partial}{\partial \mathbf{q}_i} \tag{16.27}$$

and

$$\hat{\Theta}_{ij} = -\frac{\partial V_{ij}}{\partial \mathbf{q}_i}\cdot\left(\frac{\partial}{\partial \mathbf{p}_i} - \frac{\partial}{\partial \mathbf{p}_j}\right). \tag{16.28}$$

If we tag particle 1, the probability density of finding it at point $\mathbf{r}$ at time t is

$$\rho_1(\mathbf{r}, t) = \langle \delta(\mathbf{q}_1(t) - \mathbf{r})\rangle = \int\cdots\int d\mathbf{X}_1\cdots d\mathbf{X}_N \rho_{eq}(\mathbf{X}^N)\, e^{\hat{\mathscr{H}}^N t}\, \delta(\mathbf{q}_1 - \mathbf{r}), \tag{16.29}$$

where $\mathbf{q}_1$ is the position at time $t = 0$, $\rho_{eq}(\mathbf{X}^N)$ is the equilibrium probability density

$$\rho_{eq}(\mathbf{X}^N) = (Z'^N)^{-1}\, e^{-\beta H^N}, \tag{16.30}$$

and Z'^N is proportional to the partition function

$$Z'^N = \int\cdots\int d\mathbf{X}_1\cdots d\mathbf{X}_N\, e^{-\beta H^N}. \tag{16.31}$$

The velocity autocorrelation function is given by

$$C_{\mathbf{vv}}(\mathbf{r}, t) = \langle \mathbf{v}(\mathbf{0}, 0)\mathbf{v}(\mathbf{r}, t)\rangle = m^{-2}\langle \mathbf{p}_1\, \delta(\mathbf{q}_1)\mathbf{p}_1(t)\, \delta(\mathbf{q}_1(t) - \mathbf{r})\rangle$$
$$= m^{-2}\int\cdots\int d\mathbf{X}_1\cdots d\mathbf{X}_N \rho_{\mathrm{eq}}(\mathbf{X}^N)\mathbf{p}_1\, \delta(\mathbf{q}_1)\, e^{\hat{\mathscr{H}}^N t}\mathbf{p}_1\, \delta(\mathbf{q}_1 - \mathbf{r}). \tag{16.32}$$

It is most convenient to work with the Fourier-Laplace transform

$$\bar{S}_{\mathbf{vv}}(\mathbf{k}, z) = \int_0^\infty dt \int d\mathbf{r}\, e^{-i(\mathbf{k}\cdot\mathbf{r} - zt)} C_{\mathbf{vv}}(\mathbf{r}, t) \tag{16.33}$$

where $z = \omega + i\delta$ and δ is a positive number. If we expand the time evolution operator $e^{\hat{\mathscr{H}}t}$ in a power series, integrate each term by parts, and resum it, we obtain an expression in which the time evolution operator acts on the phase function to the left. We then find, after some simplification,

$$\bar{S}_{\mathbf{vv}}(\mathbf{k}, z) = \frac{m^{-2}}{(2\pi)^3}\int_0^\infty dt\, e^{+izt}\int d\mathbf{k}_1\int\cdots\int d\mathbf{X}_1\cdots d\mathbf{X}_N\mathbf{p}_1$$
$$\times\, e^{-i\mathbf{k}\cdot\mathbf{q}_1}\, e^{-\hat{\mathscr{H}}^N t}\rho_{\mathrm{eq}}^N(\mathbf{X}^N)\, e^{i\mathbf{k}_1\cdot\mathbf{q}_1}\mathbf{p}_1 \tag{16.34}$$

where the minus sign in the evolution operator comes from the integration by parts. In Eq. (16.34) we have introduced the Fourier decomposition of the delta function $\delta(\mathbf{q})$ and we have approximated it by its value for an infinite volume system.

Let us now rewrite $\bar{S}_{\mathbf{vv}}(\mathbf{k}, z)$ in the form

$$\bar{S}_{\mathbf{vv}}(\mathbf{k}, z) = \frac{1}{m^2}\int d\mathbf{p}_1\mathbf{p}_1\chi(\mathbf{p}_1; \mathbf{k}, z)\mathbf{p}_1 \tag{16.35}$$

where

$$\chi(\mathbf{p}_1; \mathbf{k}, z) = \left(\frac{1}{2\pi}\right)^3\int_0^\infty dt\, e^{izt}\int d\mathbf{k}_1\int d\mathbf{q}_1\int\cdots\int d\mathbf{X}_2\cdots d\mathbf{X}_N$$
$$\times\, e^{-i\mathbf{k}\cdot\mathbf{q}_1}\, e^{-\mathscr{H}^N t}\rho_{\mathrm{eq}}^N(\mathbf{X}^N)\, e^{+i\mathbf{k}_1\cdot\mathbf{q}_1}. \tag{16.36}$$

We can write the operator $\chi(\mathbf{p}_1; \mathbf{k}, z)$ as a virial expansion:

$$\chi(\mathbf{p}_1; \mathbf{k}, z) = \sum_{l=0}^{\infty} n^l\chi_l(\mathbf{p}_1; \mathbf{k}, z). \tag{16.37}$$

To find a microscopic expression for the virial coefficients, we first introduce a cumulant expansion for the evolution operator:

$$S_{-t}(1, \ldots, N) \equiv e^{-t\hat{\mathscr{H}}^N} = \exp\left(\sum_{l=1}^{\infty} U(1, \ldots, l; t)\right) \tag{16.38}$$

where $U_l(1, \ldots, l; t)$ is an l-body cluster operator. The cluster operators can be expressed in terms of the evolution operator through the following hierarchy of equations:

$$U(1;t) = S_{-t}(1), \tag{16.39a}$$

$$U(1,2;t) = S_{-t}(1,2) - S_{-t}(1)S_{-t}(2), \tag{16.39b}$$

$$U(1,2,3;t) = S_{-t}(1,2,3) - \sum_{p}^{(3)} S_{-t}(1)S_{-t}(2,3) + 2!\prod_{i=1}^{3} S_{-t}(i), \tag{16.39c}$$

$$U(1,2,3,4;t) = S_{-t}(1,2,3,4) - \sum_{p}^{(4)} S_{-t}(1)S_{-t}(2,3,4) - \sum_{p}^{(3)} S_{-t}(1,2)S_{-t}(3,4)$$
$$+ 2!\sum_{p}^{(6)} S_{-t}(1)S_{-t}(2)S_{-t}(3,4) - 3!\prod_{i=1}^{(4)} S_{-t}(i), \tag{16.39d}$$

etc., where $\sum_p^{(j)}$ denotes the sum over all j distinct permutations of the particles. If we isolate those cluster operators which depend on the coordinates of particle 1, we can write the evolution operator in terms of the following hierarchy:

$$S_{-t}(1) = U(1;t), \tag{16.40a}$$

$$S_{-t}(1,2) = U(1;t)S_{-t}(2) + U(1,2;t), \tag{16.40b}$$

$$S_{-t}(1,2,3) = U(1;t)S_{-t}(2,3) + \sum_{p}^{(2)}{}' U(1,2;t)S_{-t}(3) + U(1,2,3;t), \tag{16.40c}$$

$$S_{-t}(1,2,3,4) = U(1;t)S_{-t}(2,3,4) + \sum_{p}^{(3)}{}' U(1,2;t)S_{-t}(3,4)$$
$$+ \sum_{p}^{(3)}{}' U(1,2,3;t)S_{-t}(4) + U(1,2,3,4;t), \tag{16.40d}$$

etc. The prime on the summation $\sum_p'^{(j)}$ means that particle 1 is not to be included in the j distinct permutations. We now substitute Eqs. (16.40) for the case of N-particles into Eq. (16.36), and we obtain

$$\chi(\mathbf{p}_1;\mathbf{k},z) = \int \frac{d\mathbf{k}_1}{(2\pi)^3}\int_0^\infty dt\, e^{izt}\int d\mathbf{q}_1 \int\cdots\int d\mathbf{X}_2\cdots d\mathbf{X}_N\, e^{-i\mathbf{k}\cdot\mathbf{q}_1}$$
$$\times \Bigg\{ U_t(1;t)S_{-t}(2,\ldots,N) + \sum_{p}^{(N-1)}{}' U_t(1,2;t)S_{-t}(3,\ldots,N)$$
$$+ \sum_{p}^{(1/2!)(N-1)(N-2)} U_t(1,2,3;t)S_{-t}(4,\ldots,N) + \cdots \Bigg\}$$
$$\times \rho_{\text{eq}}^N(\mathbf{X}^N)\, e^{+i\mathbf{k}_1\cdot\mathbf{q}_1}. \tag{16.41}$$

We can simplify Eq. (16.41) if we note that

$$\frac{\phi(1)}{V} \equiv \int\cdots\int d\mathbf{X}_2\cdots d\mathbf{X}_N S_{-t}(2,\cdots,N)\rho_{eq}(\mathbf{X}^N)$$

$$= \int\cdots\int d\mathbf{X}_2\cdots d\mathbf{X}_N\rho_{eq}(\mathbf{X}^N)S_t(2,\ldots,N)$$

$$= \frac{1}{V}\left(\frac{\beta}{2\pi m}\right)^{3/2} e^{-\beta p_1^2/2m}, \tag{16.42}$$

where we have integrated by parts and have used the fact that $S_{-t}(1,\ldots,N)1 = 1$. We can now write Eq. (16.41) in the form

$$\chi(\mathbf{p}_1;\mathbf{k},z) = \int \frac{d\mathbf{k}_1}{(2\pi)^3}\int d\mathbf{q}_1\int_0^\infty dt\, e^{izt}$$

$$\times\left\{e^{-i\mathbf{k}\cdot\mathbf{q}_1}U(1;t)\,e^{+i\mathbf{k}_1\cdot\mathbf{q}_1}\phi(1)\right.$$

$$+\frac{(N-1)}{V}\int d\mathbf{X}_2\, e^{-i\mathbf{k}\cdot\mathbf{q}_1}U(1,2;t)$$

$$\times\, e^{+i\mathbf{k}_1\cdot\mathbf{q}_1}V^2\int\cdots\int d\mathbf{X}_3\times\cdots\times d\mathbf{X}_N\rho_{eq}(\mathbf{X}^N)$$

$$+\frac{(N-1)(N-2)}{2!\,V^2}\int\cdots\int d\mathbf{X}_2\, d\mathbf{X}_3\, e^{-i\mathbf{k}\cdot\mathbf{q}_1}U(1,2,3;t)$$

$$\times\, e^{+i\mathbf{k}_1\cdot\mathbf{q}_1}V^3\int\cdots\int d\mathbf{X}_3\times\cdots\times d\mathbf{X}_N\rho_{eq}(\mathbf{X}^N)$$

$$+\frac{(N-1)(N-2)(N-3)}{3!\,V^3}\int\cdots\int d\mathbf{X}_2\, d\mathbf{X}_3\, d\mathbf{X}_4$$

$$\times\, e^{-i\mathbf{k}\cdot\mathbf{q}_1}U(1,2,3,4;t)\,e^{+i\mathbf{k}_1\cdot\mathbf{q}_1}V^4\int\cdots\int d\mathbf{X}_4\times\cdots$$

$$\left.\times\, d\mathbf{X}_N\rho_{eq}(\mathbf{X}^N)+\cdots\right\}. \tag{16.43}$$

For simplicity in subsequent calculations we will neglect contributions from the equilibrium probability densities and partition function due to interactions. Thus, we approximate $e^{-\beta H}$ by its value for a noninteracting gas. It can be shown that the terms we neglect give contributions to the transport coefficients which are less divergent than the ones we retain.[6] If we make this approximation and anticipate the fact that we will take the thermodynamic limit, we can write

$\chi(\mathbf{p}_1; \mathbf{k}, z)$ as

$$\chi(\mathbf{p}_1; \mathbf{k}, z) = \int \frac{d\mathbf{k}_1}{(2\pi)^3} \int d\mathbf{q}_1 \int_0^\infty dt\, e^{izt} \times \left\{ e^{-i\mathbf{k}\cdot\mathbf{q}_1} U(1; t)\, e^{+i\mathbf{k}_1\cdot\mathbf{q}_1}\phi(1) + n \int d\mathbf{X}_2\, e^{-i\mathbf{k}\cdot\mathbf{q}_1} U(1, 2; t)\, e^{+i\mathbf{k}_1\cdot\mathbf{q}_1}\phi(1)\phi(2) + \frac{n^2}{2!}\int d\mathbf{X}_2\, d\mathbf{X}_3\, e^{-i\mathbf{k}\cdot\mathbf{q}_1} U(1, 2, 3; t) \times e^{+i\mathbf{k}_1\cdot\mathbf{q}_1}\phi(1)\phi(2)\phi(3) + \cdots \right\} \tag{16.44}$$

where we have neglected terms of order $1/V$. Thus, we obtain for the virial coefficients

$$\chi_0(\mathbf{p}_1; \mathbf{k}, z) = \frac{i}{z - \frac{1}{m}\mathbf{k}\cdot\mathbf{p}_1}\phi(1), \tag{16.45a}$$

$$\chi_1(\mathbf{p}_1; \mathbf{k}, z) = \int d\mathbf{q}_1 \int d\mathbf{X}_2 \int \frac{d\mathbf{k}_1}{(2\pi)^3}\int_0^\infty dt\, e^{izt} \times \{e^{-i\mathbf{k}\cdot\mathbf{q}_1} U(1, 2; t)\, e^{+i\mathbf{k}_1\cdot\mathbf{q}_1}\phi(1)\phi(2)\}, \tag{16.45b}$$

$$\chi_2(\mathbf{p}_1; \mathbf{k}, z) = \frac{1}{2!}\int d\mathbf{q}_1 \iint d\mathbf{X}_2\, d\mathbf{X}_3 \int \frac{d\mathbf{k}_1}{(2\pi)^3}\int_0^\infty dt\, e^{izt} \times \{e^{-i\mathbf{k}\cdot\mathbf{q}_1} U(1, 2, 3; t)\, e^{+i\mathbf{k}_1\cdot\mathbf{q}_1}\phi(1)\phi(2)\phi(3)\}, \tag{16.45c}$$

etc. In the long-wavelength ($\mathbf{k} \to 0$), long time ($z \to 0$) limit, the virial coefficients diverge. This is easy to see from Eq. (16.45*a*). Part of the divergence can be removed if we resum Eq. (16.37). The procedure for resummation was first introduced by Zwanzig.[7] Let us define a new set of operators, $B_m(\mathbf{p}_1; \mathbf{k}, z)$, which satisfy the following equation:

$$\chi(\mathbf{p}_1; \mathbf{k}, z)\phi^{-1}(1) = \frac{i}{z - \frac{1}{m}\mathbf{k}\cdot\mathbf{p}_1} + \sum_{l=1}^{\infty} n^l \chi_l(\mathbf{p}_1; \mathbf{k}, z)\phi^{-1}(1) = \frac{i}{z - \frac{1}{m}\mathbf{k}\cdot\mathbf{p}_1 + \sum_{m=1}^{\infty} n^m B_m(\mathbf{p}_1; \mathbf{k}, z)}. \tag{16.46}$$

If we expand the summations in Eq. (16.46) and equate coefficients of equal powers of the density, we obtain the following relation between $B_m(\mathbf{p}_1; \mathbf{k}, z)$ and $\chi_l(\mathbf{p}_1; \mathbf{k}, z)$:

$$B_1(\mathbf{p}_1; \mathbf{k}, z) = i\left(z - \frac{1}{m}\mathbf{k}\cdot\mathbf{p}_1\right)\chi_1(\mathbf{p}_1; \mathbf{k}, z)\left(z - \frac{1}{m}\mathbf{k}\cdot\mathbf{p}_1\right)\phi^{-1}(1), \tag{16.47a}$$

$$\begin{aligned} B_2(\mathbf{p}_1; \mathbf{k}, z) = {} & i\left(z - \frac{1}{m}\mathbf{k}\cdot\mathbf{p}_1\right)\chi_2(\mathbf{p}_1; \mathbf{k}, z)\left(z - \frac{1}{m}\mathbf{k}\cdot\mathbf{p}_1\right)\phi^{-1}(1) \\ & - \left(z - \frac{1}{m}\mathbf{k}\cdot\mathbf{p}_1\right)\chi_1(\mathbf{p}_1; \mathbf{k}, z)\left(z - \frac{1}{m}\mathbf{k}\cdot\mathbf{p}_1\right) \\ & \times \chi_1(\mathbf{p}_1; \mathbf{k}, z)\left(z - \frac{1}{m}\mathbf{k}\cdot\mathbf{p}_1\right)\phi^{-1}(1), \end{aligned} \tag{16.47b}$$

etc. (Note that $\mathbf{p}_1$ and $\chi(\mathbf{p}_1; \mathbf{k}, z)$ do not commute.) We can now combine Eqs. (16.35) and (16.46) and write the velocity autocorrelation function in the form

$$C_{vv}(\mathbf{k}, z) = \frac{1}{m^2}\int d\mathbf{p}_1\mathbf{p}_1 \frac{i}{\left(z - \frac{1}{m}\mathbf{k}\cdot\mathbf{p}_1 + \sum_{m=1}^{\infty} n^m B_m(\mathbf{p}_1; \mathbf{k}, z)\right)} \mathbf{p}_1\phi(1). \tag{16.48}$$

It is of interest to note that this resummation is very much like the Dyson resummation of the exact propagator in Chap. 12. One may think of the function, $B(\mathbf{p}_1; \mathbf{k}, z)$, as a "self-energy" effect. Microscopic expressions for coefficients $B_m(\mathbf{p}_1; \mathbf{k}, z)$ are obtained from the binary collision expansion, as we shall see in the next section.

D. MICROSCOPIC EXPRESSION FOR THE VIRIAL COEFFICIENTS

From Eqs. (16.45) the lth-order virial coefficient can be expressed

$$\begin{aligned} \chi_l(\mathbf{p}_1; \mathbf{k}, z) = {} & \frac{1}{l!}\int_0^\infty dt\, e^{izt}\int \frac{d\mathbf{k}_1}{(2\pi)^3}\int d\mathbf{q}_1 \int\cdots\int d\mathbf{X}_2 \cdots d\mathbf{X}_l \\ & \times e^{-i\mathbf{k}\cdot\mathbf{q}_1} U_l(1, 2, \ldots, l; t)\, e^{i\mathbf{k}_1\cdot\mathbf{q}_1}\phi(1) \times \cdots \times \phi(l), \end{aligned} \tag{16.49}$$

where $U_l(1, 2, \ldots, l; t)$ is the contribution to the time evolution operator due to an l-body dynamical cluster (cf. Chap. 11). From Eqs. (16.39) and (16.45) we see that the virial coefficients depend on the Laplace transform of the evolution operators,

$$G(1,\ldots,N;z) = \int_0^\infty dt\, e^{izt}\, e^{-t\hat{\mathscr{H}}^N(1,\ldots,N)} = \frac{i}{z + i\mathscr{H}^N(1,\ldots,N)} \tag{16.50}$$

and

$$G_0(1,\ldots,N;z) = \int_0^\infty dt\, e^{izt}\, e^{-t\hat{\mathscr{H}}^N(1,\ldots,N)} = \frac{i}{z + i\hat{\mathscr{H}}_0(1,\ldots,N)}, \tag{16.51}$$

and on evolution operators which depend on the kinetic contribution of all N-particles and the potential energy of only a fraction of the N-particles.

1. Binary Collision Expansion

The operator $G(1,\ldots,N;z)$ can be expanded in terms of the following binary collision expansion:

$$\begin{aligned} G(1,\ldots,N;z) = G_0(1,\ldots,N;z) &+ \sum_{\mu}^{\text{all pairs}} G_0(1,\ldots,N;z)T_\mu(z)G_0(1,\ldots,N;z) \\ &+ \sum_{\mu}^{\text{all pairs}}\sum_{\nu\neq\mu} G_0(1,\ldots,N;z)T_\mu(z)G_0(1,\ldots,N;z)T_\nu(z) \\ &\times G_0(1,\ldots,N;z) + \cdots \end{aligned} \tag{16.52}$$

where

$$T_\mu(z) = -\hat{\Theta}_\mu - \hat{\Theta}_\mu G_0(1,\ldots,N;z)T_\mu(z) \tag{16.53}$$

and μ denotes a given pair of particles. It is important to note that no two operators T_μ which depend on the same pair of particles can neighbor one another.

We will want to Fourier transform the position space dependence of the above quantities. Therefore, it is of interest to define the following matrix elements:

$$\begin{aligned} &\langle \mathbf{k}_1,\ldots,\mathbf{k}_N|G_0(1,\ldots,N;z)|\mathbf{k}_1',\ldots,\mathbf{k}_N'\rangle \\ &\equiv \int d\mathbf{q}_1\cdots d\mathbf{q}_N\, e^{-i\mathbf{k}^N\cdot\mathbf{q}^N}\left(\frac{i}{z + i\sum_{j=1}^{N}\frac{\mathbf{p}_j}{m}\cdot\frac{\partial}{\partial\mathbf{q}_j}}\right)e^{+i\mathbf{k}'^N\cdot\mathbf{q}^N} \\ &= (2\pi)^{3N}\,\delta(\mathbf{k}_1-\mathbf{k}_1')\cdots\delta(\mathbf{k}_N'-\mathbf{k}_N)\left(\frac{i}{z - \mathbf{k}^N\cdot\frac{\mathbf{p}^N}{m}}\right) \end{aligned} \tag{16.54}$$

where $\mathbf{k}^N \cdot \mathbf{q}^N \equiv \sum_{j=1}^N \mathbf{k}_j \cdot \mathbf{q}_j$ and $\mathbf{k}'^N \cdot \mathbf{p}^N = \sum_{j=1}^N \mathbf{k}'_j \cdot \mathbf{p}_j$. Similarly, matrix elements of $T_{12}(1, \ldots, N; z)$ are given by

$$\begin{aligned}\langle \mathbf{k}_1, \ldots, \mathbf{k}_N | T_{12}(z) | \mathbf{k}'_1, \ldots, \mathbf{k}'_N \rangle \\ &= (2\pi)^{3(N-1)} \delta(\mathbf{k}_3 - \mathbf{k}'_3) \cdots \delta(\mathbf{k}_N - \mathbf{k}'_N)\, \delta(\mathbf{k}_1 + \mathbf{k}_2 - \mathbf{k}'_1 - \mathbf{k}'_2) \\ &\quad \times \langle \mathbf{k}_1, \ldots, \mathbf{k}_N | \tilde{T}_{12}(z) | \mathbf{k}'_1, \mathbf{k}_1 + \mathbf{k}_2 - \mathbf{k}'_1, \mathbf{k}_3, \ldots, \mathbf{k}_N \rangle \qquad (16.55)\end{aligned}$$

where

$$\begin{aligned}\langle \mathbf{k}_1, \ldots, \mathbf{k}_N | \tilde{T}_{12}(z) | \mathbf{k}'_1, \mathbf{k}'_2; \mathbf{k}_3, \ldots, \mathbf{k}_N \rangle \\ &= -\tilde{\Theta}(\mathbf{k}_1 - \mathbf{k}'_1) - i \sum_{\mathbf{k}''_1, \mathbf{k}''_2} (2\pi)^3\, \delta(\mathbf{k}''_1 + \mathbf{k}''_2 - \mathbf{k}'_1 - \mathbf{k}'_2) \tilde{\Theta}(\mathbf{k}_1 - \mathbf{k}''_1) \\ &\quad \times \left(z - \mathbf{k}''_1 \cdot \frac{\mathbf{p}_1}{m} - \mathbf{k}''_2 \cdot \frac{\mathbf{p}_2}{m} - \sum_{j=3}^{N} \mathbf{k}_j \cdot \frac{\mathbf{p}_j}{m} \right)^{-1} \\ &\quad \times \langle \mathbf{k}''_1, \mathbf{k}''_2, \mathbf{k}_3, \ldots, \mathbf{k}_N | \tilde{T}_{12}(z) | \mathbf{k}'_1, \mathbf{k}'_2, \mathbf{k}_3, \ldots, \mathbf{k}_N \rangle \qquad (16.56)\end{aligned}$$

and $\tilde{\Theta}(\mathbf{k}_1 - \mathbf{k}'_1)$ is the Fourier transform of $\hat{\Theta}_{12}$. We can now use these definitions to compute various contributions to the density expansion of the self-energies, $B_m(\mathbf{p}_1; \mathbf{k}, z)$.

2. Ring Approximation to the Self-Energy

Let us consider the first-order contribution to the self-energy $B_1(\mathbf{p}_1; \mathbf{k}, z)$. We first must find $\chi_1(\mathbf{p}_1; \mathbf{k}, z)$. From Eq. (16.45*b*) we can write

$$\begin{aligned}\chi_1(\mathbf{p}_1; \mathbf{k}, z) &= \int \frac{d\mathbf{k}_1}{(2\pi)^3} \int d\mathbf{q}_1\, d\mathbf{X}_2\, e^{-i\mathbf{k}\cdot\mathbf{q}_1} \\ &\quad \times [G(1, 2; z) - G_0(1, 2; z)]\, e^{+i\mathbf{k}_1\cdot\mathbf{q}_1} \phi(1)\phi(2) \\ &= \int \frac{d\mathbf{k}_1}{(2\pi)^3} \int d\mathbf{q}_1\, d\mathbf{X}_2\, e^{-i\mathbf{k}\cdot\mathbf{q}_1} \\ &\quad \times [G_0(1, 2; z) T_{12}(z) G_0(1, 2; z)]\, e^{+i\mathbf{k}_1\cdot\mathbf{q}_1} \phi(1)\phi(2), \qquad (16.57)\end{aligned}$$

and from Eq. (16.55) obtain

$$\begin{aligned}\chi_1(\mathbf{p}_1; \mathbf{k}, z) &= \int \frac{d\mathbf{k}_1}{(2\pi)^3} \int d\mathbf{p}_2 \\ &\quad \times \langle \mathbf{k}, 0 | G_0(1, 2; z) T_{12}(z) G_0(1, 2; z) | \mathbf{k}_1, 0 \rangle \phi(1)\phi(2) \\ &= \left(\frac{i}{z - \frac{\mathbf{p}_1}{m}\cdot\mathbf{k}} \right) \int d\mathbf{p}_2 \langle \mathbf{k}, 0 | \tilde{T}_{12}(z) | \mathbf{k}, 0 \rangle \left(\frac{i}{z - \frac{\mathbf{p}_1}{m}\cdot\mathbf{k}} \right) \phi(1)\phi(2) \qquad (16.58)\end{aligned}$$

where we have used the identity

$$\iint \frac{d\mathbf{k}_a}{(2\pi)^3}\frac{d\mathbf{k}_b}{(2\pi)^3}\,|\mathbf{k}_a\mathbf{k}_b\rangle\langle\mathbf{k}_a\mathbf{k}_b| = 1. \tag{16.59}$$

If we now use Eq. (16.47*a*), the first-order contribution to the self-energy becomes

$$B_1(\mathbf{p}_1; \mathbf{k}, z) = -i\int d\mathbf{p}_3\langle\mathbf{k}, 0|\tilde{T}_{12}(z)|\mathbf{k}, 0\rangle\phi(2). \tag{16.60}$$

Thus, if the matrix element $\langle\mathbf{k}, 0|T_{12}(z)|\mathbf{k}, 0\rangle$ is well behaved in the limit $\mathbf{k} \to 0$ and $z \to 0$, then so is $B_1(\mathbf{p}_1; \mathbf{k}, z)$.

The expression for $B_l(\mathbf{p}_1; \mathbf{k}, z)$ for $l \geqslant 2$ is more complicated than for $l = 1$ because for $l \geqslant 2$ the expansion of $G(1, \ldots, l; z)$ is given by an infinite series. We will only try to give the flavor of what happens. Let us consider a typical term in the expansion of $B_2(\mathbf{p}_1; \mathbf{k}, z)$:

$$\begin{aligned} B_2(\mathbf{p}_1; \mathbf{k}, z) = {} & i\left(z - \frac{1}{m}\mathbf{p}_1\cdot\mathbf{k}\right)\int\frac{d\mathbf{k}_1}{(2\pi)^3}\int d\mathbf{q}_1\iint d\mathbf{X}_2\,d\mathbf{X}_3\,e^{-i\mathbf{k}\cdot\mathbf{q}_1} \\ & \times [G_0(1, 2. 3; z)T_{12}(z)G_0(1, 2, 3; z)T_{23}(z) \\ & \quad \times G_0(1, 2, 3; z)T_{12}(z)G_0(1, 2, 3; z)] \\ & \times \left(z - \frac{1}{m}\mathbf{p}_1\cdot\mathbf{k}\right) e^{+i\mathbf{k}_1\cdot\mathbf{q}_1}\phi(2)\phi(3) \\ & + \text{other terms.} \end{aligned} \tag{16.61}$$

If we again make use of Eqs. (16.54) and (16.55) and insert an identity similar to Eq. (16.59) for three particles, we find

$$\begin{aligned} B_2(\mathbf{p}_1; \mathbf{k}, z) = {} & -i\int\frac{d\mathbf{k}'}{(2\pi)^3}\iint d\mathbf{p}_2\,d\mathbf{p}_3\langle\mathbf{k}, 0, 0|\tilde{T}_{12}|\mathbf{k}', \mathbf{k} - \mathbf{k}', 0\rangle \\ & \times\left(\frac{i}{z - \dfrac{\mathbf{k}'\cdot\mathbf{p}_1}{m} - \dfrac{(\mathbf{k} - \mathbf{k}')\cdot\mathbf{p}_2}{m}}\right)\langle\mathbf{k}', \mathbf{k} - \mathbf{k}', 0|\tilde{T}_{23}|\mathbf{k}', \mathbf{k} - \mathbf{k}', 0\rangle \\ & \times\left(\frac{i}{z - \dfrac{\mathbf{k}'\cdot\mathbf{p}_1}{m} - \dfrac{(\mathbf{k} - \mathbf{k}')\cdot\mathbf{p}_2}{m}}\right) \\ & \times\langle\mathbf{k}', \mathbf{k} - \mathbf{k}', 0|\tilde{T}_{12}|\mathbf{k}, 0, 0\rangle\phi(2)\phi(3). \end{aligned} \tag{16.62}$$

Let us now consider Eq. (16.62) in the limit $\mathbf{k} \to 0$ (long-wavelength or hydrodynamic regime). If we expand matrix elements about $\mathbf{k}' = 0$ and integrate over $\mathbf{k}'$, we find that in two dimensions the integral diverges as $\ln(z)$ as $z \to 0$ and in three dimensions it is well behaved. This indicates that there is a problem with the correlation function in the limit $\mathbf{k} \to 0$ and $z \to 0$. The origin of this problem is the repeated propagators which give rise to secular effects in the time-dependent correlation function. In the limit $t \to 0$, the correlation function becomes infinite. The effect is mathematically analogous to that which occurs in the low-temperature limit for equilibrium systems except that now the variable that becomes infinite is the time and not the inverse temperature. In three dimensions, higher order terms diverge for similar reasons. Just as for the equilibrium case, we must resum the terms containing repeated propagators to remove the divergences.

Higher order contributions were studied by Zwanzig and by Kawasaki and Oppenheim. Zwanzig found that in three dimensions $B_2(\mathbf{p}_1, z)$ is well defined as $z \to 0$. Kawasaki and Oppenheim found that $B_2(\mathbf{p}_1, z)$ is well defined for three dimensions and diverges as $\ln(z)$ for two dimensions:

$$B_2(\mathbf{p}_1, z) \sim \begin{cases} \text{well behaved} & \text{(3 dimensions)} \\ \ln z & \text{(2 dimensions)}; \end{cases} \tag{16.63}$$

for all higher powers they found that

$$B_l(\mathbf{p}_1, z) \sim \begin{cases} \dfrac{1}{z^l} & \text{(3 dimensions)} \\ \dfrac{1}{z^{l-1}} & \text{(2 dimensions).} \end{cases} \tag{16.64}$$

Kawasaki and Oppenheim showed that the most divergent terms in each order of the density expansion involved combinations of the type $G_0T_{12}G_0T_{1j}G_0T_{12}G_0$, $G_0T_{12}G_0T_{2j}G_0T_{12}G_0$, and $G_0T_{12}G_0T_{2j}G_0T_{1j}G_0$ for the case of self-diffusion, because these terms involve the largest number of repeated propagators in a given order in the binary expansion and therefore the largest secular dependence. As we have just noted, such terms become highly divergent in the limit $z \to 0$ or, equivalently, $t \to \infty$. They were able to resum these terms by introducing the following operator:

$$\Lambda(1, 2) = \int d\mathbf{p}_j[T_{1j} + T_{2j}(1 + P_{2j})]\phi(j) \tag{16.65}$$

where P_{2j} is the particle exchange operator which acts on functions of the phase space variables $\mathbf{p}_1$ and $\mathbf{p}_j$, which appear to the right of it, and interchanges $\mathbf{p}_1$ and $\mathbf{p}_j$. (The index j refers to particles other than 1 or 2.) In three dimensions the most divergent part of $B_3(\mathbf{p}_1; \mathbf{k}, z)$ is

$$\begin{aligned} B_3(\mathbf{p}_1; \mathbf{k}, z) \sim i\left(z - \frac{\mathbf{k}\cdot\mathbf{p}_1}{m}\right) \int \frac{d\mathbf{k}_1}{(2\pi)^3} \int d\mathbf{q}_1\, d\mathbf{X}_2 \\ \times e^{-i\mathbf{k}\cdot\mathbf{q}_1} G_0 T_{12} G_0 (\Lambda(1, 2) G_0)^2 T_{12} G_0\left(z - \frac{\mathbf{k}\cdot\mathbf{p}_1}{m}\right) \\ \times e^{+i\mathbf{k}_1\cdot\mathbf{q}_1}\phi(2), \end{aligned} \tag{16.66}$$

while the most divergent part of $B_l(\mathbf{p}_1; \mathbf{k}, z)$ is

$$B_l(\mathbf{p}_1; \mathbf{k}, z) \sim i\left(z - \frac{\mathbf{k}\cdot\mathbf{p}_1}{m}\right)\int \frac{d\mathbf{k}_1}{(2\pi)^3}\int d\mathbf{q}_1\, d\mathbf{X}_2$$

$$\times\, e^{-i\mathbf{k}\cdot\mathbf{q}_1} G_0 T_{12} G_0 (\Lambda(1, 2) G_0)^{l-1} T_{12} G_0 \left(z - \frac{\mathbf{k}\cdot\mathbf{p}_1}{m}\right)$$

$$\times\, e^{+i\mathbf{k}_1\cdot\mathbf{q}_1}\phi(2). \tag{16.67}$$

If we now resum the most divergent terms in the denominator of Eq. (16.48), we obtain

$$z - \frac{\mathbf{k}\cdot\mathbf{p}_1}{m} + \sum_{l=1}^{\infty} n^l B_l(\mathbf{p}_1; \mathbf{k}, z)$$

$$= z - \frac{\mathbf{k}\cdot\mathbf{p}_1}{m} + nB_1(\mathbf{p}_1; \mathbf{k}, z) + i\left(z - \frac{1}{m}\mathbf{k}\cdot\mathbf{p}_1\right)n\int \frac{d\mathbf{k}_1}{(2\pi)^3}\int d\mathbf{q}_1\, d\mathbf{X}_2$$

$$\times\, e^{-i\mathbf{k}\cdot\mathbf{q}_1} G_0 T_{12}\left(\frac{1}{G_0^{-1} - G_0^{-1}\Lambda(1, 2) n G_0}\right) T_{12} G_0 \left(z - \frac{1}{m}\mathbf{k}\cdot\mathbf{p}_1\right) e^{i\mathbf{k}_1\cdot\mathbf{q}_1}\phi(2) \tag{16.68}$$

(cf. Prob. 16.2) and we can write the velocity autocorrelation function in the form

$$\bar{S}_{vv}(\mathbf{k}, z) = \frac{1}{m^2}\int d\mathbf{p}_1 \mathbf{p}_1 \left(\frac{i}{z - \frac{1}{m}\mathbf{k}\cdot\mathbf{p}_1 - iC^-(\mathbf{p}_1; \mathbf{k}, z) - iR(\mathbf{p}_1; \mathbf{k}, z)}\right)\mathbf{p}_1\phi(1), \tag{16.69}$$

where

$$C^-(\mathbf{p}_1; \mathbf{k}, z) = \int d\mathbf{p}_2 \langle \mathbf{k}, 0|\tilde{T}_{12}(z)|\mathbf{k}, 0\rangle f^0(2); \tag{16.70}$$

$f^0(2) = n\phi(2)$ is the Maxwell-Boltzmann distribution (cf. Eq. (9.96)), and $R(\mathbf{p}_1; \mathbf{k}, z)$ is called the ring operator and is defined

$$R(\mathbf{p}_1; \mathbf{k}, z) = \int d\mathbf{p}_2 \int \frac{d\mathbf{k}'}{(2\pi)^3} \langle \mathbf{k}, 0|\tilde{T}_{12}|\mathbf{k}', \mathbf{k} - \mathbf{k}'\rangle$$

$$\times \left(\frac{i}{z - \frac{\mathbf{k}'\cdot\mathbf{p}_1}{m} - \frac{(\mathbf{k} - \mathbf{k}')\cdot\mathbf{p}_2}{m} - i\langle \mathbf{k}', \mathbf{k} - \mathbf{k}'|\tilde{\lambda}(1, 2)|\mathbf{k}', \mathbf{k} - \mathbf{k}'\rangle}\right)$$

$$\times \langle \mathbf{k}', \mathbf{k} - \mathbf{k}'|\tilde{T}_{12}|\mathbf{k}, 0\rangle f^0(2) \tag{16.71}$$

where

$$\langle \mathbf{k}', \mathbf{k} - \mathbf{k}'|\tilde{\lambda}(1, 2)|\mathbf{k}', \mathbf{k} - \mathbf{k}'\rangle$$

$$= \int d\mathbf{p}_3 \langle \mathbf{k}', \mathbf{k} - \mathbf{k}', 0|G_0^{-1}(\tilde{T}_{13} + \tilde{T}_{23}(1 + P_{23}))G_0|\mathbf{k}', \mathbf{k} - \mathbf{k}', 0\rangle f^0(3). \tag{16.72}$$

Note that we have included a term of the form $G_0 T_{12} G_0 T_{12} G_0$, which is forbidden. However, it does not contribute to the long time behavior. As we shall now see, the binary collision operator is closely related to the Boltzmann and Lorentz-Boltzmann collision operators and therefore we can use the hydrodynamic eigenfunctions of the collision operators to evaluate Eq. (16.69).

3. Binary Collision Operator

Let us consider the operator $C^-(\mathbf{p}_1; \mathbf{0}, z)$ acting on some function $f^0(1)F(\mathbf{p}_1)$:

$$\begin{aligned} C^-(\mathbf{p}_1; \mathbf{0}, z) f^0(1) F(\mathbf{p}_1) &= \lim_{\mathbf{k}\to 0} C^-(\mathbf{p}_1; \mathbf{k}, z) f^0(1) F(\mathbf{p}_1) \\ &= \lim_{\mathbf{k}\to 0} \int d\mathbf{p}_2 \langle \mathbf{k}, 0 | \tilde{T}_{12} | \mathbf{k}, 0 \rangle f^0(1) f^0(2) F(\mathbf{p}_1) \\ &= -z^2 \int d\mathbf{p}_2 \int d\mathbf{q}_{12} \int_0^\infty dt\, e^{izt} (e^{-t\hat{\mathscr{H}}(1,2)} - e^{-t\hat{\mathscr{H}}_0(1,2)}) \\ &\quad \times f^0(\mathbf{p}_1) f^0(\mathbf{p}_2) F(\mathbf{p}_1). \end{aligned} \tag{16.73}$$

We first note that the evolution operator $e^{-t\hat{\mathscr{H}}_0(1,2)}$ has no effect on the function $F(\mathbf{p}_1)$, while $e^{-t\hat{\mathscr{H}}(1,2)}$ transforms the momenta $\mathbf{p}_1$ and $\mathbf{p}_2$ to their values $\mathbf{p}_1(-t)$ and $\mathbf{p}_2(-t)$ at time $-t$. Thus,

$$\begin{aligned} &C^-(\mathbf{p}_1; \mathbf{0}, z) f^0(1) F(\mathbf{p}_1) \\ &\quad = -z^2 \int d\mathbf{p}_2 \int d\mathbf{q}_{12} \int_0^\infty dt\, e^{izt} f^0(1) f^0(2) (F(\mathbf{p}_1(-t)) - F(\mathbf{p}_1)), \end{aligned} \tag{16.74}$$

where we have used the fact that for elastic collisions $f^0(\mathbf{p}_1) f^0(\mathbf{p}_2) = f^0(\mathbf{p}_1(-t)) f^0(\mathbf{p}_2(-t))$. We next note that $\mathbf{p}_1(-t)$ and $\mathbf{p}_2(-t)$ will only differ from $\mathbf{p}_1$ and $\mathbf{p}_2$, respectively, if a collision has taken place in the time t. Therefore, the only contribution to $C^-(\mathbf{p}_1; \mathbf{0}, z)$ will come from the volume of configuration space for which a collision can take place in time t.

Let us assume that the particles are hard spheres of diameter d. We can write the volume element $d\mathbf{q}_{12}$ in terms of cylindrical coordinates such that the direction of the z-axis coincides with the direction of the $\mathbf{p}_{12}$-axis (cf. Fig. 16.2). A collision can only take place in time $-t$ if the particle lies within a cylinder of length $p_{12}t/m$ and radius d. Therefore, we find

$$\begin{aligned} &C^-(\mathbf{p}_1; \mathbf{0}, z) f^0(1) F(\mathbf{p}_1) \\ &\quad = -z^2 \int d\mathbf{p}_2 \int_0^{2\pi} d\phi \int_0^d b\, db \int_0^\infty dt\, e^{izt} \frac{p_{12}t}{m} (F(\mathbf{p}_1') - F(\mathbf{p}_1)) f^0(1) f^0(2) \end{aligned} \tag{16.75}$$

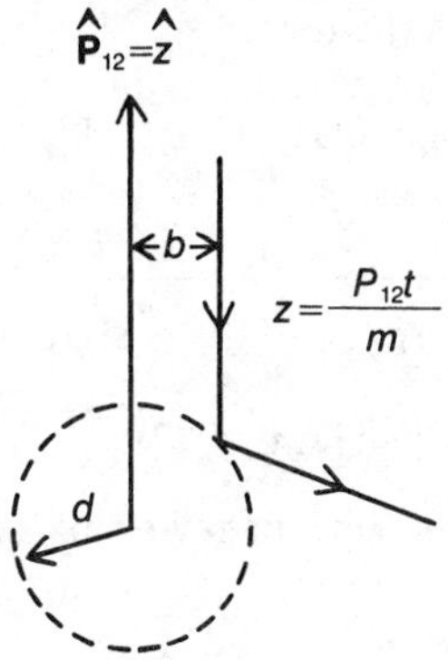

Fig. 16.2. Geometry of a collision: the dashed circle is the sphere of influence.

where $\mathbf{p}_1'$ is the momentum after the collision and depends on $\mathbf{p}_1$, the impact parameter b, and the type of force between the particles. If we now note that $\int_0^\infty dt\, e^{izt} t = -z^{-2}$, we finally obtain

$$C^-(\mathbf{p}_1; \mathbf{0}, z) \equiv C^-(\mathbf{p}_1) = \int d\mathbf{p}_2 \int_0^{2\pi} d\phi \int_0^d b\, db \frac{p_{12}}{m} (F(\mathbf{p}_1') - F(\mathbf{p}_1)) f^0(2). \tag{16.76}$$

Thus, $C^-(\mathbf{p}_1; \mathbf{0}, z)$ is just the Lorentz-Boltzmann collision operator (cf. Sec. 13.D) and is independent of z. Note, however, that it must operate on a function which is multiplied by $f^0(1)$ or $\phi(1)$.

By a similar calculation, we can obtain the Boltzmann collision operator. It is given by

$$\begin{aligned} C^+(\mathbf{p}_1) &= \lim_{\mathbf{k}\to 0} C^+(\mathbf{p}_1; \mathbf{k}, z) = \lim_{\mathbf{k}\to 0} \int d\mathbf{p}_2 \langle \mathbf{k}, 0 | \tilde{T}_{12}(1 + P_{12}) | \mathbf{k}, 0 \rangle f^0(2) \\ &= \int d\mathbf{p}_2 \int_0^{2\pi} d\phi \int_0^d b\, db \frac{p_{12}}{m} \\ &\quad \times (F(\mathbf{p}_1') + F(\mathbf{p}_2') - F(\mathbf{p}_1) - F(\mathbf{p}_2)) f^0(2) \end{aligned} \tag{16.77}$$

and is also independent of z.

It is now possible to obtain a simple approximation to the ring operator, which contains its leading contribution in the hydrodynamic regime (long wavelengths). If we replace the matrix element $\langle \mathbf{k}', \mathbf{k} - \mathbf{k}' | \tilde{\lambda}(1, 2) | \mathbf{k}', \mathbf{k} - \mathbf{k}' \rangle$ by its infinite wavelength limit, we find

$$\lim_{\substack{\mathbf{k}\to 0 \\ \mathbf{k}'\to 0}} \langle \mathbf{k}', \mathbf{k} - \mathbf{k}' | \tilde{\lambda}(1, 2) | \mathbf{k}', \mathbf{k} - \mathbf{k}' \rangle \to C^-(\mathbf{p}_1) + C^+(\mathbf{p}_2) \tag{16.78}$$

The ring operator then takes the form

$$R(\mathbf{p}_1; \mathbf{k}, z) \approx \int d\mathbf{p}_2 \int \frac{d\mathbf{k}'}{(2\pi)^3} \langle \mathbf{k}, 0|\tilde{T}_{12}|\mathbf{k}', \mathbf{k} - \mathbf{k}'\rangle$$
$$\times \left(\frac{i}{z - \frac{1}{m}\mathbf{k}'\cdot\mathbf{p}_1 - \frac{1}{m}(\mathbf{k} - \mathbf{k}')\cdot\mathbf{p}_2 - iC^-(\mathbf{p}_1) - iC^+(\mathbf{p}_2)}\right)$$
$$\times \langle \mathbf{k}', \mathbf{k} - \mathbf{k}'|\tilde{T}_{12}|\mathbf{k}, 0\rangle f^0(2). \tag{16.79}$$

We can now study the long-time behavior of the velocity autocorrelation function.

E. MICROSCOPIC DERIVATION OF THE LONG TIME TAILS

The relation between the divergences in the virial expansion of the transport coefficients and the long time tails found by Alder and Wainwright has been shown by Dorfmann and Cohen and by Dufty. The Boltzmann equation is unable to account for the long time tails observed in the correlation functions. They, in fact, result from long-wavelength hydrodynamic effects in the system. We will now show that, when the most divergent terms in the virial expansion of the transport coefficients are summed together, they give contributions to the current-current correlation functions which decay as $t^{d/2}$ (where d is the number of dimensions), even for low density. Therefore, it appears that the Boltzmann equation in principle is only capable of describing short time effects in a dilute gas.

We will be interested in the hydrodynamic limit of the velocity autocorrelation function, that is, the limit $\mathbf{k} \to 0$. If we take this limit in Eq. (16.69), we obtain

$$\lim_{\mathbf{k}\to 0} C_{\mathbf{vv}}(\mathbf{p}_1; \mathbf{k}, t)$$
$$= \frac{1}{m^2}\int_{-\infty}^{\infty} \frac{dz}{(2\pi)} e^{-izt} \int d\mathbf{p}_1 \mathbf{p}_1 \left(\frac{i}{z - iC^-(\mathbf{p}_1) - iR(\mathbf{p}_1; 0, z)}\right) \mathbf{p}_1 \phi(1) \tag{16.80}$$

where

$$R(\mathbf{p}_1; 0, z) \approx \int d\mathbf{p}_2 \int \frac{d\mathbf{k}'}{(2\pi)^3} \langle 0, 0|\tilde{T}_{12}(z)|\mathbf{k}', -\mathbf{k}'\rangle$$
$$\times \left(\frac{i}{z - \frac{\mathbf{k}'\cdot\mathbf{p}_1}{m} + \frac{\mathbf{k}'\cdot\mathbf{p}_2}{m} - iC^-(\mathbf{p}_1) - iC^+(\mathbf{p}_2)}\right)$$
$$\times \langle \mathbf{k}', -\mathbf{k}'|\tilde{T}_{12}(z)|0, 0\rangle f^0(2). \tag{16.81}$$

The hydrodynamic poles of the correlation function give its long time behavior.

We will compute the velocity autocorrelation function retaining only terms of first order in the ring operator, $R(\mathbf{p}_1; \mathbf{0}, z)$. This approximation restricts us to low-density systems and restricts the time interval over which the results are valid because more complicated dynamical events are neglected. The time for which it is valid should be comparable with the time over which the Alder and Wainwright calculation is done and therefore we can compare the two results.[6]

Let us write the velocity autocorrelation function as an expansion,

$$\lim_{\mathbf{k}\to 0} C_{\mathbf{vv}}(\mathbf{k}, t) = C_{\mathbf{vv}}(0, t)_{(0)} + C_{\mathbf{vv}}(0, t)_{(1)} + \cdots \tag{16.82}$$

where

$$C_{\mathbf{vv}}(0, t)_{(0)} = \frac{1}{m^2}\int_{-\infty}^{\infty} \frac{dz}{(2\pi)}\, e^{-izt} \int d\mathbf{p}_1 \mathbf{p}_1 \left(\frac{i}{z - iC^-(\mathbf{p}_1)}\right) \mathbf{p}_1 \phi(1) \tag{16.83}$$

and

$$C_{\mathbf{vv}}(0, t)_{(1)} = \frac{1}{m^2}\int_{-\infty}^{\infty} \frac{dz}{(2\pi)}\, e^{-izt} \times \int d\mathbf{p}_1 \mathbf{p}_1 \left(\frac{i}{z - iC^-(\mathbf{p}_1)}\right) R(\mathbf{p}_1; 0, z) \left(\frac{i}{z - iC^-(\mathbf{p}_1)}\right) \mathbf{p}_1 \phi(1). \tag{16.84}$$

Eq. (16.83) represents contributions to the self-diffusion coefficient which one would obtain starting from the Lorentz-Boltzmann equation taking only two-body interactions into account. We know that it will decay exponentially. Eq. (16.84) contains long-lived contributions from higher order collision processes. In order to evaluate Eq. (16.84) we must use the eigenfunctions and eigenvalues of the operators $[\mathbf{k}'\cdot\mathbf{p}_1/m + iC_0^+(\mathbf{p}_1)]$ and $[-\mathbf{k}'\cdot\mathbf{p}_2/m + iC^-(\mathbf{p}_2)]$. These have been obtained in Chap. 13.

We will introduce a projection operator, P_H, which projects onto the hydrodynamic subspace. If $F(\mathbf{p}_1, \mathbf{p}_2)$ is a function of $\mathbf{p}_1$ and $\mathbf{p}_2$, we can write

$$P_H F(\mathbf{p}_1, \mathbf{p}_2)\phi(1)\phi(2) = \sum_{\alpha=1}^{5} \Psi^*(\mathbf{p}_1 . \mathbf{k}')_{(-)} \Psi_\alpha^*(\mathbf{p}_2, -\mathbf{k}')_{(+)} \phi(1)\phi(2) \times \iint d\mathbf{p}_1'\, d\mathbf{p}_2' \Psi(\mathbf{p}_1', \mathbf{k}')_{(-)} \Psi_\alpha(\mathbf{p}_2', -\mathbf{k}')_{(+)} F(\mathbf{p}_1', \mathbf{p}_2')\phi(1')\phi(2'), \tag{16.85}$$

where $\Psi(\mathbf{p}_1, \mathbf{k}')_{(-)}$ is the hydrodynamic eigenfunction of the Lorentz-Boltzmann equation and $\Psi_\alpha(\mathbf{p}_1, \mathbf{k}')_{(+)}$ is one of the five hydrodynamic eigenfunctions of the Boltzmann operator. We may write $C_{\mathbf{vv}}(0, t)_{(1)}$ in terms of a hydrodynamic part, $C_{\mathbf{vv}}^H(0, t)_{(1)}$, and a nonhydrodynamic part, $C_{\mathbf{vv}}^{NH}(0, t)_{(1)}$:

$$C_{\mathbf{vv}}(0, t)_{(1)} = C_{\mathbf{vv}}^H(0, t)_{(1)} + C_{\mathbf{vv}}^{NH}(0, t)_{(1)} \tag{16.86}$$

where

$$C^{H}_{\mathbf{vv}}(0, t)_{(1)} = \frac{n}{m^2}\int_{-\infty}^{\infty}\frac{dz}{(2\pi)}\,e^{-izt}$$

$$\times \int \frac{d\mathbf{k}'}{(2\pi)^3}\iint d\mathbf{p}_1\,d\mathbf{p}_2\mathbf{p}_1\left(\frac{i}{z - iC^-(\mathbf{p}_1)}\right)\langle 0, 0|\hat{T}_{12}|\mathbf{k}', -\mathbf{k}'\rangle$$

$$\times \left(\frac{i}{z - \dfrac{\mathbf{k}'\cdot\mathbf{p}_{12}}{m} - iC^-(\mathbf{p}_1) - iC^+(\mathbf{p}_1)}\right)$$

$$\times P_H\langle\mathbf{k}', -\mathbf{k}'|\tilde{T}_{12}|0, 0\rangle\left(\frac{i}{z - iC^-(\mathbf{p}_1)}\right)\phi(1)\phi(2)\mathbf{p}_1 \qquad (16.87)$$

and $C^{NH}_{\mathbf{vv}}(0, t)_{(1)}$ is defined in a similar manner except that P_H is replaced by $(1 - P_H)$. To evaluate Eq. (16.87) we approximate $\langle 0, 0|\tilde{T}_{12}|\mathbf{k}', -\mathbf{k}'\rangle$ and $\langle\mathbf{k}', -\mathbf{k}'|\tilde{T}_{12}|0, 0\rangle$ by $\langle 0, 0|\tilde{T}_{12}|0, 0\rangle$ and use Eq. (16.85). We then obtain

$$C^{H}_{\mathbf{vv}}(0, t)_{(1)} = \frac{n}{m^2}\int_{-\infty}^{\infty}\frac{dz}{2\pi}\,e^{-izt}\int d\mathbf{p}_1\int d\mathbf{p}_2\sum_{\alpha=1}^{5}\int\frac{d\mathbf{k}'}{(2\pi)^3}\,\mathbf{p}_1$$

$$\times \left(\frac{i}{z - \omega^-(\mathbf{k}') - \omega^+(-\mathbf{k}')}\right)\left(\frac{i}{z - iC^-(\mathbf{p}_1)}\right)\langle 0, 0|\tilde{T}_{12}|0, 0\rangle$$

$$\times \Psi^*_1(\mathbf{p}_1, \mathbf{k}')_{(-)}\Psi_\alpha(\mathbf{p}_2, -\mathbf{k}')_{(+)}\phi(1)\phi(2)\iint d\mathbf{p}_1'\,d\mathbf{p}_2'\Psi(\mathbf{p}_1', \mathbf{k}')_{(-)}$$

$$\times \Psi_\alpha(\mathbf{p}_2', -\mathbf{k}')_{(+)}\langle 0, 0|\tilde{T}_{12}|0, 0\rangle\left(\frac{i}{z - iC^-(\mathbf{p}_1')}\right)\mathbf{p}_1'\phi(1')\phi(2'). \qquad (16.88)$$

We can now replace $\Psi(\mathbf{p}_1, \mathbf{k}')_{(-)}$ by $\Psi(\mathbf{p}_1, 0)_{(-)}$ and replace $\Psi_\alpha(\mathbf{p}_2, -\mathbf{k}')_{(+)}$ by $\Psi_\alpha(\mathbf{p}_1, 0)_{(+)}$. In replacing the exact eigenfunctions by their zeroth-order approximation, we are neglecting terms of order $\mathbf{k}$ and higher. These terms will, however, be less divergent than the terms we are retaining. Furthermore, we make use of the following relationships:

$$n\iint d\mathbf{p}_1\,d\mathbf{p}_2\mathbf{p}_1\left(\frac{i}{z - iC^-(\mathbf{p}_1)}\right)\langle 0, 0|\tilde{T}_{12}|0, 0\rangle\Psi_\alpha(\mathbf{p}_2)_{(+)}\phi(2)\phi(1)$$

$$= -\int d\mathbf{p}_1\mathbf{p}_1\left(\frac{i}{z - iC^-(\mathbf{p}_1)}\right)C^-(\mathbf{p}_1)\Psi_\alpha(\mathbf{p}_1)_{(+)}\phi(1) \qquad (16.89)$$

where we have used the fact that $\Psi_\alpha(\mathbf{p}_2)$ depends only on quantities which are invariant during the collision so that $\Psi_\alpha(\mathbf{p}_2') - \Psi_\alpha(\mathbf{p}_2) = -(\Psi_\alpha(\mathbf{p}_1') - \Psi_\alpha(\mathbf{p}_1))$ and the fact that $\phi(1)\phi(2) = \phi(1')\phi(2')$. Similarly, we may write

$$n\iint d\mathbf{p}_1\,d\mathbf{p}_2\Psi_\alpha(\mathbf{p}_2)_{(+)}\langle 0, 0|\tilde{T}_{12}|0, 0\rangle\left(\frac{i}{z - iC^-(\mathbf{p}_1)}\right)\phi(1)\phi(2)\mathbf{p}_1$$

$$= -\int d\mathbf{p}_1\Psi_\alpha(\mathbf{p}_1)_{(+)}C^-(\mathbf{p}_1)\left(\frac{i}{z - iC^-(\mathbf{p}_1)}\right)\phi(1)\mathbf{p}_1. \qquad (16.90)$$

Combining Eqs. (16.88), (16.89), and (16.90), we obtain

$$C^H_{\mathbf{vv}}(0, t)_{(1)} = \frac{1}{n}\frac{1}{m^2}\int_{-\infty}^{\infty}\frac{dz}{(2\pi)}e^{-izt}\sum_{\alpha=1}^{5}\int\frac{d\mathbf{k}'}{(2\pi)^3}\left(\frac{i}{z-\omega^-(\mathbf{k}')-\omega_\alpha^{(+)}(-\mathbf{k}')}\right)$$
$$\times\left[\iint d\mathbf{p}_1\mathbf{p}_1\Psi'_\alpha(\mathbf{p}_1)_+C^-(\mathbf{p}_1)\left(\frac{i}{z-iC^-(\mathbf{p}_1)}\right)\phi(1)\right]^2 \tag{16.91}$$

for the hydrodynamic part of the correlation function.

The leading contribution to $C^H_{\mathbf{vv}}(0, t)_{(1)}$ for long times comes from the pole $(z - \omega_1^-(\mathbf{k}') - \omega_\alpha^+(-\mathbf{k}'))^{-1}$. The other poles give contributions which decay exponentially after a few collision times. If we take into account the fact that $\omega_1^-(\mathbf{k}') \xrightarrow[\mathbf{k}\to 0]{} 0$ and $\omega_\alpha^+(-\mathbf{k}') \xrightarrow[\mathbf{k}\to 0]{} 0$, we can expand the $\mathbf{k}'$ dependence of the squared term in Eq. (16.91) (it gains $\mathbf{k}'$ dependence from the pole $(z - \omega_1^-(\mathbf{k}') - \omega_\alpha^+(-\mathbf{k}'))^{-1}$) and neglect terms of order $\mathbf{k}'$. Eq. (16.91) then becomes

$$C^H_{\mathbf{vv}}(0, t) \approx \frac{1}{n}\frac{1}{m^2}\sum_{\alpha=1}^{5}\int\frac{d\mathbf{k}'}{(2\pi)^3}e^{-i(\omega_1^-(\mathbf{k}')+\omega_\alpha^+(-\mathbf{k}'))t}\left[\int d\mathbf{p}_1\mathbf{p}_1\Psi'_\alpha(\mathbf{p}_1)_{(+)}\phi(1)\right]^2. \tag{16.92}$$

We now notice that not all hydrodynamic modes can contribute to the velocity autocorrelation function. Because of the integration $\int d\mathbf{p}_1\mathbf{p}_1\Psi'_\alpha(\mathbf{p}_1)\phi(1)$ in Eq. (16.92), only contributions from the sound modes ($\alpha = 1$ and 2) and the shear modes will be nonzero (only these modes depend on components of the vector $\mathbf{p}_1$). Eq. (16.92) becomes

$$C^H_{\mathbf{vv}}(0, t) \approx \frac{1}{n}\frac{1}{m^2}\sum_{\alpha=1}^{4}\int\frac{d\mathbf{k}'}{(2\pi)^3}e^{-i(\omega^-(\mathbf{k})+\omega_\alpha^+(-\mathbf{k}'))t}\left[\int d\mathbf{p}_1\mathbf{p}_1\Psi'_\alpha(\mathbf{p}_1)_{(+)}\phi(1)\right]^p. \tag{16.93}$$

Let us now look at the term $\alpha = 1$ in Eq. (16.93). From Eqs. (13.112) and (13.165), we have

$$\frac{1}{m^2}\int\frac{d\mathbf{k}'}{(2\pi)^3}e^{-i(\omega^-(\mathbf{k}')+\omega_1^+(-\mathbf{k}'))t}\left[\int d\mathbf{p}_1\mathbf{p}_1\Psi'^0_1(\mathbf{p}_1)_{(+)}\phi(1)\right]^2$$
$$= \frac{1}{m^2}\int\frac{d\mathbf{k}'}{(2\pi)^3}e^{-(ik'c_0+k'^2A)t}\left[\int d\mathbf{p}_1\mathbf{p}_1\Psi'^0_1(\mathbf{p}_1)_{(+)}\phi(1)\right]^2 \tag{16.94}$$

where

$$A = D + \frac{1}{2n_0}\left[\tfrac{4}{3}\eta + \zeta + K\left(\frac{1}{C_V} - \frac{1}{C_p}\right)\right],$$

D is the Lorentz-Boltzmann diffusion coefficient, η and ζ are the Boltzmann viscosities, K is the Boltzmann thermal conductivity, and c_0 is the speed of sound. Let us now perform the integration over k':

$$\int dk'\, e^{-(ik'c_0+k'^2A)t} = \frac{4\pi}{A\sqrt{A}}\, e^{-c_0^2t/4A} \int_{c_0i/2\sqrt{A}}^{\infty} \left(x - \frac{c_0 i}{2\sqrt{A}}\right)^2 dx\, e^{-x^2t} \tag{16.95}$$

where $x = (\sqrt{A}k' + ic_0/2\sqrt{A})$. From Eq. (16.95) we see that the presence of the imaginary term, ic_0k, in the dispersion relation for the sound modes causes an exponential decay of both the first and second terms in Eq. (16.93). Thus, we are left only with the contribution for the shear modes as the source of the long time tails.

Let us now look at the shear contributions, $\alpha = 3$ and 4. We obtain

$$\begin{aligned}
&\sum_{\alpha=3,4} \int \frac{d\mathbf{k}'}{(2\pi)^3}\, e^{-i(\omega^{-}(\mathbf{k}')+\omega_\alpha^{+}(-\mathbf{k}'))t} \left[\int d\mathbf{p}_1 \mathbf{p}_1 \Psi_\alpha{}^0(\mathbf{p}_1)_{(+)}\phi(1)\right]^2 \\
&= \int \frac{d\mathbf{k}'}{(2\pi)^3}\, e^{-k^2(D+(\eta/mn_0))t}\, \frac{\beta}{m} \left\{\left[\int d\mathbf{p}_1 p_1 p_{1y}\phi(1)\right]^2 + \left[\int d\mathbf{p}_1 p_1 p_{1x}\phi(1)\right]^2\right\} \\
&= \frac{1}{8\sqrt{\pi}}\frac{\beta}{m}\left(\frac{1}{\left(D + \dfrac{\eta}{mn_0}\right)t}\right)^{3/2} \\
&\quad \times \left\{\left[\int d\mathbf{p}_1 \mathbf{p}_1 p_{1y}\phi(1)\right]^2 + \left[\int d\mathbf{p}_1 \mathbf{p}_1 p_{1z}\phi(1)\right]^2\right\}
\end{aligned} \tag{16.96}$$

where we have assumed that for the $\mathbf{p}_1$ integration k' is directed along the x-axis. In Eq. (16.96), the integration over $\mathbf{p}_1$ simply gives a number (it is a worthwhile exercise to do it). We see that for long times the velocity autocorrelation function decays as $t^{-3/2}$. To obtain this result, we have had to retain contributions from collision processes involving all possible numbers of particles even for low density. In practice, at very low density, the many-body effects give only a very small contribution to the transport coefficients and the Boltzmann equation is adequate. But, in principle, they are there and become more important as the density is raised.

F. IMPLICATIONS FOR HYDRODYNAMICS

In Sec. 16.D we found that many-body effects must be taken into account in calculations of time-dependent correlation functions; otherwise, we obtain divergent expressions for the correlation functions of moderately dense gases.

As we mentioned before, the origin of these divergences is mathematically the same sort that we saw in the equilibrium expansions at low temperature (cf. Sec. 12.D). A simple density expansion leads to an expression for the correlation function involving repeated free propagators which give rise to secular terms in the long time limit. An expansion in terms of free propagators allows the possibility of a particle to move as if it were in free space, without any effects from the medium. This, of course, cannot happen. In a real gas, a particle, on the average, undergoes a collision each time interval of length τ, where τ is the collision time. Resummation of the most divergent terms in Sec. 16.D has the effect of including interactions of the particle with the medium in the propagators.

The fact that we must include many-body effects in our expressions for the time correlation function has important consequences for hydrodynamics. Until now we have only discussed the Navier-Stokes limit of the hydrodynamic equations. That is, we have kept terms to order k^2 in the linearized hydrodynamic equations. However, it is also of interest to go to higher orders in k and by so doing introduce new transport coefficients. For example, we may wish to write the diffusion equation in the form

$$\frac{\partial n}{\partial t} = Dk^2 n + D_2 k^4 n + \cdots. \tag{16.97}$$

Corrections of this type were first introduced by Burnett[8] for the linearized hydrodynamic equations.

We can obtain a microscopic expression for D_2 using the same methods as in Chap. 13 except that we have to go to higher order in k and, in view of the divergences, the eigenvalue problem is no longer given by Eq. (13.113) but rather by the equation

$$\left(\mathbf{k}\cdot\frac{\mathbf{p}_1}{m} + iC^-(\mathbf{k}) + iR(\mathbf{k}, z)\right)\Psi(\mathbf{k}, z) = \omega(\mathbf{p}_1, \mathbf{k})\Psi(\mathbf{k}, z) \tag{16.98}$$

where $R(\mathbf{k}, z)$ is the ring operator derived in Sec. 16.D. By obtaining an expression for $\omega(\mathbf{k}, \mathbf{p}_1)$ to fourth order in k, Dufty and McLennan[9] have shown for diffusion processes that the transport coefficient D_2 does not exist (is infinite) and that, in fact, simple corrections of the linearized hydrodynamic equations involving higher order gradients are not possible. Similar results have been obtained by Dorfmann and Ernst[10] and for the corrections to the Navier-Stokes equations. Thus, any k-dependent corrections to the Navier-Stokes equations appear to be nonanalytic in k, and simple corrections in terms of higher order gradients are not possible.

REFERENCES

1. B. J. Alder and T. E. Wainwright, Phys. Rev. Let. *18* 988 (1967).
2. M. H. Ernst, E. H. Hauge, and J. M. J. van Leeuven, Phys. Rev. *A4* 2055 (1971).
3. N. N. Bogoliubov in *Studies in Statistical Mechanics*, Vol. I, ed. J. de Boer and G. E. Uhlenbeck (North-Holland Publishing Co., Amsterdam, 1961).

4. S. T. Choh and G. E. Uhlenbeck, "The Kinetic Theory of Dense Gases," dissertation, University of Michigan, 1958.
5. K. Kawasaki and I. Oppenheim, Phys. Rev. *A139* 1763 (1965).
6. J. R. Dorfmann and E. D. G. Cohen, Phys. Rev. *A6* 776 (1972).
7. R. Zwanzig, Phys. Rev. *129* 486 (1963).
8. D. Burnett, Proc. London Math. Soc. *40* 382 (1935).
9. J. W. Dufty and J. A. McLennan, Phys. Rev. *A9* 1266 (1974).
10. E. H. Ernst and J. R. Dorfmann, Physica *61* 157 (1972).

PROBLEMS

1. Using the method of Sec. 16.B, compute the longitudinal contribution to the correlation function $C_\eta(t)$.

2. Prove that the operators $G(1, \ldots, N; z)$ given in Eq. (16.49) may be written in the terms of the binary collision expansion. (*Hint:* first obtain the time-dependent binary expansion.)

3. Prove that the quantity $\int d\mathbf{p}_2 \langle 0, 0, 0|\tilde{T}_{12}(1 + P_{12})|0, 0, 0\rangle \phi(1)\phi(2)$ is the Boltzmann collision operator.

4. Obtain the ring approximation for the correlation function $\langle J_{xy}(\mathbf{0}, 0) J_{xy}(\mathbf{r}, t)\rangle$, where $J_{xy}(\mathbf{r}, t) = \sum_{i=1}^{N} p_{x_i} p_{y_i}\, \delta(\mathbf{q}_i - \mathbf{r})$ is the microscopic momentum current. (Note that this correlation function may be used to compute the coefficient of shear viscosity.)

5. Prove that the term $G_0(1, 2, 3; z)T_{12}(z)G_0(1, 2, 3; z)T_{13}(z)G_0(1, 2, 3; z)$ does not contribute to $B_2(\mathbf{p}_1; \mathbf{k}, z)$ for the velocity autocorrelation function (it is canceled by other terms).

17
Nonequilibrium Phase Transitions

A. INTRODUCTORY REMARKS

In previous chapters, we considered phase transitions that occur in systems in thermodynamic equilibrium. In this chapter, we will give examples of phase transitions which can occur in chemical and hydrodynamic systems far from absolute thermodynamic equilibrium. The systems we will consider are locally in thermodynamic equilibrium, so the state of the system can be described in terms of thermodynamic densities which vary in space. However, they are far enough from absolute equilibrium that nonlinear effects must be included in the chemical rate equations or hydrodynamic equations. As we shall see, nonlinear effects open a whole new world of behavior for such systems. Nonlinear equations allow the possibility of multiple solutions, each with different regions of stability. Thus, as we change the parameters of a nonlinear system, it can exhibit phase transitions from one state to another.

Prigogine was first to show that near equilibrium (in the linear regime), if one of the thermodynamic forces is held fixed, the stable state of the system is a steady state characterized by a minimum entropy production and is unique (cf. Sec. 14.E). This state is said to lie on the thermodynamic branch. However, as we move away from the linear regime, nonlinearities in the hydrodynamic equations or chemical rate equations become more important and at some point the thermodynamic branch becomes unstable and a nonequilibrium phase transition occurs. The system then changes to a new state which is characterized by an order parameter. Often the change is dramatic. Even if the boundary is held fixed and the steady state in the linear regime is homogeneous or has constant gradients, the new state that appears in the nonlinear regime can oscillate in space and/or in time and often exhibits nonlinear wave motion. Thus, the symmetry of such systems in space and/or in time is broken at the phase transition, and the new stable nonequilibrium state exhibits much more structure than the state on the thermodynamic branch or the state of thermodynamic equilibrium. These structured states (called dissipative structures by

Prigogine) require a flow of energy and sometimes a flow of matter and, therefore, a production of entropy to maintain them.

In this chapter we will illustrate nonequilibrium phase transitions with some classic examples taken from hydrodynamics and chemistry. Before we begin to consider examples, we will discuss how equilibrium stability arguments determine the stability of systems far from equilibrium. We will then consider two important nonlinear chemical reactions. In the first, the Schlögl model, we show that chemical systems can exhibit nonequilibrium phase transitions similar to the first-order transitions in the van der Waals system. That is, abrupt transitions in the concentration of one of the chemicals can occur. As a second example, we discuss the so-called Brusselator, which was first introduced by Prigogine and Lefever. This is one of the simplest chemical models which can exhibit spatial and temporal dissipative structures, that is, oscillations in space and time maintained by the flow of chemicals through the system.

Many of the ideas of nonlinear chemistry can also be applied to population dynamics and ecological systems because, much like chemical systems, individuals continually interact and are changed by their interactions, thus causing dynamical evolution of the populations. One of the most famous and historically most important nonlinear models in population dynamics is the Lotka-Volterra, or predator-prey, model. Although this model is too simple to exhibit the essential features of the evolution of real populations, it is still widely discussed and we will consider it here.

As a final example, we consider a purely hydrodynamic system and show that the nonlinearities in the Navier-Stokes equations can lead to instabilities in the flow of real systems. Turbulence is one example of such an instability. The example we will consider is simpler and involves an instability (the Bénard instability) which occurs in a fluid layer when it is heated from below.

B. THERMODYNAMIC STABILITY CRITERIA FAR FROM EQUILIBRIUM

1. Entropy Production

It is possible to create a nonequilibrium steady state by holding some thermodynamic forces, such as gradients of various densities or the affinity, fixed (cf. Sec. 14.I). In the linear regime, the steady state which results is a state of minimum entropy production and is always stable. The proof that steady states in the linear regime are always stable depends on the fact that the thermodynamic currents are linear functions of the forces and uses the equilibrium stability criteria derived in Chap. 2. In the nonlinear regime, we can no longer write the currents as linear functions of the forces and it is no longer possible to prove that the steady state which is stable in the linear regime will continue to be stable in the nonlinear regime. Thus, we have the possibility of creating nonequilibrium steady states which become unstable as we vary certain parameters in the system and the system can evolve to a new state which does not exist in the linear regime. That is, it can undergo a nonequilibrium phase transition.

A thermodynamic stability theory for systems far from equilibrium has been developed by Prigogine and Glansdorff.[1] We shall illustrate the basic ideas of this theory for the case of chemically reacting systems.

When we consider systems for which the linear approximation is no longer valid, we can no longer show that the time rate of change of the entropy production has a definite sign, but we can show that a part of it has a definite sign.

If we write the entropy production in the form

$$P = -\int dV \sum_i J_i X_i, \tag{17.1}$$

its time derivative can be written

$$\frac{\partial P}{\partial t} = \frac{\partial_X P}{\partial t} + \frac{\partial_J P}{\partial t} \tag{17.2}$$

where $\partial_X P/\partial t$ is the contribution to the rate of change of the entropy production due to forces which change in time

$$\frac{\partial_X P}{\partial t} = -\int dV \sum_i J_i \frac{\partial X_i}{\partial t} \tag{17.3}$$

and $\partial_J P/\partial t$ is the rate of change due to currents which change in time

$$\frac{\partial_J P}{\partial t} = -\int dV \sum_i X_i \frac{\partial J_i}{\partial t}. \tag{17.4}$$

In the linear regime we have seen that

$$\frac{\partial_J P}{\partial t} = \frac{\partial_X P}{\partial t} = \frac{1}{2}\frac{\partial P}{\partial t} \leqslant 0. \tag{17.5}$$

In the nonlinear regime, *for systems in local equilibrium*, it is possible to prove that the relation

$$\frac{\partial_X P}{\partial t} \leqslant 0 \tag{17.6}$$

is always true because of thermodynamic stability relations which hold in each small region of the system, but it is not in general possible to say anything about the sign of $\partial_J P/\partial t$. However, under certain conditions, if we can establish one additional property for the system, Eq. (17.6) enables us to say something about the stability of a nonequilibrium steady state.

To illustrate the use of thermodynamic stability theory for nonequilibrium steady states, we shall consider the case of a chemically reacting system of n

types of molecules which is held far from equilibrium (it has large affinity). We shall assume that the effects of diffusion, viscosity, and thermal conductivity can be neglected and that the pressure and temperature are uniform throughout the system. The entropy production for such a system can be written

$$P = -\frac{1}{T}\int dV \sum_{j=1}^{r} \mathscr{J}_j \mathscr{A}_j \tag{17.7}$$

(cf. Eq. (14.152)), where r is the number of reactions. Therefore,

$$\frac{\partial_x P}{\partial t} = -\frac{1}{T}\int dV \sum_{k=1}^{r} \mathscr{J}_k \frac{\partial \mathscr{A}_k}{\partial t}. \tag{17.8}$$

From Eq. (14.132) we can write the balance equation for the jth component in the form

$$\rho \frac{\partial c_j}{\partial t} = \sum_{k=1}^{r} \nu_{jk} \mathscr{J}_k. \tag{17.9}$$

If we now use the definition of the affinity given in Eq. (3.54), we find

$$\frac{\partial_x P}{\partial t} = -\frac{1}{T}\int dV \left\{ \sum_{j=1}^{n} \rho \frac{\partial c_j}{\partial t} \left[\frac{\partial \mu_j}{\partial t}\right] \right\}. \tag{17.10}$$

Let us next note that since pressure and temperature are held constant

$$\frac{\partial \mu_j}{\partial t} = \sum_{j=1}^{n} \left(\frac{\partial \mu_j}{\partial c_i}\right)_{P,T} \frac{\partial c_i}{\partial t}, \tag{17.11}$$

and Eq. (17.10) becomes

$$\frac{\partial_x P}{\partial t} = -\int dV \left\{ \sum_{i=1}^{n} \sum_{j=1}^{n} \rho \left(\frac{\partial \mu_i}{\partial c_j}\right)_{T,P,\{c_i \neq c_j\}} \frac{\partial c_i}{\partial t} \frac{\partial c_j}{\partial t} \right\} \leqslant 0. \tag{17.12}$$

To establish the inequality we have used the stability condition in Prob. 2.19. Since we assume that the system is in local equilibrium, equilibrium stability conditions ensure that $\partial_x P/\partial t$ always has a well-defined sign.

2. Nonlinear Chemical Reactions

Thermodynamic stability theory for systems undergoing nonlinear chemical reactions is of considerable interest at present because of the importance of such reactions in biological systems. The approximation in which the rates of reaction, $\mathscr{J}_k$, are driven by terms linear in the affinities is true only near equilibrium and in general is not of great interest. We can find the exact dependence of the reaction rates on the affinities as follows. Let us consider a general chemical reaction of the form

$$-\nu_1 A_1 - \nu_2 A_2 \underset{k_2}{\overset{k_1}{\rightleftharpoons}} \nu_3 A_3 + \nu_4 A_4. \tag{17.13}$$

The rate equation can then be written

$$\frac{\partial N_3}{\partial t} = k_1 N_1^{|\nu_1|} N_2^{|\nu_2|} - k_2 N_3^{\nu_3} N_4^{\nu_4}$$

$$= k_1 N_1^{|\nu_1|} N_2^{|\nu_2|} \left[1 - \frac{k_2}{k_1} \frac{N_3^{\nu_3} N_4^{\nu_4}}{N_1^{|\nu_1|} N_2^{|\nu_2|}} \right] \tag{17.14}$$

where N_i is the molar density. The rate of reaction $\mathscr{J}$, is defined in terms of $\partial c_i/\partial t$ by the equation

$$\rho \frac{\partial c_i}{\partial t} = \nu_i \mathscr{J} = M_i \frac{\partial N_i}{\partial t} \tag{17.15}$$

where M_i is the molecular weight of constituent i. For a system of ideal gases, we can write the affinity in the form

$$\mathscr{A} = RT\left[F(T, P) + \ln \frac{N_3^{\nu_3} N_4^{\nu_4}}{N_1^{|\nu_1|} N_2^{|\nu_2|}} \right] \tag{17.16}$$

(cf. Eq. (3.57)) or

$$\mathscr{A} = RT \ln \left[K(T, P) \frac{N_3^{\nu_3} N_4^{\nu_4}}{N_1^{|\nu_1|} N_2^{|\nu_2|}} \right] \tag{17.17}$$

where $K(T, P) = e^{F(T,P)}$ and $F(T, P)$ and $K(T, P)$ are functions of temperature and pressure only. At equilibrium, $\mathscr{A} = 0$ and $\partial N_i/\partial t = 0$. If we compare Eqs. (17.14) and (17.17) we see that

$$K(T, P) = \frac{k_2}{k_1} \tag{17.18}$$

and

$$\frac{\partial N_3}{\partial t} = \nu_3 \mathscr{J} = k_1 N_1^{|\nu_1|} N_2^{|\nu_2|} [1 - e^{\mathscr{A}/RT}]. \tag{17.19}$$

Thus, Eq. (17.19) relates the reaction rate, $\mathscr{J}$, to the affinity $\mathscr{A}$. In general, $\mathscr{J}$ is a nonlinear function of $\mathscr{A}$. For small $\mathscr{A}$, $\mathscr{J}$ becomes linear in $\mathscr{A}$.

Let us now look at the stability problem for a case in which only chemical reactions take place. To determine the stability of a given steady state we can expand the local entropy around that state. The calculation is similar to that for the equilibrium case except that now all partial derivatives are evaluated at the steady state rather than the equilibrium state. The result, schematically, is of the form

$$S_{\text{local}} = S_{\text{local}}^0 + \delta S_{\text{local}} + \tfrac{1}{2}\delta^2 S_{\text{local}} + \cdots \tag{17.20}$$

The various terms in Eq. (17.20) are identical *in form* to those obtained for the Taylor expansion about absolute equilibrium. However, it differs from the equilibrium case because we expand about a nonequilibrium steady state where entropy is produced at a constant rate rather than about the absolute equilibrium state for which entropy has fixed value. The second-order term is of the form

$$\tfrac{1}{2}\delta^2 S_{\text{local}} = -\frac{\rho}{2T}\sum_{i=1}^{n}\sum_{j=1}^{n}\left(\frac{\partial \mu_i}{\partial c_j}\right)^0_{P,T\{c_i \neq c_j\}} \delta c_i\, \delta c_j \leqslant 0. \tag{17.21}$$

Since we assume that the system is in local equilibrium, the equilibrium stability conditions tell us that $\tfrac{1}{2}\delta^2 S_{\text{local}}$ will be negative or zero at each point.

We can now find the time rate of change of the total entropy due to terms which are quadratic in the local fluctuations about the steady state. If we take the derivative of Eq. (17.21), we find

$$\frac{1}{2}\frac{\partial}{\partial t}\delta^2 S_{\text{tot}} = -\int dV\left[\frac{1}{T}\rho\sum_{i=1}^{n}\sum_{j=1}^{n}\left(\frac{\partial \mu_i}{\partial c_j}\right)^0_{P,T,\{c_i \neq c_j\}} \delta c_i \frac{\partial\, \delta c_j}{\partial t}\right]. \tag{17.22}$$

We can use Eq. (17.9) to write

$$\frac{\partial}{\partial t}\delta c_i = \frac{1}{\rho}\sum_{k=1}^{r} \nu_{ik}\, \delta \mathscr{J}_k. \tag{17.23}$$

Furthermore, if we note that

$$\delta\mu_i \equiv \sum_{j=1}^{n}\left(\frac{\partial \mu_i}{\partial c_j}\right)_{T,P,\{c_i \neq c_j\}} \delta c_j \tag{17.24}$$

(pressure and temperature are assumed to be fixed) and that

$$\delta \mathscr{A}_k = \sum_{i=1}^{n} \nu_{ik}\, \delta\mu_i, \tag{17.25}$$

we obtain

$$\frac{\partial}{\partial t}(\tfrac{1}{2}\delta^2 S_{\text{tot}}) = -\int_V dV \sum_{k=1}^{r} \frac{\delta \mathscr{J}_k\, \delta \mathscr{A}_k}{T}. \tag{17.26}$$

Thus, the time variation of the second-order contribution to the entropy production due to fluctuations about the nonequilibrium steady state is given by a product of fluctuations in the current and the forces. The function $\delta^2 S_{\text{tot}}$ may be viewed as a Lyapounov function (cf. App. D). If we can show that $\delta^2 S_{\text{tot}} \leqslant 0$ and $\partial/\partial t\, \delta^2 S_{\text{tot}} > 0$, then the steady state we are considering is stable. Since the condition $\delta^2 S_{\text{tot}} \leqslant 0$ is always true, the condition that must be established is $\partial(\tfrac{1}{2}\delta^2 S_{\text{tot}})/\partial t > 0$.

Let us now compare Eq. (17.26) with the excess entropy production $\delta_x P$. The entropy production itself is given by

$$P = -\int dV\left\{\frac{1}{T}\sum_{k=1}^{r} \mathscr{J}_k \mathscr{A}_k\right\}, \tag{17.27}$$

and the contribution to the entropy production due to fluctuations of the independent forces about their equilibrium value is

$$\delta_x P = -\int dV \left[\frac{1}{T} \sum_{k=1}^{r} \mathcal{J}_k \, \delta\mathcal{A}_k\right]. \tag{17.28}$$

In the steady state, currents corresponding to the independent force vanish. Therefore, we can make the following expansion of $\mathcal{J}_j$ about the steady state:

$$\mathcal{J}_k = 0 + \delta\mathcal{J}_k + \tfrac{1}{2}\delta^2\mathcal{J}_k + \cdots \tag{17.29}$$

For very small deviations from the steady state, Eq. (17.28) becomes

$$\delta_x P = -\int dV \left[\frac{1}{T} \sum_j \delta\mathcal{J}_j \, \delta\mathcal{A}_j\right] = \frac{\partial}{\partial t}(\tfrac{1}{2}\delta^2 S_{\text{tot}}). \tag{17.30}$$

If the steady state is to be stable with respect to small fluctuations, then the fluctuations must cause the entropy production to increase.

It is of interest to apply these stability arguments to some simple chemical reactions.[4] Let us first consider the unimolecular reaction,

$$X \underset{k_2}{\overset{k_1}{\rightleftharpoons}} A. \tag{17.31}$$

The rate equation for constituent X is

$$\frac{dN_x}{dt} = k_2 N_A - k_1 N_x. \tag{17.32}$$

The unimolecular reaction is a linear chemical reaction. That is, the time rate of change of N_x depends only on terms linear in N_x. There is one steady state which occurs when

$$\frac{dN_x}{dt} = 0 \tag{17.33}$$

and, therefore, when $\bar{N}_x = k_2 N_A / k_1$. The thermodynamic flux of X for this system is given by

$$\mathcal{J} = -k_1 N_x + k_2 N_A, \tag{17.34}$$

and the affinity is given by

$$\mathcal{A}(N_x) = RT \ln \frac{k_1 N_x}{k_2 N_A}. \tag{17.35}$$

Since $\mathcal{A}(\bar{N}_x) = 0$, the steady state is also a state of thermodynamic equilibrium.

Let us now assume that the molar density N_A is held fixed away from its equilibrium value and study the entropy production that results due to fluctuations about the steady state. Variations in $\mathcal{J}$ and $\mathcal{A}$ due to fluctuations in N_x are given by

$$\delta\mathcal{J} = -k_1 \, \delta N_x \tag{17.36}$$

and

$$\delta\mathscr{A} = RT\frac{\delta N_x}{\bar{N}_x}. \tag{17.37}$$

Thus, the excess entropy production is

$$d_x P = -\delta\mathscr{A}\,\delta\mathscr{J} = Rk_1\frac{(\delta N_x)^2}{\bar{N}_x} > 0 \tag{17.38}$$

and the steady state must be stable.

As a final example, let us now consider the autocatalytic (nonlinear) reaction

$$A + X \underset{k_2}{\overset{k_1}{\rightleftharpoons}} 2X. \tag{17.39}$$

The reaction is nonlinear because it requires X to produce X. The rate equation is given by

$$\mathscr{J} = \frac{dN_x}{dt} = k_1 N_A N_x - k_2 N_x^2. \tag{17.40}$$

There are two steady states, that is, states for which $dN_x/dt = 0$. These are $\bar{N}_x = 0$ and $\bar{N}_x = k_1 N_A/k_2$. The affinity for this reaction is

$$\mathscr{A} = RT\ln\frac{k_1 N_x^2}{k_2 N_A N_x} \tag{17.41}$$

and, therefore, the equilibrium steady state (the state with zero affinity) is the state $\bar{N}_x = k_1 N_A/k_2$.

Let us now fix the value of N_A by allowing particles to flow in and out of the system. The change in the flux due to fluctuations in N_x about one of the steady states is

$$\delta\mathscr{J} = (k_1 N_A - 2k_2\bar{N}_x)\,\delta N_x, \tag{17.42}$$

and the change in the affinity is

$$\delta\mathscr{A} = RT\frac{\delta N_x}{\bar{N}_x}. \tag{17.43}$$

The excess entropy production for this reaction becomes

$$dP_x = -\frac{R}{\bar{N}_x}(k_1 N_A - 2k_2\bar{N}_x)(\delta N_x)^2. \tag{17.44}$$

Thus, the state $\bar{N}_x = k_1 N_A/k_2$ is stable and the state $\bar{N}_x = 0$ is unstable. In fact, $dP_x = +\infty$ when $\bar{N}_x = 0$, so this example is a little too simple. Nevertheless it illustrates the point that a nonlinear chemical reaction can have multiple steady states and we can use thermodynamic arguments to determine the stability of the steady states.

In general, as we increase the affinity of a chemical system in which autocatalytic (nonlinear) reactions occur, the system will undergo a series of phase transitions. The first transition occurs from the near equilibrium steady state to a new state which cannot exist near equilibrium; then, in general, there will be transitions between a whole series of states, each of which becomes stable and then unstable as the parameters of the system are changed.

The possibility of having phase transitions to more ordered states far from equilibrium has opened a new field of research in the chemical and biological sciences. For example, living systems may be thought of as chemical systems (although of *huge* complexity) maintained far from equilibrium. There is now some evidence that the formation and maintenance of ordered states far from equilibrium in such systems is important for the maintenance of life processes. In 1977, I. Prigogine received the Nobel Prize in chemistry for laying the foundations of this field of chemistry and for his many contributions to its development.

In the subsequent sections, we shall consider some more interesting chemical reactions than those which were considered here. We will generally use the linear stability theory discussed in App. D, because it is simpler to apply than the Lyapounov approach discussed in this section and it is sufficient for the purposes of this chapter. However, the stability conditions for the reactions discussed in the next two sections can be obtained using the methods of this section and therefore can be fitted into the framework of thermodynamics.

C. THE SCHLÖGL MODEL[4,5]

The first nonlinear chemical reaction we will consider was introduced by Schlögl[5] and exhibits a phase transition which is analogous to the first-order phase transition predicted by the van der Waals equation. The Schlögl model is given by the nonlinear reaction scheme

$$\begin{aligned} A + 2X &\underset{k_2}{\overset{k_1}{\rightleftharpoons}} 3X. \\ X &\underset{k_4}{\overset{k_3}{\rightleftharpoons}} B \end{aligned} \tag{17.45}$$

Thermodynamic equilibrium occurs when

$$k_1 N_A^{\text{eq}}(N_x^{\text{eq}})^2 = k_2 (N_x^{\text{eq}})^3 \tag{17.46}$$

and

$$k_3 N_x^{\text{eq}} = k_4 N_B^{\text{eq}}. \tag{17.47}$$

Thus, at thermodynamic equilibrium, the ratio of the equilibrium densities is fixed,

$$\mathscr{R}_{\text{eq}} = \frac{N_A^{\text{eq}}}{N_B^{\text{eq}}} = \frac{k_4 k_2}{k_1 k_3}, \tag{17.48}$$

and is determined by the rate constants.

If we now hold N_A and N_B fixed at some nonequilibrium value and allow N_x to vary, the rate equation for N_x can be written

$$\frac{dN_x}{dt} = k_1 N_A N_x^2 - k_2 N_x^3 - k_3 N_x + k_4 N_B. \tag{17.49}$$

The steady states occur when $dN_x/dt = 0$ and therefore for values of N_x which satisfy the equation

$$\bar{N}_x^3 - a\bar{N}_x^2 + k\bar{N}_x - b = 0 \tag{17.50}$$

where $a = k_1 N_A/k_2$, $b = k_4 N_B/k_2$, and $k = k_3/k_2$. If we fix the ratio $\mathscr{R} = N_A/N_B \neq \mathscr{R}_{eq}$, the steady state need not be a state of thermodynamic equilibrium.

Eq. (17.50) is a cubic equation as is the van der Waals equation and has similar properties. In Fig. 17.1 we sketch the steady state solutions for the concentration $\bar{N}_x$ (solutions to Eq. (17.50)) as a function of b for fixed values of a and k. The points b_1 and b_2 are easily found. They are simply the extrema of the curve $b(\bar{N}_x)$ and are obtained from the roots of the equation

$$\frac{db}{d\bar{N}_x} = 3\bar{N}_x^2 - 2a\bar{N}_x + k = 0. \tag{17.51}$$

The two roots of Eq. (17.51) are given by $\bar{N}_x^{\pm} = \frac{1}{3}(a \pm (a^2 - 3k)^{1/2})$. The maximum must satisfy the condition $(d^2b/d\bar{N}_x^2) < 0$. Thus, from the equation

$$\frac{d^2b}{d\bar{N}_x^2} = 6\bar{N}_x - 2a \tag{17.52}$$

we can easily show that $b_1 = b(\bar{N}_x^+)$ and $b_2 = b(\bar{N}_x^-)$. Explicit expressions for b_1 and b_2 can be found from Eq. (17.50). When $a = \sqrt{3k}$, $b_1 = b_2$ and all the roots of Eq. (17.50) coalesce. For $a > \sqrt{3k}$ there is a range of values for $b(\bar{N}_x)$ for which the system can have three different steady states. For $a < \sqrt{3k}$, it can only have one steady state for any value of b.

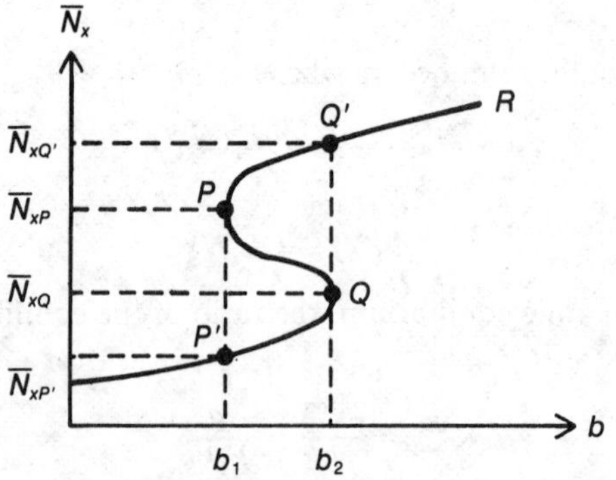

Fig. 17.1. The line of steady state solutions for the Schlögl model.

From linear stability theory (cf. App. D) it is possible to show that the region QP in Fig. 17.1 is unstable while regions $P'Q$ and PQ' are stable. If we let $N_x(t) = \bar{N}_x + x(t)$, where $x(t)$ is a small parameter, and solve the linearized equation for $x(t)$, we find that, in the regions $P'Q$ and PQ', $x(t)$ decays exponentially while in the region QP it grows exponentially. Since fluctuations are always present in the system to test a steady state solution, the system will be driven away from the steady states in the region Q to P.

As we increase the parameter b, the steady state value of N_x increases until we reach point Q. At this point the system must jump from a steady state with density N_{xQ} to a steady state with density $N_{xQ'}$ in order to exist in a stable steady state for values of $b > b_2$. Thus, the chemical system appears to undergo an abrupt transition at the point b_2. The system also exhibits hysteresis behavior. As we decrease b we remain on the curve PQ' until we reach the value b_1. Then the system undergoes an abrupt transition from a steady state with density N_{xp} to a steady state with density $N_{xp'}$. In Fig. 17.2, we have plotted the hysteresis region as a function of a and b. Such behavior has been used as a model for explosive reactions.

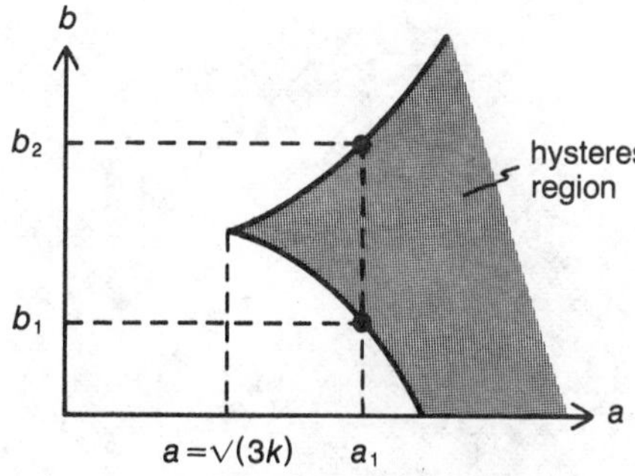

Fig. 17.2. Hysteresis region for the Schlögl model plotted as a function of a and b.

The rate equation (17.49) is a macroscopic equation for the density of the intermediate X. It is possible, using stochastic methods similar to those discussed in Chap. 5, to do a stochastic analysis of this reaction and to obtain some of the critical exponents for the reaction.[6]

D. BRUSSELATOR[1–4,7]

Chemical systems which are locally in thermodynamic equilibrium but held far from chemical equilibrium (constrained to have a large affinity) can undergo phase transitions to new stable states with striking behavior. The new stable states may be steady states in which the relative concentrations of the constituents vary in space, or they may be spatially homogeneous states in which the concentrations of some constituents vary in time (chemical clocks), or they may be states with nonlinear traveling waves in the concentrations of some constituents. The classic example of such behavior is the Belousov[8]-Zhabotinski[9] reaction. This reaction, which is too complicated to discuss in detail here (cf. Ref. 17.10), involves the cerium ion catalyzed oxidation of malonic acid by bromate in a sulfuric acid medium. The reaction is nonlinear because it contains autocatalytic steps. The system, when it is well stirred, can behave like a chemical clock. That is, there is a periodic change in the concentration of Br^- and of the relative concentration Ce^{4+}/Ce^{3+}. The system oscillates between red and blue color with a period of the order of a minute. The system is also thought to exhibit a symmetry-breaking spatial ordering as well, that is, periodic oscillations of concentration of Br^- and of the relative concentration Ce^{4+}/Ce^{3+} in space, although the evidence is more controversial. Perhaps the most fascinating behavior of this system is the traveling waves in the concentration of Br^- and the relative concentration Ce^{4+}/Ce^{3+} which are observed in shallow unstirred dishes (cf. Fig. 17.3).

Fig. 17.3. Traveling waves in the chemical concentrations for the Belousov-Zhabotinski reaction (reprinted, by permission, from A. T. Winfree, Adv. Bio. Med. Phys. *16* 115 [1977]).

The Belousov-Zhabotinski reaction appears to be well described by a model which contains three variable intermediates. This model is called the Oregonator and was first introduced by Field and Noyes.[11] However, the qualitative behavior of the type appearing in the Belousov-Zhabotinski reaction also occurs in a simpler model called the Brusselator, first introduced by Prigogine and Lefever,[7] which contains two variable intermediates. In this section, we shall discuss some properties of the Brusselator, using linear stability theory (cf. App. D).

The Brusselator is one of the simplest models of a nonlinear chemical system for which the relative concentration of the constituents can oscillate in time, or can vary in space, or can exhibit nonlinear traveling waves. It has six different components, four of which are held fixed and two others whose concentrations can vary in space and time. The chemical reaction takes place in four steps and is held far from equilibrium by allowing the reactions to go in one direction only. The four steps are

$$A \xrightarrow{k_1} X,$$

$$B + X \xrightarrow{k_2} Y + D,$$

$$2X + Y \xrightarrow{k_3} 3X, \tag{17.53}$$

and

$$X \xrightarrow{k_4} E.$$

In practice, A and B are present in excess and D and E are removed as soon as they appear. The rate equations for X and Y can be written

$$\frac{dN_x}{dt'} = k_1 N_A - (k_2 N_B + k_4) N_x + k_3 N_x^2 N_Y + D_1' \nabla_{\mathbf{r}}^2 N_x \tag{17.54}$$

and

$$\frac{dN_y}{dt'} = k_2 N_B N_x - k_3 N_x^2 N_Y + D_2' \nabla_{\mathbf{r}}^2 N_y. \tag{17.55}$$

We have allowed the densities to vary in space and have allowed for the possibility of diffusion (D_1' and D_2' are the coefficients of diffusion). If we now introduce a change of variables $t = k_4 t'$, $X = (k_3/k_4)^{1/2} N_x$, $Y = (k_3/k_4)^{1/2} N_y$, $A = (k_1^2 k_3/k_4^3)^{1/2} N_A$, $B = (k_2/k_4) N_B$, and $D_i = D_i'/k_4$ then Eqs. (17.54) and (17.55) take the form

$$\frac{dX}{dt} = A - (B + 1)X + X^2 Y + D_1 \nabla_{\mathbf{r}}^2 X \tag{17.56}$$

and

$$\frac{dY}{dt} = BX - X^2 Y + D_2 \nabla_{\mathbf{r}}^2 Y. \tag{17.57}$$

Eqs. (17.56) and (17.57) have a uniform steady state solution of the form

$$X_0 = A \quad \text{and} \quad Y_0 = B/A, \tag{17.58}$$

which is the continuation far from equilibrium of the steady state solution which occurs at chemical equilibrium when the reverse reactions in Eq. (17.53) are allowed to occur. That is, the steady state (Eq. (17.58)) lies on the "thermodynamic branch" of steady state solutions. We wish to look for conditions under which the thermodynamic branch becomes unstable and a bifurcation (nonequilibrium phase transition) occurs to a state which may oscillate in space or time.

The first step is to specify the boundary conditions of our system. We shall consider two cases:

Case I: The concentrations X and Y are constant on the boundaries and take on the values

$$X_{\text{bound}} = A \quad \text{and} \quad Y_{\text{bound}} = B/A.$$

These boundary conditions can be maintained by allowing a flow of X and Y through the surface.

Case II: There is no flux of X and Y perpendicular to the surface

$$\hat{\mathbf{n}} \cdot \nabla_{\mathbf{r}} X = \hat{\mathbf{n}} \cdot \nabla_{\mathbf{r}} Y = 0$$

on the boundaries, where $\hat{\mathbf{n}}$ is the normal to the boundaries.

We now use linear stability theory to look for instabilities in the steady state homogeneous solution in Eq. (17.58). We write X and Y as

$$X(\mathbf{r}, t) = A + \alpha(\mathbf{r}, t) \tag{17.59}$$

and

$$Y(\mathbf{r}, t) = B/A + \beta(\mathbf{r}, t) \tag{17.60}$$

where α and β are small space- and time-dependent perturbations. We shall introduce Eqs. (17.59) and (17.60) into Eqs. (17.56) and (17.57) and linearize them with respect to $\alpha(\mathbf{r}, t)$ and $\beta(\mathbf{r}, t)$. We then obtain

$$\frac{d\alpha}{dt} = ((B - 1) + D_1 \nabla_{\mathbf{r}}^2)\alpha + A^2\beta \tag{17.61}$$

and

$$\frac{d\beta}{dt} = -B\alpha + (-A^2 + D_2 \nabla_{\mathbf{r}}^2)\beta. \tag{17.62}$$

Since Eqs. (17.61) and (17.62) are linear, it is sufficient to consider one Fourier component of $\alpha(\mathbf{r}, t)$ and $\beta(\mathbf{r}, t)$. For simplicity, let us assume that the system is contained in a cubic container with sides of length L. For Case I boundary conditions, a given Fourier component of $\alpha(\mathbf{r}, t)$ and $\beta(\mathbf{r}, t)$ with frequency ω and wave vector

$$\mathbf{k} = \frac{n_x \pi}{L} \hat{\imath} + \frac{n_y \pi}{L} \hat{\jmath} + \frac{n_z \pi}{L} \hat{z}$$

has the form

Case I:

$$\alpha(\mathbf{r}, t) = \tilde{\alpha}(\mathbf{k}, \omega) \sin k_x x \sin k_y y \sin k_z z \; e^{\omega(\mathbf{k})t} \tag{17.63a}$$

and

$$\beta(\mathbf{r}, t) = \tilde{\beta}(\mathbf{k}, \omega) \sin k_x x \sin k_y y \sin k_z z \, e^{\omega(\mathbf{k})t} \quad (17.63b)$$

where n_x, n_y, and n_z can each have values $(1, 2, \ldots, \infty)$. For the Case II boundary conditions a given Fourier component is of the form

Case II:

$$\alpha(\mathbf{r}, t) = \tilde{\alpha}(\mathbf{k}, \omega) \cos k_x x \cos k_y y \cos k_z z \, e^{\omega(\mathbf{k})t} \quad (17.64a)$$

and

$$\beta(\mathbf{r}, t) = \tilde{\beta}(\mathbf{k}, \omega) \cos k_x x \cos k_y y \cos k_z z \, e^{\omega(\mathbf{k})t} \quad (17.64b)$$

where n_x, n_y, and n_z can each have values $(0, 1, 2, \ldots, \infty)$.

To determine if a given mode is unstable we must substitute the solutions Eqs. (17.63*a*) and (17.63*b*) or (17.64*a*) and (17.64*b*) into Eqs. (17.61) and (17.62) and find an equation for the frequency $\omega(\mathbf{k})$. In general, $\omega(\mathbf{k})$ can be complex. If it is complex, the component of the perturbations $\alpha(\mathbf{r}, t)$ and $\beta(\mathbf{r}, t)$ with wave vector **k** and frequency $\omega(\mathbf{k})$ will oscillate about the steady state (17.58) with frequency given by the imaginary part of $\omega(\mathbf{k})$. The real part of $\omega(\mathbf{k})$ determines if the steady state is stable with respect to the perturbation. If the real part of $\omega(\mathbf{k})$ is negative, then the perturbation will decay away and the steady state is stable with respect to fluctuations with wave vector **k** and frequency $\omega(\mathbf{k})$. However, if the real part of $\omega(\mathbf{k})$ is positive the perturbations $\alpha(\mathbf{r}, t)$ and $\beta(\mathbf{r}, t)$ grow exponentially. Thus, when the real part of $\omega(\mathbf{k})$ is positive, a bifurcation or phase transition occurs. Linear stability analysis cannot tell us the form of the new state, but it can tell us where the bifurcation occurs.

If we substitute either of the solutions Eqs. (17.63*a*) and (17.63*b*) or (17.64*a*) and (17.64*b*) into Eqs. (17.61) and (17.62), we obtain the following set of equations:

$$\omega(\mathbf{k})\tilde{\alpha} = (B - 1 - D_1 k^2)\tilde{\alpha} + A^2\tilde{\beta} \quad (17.65)$$

and

$$\omega(\mathbf{k})\tilde{\beta} = -B\tilde{\alpha} + (-A^2 + D_2 k^2)\tilde{\beta} \quad (17.66)$$

where $k^2 = (\pi^2/L^2)(n_x^2 + n_y^2 + n_z^2) = (\pi^2/L^2)n^2$. We can write Eqs. (17.65) and (17.66) in the form of a matrix acting on the vector $\begin{pmatrix}\tilde{\alpha}\\ \tilde{\beta}\end{pmatrix}$. An equation for $\omega(\mathbf{k})$ is found by setting the determinant of the matrix to zero as usual. We then have

$$\omega^2(\mathbf{k}) + (c_1 - c_2)\omega(\mathbf{k}) + A^2B - c_1c_2 = 0 \quad (17.67)$$

where

$$c_1 = B - 1 - k^2 D_1 \quad (17.68)$$

and

$$c_2 = A^2 + k^2 D_2. \quad (17.69)$$

Eq. (17.67) has two solutions:

$$\omega^{\pm}(\mathbf{k}) = \tfrac{1}{2}\{c_1 - c_2 \pm \sqrt{(c_1 + c_2)^2 - 4A^2B}\}. \tag{17.70}$$

From Eq. (17.70) it is easy to see that, depending on the variables D_1, D_2, A, and B, $\omega(\mathbf{k})$ can be either complex or real and the real part can be either positive or negative.

Let us consider the cases of real and complex frequency, $\omega(\mathbf{k})$, separately. We first consider the case of real frequency.

1. Real $\omega(\mathbf{k})$

For $\omega(\mathbf{k})$ real, that is,

$$(c_1 + c_2)^2 - 4A^2B > 0, \tag{17.71}$$

$\omega(\mathbf{k})$ will be positive if

$$c_1c_2 - A^2B > 0 \tag{17.72}$$

or

$$B > B_n = 1 + \frac{D_1}{D_2}A^2 + \frac{A^2}{D_2n^2\pi^2}L^2 + \frac{D_1n^2\pi^2}{L^2}. \tag{17.73}$$

In this case, there will be no oscillations in time but only in space. The linear stability diagram for the onset of time-independent spatial oscillations of the concentrations of X and Y is given in Fig. 17.4. The curved line is B_n. The bifurcation first occurs for $B = B_c$, where B_c is the lowest value of B corresponding to an integer value of n. Linear stability analysis can only tell us that a bifurcation to a new spatially oscillating steady state is possible for a particular set of values for the parameters A, B, D_1, D_2, and L. It cannot give us information about the form of the new state. However, for the Brusselator the spatially

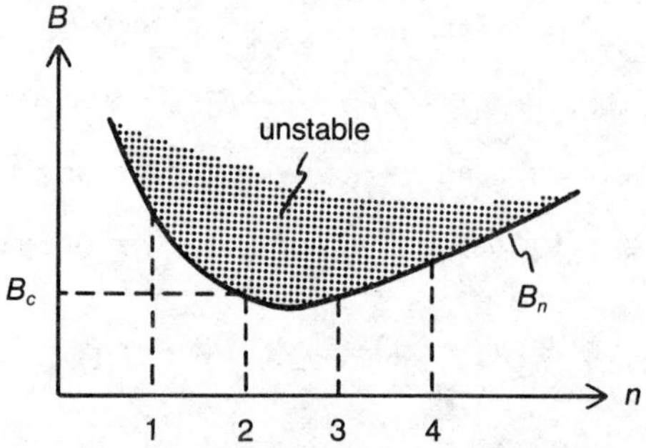

Fig. 17.4. Linear stability diagram for the onset of a time-independent dissipative structures: for the case pictured here, as B increases, the first bifurcation occurs at $n = 2$ (based on Ref. 17.4).

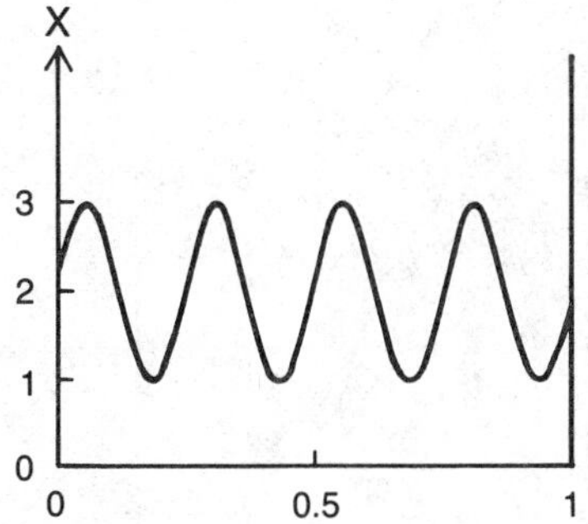

Fig. 17.5. A steady state spatial dissipative structure for $B = 4.6$, $A = 2$, $L = 1$, $D_1 = 1.6 \times 10^{-3}$, and $D_2 = 6.0 \times 10^{-3}$: the boundary conditions are fixed at $X = A$ and $Y = B/A$ (based on Ref. 17.4).

varying states can be obtained numerically. In Fig. 17.5, we give an example for a one-dimensional system for the following values of the parameters: $B = 4.6$, $A = 2$, $L = 1$, $D_1 = 1.6 \times 10^{-3}$, $D_2 = 6.0 \times 10^{-3}$, and fixed boundary conditions $X_{\text{bound}} = A$ and $Y_{\text{bound}} = B/A$. The figure shows a state with oscillations in the concentration of X as a function of position.

2. Complex $\omega(\mathbf{k})$

The condition that $\omega(\mathbf{k})$ be complex is

$$(c_1 + c_2)^2 - 4A^2B < 0. \tag{17.74}$$

From Eq. (17.74), we can show that $\omega(\mathbf{k})$ will be complex if

$$D_2 - D_1 \leqslant \frac{L^2}{n^2\pi^2}. \tag{17.75}$$

In addition, $\omega(\mathbf{k})$ will have a positive real part if $c_1 - c_2 < 0$ or if

$$B > A^2 + 1 + \frac{n^2\pi^2}{L^2}(D_1 + D_2). \tag{17.76}$$

Thus, the curve

$$B_n = A^2 + 1 + \frac{n^2\pi^2}{L^2}(D_1 + D_2) \tag{17.77}$$

denotes the boundary between the region where the steady state (Eq. (17.58)) is stable and where it is unstable as a function of n.

We note that for the Case II boundary conditions we can have $n = 0$ and, therefore, the possibility of spatially homogeneous oscillations in time of the relative concentrations of X and Y, that is, a chemical clock. For all finite values of n, we have the possibility of a new state which varies in both space and

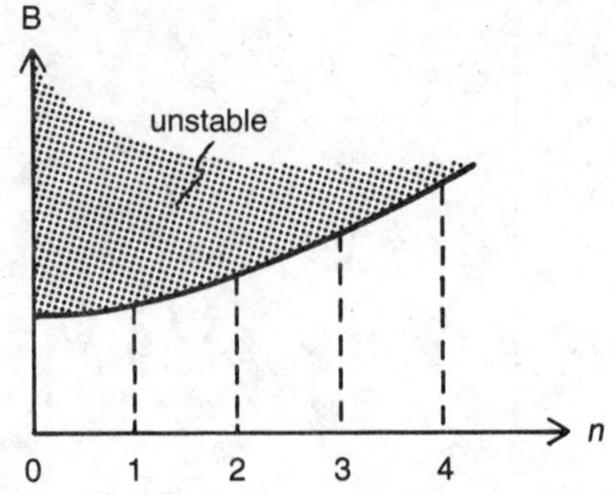

Fig. 17.6. Linear stability diagram for the onset of time-dependent dissipative structures: as B is increased, the first bifurcation occurs at $n = 0$ and yields a spatially homogeneous chemical clock (based on Ref. 17.4).

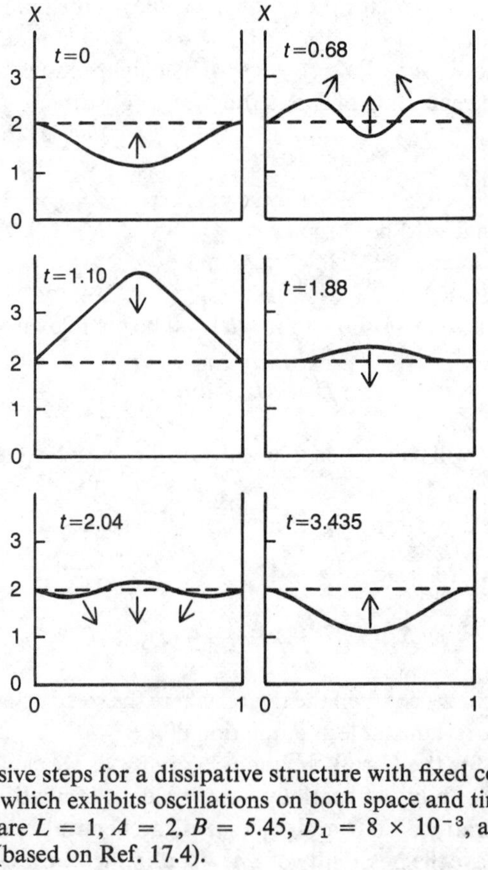

Fig. 17.7. Successive steps for a dissipative structure with fixed concentrations on the boundary which exhibits oscillations on both space and time: the parameters used are $L = 1$, $A = 2$, $B = 5.45$, $D_1 = 8 \times 10^{-3}$, and $D_2 = 4 \times 10^{-3}$ (based on Ref. 17.4).

time and therefore may exhibit wave motion. The linear stability diagram for the transition to time-dependent states is given in Fig. 17.6. In Fig. 17.7, we give the results of a computer simulation, of a one-dimensional system, for a state which oscillates in space and time. The solution is reminiscent of a standing wave on a string, but, due to the nonlinearities, it has a much more complicated structure. In two dimensions, computer simulations have shown that traveling waves can exist which are similar to the waves shown in the Belousov-Zhabotinski reaction in Fig. 17.3 (cf. Ref. 17.4).

As we mentioned earlier, spatially varying steady states, temporally oscillating homogeneous states, and nonlinear traveling waves have been observed in chemical systems. These states, which only become stable far from equilibrium, have been called "dissipative structures" by I. Prigogine[1,4] because their very existence depends upon dissipative processes, such as chemical reactions far from equilibrium or diffusion (if spatial structures occur). As we have seen in Sec. 17.B, they are maintained by production of entropy and by a related flow of energy and matter from the outside world. It is curious that the type of autocatalytic reactions which produce them are abundant in living systems and yet are rare in nonliving systems. It is possible that dissipative structures in living systems play an important role in the maintenance of life processes.[12]

E. LOTKA-VOLTERRA MODEL[4,13–15]

The ideas and techniques of nonlinear chemistry are being applied more and more to fields outside chemistry, such as population dynamics, urban growth, and ecology. We shall consider a model which has often been used in ecology and population dynamics and has had great historical impact, although it is now known to be too simple to describe the qualitative behavior of such systems well. The model, called the Lotka-Volterra model (or predator-prey model) is rather special because it is a conservative nonlinear system and the meaningful fixed point (steady state solution) is a center rather than a focus as one usually finds in dissipative systems (cf. App. D). Since the Lotka-Volterra model is conservative, we can define a single-valued first integral of the motion which allows us to determine the structure of the entire phase space.

Let us consider a small lake whose bottom is covered with an unlimited amount of grass. We shall assume that the lake contains N_A small fish of type A, which only eat grass, and N_B large fish of type B, which only eat the small fish A but not grass (cf. Fig. 17.8). If there were no type B fish present, then the equation of motion for the type A fish would be

$$\frac{dN_A}{dt} = \epsilon_A N_A. \tag{17.78}$$

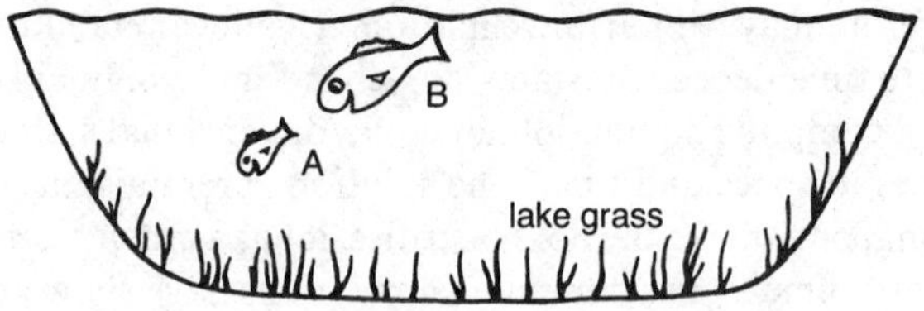

Fig. 17.8. Predator-prey model: the small fish, A, eats grass, and the big fish, B, eats the small fish.

They would simply multiply exponentially. If there were no type A fish present, then the type B fish would die out exponentially:

$$\frac{dN_B}{dt} = -\epsilon_B N_B. \tag{17.79}$$

We now allow the fish to interact. Whenever a fish B encounters a fish A, it will eat it. This will contribute to an increase in the number of type B fish and a decrease in the number of type A fish. The rate of increase of type B fish and the rate of decrease of type A fish will each be proportional to $N_A N_B$ but with different proportionality constants. We therefore modify Eqs. (17.78) and (17.79) to include interactions as follows:

$$\frac{dN_A}{dt} = \epsilon_A N_A - \gamma_A N_A N_B \tag{17.80}$$

and

$$\frac{dN_B}{dt} = -\epsilon_B N_B + \gamma_B N_A N_B. \tag{17.81}$$

Eqs. (17.80) and (17.81) are nonlinear equations which describe the rate of change of populations A and B. The quantities N_A and N_B can only change by integer amounts. Let us assume, however, that N_A and N_B are large and approximate them by continuous variables.

We now wish to find the steady state solutions of Eqs. (17.80) and (17.81)—that is, those values of N_A and N_B for which the populations remain constant in time. The steady state solutions are determined by the condition

$$\frac{d\overline{N}_A}{dt} = 0 = \epsilon_A \overline{N}_A - \gamma_A \overline{N}_A \overline{N}_B \tag{17.82}$$

and

$$\frac{d\overline{N}_B}{dt} = 0 = -\epsilon_B \overline{N}_B + \gamma_B \overline{N}_A \overline{N}_B. \tag{17.83}$$

We see that there are two solutions, ($\overline{N}_A = 0$, $\overline{N}_B = 0$) and ($\overline{N}_A = \epsilon_B/\gamma_B$; $\overline{N}_B = \epsilon_A/\gamma_A$). One steady state solution occurs at the origin for zero values of both populations and the other occurs for finite values of both populations.

Let us first examine the stability of the steady state at the origin. We let $N_A = x_A$ and $N_B = x_B$ and keep only terms linear in x_A and x_B. We then obtain

$$\dot{x}_A = \epsilon_A x_A \tag{17.84}$$

and

$$\dot{x}_B = -\epsilon_B x_B. \tag{17.85}$$

The coordinates are already in normal form. The eigenvalues are real but one is positive, so the steady state at the origin is unstable and corresponds to a saddle point or hyperbolic fixed point (cf. App. D).

Let us next consider the steady state at $(\bar{N}_A = \epsilon_B/\gamma_B;\ \bar{N}_B = \epsilon_A/\gamma_A)$. We write $N_A = \epsilon_B/\gamma_B + x_A$ and $N_B = \epsilon_A/\gamma_A + x_B$ and obtain the following linearized equations for x_A and x_B:

$$\frac{dx_A}{dt} = -\frac{\gamma_A \epsilon_B}{\gamma_B} x_B \tag{17.86}$$

and

$$\frac{dx_B}{dt} = \frac{\gamma_B \epsilon_A}{\gamma_A} x_A. \tag{17.87}$$

In matrix form, they become

$$\begin{pmatrix} \dot{x}_A \\ \dot{x}_B \end{pmatrix} = \begin{pmatrix} 0 & \dfrac{-\gamma_A \epsilon_B}{\gamma_B} \\ \dfrac{\gamma_B \epsilon_A}{\gamma_A} & 0 \end{pmatrix} \begin{pmatrix} x_A \\ x_B \end{pmatrix}. \tag{17.88}$$

The eigenvalues of the matrix in Eq. (17.88) are of the form $\pm i\sqrt{\epsilon_A \epsilon_B}$ and the nonzero steady state is a center.

The linear stability analysis only tells us something about the behavior of orbits in the neighborhood of the steady states. In general, it is difficult to say anything about the stability of the entire phase space. However, for the Lotka-Volterra model it is possible to find an invariant of the motion (analogous to the energy for dynamical systems). Therefore, we can determine the structure of the phase space.

Let us now multiply Eq. (17.80) by ϵ_B/N_A and Eq. (17.81) by ϵ_A/N_B and add the equations together. We then obtain

$$\frac{\epsilon_B}{N_A}\frac{dN_A}{dt} + \frac{\epsilon_A}{N_B}\frac{dN_B}{dt} = -\gamma_A \epsilon_B N_B + \gamma_B \epsilon_A N_A. \tag{17.89}$$

If we next substitute Eqs. (17.80) and (17.81) into Eq. (17.89), we obtain

$$\epsilon_B \frac{d}{dt} \ln N_A + \epsilon_A \frac{d}{dt} \ln N_B - \gamma_A \frac{dN_B}{dt} - \gamma_B \frac{dN_A}{dt} = 0. \tag{17.90}$$

If we integrate Eq. (17.90) we find

$$\epsilon_B \ln N_A + \epsilon_A \ln N_B - \gamma_A N_B - \gamma_B N_A = \text{const.} \tag{17.91}$$

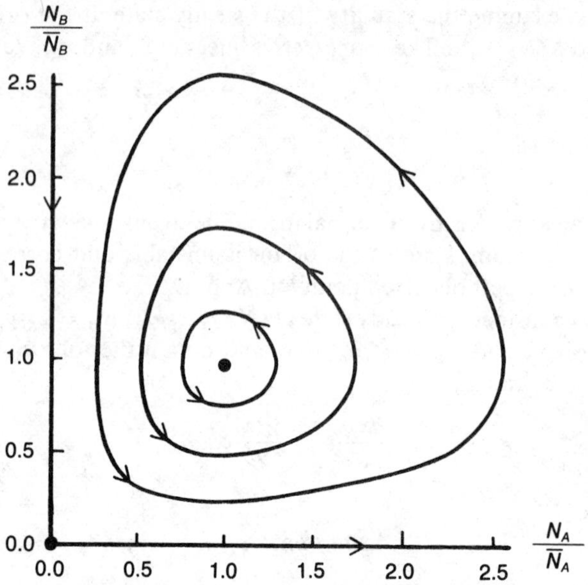

Fig. 17.9. Phase space for Lotka-Volterra (predator-prey) model.

Knowledge of Eq. (17.91) allows one to determine the structure of the entire phase plane. We see that the population oscillates about the steady state. However, if $N_A = 0$ then $N_B \to 0$ and the big fish die away, while if $N_B = 0$ then $N_A \to \infty$ and the population of small fish grows exponentially (cf. Fig. 17.9).

F. BÉNARD INSTABILITY[16–20]

For a change of pace, let us now consider a hydrodynamic system. Nonequilibrium phase transitions are abundant in hydrodynamic systems because they too are governed by nonlinear equations. For example, if we let fluid flow in a pipe, then we find that for low velocities or high viscosities the flow is smooth and steady. However, as we increase the velocity or decrease the viscosity, we get a transition to turbulent flow. The smooth steady state becomes unstable and a turbulent state becomes favored. As another example, let us consider a fluid at rest and place it between horizontal parallel plates in the gravitational field. If we put a temperature gradient across the plates (with the hottest plate below) and vary the viscosity or thermal conductivity, we find that at some point the rest state becomes unstable and the fluid breaks into convective flow cells which occur periodically in space. In each cell, fluid rises in one part of the cell and descends in another part. The circulation of fluid repeats itself in each cell. This instability is called the Bénard instability and is the one we shall study in this section. We will follow closely the classic presentation of Chandrasekhar.[16]

Let us first write the hydrodynamic equations in a form which will be useful to us. The full nonlinear Navier-Stokes equations can be written

$$\frac{\partial \rho}{\partial t} + \nabla_{\mathbf{r}}\cdot \rho\mathbf{v} = 0, \tag{17.92}$$

$$\frac{\partial \rho\mathbf{v}}{\partial t} + \nabla_{\mathbf{r}}\cdot(\rho\mathbf{v}\mathbf{v}) = -\nabla_{\mathbf{r}}P + \rho\mathbf{F} + \eta\nabla_{\mathbf{r}}^2\mathbf{v} + (\zeta + \tfrac{1}{3}\eta)\nabla_{\mathbf{r}}(\nabla_{\mathbf{r}}\cdot\mathbf{v}), \tag{17.93}$$

and

$$\frac{\partial \rho u}{\partial t} + \nabla_{\mathbf{r}}\cdot(\rho u\mathbf{v} - K\nabla_{\mathbf{r}}T) = \overline{\overline{\mathbf{\Pi}}}:\nabla_{\mathbf{r}}\mathbf{v}, \tag{17.94}$$

where ρ is the mass density and all quantities have been defined in Sec. 13.F. We will consider the case of a fluid at rest constrained to lie between two infinitely large parallel plates which extend in the x- and y-directions. The distance between the plates is d. We will put a temperature gradient across the plates so that the temperature of the bottom plate is greater than that of the top, and we will assume that a gravitational field acts in the negative z-direction (cf. Fig. 17.10). If we have a thin enough fluid layer, any density variations in the system will be due primarily to the temperature gradient and we can write

$$\rho = \rho_0[1 + \alpha_p(T_0 - T)] \tag{17.95}$$

where $\alpha_p = -(1/\rho)(\partial\rho/\partial T)_p$ is the thermal expansivity. Heating the bottom more than the top causes the fluid at the top to be denser and heavier and creates the possibility for an instability to occur.

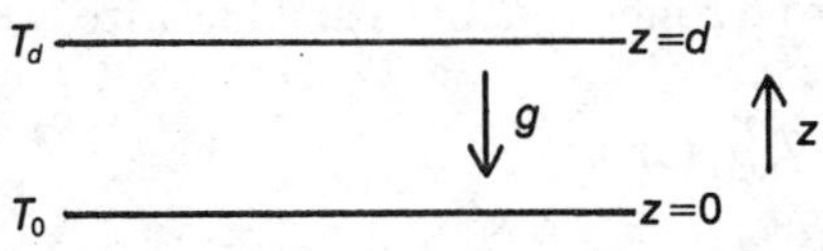

Fig. 17.10. Geometrical configuration for the Bénard problem.

Let us now write the equations for the steady state in the absence of any macroscopic flow; that is, $\mathbf{v} = 0$. We obtain

$$\nabla_{\mathbf{r}}\cdot \rho\mathbf{v} = 0, \tag{17.96}$$

$$\nabla_{\mathbf{r}}P = -\rho g, \tag{17.97}$$

and

$$\nabla_{\mathbf{r}}^2 T = 0, \tag{17.98}$$

where we have let $\mathbf{F} = -g\hat{\mathbf{z}}$ and g is the acceleration of gravity. If we note the boundary conditions $T(0) = T_0$ and $T(d) = T_d$, we can solve Eqs. (17.97) and (17.98) to obtain the steady state solutions. We find

$$T(z) = T_0 - az, \tag{17.99}$$

$$\rho(z) = \rho_0[1 + \alpha_p(T_0 - T)] = \rho_0[1 + \alpha_p az], \tag{17.100}$$

and

$$P(z) = P_0 - g\rho_0[z + \tfrac{1}{2}\alpha_p az^2], \tag{17.101}$$

where a is the temperature gradient, $a = (T_0 - T_d)/d$, and P_0 is the pressure at $z = 0$.

Eqs. (17.99)–(17.101) are the steady state solutions in the absence of flow. We now wish to determine conditions for which they are stable. As usual, we will perturb the steady state solutions slightly and study the linearized equations for the perturbations. Thus, we write

$$T(\mathbf{r}, t) = T(z) + \delta T(\mathbf{r}, t), \tag{17.102}$$

$$P(\mathbf{r}, t) = P(z) + \delta P(\mathbf{r}, t), \tag{17.103}$$

$$\rho(\mathbf{r}, t) = \rho(z) + \delta\rho(\mathbf{r}, t), \tag{17.104}$$

and

$$\mathbf{v}(\mathbf{r}, t) = 0 + \mathbf{v}(\mathbf{r}, t). \tag{17.105}$$

Let us further note that the dominant contribution to internal energy variations will come from temperature variations. Thus, we write

$$\delta u = c_V\, \delta T \tag{17.106}$$

where c_V is the constant volume specific heat. If we substitute Eqs. (17.102)–(17.105) into the hydrodynamic equations and linearize, we find

$$\frac{\partial}{\partial t}\delta\rho = -\nabla_{\mathbf{r}}\cdot(\rho(z)\mathbf{v}), \tag{17.107}$$

$$\frac{\partial}{\partial t}\rho(z)\mathbf{v} = -\nabla_{\mathbf{r}}(P(z) + \delta P) - \rho g\hat{\mathbf{z}} + \eta\nabla_{\mathbf{r}}^2\mathbf{v} + (\zeta + \tfrac{1}{3}\eta)\nabla_{\mathbf{r}}(\nabla_{\mathbf{r}}\cdot\mathbf{v}), \tag{17.108}$$

and

$$\frac{\partial}{\partial t}\rho(z)c_V\,\delta T = -\nabla_{\mathbf{r}}\cdot(\rho(z)c_V T(z)\mathbf{v} - K\nabla_{\mathbf{r}}(T(z) + \delta T)). \tag{17.109}$$

Let us now note that

$$\delta\rho(\mathbf{r}, t) = -\rho_0\alpha_p\,\delta T(\mathbf{r}, t) \tag{17.110}$$

(cf. Eq. (17.95)) and, therefore, $\nabla_{\mathbf{r}}\cdot\mathbf{v} \approx \alpha_p$. If we neglect all terms which depend on α_p except those that appear in the term involving the external field, we find

$$\nabla_{\mathbf{r}}\cdot\mathbf{v} = 0, \tag{17.111}$$

$$\rho_0 \frac{\partial \mathbf{v}}{\partial t} = -\nabla_{\mathbf{r}}\, \delta P + \rho_0\alpha_p\, \delta T g\hat{\mathbf{z}} + \eta\nabla_{\mathbf{r}}^2\mathbf{v}, \tag{17.112}$$

and

$$\frac{\partial}{\partial t}\,\delta T = av_z + \frac{K}{\rho_0 c_V}\,\nabla_{\mathbf{r}}^2\,\delta T. \tag{17.113}$$

The above approximation is called the Boussinesq approximation. The consistency of the Boussinesq approximation has been demonstrated by Mihaljan.[20] Eqs. (17.111)–(17.113) form the starting point of our linear stability analysis.

We can simplify these equations somewhat through the following steps. We first take the curl of Eq. (17.112) to obtain

$$\frac{\partial \boldsymbol{\omega}}{\partial t} = g\alpha_p\nabla_{\mathbf{r}} \times (\delta T\hat{\mathbf{z}}) + \nu\nabla_{\mathbf{r}}^2\boldsymbol{\omega}, \tag{17.114}$$

where

$$\boldsymbol{\omega} = \nabla_{\mathbf{r}} \times \mathbf{v} \tag{17.115}$$

is the vorticity of the fluid and $\nu = \eta/\rho_0$. We next take the curl of Eq. (17.114) and note that $\nabla_{\mathbf{r}} \times (\nabla_{\mathbf{r}} \times \mathbf{v}) = \nabla_{\mathbf{r}}(\nabla_{\mathbf{r}}\cdot\mathbf{v}) - \nabla_{\mathbf{r}}^2\mathbf{v}$ (we have used Eq. (17.111)). We then find

$$\frac{\partial}{\partial t}\,\nabla_{\mathbf{r}}^2\mathbf{v} = -g\alpha_p\nabla_{\mathbf{r}}\left(\frac{\partial T}{\partial z}\right) + g\alpha_p\nabla_{\mathbf{r}}^2(\delta T\hat{\mathbf{z}}) + \nu\nabla_{\mathbf{r}}^4\mathbf{v}. \tag{17.116}$$

In our stability analysis, we shall be interested in instabilities in the z-components of velocity and vorticity. The equations of motion for the z-components of velocity and vorticity are given by

$$\frac{\partial}{\partial t}\,\nabla_{\mathbf{r}}^2 v_z = g\alpha_p\left(\frac{d^2}{dx^2} + \frac{d^2}{dy^2}\right)\,\delta T + \nu\nabla^4 v_z \tag{17.117}$$

and

$$\frac{\partial \omega_z}{\partial t} = \nu\nabla_{\mathbf{r}}^2\omega_z. \tag{17.118}$$

If an instability occurs in which flow develops in the z-direction, then $v_z\hat{\mathbf{z}}$ must change sign as we move in the x- or y-direction (what goes up must come down).

As for the case of the Brusselator, we must specify boundary conditions before we apply the stability analysis. We first note the general boundary conditions (valid for all surfaces) that

$$\delta T(x, y, 0; t) = \delta T(x, y, d; t) = 0 \tag{17.119}$$

and

$$v_z(x, y, 0; t) = v_z(x, y, d; t) = 0. \tag{17.120}$$

In addition to the above general boundary conditions, we have additional constraints on the surfaces at $z = 0$ and $z = d$. We can have either rigid surfaces for which there can be no tangential components of velocity or smooth surfaces where we can have a tangential flow.

Case I: For a rigid surface we have the boundary conditions

$$\mathbf{v}(x, y, 0; t) = \mathbf{v}(x, y, d; t) = 0. \tag{17.121}$$

Since $\mathbf{v} = 0$ for all x and y, we find $dv_x/dx = dv_y/dy = 0$, and thus from Eq. (17.111) we have the additional condition $dv_z/dz = 0$. The vorticity satisfies the boundary conditions

$$(\omega_z)_{\text{bound}} = \left(\frac{\partial v_x}{\partial y} - \frac{\partial v_y}{\partial x}\right)_{\text{bound}} = 0 \tag{17.122}$$

on the rigid surface.

Case II: On a smooth surface there can be horizontal flow but no transport of the horizontal components of velocity in the z-direction. Thus, the components of the stress tensor satisfy the condition

$$\pi_{xz}(x, y, 0; t) = \pi_{xz}(x, y, d; t) = \pi_{yz}(x, y, 0; t) = \pi_{yz}(x, y, d; t) = 0 \tag{17.123}$$

or

$$\left(\frac{\partial v_x}{\partial z} + \frac{\partial v_z}{\partial x}\right)_{\text{bound}} = 0 \tag{17.124}$$

and

$$\left(\frac{\partial v_y}{\partial z} + \frac{\partial v_z}{\partial y}\right)_{\text{bound}} = 0. \tag{17.125}$$

Since $v_z = 0$ for all x and y on the surface, we have

$$\left(\frac{\partial v_z}{\partial x}\right)_{\text{bound}} = \left(\frac{\partial v_z}{\partial y}\right)_{\text{bound}} = 0 \quad \text{and} \quad \left(\frac{\partial v_x}{\partial z}\right)_{\text{bound}} = \left(\frac{\partial v_y}{\partial z}\right)_{\text{bound}} = 0 \tag{17.126}$$

on the smooth surface. The vorticity satisfies the boundary condition

$$\left(\frac{\partial \omega_z}{\partial z}\right)_{\text{bound}} = 0 \tag{17.127}$$

on the smooth surface.

We shall now look for instabilities in v_z, ω_z, and T for fluctuations which vary in space in the x- and y-directions. Since in performing the linear stability analysis we are working with linear equations, we only need consider one Fourier component. Thus, we shall write

$$v_z(\mathbf{r}, t) = \tilde{V}_z(z)\, e^{i(k_x x + k_y y)}\, e^{\omega t}, \tag{17.128}$$

$$\omega_z(\mathbf{r}, t) = \tilde{W}_z(z)\, e^{i(k_x x + k_y y)}\, e^{\omega t}, \tag{17.129}$$

and

$$T(\mathbf{r}, t) = \tilde{T}(z)\, e^{i(k_x x + k_y y)}\, e^{\omega t}, \tag{17.130}$$

where $\tilde{V}_z(z)$, $\tilde{W}_z(z)$, and $\tilde{T}(z)$ are assumed to depend on $\mathbf{k} = k_x\hat{\imath} + k_y\hat{\jmath}$ and on ω. Substitution into Eqs. (17.113), (17.117), and (17.118) yields

$$\omega\tilde{T} = a\tilde{V}_z + \frac{K}{\rho_0 c_V}\left(\frac{d^2}{dz^2} - k^2\right)\tilde{T}, \tag{17.131}$$

$$\omega\left(\frac{d^2}{dz^2} - k^2\right)\tilde{V}_z = -g\alpha_p k^2 \tilde{T} + \nu\left(\frac{d^2}{dz^2} - k^2\right)^2 \tilde{V}_z, \tag{17.132}$$

and

$$\omega\tilde{W}_z = \nu\left(\frac{d^2}{dz^2} - k^2\right)\tilde{W}_z. \tag{17.133}$$

From the general boundary conditions we must have $\tilde{T}(0) = \tilde{T}(d) = 0$ and $\tilde{V}_z(0) = \tilde{V}_z(d) = 0$. On a rigid surface we have $\tilde{W}_z(0) = \tilde{W}_z(d) = 0$ and

$$\left.\frac{d\tilde{V}_z}{dz}\right|_{z=0} = \left.\frac{d\tilde{V}_z}{dz}\right|_{z=d} = 0.$$

On a smooth surface we have

$$\left.\frac{d\tilde{W}_z}{dz}\right|_{z=0} = \left.\frac{d\tilde{W}_z}{dz}\right|_{z=d} = 0$$

and

$$\left.\frac{d^2{}_z\tilde{V}}{dz^2}\right|_{z=0} = \left.\frac{d^2{}_z\tilde{V}}{dz^2}\right|_{z=d} = 0.$$

It is convenient to introduce a change of length scale. Thus, we let $\mathscr{Z} = z/d$, $s = \omega d^2/2$, $\alpha = kd$, and $P = \nu\rho_0 c_V/K$ (P is the Prandl number). Then Eqs. (17.131) and (17.132) can be written

$$\left(\frac{\partial^2}{\partial\mathscr{Z}^2} - \alpha^2\right)\left(\frac{\partial^2}{\partial\mathscr{Z}^2} - \alpha^2 - s\right)\left(\frac{\partial^2}{\partial\mathscr{Z}^2} - \alpha^2 - Ps\right)\tilde{V}_z(z) = -R\alpha^2\tilde{V}_z(z) \tag{17.134}$$

and an identical equation holds for $\tilde{T}(z)$. In Eq. (17.134), R is the Rayleigh number, $R = g\alpha_p d^4 \rho_0 c_V/\nu K$.

Following Chandrasekhar, we now show that s, the rescaled frequency, must be real and, therefore, an instability will occur when $s = 0$ (that is, when s

changes from negative to positive). We first define two functions, $G(\mathscr{Z})$ and $F(\mathscr{Z})$, such that

$$G(\mathscr{Z}) = \left(\frac{\partial^2}{\partial \mathscr{Z}^2} - \alpha^2\right) \tilde{V}_z(\mathscr{Z}) \tag{17.135}$$

and

$$F(\mathscr{Z}) = \left(\frac{\partial^2}{\partial \mathscr{Z}^2} - \alpha^2 - s\right) G(\mathscr{Z}) = \left(\frac{\partial^2}{\partial \mathscr{Z}^2} - \alpha^2 - s\right)\left(\frac{\partial^2}{\partial \mathscr{Z}^2} - \alpha^2\right) \tilde{V}_z(\mathscr{Z}). \tag{17.136}$$

Then Eq. (17.134) can be written

$$\left(\frac{\partial^2}{\partial \mathscr{Z}^2} - \alpha^2 - Ps\right) F(\mathscr{Z}) = -R\alpha^2 \tilde{V}_z(\mathscr{Z}). \tag{17.137}$$

We next multiply by the complex conjugate of $F(\mathscr{Z})$ and integrate $\mathscr{Z}$ from 0 to 1:

$$\int_0^1 F^* \left(\frac{\partial^2}{\partial \mathscr{Z}^2} - \alpha^2 - Ps\right) F \, d\mathscr{Z} = -R\alpha^2 \int_0^1 F^* \hat{V}_z \, d\mathscr{Z}. \tag{17.138}$$

An integration by parts, together with the boundary conditions in Eqs. (17.119) and (17.120), yields for the first term on the left

$$\int_0^1 F^* \frac{\partial^2}{\partial \mathscr{Z}^2} F \, d\mathscr{Z} = -\int_0^1 \left|\frac{\partial F}{\partial \mathscr{Z}}\right|^2 d\mathscr{Z}. \tag{17.139}$$

If we now use the definition of $F^*(\mathscr{Z})$ obtained from Eq. (17.136) and integrate by parts to rearrange the right-hand side of Eq. (17.138), we find, after some algebra,

$$\int_0^1 \left\{\left|\frac{\partial F}{\partial \mathscr{Z}}\right|^2 + \alpha^2 |F|^2 + Ps|F|^2\right\} d\mathscr{Z}$$
$$= R\alpha^2 \int_0^1 \left\{|G|^2 + s^* \left[\left|\frac{\partial V_z}{\partial \mathscr{Z}}\right|^2 + \alpha^2 |V_z|^2\right]\right\} d\mathscr{Z}. \tag{17.140}$$

The real and imaginary parts of Eq. (17.140) must vanish separately. Thus, we find

$$(\mathrm{Im}\, s)\left[\int_0^1 P|F|^2 \, d\mathscr{Z} + R\alpha^2 \int_0^1 \left[\left|\frac{\partial V_z}{\partial \mathscr{Z}}\right|^2 + \alpha^2 |V_z|\right] d\mathscr{Z}\right] = 0 \tag{17.141}$$

where Im s is the imaginary part of s. Since the term in brackets is positive and finite, the imaginary part of s must vanish. Thus, s is a real frequency as long as R and P are positive numbers (R and P must be positive from thermodynamic stability considerations). The transition from the steady state in Eqs. (17.99)–(17.101) to a new steady state (often called an "exchange of stabilities" or soft mode transition when the frequency is real and no temporal oscillations are allowed) occurs when $s = 0$. To find conditions under which the transition occurs, we set $s = 0$ in Eq. (17.134) and find values of R which satisfy the equation

$$\left(\frac{\partial^2}{\partial \mathscr{Z}^2} - \alpha^2\right)^3 \tilde{V}_z(\mathscr{Z}) = -R\alpha^2 \tilde{V}_z(\mathscr{Z}). \tag{17.142}$$

Thus, the whole problem has been reduced to an eigenvalue problem.

Let us now find the conditions for a phase transition to a new steady state for the case of smooth boundaries. We will then give the results for the other cases. The boundary conditions for the case of two smooth boundaries are such that

$$\left(\frac{\partial^{2n}}{\partial \mathscr{Z}^{2n}} \tilde{V}_z(\mathscr{Z})\right)_{\text{bound}} = 0 \tag{17.143}$$

for all integer n. Thus, $\tilde{V}_z(\mathscr{Z})$ must have the form

$$\tilde{V}_z(\mathscr{Z}) = A \sin n\pi\mathscr{Z} \tag{17.144}$$

where $n = (1, 2, \ldots, \infty)$. Substitution into Eq. (17.142) leads to the eigenvalue equation

$$R = \frac{(n^2\pi^2 + \alpha^2)^3}{\alpha^2}. \tag{17.145}$$

The smallest value of R for which an instability can occur is for $n = 1$ or

$$R = \frac{(\pi^2 + \alpha^2)^3}{\alpha^2}. \tag{17.146}$$

If we plot R as a function of α (cf. Fig. 17.11), we find a critical value of α (α is related to the size of cells in the x and y directions) for which R takes on its lowest value. To find it we must find the position of the minimum of R. Thus, we set

$$\frac{\partial R}{\partial \alpha^2} = 3\frac{(\pi^2 + \alpha^2)^2}{\alpha^2} - \frac{(\pi^2 + \alpha^2)^3}{\alpha^4} = 0 \tag{17.147}$$

and obtain

$$\alpha_c = \frac{\pi}{\sqrt{2}} = 2.22. \tag{17.148}$$

The critical wavelength (the distance across a cell in the x-y plane) for the onset of instabilities is

$$\lambda_c = \frac{2\pi d}{\alpha} = 2^{3/2} d \tag{17.149}$$

and the critical Rayleigh number is

$$R_c = \tfrac{27}{4}\pi^4 = 657.51. \tag{17.150}$$

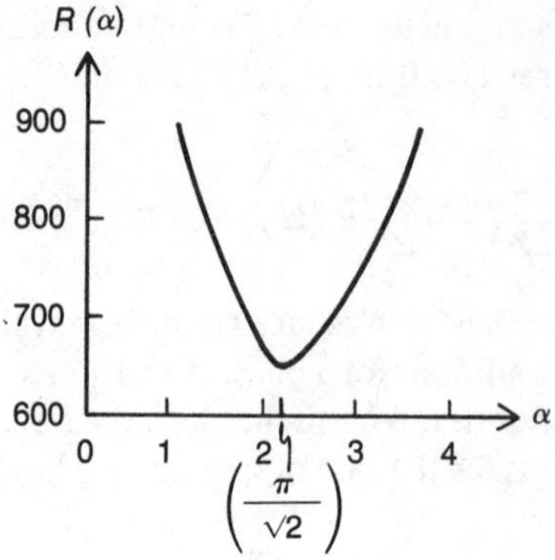

Fig. 17.11. Plot of $R(\alpha) = (\pi^2 + \alpha^2)^3/\alpha^2$.

Thus, the critical wavelength depends on the size of the container. The case of two smooth surfaces is difficult to realize in practice. Generally, for experiments, one has one smooth surface and one rigid surface or two rigid surfaces. For those boundary conditions the analysis proceeds along similar lines but is slightly more complicated[16] and will not be included here. For the case of one rigid and one free surface, we find $R_c = 1100.65$ and $\alpha_c = 2.68$. For the case of two rigid boundaries, we find $R_c = 1707.76$ and $\alpha_c = 3.12$.

Let us now summarize our results. For the region in which the steady state in Eqs. (17.99)–(17.101) is stable, the system is homogeneous. However, at the transition point convection begins and the system breaks into cells in the x- and y-directions. The cells will have a periodicity in the x- and y-directions which is proportional to d, distance between the plates. We expect the cell walls to be vertical and that at the cell walls the normal gradient of velocity $v_z(\mathbf{r}, t)$ is zero. Thus, at the cell walls

$$(\hat{\mathbf{n}} \cdot \nabla_{\mathbf{r}_\perp} v_z(\mathbf{r}, t)) = 0 \tag{17.151}$$

where $\nabla_{\mathbf{r}_\perp} = \hat{\mathbf{i}}(\partial/\partial x) + \hat{\mathbf{j}}(\partial/\partial y)$ and $\hat{\mathbf{n}}$ is normal to the walls of the cell.

The simplest type of cell pattern that can develop is that of a roll, that is, a cell pattern which depends only on one horizontal coordinate—say, x. Then we have

$$v_z(\mathbf{r}, t) = \tilde{V}_z(z) \cos kx \tag{17.152}$$

where k is the wave vector of the rolls.

The theory we have developed here assumes that the planes are of infinite extent in the x- and y-directions. In practice the experiments must be done in finite containers. Then the cell pattern that emerges is strongly dependent on the shape of the container. For rectangular containers and circular covered

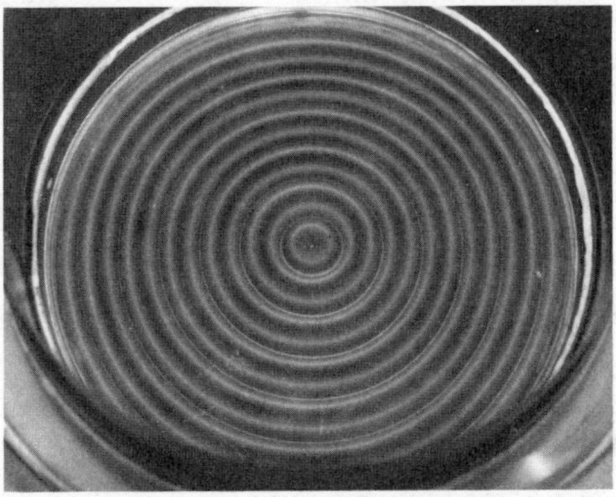

Fig. 17.12. Bénard instability: in a circular covered container, rolls usually form; the bright areas represent horizontal motion; the fine white lines are settled aluminum powder (reprinted, by permission, from E. L. Koschmieder, Adv. Chem. Phys. *26* 177 [1974]).

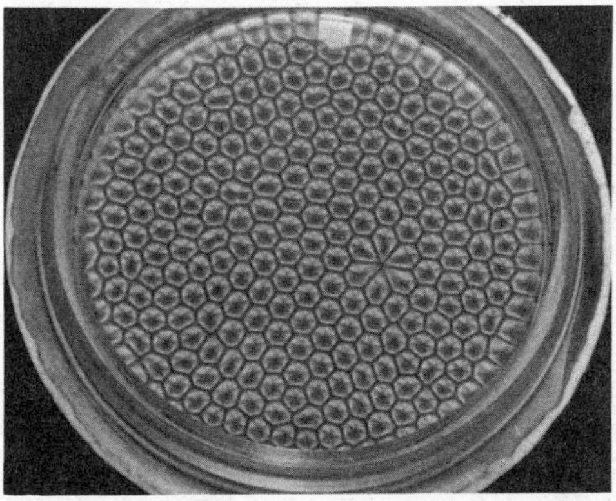

Fig. 17.13. Bénard instability: in an open container, where the effects of surface tension are important, hexagonal cells usually form; the dark lines indicate vertical motion; the bright areas indicate predominantly horizontal motion (reprinted, by permission, from E. L. Koschmieder, Adv. Chem. Phys. *26* 177 [1974]).

containers, rolls usually form. Square cells can be formed in square containers. In uncovered circular containers, where the effects of surface tension are important, hexagonal cells usually form.[21]. In Figs. 17.12 and 17.13, we give some examples of rolls and hexagons, respectively.

REFERENCES

1. P. Glansdorff and I. Prigogine, *Structure, Stability, and Fluctuations* (Wiley-Interscience, New York, 1971).
2. G. Nicolis, Adv. Chem. Phys. *19* 209 (1971).
3. J. Turner in *Lectures in Statistical Physics*, Vol. 28, ed. W. Schieve and J. Turner (Springer-Verlag, Berlin, 1974).
4. G. Nicolis and I. Prigogine, *Self-Organization in Nonequilibrium Systems* (Wiley-Interscience, New York, 1977).
5. F. Schlögl, Z. Physik *248* 446 (1971); *253* 147 (1972).
6. G. Nicolis and J. W. Turner, Physica *89A* 326 (1977).
7. I. Prigogine and R. Lefever, J. Chem. Phys. *48* 1695 (1968).
8. B. B. Belousov, Sb. Ref. Radiats. Med. Moscow (1958).
9. A. M. Zhabotinski, Biofizika *9* 306 (1964).
10. R. J. Field and R. M. Noyes, J. Chem. Phys. *60* 1877 (1974).
11. R. J. Field, J. Chem. Phys. *63* 2289 (1975).
12. A. D. Nazarea, Adv. Chem. Phys. *38* 415 (1978).
13. A. J. Lotka, J. Amer. Chem. Soc. *42* 1595 (1920).
14. V. Volterra, *Théorie mathématique de la lutte pour la vie* (Gauthier-Villars, Paris, 1931).
15. N. Minorsky, *Nonlinear Oscillations* (Van Nostrand Co., Princeton, 1962).
16. S. Chandrasekhar, *Hydrodynamic and Hydromagnetic Stability* (Clarendon Press, Oxford, 1961).
17. G. Normand, Y. Pomeau, and M. G. Velarde, Rev. Mod. Phys. *49* 581 (1977).
18. J. Swift and P. C. Hohenberg, Phys. Rev. *A15* 319 (1977).
19. E. L. Koschmieder, Adv. Chem. Phys. *26* 177 (1974).
20. J. M. Mihaljan, Astrophysics J. *136* 1126 (1962).
21. The author wishes to thank Professor Koschmieder for pointing this out.

PROBLEMS

1. Use linear stability theory to show that the region QP in Fig. 17.1 for the Schlögl model is unstable.

2. For the Schlögl model, show that multiple steady states can only occur far from thermodynamic equilibrium.

3. A one-dimension chemically reacting system has a variable constituent X whose rate equation is linear:

$$\frac{\partial N_x}{\partial t} = KN_x - D\nabla_{\mathbf{r}}^2 N_x.$$

Assume that the concentration N_x is held fixed on the boundaries at $N_x = N_0$. Find the steady state value of N_x as a function of position (D is the coefficient of diffusion). Sketch the steady state solution as a function of position.

4. Write the rate equations for the Brusselator near equilibrium and find the equilibrium values of N_x and N_y. Find the steady state solutions for the system near equilibrium (assume it is homogeneous) and show that far from equilibrium (A and B large and D and E eliminated as soon as they are produced) they reduce to the expressions $X_0 = A$ and $Y_0 = B/A$ given in Eq. (17.58).

5. Show that the Brusselator admits traveling wave solutions $X(r\theta \pm vt)$ and $Y(r\theta \pm vt)$ for linear motion in a circle, assuming periodic boundary conditions. Find the condition for a bifurcation from the thermodynamic branch to a traveling wave solution using linear stability theory. Show that the velocity v of the waves is a decreasing function of the wave number n.

6. Write a set of chemical reactions which gives rise to the Lotka-Volterra rate equations. Assume that the system can be treated as an ideal mixture. Find an expression for the excess entropy, $\frac{1}{2}\delta^2 S$, due to fluctuations about each of the steady states.

7. A fluid is constrained between two smooth surfaces a distance d apart in a gravitational field with a temperature gradient across the fluid. If a Bénard instability develops, show that it is possible to have square cells. Find the horizontal width of the cells in terms of d. Find an expression for the velocity, $\mathbf{v}$, of the fluid and sketch the flow in each cell. Assume that, at the cell walls, the fluid flows upward and has its maximum speed.

Appendices

A. BALANCE EQUATIONS

1. General Fluid Flow

Let us consider a continuous medium which moves with velocity $\mathbf{v}(\mathbf{x}, t)$ at time t and position $\mathbf{x}$. The position $\mathbf{x}$ is measured with respect to a fixed frame of reference whose origin lies at $O_{\mathbf{x}}$. In addition, it is convenient to introduce a second coordinate system whose origin lies at $O_{\mathbf{z}}$ and which moves with the fluid. We assume that at time $t = 0$ the two coordinate systems coincide; and at time $t = 0$ we select a given fluid "particle" (small volume element of the fluid) which has position $\mathbf{x}(0)$ with respect to frame $O_{\mathbf{x}}$ and position $\mathbf{z}$ with respect to frame $O_{\mathbf{z}}$. Because the coordinate systems coincide, $\mathbf{z} = \mathbf{x}(0)$.

Let us assume that both the frame $O_{\mathbf{z}}$ and the fluid particle move freely with the fluid, and that the fluid particle always has position $\mathbf{z}$ with respect to the frame $O_{\mathbf{z}}$. The position of the fluid particle with respect to the frame $O_{\mathbf{x}}$ at time t is

$$\mathbf{x}(t) = \mathbf{x}(\mathbf{z}, t) \tag{A.1}$$

where $\mathbf{z} = \mathbf{x}(0)$ is the initial position of the particle. Eq. (A.1) relates the coordinates of the fluid particle as seen by observers in the two frames (cf. Fig. A.1).

We now introduce a density $\mathscr{D}(\mathbf{x}, t)$, where $\mathscr{D}(\mathbf{x}, t)$ could be a probability per unit volume, internal energy per unit volume, mass per unit volume, angular momentum per unit volume, etc. At time $t = 0$, an observer in frame $O_{\mathbf{x}}$ sees a density $\mathscr{D}(\mathbf{x}(0), 0)$ and an observer in frame $O_{\mathbf{z}}$ sees a density $\mathscr{D}'(\mathbf{z}, 0)$ in the neighborhood of the fluid particle, where $\mathscr{D}(\mathbf{x}(0), 0) = \mathscr{D}'(\mathbf{z}, 0)$. At time t, an observer in $O_{\mathbf{x}}$ sees a density $\mathscr{D}(\mathbf{x}(\mathbf{z}, t), t)$ and an observer in $O_{\mathbf{z}}$ sees a density $\mathscr{D}'(\mathbf{z}, t)$ in the neighborhood of the fluid particle, where $\mathscr{D}(\mathbf{x}(\mathbf{z}, t), t) = \mathscr{D}'(\mathbf{z}, t)$. (Once the density is given in one coordinate system, it can be found in the other using Eq. (A.1).)

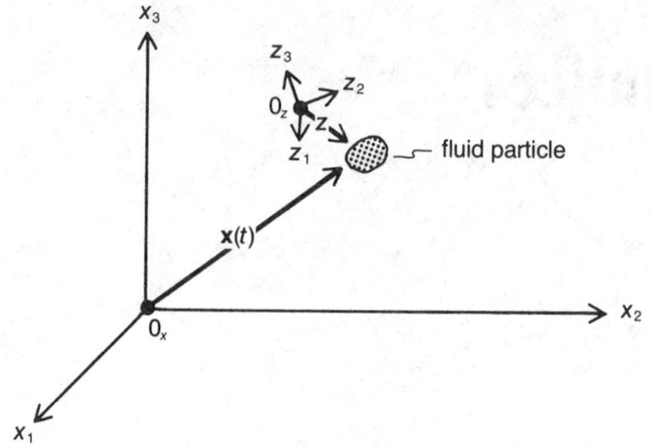

Fig. A.1. Coordinates of a fluid particle at time *t*.

We shall denote the total time rate of change of the density in the neighborhood of the fluid particle by $d\mathscr{D}/dt$. Then

$$\frac{d\mathscr{D}}{dt} = \left(\frac{\partial \mathscr{D}'}{\partial t}\right)_{\mathbf{z}} = \left(\frac{\partial \mathscr{D}}{\partial t}\right)_{\mathbf{x}} + \left(\frac{\partial \mathscr{D}}{\partial \mathbf{x}}\right)_t \cdot \left(\frac{\partial \mathbf{x}}{\partial t}\right)_{\mathbf{z}}. \tag{A.2}$$

However, $(\partial \mathbf{x}/\partial t)_{\mathbf{z}}$ is the velocity of the fluid particle:

$$\mathbf{v} \equiv \frac{d\mathbf{x}}{dt} = \left(\frac{\partial \mathbf{x}}{\partial t}\right)_{\mathbf{z}}. \tag{A.3}$$

From Eq. (A.2), we see that $d\mathscr{D}/dt = (\partial \mathscr{D}'/\partial t)_{\mathbf{z}}$ is the time rate of change of the density as seen by an observer moving with the fluid particle, and $(\partial \mathscr{D}/\partial t)_{\mathbf{x}}$ is the time rate of change of the density at a fixed point in space (not moving with the fluid). Combining Eqs. (A.2) and (A.3), we may write the total derivative d/dt in the form

$$\frac{d}{dt} = \frac{\partial}{\partial t} + \mathbf{v}\cdot\nabla_{\mathbf{x}}. \tag{A.4}$$

Eq. (A.4) is often called the convective time derivative. It allows one to find the time rate of change at a moving point in terms of coordinates which are fixed in space.

Let us now consider a given set of fluid particles contained in a volume element dV_0 at time $t = 0$. We wish to follow the same set of fluid particles with time and see how their volume changes with time. If we remember that $\mathbf{x}(0) = \mathbf{z}$, we may write

$$dV_t = \mathscr{J}\begin{pmatrix} x_1, x_2, x_3 \\ z_1, z_2, z_3 \end{pmatrix} dV_0 \tag{A.5}$$

where

$$dV_0 = d\mathbf{z} = d\mathbf{x}(0) \tag{A.6}$$

and

$$dV_t = d\mathbf{x}(t). \tag{A.7}$$

In Eq. (A.5), x_1, x_2, and x_3 are the components of the vector $\mathbf{x}(t)$, and z_1, z_2, and z_3 are the components of the vector $\mathbf{z}$. The Jacobian is defined

$$\mathscr{J}\begin{pmatrix} x_1 & x_2 & x_3 \\ z_1 & z_2 & z_3 \end{pmatrix} = \det \begin{vmatrix} \frac{\partial x_1}{\partial z_1} & \frac{\partial x_1}{\partial z_2} & \frac{\partial x_1}{\partial z_3} \\ \frac{\partial x_2}{\partial z_1} & \frac{\partial x_2}{\partial z_2} & \frac{\partial x_2}{\partial z_3} \\ \frac{\partial x_3}{\partial z_1} & \frac{\partial x_3}{\partial z_2} & \frac{\partial x_3}{\partial z_3} \end{vmatrix} \tag{A.8}$$

and may be evaluated from Eq. (A.1). From Eq. (A.5) we see that the Jacobian is a measure of the amount by which the volume of a given set of fluid particles changes as a function of time. The change in the Jacobian with time, as seen by an observer moving with the volume element, is $d\mathscr{J}/dt$. If we note that

$$\frac{d}{dt}\frac{\partial x_i}{\partial z_j} = \frac{\partial}{\partial z_j}\frac{dx_i}{dt} = \frac{\partial v_i}{\partial z_j} \tag{A.9}$$

($\mathbf{z}$ is held constant when the differentiation d/dt is carried out), we may write

$$\frac{d\mathscr{J}}{dt} = \det \begin{vmatrix} \frac{\partial v_1}{\partial z_1} & \frac{\partial v_1}{\partial z_2} & \frac{\partial v_1}{\partial z_3} \\ \frac{\partial x_2}{\partial z_1} & \frac{\partial x_2}{\partial z_2} & \frac{\partial x_2}{\partial z_3} \\ \frac{\partial x_3}{\partial z_1} & \frac{\partial x_3}{\partial z_2} & \frac{\partial x_3}{\partial z_3} \end{vmatrix} + \det \begin{vmatrix} \frac{\partial x_1}{\partial z_1} & \frac{\partial x_1}{\partial z_2} & \frac{\partial x_1}{\partial z_3} \\ \frac{\partial v_2}{\partial z_1} & \frac{\partial v_2}{\partial z_2} & \frac{\partial v_2}{\partial z_3} \\ \frac{\partial x_3}{\partial z_1} & \frac{\partial x_3}{\partial z_2} & \frac{\partial x_3}{\partial z_3} \end{vmatrix} + \det \begin{vmatrix} \frac{\partial x_1}{\partial z_1} & \frac{\partial x_1}{\partial z_2} & \frac{\partial x_1}{\partial z_3} \\ \frac{\partial x_2}{\partial z_1} & \frac{\partial x_2}{\partial z_2} & \frac{\partial x_2}{\partial z_3} \\ \frac{\partial v_3}{\partial z_1} & \frac{\partial v_3}{\partial z_2} & \frac{\partial v_3}{\partial z_3} \end{vmatrix}. \tag{A.10}$$

However,

$$\frac{\partial v_i}{\partial z_j} = \sum_{k=1}^{3} \frac{\partial v_i}{\partial x_k}\frac{\partial x_k}{\partial z_j}. \tag{A.11}$$

If we substitute Eq. (A.11) into Eq. (A.10), then, in the first determinant on the right-hand side, only the term in the sum with $k = 1$ is nonzero, because for $k = 2$ or 3 two rows of the determinant become identical and therefore give zero. The second and third determinants are only nonzero for $k = 2$ and 3, respectively. We therefore obtain

$$\frac{d\mathscr{J}}{dt} = \left(\frac{\partial v_1}{\partial x_1} + \frac{\partial v_2}{\partial x_2} + \frac{\partial v_3}{\partial x_3}\right)\mathscr{J} = \nabla_{\mathbf{x}} \cdot \mathbf{v}\mathscr{J}. \tag{A.12}$$

If $\nabla_{\mathbf{x}} \cdot \mathbf{v} = 0$, then $d\mathcal{J}/dt = 0$ and the size of the volume element remains constant with time; that is, the fluid is incompressible. If $\nabla_{\mathbf{x}} \cdot \mathbf{v} \neq 0$, the volume element may expand and contract with time.

2. General Balance Equation

We now wish to find the equation of motion for an arbitrary density $\mathcal{D}(\mathbf{x}, t)$. Let us consider a finite volume element V_0 at time $t = 0$ containing a given set of fluid particles. At time t, that same set of fluid particles will occupy a volume $V(t)$. The total amount of quantity D contained in $V(t)$ is found by integrating $\mathcal{D}(\mathbf{x}, t)$ over the volume $V(t)$:

$$D(t) = \iiint_{V(t)} \mathcal{D}(\mathbf{x}, t)\, dV_t. \tag{A.13}$$

The rate of change of the amount of D in $V(t)$ as seen by an observer moving with $V(t)$ is given by

$$\frac{dD}{dt} = \frac{d}{dt} \iiint_{V(t)} \mathcal{D}(\mathbf{x}, t)\, dV_t. \tag{A.14}$$

Since the limits of integration depend on time, the derivative cannot be taken inside the integral. However, if we write the integral in terms of the coordinates at time $t = 0$, we can take the derivative inside:

$$\frac{dD}{dt} = \frac{d}{dt} \iiint_{V_0} \mathcal{D}(\mathbf{x}, t)\mathcal{J}\, dV_0 = \iiint_{V_0} \left(\frac{d\mathcal{D}}{dt}\mathcal{J} + \mathcal{D}\frac{d\mathcal{J}}{dt}\right) dV_0. \tag{A.15}$$

In Eq. (A.15), $\mathcal{J}$ is the Jacobian. Using Eq. (A.12), we obtain

$$\frac{dD}{dt} = \iiint_{V_0} \left(\frac{d\mathcal{D}}{dt} + \mathcal{D}\nabla_{\mathbf{x}} \cdot \mathbf{v}\right)\mathcal{J}\, dV_0 = \iiint_{V(t)} \left(\frac{d\mathcal{D}}{dt} + \mathcal{D}\nabla_{\mathbf{x}} \cdot \mathbf{v}\right) dV_t. \tag{A.16}$$

The total amount of the quantity D in $V(t)$ can change if there is a flow of D through the sides of $V(t)$ or a source of D in $V(t)$. We can imagine a flow of D through the sides. An example would be the case when $\mathcal{D}(\mathbf{x}, t)$ is the heat density at position $\mathbf{x}$ and time t, and the surrounding medium is at a different temperature. If we let $\mathbf{J}_D$ represent a current of D through the sides of $V(t)$ and σ_D represent a source of D inside $V(t)$, then we can write

$$\frac{dD}{dt} = \iiint_{V(t)} \left(\frac{d\mathcal{D}}{dt} + \mathcal{D}\nabla_{\mathbf{x}} \cdot \mathbf{v}\right) dV_t = \iiint_{V(t)} \sigma_D\, dV_t - \iint_{S(t)} \mathbf{J}_D \cdot d\mathbf{S}_t. \tag{A.17}$$

(The surface area element $d\mathbf{S}_t$ is assumed to point out of $V(t)$.) If we first change the surface integral to a volume integral ($\iint_{S(t)} \mathbf{J}_D \cdot d\mathbf{S}_t = \iiint_{V(t)} \nabla_{\mathbf{x}} \cdot \mathbf{J}_D \, dV_t$), then, since $V(t)$ is arbitrary, we may equate integrands to obtain

$$\frac{d\mathscr{D}}{dt} = -\mathscr{D}\nabla_{\mathbf{x}} \cdot \mathbf{v} + \sigma_D - \nabla_{\mathbf{x}} \cdot \mathbf{J}_D. \tag{A.18}$$

Eq. (A.18) is one form of the Balance Equation. If we use the definition of the convective time derivative given in Eq. (A.4), we may write the Balance Equation in the more usual form:

$$\frac{\partial \mathscr{D}}{\partial t} = -\nabla_{\mathbf{x}} \cdot (\mathscr{D}\mathbf{v} + \mathbf{J}_D) + \sigma_D. \tag{A.19}$$

The quantity $\mathscr{D}\mathbf{v}$ is a convective current of D. It is a current which occurs simply because the fluid is in motion. Eq. (A.18) gives the time rate of change of the density in the neighborhood of a point moving with the fluid. Eq. (A.19) gives the time rate of change of the density at a fixed point in the fluid.

We can get a better understanding of the meaning of various terms in Eq. (A.19) if we integrate it over a fixed volume, V_f, in the fluid. Then

$$\iiint_{V_f} \frac{\partial \mathscr{D}}{\partial t} \, dV_f = -\iiint_{V_f} \nabla_{\mathbf{x}} \cdot (\mathscr{D}\mathbf{v} + \mathbf{J}_D) \, dV_f + \iiint \sigma_D \, dV_f. \tag{A.20}$$

Since the volume V_f is fixed, the limits of integration are time independent and the time derivative on the left-hand side can be taken outside the integral. Also, we can use Stoke's theorem to change the volume integral of the divergence on the right-hand side to a surface integral. We then obtain

$$\frac{\partial}{\partial t} \iiint_{V_f} \mathscr{D} \, dV_f = -\iint_{S_f} (\mathscr{D}\mathbf{v} + \mathbf{J}_D) \cdot d\mathbf{S}_f + \iiint_{V_f} \sigma_D \, dV_f \tag{A.21}$$

where S_f is the surface of V_f and $d\mathbf{S}_f$ is an infinitesimal surface area segment whose direction is along the outward normal to the surface S_f. From Eq. (A.21), we see that the rate of change of the amount of D in V_f is due, in part, to a flow of D through the surface and to the creation of D in the fluid.

B. REPRESENTATIONS FOR MANY-BODY QUANTUM SYSTEMS

The density operator, $\hat{\rho}$, introduced in Sec. 7.E, is completely abstract and can be evaluated in any desired representation. In dealing with degenerate many-body systems of Fermi-Dirac or Bose-Einstein particles, as we often do in this book, it is easiest to use field operators (although not always possible to do so). For classical many-body systems and nondegenerate and sometimes degenerate quantum systems, phase space variables are often used. In this appendix, we shall review the quantum theory of many-body systems and discuss in a rather simple way the relation between the coordinate and number representations.

1. Representations of Position and Momentum Operators[2]

We will denote the one-body position operator by $\hat{\mathbf{q}}$ (hat always will denote an operator) and the complete set of orthonormal eigenstates of $\hat{\mathbf{q}}$ by $|\mathbf{r}\rangle$. Orthonormality of states $|\mathbf{r}\rangle$ gives

$$\langle \mathbf{r}|\mathbf{r}'\rangle = \delta(\mathbf{r} - \mathbf{r}') \tag{B.1}$$

where $\langle \mathbf{r}|$ is the dual vector to $|\mathbf{r}\rangle$ and $\delta(\mathbf{r} - \mathbf{r}')$ denotes the Dirac delta function. Completeness of states $|\mathbf{r}\rangle$ gives

$$\int d\mathbf{r}|\mathbf{r}\rangle\langle \mathbf{r}| = 1 \tag{B.2}$$

where the integral ranges over the entire volume of the system. A matrix element of $\hat{\mathbf{q}}$ in the position representation is given by

$$\langle \mathbf{r}|\hat{\mathbf{q}}|\mathbf{r}'\rangle = \mathbf{r}'\,\delta(\mathbf{r} - \mathbf{r}'). \tag{B.3}$$

For a particle in a finite cubic box of volume $V = L^3$ (L the length of a side), the momentum operator $\hat{\mathbf{p}}$ may be written

$$\hat{\mathbf{p}} = \hbar\hat{\mathbf{k}} \tag{B.4}$$

where $\hat{\mathbf{k}}$ is the wave vector operator (not a unit vector) and is given by

$$\hat{\mathbf{k}} = \frac{2\pi}{L}\hat{\mathbf{n}}. \tag{B.5}$$

In Eq. (B.5) $\hat{\mathbf{n}}$ is a vector; $\hat{\mathbf{n}} = \hat{n}_x\mathbf{i} + \hat{n}_y\mathbf{j} + n_z\mathbf{k}$; $\hat{n}_x$, $\hat{n}_y$, and $\hat{n}_z$ are number operators whose eigenvalues range over all integers from $-\infty$ to $+\infty$; and $\mathbf{i}$, $\mathbf{j}$, and $\mathbf{k}$ are unit vectors in the x-, y-, and z-directions, respectively. The complete set of orthonormal eigenstates of $\hat{\mathbf{k}}$ will be denoted $|\mathbf{k}\rangle$. For a finite box, they obey the orthonormality condition

$$\langle \mathbf{k}|\mathbf{k}'\rangle = \delta^{\mathrm{Kr}}_{\mathbf{k},\mathbf{k}'} \tag{B.6}$$

where $\delta^{\mathrm{Kr}}_{\mathbf{k},\mathbf{k}'}$ is the Kronecker delta (for an infinite box, the Kronecker delta is replaced by a Dirac delta). Completeness of the states $|\mathbf{k}\rangle$ yields

$$\sum_{\mathbf{k}} |\mathbf{k}\rangle\langle \mathbf{k}| = 1 \tag{B.7}$$

for a finite box, while for an infinite box the summation over $\mathbf{k}$ is replaced by an integration over $\mathbf{k}$. Matrix elements of the momentum operator in the momentum representation (we shall often call $\mathbf{k}$ the momentum) are

$$\langle \mathbf{k}|\hat{\mathbf{p}}|\mathbf{k}'\rangle = \hbar\mathbf{k}'\,\delta^{\mathrm{Kr}}_{\mathbf{k},\mathbf{k}'}. \tag{B.8}$$

for a finite box. For an infinite box the Kronecker delta is replaced by a Dirac delta.

The vectors $|\mathbf{r}\rangle$ and $|\mathbf{k}\rangle$ are abstract eigenstates of position and momentum, respectively. We can evaluate them in various representations. The state of momentum, $|\mathbf{k}\rangle$, in the momentum representation will be written

$$\phi_{\mathbf{k}}(\mathbf{k}_0) = \langle \mathbf{k}_0|\mathbf{k}\rangle = \begin{cases} \delta(\mathbf{k} - \mathbf{k}_0) & \text{infinite box} \\ \delta^{Kr}_{\mathbf{k},\mathbf{k}_0} & \text{finite box} \end{cases}. \tag{B.9}$$

The state of momentum **k** in the position representation is written

$$\phi_{\mathbf{k}}(\mathbf{r}) = \langle \mathbf{r}|\mathbf{k}\rangle = \begin{cases} \dfrac{1}{\sqrt{V}} e^{i\mathbf{k}\cdot\mathbf{r}} & \text{finite box} \\ \left(\dfrac{1}{2\pi}\right)^{3/2} e^{i\mathbf{k}\cdot\mathbf{r}} & \text{infinite box} \end{cases}. \tag{B.10}$$

Note that

$$\phi^*_{\mathbf{k}}(\mathbf{r}) = \langle \mathbf{r}|\mathbf{k}\rangle^+ = \langle \mathbf{k}|\mathbf{r}\rangle. \tag{B.11}$$

Similarly, the vector $|\mathbf{r}\rangle$ in the position representation is

$$\phi_{\mathbf{r}}(\mathbf{r}_0) = \langle \mathbf{r}_0|\mathbf{r}\rangle = \delta(\mathbf{r} - \mathbf{r}_0) \tag{B.12}$$

and in the momentum representation is

$$\phi_{\mathbf{r}}(\mathbf{k}_0) = \langle \mathbf{k}_0|\mathbf{r}\rangle = \phi^*_{\mathbf{k}_0}(\mathbf{r}). \tag{B.13}$$

With the above definitions, Eq. (B.6) is satisfied. For a finite box

$$\langle \mathbf{k}_0|\mathbf{k}\rangle = \int d\mathbf{r}\langle \mathbf{k}_0|\mathbf{r}\rangle\langle \mathbf{r}|\mathbf{k}\rangle = \frac{1}{V}\int d\mathbf{r}\, e^{i\mathbf{r}\cdot(\mathbf{k}-\mathbf{k}_0)} = \delta^{Kr}_{\mathbf{k},\mathbf{k}_0}, \tag{B.14}$$

while for an infinite box the proof is similar.

We can write the operators $\hat{\mathbf{p}}$ and $\hat{\mathbf{r}}$ in various representations. Let us first find an expression for $\hat{\mathbf{p}}$ in the position representation. For infinite volume box,

$$\langle \mathbf{r}|\hat{\mathbf{p}}|\mathbf{r}'\rangle = \int d\mathbf{k}\hbar\mathbf{k}\langle \mathbf{r}|\mathbf{k}\rangle\langle \mathbf{k}|\mathbf{r}'\rangle = \left(\frac{1}{2\pi}\right)^3 \int d\mathbf{k}\hbar\mathbf{k}\, e^{i\mathbf{k}\cdot(\mathbf{r}-\mathbf{r}')}$$

$$= \frac{i\hbar}{(2\pi)^3}\nabla_{\mathbf{r}}\int d\mathbf{k}\, e^{i\mathbf{k}\cdot(\mathbf{r}-\mathbf{r}')} = i\hbar[\nabla_{\mathbf{r}}\,\delta(\mathbf{r} - \mathbf{r}')]. \tag{B.15}$$

In the last term, $\nabla_{\mathbf{r}}$ operates only on the delta function. We can write $\langle \mathbf{r}|\hat{\mathbf{p}}|\mathbf{r}'\rangle$ in a more convenient form. Let us take the matrix element of $\hat{\mathbf{p}}$ with respect to the abstract vectors $|\phi_\alpha\rangle$:

$$\langle \phi_\beta|\hat{\mathbf{p}}|\phi_\alpha\rangle = \iint d\mathbf{r}_1\, d\mathbf{r}_2\langle \phi_\beta|\mathbf{r}_1\rangle\langle \mathbf{r}_1|\hat{\mathbf{p}}|\mathbf{r}_2\rangle\langle \mathbf{r}_2|\phi_\alpha\rangle$$

$$= i\hbar\iint d\mathbf{r}_1\, d\mathbf{r}_2\phi_\beta(\mathbf{r}_1)\phi_\alpha(\mathbf{r}_2)[\nabla_{\mathbf{r}_2}\,\delta(\mathbf{r}_1 - \mathbf{r}_2)]. \tag{B.16}$$

If we now integrate by parts, we find

$$\langle \phi_\beta|\hat{\mathbf{p}}|\phi_\alpha\rangle = -i\hbar\iint d\mathbf{r}_1\, d\mathbf{r}_2\phi_\beta(\mathbf{r}_1)\,\delta(\mathbf{r}_1 - \mathbf{r}_2)\nabla_{\mathbf{r}_2}\phi_\alpha(\mathbf{r}_2). \tag{B.17}$$

From Eq. (B.17), we can also write the matrix element of $\hat{\mathbf{p}}$ as

$$\langle \mathbf{r}|\hat{\mathbf{p}}|\mathbf{r}'\rangle = -i\hbar\, \delta(\mathbf{r} - \mathbf{r}')\nabla_{\mathbf{r}'} \tag{B.18}$$

where the gradient $\nabla_{\mathbf{r}'}$ operates on everything to its right. By a similar argument, we obtain for the matrix element $\langle \mathbf{k}|\hat{\mathbf{q}}|\mathbf{k}'\rangle$

$$\langle \mathbf{k}|\hat{\mathbf{q}}|\mathbf{k}'\rangle = i\, \delta(\mathbf{k} - \mathbf{k}')\nabla_{\mathbf{k}'}. \tag{B.19}$$

We now can write the Schrödinger equation for a many-body system.

2. *N*-body Schrödinger Equation: General Form

If we denote the wave vector of an N-body interacting system by $|\Psi^N(t)\rangle$, its time evolution is given by the N-body Schrödinger equation:

$$i\hbar \frac{\partial}{\partial t} |\Psi^N(t)\rangle = \hat{H}^N |\Psi^N(t)\rangle \tag{B.20}$$

where $\hat{H}^N$ is the N-body Hamiltonian of the system. Knowledge of $|\Psi^N(t)\rangle$ enables us to obtain expectation values of various observables (such as energy, momentum, etc.) by taking matrix elements of the appropriate operators with respect to the state $|\Psi^N(t)\rangle$.

In this book, we often consider systems containing N identical particles with spin, confined to a finite or infinite volume. We can then decompose the total Hamiltonian $\hat{H}^N$ into kinetic energy $\hat{T}^N$ and potential energy $\hat{V}^N$:

$$\hat{H}^N = \hat{T}^N + \hat{V}^N \tag{B.21}$$

where

$$\hat{T}^N = \sum_{i=1}^{N} \frac{\hbar^2 |\hat{\mathbf{k}}_i|^2}{2m} \tag{B.22}$$

and

$$\hat{V}^N = \sum_{i<j=1}^{(1/2)N(N-1)} V(\hat{\mathbf{q}}_i, \hat{\mathbf{q}}_j, \hat{\boldsymbol{\sigma}}_i, \hat{\boldsymbol{\sigma}}_j) \tag{B.23}$$

(m is the mass of the particles). For fermions with spin $\frac{1}{2}$, the potential will, in general, depend on the Pauli spin matrices $\hat{\boldsymbol{\sigma}}_i$ and $\hat{\boldsymbol{\sigma}}_j$ for particles i and j. For spin zero bosons there will be no dependence on spin matrices.

For particles with spin we must specify position or momentum and also z-component of spin. We will let $|r\rangle$ denote both position and spin components,

$$|r\rangle = |\mathbf{r}, \sigma_z\rangle, \tag{B.24}$$

and let k denote both the momentum and the spin component,

$$|k\rangle = |\mathbf{k}, \sigma_z\rangle. \tag{B.25}$$

The position eigenstates of the N-body system may be written as the direct product of the position eigenstates of the constituent particles:

$$|r_a, r_b, \ldots, r_l\rangle \equiv |r_a\rangle_1 |r_b\rangle_2 \times \cdots \times |r_l\rangle_N. \tag{B.26}$$

On the left-hand side of Eq. (B.26) we use the convention that the positions of the particles labeled from 1 to N are ordered from left to right from 1 to N. Thus, particle 1 has position r_a, particle 2 has position $r_b, \ldots$ particle N has position r_l. On the right-hand side of Eq. (B.26), the ket$|r_a\rangle_i$ is a position eigenstate of particle i. (When we say "position" we include spin if the particle has spin.) A momentum (and spin) eigenstate of the N-body system is written

$$|k_a, k_b, \ldots, k_l\rangle \equiv |k_a\rangle_1 |k_b\rangle_2 \times \cdots \times |k_l\rangle_N \tag{B.27}$$

where particle 1 has momentum k_a, particle 2 has momentum k_b, etc.

In the position representation, the Hamiltonian takes the form

$$\begin{aligned} H(r_1, \ldots, r_N) &\equiv \langle r_1, \ldots, r_N | \hat{H}^N | r_1, \ldots, r_N \rangle \\ &= \sum_{i=1}^{N} \frac{-\hbar^2 \nabla_{\mathbf{r}_i}^2}{2m} + \sum_{i<j=1}^{(1/2)N(N-1)} V(\mathbf{r}_i, \mathbf{r}_j, \boldsymbol{\sigma}_i, \boldsymbol{\sigma}_j) \end{aligned} \tag{B.28}$$

and the Schrödinger equation takes the form

$$\begin{aligned} i\hbar \frac{\partial}{\partial t} \langle r_1, \ldots, r_N | \Psi^N(t) \rangle = \int \cdots \int dr_1' \cdots dr_N' \langle r_1 \cdots r_N | \hat{H}^N | r_1' \cdots r_N' \rangle \\ \times \langle r_1' \cdots r_N' | \Psi^N(t) \rangle \end{aligned} \tag{B.29}$$

where

$$\int dr \equiv \int d\mathbf{r} \sum_{\sigma_z}. \tag{B.30}$$

From Eqs. (B.3) and (B.18) we may write

$$i\hbar \frac{\partial}{\partial t} \Psi^N(r_1, \ldots, r_N, t) = H^N(r_1, \ldots, r_N) \Psi^N(r_1, \ldots, r_N, t) \tag{B.31}$$

where $\Psi^N(r_1, \ldots, r_N, t) \equiv \langle r_1, \ldots, r_N | \Psi^N(t) \rangle$.

In the momentum representation, we would write

$$i\hbar \frac{\partial}{\partial t} \Psi^N(k_1, \ldots, k_N; t) = H^N(k_1, \ldots, k_N) \Psi^N(k_1, \ldots, k_N; t) \tag{B.32}$$

where

$$H^N(k_1, \ldots, k_N) \equiv \sum_{i=1}^{N} \frac{-\hbar^2 |\hat{\mathbf{k}}_i|^2}{2m} + \sum_{l<m=1}^{(1/2)N(N-1)} V(i\hbar \nabla_{\mathbf{k}_l}, i\hbar \nabla_{\mathbf{k}_m}, \boldsymbol{\sigma}_l, \boldsymbol{\sigma}_m) \tag{B.33}$$

and

$$\Psi^N(k_1, \ldots, k_N; t) \equiv \langle k_1, \ldots, k_N | \Psi^N(t) \rangle.$$

3. Noninteracting Particles

(*a*) *States for Noninteracting Distinguishable Particles.* The Schrödinger equation for a system of N noninteracting particles takes the form

$$i\hbar \frac{\partial}{\partial t} |\Psi_0{}^N(t)\rangle = \hat{T}^N |\Psi_0{}^N(t)\rangle \tag{B.34}$$

where

$$\hat{T}^N = \sum_{i=1}^{N} \hat{T}_i. \tag{B.35}$$

$\hat{T}_i$ is the kinetic energy operator for the ith particle. Because $\hat{T}^N$ can be written as a sum of one-body operators, the state $|\Psi_0{}^N(t)\rangle$ separates into a direct product of states for each particle:

$$|\Psi_0{}^N(t)\rangle = |\psi(t)\rangle_1 |\psi(t)\rangle_2 \times \cdots \times |\psi(t)\rangle_N. \tag{B.36}$$

The single-particle states $|\psi(t)\rangle_j$ obey the equation

$$i\hbar \frac{\partial}{\partial t} |\psi(t)\rangle_j = \hat{T}_j |\psi(t)\rangle_j, \tag{B.37}$$

which has the solution

$$|\psi(t)\rangle_j = e^{-i\hat{T}_j t/\hbar} |\psi\rangle_j \tag{B.38}$$

(we have let $|\psi(0)\rangle_j \equiv |\psi\rangle_j$). If we now expand $|\psi\rangle_j$ in momentum eigenstates $|k\rangle_j$, we obtain

$$|\psi\rangle_j = \sum_k {}_j\langle k|\psi\rangle_j |k\rangle_j = \sum_k \alpha_k{}^j |k\rangle_j \tag{B.39}$$

where $\alpha_k{}^j = {}_j\langle k|\psi\rangle_j$ and $\sum_k = \sum_{\sigma_z} \sum_{\mathbf{k}}$ or $\sum_{\sigma_z} \int d\mathbf{k}$ (if $\hat{H}_0{}^N$ were not kinetic energy then we would expand $|\psi\rangle_j$ in the appropriate eigenstates of $\hat{H}_0{}^N$). Substitution of Eq. (B.39) into Eq. (B.38) gives

$$|\psi(t)\rangle_j = \sum_k \alpha_k{}^j \, e^{-(i\hbar|\mathbf{k}|^2/2m)t} |k\rangle_j. \tag{B.40}$$

The N-body state $|\Psi_0{}^N(t)\rangle$ can now be written

$$|\Psi_0{}^N(t)\rangle = \sum_{k_1,\ldots,k_N} A_{k_1,\ldots,k_N} \exp\left\{-\frac{i\hbar t}{2m} \sum_{j=1}^{N} |\mathbf{k}_j|^2\right\} |k_1, k_2, \ldots, k_N\rangle \tag{B.41}$$

where $|k_1, \ldots, k_N\rangle$ is a direct product of single-particle momentum states and the convention for ordering the states is the same as in Eq. (B.27).

Eq. (B.41) would be the correct form for $|\Psi_0(t)\rangle$ if we were dealing with distinguishable particles. However, we are often interested in identical particles at low temperature where the effect of the statistics of the particles plays a major role.

(*b*) *States for Noninteracting Bose-Einstein Particles.*[3] By definition, any wave function describing the state of a system containing N identical Bose-Einstein particles must be symmetric under interchange of coordinates of any two particles. Furthermore, any number of bosons can have the same quantum numbers.

Let us first introduce a permutation operator P_{ij} such that P_{ij} interchanges the momenta of particles i and j:

$$P_{ij}|k_1, k_2, \ldots, k_i, k_j, \ldots, k_N\rangle = |k_1, k_2, \ldots, k_j, k_i, \ldots, k_N\rangle. \quad \text{(B.42)}$$

We shall use the following notation for the sum of all possible permutations:

$$\sum_P P|k_1, \ldots, k_N\rangle \equiv \sum (\text{all } N! \text{ permutations of momenta in } |k_1, k_2, \ldots, k_N\rangle). \quad \text{(B.43)}$$

For example,

$$\sum_P P|k_1, k_2, k_3\rangle = \{|k_1, k_2, k_3\rangle + |k_2, k_1, k_3\rangle + |k_1, k_3, k_2\rangle + |k_3, k_2, k_1\rangle + |k_3, k_1, k_2\rangle + |k_2, k_3, k_1\rangle\}. \quad \text{(B.44)}$$

Note that in $|k_2, k_3, k_1\rangle$, particle 1 has momentum k_2, particle 2 has momentum k_3, and particle 3 has momentum k_1.

If there are n_1 particles with momentum k_1, n_2 particles with momentum k_2, n_3 particles with momentum k_3, etc., then $\sum_P P|k_1, \ldots, k_N\rangle$ contains $N!$ terms but only $N!/\prod_{\alpha=1}^{\infty} n_\alpha!$ of them are different. In $\sum_P P|k_1, \ldots, k_N\rangle$ each term appears $\prod_{\alpha=1}^{\infty} n_\alpha!$ times. Hence, $(\prod_{\alpha=1}^{\infty} n_\alpha!)^{-1} \sum_P P|k_1, \ldots, k_N\rangle$ has precisely $N!(\prod_{\alpha=1}^{\infty} n_\alpha!)^{-1}$ different terms. If we now use the orthonormality of momentum states,

$$\langle k_a, k_b, \ldots, k_l|k'_a, k'_b, \ldots, k'_l\rangle = \delta_{k_a k'_a}\, \delta_{k_b k'_b} \times \cdots \times \delta_{k_l, k'_l}, \quad \text{(B.45)}$$

we easily see that

$$|k_1, k_2, \ldots, k_N\rangle^{(S)} = \frac{1}{\sqrt{N! \prod_{\alpha=0}^{\infty} n_\alpha!}} \sum_P P|k_1, \ldots, k_N\rangle \quad \text{(B.46)}$$

is a symmetrized orthonormal N-body momentum eigenstate. It is useful to write Eq. (B.46) as follows:

$$\begin{aligned} |k_1, k_2, \ldots, k_N\rangle^{(S)} &= \frac{1}{\sqrt{N! \prod_{\alpha=0}^{\infty} n_\alpha!}} \left[\sum (\text{all } N! \text{ permutations of } |k_1, \ldots, k_N\rangle\right] \\ &= \sqrt{\frac{\prod_{\alpha=1}^{\infty} n_\alpha!}{N!}} \left[\sum (\text{all \textit{different} permutations of } |k_1, \ldots, k_N\rangle\right]. \end{aligned} \quad \text{(B.47)}$$

The symmetrized eigenstates have the normalization

$$^{(S)}\langle k_1, \ldots, k_N | k_1, \ldots, k_N\rangle^{(S)} = 1 \tag{B.48}$$

and satisfy the completeness relation

$$1^s = \left(\frac{1}{N!}\right) \sum_{k_1 \ldots k_N} \prod_{\alpha=0}^{\infty} n_\alpha |k_1, k_2, \ldots, k_N\rangle^{(S)} \, {}^{(S)}\langle k_1, k_2, \ldots, k_N|. \tag{B.49}$$

The factor of $(N!|\prod_\alpha n_\alpha!)^{-1}$ in Eq. (B.49) comes from the fact that the summation counts a given state $N!/\prod_\alpha n_\alpha!$ times.

We can now obtain the following symmetrized N-body wave vector:

$$|\Psi_0{}^N(t)\rangle^{(S)} = \sum_{k_1, \ldots, k_N} A_{k_1, \ldots, k_N} \exp\left\{\frac{-i\hbar t}{2m} \sum_{j=1}^{N} |\mathbf{k}_j|^2\right\} |k_1, k_2, \ldots, k_N\rangle^{(S)}. \tag{B.50}$$

In the coordinate representation we obtain

$$\begin{aligned} \Psi_0{}^N(r_1, \ldots, r_N, t)^{(S)} &\equiv \langle r_1, \ldots, r_N | \Psi_0{}^N(t)\rangle^{(S)} \\ &= \sum_{k_1 \ldots k_N} A_{k_1, \ldots, k_N} \exp\left\{\frac{-i\hbar t}{2m} \sum_{j=1}^{N} |\mathbf{k}_j{}^2|\right\} \\ &\quad \times \langle r_1, \ldots, r_N | k_1, \ldots, k_N\rangle^{(S)}. \end{aligned} \tag{B.51}$$

Note that since $\langle r_1, \ldots, r_N | k_1, \ldots, k_N\rangle^{(S)} = {}^{(S)}\langle r_1, \ldots, r_N | k_1, \ldots, k_N\rangle$, the wave vector $\Psi_0{}^N(r_1, \ldots, r_N, t)^{(S)}$ is symmetric under interchange of position variables r_i and r_j for any two particles.

(*c*) *States for Noninteracting Fermi-Dirac Particles.* Any wave function describing the state of a system containing N identical Fermi-Dirac particles must be antisymmetric under interchange of coordinates of any two particles. Because of the Pauli exclusion principle, no more than one fermion can have a given set of quantum numbers.

An antisymmetrized momentum eigenstate may be written

$$|k_1, \ldots, k_N\rangle^{(A)} \equiv \frac{1}{\sqrt{N!}} \sum_P (-1)^P P |k_1, \ldots, k_N\rangle, \tag{B.52}$$

where $(-1)^P = +1$ for an even number of permutations and -1 for an odd number of permutations. For example,

$$\begin{aligned} \sum_P (-1)^P P |k_1, k_2, k_3\rangle = [&|k_1, k_2, k_3\rangle - |k_2, k_1, k_3\rangle - |k_1, k_3, k_2\rangle \\ &- |k_3, k_2, k_1\rangle + |k_3, k_1, k_2\rangle + |k_2, k_3, k_1\rangle]. \end{aligned} \tag{B.53}$$

The antisymmetrized states have the normalization

$$^{(A)}\langle k_1, k_2, \ldots, k_N | k_1, k_2, \ldots, k_N\rangle^{(A)} = 1 \tag{B.54}$$

and obey a completeness relation,

$$1^{(A)} = \frac{1}{N!} \sum_{k_1 \ldots k_N} |k_1, \ldots, k_N\rangle^{(A)} \, {}^{(A)}\langle k_1, \ldots, k_N|. \tag{B.55}$$

Note that if any two particles have the same momentum and spin, $|k_1, \ldots, k_N\rangle^{(A)} \equiv 0$.

We can write a general antisymmetrized N-body wave vector as

$$|\Psi_0^N(t)\rangle^{(A)} = \sum_{k_1, \ldots, k_N} A_{k_1, \ldots, k_N} \exp\left\{\frac{-i\hbar t}{2m} \sum_{j=1}^{N} |\mathbf{k}_j|^2\right\} |k_1, \ldots, k_N\rangle^{(A)}. \tag{B.56}$$

In the position representation, Eq. (B.56) becomes

$$\begin{aligned} \Psi_0^N(r_1, \ldots, r_N; t)^{(A)} &\equiv \langle r_1, \ldots, r_N|\Psi_0^N(t)\rangle^{(A)} \\ &= \sum_{k_1, \ldots, k_N} A_{k_1, \ldots, k_N} \exp\left\{\frac{-i\hbar t}{2m} \sum_{j=1}^{N} |\mathbf{k}_j|^2\right\} \\ &\quad \times \langle r_1, \ldots, r_N|k_1, \ldots, k_N\rangle^{(A)}. \end{aligned} \tag{B.57}$$

Since $\langle r_1, \ldots, r_N|k_1, \ldots, k_N\rangle^{(A)} = {}^{(A)}\langle r_1, \ldots, r_N|k_1, \ldots, k_N\rangle$, the wave function $\Psi_0^N(r_1, \ldots, r_N; t)^{(A)}$ will be antisymmetric under interchange of coordinates for any two particles. The wave function $\langle r_1, \ldots, r_N|k_1, \ldots, k_N\rangle^{(A)}$ can be written in the form of a determinant (known as the Slater determinant):

$$\langle r_1, \ldots, r_N|k_1, \ldots, k_N\rangle^{(A)} = \frac{1}{\sqrt{N!}} \begin{vmatrix} \langle r_1|k_1\rangle & \langle r_2|k_1\rangle & \cdots & \langle r_N|k_1\rangle \\ \langle r_1|k_2\rangle & & \cdots & \vdots \\ \vdots & & & \vdots \\ \langle r_1|k_N\rangle & & \cdots & \langle r_N|k_N\rangle \end{vmatrix}. \tag{B.58}$$

(*Note:* when using symmetrized or antisymmetrized wave functions to take matrix elements of operators, the operators must themselves be symmetric under interchange of particle indices.)

The wave functions $|k_1, \ldots, k_N\rangle$ tell us which momentum state each particle is in. However, the states $|k_1, \ldots, k_N\rangle^{(S)(A)}$ contain no such information because they contain permutations of particles among all momentum states listed in the ket. Also, they can be clumsy to work with. Therefore, it is convenient to change representations from one which tells the momentum state of each particle (the momentum representation) to one which tells us the number of particles in each momentum state (the number representation).

4. Number Representation for Bosons

The basis states in the number representation are written $|n_1, n_2, \cdots, n_\infty\rangle$, where n_α is the number of particles in the state of momentum and spin k_α.

Operators in the number representation are written $\hat{a}_k{}^+$ and $\hat{a}_k$ and are called creation and annihilation operators, respectively. They are Hermitian conjugates of one another.

The states and operators of the number representation are related to the N-particle states $|k_1, \ldots, k_m\rangle^{(S)}$ as follows:

$$|k_1, \ldots, k_m\rangle^{(S)} = |0, \ldots, n_1, \ldots, n_m, 0 \to 0\rangle$$
$$\equiv \frac{1}{\sqrt{\prod_{\alpha=1}^{\infty} n_\alpha!}} (\hat{a}_{k_1}^+)^{n_1}(\hat{a}_{k_2}^+)^{n_2} \times \cdots \times (\hat{a}_{k_m}^+)^{n_m}|0\rangle \quad \text{(B.59)}$$

where $\sum_{i=1}^{m} n_i = N$. We have labeled the momenta in $|k_1, \ldots, k_m\rangle^{(S)}$ from 1 to m. However, it is understood that there are N positions in the ket. Several particles may have the same momentum. The notation $0 \to 0$ means that all occupation numbers after n_m are zero. Note that zeros may be interspersed between occupation numbers $n_1, \ldots, n_m$. If we interchange momenta of two particles, the wave functions should not change:

$$|k_1, k_2, \ldots, k_m\rangle^{(S)} = |k_2, k_1, \ldots, k_m\rangle^{(S)}$$
$$= \frac{1}{\sqrt{\prod_{\alpha}^{\infty} n_\alpha!}} (\hat{a}_{k_1}^+)^{n_1}(\hat{a}_{k_2}^+)^{n_2} \times \cdots \times (\hat{a}_{k_m}^+)^{n_m}|0\rangle$$
$$= \frac{1}{\sqrt{\prod_{\alpha}^{\infty} n_\alpha!}} (\hat{a}_{k_1}^+)^{n_1-1}\hat{a}_{k_2}^+\hat{a}_{k_1}^+(\hat{a}_{k_1}^+)^{n_2-1} \times \cdots \times (\hat{a}_{k_m}^+)^{n_m}|0\rangle. \quad \text{(B.60)}$$

Thus, we find that symmetrization of the wave functions gives

$$\hat{a}_{k_1}^+\hat{a}_{k_2}^+ = \hat{a}_{k_2}^+\hat{a}_{k_1}^+ \quad \text{(B.61)}$$

and, therefore, creation operators commute. If we take the Hermitian conjugate of Eq. (B.60), we find

$${}^{(S)}\langle k_1, k_2, \ldots, k_m| = {}^{(S)}\langle k_2, k_1, \ldots, k_m|$$
$$= \frac{1}{\sqrt{\prod_{\alpha}^{\infty} n_\alpha!}} \langle 0|(\hat{a}_{k_m})^{n_m} \times \cdots \times (\hat{a}_{k_2})^{n_2}(\hat{a}_{k_1})^{n_1}$$
$$= \frac{1}{\sqrt{\prod_{\alpha}^{\infty} n_\alpha!}} \langle 0|(\hat{a}_{k_m})^{n_m} \times \cdots \times (\hat{a}_{k_2})^{n_2-1}\hat{a}_{k_1}\hat{a}_{k_2}(\hat{a}_{k_1})^{n_1-1}. \quad \text{(B.62)}$$

Thus, annihilation operators also commute:

$$\hat{a}_{k_1}\hat{a}_{k_2} = \hat{a}_{k_2}\hat{a}_{k_1}. \quad \text{(B.63)}$$

The wave functions must satisfy the following condition:

$$
\begin{aligned}
&{}^{(S)}\langle k_1, k_2, \ldots, k_m | k'_1, k'_2, \ldots, k'_m\rangle^{(S)} \\
&\quad \equiv \frac{1}{\prod_{\alpha}^{\infty} n_\alpha!} \langle 0|\hat{a}_{k_m} \times \cdots \times \hat{a}_{k_1}\hat{a}^+_{k'_1} \times \cdots \times \hat{a}^+_{k'_m}|0\rangle \\
&\quad = \begin{cases} 0 & \text{if the sets } [k_1, \ldots, k_m] \text{ and } [k'_1, \ldots, k'_m] \text{ are not the same.} \\ 1 & \text{if the sets } [k_1, \ldots, k_m] \text{ and } [k'_1, \ldots, k'_m] \text{ are the same.} \end{cases}
\end{aligned}
\tag{B.64}
$$

Eq. (B.64) will be satisfied if the state $|0\rangle$ and the creation and annihilation operators have the following properties:

$$\langle 0|0\rangle = 1, \tag{B.65}$$

$$\hat{a}_k|0\rangle = 0 \qquad \text{for all } k, \tag{B.66}$$

and

$$[\hat{a}_k, \hat{a}^+_{k'}]_- = \hat{a}_k\hat{a}^+_{k'} - \hat{a}^+_{k'}\hat{a}_k = \delta_{k,k'}. \tag{B.67}$$

One can easily show that Eq. (B.64) follows from Eqs. (B.65)–(B.67) by doing some examples. One simply commutes each annihilation operator to the right using Eqs. (B.65)–(B.67) as they are needed. The result is Eq. (B.64).

Again, by using Eqs. (B.59), (B.66), and (B.67), we can show that

$$\hat{a}_{k_j}|n_1, \ldots, n_j, \ldots, n_\infty\rangle = \sqrt{n_j}|n_1, \ldots, n_j - 1, \ldots, n_\infty\rangle, \tag{B.68}$$

$$\hat{a}^+_{k_j}|n_1, \ldots, n_j, \ldots, n_\infty\rangle = \sqrt{n_j + 1}|n_1, \ldots, n_j + 1, \ldots, n_\infty\rangle, \tag{B.69}$$

and

$$\langle n_1, \ldots, n_j, \ldots, n_\infty|\hat{a}^+_{k_j}\hat{a}_{k_j}|n_1, \ldots, n_j, \ldots, n_\infty\rangle = n_j. \tag{B.70}$$

Eq. (B.70) follows from the normalization of the states:

$$\langle n_1, \ldots, n_j, \ldots, n_\infty|n_1, \ldots, n_j, \ldots, n_\infty\rangle = 1. \tag{B.71}$$

When working with N-body systems, we commonly must evaluate one-body operators of the type

$$\hat{O}^N_{(1)} = \sum_{i=1}^{N} O_i(\hat{\mathbf{q}}_i, \hat{\mathbf{p}}_i, \hat{\boldsymbol{\sigma}}_i). \tag{B.72}$$

(Note that these operators are symmetric under interchange of particle labels.) Some examples are the kinetic energy, the number density, and the particle current. We wish to express these operators in the number representation.

The correspondence between $\hat{O}^N_{(1)}$ and its analog in the number representation is

$$\hat{O}^N_{(1)} \to \hat{O}^{nb}_{(1)} = \sum_{k_a k_b} \langle k_a|\hat{O}_1(\hat{\mathbf{q}}_1, \hat{\mathbf{p}}_1, \hat{\boldsymbol{\sigma}}_1)|k_b\rangle \hat{a}^+_{k_a}\hat{a}_{k_b}. \tag{B.73}$$

To show that Eq. (B.73) is correct, we must show that

$$
\begin{aligned}
&{}^{(S)}\langle k_1, \ldots, k_m | \hat{O}^N_{(1)} | k'_1, \ldots, k'_m \rangle^{(S)} \\
&\qquad = \langle 0, \ldots, n_1, \ldots, n_m, 0 \to 0 | \hat{O}^{nb}_{(1)} | 0, \ldots, n'_1, \ldots, n'_m, 0 \to 0 \rangle.
\end{aligned} \tag{B.74}
$$

The proof of Eq. (B.74) is given in Sec. 7 of this appendix.

In going from the momentum representation to the number representation, we thus make the correspondence

$$|k_1, k_2, \ldots, k_m\rangle^{(S)} \to |0, \ldots, n_1, n_2, \ldots, n_m, 0 \to 0\rangle \tag{B.75}$$

and

$$\sum_{i=1}^{N} O_i(\hat{\mathbf{q}}_i, \hat{\mathbf{p}}_i, \hat{\boldsymbol{\sigma}}_i) \to \sum_{k_a, k_b} \langle k_a | O_1(\hat{\mathbf{q}}_1, \hat{\mathbf{p}}_1, \hat{\boldsymbol{\sigma}}_1) | k_b \rangle \hat{a}^+_{k_a} \hat{a}_{k_b}. \tag{B.76}$$

In practice, the only other type of N-body operators we encounter are two-body operators of the form

$$\hat{O}^N_{(2)} = \sum_{i<j}^{(1/2)N(N-1)} O_{ij}(\hat{\mathbf{q}}_i, \hat{\mathbf{q}}_j, \hat{\mathbf{p}}_i, \hat{\mathbf{p}}_j, \hat{\boldsymbol{\sigma}}_i, \hat{\boldsymbol{\sigma}}_j). \tag{B.77}$$

Some examples are the potential energy and the momentum currents (stress tensor). By a proof completely analogous to that given in App. A we can show that

$$
\begin{aligned}
\sum_{i<j}^{(1/2)N(N-1)} O_{ij}(\hat{\mathbf{q}}_i\hat{\mathbf{q}}_j\hat{\mathbf{p}}_i\hat{\mathbf{p}}_j\hat{\boldsymbol{\sigma}}_i\hat{\boldsymbol{\sigma}}_j) \to \frac{1}{2} \sum_{k_a k_b k'_a k'_b} &\langle k_a k_b | O_{12}(\hat{\mathbf{q}}_1\hat{\mathbf{q}}_2\hat{\mathbf{p}}_1\hat{\mathbf{p}}_2\hat{\boldsymbol{\sigma}}_1\hat{\boldsymbol{\sigma}}_2) | k'_a k'_b \rangle \\
&\times \hat{a}^+_{k_a}\hat{a}^+_{k_b}\hat{a}_{k'_b}\hat{a}_{k'_a}
\end{aligned} \tag{B.78}
$$

when going from the coordinate to number representation.

5. Number Representation for Fermions [4]

The basis states for fermions are also written $|n_1, n_2, \ldots, n_\infty\rangle$, but because of the Pauli exclusion principle the occupation numbers take on values $n_\alpha = 0$ or 1 only. We also introduce creation and annihilation operators $\hat{a}_k{}^+$ and $\hat{a}_k$, respectively. But they obey different commutation relations.

The correspondence between states in the number representation and antisymmetrized momentum states is

$$
\begin{aligned}
|0, \ldots, 1_1, 1_2, \ldots, 1_N, 0 \to 0) &\equiv \hat{a}^+_{k_1}\hat{a}^+_{k_2} \times \cdots \times \hat{a}^+_{k_N}|0\rangle \\
&= |k_1, k_2, \ldots, k_N\rangle^{(A)},
\end{aligned} \tag{B.79}
$$

where we have used the convention $n_\alpha = 1_\alpha$ when the state k_α is filled and zero otherwise. In general, there will be zeros interspersed between the occupation numbers $1_1, 1_2, \ldots, 1_N$ on the right-hand side of Eq. (B.79). If we interchange momenta of two particles, the states must change sign:

$$\begin{aligned}|k_1, k_2, \ldots, k_N\rangle^{(A)} &= -|k_2, k_1, \ldots, k_N\rangle^{(A)} \\ &= \hat{a}^+_{k_1}\hat{a}^+_{k_2} \times \cdots \times \hat{a}^+_{k_N}|0\rangle = -\hat{a}^+_{k_2}\hat{a}^+_{k_1} \times \cdots \times \hat{a}^+_{k_N}|0\rangle.\end{aligned} \tag{B.80}$$

Thus, the creation and annihilation operators must obey the relation

$$\hat{a}^+_{k_1}\hat{a}^+_{k_2} = -\hat{a}^+_{k_2}\hat{a}^+_{k_1}. \tag{B.81}$$

The creation operators anticommute. If we take the Hermitian adjoint of Eq. (B.80), we find

$$\hat{a}_{k_2}\hat{a}_{k_1} = -\hat{a}_{k_1}\hat{a}_{k_2}. \tag{B.82}$$

The annihilation operators also anticommute.

The wave functions must satisfy the condition

$$\begin{aligned}&{}^{(A)}\langle k_1, k_2, \ldots, k_N|k'_1, k'_2, \ldots, k'_N\rangle^{(A)} \\ &\equiv \langle 0|\hat{a}_{k_N} \times \cdots \times \hat{a}_{k_1}\hat{a}^+_{k'_1} \times \cdots \times \hat{a}^+_{k'_N}|0\rangle \\ &= \begin{cases} 0 & \text{if the sets } \{k_i\} \text{ and } \{k'_i\} \text{ differ,} \\ +1 & \text{if the sets } \{k_i\} \text{ and } \{k'_i\} \text{ are the same but differ by} \\ & \text{an even permutation,} \\ -1 & \text{if the sets } \{k_i\} \text{ and } \{k'_i\} \text{ are the same but differ by} \\ & \text{an odd permutation.}\end{cases}\end{aligned} \tag{B.83}$$

Eq. (B.83) will be satisfied if

$$\langle 0|0\rangle = 1, \tag{B.84}$$

$$\hat{a}_k|0\rangle = 0 \qquad \text{for all } k, \tag{B.85}$$

and

$$[\hat{a}_k, \hat{a}_{k'}{}^+]_+ \equiv \hat{a}_k\hat{a}^+_{k'} + \hat{a}^+_{k'}\hat{a}_k = \delta_{k,k'}. \tag{B.86}$$

To show Eq. (B.83) for the case of identical states, we use Eq. (B.81) to permute operators until

$$\hat{a}^+_{k'_1} \times \cdots \times \hat{a}^+_{k'_N} = \pm\hat{a}^+_{k_1} \times \cdots \times \hat{a}^+_{k_N}. \tag{B.87}$$

There will be a plus sign if an even number of permutations are needed and a minus sign if an odd number are needed. One can then show that

$$\langle 0|\hat{a}_{k_N} \times \cdots \times \hat{a}_{k_1}\hat{a}^+_{k_1} \times \cdots \times a^+_{k_N}|0\rangle = 1 \tag{B.88}$$

by permuting each annihilation operator all the way to the right using Eqs. (B.85) and (B.86).

For Fermi systems there can be an ambiguity in the overall sign of the states. Therefore, *one uses the convention that in the states $|n_1, n_2, \ldots, n_\infty\rangle$ the occupa-*

tion numbers are ordered from left to right in order of increasing energy. We then find that

$$\hat{a}_{k_j}|n_1, n_2, \ldots, n_j, \ldots, n_\infty\rangle = (-1)^{\sum_{l=1}^{j-1} n_l}\sqrt{n_j}|n_1, n_2, \ldots, (1 - n_j), \ldots, n_\infty\rangle, \quad \text{(B.89)}$$

$$\hat{a}^+_{k_j}|n_1, n_2, \ldots, n_j, \ldots, n_\infty\rangle = (-1)^{\sum_{l=1}^{j-1} n_l}\sqrt{1 - n_j}|n_1, n_2, \ldots, (1 + n_j), \ldots, n_\infty\rangle, \quad \text{(B.90)}$$

and

$$\langle n_1, n_2, \ldots, n_j, \ldots, n_\infty|\hat{a}^+_{k_j}\hat{a}_{k_j}|n_1, n_2, \ldots, n_j, \ldots, n_\infty\rangle = n_j, \quad \text{(B.91)}$$

where we have used Eqs. (B.81), (B.82), and (B.84)–(B.86). In Eqs. (B.89)–(B.91), the occupation numbers may equal 0 or 1.

For systems of identical fermions, the same correspondence holds between operators in the coordinate representation and number representation (cf. Eqs. (B.76) and (B.77)) as was found to hold for systems of identical bosons. However, for fermions the creation and annihilation operators anticommute and for bosons they commute.

To establish this correspondence, we must show that

$$\begin{aligned}&{}^{(A)}\langle k_1, \ldots, k_N|\hat{O}^N_{(1)}|k'_1, \ldots, k'_N\rangle^{(A)} \\ &\quad = \langle 0, \ldots, 1_1, \ldots, 1_N, 0 \to 0|\hat{O}^{nb}_{(1)}|0, \ldots, 1_1, \ldots, 1_N, 0 \to 0\rangle.\end{aligned} \quad \text{(B.92)}$$

In Eq. (B.92) we have used the ordering convention discussed above. We will assume that the occupation numbers and momenta are labeled in order of increasing energy, and that the states $k_1, \ldots, k_N$ are all filled. The proof of Eq. (B.92) is given in Sec. 8 of this appendix.

A similar correspondence holds for the operator $O^N_{(2)}$. However, we must make a brief comment about sign conventions. The correspondence is

$$\begin{aligned}\hat{O}^N_{(2)} &= \sum_{i<j}^{(1/2)N(N-1)} O_{ij}(\hat{\mathbf{q}}_i\hat{\mathbf{q}}_j\hat{\mathbf{p}}_i\hat{\mathbf{p}}_j\hat{\boldsymbol{\sigma}}_i\hat{\boldsymbol{\sigma}}_j) \\ &\to \frac{1}{2}\sum_{k_ak_bk'_ak'_b} \langle k_a, k_b|O_{12}(\hat{\mathbf{q}}_1\hat{\mathbf{q}}_2\hat{\mathbf{p}}_1\hat{\mathbf{p}}_2\hat{\boldsymbol{\sigma}}_1\hat{\boldsymbol{\sigma}}_2)|k'_a, k'_b\rangle \hat{a}^+_{k_a}\hat{a}^+_{k_b}\hat{a}_{k'_b}\hat{a}_{k'_a}.\end{aligned} \quad \text{(B.93)}$$

Note that the order of the operators $\hat{a}_{k'_a}$ and $\hat{a}_{k'_b}$ is opposite to the order of k'_a and k'_b in the ket $|k'_ak'_b\rangle$. The reason is most easily seen by evaluating a matrix element of the operator on the right-hand side of Eq. (B.93) with respect to a state containing only two particles. We find

$$\begin{aligned}&\langle 0, \ldots, 0, 1_2, 0, \ldots, 1_1, 0, \ldots|\hat{a}^+_{k_a}\hat{a}^+_{k_b}\hat{a}_{k'_b}\hat{a}_{k'_a}|0, \ldots, 0, 1_3, \ldots, 1_4, \ldots, 0\rangle \\ &\quad = \langle 0|\hat{a}_{k_2}\hat{a}_{k_1}\hat{a}^+_{k_4}\hat{a}^+_{k_b}\hat{a}_{k'_b}\hat{a}_{k'_a}\hat{a}^+_{k_3}\hat{a}^+_{k_4}|0\rangle \\ &\quad = +\delta_{1,a}\delta_{2,b}\,\delta_{3,a'}\,\delta_{4,ba'} - \delta_{2,a}\,\delta_{1,b}\,\delta_{4,a'}\,\delta_{3,b'} - \delta_{1,a}\,\delta_{2,b}\,\delta_{4,a'}\,\delta_{3,b'} \\ &\qquad + \delta_{2,a}\,\delta_{1,b}\,\delta_{3,a'}\,\delta_{4,b'}\end{aligned}$$

and, hence,

$$\langle 0, \ldots, 0, 1_2, 0, \ldots, 1_1, 0, \ldots |\hat{O}^{nb}_{(2)}| 0, \ldots, 0, 1_3, \ldots, 1_4, \ldots, 0\rangle = \langle k_1 k_2|\hat{O}_{12}|k_3 k_4\rangle - \langle k_1 k_2|\hat{O}_{12}|k_4 k_3\rangle.$$

Thus, with this convention, the matrix elements whose momenta appear in the same order (of increasing or decreasing energy) on each side have a positive sign.

6. Field Operators

In Secs. B.4 and B.5, we have written general one- and two-body operators in the number representation using momentum and spin as coordinates. We can also write them using position and spin as coordinates.

Let us first consider the single-particle operators in Eq. (B.72). If we insert a complete set of position (and spin) states $|r\rangle$ into the matrix element, we find

$$\begin{aligned}\hat{O}^{nb}_{(1)} &= \sum_{k_a k_b} \langle k_a|O(\hat{\mathbf{q}}, \hat{\mathbf{p}}, \hat{\boldsymbol{\sigma}})|k_b\rangle \hat{a}^{+}_{k_a}\hat{a}_{k_b} \\ &= \iint dr_1\, dr_2 \langle r_1|O(\hat{\mathbf{q}}, \hat{\mathbf{p}}, \hat{\boldsymbol{\sigma}})|r_2\rangle \hat{\psi}^{+}(r_1)\hat{\psi}(r_2)\end{aligned} \tag{B.94}$$

where

$$\hat{\psi}(r) = \sum_k \langle r|k\rangle \hat{a}_k \tag{B.95}$$

and

$$\hat{\psi}^{+}(r) = \sum_k \langle k|r\rangle \hat{a}_k{}^{+}. \tag{B.96}$$

Similarly, for two-particle operators we find

$$\begin{aligned}\hat{O}^{nb}_{(2)} &= \frac{1}{2}\sum_{k_a k_b k_c k_d} \langle k_a, k_b|O(\hat{\mathbf{q}}_1, \hat{\mathbf{q}}_2, \hat{\mathbf{p}}_1, \hat{\mathbf{p}}_2, \hat{\boldsymbol{\sigma}}_1, \hat{\boldsymbol{\sigma}}_2)|k_c, k_d\rangle \hat{a}^{+}_{k_a}\hat{a}^{+}_{k_b}\hat{a}_{k_d}\hat{a}_{k_c} \\ &= \frac{1}{2}\int\cdots\int dr_1\cdots dr_4 \\ &\quad\times \langle r_1 r_2|O(\hat{\mathbf{q}}_1, \hat{\mathbf{q}}_2, \hat{\mathbf{p}}_1, \hat{\mathbf{p}}_2, \hat{\boldsymbol{\sigma}}_1, \hat{\boldsymbol{\sigma}}_2)|r_3 r_4\rangle \hat{\psi}^{+}(r_1)\hat{\psi}^{+}(r_2)\hat{\psi}(r_4)\hat{\psi}(r_3).\end{aligned} \tag{B.97}$$

It is easy to show that the momentum-dependent commutation relations

$$[\hat{a}_{\mathbf{k},\lambda}\hat{a}^{+}_{\mathbf{k}',\lambda'}]_{\pm} = \delta_{\mathbf{k},\mathbf{k}'}\,\delta_{\lambda,\lambda'} \tag{B.98}$$

are equivalent to position-dependent commutation relations

$$[\hat{\psi}_{\lambda}(\mathbf{r}), \hat{\psi}^{+}_{\lambda'}(\mathbf{r}')]_{\pm} = \delta(\mathbf{r} - \mathbf{r}')\,\delta_{\lambda,\lambda'} \tag{B.99}$$

for bosons and fermions (+ for fermions and − for bosons).

It is worthwhile to give a few examples of commonly used operators. The number operator may be written

$$N^N = \sum_{i=1}^{N} 1_i \to \hat{N}_{nb} = \sum_{\lambda} \int d\mathbf{r} \psi_{\lambda}{}^{+}(\mathbf{r}) \psi_{\lambda}(\mathbf{r}) = \sum_{\lambda} \sum_{\mathbf{k}} \hat{a}^{+}_{\mathbf{k},\lambda} \hat{a}_{\mathbf{k},\lambda}. \tag{B.100}$$

The number density operator may be written

$$\hat{n}^N(\hat{\mathbf{q}}_1, \ldots, \hat{\mathbf{q}}_N; \mathbf{R}) = \sum_{i=1}^{N} \delta(\hat{\mathbf{q}}_i - \mathbf{R}) \to \hat{n}_{nb}(\mathbf{R}) = \sum_{\lambda} \hat{\psi}_{\lambda}{}^{+}(\mathbf{R}) \psi_{\lambda}(\mathbf{R})$$

$$= \frac{1}{V} \sum_{\lambda} \sum_{\mathbf{k}_1 \mathbf{k}_2} e^{-i\mathbf{R}\cdot(\mathbf{k}_1 - \mathbf{k}_2)} \hat{a}^{+}_{\mathbf{k}_1,\lambda} \hat{a}_{\mathbf{k}_2,\lambda}. \tag{B.101}$$

The spin density operator is given by

$$\hat{\mathbf{S}}^N(\hat{\mathbf{q}}_1, \ldots, \hat{\mathbf{q}}_N; \hat{\boldsymbol{\sigma}}_1, \ldots, \hat{\boldsymbol{\sigma}}_N; \mathbf{R}) = \sum_{i=1}^{N} \boldsymbol{\sigma}_i \, \delta(\hat{\mathbf{q}}_i - \mathbf{R}) \to \hat{\mathbf{S}}_{nb}(\mathbf{R})$$

$$= \sum_{\lambda_1 \lambda_2} \langle \lambda_1 | \boldsymbol{\sigma} | \lambda_2 \rangle \hat{\psi}^{+}_{\lambda_1}(\mathbf{R}) \psi_{\lambda_2}(\mathbf{R})$$

$$= \frac{1}{V} \sum_{\lambda_1 \lambda_2} \sum_{\mathbf{k}_1, \mathbf{k}_2} \langle \lambda_1 | \boldsymbol{\sigma} | \lambda_2 \rangle \, e^{-i\mathbf{R}\cdot(\mathbf{k}_1 - \mathbf{k}_2)} \hat{a}_{\mathbf{k}_1,\lambda_1} \hat{a}^{+}_{\mathbf{k}_2,\lambda_2}. \tag{B.102}$$

The kinetic energy operator is written

$$T^N = \sum_{i=1}^{N} \frac{\hbar^2 k_i{}^2}{2m} \to \hat{T}_{nb} = \sum_{\lambda} \sum_{\mathbf{k}} \frac{\hbar^2 \mathbf{k}^2}{2m} \hat{a}^{+}_{\mathbf{k}\lambda} \hat{a}_{\mathbf{k},\lambda}$$

$$= \sum_{\lambda} \int d\mathbf{r} \hat{\psi}_{\lambda}{}^{+}(\mathbf{r}) \left(\frac{-\hbar^2 \nabla_{\mathbf{r}}{}^2}{2m} \right) \hat{\psi}_{\lambda}(\mathbf{r}). \tag{B.103}$$

The potential energy operator takes the form

$$V^N = \sum_{i<j=1}^{(1/2)N(N-1)} V_{ij} \to \hat{V}_{nb}$$

$$= \frac{1}{2} \sum_{\lambda_1, \ldots, \lambda_4} \iint d\mathbf{r}_1 \, d\mathbf{r}_2 \langle \mathbf{r}_1, \lambda_1; \mathbf{r}_2, \lambda_2 | V_{12} | \mathbf{r}_1, \lambda_3; \mathbf{r}_2, \lambda_4 \rangle$$

$$\times \hat{\psi}^{+}_{\lambda_1}(\mathbf{r}_1) \hat{\psi}^{+}_{\lambda_2}(\mathbf{r}_2) \hat{\psi}_{\lambda_4}(\mathbf{r}_2) \hat{\psi}_{\lambda_3}(\mathbf{r}_1)$$

$$= \frac{1}{2} \sum_{\substack{\mathbf{k}_1, \ldots, \mathbf{k}_4 \\ \lambda_1, \ldots, \lambda_4}} \langle \mathbf{k}_1, \lambda_1; \mathbf{k}_2, \lambda_2 | V | \mathbf{k}_3, \lambda_3; \mathbf{k}_4, \lambda_4 \rangle \hat{a}^{+}_{\mathbf{k}_1,\lambda_1} \hat{a}^{+}_{\mathbf{k}_2,\lambda_2} \hat{a}_{\mathbf{k}_4,\lambda_4} \hat{a}_{\mathbf{k}_3,\lambda_3}. \tag{B.104}$$

7. Proof of Eq. (B.74)

We wish to show that

$$^{(S)}\langle k_1, \ldots, k_M | O^N_{(1)} | k'_1, \ldots, k'_M \rangle^{(S)}$$

$$= \langle 0, \ldots, n_1, \ldots, n_M, 0 \to 0 | O^{nb}_{(1)} | 0, \ldots, n'_1, \ldots, n'_M, 0 \to 0 \rangle \tag{B.74}$$

for Bose-Einstein particles. Let us first evaluate the left-hand side of Eq. (B.74). We can write

$$\hat{O}^N_{(1)} = \sum_{j=1}^{N} \hat{O}_j = \sum_{i=1}^{N} \sum_{k_a} \sum_{k_b} |k_a\rangle_j\,{}_j\langle k_a|O_j|k_b\rangle_j\,{}_j\langle k_b|$$

$$= \sum_{k_a k_b} \langle k_a|O_1|k_b\rangle \sum_{j=1}^{N} (|k_a\rangle\langle k_b|)_j \tag{B.105}$$

where the operator $(|k_a\rangle\langle k_b|)_j$ only operates on the state of particle j. Let us now consider matrix elements of $(|k_a\rangle\langle k_b|)_j$:

$${}^{(S)}\langle k_1, \ldots, k_M|(|k_a\rangle\langle k_b|)_j|k_1', \ldots, k_M'\rangle^{(S)} = 0 \tag{B.106}$$

if two or more labels differ in the sets $(k_1, \ldots, k_M)$ and $(k_1', \ldots k_M')$. If only one label is different in each of the sets, then we obtain

$${}^{(S)}\langle k_1, \ldots, k_l, \ldots, k_M|(|k_a\rangle\langle k_b|)_1|k_1, \ldots, k_m, \ldots, k_M\rangle^{(S)}$$

$$= \sqrt{\frac{\left(\prod_\alpha^{\infty}{}' n_\alpha!\right)(n_l + 1)!\,(n_m - 1)!}{N!}} \sqrt{\frac{\left(\prod_\alpha^{\infty}{}' n_\alpha!\right)n_l!\,n_m!}{N!}}$$

$$\times \left(\sum \begin{matrix}\text{all different}\\ \text{permutations of}\end{matrix} \langle \underbrace{k_l \cdots k_l}_{n_l+1}, \underbrace{k_m \cdots k_m}_{n_m-1}, k_1, \ldots, k_M|\right)(|k_a\rangle\langle k_b|)_1$$

$$\times \left(\sum \begin{matrix}\text{all different}\\ \text{permutations of}\end{matrix} |\underbrace{k_m \cdots k_m}_{n_m}, \underbrace{k_l \cdots k_l}_{n_l}, k_1, \ldots, k_M\rangle\right). \tag{B.107}$$

(We have chosen $j = 1$. The result is independent of j.) We next note that in the bra

$$\langle \underbrace{k_l \cdots k_l}_{n_l+1}, \underbrace{k_m \cdots k_m}_{n_m-1}, k_1, \ldots, k_M|$$

there are $(N-1)!/(n_l!\,(n_m-1)!\prod_{\alpha=1}^{\sigma'} n_\alpha!)$ different permutations of the variables appearing to the right of the first momentum k_l. Similarly, in the ket

$$|\underbrace{k_m \cdots k_m}_{n_m}, \underbrace{k_l \cdots k_l}_{n_l}, k_1, \ldots, k_M\rangle,$$

there are the same number of different permutations of variables to the right of the first momentum k_m. Thus, in Eq. (B.107) the term

$$\left(\sum \vdots\vdots\vdots \langle\ \ |\right)(|k_a\rangle\langle k_b\rangle)_1\left(\sum \vdots\vdots\vdots\ |\ \ \rangle\right)$$

$$= \frac{(N-1)!}{n_l!\,(n_m-1)!\prod_{\alpha=1}^{\infty}{}' n_\alpha!} \langle k_l|k_a\rangle\langle k_b|k_m\rangle. \tag{B.108}$$

Combining Eqs. (B.107) and (B.108), we find

$$
{}^{(S)}\langle k_1, \ldots, k_l, \ldots, k_M | (|k_a\rangle\langle k_b|)_1 | k_1, \ldots, k_m, \ldots, k_M\rangle^{(S)} = \frac{1}{N}\sqrt{(n_l + 1)n_m}\, \delta_{k_l,k_a}\, \delta_{k_b,k_m} \tag{B.109}
$$

and

$$
\sum_{i=1}^{N} {}^{(S)}\langle k_1, \ldots, k_l, \ldots, k_M | (|k_a\rangle\langle k_b|)_i | k_1, \ldots, k_m, \ldots, k_M\rangle^{(S)} = \sqrt{(n_l + 1)n_m}\, \delta_{k_l,k_a}\, \delta_{k_b,k_m}. \tag{B.110}
$$

If we now combine Eqs. (B.105) and (B.110), we find

$$
{}^{(S)}\langle k_1, \ldots, k_l, \ldots, k_M | \hat{O}_{(1)}^N | k_1, \ldots, k_m, \ldots, k_M\rangle^{(S)} = \langle k_l | \hat{O}_1 | k_m\rangle \sqrt{n_m(n_l + 1)}. \tag{B.111}
$$

On deriving Eq. (B.111), we have used $\prod_{\alpha=1}^{\infty\prime} n_\alpha!$ to denote a product of all occupation numbers except those for states k_l and k_m.

We now have to evaluate the right side of Eq. (B.74). First, we see that

$$
\langle 0, \ldots, n_1, \ldots, n_M, 0 \to 0 | \hat{O}_{(1)}^{nb} | 0, \ldots, n_1', \ldots, n_M', 0 \to 0\rangle \equiv 0 \tag{B.112}
$$

if more than two momentum states in the sets $\{n_i\}$ and $\{n_i'\}$ differ in occupation number. This agrees with Eq. (B.106). If we finally note that

$$
\langle 0, \ldots, n_1, \ldots, n_l + 1, n_m - 1, \ldots, n_M, 0 \to 0 | \hat{O}_{(1)}^{nb} | 0, \ldots, n_1, \ldots, n_l, n_m, \ldots, n_M, 0 \to 0\rangle, = \sqrt{(n_l + 1)n_m}\langle k_l | \hat{O}_1 | k_m\rangle \tag{B.113}
$$

where we have used Eqs. (B.67), (B.68), (B.70), and (B.72), then we have established Eq. (B.77).

8. Proof of Eq. (B.92)

In this section, we prove that

$$
{}^{(A)}\langle k_1, \ldots, k_N | \hat{O}_{(1)}^N | k_1', \ldots, k_N'\rangle^{(A)} = \langle 0, \ldots, 1_1, \ldots, 1_N, 0 \to 0 | \hat{O}_{(1)}^{nb} | 0, \ldots, 1_{1'}, \ldots, 1_{N'}, 0 \to 0\rangle \tag{B.114}
$$

for Fermi-Dirac particles. We first consider the left-hand side of Eq. (B.114). As before, we can write

$$
\hat{O}_{(1)}^N = \sum_{k_a k_b} \langle k_a | \hat{O}_1 | k_b\rangle \sum_{j=1}^{N} (|k_a\rangle\langle k_b|)_j \tag{B.115}
$$

(cf. Eq. (B.105)). We first consider the matrix element

$$^{(A)}\langle k_1, \ldots, k_N|(|k_a\rangle\langle k_b|)_1|k_1', \ldots, k_N'\rangle^{(A)}$$

and note that

$$^{(A)}\langle k_1, \ldots, k_N|(|k_a\rangle\langle k_b|)_1|k_1', \ldots, k_N'\rangle^{(A)} = 0 \tag{B.116}$$

if two or more elements of the sets $\{k\}$ and $\{k'\}$ are different. If only one element in the sets $\{k\}$ and $\{k'\}$ is different, we find

$$\begin{aligned} &^{(A)}\langle k_1, \ldots, k_l, \ldots, k_N|(|k_a\rangle\langle k_b|)_1|k_1, \ldots, k_m', \ldots, k_N\rangle^{(A)} \\ &\quad = \pm\, ^{(A)}\langle k_l, k_1, \ldots, k_N|(|k_a\rangle\langle k_b|)_1|k_m', k_1, \ldots, k_N\rangle^{(A)}. \end{aligned} \tag{B.117}$$

The right-hand side of Eq. (B.117) has a $\pm$ sign if it takes an $\begin{pmatrix}\text{even}\\ \text{odd}\end{pmatrix}$ number of permutations to move k_l and k_m' to the extreme left. It is easy to see that only $(N-1)!$ terms can contribute to Eq. (B.117). These correspond to the $(N-1)!$ permutations of the $(N-1)$ labels $k_1, \ldots, k_N$ that appear to the right of k_l and k_m' on the right-hand side of Eq. (B.117). Thus, we find

$$\begin{aligned} &^{(A)}\langle k_1, \ldots, k_l, \ldots, k_N|(|k_a\rangle\langle k_b|)_1|k_1, \ldots, k_m', \ldots, k_N\rangle^{(A)} \\ &\quad = \pm\frac{1}{N}\langle k_l|k_a\rangle\langle k_b|k_m'\rangle = \pm\frac{1}{N}\delta_{k_l,k_a}\,\delta_{k_b,k_m'} \end{aligned} \tag{B.118}$$

and

$$\begin{aligned} &\sum_{i=1}^{N} {}^{(A)}\langle k_1, \ldots, k_l, \ldots, k_N|(|k_a\rangle\langle k_b|)_1|k_1, \ldots, k_m', \ldots, k_N\rangle^{(A)} \\ &\quad = \pm\,\delta_{k_l,k_a}\,\delta_{k_b,k_m'}. \end{aligned} \tag{B.119}$$

Combining Eqs. (B.119) and (B.115), we obtain

$$^{(A)}\langle k_1, \ldots, k_l, \ldots, k_N|\hat{O}^N_{(1)}|k_1, \ldots, k_m', \ldots, k_N\rangle^{(A)} = \pm\langle k_l|\hat{O}_1|k_m'\rangle. \tag{B.120}$$

(Note that in Eq. (B.118), the overall sign is independent of j because the number of permutations needed to move k_l and k_m' to position j rather than position 1 differs by an even number.)

Let us now consider the right-hand side of Eq. (B.114). We may write

$$\begin{aligned} &\langle 0, \ldots, 1_1, \ldots, 1_N, 0 \to 0|\hat{O}^{nb}_{(1)}|0, \ldots, 1_{1'}, \ldots, 1_{N'}, 0 \to 0\rangle \\ &\quad = \sum_{k_a k_b}\langle k_a|\hat{O}_1|k_b\rangle\langle 0|\hat{a}_{k_N}\cdots\hat{a}_{k_1}\hat{a}^+_{k_a}\hat{a}_{k_b}\hat{a}^+_{k_1'}\cdots\hat{a}^+_{k_N'}|0\rangle. \end{aligned} \tag{B.121}$$

Thus, in Eq. (B.121), if two or more labels in the set $\{k_i\}$ differ from those in $\{k_i'\}$ then one of the set $\{k_i\}$ can be anticommuted through to the right to give zero. Thus, we have agreement with Eq. (B.116). If k_i is not in the set $\{k_i'\}$ then it must equal k_a. Otherwise, $\hat{a}_{k_l}$ could be anticommuted through to give zero. Thus,

$$\langle 0|\hat{a}_{k_N}\cdots\hat{a}_{k_l}\cdots\hat{a}_{k_1}\hat{a}^+_{k_a}\hat{a}_{k_b}\hat{a}^+_{k_1}\cdots\hat{a}^+_{k'_l}\cdots\hat{a}^+_{k_N}|0\rangle = \pm\,\delta_{k_l,k_a}\,\delta_{k_b,k'_l}. \tag{B.122}$$

There is a $\begin{Bmatrix}+\\-\end{Bmatrix}$ sign if it takes an $\begin{Bmatrix}\text{even}\\\text{odd}\end{Bmatrix}$ number of permutations to bring $\hat{a}_{k_l}$ to the left of $\hat{a}^+_{k_a}$ and $\hat{a}^+_{k'_l}$ to the right of $\hat{a}_{k_b}$. This again is in agreement with the signs in Eq. (B.120). Thus,

$$\begin{aligned}\langle 0,\ldots,1_1,\ldots,1_l,\ldots,1_N,0\to 0|\hat{O}^{nb}_{(1)}|0,\ldots,1_1,\ldots,1_{m'},\ldots,1_N,0\to 0\rangle \\ = \pm\langle k_l|\hat{O}_1|k_m\rangle\end{aligned} \tag{B.123}$$

and Eq. (B.114) is established.

C. ISOTROPIC SYSTEMS: CURIE PRINCIPLE[5,6]

For systems near equilibrium, we assume that there is a linear relation between the generalized forces and the generalized currents which result from these forces. The constants of proportionality are the transport coefficients. Symmetry properties reduce the number of independent transport coefficients needed to describe a system. In this appendix we will demonstrate this for the case of an isotropic system. Before we begin our discussion, we will list some useful mathematical properties of tensors.

1. Some Mathematical Properties of Tensors

An arbitrary second-order tensor $\bar{\bar{\mathbf{T}}}$ can be decomposed into three orthogonal components:

$$\bar{\bar{\mathbf{T}}} = \tfrac{1}{3}(\mathrm{Tr}\,\bar{\bar{\mathbf{T}}})\bar{\bar{\mathbf{U}}} + \bar{\bar{\mathbf{T}}}^s + \bar{\bar{\mathbf{T}}}^a \tag{C.1}$$

where $\bar{\bar{\mathbf{U}}}$ is the unit tensor; $\mathrm{Tr}\,\bar{\bar{\mathbf{T}}}$ is a scalar,

$$\mathrm{Tr}\,\bar{\bar{\mathbf{T}}} = T_{11} + T_{22} + T_{33}; \tag{C.2}$$

$\bar{\bar{\mathbf{T}}}^s$ is a symmetric tensor with zero trace,

$$\bar{\bar{\mathbf{T}}}^s = \tfrac{1}{2}(\bar{\bar{\mathbf{T}}} + \bar{\bar{\mathbf{T}}}^T) - \tfrac{1}{3}(\mathrm{Tr}\,\bar{\bar{\mathbf{T}}})\bar{\bar{\mathbf{U}}}; \tag{C.3}$$

and $\bar{\bar{\mathbf{T}}}^a$ is an antisymmetric tensor with zero trace,

$$\bar{\bar{\mathbf{T}}}^a = \tfrac{1}{2}(\bar{\bar{\mathbf{T}}} - \bar{\bar{\mathbf{T}}}^T). \tag{C.4}$$

For example, the tensors $\bar{\bar{\mathbf{T}}}^s$ and $\bar{\bar{\mathbf{T}}}^a$ have the following elements:

$$(\bar{\bar{\mathbf{T}}}^s)_{11} = \tfrac{2}{3}T_{11} - \tfrac{1}{3}T_{22} - \tfrac{1}{3}T_{33}, \qquad (\bar{\bar{\mathbf{T}}}^s)_{12} = \tfrac{1}{2}(T_{12} + T_{21}),$$
$$(\bar{\bar{\mathbf{T}}}^a)_{11} = 0 \quad \text{and} \quad (\bar{\bar{\mathbf{T}}}^a)_{12} = \tfrac{1}{2}(T_{12} - T_{21}).$$

These tensors have the following useful properties:

$$\bar{\bar{\mathbf{U}}}:\bar{\bar{\mathbf{T}}}^a = 0, \tag{C.5}$$

$$\bar{\bar{\mathbf{U}}}:\bar{\bar{\mathbf{T}}}^s = 0, \tag{C.6}$$

and

$$\bar{\bar{\mathbf{T}}}^s:\bar{\bar{\mathbf{V}}}^a = 0, \tag{C.7}$$

where $\bar{\bar{\mathbf{V}}}^a$ is an arbitrary antisymmetric tensor with zero trace. Two arbitrary tensors $\bar{\bar{\mathbf{T}}}$ and $\bar{\bar{\mathbf{V}}}$, when contracted, decompose as follows:

$$\bar{\bar{\mathbf{T}}}:\bar{\bar{\mathbf{V}}} = \tfrac{1}{3}(\mathrm{Tr}\,\bar{\bar{\mathbf{T}}})(\mathrm{Tr}\,\bar{\bar{\mathbf{V}}}) + \bar{\bar{\mathbf{T}}}^s:\bar{\bar{\mathbf{V}}}^s + \bar{\bar{\mathbf{T}}}^a:\bar{\bar{\mathbf{V}}}^a. \tag{C.8}$$

An arbitrary antisymmetric tensor $\bar{\bar{\mathbf{T}}}^a$ can be written as an axial vector (or pseudo vector) $\mathbf{t}^a$ according to the relation

$$t_i^a = \frac{1}{2}\sum_{jk} \epsilon_{ijk} T_{jk}^a, \tag{C.9}$$

where i, j, and k range from 1 to 3 and $\epsilon_{ijk} = +1$ for even permutations of i, j, and k; $\epsilon_{ijk} = -1$ for odd permutations; and $\epsilon_{ijk} = 0$ if any two of the indices i, j, or k are equal.

There are also some transformation properties of tensors which we will find useful. First, we note that under an orthogonal transformation $\bar{\bar{\mathbf{A}}}$ (where the transformation matrix has the property $\bar{\bar{\mathbf{A}}}^T = \bar{\bar{\mathbf{A}}}^{-1}$ and ^{T} denotes transpose), such as a rotation, an nth-order tensor transforms according to the equation

$$T'_{i_1,i_2,\ldots,i_n} = \det|\bar{\bar{\mathbf{A}}}|^\epsilon \sum_{j_1,j_2,\ldots,j_n} A_{i_1j_1}A_{i_2j_2} \times \cdots \times A_{i_nj_n}T_{j_1j_2\times\cdots\times j_n}. \tag{C.10}$$

Here $\det|\bar{\bar{\mathbf{A}}}|$ denotes the determinant of $\bar{\bar{\mathbf{A}}}$ and $\epsilon = 0$ for polar tensors and $\epsilon = 1$ for axial tensors. For a rotation $|\bar{\bar{\mathbf{A}}}| = 1$ and the axial and polar tensors transform in the same way. For an inversion in three-dimensional space

$$\bar{\bar{\mathbf{A}}} = \begin{pmatrix} -1 & 0 & 0 \\ 0 & -1 & 0 \\ 0 & 0 & -1 \end{pmatrix} \tag{C.11}$$

and $|\bar{\bar{\mathbf{A}}}| = -1$. Thus, there is a change in sign for axial vectors.

The local entropy production can be written, in general, in the form

$$\sigma_s = -J^0X^0 - \mathbf{J}^v\cdot\mathbf{X}^v - \mathbf{J}^a\cdot\mathbf{X}^a - \bar{\bar{J}}^s:\bar{\bar{X}}^s \tag{C.12}$$

where J^0 and X^0 are scalars, $\mathbf{J}^a$ and $\mathbf{X}^a$ are axial vectors, $\mathbf{J}^v$ and $\mathbf{X}^v$ are ordinary or polar vectors, and $\overline{\mathbf{J}}^s$ and $\overline{\mathbf{X}}^s$ are symmetric traceless tensors. If we wish to write a linear relation between the forces and the fluxes, we must decide which forces can drive the various fluxes. In general, all combinations are possible so that

$$J^0 = -L^{00}X^0 - \mathbf{L}^{0v}\cdot\mathbf{X}^v - \mathbf{L}^{0a}\cdot\mathbf{X}^a - \overline{L}^{0s}:\overline{X}^s, \tag{C.13}$$

$$\mathbf{J}^v = -\mathbf{L}^{v0}X^0 - \overline{L}^{vv}\cdot\mathbf{X}^v - \overline{L}^{va}\cdot\mathbf{X}^a - \overline{\overline{L}}^{vs}:\overline{X}^s, \tag{C.14}$$

$$\mathbf{J}^a = -\mathbf{L}^{a0}X^0 - \overline{L}^{av}\cdot\mathbf{X}^v - \overline{L}^{aa}\cdot\mathbf{X}^a - \overline{\overline{L}}^{as}:\overline{X}^s, \tag{C.15}$$

and

$$\overline{J}^s = -\overline{L}^{s0}X^0 - \overline{\overline{L}}^{sv}\cdot\mathbf{X}^v - \overline{\overline{L}}^{sa}\cdot\mathbf{X}^a - \overline{\overline{\overline{L}}}^{ss}:\overline{X}^s. \tag{C.16}$$

We see that the phenomenological coefficients can themselves have a variety of tensor properties. The actual form of the phenomenological coefficients will depend on the symmetry properties of the system being considered. We shall now deduce their form for an isotropic system.

2. Phenomenological Coefficients for an Isotropic System

An isotropic system is one whose phenomenological coefficients remain invariant under inversion and rotation of coordinates. Let us first consider inversion (cf. Eq. (C.11)). Then Eq. (C.10) can be written

$$T'_{i_1,i_2,\ldots,i_n} = (-1)^{\epsilon}(-1)^n \sum_{j_1,\ldots,j_n} |A_{i_1j_1}| \times \cdots \times |A_{i_nj_n}| T_{j_1,\ldots,j_n} \tag{C.17}$$

where $|A_{ij}| = 1$ for $i = j$ and zero otherwise.

Polar tensors of odd rank ($n = 1, 3, \ldots$) and all axial tensors of even rank ($n = 2, 4, \ldots$) must vanish since they do not remain invariant under inversion. Thus, in Eqs. (C.11)–(C.14) the phenomenological coefficients $\mathbf{L}^{0v}$, $\mathbf{L}^{v0}$, $\overline{\overline{\mathbf{L}}}^{vs}$, $\overline{\overline{\mathbf{L}}}^{sv}$, and $\overline{\mathbf{L}}^{va}$, $\overline{\mathbf{L}}^{av}$ vanish identically.

Let us now itemize the tensors which remain invariant under rotation. In zeroth order (scalars) all tensors remain invariant. In first order (vectors) no tensor remains invariant. In second order only the polar tensor, the delta function $\delta_{i,j}$ remains invariant. In third order only the pseudo tensor ϵ_{ijk} remains invariant. In fourth order there are three tensors which remain invariant under rotation: $\delta_{i,j}\,\delta_{p,q}$; $(\delta_{i,p}\,\delta_{j,q} + \delta_{i,q}\,\delta_{j,p})$; and $(\delta_{i,p}\,\delta_{j,q} - \delta_{i,q}\,\delta_{j,p})$.

We can use this fact to eliminate more phenomenological coefficients for an isotropic system. In first order no tensors remain invariant under rotation, so $\mathbf{L}^{0a}$ and $\mathbf{L}^{a0}$ are zero. In second order all traceless tensors vanish, so $\overline{\mathbf{L}}^{va}$, $\overline{\mathbf{L}}^{av}$, $\overline{\mathbf{L}}^{0s}$, and $\overline{\mathbf{L}}^{s0}$ are zero. In third order any nonzero phenomenological coefficient must be antisymmetric with respect to interchange of any two indices since it is proportional to ϵ_{ijk}. Since $\overline{\overline{\mathbf{L}}}^{vs}$, $\overline{\overline{\mathbf{L}}}^{sv}$, $\overline{\overline{\mathbf{L}}}^{as}$, and $\overline{\overline{\mathbf{L}}}^{sa}$ are symmetric with respect to at least one pair of indices, they must be zero. Thus, for an isotropic system, Eqs. (C.13)–(C.16) reduce to the following set:

$$J^0 = -L^{00}X^0 \tag{C.18}$$

$$\mathbf{J}^v = -\overline{\overline{\mathbf{L}}}^{vv}\cdot\mathbf{X}^v, \tag{C.19}$$

$$\mathbf{J}^a = -\overline{\overline{\mathbf{L}}}^{aa}\cdot\mathbf{X}^a, \tag{C.20}$$

and

$$\overline{\overline{\mathbf{J}}}^s = -\overline{\overline{\overline{\overline{\mathbf{L}}}}}^{ss}:\overline{\overline{\mathbf{X}}}^s. \tag{C.21}$$

As pointed out by Prigogine,[5] Eqs. (C.18)–(C.21) give an illustration of the so-called Curie principle,[7] which states that a macroscopic cause cannot produce an effect with a lower symmetry in an isotropic system.

We can reduce Eqs. (C.18)–(C.21) still further. To be invariant under rotation, $\overline{\overline{\mathbf{L}}}^{vv}$ and $\overline{\overline{\mathbf{L}}}^{aa}$ must be proportional to the unit tensors. Thus,

$$\overline{\overline{\mathbf{L}}}^{vv} = c_v\overline{\overline{\mathbf{U}}} \tag{C.22}$$

and

$$\overline{\overline{\mathbf{L}}}^{aa} = c_a\overline{\overline{\mathbf{U}}} \tag{C.23}$$

where c_v and c_a are constants. The tensor $\overline{\overline{\overline{\overline{\mathbf{L}}}}}^{ss}$ must be symmetric and traceless in its two leftmost indices and in its two rightmost indices. The tensor $\overline{\overline{\overline{\overline{\mathbf{L}}}}}^{ss}$, in general, has the form

$$(\overline{\overline{\overline{\overline{\mathbf{L}}}}}^{ss})_{\alpha\beta,\delta\gamma} = a\,\delta_{\alpha\beta}\,\delta_{\delta\gamma} + b(\delta_{\alpha\gamma}\,\delta_{\beta\delta} + \delta_{\alpha\delta}\,\delta_{\beta\gamma}) + c(\delta_{\alpha\gamma}\,\delta_{\beta\delta} - \delta_{\alpha\delta}\,\delta_{\beta\gamma}) \tag{C.24}$$

where a, b, and c are arbitrary constants. However, because of the properties of tensors discussed in Sec. C.1, it must be symmetric and traceless in its two leftmost indices and in its two rightmost indices. This leads to the expression

$$(\overline{\overline{\overline{\overline{\mathbf{L}}}}}^{ss})_{\alpha\beta,\delta\gamma} = c_s(\tfrac{1}{2}(\delta_{\alpha\delta}\,\delta_{\beta\gamma} + \delta_{\alpha\gamma}\,\delta_{\beta\delta}) - \tfrac{1}{3}\delta_{\alpha\beta}\,\delta_{\gamma\delta}). \tag{C.25}$$

Thus, for an isotropic system, each of the phenomenological coefficients in Eqs. (C.18)–(C.21) depends only on a single scalar parameter.

D. STABILITY OF SOLUTIONS TO NONLINEAR EQUATIONS

1. Linear Stability Theory[8,9]

Any set of nonlinear differential equations of arbitrary order can be reduced to a larger number of first-order nonlinear differential equations. Therefore, to study the stability of a system, it is sufficient to study the stability of the set of first-order nonlinear equations governing the behavior of the system.

For simplicity, we will consider a system of two nonlinear first-order differential equations:

$$\frac{dy_1}{dt} \equiv \dot{y}_1 = f_1(y_1, y_2) \tag{D.1}$$

and

$$\frac{dy_2}{dt} = \dot{y}_2 = f_2(y_1, y_2). \tag{D.2}$$

If there exist positive numbers k and δ such that for given values of coordinates $y_1{}^0$ and $y_2{}^0$ the functions $f_i(y_1, y_2)$ satisfy the condition

$$|f_i(y_1{}^0, y_2{}^0) - f_i(y_1, y_2)| \leqslant k \sum_{j=1}^{2} |y_j{}^0 - y_j|$$

for $|y_j{}^0 - y_j| < \delta$, and if the partial derivatives of $f_i(y_1, y_2)$ exist and are continuous, then the functions $f_i(y_1, y_2)$ satisfy Lipschitz conditions and the differential equations (D.1) and (D.2) have unique solutions in the neighborhood $y_1{}^0$, $y_2{}^0$. We shall always assume that the above conditions are satisfied.

In general, the set of Eqs. (D.1) and (D.2) will have a number of steady state solutions (which are often called fixed or singular or critical points). These are solutions $\bar{y}_1$, $\bar{y}_2$ which satisfy the conditions

$$\dot{\bar{y}}_1 = f_1(\bar{y}_1, \bar{y}_2) = 0 \tag{D.3}$$

and

$$\dot{\bar{y}}_2 = f_2(\bar{y}_1, \bar{y}_2) = 0. \tag{D.4}$$

We may plot the solutions of Eqs. (D.1) and (D.2) in a two-dimensional phase plane. The time-independent steady state solutions correspond to points in the phase plane which do not change their position with time. The time-dependent solutions form trajectories in the phase plane. The equation for the trajectories can be found by dividing Eq. (D.1) by Eq. (D.2):

$$\frac{dy_1}{dy_2} = \frac{f_1(y_1, y_2)}{f_2(y_1, y_2)}. \tag{D.5}$$

Eq. (D.5) uniquely relates y_1 to y_2 for a given value of y_2.

In general, a given steady state solution will occur at some point $\bar{y}_1 = A$ and $\bar{y}_2 = B$ in the phase plane. It can always be moved to the origin by the change of variables $y_1' = y_1 - A$ and $y_2' = y_2 - B$. We therefore wish to study the stability of a given steady state solution to Eqs. (D.1) and (D.2) which occurs at the origin.

Consider a system which is described by Eqs. (D.1) and (D.2) and which is in the steady state $\bar{y}_1$, $\bar{y}_2$. We often want to know how the system behaves under the influence of a small perturbation. Will the system leave the steady state and move to a different state? Will it decay back to the steady state? Or will the system remain in the neighborhood of the steady state and not decay back?

With linear stability theory, we can say something about the stability of a very small neighborhood of the steady state. We can say nothing about the global stability of the phase space. To examine the stability of solutions in the neighborhood of the steady state, let us expand y_1 and y_2 about the steady state,

$$y_1 = \bar{y}_1 + x_1 \tag{D.6}$$

and

$$y_2 = \bar{y}_2 + x_2, \tag{D.7}$$

and let us expand both sides of Eqs. (D.1) and (D.2) about the steady state. We then find

$$\dot{x}_1 = \frac{dx_1}{dt} = \left(\frac{\partial f_1}{\partial y_1}\right)_{\bar{y}_1,\bar{y}_2} x_1 + \left(\frac{\partial f_1}{\partial y_2}\right)_{\bar{y}_1,\bar{y}_2} x_2 + O(x^2) \tag{D.8}$$

and

$$\dot{x}_2 = \frac{dx_2}{dt} = \left(\frac{\partial f_2}{\partial y_1}\right)_{\bar{y}_1,\bar{y}_2} x_1 + \left(\frac{\partial f_2}{\partial y_2}\right)_{\bar{y}_1,\bar{y}_2} x_2 + O(x^2). \tag{D.9}$$

For very small values of x_i (for solutions very close to the steady state), we can linearize Eqs. (D.8) and (D.9) and obtain

$$\begin{pmatrix}\dot{x}_1\\ \dot{x}_2\end{pmatrix} = \bar{\bar{\mathbf{A}}}\begin{pmatrix}x_1\\ x_2\end{pmatrix} \tag{D.10}$$

where

$$\bar{\bar{\mathbf{A}}} = \begin{pmatrix}a & b\\ c & d\end{pmatrix} \tag{D.11}$$

and

$$a = \left(\frac{\partial f_1}{\partial y_1}\right)_{\bar{y}_1,\bar{y}_2}, \qquad b = \left(\frac{\partial f_1}{\partial y_2}\right)_{\bar{y}_1,\bar{y}_2},$$

$$c = \left(\frac{\partial f_2}{\partial y_1}\right)_{\bar{y}_1,\bar{y}_2}, \quad \text{and} \quad d = \left(\frac{\partial f_2}{\partial y_2}\right)_{\bar{y}_1,\bar{y}_2}.$$

We next transform to normal coordinates of Eq. (D.10). If we find the eigenvalues of $\bar{\bar{\mathbf{A}}}$ and expand Eq. (D.10) in terms of the right eigenvectors of $\bar{\bar{\mathbf{A}}}$, which we denote z_1 and z_2 (cf. Sec. 6.C), Eq. (D.10) then takes the form

$$\begin{pmatrix}\dot{z}_1\\ \dot{z}_2\end{pmatrix} = \begin{pmatrix}\lambda_1 & 0\\ 0 & \lambda_2\end{pmatrix}\begin{pmatrix}z_1\\ z_2\end{pmatrix} \tag{D.12}$$

where the eigenvalues are determined by the condition that

$$\det\begin{vmatrix}a-\lambda & b\\ c & d-\lambda\end{vmatrix} = 0 \tag{D.13}$$

or

$$\lambda_{\binom{1}{2}} = \tfrac{1}{2}\{(a+d) \pm \sqrt{(a+d)^2 - 4(ad-bc)}\}. \tag{D.14}$$

For a conservative system, which is area preserving, Eq. (D.14) simplifies because $\det|\bar{\bar{\mathbf{A}}}| = ad - cb = 1$. If the coordinates initially have values $z_1{}^0$ and $z_2{}^0$, at a later time they will have values

$$z_1(t) = e^{\lambda_1 t} z_1{}^0 \quad \text{and} \quad z_2(t) = e^{\lambda_2 t} z_2{}^0. \tag{D.15}$$

The eigenvalues, in general, are complex. We can consider several cases which can occur depending on values of the real and imaginary parts of λ_1 and λ_2:

(*a*) Fig. D.1*a*: Both λ_1 and λ_2 are real and $\lambda_1 < \lambda_2 < 0$. This is a completely stable case. A solution displaced from the steady state will decay back to the steady state, exponentially.

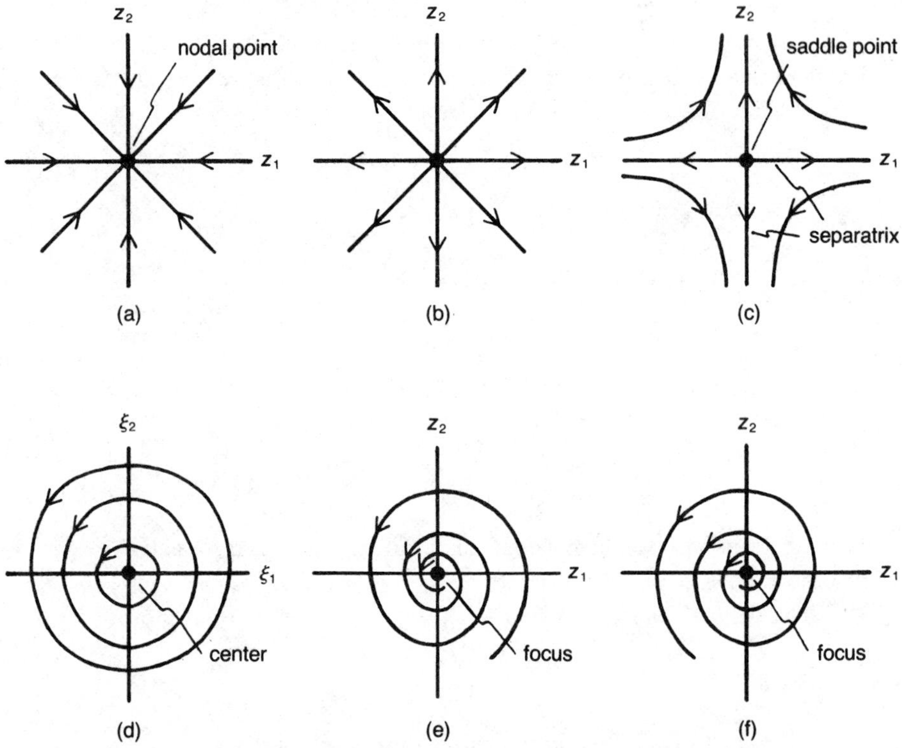

Fig. D.1. Different types of fixed points: (*a*) both λ_1 and λ_2 are real and $\lambda_1 < \lambda_2 < 0$; (*b*) both λ_1 and λ_2 are real and $0 < \lambda_1 < \lambda_2$; (*c*) both λ_1 and λ_2 are real and $\lambda_1 < 0 < \lambda_2$; (*d*) both λ_1 and λ_2 are pure imaginary with $\lambda_1 = i\beta$ and $\lambda_2 = -i\beta$ ($z_1 = \zeta + i\zeta_2$ and $z_2 = \zeta_1 - i\zeta_2$); (*e*) both λ_1 and λ_2 are complex, $\lambda_1 = \alpha + i\beta$, $\lambda_2 = \alpha - i\beta$, and $\alpha < 0$; (*f*) both λ_1 and λ_2 are complex, $\lambda_1 = \alpha + i\beta$, $\lambda_2 = \alpha - i\beta$, and $\alpha > 0$.

(*b*) Fig. D.1*b*: Both λ_1 and λ_2 are real and $0 < \lambda_1 < \lambda_2$. This is a completely unstable case. A solution displaced from the steady state will move away exponentially.

(*c*) Fig. D.1*c*: Both λ_1 and λ_2 are real and $\lambda_1 < 0 < \lambda_2$. This is an unstable case. A solution which starts with a nonzero value of z_2 will move away from the steady state. (Note that the trajectories moving along the z_1-axis never reach the steady state. It is a separate solution.)

(*d*) Fig. D.1*d*: Both λ_1 and λ_2 are pure imaginary; that is, $\lambda_1 = i\beta$ and $\lambda_2 = -i\beta$. To find the stability in this case, one must consider nonlinear terms. The solution neither decays nor moves away from the steady state. It simply oscillates around it. In Fig. D.1*d* we have let $z_1 = \zeta_1 + i\zeta_2$ and $z_2 = \zeta_1 - i\zeta_2$; therefore,

$$\begin{pmatrix} \dot{\zeta}_1 \\ \dot{\zeta}_2 \end{pmatrix} = \begin{pmatrix} \cos\beta & \sin\beta \\ -\sin\beta & \cos\beta \end{pmatrix} \begin{pmatrix} \zeta_1 \\ \zeta_2 \end{pmatrix}$$

(this is a case of marginal stability). For this case, the fixed point is often called a center.

(*e*) Fig. D.1*e*: Both λ_1 and λ_2 are complex, so $\lambda_1 = \alpha + i\beta$ and $\lambda_2 = \alpha - i\beta$ where $\alpha < 0$. This is a stable case in which the solution spirals toward the steady state (which is called a focus).

(*f*) Fig. D.1*f*: Both λ_1 and λ_2 are complex, so that $\lambda_1 = \alpha + i\beta$ and $\lambda_2 = \alpha - i\beta$ and $\alpha > 0$. This is an unstable case in which the solution spirals away from the steady state.

The linear stability analysis provides a very simple tool for studying the stability of a steady state. For example, if the real parts of λ_1 and λ_2 are both negative, the steady state is stable. It can be shown that the nonlinear terms which were neglected in the stability analysis do not change this fact (Ref. A.8, p. 32). Similarly, if at least one of the eigenvalues has a positive real part, the steady state will be unstable. Inclusion of the nonlinear terms cannot make it stable.

When linear stability analysis reveals a center (*d*), then one must look at the effect of nonlinear terms on the stability. They will, usually (but not always), change the center to a stable or unstable focus.

2. Limit Cycles

In Sec. D.1 we studied conditions under which steady state solutions might be stable. We found that, if a steady state solution was stable, then all states in the neighborhood of the steady state would decay to it after long times.

For nonconservative nonlinear equations, it is also possible to find stable *periodic* solutions. Such solutions are called limit cycles. If a periodic solution is stable, then all solutions in its neighborhood will decay to it after long times.

An example of a set of equations which admits a limit cycle is given by

$$\frac{dy_1}{dt} = \dot{y}_1 = +y_2 + \frac{y_1}{\sqrt{y_1^2 + y_2^2}}\,[1 - (y_1^2 + y_2^2)] \tag{D.16}$$

and

$$\frac{dy_2}{dt} = \dot{y}_2 = -y_1 + \frac{y_2}{\sqrt{y_1^2 + y_2^2}}\,[1 - (y_1^2 + y_2^2)]. \tag{D.17}$$

If we change to polar coordinates $y_1 = r\cos\theta$ and $y_2 = r\sin\theta$, these equations become

$$\frac{dr}{dt} = \dot{r} = 1 - r^2 \tag{D.18}$$

and

$$\frac{d\theta}{dt} = \dot{\theta} = 1. \tag{D.19}$$

We can solve Eq. (D.18) to obtain

$$\int_{R_0}^{R(t)} \frac{dr}{1 - r^2} = \int_0^t dt = t = \tfrac{1}{2}\ln\left|\frac{1 + r}{1 - r}\right|_{R_0}^{R}$$

or

$$R(t) = \frac{A\,e^{2t} - 1}{A\,e^{2t} + 1} \tag{D.20}$$

where $A = (1 + R_0)/(1 - R_0)$. From Eq. (D.20) we see that regardless of the value of R_0 (unless $R_0 = 1$) we have

$$R(t) \rightarrow 1. \tag{D.21}$$

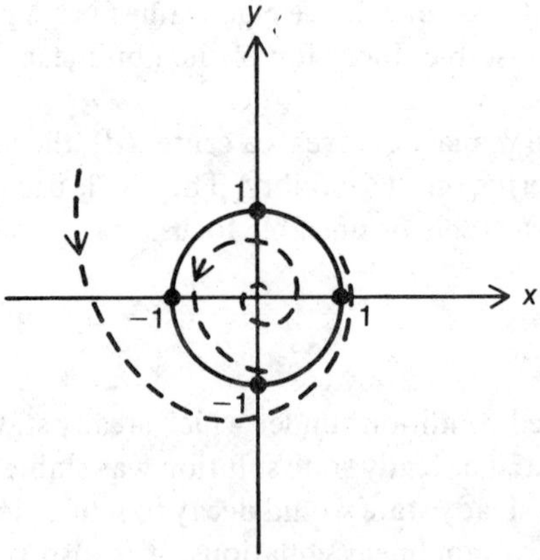

Fig. D.2. All trajectories in the neighborhood of a stable limit cycle decay to it after long times.

If $R_0 = 1$, then $R(t) = 1$ for all t. Therefore, Eqs. (D.18) and (D.19) admit a periodic solution with radius $R = 1$ and all trajectories decay to it after long times. If a solution starts with $R_0 = 1$, then it will keep that value for all times (cf. Fig. D.2). Not all limit cycles are stable. One can also find limit cycles for which the neighboring trajectories move away in time.

3. Liapounov Functions and Global Stability

In Sec. D.1, we discussed methods for determining the infinitesimal stability of stationary solutions. Liapounov introduced a means of determining the stability of a finite region of the phase space surrounding a stationary state.

Let us again consider Eqs. (D.1) and (D.2) and assume that a steady state occurs at $\bar{y}_1 = \bar{y}_2 = 0$. If we can find a function $V(y_1, y_2)$ such that for a region of space (D: $|y_1| < h$, $|y_2| < h$) around the steady state, $V(y_1, y_2) > 0$ (assume $V(0, 0) = 0$), and $dV/dt \leqslant 0$, then the steady state will be stable. In other words, if $V > 0$ (except at the origin) and $dV/dt \leqslant 0$, a trajectory which starts in D remains in D for all time.

The region D forms a square with sides $2h$ centered at the origin. Consider two more squares with sides 2ϵ and 2δ in D such that $\delta < \epsilon < h$ (cf. Fig. D.3). Let B^h denote the boundary of square $2h$, B^ϵ the boundary of square 2ϵ, and B^δ the boundary of square 2δ. We now let α denote the lower bound of $V(y_1, y_2)$ on B^ϵ. Thus, on B^ϵ, $V \geqslant \alpha > 0$.

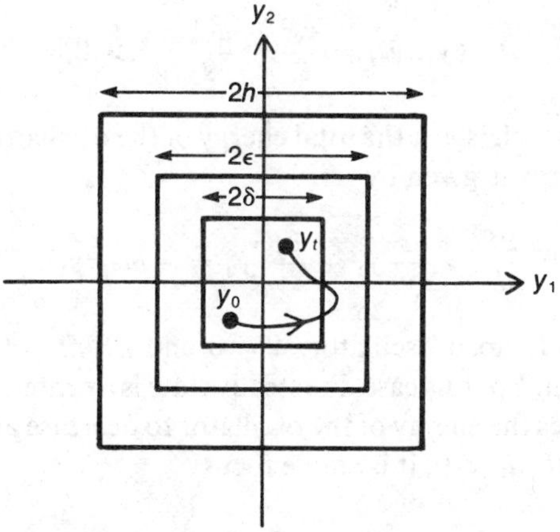

Fig. D.3. Regions of phase space used to prove stability of the steady state if Liapounov's criteria are satisfied.

Let us next consider a point $y_0 = (y_1(t_0), y_2(t_0))$ contained in the square B^δ at time t_0 and let us follow its motion with time. At time t, the point will have coordinates $y_t = (y_1(t), y_2(t))$. We shall choose B^δ so that $V(y_1(t_0), y_2(t_0)) < \alpha$. Then since $dV/dt \leqslant 0$ we must have

$$V(y_1(t), y_2(t)) \leqslant V(y_1(t_0), y_2(t_0)) < \alpha.$$

Therefore, for all times $t > t_0$, the point $y_t = (y_1(t), y_2(t))$ must remain inside the boundary B^ϵ and the steady state is stable.

An identical argument can be made if $V(y_1, y_2) < 0$ in D and $dV/dt \geqslant 0$. If such a function can be found, then, again, the steady state is stable.

The simplest example of the above criteria is the harmonic oscillator, whose equation of motion can be written

$$\frac{d^2y_1}{dt^2} + \alpha\frac{dy}{dt} + \omega_0{}^2y_1 = 0 \tag{D.22}$$

if the oscillator is damped. If we introduce the variable $dy_2/dt = -\omega_0{}^2y_1$, then we can write Eq. (D.22) as a set of linear equations:

$$\frac{dy_1}{dt} = y_2 - \alpha y_1 \tag{D.23}$$

and

$$\frac{dy_2}{dt} = -\omega_0{}^2y_1. \tag{D.24}$$

(Note that for this example the exact equations of motion are linear.) We may now introduce the Liapounov function,

$$F(y_1, y_2) = \frac{y_2{}^2}{2} + \frac{\omega_0{}^2}{2}\,y_1{}^2 > 0, \tag{D.25}$$

for all y_1 and y_2, which is just the total energy of the oscillator. The time rate of change of the energy is given by

$$\frac{dF}{dt} = \frac{\partial F}{\partial y_1}\dot{y}_1 + \frac{\partial F}{\partial y_2}\dot{y}_2 = -\alpha\omega_0{}^2y_1{}^2. \tag{D.26}$$

Hence, for an undamped oscillator, $\alpha = 0$ and $dF/dt = 0$. The energy is a constant of motion. For this case, the steady state is a center. When $\alpha > 0$, then the damping causes the energy of the oscillator to decrease and the steady state is a focus. Since $dF/dt < 0$, it is stable focus.

REFERENCES

1. R. Aris, *Vectors, Tensors, and the Basic Equations of Fluid Mechanics* (Prentice-Hall, Englewood Cliffs, N.J., 1962).
2. A. S. Davidov, *Quantum Mechanics* (NEO Press, Ann Arbor, Mich., 1966).
3. A. Messiah, *Quantum Mechanics* (John Wiley & Sons, New York, 1962).

4. B. R. Easlea, "The Many-Body Problem and Applications to Nuclear Physics," lecture notes, University of Pittsburgh, 1963.
5. I. Prigogine, *Introduction to Thermodynamics of Irreversible Processes* (Wiley-Interscience, New York, 1967).
6. S. R. de Groot and P. Mazer, *Nonequilibrium Thermodynamics* (North-Holland Publishing Co., Amsterdam, 1951).
7. P. Curie, *Oeuvres* (Société Française de Physique, Paris, 1908).
8. N. Minorsky, *Nonlinear Oscillations* (Van Nostrand Co., Princeton, 1962).
9. T. Saarty and J. Bram, *Nonlinear Mathematics* (McGraw-Hill Book Co., New York, 1964).
10. D. H. Sattinger, *Topics in Stability and Bifurcation Theory* (Springer-Verlag, Berlin, 1973).

Author Index

Subject Index